D0086650

Arguing for Evolution

Arguing for Evolution

AN ENCYCLOPEDIA FOR UNDERSTANDING SCIENCE

Sehoya Cotner and Randy Moore

AN IMPRINT OF ABC-CLIO, LLC
Santa Barbara, California • Denver, Colorado • Oxford, England

Library of Congress Cataloging-in-Publication Data

Cotner, Sehoya.
  Arguing for evolution : an encyclopedia for understanding science / Sehoya Cotner and Randy Moore.
    p. cm.
  Includes index.
  ISBN 978-0-313-35947-7 (hardcopy : alk. paper) —
ISBN 978-0-313-35948-4 (ebook)
  1. Evolution (Biology) 2. Natural selection. I. Moore, Randy.
II. Title.
QH366.2.C735 2011
576.8—dc22 2011009608

ISBN: 978-0-313-35947-7
EISBN: 978-0-313-35948-4

15 14 13 12 11 1 2 3 4 5

This book is also available on the World Wide Web as an eBook.
Visit www.abc-clio.com for details.

Greenwood
An Imprint of ABC-CLIO, LLC

ABC-CLIO, LLC
130 Cremona Drive, P.O. Box 1911
Santa Barbara, California 93116-1911

This book is printed on acid-free paper ∞

Manufactured in the United States of America

For Jim, with love and thanks . . . for the patience,
and for the fitness!—SC

For Maryuri, Javier, and Alfredo, and the good times
we've had through the years in Galápagos.—RM

Contents

Quick A–Z Guide to the Evidence

Acknowledgments

Many talented colleagues and supportive friends and family have assisted us as we collected evidence for evolution. We especially thank the following people for their help with our book: Tom Linley, Cinzia Cervato, Robert Wallace, Diane Seward, Bette Englund, Kara Witt, Angus Miller, Sue Hendrickson, Larry Shaffer, Lachlan and Flannery Cotner, Carolyn Belardo, Eileen Mathias, Mindy McNaugher, Norman Butcher, Janice Moore, Charity Putuhena, Gigi Boyer, Kyle Hammond, Rebecca Forman, Leah Johnson, Meghan Gordon, Lloyd Patton, Örn Óskarsson, Michael Wilson, Susan Lundy, Mike Caputo, Csaba Moskat, Miklós Bán, Arthur Chapman, Alan Cressler, Andrew Hoffman, Antony Dean, Susan Harding, Lynn Dion, Deus Mjungu, Mark Decker, Deena Wassenberg, and Frank Barnwell. We are also especially grateful to our editor, David Paige, for his guidance and patience.

Introduction: Evolution as a Predictive Science

"What insect could suck it?"

These were the words Charles Darwin used when, in 1862, he was confronted with the Madagascan orchid, *Angraecum sesquipedale*. A specimen had been sent to Darwin by James Bateman, a British naturalist and botanical illustrator who was helping Darwin with his forthcoming manuscript, *On the Various Contrivances by which British and Foreign Orchids are Fertilised by Insects* (1862). The orchid amazed Darwin, largely because it possessed "a whip-like green nectary of astonishing length." Because the flower's nectar sat at the base of a foot-long spur, Darwin could not imagine what creature might be able to pollinate it. Yet based on his decades of observation, including his experiences on his famous voyage aboard HMS *Beagle* (1831–1836), Darwin was able to make a logical prediction: "[I]n Madagascar there must be moths with proboscises capable of extension to a length of between ten and eleven inches!" Nobody had ever reported on such a creature; moreover, the same group of orchids had been used earlier by the Duke of Argyll to cite the existence of a creator. Yet five years later, Darwin's correspondent—and co-discoverer of natural selection, Alfred Russel Wallace—refuted the Duke's claim, saying "[t]hat such a moth exists in Madagascar may be safely predicted; and naturalists who visit that island should search for it with as much confidence as astronomers searched for the planet Neptune, and they will be equally successful!" Sure enough, in 1903—21 years after Darwin's death—scientists reported the existence of a moth in Madagascar with a proboscis long enough to reach the unusual orchid's nectar. The aptly named *Xanthopan morgani praedicta* is rare and nocturnal, thus explaining why it took decades to confirm its existence.

Science, a process by which we interpret the natural world, involves making observations and testing predictions. Charles Darwin, best known for his contributions to evolutionary biology, made several predictions, such as the existence of the unusual orchid-pollinating moth in Madagascar. For decades, this prediction was neither supported nor falsified.

However, Darwin's claim was logical and theoretically possible to confirm, as are all scientific predictions.

Although some of Darwin's predictions (e.g., his claims about inheritance) turned out to be wrong, his most famous prediction—namely, that evolution is driven by natural selection—was remarkably insightful. Darwin wasn't the first person to propose that evolution occurs, but he knew that such a claim "would be unsatisfactory until it could be shown *how* the innumerable species inhabiting this world have been modified so as to acquire that perfection of structure and coadaptation that most justly excites our admiration" (italics added). Darwin described his 1859 masterpiece, *On the Origin of Species,* as "one long argument" detailing this mechanism with numerous specific examples and an outline for future discoveries. Darwin understood the value of testing predictions to the scientific process: "as modern geology has almost banished such views as the excavation of a great valley by a single diluvial wave, so will natural selection, if it be a true principle, banish the belief of the continued creation of new organic beings, or of any great and sudden modification in their structure." Darwin's proposed mechanism (i.e., natural selection—the principle "by which each slight variation, if useful, is preserved") has been repeatedly confirmed by evidence gathered from paleontology, biogeography, biochemistry, medicine, comparative anatomy, embryology, molecular biology, behavioral ecology, anthropology, and other scientific disciplines. As American paleontologist Niles Eldredge noted, "Nothing that we have learned [since Darwin] has contravened Darwin's basic description of how natural selection works."

In *Arguing for Evolution,* we discuss natural selection and associated predictions regarding evolution. To model evolution as "science in action," we have organized each chapter around a series of confirmed predictions arising from basic tenets of evolution, and loosely arranged by discipline (geology, paleontology, molecular biology, etc.).

Coverage

Libraries, museums, hospitals, universities, and other institutions throughout the world are filled with books, journals, monographs, videos, and other materials that document the evidence for evolution. We cannot possibly include all of this evidence in a one-volume book such as *Arguing for Evolution*, but we have presented the basic lines of evidence for Darwin's great idea. Although *Arguing for Evolution* covers general information and iconic topics about evolution (e.g., Galápagos, Darwin's finches), we have also included numerous fresh examples that illustrate these basic principles. Furthermore, we have dedicated entire chapters to topics that are often absent (or included parenthetically) in other treatments of evolution, such as coevolution, behavior, biogeography, and the age of Earth. For example, whereas many books simply declare Earth's age to be ~4.5 billion years, we discuss *how* scientists first came to know that Earth is old, and then *how* they determined that it is ~4.5 billion years old. These arguments provide excellent examples of how to use an understanding of evolution to illustrate "science in action." They are also important for countering the claims of evolution's most vocal, persistent, and well-funded critics—the young-Earth creationists—who, as their name suggests, base their rejection of evolution, geology, physics, and other sciences on an age of Earth that they derive from their particular religious beliefs. As John Morris—president of the young-Earth organization Institute for Creation Research—claimed in 2010, "any system that includes great ages [for Earth] is not compatible with Scripture."

Boxed Inserts

Each chapter of *Arguing for Evolution* also includes features that elaborate on topics ranging from the icons of evolution (e.g., Darwin's finches) to related issues (e.g., the business of fossil collecting) that supplement the chapter's arguments and predictions.

Accessibility

Arguing for Evolution presents the evidence for evolution in a single, concise, accessible, and easy-to-use volume. We have tried to eliminate the jargon that impedes understanding of many science books, and in its place have substituted easily understood explanations of various evolutionary principles.

Illustrations

Throughout *Arguing for Evolution,* you'll encounter artwork and photographs that help explain the evidence for evolution. Many of these photographs have never been published before.

Supplemental Information

Arguing for Evolution includes three appendices, one of which includes two key chapters of Charles Darwin's *On the Origin of Species*—Chapter IV, "Natural Selection," and Chapter XIV, "Recapitulation and Conclusion"—which we discuss throughout the first nine chapters. Darwin's book revolutionized biology; Sir Joseph Dalton Hooker noted in 1859 that he "would

Figure I.1 Alfred Russel Wallace's conception of the pollination of *A. sesquipedale,* published alongside his article "Creation by Law" in 1867. This drawing, by Thomas William Wood, is based on Wallace's description because the actual pollinator would not be described for another 36 years. (Alfred Russel Wallace, "Creation by Law" [October 1867], *Quarterly Journal of Science* 4 (16): 470. London: John Churchill & Sons.)

rather be author of [*Origin*] than of any other on Natural History Science," and in 1964—more than a century later—Sir Julian Huxley described Darwin's idea as "the most powerful and the most comprehensive idea that has ever arisen on Earth." Other appendices summarize two important topics: (1) the geologic history of Earth, and (2) the evolution of humans. We conclude with a Glossary; terms in the text that are defined in the Glossary are formatted in bold.

We hope you'll be convinced by our arguments.

Sehoya Cotner and Randy Moore, June 2011

1

Understanding the Natural World: Evolution and the Process of Science

> The publication of the Darwin and Wallace papers in 1858, and still more that of the "Origin" in 1859, had the effect upon them of the flash of light, which to a man who has lost himself in a dark night, suddenly reveals a road which, whether it takes him straight home or not, certainly goes his way. That which we were looking for, and could not find, was a hypothesis respecting the origin of known organic forms, which assumed the operation of no causes but such as could be proved to be actually at work. We wanted, not to pin our faith to that or any other speculation, but to get hold of clear and definite conceptions which could be brought face to face with facts and have their validity tested. The "Origin" provided us with the working hypothesis we sought.
>
> —*Thomas Henry Huxley, 1888*

> [T]he alternative to thinking in evolutionary terms is not to think at all.
>
> —*Sir Peter Medawar, 1980*

Definitions of Science and Scientific Theory

It is difficult to argue for the scientific validity of evolution without clarifying what "science" is, and what it means to accrue scientific evidence. Science is a process by which we interpret the natural world. This process relies on observation, generating hypotheses (or logical explanations), making and testing predictions, and constantly using our observations and data to revise our understanding of natural phenomena. Scientific evidence must be at least theoretically attainable through interactions with the natural world. Nonnatural, or *supernatural* explanations, are neither scientific nor evidentiary (i.e., they cannot be supported by scientific evidence).

Scientific hypotheses must be *testable* and *falsifiable*. Making conclusions that cannot be tested through experimentation and observation is not scientifically valid. Scientific hypotheses can involve either deductive reasoning, whereby logic flows from the general to the specific (e.g., "all bats can fly; therefore, this bat can fly"), or inductive reasoning, whereby logic flows from the specific to the general (e.g., "these ten snakes are poisonous; therefore, all snakes are poisonous"). In any case, testing a hypothesis involves making *predictions* that logically flow from the hypothesis, and then testing these individual predictions.

A *scientific theory* is a related set of hypotheses that together form a broad, testable explanation about some fundamental aspect of nature. That is, a **theory** is a well-supported broad idea; biological examples include the cell theory ("all living organisms are composed of, and arise from, cells and cell products") and the theory of evolution. Although scientific knowledge is provisional and cannot be proven, many theories are supported by so much evidence that they are not questioned by serious scientists. These theories, in the words of the late Stephen Jay Gould, have been "confirmed to such a degree that it would be perverse to withhold provisional assent." Evolution is one such theory.

Definition of Evolution

Most simply, evolution is any change in a population's genetic composition over time. There are several key features of this definition:

1. Evolution is a population-level phenomenon, rather than something that can be measured in individuals. Populations evolve; individuals do not.
2. Evolution has occurred when *any* genetic change—even something that seems insignificant—happens to any number of individuals in a population.
3. Evolution can be measured in generational time.

To appreciate this, consider a population of rock pocket mice living on a lava outcrop in the southwestern United States. At an initial observation, 40% of the mice possessed an **allele** (or genetic variant) that produces lightly colored fur when inherited from each parent. A few generations later, only 34% of the mice possessed this allele. This change in the genetic composition of the population means that evolution has occurred.

For evolutionary biologists, the fact that evolution has occurred is often not as exciting as the questions that arise as a result of this observation. For example, *why* did evolution occur? Were the changes entirely random, or can they tell us something about the organisms or their environment?

How Evolution Occurs

A population can evolve in a number of ways. Random genetic changes, or **mutations,** can arise in any gene in any member of the population. The random, rare, and recurring nature of mutations means that all populations are undergoing evolution all the time. These mutations can have no effect, a negative effect, or a positive effect on the bearer. Most mutations are *silent;* that is, they have no immediate discernable impact on the individual or population. However, mutations ultimately provide the raw material for evolutionary change. (Mutations, genes, and alleles are discussed further in Chapter 5.)

Random changes in genetic frequencies can occur in a population, a type of evolution known as **genetic drift.** Genetic drift often typifies small populations (e.g., as with endangered species), in which by chance some individuals do not reproduce. By definition this type of evolution is not a result of environmental pressures.

Migratory events often produce genetic changes, either in the parent population, the population on the move, or the new, target population. Evolution that occurs when individuals move, taking their alleles with them to a new population, is known as **gene flow.**

When most people speak of evolution, they are talking about **natural selection,** one of the ways that a population can change over time. Unlike other mechanisms for evolution (e.g., mutation, genetic drift, gene flow), natural selection is the only scientifically supported explanation for *adaptive* change.

Adaptation and Natural Selection

Any feature that confers a benefit on an individual in its present environment is considered adaptive, and is referred to as an **adaptation.** On average, individuals with this feature are more likely to survive and reproduce than are unaffected members of the population. If the adaptation is inherited, then we would expect affected individuals to survive and pass the feature to their offspring in greater numbers than those who are unaffected.

Natural selection refers to adaptive evolutionary change that occurs when some heritable genetic variants—that is, adaptations—confer a survival and reproductive benefit that, in turn, alters a population's genetic structure. In the short- or long-term, natural selection can alter populations, species, and entire groups of organisms. Given the amazing ability of organisms to procreate, and the fact that resources are limited, we know there must be some culling mechanism (i.e., a competition) in place. That mechanism is natural selection. Populations do not consciously choose what is selected; instead, individuals that inherit adaptive traits simply out-compete individuals that do not. The "winners" of the competition are

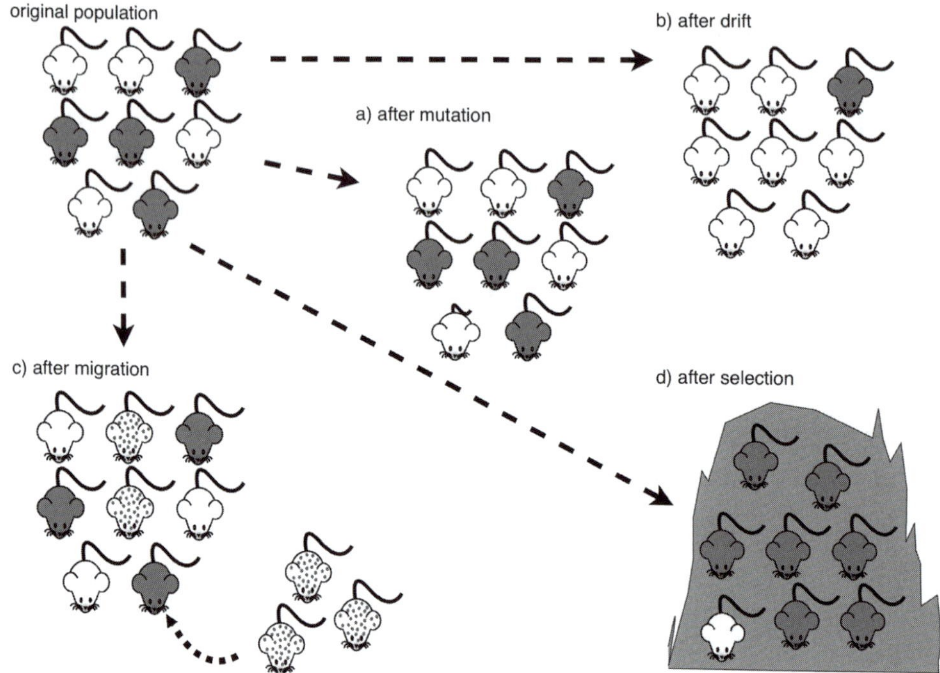

Figure 1.1 Evolution via mutation, genetic drift, gene flow, and natural selection. Populations evolve as a consequence of (a) mutation, the formation of new gene variants; (b) genetic drift, the generational changes that occur due to random events; (c) gene flow, the movement of individuals—and their genes—from one population to another; and (d) natural selection, the differential reproductive success due to adaptation (in this case, dark coloration on a dark background conceals prey from an overhead predator and gives the mice a survival advantage). (Sehoya Cotner)

those individuals that survive, reproduce, and pass on the adaptations to their offspring.

It can be tempting to imagine that large, complex organisms (like ourselves) are somehow "better" and "more advanced" than smaller, simpler organisms that appeared on Earth long before us. This is a pervasive misconception. Indeed, bacteria have existed for *billions* of years; they continue to evolve and adapt to new environments. Similarly, much larger organisms such as *Tyrannosaurus* and *Apatosaurus* have gone extinct because they could not adapt. Indeed, *all* nonavian dinosaurs—such as the iconic *T. rex*, *Stegosaurus*, *Triceratops*, and *Apatosaurus*—are now extinct, despite the fact that they dominated life on every continent for more than 160 million years. Although these nonavian dinosaurs were the most successful terrestrial vertebrate ever, they, too, went extinct 65 million years ago at the end of the Cretaceous. (Today, the only remaining dinosaurs are birds, as will be discussed in Chapter 3.) This is because evolution never produces perfection, nor does it necessarily produce "progress." It simply continues to operate.

Natural Selection and Fitness

The effects of natural selection can be measured simply, and empirically, by determining an individual's reproductive success, or **fitness.** That is, the focus of natural selection is not on survival, but rather on reproductive success. Individuals that possess an adaptation are more likely to survive to reproduce. Alternatively, the adaptation may serve a strictly reproductive

SIDEBAR 1.1 The Inevitability of Natural Selection

Natural selection is a simple idea that logically flows from a few observations and their associated conclusions:

1. Organisms produce many offspring.
2. There are finite resources to support these offspring.

   Reproducing at full capacity, any population of any organism will outstrip the available resources within a few generations. This produces a competition for survival and reproduction. This competition is often termed a struggle for existence, yet the struggle is not necessarily overt. For example, two birds could actually fight to secure an ideal nesting site, or one of them might just fly faster and get to the site first. Either way, there are winners and losers in the unavoidable competition.
3. Competing organisms exhibit individual variability.

   This variability (e.g., some birds are stronger or faster than others) may play a role in an individual's success, allowing some individuals to survive to reproductive age. Thus, individuals in a population exhibit differential reproduction due to their varying features, be they structural, physiological, or behavioral.
4. Some of the variability among organisms is heritable.

   Winners are those organisms that survive to reproduce. In doing so, they pass their winning features on to the next generation, altering the genetic composition of the population. This change in gene frequency is evolution by natural selection.

Therefore, if there is competition for resources and differential reproductive success due to heritable variations, natural selection is inevitable.

function. For example, a dark-colored mouse on a dark, lava-rock background will be less susceptible to predation from owls; this means that the mouse's dark coloration is an adaptation that confers a survival benefit. A damselfly with a sperm-gouging device on his genitalia can evict the sperm of rival males from the genitalia of a female damselfly, increasing the male's chance of paternity; this means that the sperm-gouging device is an adaptation that confers a reproductive benefit. In either case, we infer an adaptive advantage only if the individual mouse or dragonfly has more offspring (i.e., is more "fit") relative to others in the population. Differential reproductive success sets the stage for natural selection.

Fitness can be calculated in several ways; as a result, it is not always immediately obvious how a given feature could have resulted from natural selection. A response that increases an individual's own survival and reproductive potential provides a *direct* fitness benefit, while one that increases the survival and reproduction of an individual's relatives provides an *indirect* fitness benefit. The combination of fitness-enhancing features, direct and indirect, combines to form an individual's inclusive fitness.

It is often easy to be confused by features that are adaptive due to an indirect, rather than direct, fitness gain. A parent that allows one of its offspring to die (or that kills its progeny outright) may be increasing the odds of survival for its higher-quality offspring. In this way, infanticide of low-quality individuals leads to an increase in the parent's inclusive fitness. Inclusive fitness is discussed further in Chapters 7 and 9.

Proximate versus Ultimate Causation

In scientific investigations, we distinguish between proximate and ultimate questions and explanations. *Proximate questions* focus on how a phenomenon works. Take the case of the arctic tern, which migrates over 65,000 kilometers each year. *How* does the tern know when to migrate? *How* does it know where to go? *How* do its muscles make its wings move, and *how* can this flight be sustained metabolically in such a small bird? These are proximate questions. The goal of proximate-level investigations is to understand mechanisms and immediate reinforcements. How does something work? What is the immediate reward?

Ultimate questions focus on why a trait exists, rather than one of other plausible alternatives. *Why* do birds migrate? *What* advantage do migratory birds enjoy? *What* fitness gains are associated with this phenomenon? These are ultimate questions, and the goal of ultimate-level investigations is to understand evolution. Why does a trait exist? How does a behavior help an organism exist, reproduce, or both? In a nutshell, how does an adaptation increase an individual's fitness?

Products of Evolution

Although the mechanics of evolution involve mutation, genetic drift, migration, and natural selection, the products of evolution are the seemingly limitless variety of life on earth. Evolution constrains diversity through genetic drift and natural selection against harmful features, and it enhances diversity through mutation and natural selection for beneficial features. Logic and the evidence accumulated thus far indicate that all organisms share common ancestry, and that evolution creates new features, new species, and even entirely new groups of organisms. Many of the predictions associated with evolutionary theory involve shared traits within a lineage of organisms, and the adaptive significance of these shared traits. We develop these predictions in Chapters 2 through 9.

An organism's environment generates selection pressures that favor certain traits over others. Environmental constraints on adaptability can be biotic (involving a community of living organisms) or abiotic (consisting of nonliving variables such as climate). Regardless, the environment is constantly changing in ways both predictable and not. These changes, coupled with the randomness of mutation and genetic drift, as well as the tremendous time-span of life on Earth, contribute to our understanding of evolutionary change. And in spite of the stochastic nature of several tenets of evolution (e.g., mutation, environmental perturbations), several themes emerge that will reappear throughout this book. These themes are discussed in the sections that follow.

Transitional Forms

Transitional forms represent intermediate stages in the evolution of a lineage. The most well-known examples of a transitional form are fossils such as *Archaeopteryx,* which is actually just one of many linkages between reptilian dinosaurs and modern-day birds. More recently, the discovery of *Tiktaalik* in northern Canada was hailed as a **"missing link"** in the transition of fish-like tetrapods from water to land. The term missing link creates an unfortunate demand for evidence, for given the nature of fossilization and the fossil record, any lineage is likely to be riddled with holes in its past progression. However, many lineages—avian, equine, and tetrapod, to name a few—have been well documented with the help of transitional forms. Some of these examples are discussed in Chapter 3.

Complexity

The origin of complexity, including the evolution of sophisticated structures such as the vertebrate eye, or sophisticated behaviors such as learning to play a musical instrument, poses an exciting challenge for biologists. Constructing a picture of, for example, eye evolution, has involved comparative anatomy as well as developmental and molecular biology. Using sophisticated tools from diverse fields creates a lot of room for misunderstanding, as evidenced by the large number of otherwise educated individuals who cite the "argument from design" (i.e., creationism) when confronted with explaining complicated features. The origin of complexity is discussed throughout the book, and is emphasized in Chapters 5 and 8.

Extinction

Extinction, or the loss of an entire taxonomic group, is one of the criteria for the evolution of new forms and serves as a testament to the noncalculating nature of natural selection. Mass extinctions, such as the end-Cretaceous event that included the demise of the nonavian dinosaurs, have been important for the subsequent emergence of many new forms. In addition to the mass extinctions, there is also a baseline rate of ongoing extinctions. Extinction is discussed in Chapter 3.

Artificial Selection

Artificial selection, whereby humans manipulate individual reproductive success to achieve a desired result, has population-level impacts and informs our understanding of natural selection. In Chapter 4 of *On the Origin of Species,* Darwin asks "Can the principle of selection, which we have seen is so potent in the hands of man, apply in nature?" If we can manipulate *Brassica oleracea* to create cauliflower, broccoli, Brussels sprouts, and kale, then it logically follows that natural selection—by operating similarly via differential reproductive success—can alter populations in nature. Examples of artificial selection are addressed in Chapter 7.

Adaptive Radiation

Adaptive radiation describes the origin of new species from a common stock as a result of migration and the colonization of novel habitats. The hundreds of species of cichlid fish in the African Great Lakes are the result of adaptive radiation made possible by extreme specialization. The finches of Galápagos, the honeycreepers of Australia, the fruit flies of Hawaii, and the *Anolis* lizards of the neotropics are well-studied examples of adaptive radiation. Adaptive radiations are discussed in Chapters 3 and 8.

Homologous and Analogous Features

Shared derived-traits (or **homologies**) are features of related organisms that serve as clues to the organisms' past. For example, all mammals possess fur, a shared derived-trait that characterizes this lineage (specifically, the Order Mammalia). A more recent branch on the mammalian tree includes the placental mammals, which share the adaptation of a placenta and are presumed to have split from the older mammal groups (egg-laying and marsupial mammals) millions of years ago. A placenta is a derived trait that is not shared by all mammals. Shared derived-traits are discussed throughout the remaining chapters.

While *homologies* are structures or behaviors that organisms share because of their shared ancestry, *analogies* are structures or behaviors that are similar but *not* due to shared ancestry. Analogous features, such as the mobbing behaviors that disparate prey species exhibit when threatened by a predator, are typically similar solutions to similar environmental pressures and are therefore the result of convergent evolution rather than shared ancestry.

Exaptations

Exaptations are features that evolved in one context but have become used in another. Bird feathers are presumed to be an exaptation—initially they may have been used for insulation or ornamentation, but over time were commandeered for flight. Likewise, the fish's swim bladder, an organ that controls buoyancy in water, was derived from the lung. Exaptations are discussed in Chapter 6.

Vestigial Traits

Vestigial traits are features that are no longer functioning, or exhibit a very reduced or novel function. Vestigial traits, such as the whale's pelvis, hind limbs of many snakes, and body hair of humans, are relicts of the past that tell us what features were adaptive in the history of a lineage. Vestigial traits are discussed in Chapters 3, 5, 6, and 9.

Evolution in Action: The Modern Threat of Antibiotic Resistance

When Charles Darwin proposed evolution by natural selection in *On the Origin of Species,* he believed that evolution typically occurs very slowly, as "the hand of time has marked the lapse of ages." We now know better. Indeed, a variety of studies—such as those by Peter and Rosemary Grant of finches' beaks in Galápagos (Chapter 4)—have shown that evolution can occur remarkably fast, so fast that we can watch it happen. For example,

- Cotton plants in the southeastern United States are eaten by the moth *Heliothis*. A pesticide introduced in 1986 killed 94% of the moths, but within two years (i.e., several moth-generations later) the pesticide killed fewer than 40% of the moths. The moths had evolved resistance to the pesticide. Similar resistance has developed to all major pesticides; this is why farmers must change pesticides every few years.

- For several decades, hunters in Alberta, Canada, killed the biggest bighorn rams. This selection against large bighorn rams would be predicted to favor smaller rams. This is what happened: the frequency of small-horned, small-bodied adult males has increased.

The development of **antibiotic resistance** by bacterial pathogens presents an example of evolution in action with tremendous consequences for humans. To appreciate this, consider penicillin, the first antibiotic to be discovered and made available to the public. When it was introduced in 1943, penicillin was hailed as a "miracle drug" because of its effectiveness in killing pathogens and saving lives. Indeed, the U.S. Surgeon General, believing that penicillin would end infectious disease, proclaimed that it was time "to close the book on infectious diseases." However, it took only three years for reports of penicillin resistance to appear; this resistance resulted from the bacteria's ability to produce enzymes such as β-lactamase that inactivated the drug. By 1992, infections by *Staphylococcus aureus*—which are common after hospital-based surgery—had returned. Today, more than 95% of all strains of *S. aureus* worldwide are resistant to penicillin and related drugs. Yesterday's "miracle drug" is worthless for treating many infections for which it was once effective.

Bacterial resistance can develop quickly as genes for resistance spread in a population. To appreciate this, consider a population of 10,000 bacterial cells, only one of which is resistant to a particular antibiotic. This means that the frequency of the resistant bacterium is 1/10,000 = 0.01%, and that the frequency of susceptible bacteria is 99.99%. If, in the presence of the antibiotic, reproduction of the wild-type (i.e., antibiotic-susceptible) cells is slowed from two divisions per hour to one division per hour, then the wild-type cells will go through 24 generations during the next 24 hours and produce 1.7×10^{11} offspring that are susceptible to the antibiotic. However, if the reproductive rate of the resistant bacterium is unaffected and remains at two divisions per hour, then the resistant bacterium will go through 48 generations and, in the process, produce 2.8×10^{14} resistant cells. As a result, in only one day, natural selection alters the antibiotic-resistant bacteria from 0.01% of the population to 99.99% of the population. Thanks to short generation-times, bacterial resistance to antibiotics can develop quickly.

The evolution-driven uselessness of penicillin is not an isolated example; indeed, similar types of resistance have evolved in response to the introduction of *every* antibiotic. For example,

- Methicillin was introduced to kill *Staphylococcus aureus* in 1959; resistance strains appeared two years later. These were the first cases of methicillin-resistant *Staphylococcus aureus* (MRSA). Whereas in 1974 only 2% of *S. aureus* strains was resistant to methicillin, today almost 70% of *S. aureus* in hospitals is resistant to methicillin and other antibiotics. Treating drug-resistant strains of pathogens is expensive; *S. aureus* infections alone cost more than $30 billion per year.

- Streptomycin was approved by the FDA in 1947; resistant strains appeared the same year.
- Tetracycline was approved in 1952; resistant strains appeared four years later.
- Cephalothin (the first antibiotic in the cephalosporin class) was introduced in 1964; resistance appeared two years later.
- Cefotaxime was introduced in 1981; resistant strains appeared two years later.
- Gentamicin was approved by the FDA in 1967; resistance appeared in 1970.
- Linezolid (the first antibiotic in the oxazolidinone class) was approved by the FDA in 2000; resistant strains appeared the following year.
- In 2010 researchers reported that urinary infections caused by *Escherichia coli* will soon resist *all* known antibiotics.
- The gene for the enzyme New Delhi metallo-β-lactamase (NDM-1), which makes bacteria resistant to a broad range of antibiotics, was detected in *Klebsiella pneumoniae* isolated from a medical patient in 2001. In 2010 a patient hospitalized with a leg injury died from an infection with bacteria expressing this gene. The next year, the gene was found in 1/4 of the water samples taken from drinking supplies and puddles in the streets of New Delhi and it has now spread to Australia, Canada, the United States, Britain, Sweden, and elsewhere. Patients infected with resistant bacteria can be treated only with expensive, highly toxic antibiotics.

The list goes on and on, as documented by such names as VRSA (vanomycin-resistant *Staphylococcus aureus*), VISA (vanomycin-insensitive *Staphylococcus aureus*), and extremely drug-resistant tuberculosis (XDR-TB). These products of evolution pose an increasingly serious public health concern. For example, every year in the United States, MRSA kills more than 20,000 people, a number that exceeds the deaths caused by HIV/AIDS.

The development of antibiotic resistance is evolution in action: the genetic composition of the population of the bacteria changes over time in response to changes in the population's environment (i.e., the introduction of the antibiotics). The antibiotic does not cause resistance to appear; that occurs spontaneously. Moreover, the appearance of resistance is neither good nor bad—certain bacteria are favored only because of the presence of the antibiotic. As evolution would predict, the most resistance is found in places such as hospitals, where antibiotics are used intensively.

And as evolution would also predict, our best efforts to fight pathogens have made the pathogens stronger.

Summary

Evolution in general, and natural selection specifically, is supported by a tremendous amount of evidence from multiple fields of study, including geology, paleontology, anatomy and physiology, geography, molecular biology and genetics, ecology, behavior, and anthropology. Within each discipline, ultimate-level investigations have included predictions that flow logically from the claims of evolutionary theory. For example, within the field of behavior, we can test the predictions that individuals will exhibit behaviors that increase their chances of survival, behave preferentially toward relatives, and exhibit behaviors that help them acquire mates. Within the field of molecular biology, we can test whether ubiquitous proteins exist in disparate taxa, and whether shared ancestry is reflected by the occurrence of vestigial molecular elements (i.e., pseudogenes).

Although it is impossible for us to present all of the major evidence for evolution in this one volume, the following chapters in this book use an evidentiary basis to argue the validity of evolution. For each category of evidence for evolution, we present predictions along with scientific support for each prediction. We hope that this approach will help you appreciate the explanatory power of evolutionary theory.

2

Age of Earth

It is perhaps a little indelicate to ask of our Mother Earth her age, but Science acknowledges no shame and from time to time has boldly attempted to wrest from her a secret which is proverbially well guarded.

—Arthur Holmes, 1913

The [Darwinian] revolution began when it became obvious that the earth was very ancient rather than having been created only 6,000 years ago. This finding was the snowball that started the whole avalanche.

—Ernst Mayr, 1972

Any event that is not absolutely impossible . . . becomes probable if enough time passes.

—George Gaylord Simpson, 1952

PREDICTION

If evolution accounts for the diversity of life on Earth, then we would predict that Earth has a vast age.

All definitions of evolution—for example, that evolution is a change in allelic frequencies in a population over time (Chapter 1)—involve time, and therefore involve history. Although Earth's age is not evidence for evolution *per se,* it does provide a limiting context for all other evidence for evolution because it constrains the development of life on Earth. Few people understood this better than Charles Darwin did as he was developing his ideas for *On the Origin of Species* (1859; see Appendix 1). In Darwin's day, the most widely accepted estimates of Earth's age were those of William Thomson (later Lord Kelvin), who argued that Earth is only 20–100 million years old—an age far too brief to accommodate evolution by natural selection. Darwin, who referred to Thomson as an "odious spectre," understood that the history of life on Earth ultimately relies on geology. Consequently, Thomson's claim about Earth's age was one of Charles Darwin's "sorest troubles," for it did not provide sufficient time for Darwin's theory of evolution by natural selection to produce life's diversity.

Darwin suspected that Earth was much older than Thomson claimed, but Thomson's enormous stature as a scientist obliged Darwin to try to reconcile his claims with Thomson's data. To accommodate Thomson's timeline, Darwin proposed **pangenesis** as an explanation of inheritance (i.e., every sperm and egg contained "gemmules thrown off from each different unit throughout the body"). Although Darwin's explanation sped evolution while avoiding Jean-Baptiste Lamarck's (1744–1829) quasi-spiritual sources of acquired traits, Darwin's explanation of inheritance was wrong. Nevertheless, Darwin continued to suspect that Kelvin was wrong. Even Lamarck suspected that evolution required an old Earth; as he noted in *Hydrogéologie* (1802[1964]), "The great age of the earth will appear greater to man when he understands the origin of living organisms and the reasons for the gradual development and improvement of their organization. This antiquity will appear even greater when he realizes the length of time and the particular conditions which were necessary to bring all the living species into existence."

Our understanding of Earth's age is an interesting story that showcases "science in action"—namely, how new ideas and techniques continually enable scientists to discover new things and refine established principles. The story involves scriptural claims, cooling rates, sedimentation rates, salinity studies, and radioactivity (Table 2.1).

Theories of Earth's Age

Scriptural Claims about Earth's Age

Before the 18th century, claims about Earth's age were based on gods and religious texts. For example, ancient Hindus believed that the universe developed from a golden egg (Hiranyagarbha) that contained the universe and humanity, and produced the supreme god (the Brahman). The ancient Chinese believed that Earth's history is cyclical, whereas the Mayans believed that there have been several creations, with the most recent being several thousand years ago. These creation stories, like more modern ones, allowed the inexplicable to be reasoned without much thought. Throughout history, however, many people have questioned such claims about Earth's age; as Carus Titus Lucretius (ca. 95–55 B.C.E.) noted just two years before committing suicide, "the question troubles the mind with doubts, whether there was ever a birth time of the world and whether likewise there is to be any end."

Table 2.1 Estimates of Earth's Age

Claimed Age of Earth (in years)	Date of Claim	Advocate	Basis of Claim
1,972,949,091	120–150 B.C.E.	Hindu priests	Religion
5,698	169	Theophilus	Biblical chronology
6,321	Fifth century	St. Augustine	Biblical chronology
5,918	1644	John Lightfoot	Biblical chronology
5,994	1650	James Ussher	Biblical chronology
5,983	ca. 1620	Johannes Kepler	Movement of solar apogee
>2 billion	1748	Benoît de Maillet	Decline of sea level
75,000	1774	Comte de Buffon	Cooling of Earth
38–96 million	1860	John Phillips	Sediment accumulation
20–400 million	1862	William Thomson	Cooling of Earth
10–500 million	1862	William Thomson	Cooling of Sun
100 million	1869	Thomas Huxley	Sediment accumulation
>56 million	1879	George Darwin	Orbital period of the moon
28 million	1892	Alfred Wallace	Sediment accumulation
80–90 million	1899	John Joly	Sodium accumulation in ocean
>1.3 billion	1917	Arthur Holmes	Cooling of Earth
1.6-3.9 billion	1927	Arthur Holmes	Radiometric dating
3.4 billion	1929	Ernest Rutherford	Radiometric dating
4.55 billion	1956	Claire Patterson	Radiometric dating

Modified from G. B. Dalrymple, *The Age of the Earth* (1991), Stanford, CA: Stanford University.

The first person to use the Bible as a scholarly foundation for geologic chronology was Syrian Saint Theophilus of Antioch (ca. 115–180), who in 169 claimed that Earth was created in 5529 B.C.E. Since Theophilus' time, many scholars and other theologians have used the Bible to establish a date for creation. For example, Benedictine monk Venerable Bede (672–735) claimed that creation occurred in 3952 B.C.E., German astronomer Johannes Kepler (1571–1630) claimed that creation occurred in 3993 B.C.E., and Martin Luther (1483–1546)—who launched the Reformation in 1517 when he posted the *95 Theses* on the door of the Wittenberg Cathedral—in 1541 claimed that creation occurred in 3961 B.C.E. By the mid-1700s, virtually all of the more than 200 Bible-based computations of Earth's age set creation near 4,000 B.C.E.

The most famous Bible-based chronology was proposed in 1650 by Irish scholar and archbishop James Ussher (1581–1656), who used ancient texts and extensive biblical research to produce his 2,000-page *Annals of the World.* Although the Hebrew Bible describes a direct, unbroken lineage from Creation to Solomon, Ussher used additional sources for dates between Solomon and the destruction of the Temple (e.g., the death of Chaldean King Nebuchadnezzar in 562 B.C.E.). Ussher also made several assumptions, such as that creation must have occurred near the autumnal equinox because that date corresponded to the harvest time of fruits growing in the Garden of Eden. Ussher accepted the ages of biblical patriarchs listed in the Bible—for example, that Adam lived 930 years, that Seth lived 912 years, and that Methuselah (Adam's great-great-great-great-great-grandson) lived 969 years.

Ussher calculated the dates of numerous events, including the Great Flood (2348 B.C.E.), the Exodus from Egypt (1491 B.C.E.), the founding of the Temple in Jerusalem (1012 B.C.E.), the destruction of Israel by Babylon (586 B.C.E.), the birth of Jesus (4 B.C.E.), Adam and Eve being driven from Paradise (Monday, November 10, 4004 B.C.E.), and Noah's ark landing at Ararat (Wednesday, May 5, 2348 B.C.E.). However, Ussher remains most famous for his calculation of the days of creation. Apparently not a believer in building suspense, Ussher stated his famous conclusion in *Annals'* first paragraph (Figure 2.1):

> In the beginning God created Heaven and Earth, Gen. 1, v. 1. Which beginning of time, according to our chronologie, fell upon the entrance of the night preceding the twenty third day of Octob[er], in the year of the Julian [Period] 710. The year before Christ 4004. The Julian Period 710.

Ussher's calculation that creation began on the evening of Saturday, October 22, 4004 B.C.E., became the most famous creation-date when it was included in annotated versions of the King James Version of the Bible in the late 1600s. In 1701 the Church of England—at the urging of Oxford cleric John Feel and William Lloyd (Bishop of Winchester)—began listing Ussher's dates throughout Genesis in its official Bibles; Ussher's dates remained in these Bibles for almost three centuries (Figure 2.2). Later, some of Ussher's dates were also included in the influential *Scofield Reference Bible,* as well as in Bibles placed by Gideons International in hotel rooms. (Ussher's date also appeared elsewhere, including in Shakespeare's *As You Like It,* in which Rosalind laments, "The poor world is almost six thousand years old.") The Bible introduced into evidence at the famous Scopes "Monkey Trial" in 1925 included Ussher's chronology, which was cited during Clarence Darrow's questioning of William Jennings Bryan. In that testimony, Bryan famously invoked day-age creationism by claiming that "it would

Anno Mundi | I | Anno Periodi Julianæ | Anno ante æram Chriſtianam

ANNALES

VETERIS TESTAMENTI,

à primâ Mundi origine deducti.

1 a. IN PRINCIPIO creavit DEUS Cœlum & Terram. [*Geneſ.* I. 1] quod temporis principium (juxta noſtram Chronologiam) incidit in noctis illius initium, quæ XXIII. diem Octobris præceſſit, in anno Periodi Julianæ 710. 710 4004

Primo igitur ſeculi die (*Octob*.23. *feriâ* 1.) cum ſupremo Cœlo creavit Deus Angelos: deinde ſummo operis faſtigio primùm perfecto, ad ima Mundanæ hujus fabricæ fundamenta progreſſus mirandus artifex, infimum hunc globum ex Abyſſo & Terrâ conflatum conſtituit, concinentibus & collaudantibus eum ſimul omnibus ipſius An-

Figure 2.1 James Ussher's claim in *Annals of the World* (1650) that the Christian God had created heaven and Earth in 4004 B.C.E. is the most famous date associated with **young-Earth creationism**. Ussher announced his claim on the book's first page, which included three columns. The column on the left was the year of the world; these dates began at year 1. The inside-right column listed dates according to the Julian calendar, which starts at 710. The outside column, which listed "the year before Christ," listed Ussher's famous date (arrowhead). (J. Ussher, 1650, *Annals of the World: James Ussher's Classic Survey of World History*)

be just as easy for the kind of God we believe in to make the Earth in six days as in six years or in 6,000,000 years or in 600,000,000 years. I do not think it important whether we believe one or the other."

Ussher's *Annals* was one of the most influential books of the seventeenth century, and it represented some of the best scholarship of its time; Ussher had a personal library of over 10,000 books, and *Annals* included more than 10,000 footnotes. Although Ussher's chronology has been abandoned by most Judeo-Christian religions and seldom appears in modern Bibles, his creation date of 4004 B.C.E. remains the foundation of young-Earth creationism. Advocates of young-Earth creationism, who base their claims on

THE FIRST BOOK OF MOSES,

CALLED

GENESIS.

Year before the common Year of CHRIST, 4004. – Julian Period, 0710. – Cycle of the Sun, 0010. – Dominical Letter, B.
Cycle of the Moon, 0007. – Indiction, 0005. – Creation from Tisri, 0001.

[1] CHAPTER 1

1 *The creation of heaven and earth.* 14 *Of the sun, moon, and stars.*
26 *Of man in the image of God.* 29 *Also the appointment of food.*

IN the [a]beginning [b]God created the heaven and
the earth.
2 And the earth was without form, and void; and
darkness *was* upon the face of the deep: [c]and the
Spirit of God moved upon the face of the waters.
3 ¶ [d]And God said, [e]Let there be light: and there
was light.
4 And God saw the light, that *it was* good: and
God divided † the light from the darkness.
5 And God called the light [f]Day, and the darkness he called Night: † and the evening and the
morning were the first day.
6 ¶ And God said, [g]Let there be a † firmament in
the midst of the waters: and let it divide the waters
from the waters.

Before CHRIST 4004.

a John 1. 1, 2. Heb. 1. 10. *b* Ps. 8. 3. & 33. 6. & 89. 11, 12. & 102. 25. & 136. 5. & 146. 6. Isa. 44. 24. Jer. 10. 12. & 51. 15. Zech. 12. 1. Acts 14. 15. & 17. 24. Col. 1. 16, 17. Heb. 11. 3. Rev. 4. 11. & 10. 6. *c* Ps. 33. 6. Isa. 40. 13, 14. *d* Ps. 33. 9. *e* 2 Cor. 4. 6. † Heb. *between the light and*

Before CHRIST 4004.

s Jer. 31. 35. ‖ Or, *creeping.* † Heb. *soul.* † Heb. *let fowl fly.* † Heb. *face of the firmament of heaven.* *u* ch. 6. 20. & 7. 14. & 8. 19. Ps. 104. 26. *w* ch. 8. 17.

18 And to [s]rule over the day, and over the night,
and to divide the light from the darkness: and God
saw that *it was* good.
19 And the evening and the morning were the
fourth day.
20 ¶ And God said, Let the waters bring forth
abundantly the ‖ moving creature that hath † life,
and † fowl *that* may fly above the earth in the † open
firmament of heaven.
21 And [u]God created great whales, and every living
creature that moveth, which the waters brought forth
abundantly after their kind, and every winged fowl
after his kind: and God saw that *it was* good.
22 And God blessed them, saying, [w]Be fruitful
and multiply, and fill the waters in the seas, and let
fowl multiply in the earth.
23 And the evening and the morning were the
fifth day.

Figure 2.2 Ussher's date of creation became famous when it was included in Bibles. This Bible, published by Oxford University Press in the mid-1800s, included Ussher's famous dates.

religious faith rather than scientific evidence, reject not only evolution but also many basic tenets of modern geology, paleontology, physics, and other sciences. As will be shown later in this chapter and throughout this book, these claims of a "young Earth" are overwhelmingly refuted by evidence from modern science.

Using Seas and Salt to Estimate Earth's Age

The first person to produce a science-based estimate of Earth's age was French anthropologist and diplomat Benoît de Maillet (1656–1738), who used measurements of the decline of sea level to claim that Earth is more than 2 billion years old. De Maillet, who did not try to reconcile geology with the Bible, suspected from his extensive travels and geological observations that Earth must be much older than the few thousand years proposed by Biblical chronologers, but he was also aware of the Church's prosecutions (and killings) of anyone making claims contrary to religious dogma. To avoid problems with the Church, de Maillet described his ideas in a fictitious series of discussions (held over six days) between a French missionary and an Indian philosopher named Telliamed (de Maillet spelled backward). Speaking through Telliamed, de Maillet claimed that some ancient towns now as high as 1,800 meters above sea level had originated as seaports and that these towns—assuming a rise in elevation of 7.5 centimeters per century—must be approximately 2.4 billion years old.

De Maillet also described mermaids and men with tails, claimed that elephant seals are ancestors of elephants, and believed that petrified wood is the remains of ships' masts. De Maillet's manuscript—*Telliamed: or Conversations between an Indian Philosopher and a French Missionary on the Diminution of the Sea* (1748)—was advertised by a Baltimore publisher as "a very curious book" and was not published until a decade after his death. De Maillet's assumptions were wrong, as were his conclusions.

Seas were again used to estimate Earth's age in the late 1800s, when Irish geologist John Joly (1857–1933) speculated that when the oceans originally formed, they contained freshwater, and that the oceans' present salinity was due to salts leaching from rocks. Joly assumed that the rate of leaching had been constant, and therefore that the age of Earth's first ocean equaled the mass of sodium in the oceans divided by the annual rate of sodium input into the oceans. This calculation—using a method originally proposed in 1715 by English astronomer Edmond Halley to show evidence "of the Sacred Writ [that] Mankind has dwelt about 6,000 years"—produced an age of 89 million years, which Joly later adjusted to between 90 and 100 million years (in 1930, Joly again modified his estimate to 300 million years). Joly's calculation—first announced in 1899 in his classic paper "An Estimate of the Geological Age of the Earth"—was embraced by some scientists but rejected by others. Today, we know that the processes invoked by Joly were not uniform throughout geological time, and that the assumptions and numbers used in his calculations were erroneous. For example, the oceans did not form at the same time, the oceans have not become progressively saltier, and Joly's calculation did not consider salts washed into soils and sediments.

Sedimentation Rates as a Measure of Time

In the mid-1600s, Danish scientist Nicholaus Steno[1] (1638–1686) found that particles of sand, mud, and gravel settle from water according to their relative weight, and that slight changes in the size or composition of the particles produce layers in rocks. These layers, which characterize sedimentary rocks, are called strata. In undisturbed rocks, **strata** are stacked atop each other like a stack of books, and the order of the strata reflects the order in which the strata were deposited (Figures 2.3a and 2.3b). Thus, strata at the bottom were deposited before those closer to the top, thereby meaning that each layer represents a unit of time. Steno's use of geometry to infer time was a major advancement in geology, and is known as the **Principle of Superposition**.

1. Steno's actual name was Nils Stensen. His Latinized name was Nicolaus Stenonis, and his Anglicized—and most commonly cited—name was Nicholas Steno.

Figure 2.3a Geology of the Grand Canyon, which has been formed over millions of years by erosion caused by the Colorado River. Many sediments in the Grand Canyon are thin, with several being required to span the height of a U.S. quarter. (Randy Moore)

Figure 2.3b These thin sediments are often aggregated into thicker sediments, as shown in this photograph taken of the Grand Canyon's wall from the Colorado River. (Randy Moore)

Figure 2.3c On a larger scale, sediments formed by different types of rocks create the spectacular walls of the Grand Canyon. This view is from Nankoweap. (Mike Quinn, NPS, U.S. Geological Survey)

Name (top to bottom)	Era	Period	Age (mya)
Kaibab Limestone	Paleozoic	Permian	270
Toroweap Formation			
Coconino Sandstone			
Hermit Shale			
Supai Group		Pennsylvanian	310
Redwall Limestone		Mississippian	350
Muav Limestone		Cambrian	600
Bright Angel Shale			
Tapeats Sandstone			
Vishnu Schist	Proterozoic	Pre-Cambrian	>600

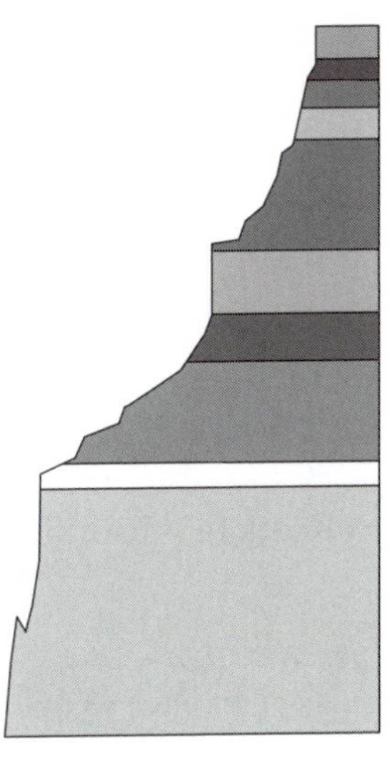

Figure 2.3d Strata at the base of the Grand Canyon are older than strata near the surface. (Jim Cotner)

Today, Steno's Principle of Superposition is a foundation of geology, and it establishes the relative—but not the absolute—ages of rocks. More recent methods of determining the absolute ages of rocks (e.g., radiometric dating; see below) have confirmed the Principle of Superposition—namely, that in undisturbed sediments, the deeper strata *are* older than those nearer Earth's surface. To appreciate this, consider the geology of the Grand Canyon and its surrounding region (Figures 2.3c and 2.3d). The oldest rocks of the canyon—the Vishnu Basement Rocks—are at the bottom of the canyon and are 1.84–1.68 billion years old. Sediments are progressively younger as you move up the sides of the canyon, and the youngest rocks—that is, the 270-million-year-old Kaibab limestone—form the canyon's rim (erosion has stripped away even younger rocks from the rim). At the nearby canyons of Antelope Valley, the Kaibab limestone forms the bottom of the canyons, and it is far below the bottoms of Zion and Bryce Canyons.

Steno's Principle of Superposition has been confirmed throughout the world, even by some young-Earth creationists. For example, in 1938, famed young-Earth creationist Harold Clark (1891–1986) visited oil fields in Texas and Oklahoma, where he examined cores mined by drillers. Clark saw first-hand that "the rocks do lie in a much more definite sequence [than young-Earth creationists] have ever allowed," and concluded that the "flood geology" that underlies young-Earth creationism (i.e., that Earth's geological features are due to a worldwide flood) "does not harmonize with the conditions in the field." The different strata also contain distinctive fossils, and can therefore be used to describe the history of life on Earth (see discussion in Chapter 3). Geologists use Steno's principle to find oil, gas, coal, uranium, and other economically important rocks and minerals.

Steno, whose work established the science of stratigraphy, was the first person to use geologic strata to deduce Earth's history. Although Steno abandoned science for religion a few years after making his monumental discovery,[2] others began using Steno's ideas to estimate Earth's age. These studies were based on the assumption that if one can estimate the thickness of a modern sediment, and if one knows the rate at which the sediment was deposited, then one can calculate the time that the sediment has been forming. The first person to do this—that is, use geologic strata as an hourglass—was English geologist John Phillips (1800–1874), who claimed that it took between 38 and 96 million years to form the Cambrian and later sediments. People who subsequently used sedimentation rates to estimate Earth's age included Thomas Huxley (100 million years in 1869), Charles Walcott (45–70 million years in 1893), Alfred Russel Wallace (28 million

2. Steno later recanted his work because it suggested that Earth is far older than 6,000 years. However, his recantation made little difference because he had already published his findings in a short abstract titled *Prodromus*. Steno is the only geologist to ever be made a saint.

years in 1880), Alexander Winchell (3 million years in 1883), and John Joly (80 million years in 1909). Today, however, we know that this method is not reliable because, among other things, one cannot know deposition rates or the original thickness of a sediment (e.g., some of the sediment might have been eroded away).

Geological Formations as Evidence of Earth's History

By the end of the 1700s, some scientists were questioning James Ussher's claims about Earth's age, and many of their questions were based on rather simple but insightful geologic observations. For example, in 1788 Scottish geologist James Hutton visited Siccar Point along the Scottish coast near the border with England (Figures 2.4a and 2.4b). Sediments at the base of Siccar Point are vertical and, because sediments are originally deposited only horizontally, Hutton knew that these sediments had been tilted vertically and raised above land by geological activity. Erosion then wore away the above-ground parts of the vertical sediments, after which they were again submerged and covered by new horizontally deposited sediments. Hutton concluded from his observations at Siccar Point and elsewhere (e.g., Hadrian's Wall) that Earth's age was vast, "indefinite," and beyond comprehension, and that geological formations were best explained by common events such as rain, erosion, and wind. This constancy of natural laws, which Hutton summarized by noting that "in examining things present we have data from which to reason with regard to what has been," became known as **uniformitarianism.**

Hutton's uniformitarianism was extended by Charles Darwin's friend Charles Lyell (1797–1875), who concluded that Earth has been in a perpetual directionless flux for an inconceivable period. Like Hutton, Lyell rejected **catastrophism** (i.e., that Earth's geology is due to catastrophic events such as worldwide floods), and instead attributed geologic formations to "the slow agency of existing causes," noting that "the present is the key to the past." In 1828, Lyell was particularly impressed when he visited the Temple of Serapis near Pozzuoli, Italy (Figure 2.5). The marble pillars of the temple were carved almost 2,000 years ago and set into the building, which was a Roman marketplace. Since then, geological events have raised and lowered the Temple. Lyell knew this when he noted that the three remaining upright pillars were marked by a dark 3-meters-wide band produced by the marine bivalve *Lithodomus* that had bored into each pillar (many of the holes still have shells in them). Clearly, the original temple had been built above sea level, but the presence of bivalves on the pillars meant that the pillars had been partially submerged in the ocean. The pillars were later raised to their present position by the geologic activity that produced Monte Nuovo just northwest of Pozzuoli. Lyell made the temple's columns an icon of uniformitarianism when he included a sketch of them as the frontispiece of his monumental *Principles of Geology.* The full title of Lyell's

Figures 2.4a and 2.4b Siccar Point, where James Hutton glimpsed the "abyss of time," is arguably the most famous geological site in the world. The vertical sediments are Silurian greywacke, a gray sedimentary rock that is 425 million years old. Approximately 80 million years later—more than 10,000-times longer than all of Ussher's proposed history of Earth—erosion of the nearby Caledonian mountains produced the reddish sandstone sediments (of the Devonian) that overlaid the vertical greywacke (of the Silurian). Hutton's observations at Siccar Point and elsewhere helped him open geology to scientific observation, remove it from the influence of Bible-based chronologies, and establish geology's most transforming contribution to human knowledge: deep time. For scale, note the two people near the center of Figure 2.4a. (Randy Moore)

Figure 2.5 The Temple of Serapis. These columns became an icon for uniformitarianism when famed geologist Charles Lyell included them in *Principles of Geology*, the most influential geology book in history. (Randy Moore)

book—*Principles of Geology, being an Attempt to Explain the Former Changes of the Earth's Surface by Reference to Causes now in Operation*—emphasized Lyell's claim that the geological changes that have shaped Earth for millennia are observable today.

Lyell, Hutton, and others knew that Earth is much older than a few thousand years. But how much older, and how do we know?

Earth's Age as a Function of Temperature

Some of the earliest estimates of Earth's age assumed that Earth was once a molten ball that was in the process of cooling. For example, in 1687, Isaac Newton suggested in his *Philosophiae naturalis principia mathematica* that "a red hot iron equal to our earth, that is, about 40,000,000 feet in diameter, would scarcely cool . . . in above 50,000 years." In the 1700s, Georges-Louis Leclerc (Comte de Buffon; 1707–1788)—who owned an iron factory—tested this idea by heating 10 iron spheres (ranging in diameter from 1 to 13 centimeters) in a forge and recording how long it took "4 or 5 lovely, sweetly complexioned ladies" to determine that the spheres had cooled to ambient temperature. Despite censures from the church, Buffon did not use the Bible as a basis for his claims about Earth's history. When Buffon extrapolated his observations to a sphere the size of Earth, he concluded that Earth was 74,832 years old.

Nearly a century later, Buffon's work was revisited by William Thomson, who rejected Lyell's claims about uniformitarianism. Although some of Thomson's estimates of Earth's age were based on the age of the sun and tidal friction, his most influential claims about Earth's age revisited Buffon's experiments involving the cooling rates of spheres. In a one-paragraph paper published in 1862, Thomson used measurements of the conductivity of various types of rocks to conclude that Earth is 20 to 400 million years old, with 98 million years the most likely age. By 1868, Thomson had concluded that Earth is less than 100 million years old, and in 1897—in a paper titled "The Age of the Earth as an Abode Fitted for Life"—he argued that Earth is 20 to 40 million years old.

Thomson's stature in the scientific community was enormous; his *Treatise on Natural Philosophy* helped define modern physics, he helped engineer the first transoceanic cable, and in 1848 he developed the Kelvin temperature scale. This stature gave immense credibility to Thomson's claims about Earth's age; even Mark Twain noted that "we must yield to [Thomson] and accept his view." However, some scientists continued to suspect that Earth was older—*much* older—than Thomson claimed. For example, in 1869, Thomas Huxley—who, like Darwin, believed that the geological and fossil record (Chapter 3) indicated that Earth was much older than 100 million years—attacked Thomson's thermodynamics-based data and assumptions, noting that "pages of formulae will not get a definite result out of loose data." Charles Darwin was also skeptical; as his son George—a pioneer in the field of geophysics—noted in a letter to Thomson in 1878 (four years before Charles Darwin's death), "[my father] cannot quite bring himself down to the period assigned by you, but does not pretend to say how long may be required."

There was something significant that Kelvin did not know: Earth and other planets formed from clouds of dust that included radioactive atoms, which release heat during radioactive decay and, in the process, heat the Earth. Kelvin did not account for this input of energy in his calculations of Earth's age. The discovery of radioactivity in 1896 (by Henri Becquerel, and so named a few years later by Marie and Pierre Curie) eventually rendered Kelvin's conductive-cooling calculations obsolete, and forever disproved claims of a young Earth.

Radioactivity and Radiometric Dating

Radioactivity

Radioactivity is caused by differences in atomic structure. Therefore, to understand radioactivity, we need to review some basic aspects of atoms. All atoms are made of the same three parts:

The number of *electrons* determines the chemical reactivity of the atom. Electrons have an inconsequential mass and have no role in the radioactivity or identity of the element.

The number of *protons* identifies the element. For example, if any atom has six protons, it is carbon, regardless of the number of other particles comprising the atom. Changing the number of protons produces a new element; for example, removing a proton from nitrogen (which normally has seven protons) transforms the atom into carbon (which has six protons). Protons have no roles in the chemical reactivity of an atom.

The number of *neutrons,* along with the number of protons, accounts for atomic mass. An atom of carbon normally has six protons and six neutrons, giving it an atomic mass of 12. These atoms are described as Carbon-12, or C-12. The number of neutrons can vary in atoms of a particular element. For example, atoms that have eight neutrons and six protons are carbon (because *any* atom having six protons is carbon), but have an atomic mass of 14 (i.e., they are C-14).

Atoms are radioactive because they contain an unstable number of protons and neutrons. For example, C-12 is not radioactive because its number of protons and neutrons is stable. However, C-14—which contains two additional neutrons—is unstable (i.e., radioactive). These unstable arrangements spontaneously change (i.e., decay) to more stable arrangements and, in the process, release energy that we call radioactivity.

Radiometric Dating

Radiometric dating uses naturally occurring radioactivity as a clock, and is based on the fact that some elements can exist in different forms called **isotopes.** Of the 339 isotopes of 84 elements found in nature, 269 are stable and 70 are unstable (i.e., radioactive). When these unstable isotopes become more stable by spontaneously releasing energetic rays or particles, they become a different element or a different isotope of the same element *at a characteristic, unalterable rate.* For example, radioactive potassium (K-40, which has 19 protons and 21 neutrons) spontaneously releases energy as it undergoes radiometric decay and becomes argon (Ar-40, which has 18 protons and 22 neutrons):

$$\text{K-40} \rightarrow \text{Ar-40} + \text{energy}$$

This transformation is called *radioactive decay.* Radioactive decay is exponential, meaning that the rate of decay does not involve fixed amounts of atoms, but instead involves fixed *proportions* of atoms. After one **half-life,** half of the original radioactive atoms are present. This rate remains constant regardless of how many atoms have already decayed. As a result, geologists can use radioactive decay as a natural clock because it is irreversible, has a starting point, has an ending point, and proceeds at a constant, measurable rate that is unaffected by chemical and physical factors such as pressure, temperature, magnetic fields, and humidity.

Radiometric dating, which was proposed by Ernest Rutherford (1871–1937) in 1905, is based on a group of techniques that use the spontaneous decay of long-lived, naturally occurring isotopes to date meteorites and rocks. Simple radiometric techniques rely on the accumulation over time of "daughter" atoms from "parent" atoms in closed systems. In the example cited above, a parent atom of K-40 decays to produce a daughter atom of Ar-40. Because it takes approximately 1.3 billion years for half of a sample of K-40 to decay to Ar-40, the half-life of K-40 is 1.3 billion years.

Radioactive decay is a statistical process; it occurs at a predictable rate, but it is not possible to determine when a particular atom will decay. Although this uncertainty may seem problematic, it is not. Indeed, if given a large number of radioactive atoms, the average decay rate is highly predictable. And the populations of atoms in samples *are* large—for example, a mere 0.00001 g of potassium contains about 150,000 *trillion* (1.5×10^{17}) atoms. These large numbers of atoms make the statistical uncertainty of radiometric dating exceedingly small.

What Isotopes Occur in Nature?

Some isotopes are produced continually by natural processes. For example, carbon-14 is continually produced by interactions of cosmic rays with atoms in the atmosphere, radium-226 is continually produced as a decay product of uranium-238, and radon-222 is continually produced as a decay product of radium-226. When we disregard those isotopes that are continually produced by natural processes, we are left with only 18 that have persisted since the formation of Earth. Each of these 18 isotopes has a half-life greater than 80 million (i.e., 8.0×10^7) years (Table 2.2). This means that none of these isotopes, in our region of the solar system, has a half-life *less* than 80 million (8.0×10^7) years (Table 2.3). To appreciate why this is important, consider that after only five half-lives, almost 97% of the radioactivity is gone. After another 5 to 15 half-lives, virtually all of the radioactivity is gone. Because an isotope decays itself out of existence (i.e., becomes undetectable) after 10 to 20 half-lives, the absence of isotopes having a half-life less than 80 million years in our region of the solar system means that the solar system must be older than 80 million years. Multiplying 80 million times 10 (or 20) gives us about 800 million (or 1.6 billion) years. That is, for these isotopes to have disappeared from nature (i.e., to have decayed through 10 to 20 half-lives), Earth must be *at least* 80 million years old. But knowing that Earth is *at least* 80 million years old doesn't answer the question: How old *is* Earth?

SIDEBAR 2.1 Radiocarbon Dating

Carbon is part of all living organisms. Most carbon in Earth's atmosphere is C-12, which is not radioactive. However, approximately one-in-a-trillion carbon atoms in the atmosphere are radioactive C-14, which form when cosmic rays hit nitrogen atoms. Organisms do not distinguish C-14 from C-12, so when they eat (e.g., animals) or photosynthesize (e.g., plants), they incorporate C-14 into their bodies. This means that as you read this, your body—as well as those of all other living organisms on Earth—contains the same proportion of C-14 as does the atmosphere. As long as organisms stay alive, that proportion remains unchanged. However, when organisms die, they stop incorporating carbon into their bodies because they are no longer eating food or absorbing carbon dioxide from the atmosphere for photosynthesis. This "zeroes" their carbon clock because, at that point, the proportion of C-14 begins to decline due to radiometric decay.

The radioactive decay of C-14 to N-14 has a half-life of 5,730 years; this means that it can be used to date objects up to about 50,000 years old (approximately 10 half-lives). For example, if the bones of a rodent have only one-fourth of the expected ratio of C-14 to C-12, the C-14 in the rodent has undergone two half-lives (i.e., $1/2 \times 1/2 = 1/4$). This means that the rodent died about 11,460 years ago (i.e., 5,730 years per half-life × 2 half-lives = 11,460 years).

Creationists often note that radiocarbon dating cannot be used to accurately date geologic sediments. They are correct; those sediments are far older than 50,000 years. This is also why geologists do not use C-14 for such measures.

Using Radioactivity to Determine the Ages of Rocks

Here's how scientists use radioactivity to determine the ages of rocks. Igneous (Latin for "fire") rocks have solidified from molten rock (e.g., granite from magma, basalt from lava). When these rocks form, they incorporate into themselves various elements from their surroundings. Some of these elements are radioactive, meaning they emit radiation (i.e., energetic rays or particles) as they decay and become more stable. For example, volcanic rocks seldom contain Ar, but often trap K-40 as they cool. After cooling, however, the rocks are closed systems that do not incorporate any more K-40. When this trapped K-40 decays into Ar-40 inside the cooled rock, the original "parent" element (K-40) is trapped in the rock, as is the "daughter" element (Ar-40). We can count both elements and, based on the ratio of K-40 to Ar-40, determine the age of the rock. We're able to do this because each parent-daughter pair is an independent clock in which atoms of the parent are transformed into atoms of its daughter at a constant rate. Table 2.4 lists some of the parent-daughter isotopes (and their known half-lives) that have been used to determine the ages of numerous moon rocks, tens

Table 2.2 Isotopes That Have Persisted Since the Formation of Earth. All of these isotopes have a half-life greater than 80 million (i.e., 8.0 x 10^7) years.

Element	Isotope	Half-Life (years)
Vanadium	^{50}V	6.0×10^{15}
Neodymium	^{144}Nd	2.4×10^{15}
Hafnium	^{174}Hf	2.0×10^{15}
Platinum	^{192}Pt	1.0×10^{15}
Indium	^{115}In	6.0×10^{14}
Gadolinium	^{152}Gd	1.1×10^{14}
Tellurium	^{123}Te	1.2×10^{13}
Platinum	^{190}Pt	6.9×10^{11}
Lanthanum	^{138}La	1.12×10^{11}
Samarium	^{147}Sm	1.06×10^{11}
Rubidium	^{87}Rb	4.88×10^{10}
Rhenium	^{187}Re	4.3×10^{10}
Lutetium	^{176}Lu	3.5×10^{10}
Thorium	^{232}Th	1.40×10^{10}
Uranium	^{238}U	4.47×10^{9}
Potassium	^{40}K	1.25×10^{9}
Uranium	^{235}U	7.04×10^{8}
Plutonium	^{244}Pu	8.2×10^{7}

Table 2.3 There Are No Isotopes in Our Region of the Solar System That Have a Half-Life Less Than 80 Million (i.e., 8.0 x 10^7) Years

Element	Isotope	Half-Life (years)
Samarium	^{146}Sm	7.0×10^{7}
Lead	^{205}Pb	3.0×10^{7}
Curium	^{247}Cm	1.6×10^{7}
Hafnium	^{182}Hf	9×10^{6}
Palladium	^{107}Pd	7×10^{6}
Cesium	^{135}Cs	3.0×10^{6}
Tellurium	^{97}Tc	2.6×10^{6}
Gadolinium	^{150}Gd	2.1×10^{6}
Zirconium	^{93}Zr	1.5×10^{6}
Tellurium	^{98}Tc	1.5×10^{6}
Dysprosium	^{154}Dy	1.0×10^{6}

of meteors, and thousands of rocks. Scientists have more than 40 different techniques for radiometric dating. Table 2.5 shows some of the most common of those methods.

The K-Ar method described previously is one of the most common methods used by geologists to radiometrically date rocks. Because Ar-40 is an inert gas that does not combine with any other element, it escapes easily from **igneous rocks** when they are heated, thereby resetting the clock;

Table 2.4 Parent–Daughter Isotopes

Parent Isotope	→	Daughter Isotope	Half-Life (millions of years)
Potassium (^{40}K)	→	^{40}Ar	1,250
Rubidium (^{87}Rb)	→	^{87}Sr	48,800
Samarium (^{147}Sm)	→	^{143}Nd	106,000
Lutetium (^{176}Lu)	→	^{176}Hf	35,900
Rhenium (^{187}Re)	→	^{187}Os	43,000
Thorium (^{232}Th)	→	^{208}Pb	14,000
Uranium (^{235}U)	→	^{207}Pb	704
Uranium (^{238}U)	→	^{206}Pb	4,470

Table 2.5 Radiometric Dating Methods

Method	Parent/ Daughter Isotopes	Half-Lives (years)	Materials Dated	Age Dating-Range
Carbon (C)/ Nitrogen (N)	C-14 → N-14	5,730	shells, limestone, organic materials	100–50,000 yrs
Potassium (K)/ Argon (Ar)	K-40 → Ar-40	1.3 billion	biotite, volcanic rocks	100,000–4.5 billion
Rubidium (Rb)/ Strontium (Sr)	Rb-87 → Sr-87	49 billion	micas	10 million–4.5 billion+
Uranium (U)/ Lead (Pb)	U-238 → Pb-206	4.5 billion	zircon	10 million–4.5 billion+
Uranium (U)/ Lead (Pb)	U-235 → Pb-207	704 million	zircon	10 million–4.5 billion+

this is why the K-Ar method is best used to date igneous rocks not heated since forming. Conversely, the K-Ar method is not well suited for estimating the ages of sedimentary and metamorphic rocks because these rocks have sometimes been heated and are composed of other rocks (or debris from older rocks). However, this concern is alleviated with **isochron** dating, which is a self-correcting dating method that involves measuring—in several different objects, such as rocks that include different minerals—the amount of a different isotope of the same element as the daughter product of radioactive decay. A statistical analysis of the data reveals an age (i.e., the slope of the best-fit line) and the uncertainty in the age (i.e., the variance in the slope).

The most direct method for estimating Earth's age is a lead/lead isochron method using meteorites and ancient sediments. Lead isotopes are ideal for radiometric dating because two different isotopes of lead (Pb-207 and Pb-206) are produced by the decay of two different isotopes of uranium (U-235 and U-238, respectively). These isotopes of lead are then normalized to Pb-204, which has remained constant through history because it is not a product of radioactive decay (i.e., Pb-204 is a stable isotope). Since the decay pathways of both isotopes of uranium contain unique intermediates and have different half-lives, scientists can use this information to make independent calculations of age.

If Earth and meteorites were not formed at the same time, there would be no reason for their Pb-isotopic compositions to lie along the same line (i.e., isochron). However, if the solar system was formed from a common pool of matter in which the lead isotopes of the meteorites and ancient sediments were distributed uniformly, then the initial plots for all objects created from that pool of matter would fall on a single line. This is, in fact, what occurs. These and other measurements have enabled scientists to know that the meteor associated with the extinction of dinosaurs hit Earth approximately 65.5 million years ago, and that the Permian-Triassic mass extinction—which occurred 252.5 million years ago—coincided with the Siberian Traps eruptions, the largest volcanic eruptions in Earth's history.

Thanks to many technological advances (e.g., the development by A. O. Nier in the late 1930s of the first modern mass spectrometer) and a variety of theoretical and experimental validations of the technique, radiometric dating has provided remarkably consistent results about Earth's history. For example, minerals from a bed of volcanic ash in Saskatchewan, Canada, have been dated as 72.5 ± 0.2 million years old with the K/Ar method (an error of ± 0.27%), 72.4 ± 0.4 million years old with the U/Pb method (an error of ± 0.55%), and 72.54 ± 0.18 million years old with the Rb/Sr method (an error of ± 0.25%). To appreciate the small amount of error in these measurements (i.e., ± 0.25 to 0.55%), note that 1% of a day is 14.4 minutes.

Other Radiometric Methods

There are several other techniques that geologists use to estimate the ages of rocks. All these techniques, when applied properly, have confirmed an ancient Earth.

Fission-Track Dating

Volcanic eruptions melt minerals and create glasslike rocks. Within these rocks, radioactive atoms often produce "tracks" that can be counted by **fission-track dating.** For example, U-238 occurs naturally within zircon, a common mineral. When U-238 decays to a more stable isotope in a zircon crystal, the decay produces tiny scratches called "tracks." Since the decay of U-238 is constant over time, the number of these tracks is directly proportional to the age of the rock.

Thermoluminescence

When rocks are heated by radioactive decay, electrons are sometimes trapped inside the rocks. If these rocks are then superheated in a laboratory, the electrons released from the rock during the heating can be measured by the amount of light produced via thermoluminescence. If one can deduce the source and rate of entrapment of the electrons, one can then use the intensity of the light to estimate the age of the rock. This method is especially useful for measuring ages between the lower limits of K-Ar dating and the upper limits of radiocarbon dating (see Sidebar 2.1, "Radiocarbon Dating").

Paleomagnetism

Occasionally throughout geological history, the magnetic poles of Earth have switched so that magnetic south becomes magnetic north, and vice versa (i.e., a compass needle would point in the opposite direction). These switches are shown in Table 2.6.

Table 2.6 Compass Orientation

0.7 million years ago to present	Current orientation
2.6 to 0.7 million years ago	Reversed orientation
3.4 to 2.6 million years ago	Current orientation
? to 3.4 million years ago	Reversed orientation

Just as Earth's magnetic field can be determined with a needle of a compass, so too can it be determined by the orientation of particles in once-molten rocks (e.g., basalt) that are iron-rich and "acquire" the magnetic field of the surroundings as the rock cools (i.e., these particles line up like tiny compass-needles). Similarly, wind-blown dust settling to Earth and mineral particles settling to the bottoms of oceans orient themselves to the current magnetic field; when the sediment hardens, this orientation is preserved, thereby leaving a record of Earth's magnetic field when the sediments were deposited. The magnetic orientations of these rocks can be matched to the history of sediments, thereby allowing scientists to estimate the age of the rocks.

Creationists' Claims about Radiometric Dating

Not surprisingly, young-Earth creationists reject radiometric dating. For example, Andrew Snelling of the antievolution organization Answers in Genesis claims that radiometric dating is inaccurate because it contradicts the Bible and because "we now have impeccable evidence that radioactive decay rates were greatly sped up at some point during the past, for example, during the global catastrophic Genesis Flood." This alleged evidence has never been published in a peer-reviewed scientific journal. Just as there is no evidence for a global flood, so too is there no evidence that rates of radiometric decay have changed significantly over time, nor is there any evidence that the laws of physics change during floods. Similarly, John Morris—the President of the Institute for Creation Research, another young-Earth organization that rejects evolution and many other foundations of modern science—claims that radiometric dating is "questionable and commonly gives erroneous, inconsistent, and sometimes bizarre dates." This claim contradicts decades of research by competent, mainstream physicists.

The Age of Earth

In 1927, Arthur Holmes claimed in *The Age of the Earth: An Introduction to Geological Ideas* that Earth "is just over 3,000 million years" old. Since then, scientists have found rocks on every continent that are 3.8 to 3.9 billion years old, indicating that Earth must be *at least* that old (some samples have included sedimentary rocks containing minerals as old as 4.2 billion years). However, Earth's earliest crust was likely in flux and later obliterated by erosion, meteorites, and plate tectonics, meaning that Earth's age cannot be computed directly from rocks found only on Earth. So how old is Earth?

When American geochemist Cameron Claire "Pat" Patterson (1922–1995) used the lead isochron method of radiometric dating to determine the ages

> **SIDEBAR 2.2 Trying to Grasp "Deep Time"**
>
> Many people can appreciate periods up to a hundred or so years because we've seen photos of, and have heard relatives' stories about, previous decades. However, we know relatively little about life during the time of Christ (about 2,000 years ago), and even less about life 5,000 years ago when Egyptians were building the pyramids. Nevertheless, if we could go back 5,000 years to watch the Egyptians work, we would have seen only about 0.00001% of Earth's history.
>
> The vastness of Earth's history is exceedingly difficult to grasp, but this may help put it in perspective: if Earth's age (i.e., 4.55 billion years) were condensed to one year, one day would represent approximately 12,300,000 years, one hour 513,000 years, one minute 8,550 years, and one second 142 years. If we began this "condensed year" on January 1, multicellular animals would not appear until November 6, vertebrates and primitive fish about November 21, vascular plants about November 27, and mammals about December 14. The first birds would appear on December 15, the first primates and grasses on Christmas Eve, and the first *Homo sapiens* just before midnight on New Year's Eve.

of five different meteorites (which he assumed to have originated when the solar system originated), he discovered that all of the meteorites, as well as a modern deep-sea marine sediment (which was presumably formed from rocks from several different continents), had the same age. In 1956, Patterson addressed hundreds of years of speculation and research by scientists and theologians such as Ussher, Kelvin, Wallace, and others about the age of Earth when he reported in a famous paper titled "Age of Meteorites and the Earth" that Earth has "an age of $4.55 \pm 0.07 \times 10^9$ years." Since then, diverse radiometric studies of moon rocks, tens of meteorites, and thousands of rocks and sediments have yielded remarkably similar results. As paleontologist Niles Eldredge noted in *The Monkey Business* (1982; p. 108), "We now have literally thousands of separate analyses using a wide variety of radiometric techniques. It is an interlocking, complex system of predictions and verified results—not a few crackpot samples with wildly varying results, as some creationists would prefer to believe."

Summary

Earth is approximately 4.55 billion years old. This age easily accommodates evolution by natural selection as a major force for the generation of life's diversity.

3

Fossils

> Such is the economy of nature, that no instance can be produced of her having permitted any one race of her animals to become extinct; or her having formed any link in her great work so weak as to be broken.
>
> —*Thomas Jefferson, 1784*

> Since Darwin's time, the number of known fossil forms has grown enormously. Researchers have discovered many hundreds of transitional fossils that document various intermediate stages in the evolution of modern species from organisms that are now extinct. Gaps remain, of course, in the fossil records of many species, although a lot of them shrink each year as new fossils are discovered. These gaps do not indicate weaknesses in the theory of evolution itself. Rather, they point out uncertainties in our understanding of exactly how some species evolved.
>
> —*Kenneth Miller and Joseph Levine, 2008*

PREDICTIONS

If evolution has occurred, we predict that

1. We will see direct evidence of the history of life on Earth.
2. The fossil record can be matched with geologic evidence to describe the history of life on Earth.
3. The fossil record will include transitional forms linking different groups of organisms.
4. Environmental changes and competition inherent in natural selection will produce extinctions, which will, in turn, open niches for other groups of organisms.

In the previous chapter, we discussed how scientists came to understand that Earth has a vast age, and how Claire Patterson in 1956 finally

Figure 3.1a Fossils, which are remnants, impressions, or traces of organisms that lived in the past, show that life has changed throughout Earth's history. *Phareodus encaustus,* a extinct fish from the Eocene, lived 50 million years ago in what is now the Green River Formation near Fossil Lake by Kemmerer, Wyoming. This lake was created by the same geological events that had earlier uplifted the Rocky Mountains. The fossil shown here is 50 centimeters long. (Randy Moore)

determined that Earth is approximately 4.55 billion years old. Subsequent observations with a variety of materials and techniques have repeatedly confirmed Patterson's findings. This evidence is why scientists accept that Earth is 4.55 billion years old.

However, knowing that Earth is so old that evolution *might* have occurred does not necessarily mean that evolution *has* occurred. Perhaps Earth and its organisms have always been like they are now, as was claimed by Aristotle and others. How could this be tested? How do scientists know that Earth and its inhabitants have changed over time? Throughout this book we present some of the overwhelming evidence and accompanying arguments for this conclusion. In this chapter, you'll learn about the most convincing evidence: fossils (Figure 3.1). A **fossil** (from Latin *fossus,* meaning "having been dug up") is a preserved remnant, impression, or trace of an organism that lived in the past. Fossils are records—usually written in stone or rock[1]—of life's history (Figure 3.1a–c).

If new species of organisms evolved from other, pre-existing species, we should see a sequence of species appearing chronologically throughout Earth's history. The paleontological record, as documented by fossils, provides extensive and detailed evidence for this prediction.

1. Some organisms have also been preserved in ice (e.g., mammoths frozen in glacial ice caps), tar, hardened resin (amber) of coniferous trees, and acidic peat bogs.

Figure 3.1b "Big Stump," a fossilized redwood (*Sequoia affinis*) 11 meters in circumference, lived 33 million years ago in what is now Florissant Fossil Beds National Monument in Colorado. When it was buried in a volcanic mudflow, this tree was 750 years old and more than 230 feet tall. Florissant Fossil Beds has yielded over 50,000 museum-quality specimens of fossils of more than 1,700 species. (Randy Moore)

Figure 3.1c *Acrocanthosaurus,* one of the largest bipedal, meat-eating dinosaurs (it was 12-meters long and weighed more than 5,400 kilograms), left these tracks approximately 107 million years ago in what is now Dinosaur Valley State Park near Glen Rose, Texas. Each of these tracks is approximately 35 centimeters wide. (Randy Moore)

We See Direct Evidence of the History of Life on Earth

Fossils puzzled people for many generations. Naturalists and others found many fossils that resembled plants and animals, but the fossils were made of stone. What could explain this? In many instances, the fossilized organisms did not resemble familiar organisms, prompting some people to speculate that the fossils represented God's first tries at creation. Some Native Americans believed that dinosaurs' bones were the remains of giant buffalos, other people claimed that fossils represented plants and animals living in foreign lands, and still others claimed that fossils were the remains of animals that had not gotten onto (or that had fallen off) Noah's ark, and had therefore died in the Genesis flood. It was a huge conceptual leap from finding an object in a solid rock to recognizing it as evidence of past life. Today, we know that fossils document life's history and, in doing so, provide strong evidence for evolution.

Contrary to popular belief, fossils had a negligible role in Charles Darwin's argument in *On the Origin of Species*. In fact, Darwin spoke almost apologetically of "difficulties" with the fossil record. For example, in Chapter VI of *On the Origin of Species*,[2] Darwin asks, "[W]hy, if species have descended from other species by fine gradations, do we not everywhere see innumerable transitional forms?" Similarly, in Chapter X, Darwin asks, "[W]hy, then, is not every geological formation and every stratum full of . . . intermediate links? Geology assuredly does not reveal any such finely-graduated organic chain; and this, perhaps, is the most obvious and serious objection which can be urged against [my] theory." However, discoveries of fossils after the publication of Darwin's book strengthened his arguments, and today, fossils provide convincing evidence that life has changed—*ramatically*—throughout Earth's long history.

Fossils provide three types of evidence for evolution: the chronology of evolutionary history, the transitional forms in evolutionary history, and the extinctions in evolutionary history. We'll examine each of these types of evidence in this chapter, but let's first examine how and where fossils form.

How and Where Do Fossils Form?

Most organisms do not leave traces of their existence. That is, few organisms become fossils; this is because when organisms die, their soft parts are usually destroyed, crushed, eaten, torn apart, or broken down by scavengers and decomposers. When organisms die, virtually all their parts are quickly recycled.

2. Darwin's most compelling evidence for evolution by natural selection was biogeography, which is the geographic distribution of organisms (see Chapter 4). Without invoking evolution, these distributions are exceedingly difficult to explain (unless, of course, one invokes an arbitrary number of special creations).

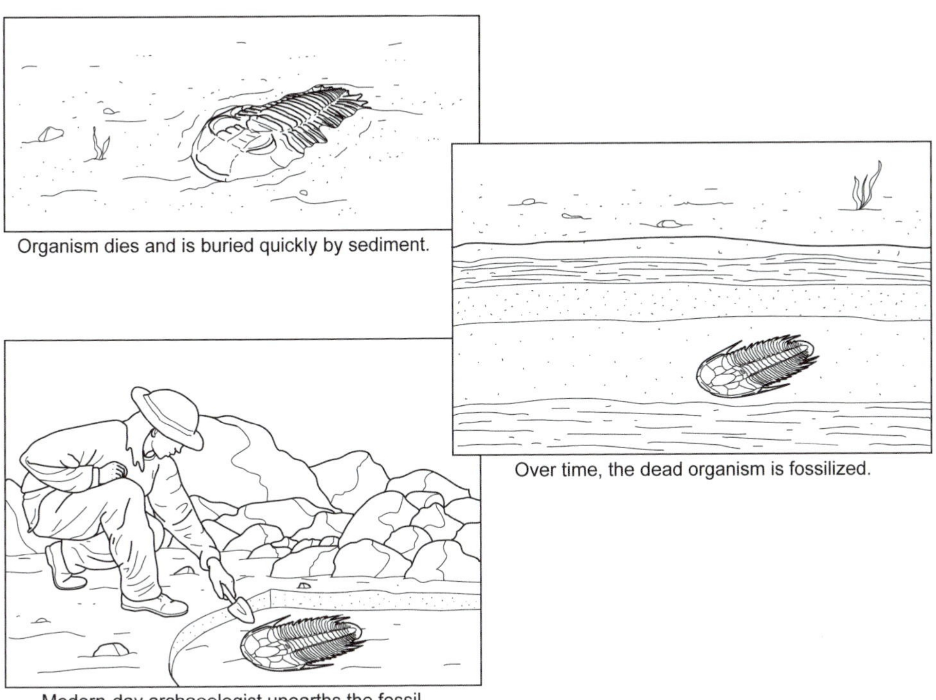

Figure 3.2 How fossils form. Fossilization is an exceedingly rare event. Only a tiny percentage of organisms that have ever lived are preserved as fossils. (Jeff Dixon)

Fossils form only in the exceedingly rare conditions in which an animal or plant is buried rapidly by volcanic ash, in a bog, or at the bottom of a river, sea, or lake where it is protected from scavengers and decomposers (Figure 3.2). Not surprisingly, soft tissues are almost always destroyed; this is why there are virtually no fossils of soft-bodied animals (e.g., worms, jellyfish). This is also why most fossils are of shelled organisms that lived in shallow seas (e.g., snails, clams, corals). However, sometimes even the hardest parts of organisms are destroyed by the pressure of sediments. For example, coal is the fossil remains of plants whose individual structures have been destroyed as the sediments were compressed.

Organisms' hard parts (e.g., teeth, bones, seeds, pollen, scales, claws, shells, lignified tissues) are sometimes preserved when the organisms are covered by sediments. Over millions of years, minerals in water percolating through the sediments slowly infiltrate the organisms' hard parts and convert these structures to sedimentary rock. As a result, most fossils are not made of the actual substances from once-living organisms. Instead, the process of fossilization slowly replaces living, organic tissues with inorganic minerals, thereby creating stone replicas of the organism. These stone replicas are what we call fossils (Figure 3.1).

There are three major types of fossils: *Compression fossils* form when organisms or parts of organisms are buried in sediments before decomposing.

Figure 3.3a How large fossils appear when they are discovered. Sue Hendrickson standing beside Sue, a 67-million-year-old *Tyrannosaurus rex* that she discovered just outside Faith, South Dakota, in the summer of 1990. After a long legal battle to determine who owned Sue, the magnificent fossil was sold at an auction. Bidding for Sue began at $500,000, but the final price was $7.6 million (the commission on the sale raised the price to $8.4 million). For a different view of Sue, see Figure 3.3b, on the following page. (© 1990, Black Hills Institute of Geological Research. Photographer: Peter L. Larson)

The weight of the sediment leaves an impression of the organism in the sediment, much like footprints in mud. *Casts* and *molds* form when organisms decay after being buried in sediment. Molds are unfilled spaces, and casts form when new materials fill the spaces and solidify.

Fossils have been found in a variety of places. For example, about 500 years ago, Leonardo da Vinci was puzzled by the fact that rocks atop Italy's northern mountains contained fossilized seashells and sharks' teeth.[3] How did fossilized shells of marine organisms get into rocks atop mountains far from the sea? One explanation was that the seashells were carried there by a flood, but that didn't make much sense; after all, water flows downhill (not up mountains), and besides, the shells would have been battered, if not destroyed, by a violent flood. A simpler and more logical argument

3. Since da Vinci's time, fossils have been found atop many mountains, including the famed North Face of Mount Everest, not far from the 12-meters high stone wall encountered by Everest's climbers at 8,700 meters (the Hillary Step).

Figure 3.3b How large fossils appear in museums. It took six fossil hunters from the Black Hills Institute 17 days to excavate Sue, and 10 museum preparators more than two years (30,000 hours of work) to prepare Sue's more than 250 teeth and bones. Sue, the product of their work, today greets visitors to the Field Museum in Chicago. Sue is 13 meters long, 4 meters high at the hips, and weighed almost seven tons when alive. (© 2000 The Field Museum, GN89860.1c, Photographer George Papdakis)

was that the shells were once in ocean sediments, where they became fossilized parts of rocks. When these rocks were lifted above sea level by geological processes such as earthquakes, the fossils were exposed for da Vinci and others to see.

Although only a tiny percentage of organisms become fossils, fossils are nevertheless abundant; humans have described more than 250,000 fossilized species from throughout the world, and more are discovered every day. However, this doesn't mean that we find *all* fossils. For us to *find* a fossil, geologic forces must raise these fossil-bearing rocks to Earth's surface (for example, as mountains), and wind or rain must erode the rocks and reveal the fossils (Figure 3.3a and 3.3b).

Consider again how fossils form (Figure 3.2). Because fossils originate as organisms buried in sediments, it is not surprising that virtually all fossils are found in sedimentary (from the Latin word *sedimentum,* meaning *to settle*) rock, which is made primarily of cemented particles of eroded rocks. Today, sedimentary rocks are the most common rocks at Earth's surface, and include sandstone, limestone, siltstone, mudstone, and shales. Fossils never occur in igneous (from the Latin word *ignis,* meaning *fire*) rocks such as basalt

Figure 3.4 A 12-million-year-old fossil of an adult female barrel-bodied rhino (*Teleoceras major*) at Ashfall Fossil Beds State Historical Park near Royal, Nebraska. This site, a National Natural Landmark, is the only known site in the world where entire three-dimensional skeletons of large prehistoric animals are preserved. (Randy Moore)

and granite, which are formed by the cooling of lava or magma (however, there are some molds of organisms incinerated by lava, such as at Pompeii and in Hawaii's Lava Tree National Park). Fossils are also common in some ash deposits (Figure 3.4), but are exceedingly rare in metamorphic (from the Greek words *meta* and *morphe,* meaning *change* and *form,* respectively) rocks (e.g., marbles, schists, diamonds), which are formed by pressure and high temperatures that destroy fossils (or alter them beyond recognition).

Is the Fossil Record Complete?

Biologists have long known that the fossil record is incomplete. In *On the Origin of Species*, Charles Darwin lamented the fact that, as a result of the process of fossil-formation (Figure 3.2), the fossil record *must* be incomplete, "like pages torn from a book." He was right: most animal phyla are not represented in the fossil record. Here's why:

Geologic sediments form episodically, and only a small percentage of the living species are included in those sediments. Some time-periods are represented by relatively few sedimentary formations worldwide.

There are systematic biases regarding which organisms become fossils. Many organisms are delicate, lack hard parts, or live where decay is rapid. Consequently, regardless of their abundance, there is only a remote chance that these organisms will become fossils (see above). Only a tiny fraction of organisms that have lived on Earth becomes fossils and, of these, only a tiny fraction has been found by humans. Most fossils are of plants and animals (e.g., shelled creatures) having hard parts that lived near water or wet, terrestrial areas.

Many organisms *must* be missing from the fossil record. To appreciate this, consider flatworms (Phylum Platyhelminthes), a large group of bilaterally symmetrical, unsegmented, soft-bodied invertebrates. There are thousands of common species of flatworms alive today, many of which have been studied intensively because of their economic importance (e.g., the flatworm *Schistosoma* causes schistosomiasis, which infects approximately 200 million people worldwide). However, not one fossilized flatworm has been discovered. This lack of flatworms in the fossil record results not from their absence or extinction, but instead because the fossilization of soft-tissue organisms is exceedingly rare. Indeed, the fossil record of soft-bodied organisms is so poor that relatively few paleontologists study these organisms or can say much about their evolution.

The incompleteness of the fossil record has also produced other problems. Once fossils are uncovered, how should they be interpreted? What differences among fossils actually denote different species? Debates about this question have prompted paleontologists to reevaluate many fossils and the associated claims about life's history. For example, in 2009, paleontologists determined that three dinosaurs—*Dracorex, Stygimoloch,* and *Pachycephalosaurus*—are actually the young, juvenile, and adult stages of the same species. Suddenly, because of a new interpretation, the fossils of three different species became different stages of just one species.

The fossil record remains incomplete; for example, biologists have not yet found fossils of an organism that clearly "links" bats and their mammalian predecessors. Nevertheless the fossil record comprises a vast library that enables us to piece together the broad patterns of life's history on Earth, much like forensic experts use incomplete historical evidence to determine what happened at a crime scene. Whereas Darwin spoke apologetically of "difficulties" with the fossil record, today's fossil record provides extensive, strong, and unequivocal evidence for evolution.

The Fossil Record Can Be Matched with Geologic Evidence to Describe the History of Life on Earth

By the 1890s, geologists knew that Earth's history is written in rocks. Recall from Chapter 2 Steno's Principle of Superposition—namely, that in undisturbed areas, younger layers of sediments always overlie older layers of sediments. This means that undisturbed sediments at Earth's surface are youngest, and that sediments are progressively older as one descends the geologic column (Figure 2.3).

When geologists began to study and name the various geologic sediments (Appendix 2), what they discovered was surprising: the layers of rock are similar and occur in the same order throughout vast areas of the world. For example, the Cretaceous is always younger than (i.e., is above) the Triassic, and the Triassic is always younger than the Cambrian. The simplest interpretation of these observations was that the sediments had been deposited over long periods of time at the bottoms of vast ancient oceans, and were later raised by geologic forces to their present positions. Although geologic sediments are most prominent in places such as the Grand Canyon (Figure 2.3), the same stratification occurs elsewhere. Today we know that different layers of rocks—which can be arranged according to their ages—are evidence of the passing of geologic time (Figure 2.3; Appendix 2). But what about the fossils in those sediments?

Fossils in the Sediments and Biostratigraphy

In 1796 English engineer and canal-digger William Smith (1769–1839) discovered that similar geologic strata could be identified by the fossils they contained—as Smith noted, there were fossils "particular to each stratum." We now know that each layer of rocks contains unique fossils that can be used to distinguish that layer from other layers.[4] Because the different strata were formed at different times in Earth's history, the fossils in those rocks must represent organisms that lived during different times in Earth's history. This finding helped establish **biostratigraphy,** which is based on the deepest (i.e., oldest) layers of rock containing fossils of the oldest, most primitive organisms. In contrast, the younger layers of sedimentary rock above them contain fossils of organisms that are similar but structurally more complex, and the youngest (i.e., uppermost) sediments contain fossils of organisms that resemble species alive today. Fossils of the earliest known forms of life (e.g., bacteria) first appear in sediments that are about 3.8 billion years old, and younger (i.e., more recent) sediments contain a progression of multicellular organisms: jawless fish (agnathans, such as lampreys, and the first vertebrates; about 500 million years

---

4. In 1815, Smith used this information to produce the first geologic map of an entire country. Smith's map—"The Map That Changed the World"—was based on fossils and minerals in various sediments, and showed that fossils are not distributed randomly as would result from a flood, but instead are arranged in a definite, predictable order.

SIDEBAR 3.1 The Business of Fossil Collecting

Fossil collecting is a booming business. For example, in 2007, a fossilized dinosaur egg sold for $97,500, a fossilized skeleton of a Siberian mammoth sold for $421,200, and a 70-million-year-old mosasaur (a 9-meters-long carnivorous underwater reptile) sold for $350,000. The following year, a mastodon skull fetched $191,000, and a 55-million-year-old lizard preserved in amber sold for $97,000. Sue, the famous *T. rex* that greets visitors to Chicago's Field Museum (Figure 3a and 3b), was bought by the museum for more than $8 million. With these potential rewards available, it's not surprising that prospectors are combing the American Southwest and elsewhere for fossils.

Although landowners can excavate and sell any fossil found on their property, the situation is far more complex when a fossil is discovered on state or federal land. These fossils often pit amateur collectors against scientists and governmental officials. Collectors are often prosecuted when they illegally collect fossils. For example, in 2006, a collector excavated and sold fossils of *Allosaurus* (an older cousin of *T. rex*) from public land in Utah; the collector was later convicted of theft of federal property. Earlier, another collector sold claws of *Therizinosaurus*—collected illegally in Utah—for $15,000, after which he was fined $15,000 and incarcerated for 10 months.

ago), then bony fish (about 410 million years ago), then amphibians (about 370 million years ago), then reptiles (about 310 million years ago) and then dinosaurs, and, finally, mammals and birds (about 225 million years ago). As one would expect, mammals appear in the fossil record just after mammal-like reptiles having features that are intermediate between reptiles and mammals. The youngest fossils, which occur in sediments nearest Earth's surface, most closely resemble organisms living on Earth today. Evolution by natural selection not only predicts this, but is alone in being able to *explain* it in a rational, testable way (i.e., that avoids deities and mystical or magical events that could be used to explain *any* observation).

Since Smith's era, geologists have repeatedly shown that the major geologic sediments represent natural units of the geologic column; as geologist Sir Roderick Impey Murchison noted in *Siluria* (1872), "Placed as the fossils are in their several tiers of burial-places the one over the other, we have in them true witnesses of successive existences." Moreover, each of these units can be defined by its characteristic (i.e., indicator) fossils (Figure 3.5). For example, Jurassic sediments and their indicator fossils are always above those of the Carboniferous, which, in turn, are always above those of the Permian. The patterns are remarkably consistent; for example, there are no trilobites (mobile, detritus-feeding arthropods that were abundant in the Paleozoic) above the Permian sediments, and there are no

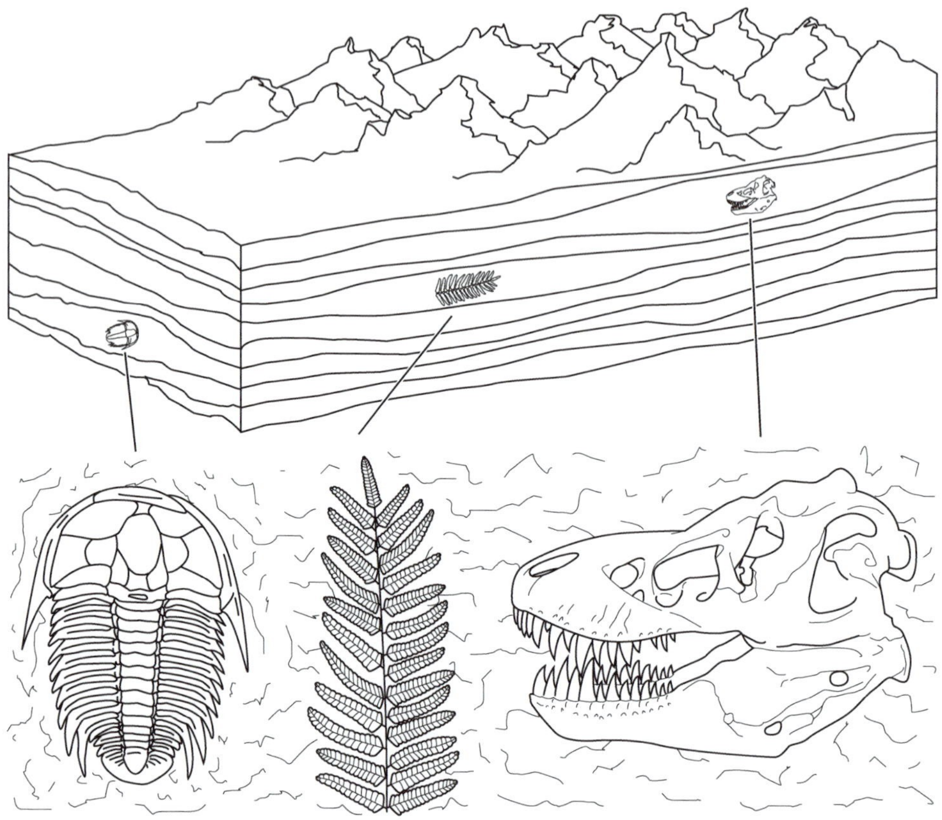

Figure 3.5 The deepest (i.e., oldest) layers of sediments contain the oldest fossils, whereas sediments closer to Earth's surface contain the younger fossils. Shown here are a trilobite (which became extinct by about 250 million years ago); a seed fern (which became extinct by about 150 million years ago); and the skull of a nonavian dinosaur, which became extinct about 65 million years ago. (Jeff Dixon)

mammals in the Devonian or earlier sediments. For many indicator species, the pattern is absolute: there has not been a single authenticated discovery of a fossilized mammal—*not one*—in the Precambrian, nor has there been an authenticated discovery of a single fossilized trilobite—*not one*—above the Permian. Similarly, there are no authenticated fossils of birds, mammals, reptiles, amphibians, flowering plants, or insects—*not one*—in Cambrian sediments.

Geologists also noted that there was a general progression of life forms as one moved from the oldest to the youngest sediments—invertebrates were abundant in the lower sediments, then amphibians appeared, then reptiles, and then mammals (including humans). Human fossils appear only in the youngest sediments; as geologist William Buckland noted in *Geology and Mineralogy* (1836), "No conclusion is more fully established, than the important fact of the total absence of any vestiges of the human species throughout the entire species of geological formations." The conclusion was inescapable: fossils are the timekeepers of sediments, and each sediment includes distinctive fossils that can be used to measure time.[5] The fact that fossils can be used to identify the ages of sediments explains why oil companies hire fossil experts (i.e., paleontologists) to help them locate underground oil. Not a single authenticated discovery of a fossil in the "wrong" sediment has ever been documented.

By the mid-1800s, naturalists had established—thanks to fossils—that life on Earth had changed throughout history, and that fossils in older strata are different from organisms alive today. In this sense, the geological column resembles a multitiered wedding cake, with each tier having a unique flavor (i.e., a unique set of distinguishing fossils). For example, Cambrian sediments contain fossilized trilobites, but not dinosaurs, and fossils from the Jurassic period contain fossilized dinosaurs, but not trilobites. Georges Cuvier (1769–1832), the most famous anatomist of his age, understood the significance of this evidence; as he noted in 1801, "the older the beds in which [fossils] are found, the more they differ from those of animals that we know today. [This is] the most remarkable and astonishing result that I have obtained from my research."

Creationists' Objections to Biostratigraphy

But not everyone today accepts modern science. For example, some young-Earth creationists claim that the fossil record "does not support an evolutionary view of the past. Rather, it is fully compatible with biblical thinking." These creationists claim that evidence for biostratigraphy is a conspiracy perpetuated by dishonest scientists to prove Darwinian evolution. These creationists forget that most of the naturalists who established these principles were religious people who lived several decades *before* Darwin, and were either ambivalent or hostile to evolution. Other young-Earth creationists reject modern science for more incredible ideas. For example, George McCready Price (1870–1963), a self-taught geologist who pioneered "flood geology," claimed that fossils were deposited in the Noachian flood described in Genesis; the helpless invertebrates were buried first, and larger land animals fleeing the flood waters (or floating to the top of the waters) were buried in higher strata. This idea was developed in 1961 by John Whitcomb, Jr., and Henry Morris in their influential *The Genesis Flood: The Biblical Record and Its Scientific Implications,* the book that became the founding document for "creation science" and the

5. The absolute ages of these sediments have been repeatedly verified with a variety of radiometric techniques (Chapter 2). The ages of fossils contained in the sediments have also been verified by applying radiometric techniques to deposits of volcanic ash above or below the fossils.

modern creationist movement.[6] Whitcomb and Morris, and many subsequent creationists, explained the fossil record with *hydraulic sorting*, in which the biblical flood buried the heavy shells of immobile marine invertebrates and fish in the lower sediments, and motile animals such as amphibians—who were fleeing the floodwaters—in higher, intermediate sediments. "Smart animals" such as humans, who were also fleeing the rising floodwaters, allegedly climbed to the highest levels before drowning and being buried. According to many creationists, this is a "perfectly satisfying explanation" for why each major sediment contains unique fossils.

There are countless problems with the hydraulic sorting model. For example, the model would have to be based on a statistical *tendency* for different groups to be sorted by floodwaters. For example, some mammals—perhaps crippled, sick, trapped, or recently deceased—would presumably have been unable to flee to higher ground, and would therefore have been trapped in the lower sediments. However, there are *no* mammals—*not one*—in the lower geologic strata. Similarly, there are *no* nonavian dinosaurs—*not one*—above the Cretaceous. Fossilized invertebrates occur in virtually *all* strata, and "dumb" animals such as marine clams and snails (that supposedly drowned in the early stages of the rising flood) are often found above "smarter, faster" reptiles and amphibians (and dinosaurs). In the famous Eocene Green River shale of the western United States, fossilized fish (Figure 3.1a) are found above dinosaurs, whereas in the lower Eocene Wasatch Formation, fossilized fish are found above mammals. In the cliffs overlooking Chesapeake Bay in Maryland, the fossils of land mammals are mixed with marine shells, and there are also beds in which marine fossils are above *and* below those containing fossils of land mammals. These are not isolated examples—there are hundreds of places throughout the world where fossilized animals that allegedly drowned in the early stages of the flood are found in higher sediments than are land animals.

The problems for young-Earth creationists who endorse flood geology and hydraulic sorting also extend to the rocks in which the fossils occur. When a flood recedes, it leaves only one kind of deposit—a single layer of mud (*not* shale, which requires burial and millions of years of compression) overlying coarse gravel and boulders. However, at the Grand Canyon (Figure 2.3)—young-Earth creationists' most-cited example of the alleged

6. At Kentucky's Creation Museum, which promotes the young-Earth creationism popularized by Whitcomb and Morris, the fossil record is explained as follows: "Fossil layers were formed by Noah's Flood (~4,350 years ago) and its aftermath." The museum, which is owned and operated by Answers in Genesis CEO Ken Ham, attracted more than 1 million visitors during its first three years in operation.

evidence for flood geology—there is no great deposit of coarse gravel and boulders at the base of the canyon. Moreover, when one examines the sides of the canyon, one sees sequences of shales, sandstones, and limestones, not just a thick layer of mud as would have been left by a flood. The lava depositions in the canyon[7] are characteristic of lava flowing across land (such as at Hawaii's Mount Kilauea), not the "pillow lava" that forms when lava solidifies in water (as it would in a flood). Finally, if the Grand Canyon had been carved rapidly by a receding flood (as young-Earth creationists claim), then the mud left behind would have slumped into the gorge. The Grand Canyon's steep sides—some more than a kilometer high—are evidence that this is not how the canyon formed. These and the myriad other inconsistencies and contradictions of flood geology and hydraulic sorting attest to why modern geologists reject this explanation for Earth's crust. In simplest terms, the results predicted by flood geology and hydraulic sorting are found nowhere.

So why do claims about flood geology and hydraulic sorting persist? Most young-Earth creationists who promote hydraulic sorting and flood geology have seldom sought advanced training in geology or paleontology, and virtually none has ever gone into the field to critically examine the actual evidence. In the rare instances in which creationists *have* critically examined the evidence, they've often questioned or rejected their earlier claims. Recall from Chapter 2 the story of young-Earth creationist Harold W. Clark's visits to oil fields of Oklahoma and northern Texas. What he saw was familiar to working geologists, but a shock to him; as he confessed in a letter to Price, "The rocks do lie in a much more definite sequence than we have ever allowed. The statements [about flood geology] do not harmonize with the conditions in the field. . . . Thousands of cores prove this. . . . The science has become a very exact one, and millions of dollars are spent in drilling, with the paleontological findings of the company geologists taken as the basis for the work. The sequence of the microscopic fossils in the strata is very remarkably uniform. . . . The same sequence is found in America, Europe, and anywhere that detailed studies have been made." Instead of accepting—or at least examining—the evidence, Price's response typified that of many young-Earth creationists: he denounced Clark and those like him as having been influenced by Satan.

---

7. During the past 2 million years, more than 150 lava flows have poured into the Grand Canyon. These flows, which account for the canyon's dark-colored basalt on the north side of the river, formed several lava dams (60–600 meters high) that impounded enormous lakes. In each instance, the Colorado River incised the dams. The Grand Canyon's challenging Lava Falls—the last major rapid encountered on most river trips—is named for one of the overflows of lava.

Fossilized Invertebrates

Today, fossils are accepted by virtually everyone as evidence of changes in life's history. We are especially intrigued by large fossils. Indeed, when most people think of fossils, they envision large, ferocious animals such as the famous *Tyrannosaurus rex* "Sue"—usually having heads full of menacing teeth—that glare at us when we enter natural history museums (Figure 3.3b). Similar fossils of *Triceratops, Stegosaurus,* pterosaurs, and other ancient creatures that fill museums throughout the world document that evolution has occurred and that life on Earth has not always been the way it is now. Fossils have also taught us that the ancestors of many familiar species of animals did not look like their modern-day descendants. For example, most fossil giraffes lacked long necks, most fossil camels lacked humps, and most fossil rhinos lacked horns.

Although we tend to focus our attention on large animals (few people visit zoos to see the "Insect House"), invertebrates are the most abundant and diverse animals alive today, as well as the best fossilized. These fossils are easy to overlook, but they are abundant—the sand of many tropical beaches is mostly microfossils of these organisms. There are hundreds of well-documented examples of fossilized invertebrates that are described expertly in a variety of books, so here we will mention only one: the evolution of horseshoe crabs.

Horseshoe crabs are easily recognized by their hard shells, five pairs of legs, and long, straight, rigid tails that can be used to turn themselves over if they are flipped upside down. Despite their name, horseshoe "crabs" are not actually crabs (a type of crustacean), but instead are marine chelicerate arthropods that are more closely related to spiders, ticks, and scorpions than to crabs. This is why, beneath the shell, they resemble a large spider. Horseshoe crabs, which have the rare ability to regenerate lost limbs, often wash ashore in the Gulf of Mexico and along the northern Atlantic coast of North America.

Horseshoe crabs predate flying insects, humans, and dinosaurs; their earliest fossils are found in strata of the late Ordovician period (450 million years ago). The ancestors of modern horseshoe crabs included primitive Cambrian aglaspids, which were arthropods having tails and segmented bodies. Throughout the Paleozoic, various species of horseshoe crabs developed larger and larger covers for their heads, fewer thoracic segments, and their unique tail spine. More recent ancestors were more specialized and more closely resembled modern-day horseshoe crabs; for example, *Mesolimulus* (from the Jurassic) was similar to the living species *Limulus.* Horseshoe crabs, which may have evolved in shallow seas with other arthropods such as trilobites, are often called "living fossils" because they have not changed much in the past 230 million years. However, they changed a lot before that.

Evolutionary Lineages for Vertebrates

Paleontologists have also produced numerous evolutionary lineages for vertebrates. Of these, human evolution attracts the most attention; that's one reason why human evolution is treated in a separate chapter of this book. However, there are many other well-described vertebrate lineages. Here, we'll discuss one: the evolution of whales (Figures 3.6 and 3.7).

Ever since biologists understood that whales are mammals, they have wondered from which group of mammals whales originated. Fossils, as well as much molecular and other evidence, now show that whales descended from a group of carnivorous, hoofed, hippo-like land mammals (hippos are more closely related to whales than to pigs, which they resemble; Figure 3.7). Whales' ancestors have included *Pakicetus* (a wolf-sized animal that lived primarily on land 47 million years ago), *Ambulocetus* ("walking whale," a 3-meter-long amphibious crocodile-like animal that could walk as well as swim), and *Rodhocetus* (which swam with an up-and-down motion similar to that of modern whales; Figure 3.7). These ancestors had characteristics now present only in whales (especially regarding the anatomy of their ears), but also had limbs like those of terrestrial mammals from which they evolved. Some of whales' ancestors—which entered water about 50 million years ago—positioned their fetuses for head-first delivery (typical of modern land mammals), suggesting that the species gave birth to its young on land. Modern

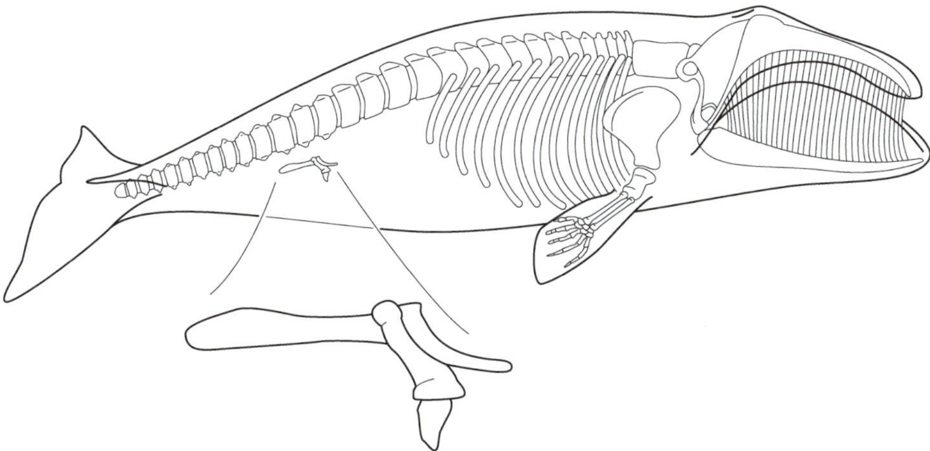

Figure 3.6 Although Ishmael of Herman Melville's *Moby-Dick* claimed that he took "the old-fashioned ground that the whale is a fish, and call upon holy Jonah to back me," he was wrong. Whales (as well as porpoises and dolphins) are not fish; they are mammals that descended from hippopotamus-like animals that lived on land. These ancestors had legs attached to their shoulders and hips, which in turn were attached to the spine. Modern whales' tiny pelvic bones are evidence of their land-based ancestors (Jeff Dixon)

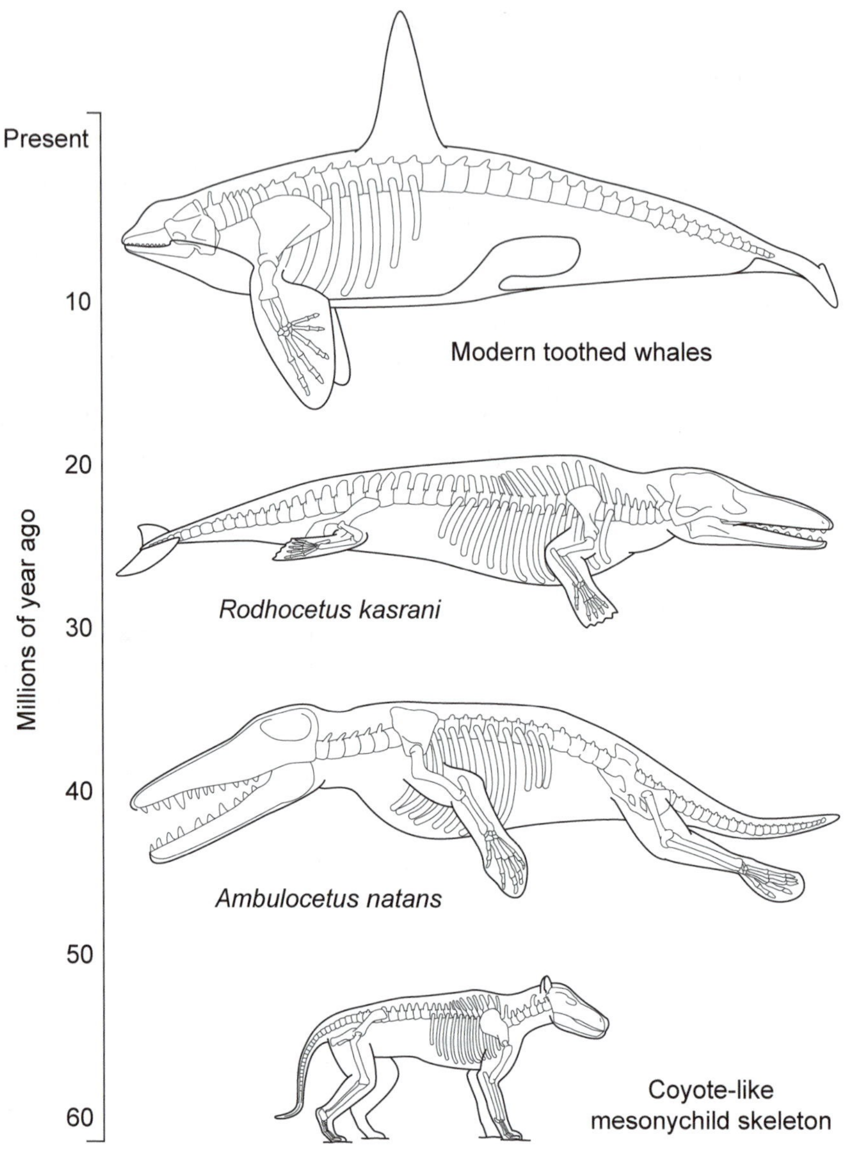

Figure 3.7 Paleontologists have discovered 10 transitional forms linking the various ancestors of whales. This figure shows three of those ancestors. (Jeff Dixon)

whales—which include baleen whales (e.g., blue whales) and toothed whales (e.g., killer whales)—still have vestiges of their ancestral land-mammals; these vestiges include hind legs, whose hip and thigh bones are buried deep in the whale's muscles (Figure 3.6). Although modern-day whales always live in water, seals and sea lions are mammals that have gone only partway back to the water.[8] These animals, as well as dolphins (a type of toothed whale), never evolved gills, thus explaining why they still breathe air into their lungs.

The Fossil Record, "Sudden Appearance," and the Cambrian Explosion

The first multicellular animals appeared in the fossil record about 565 million years ago. Fossils from this time were first discovered in the Ediacaran Hills of Australia,[9] and were later also discovered in a variety of other places. Most of these fossils are small (no larger than a few centimeters in diameter) impressions of soft-bodied animals such as jellyfish and sponges. These organisms, if they are symmetrical at all, are shaped like pies or cylinders; that is, they are *radially symmetrical,* meaning they are circular, and similar all the way around, with a top and bottom. There are also trace fossils, such as burrows and the like, that suggest that at least some animals at this time were *bilaterally symmetrical* (i.e., had a head and a tail). Because these animals left only trace fossils, these organisms probably had no shells or other hard parts with which to leave more substantial remains. Their burrows were fairly simple, suggesting that they may have faced few ecological challenges. An organism that must evade predators, for example, is likely to construct a convoluted, or otherwise impenetrable, burrow. These Ediacaran organisms flourished for 40 million years, until the start of the Cambrian.

It is the slightly younger (in geological terms) fossils at the start of the Cambrian that are truly amazing. Between 543 to 506 million years ago, an incredibly diverse group of animals appeared in the fossil record. Many of these animals were bizarre—for example, *Opabinia* had five eyes, and *Wiwaxia* was a spined, slug-like organism. These larger animals (relative to the Ediacaran fossils) were bilaterally symmetrical, and among them we see evidence of segmentation, limbs, shells, exoskeletons (the crunchy covering of insects and crustaceans), and even notochords, the flexible, rod-like signature of our own phylum, Chordata. These fossils, which are the early members of several existing phyla, exhibit many traits that characterize today's mollusks, crustaceans, worms, and primitive fish. This relatively sudden (in geologic time) appearance of diverse animals at the base

---

8. Other animals that have gone from land to water, at least part of the time, include crocodiles, otters, Galápagos flightless cormorants (Figure 4.9), Galápagos marine iguanas (Figures 4.2b, c), and penguins.

9. This site, approximately 400 kilometers north of Adelaide, is now part of the Ediacara Fossil Reserve Paleontological Site in Flinders Chase National Park.

of the Cambrian troubled Charles Darwin, who noted in *On the Origin of Species* that it "may be truly urged as a valid argument against the views here entered."[10]

A trove of these fossils was discovered in the early part of the 20th century in British Columbia, Canada, in a formation called the **Burgess Shale.** (Since then, similar fossils have been discovered in China, Greenland, and Sweden.) These remarkable animals were all new to science, and were preserved in a massive underwater mudslide that protected the animals' bodies from predators, scavengers, and decomposers. This "explosion" of fossils at the base of the Cambrian period has come to be known as the **Cambrian Explosion;** it documents the first appearance of fossils that can be assigned to many major animal phyla. But then the fossil record goes dark; insofar as we know, the actual species of the Cambrian are extinct. However, many of their descendants are still with us, and these can be recognized by the existing modifications of those Cambrian body-plans (e.g., the type of body cavity, appendages, body covering). Today, animals that bear these traits live in almost every environment, from the ice of montane glaciers to the deepest hydrothermal vents of the ocean.

The Cambrian Explosion does not account for all known animals; some animal phyla—such as sponges and radiate animals—were present in the Ediacaran period (i.e., before the Cambrian Explosion). Nevertheless, the abundance and diversity of fossils from the Cambrian Explosion have long amazed scientists; no Cambrian species exists today, and most animals from the Cambrian Explosion look nothing like today's animals. What could account for the seemingly abrupt appearance of so many animals, most having no known antecedents? Stated another way, what caused the "explosion"?

Many scientists believed that the disappearance of the Cambrian organisms was due to a mass extinction. However, in 2010, scientists discovered a trove of 1,500 480-million-year-old fossils in Morocco that had strong links to Cambrian species found in the Burgess Shale. Thus, the Burgess Shale type continued to thrive until at least the Middle Cambrian, suggesting that the sudden appearance of diverse Cambrian animals was at least partly due to the exceedingly rare fossilization of soft-bodied organisms (recall from earlier in this chapter the absence of fossilized flatworms, despite their abundance in nature). Other explanations for the Cambrian Explosion have included the evolution of eyes, an end-Ediacaran mass extinction (see below), and disruptions of food chains. However, the Cambrian Explosion

10. In *On the Origin of Species,* Darwin discussed the Silurian, which, at the time, included what we now regard as the Cambrian.

may also have resulted from changes in Earth that were occurring at that time. Expanding populations of photosynthetic algae released increasingly large amounts of oxygen into the atmosphere. With more abundant oxygen, animals that used oxygen to fuel locomotion were able to be more active and support more tissues and more complexity. The animals that left impressions in the Ediacaran Hills were either sedentary or drifted through the ocean, filtering or trapping small food particles. In contrast, the animals of the Cambrian were on the move, seeking prey, fleeing predators, and developing defenses that left marvelous fossils, some of which still bear the marks of predators' attacks.

The Fossil Record Includes Transitional Forms Linking Different Groups of Organisms

Missing Links

The term "**missing link**" was used during Charles Darwin's era to denote hypothetical organisms that linked different groups, and especially humans with anthropoid apes. For example, in 1866, when Prussian biologist Ernst Haeckel (1834–1919) published the first evolutionary trees, he claimed that human evolution consists of 22 phases, the twenty-first being a "missing link" represented by *Pithecanthropus alalus* ("speechless ape-man"). Haeckel's claim that this hypothetical ancestor would someday be found inspired Dutch paleontologist Eugène Dubois (1858–1940)—the first anthropologist to purposefully search for hominin fossils—to name his subsequent discovery in Java of this "missing link" *Pithecanthropus erectus* ("erect ape-man"; see Chapter 9 and Appendix 3). Newspapers followed suit; for example, the February 3, 1895, issue of the *Philadelphia Inquirer* announced Dubois' discovery with the headline "The Missing Link: A Dutch Surgeon in Java Unearths the Needed Specimen."

The term "missing link" became increasingly popular, and the alleged absence of these links was soon cited by creationists to denounce evolution. For example, in 1924, North Carolina Governor Cameron Morrison convinced the North Carolina Board of Education to reject biology textbooks that included evolution "because, among other things, I don't believe in any missing links. If there were any such things as missing links, why don't they keep on making them?" Similarly, in 1980, Louisiana state legislator Bill Keith convinced his colleagues to pass a bill requiring "equal time" for creationism and evolution in science classrooms by telling them that creationism is "pure science" having "no missing links." And in 2001, the discovery of the hominin *Sahelanthropus tchadensis* (which lived 6–7 million years ago in central Chad) was hailed by the *Christian Science Monitor* as "the holy grail of anthropology" that was "the missing link between humans and their ape forebears."

Transitional Forms

Although the term "missing link" became increasingly popular with the public, modern biologists seldom use the term. Here's why:

- "Missing link" implies that organisms are linked by a hierarchal chain or ladder, when, in fact, *all organisms share common ancestors. Anything* alive today has been subject to evolution by natural selection just as long as *everything* alive today has been subject to evolution by natural selection.
- "Missing link" implies that a particular fossil has not yet been found, and that somehow, this shortcoming renders evolution invalid.

Instead of "missing links," modern biologists speak of **transitional forms.** A transitional form is an organism having features intermediate between those of two major groups of organisms in an evolutionary sequence. Contrary to popular belief, transitional forms are not rare. Indeed, because all organisms share a common ancestor, *most* fossils are—to some degree or another—transitional forms.

Transitional forms show evolutionary sequences between lineages because they have characteristics of ancestral and newer lineages. Since *all* populations are in evolutionary transition, a transitional form represents a particular evolutionary stage that is recognized in hindsight. There are hundreds of well-documented transitional forms, including those between fish and amphibians, amphibians and reptiles, reptiles and mammals (the mammal-like reptiles), and extinct hominins and humans (e.g., australopithecines; see Chapter 9). Here, we'll consider two famous transitional forms: *Archaeopteryx,* a transitional form between birds and dinosaurs, and *Tiktaalik,* a transitional form between land-dwelling tetrapods and their fish ancestors.

Archaeopteryx: From Dinosaurs to Birds

In 1861—two years after the publication of *On the Origin of Species*—miners in the Jurassic limestone quarries near Solnhofen, Germany, unearthed a skeleton of an unusual bird-like creature that had several dinosaur-like traits. The fossil was named *Archaeopteryx lithographica* (*Archaeopteryx* means "ancient wing"; *lithographica* denotes the use of the Solnhofen limestone in the printing industry; Figure 3.8).

The 150-million-year-old *Archaeopteryx* had feathers like a bird, but it also had a long, bony tail, three claws on each wing, and a mouth full of teeth (like reptiles). Two years after its discovery, famed anatomist Richard Owen[11]—realizing the potential significance of *Archaeopteryx*—paid £700 for the fossil for the Natural History Museum (also known as the British

11. In 1842, Owen coined the term *dinosauria,* which means "terrible lizard." However, none of the dinosaurs were lizards and few were terrible. Many dinosaurs were herbivores no bigger than small dogs.

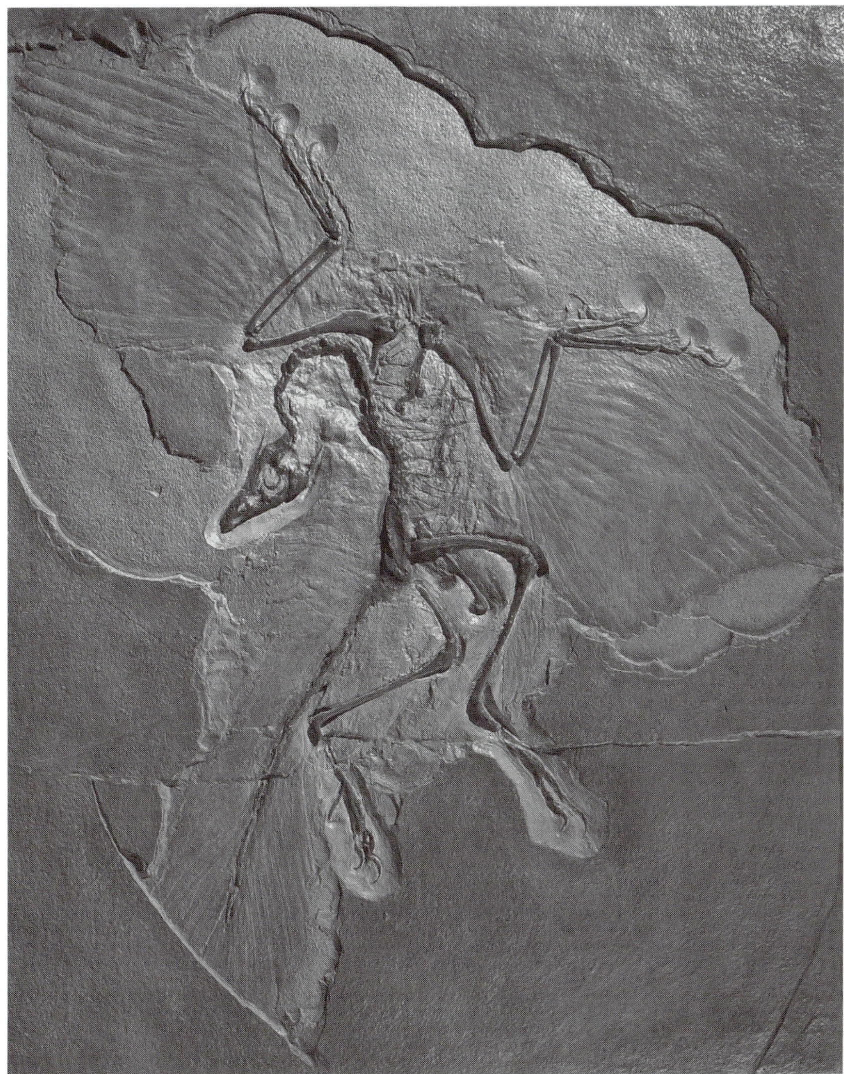

Figure 3.8 *Archaeopteryx* (in German, *Urvogel*, meaning *original bird*) was a feathered, birdlike creature having several dinosaur-like traits that provided evidence for birds having evolved from nonavian dinosaurs. Specimens of *Archaeopteryx* are named for the cities having museums that house them. Shown here is the famous "Berlin specimen," which was discovered in 1877 and which today is housed in Berlin's Museum of Natural History. (Courtesy of the Berlin Natural History Museum)

Museum of Natural History). This was double the museum's annual budget for acquisitions.

Thomas Huxley's analysis of *Archaeopteryx* prompted him to suggest that birds descended from dinosaurs. Since Huxley's time, biologists have discovered fossils of many other feathered dinosaurs, including *Sinosauropteryx* (whose body was covered by small feathers), *Protarchaeopteryx* (a flightless, turkey-sized dinosaur that had feathers on its tail and forelimbs), *Microraptor* (a four-winged dinosaur), *Caudipteryx* (a peacock-size flightless dinosaur having feathers on its hands and tail), and, in 2008, *Epidexipteryx* (a small dinosaur clad in downy plumage and sporting four long plumes from its tail). These fossils document that feathers and feather-like structures were characteristic of many different dinosaurs, thus supporting the claim that birds evolved from dinosaurs. *Archaeopteryx*—which possesses a combination of bird features and dinosaur features, just as evolutionary theory would predict—remains the most famous of these fossils. Other dinosaurs such as *Deinonychus* ("Velociraptor" in the *Jurassic Park* movies) were structurally similar to birds and had similar behaviors (e.g., *Oviraptor* was discovered in brooding postures atop eggs, as occur in birds). Similarly, there are many bird fossils from the Mesozoic that have features of more dinosaur-like animals such as *Archaeopteryx:*

Rahonavis, discovered in Cretaceous sediments in Madagascar, had a bony tail, long clawed fingers, and teeth like *Archaeopteryx,* but hips like modern birds.

Confuciusornis had fused tail vertebrae and no teeth, but had long dinosaur-like fingers.

Sinornis had teeth but shorter fingers, an opposable toe for perching, short fused tail vertebrae, and a broad breastbone.

And in 2009, paleontologists working in northeast China unearthed several specimens of *Anchiornis huxleyi,* the oldest-known feathered dinosaur. *Anchiornis* was about 25 centimeters tall at its hips, and was 1 to 11 million years older than *Archaeopteryx.*

Modern birds are so adapted to flight that it is hard to link them with any group of land-bound animals. However, thanks to *Archaeopteryx* and several other transitional forms, we now know that birds are the only surviving dinosaurs. When we eat turkey sandwiches or chicken salad, we're eating dinosaurs.

Tiktaalik: From Fish to Land Vertebrates

Tiktaalik, which lived about 375 million years ago, was unearthed by Neil Shubin and his colleagues in 2004 on Ellesmere Island in the Canadian arctic territory of Nunavut. The Canadian arctic might seem like an unlikely place to find an important fossil, but it was not. On the contrary, its discovery shows the predictive power of paleontology, for Ellesmere Island was a site where Shubin *chose* to look for the fossil. Here's why: When Shubin began his quest to find a transitional form between fish and tetrapods, he knew from the fossil record that the fish-tetrapod transition had occurred about 375 million years ago. Thus, if Shubin were to find fossils documenting this transition, he had to look in 375-million-year-old (i.e., Devonian, "The Age of Fishes") sediments from the likely environments in which the transitional forms lived (i.e., tropical rivers). Ellesmere Island fit that description: 375 million years ago the island had been part of a tropical river delta. Even with this knowledge, it took Shubin and his colleagues five trips to the area to find *Tiktaalik.*

The discovery of *Tiktaalik* confirmed theories about how land-dwelling tetrapods (the name given to all land vertebrates) evolved from fish ancestors. *Tiktaalik* was technically a fish; it had scales, gills, and thin ray bones. It also had a tetrapod-like head, a neck, elbow, thick vertebrate-like ribs, and wrist and finger bones that presumably it could use to prop itself up. Unlike fish, *Tiktaalik* also lacked gill plates, which prevent fish from being able to move their head from side to side independently of their body. Thus, *Tiktaalik* is a classic example of a transitional form that bridges the evolutionary gap between two different types of animals—in this instance, between swimming fish and their descendents, the four-legged vertebrates, which include dinosaurs, amphibians, birds, and mammals. For almost every trait, *Tiktaalik* splits the difference between fish and amphibians.

As was true for *Archaeopteryx, Tiktaalik* was merely one of many transitional forms linking these lineages (Figure 3.9). For example, the early tetrapod *Acanthostega*—a 365-million-year-old ancestor to modern land animals—was a transitional form showing the development of legs and having a mix of fish traits and land animal traits (e.g., a fish-like tail, tetrapod-like limbs with fully-formed digits, a flattened amphibian skull, and a fish-like skeleton similar to those of modern lungfishes). *Acanthostega,* which dates from the boundary between the Devonian and Carboniferous, was a fish-like amphibian that lived in water, but could breathe air and use its paddle-like limbs to walk underwater.

The skeleton of *Tiktaalik* indicated that it probably breathed through gills and through a lung-like structure, like the subsequent *Ichthyostega* (a land-dwelling salamander-like tetrapod that appeared 20 million years later). *Ichthyostega*[12] had lungs, fingers and toes, muscular shoulders, and an amphibious life style. If it moved at all on land, it probably moved like a sea lion by arching its back and advancing both forelimbs, then bringing up the rest of its body. *Eusthenopteron*—a prehistoric lobe-finned fish discovered in 1881 in a collection of Canadian fossils—looked and behaved like a modern fish, but its skull, teeth, and fins were similar to those of amphibians that appeared 50 million years later. This creature lived 385 million

---

12. *Ichthyostega* was discovered in 1932 in Greenland. When *Ichthyostega* was alive, Greenland was near the equator. The lineage of *Ichthyostega* has gone extinct; paleontologists consider *Ichthyostega* to be a "dead end" side-branch instead of a direct ancestor of tetrapods.

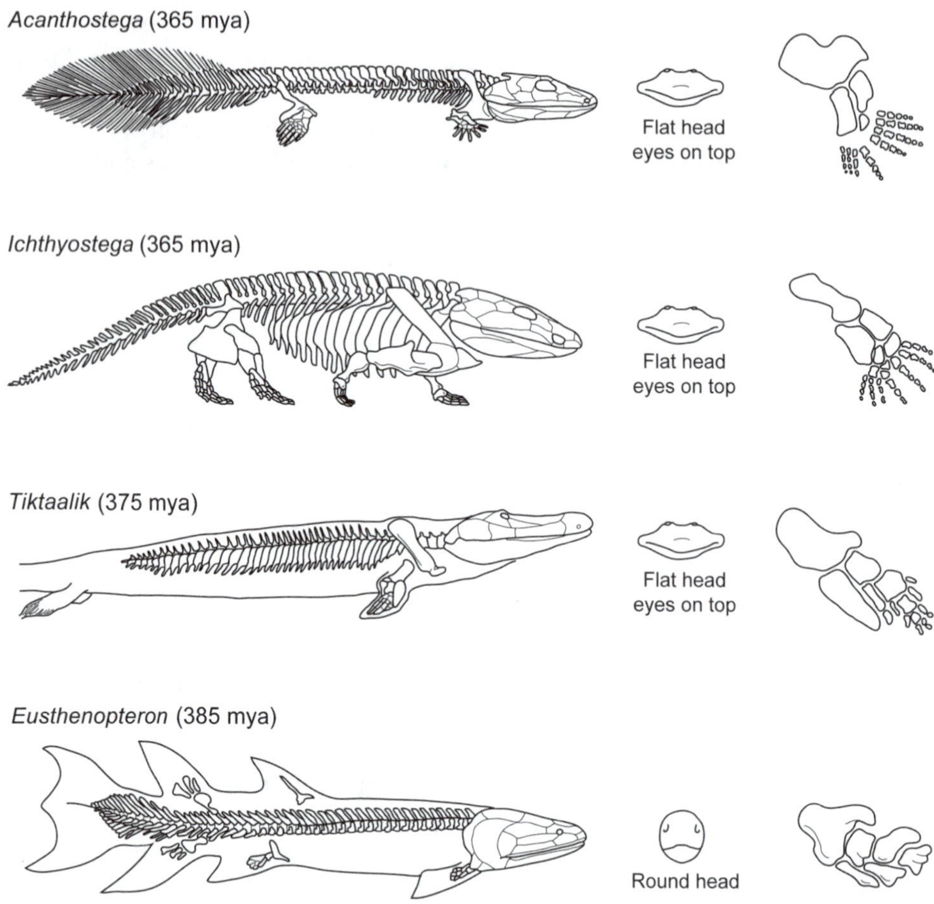

Figure 3.9 *Tiktaalik* is a transitional form bridging fish with tetrapods. Although *Tiktaalik* was not a terrestrial animal, it had muscular bony limbs and wrists that enabled it to prop itself up. It had a head like a crocodile, a body like a salamander, a tail and rear like a fish, and, unlike any fish, a neck that could turn its head. *Tiktaalik* linked primitive tetrapods such as *Ichthyostega* and *Acanthostega* with lobe-finned fishes such as *Eusthenopteron,* which had bones that compare to our upper arm and forearm. These features of *Eusthenopteron* help account for why many textbooks have referred to the *Eusthenopteron* as a "missing link" between fish and tetrapods. Although the fossil record documents the steps by which fleshy-finned vertebrates evolved four legs and, eventually, the ability to walk on land, supporting evidence also comes from comparative anatomy, molecular biology, and stratigraphy. (Jeff Dixon)

years ago, and hidden in its fins were the precursors of the arm and leg bones of tetrapods that appeared prominently in *Tiktaalik.*

Creationists claim that transitional fossils do not exist, and have dismissed *Archaeopteryx* as "just a bird," and *Tiktaalik* as "just a fish." These claims may be convenient for creationists' defenses of their religious claims, but they exhibit a willful and gross ignorance of science, evidence, and logic.

In *On the Origin of Species*, Charles Darwin predicted that "if my theory [of evolution by natural selection] be true, numberless intermediate varieties, linking most closely all the species of the same group together, must assuredly have existed." He was right. Just as evolutionary theory would predict, there are hundreds of transitional forms that document the evolution of life on Earth.

Environmental Changes and Competition Inherent in Natural Selection Produce Extinctions

Fossils as Evidence of Extinction

Before Darwin's time, many people believed that life on Earth had not changed since creation. For example, in 1686, botanist and Anglican priest John Ray (1627–1705)—the "father of natural history" and the greatest British naturalist before Charles Darwin—claimed that "the works created by God at first" had been "by Him conserved to this day in the same state and condition in which they were first made." But some naturalists soon began uncovering fossils (Chapter 2) that were unlike anything then alive. If life on Earth had not changed since creation, why did fossils make it appear that it had?

Some people dismissed fossils as accidental resemblances of rocks to living organisms; others believed that the strange organisms depicted by fossils were still alive in foreign lands. However, some people accepted the simplest explanation—namely, that many species that had lived earlier were no longer in existence. This meant that many species had become extinct.

Cuvier's Theory of Extinction

Extinctions were initially documented most thoroughly by Georges Cuvier. Cuvier worked at a museum that had a vast collection of fossils and mummified animals provided by Napoleon's plundering armies. Cuvier studied these fossils meticulously. When he compared the skulls of elephants living in India and Africa with skulls of fossilized elephant-like animals unearthed in Siberia, Europe, and North America (where no elephants live now), he concluded that elephants of India and Africa were different species, and that both were different from the ancient mammoths excavated in Siberia. As Cuvier noted, they all differed "as much as, or more than, the dog differs from the hyena." In 1796, Cuvier—for the first time—formally proposed a theory of extinction in a famous paper titled *Notes on the Species of Living and Fossil Elephants.*

Cuvier's claim was troubling for some religious adherents: for example, why would God create species only to eradicate them later? To reconcile extinctions with his religious beliefs, Cuvier claimed that the fossilized organisms had died in a series of catastrophes; as Cuvier noted in 1796, there was "a world previous to ours, destroyed by some kind of catastrophe," and that "life on Earth has been disturbed by terrible events" that killed "numberless living beings. . . . Their races even have become extinct, and have left no memorial of them except some small fragments which the naturalist can scarcely recognize." Each of these catastrophes, Cuvier claimed, eliminated some species that dated from creation, and after each catastrophe, organisms repopulated the world. According to Cuvier, the species that repopulated the world were not new species; instead, they were existing species for which scientists had not yet found fossils, and these species remained unchanged until they, too, were destroyed by the next catastrophe. Cuvier's explanation enabled him to reconcile extinctions with his religious beliefs and with geologists' discoveries of biostratigraphy and an ancient Earth. Ironically, however, Cuvier refused to accept the insights provided about life's history by his own work. Although Cuvier could offer no explanation for the catastrophes he proposed, he vehemently opposed evolution until his death in 1832. Nevertheless, his findings provide powerful support for evolution; as he himself had noted, in 1812, "Why has not anyone seen that fossils alone gave birth to a theory about the formation of the earth, that without them, no one would have ever dreamed that there were successive epochs in the formation of the globe. . . . By [fossils] we are enabled to ascertain, with the utmost certainty, that our earth has not always been covered over by the same external crust, because we are thoroughly assured that the organized bodies to which these fossil remains belong must have lived upon the surface before they came to be buried, as they are now, at a great depth."

Cuvier's evidence and arguments about extinction were rejected by many naturalists because they believed that organisms had been created perfectly by God and therefore could not have simply vanished from Earth. These naturalists noted that organisms are exquisitely adapted for the lives that they lead. For example, plants that live in deserts often have succulent, water-storing roots and stems—these organisms live where water is scarce, and their ability to store water is important for their survival. Many people believed that these and other correlations of structure and function throughout nature were evidence that the organisms were created perfectly by a deity for life in their particular environments. Many naturalists also rejected extinction because they believed that extinctions contradicted the Bible. After all, if organisms were the creations of God, how could they be so imperfect that they would disappear from the planet? And then there was the issue of the *extent* of extinctions. Although Charles Darwin famously concluded *On the Origin of Species* (Appendix 1) by noting that "endless forms most beautiful and most wonderful have been, and are being, evolved," the fossil record documents that extinction has been the fate of virtually *all* species ever to appear on Earth, because more than 99% of all species that have lived on Earth no longer exist. Stated another way, it is the fate of almost all species to go extinct. If species were created by God, how could these extinctions occur? Why have virtually all species been failures? How can the constant extinction of species throughout history be evidence of perfect design?

Many people struggled with these and related questions. For example, Thomas Jefferson's *Notes on the State of Virginia* (1781)—the only book that Jefferson published in his lifetime—famously detailed Jefferson's political philosophy, condemnation of slavery, ideas about religious freedom, and interest in fossil vertebrates. Jefferson was most interested in "the Mammoth, or big buffalo," a creature he claimed to be six times the size of an elephant. Jefferson used this organism to refute Buffon's claim that organisms in the New World were smaller, slower, and less fertile than their counterparts in the Old World. Jefferson's problem was that he had fossils, but had no living mammoths, and therefore could not prove that mammoths were still in North America. To find mammoths, Jefferson financed several expeditions to search for these and other mysterious organisms. For example, in late 1781, Jefferson asked Daniel Boone to deliver a note to Jefferson's friend George Rogers Clark asking Clark to go to Big Bone Lick, Kentucky, to retrieve mammoth bones. Jefferson viewed such bones as "the most desirable object in Natural History," and pledged that "there is no expense . . . which I will not gladly reimburse to procure them safely." Clark found giant bones but no living mammoths.

Although Cuvier was an influential scientist, other naturalists soon began to question his claims about the causes of extinctions. For example, Cuvier believed that each layer of sediment described a catastrophe, but geologists soon discovered that there are *hundreds* of distinct layers of sediments and fossils (Chapter 2). Were there hundreds of these catastrophes? Thousands of them? Moreover, some organisms appeared in only one layer of sediment, while others appeared in several layers, and human bones were found in layers much older than 6,000 years. These observations were not consistent with Cuvier's claims that all extinctions were caused by catastrophes.

Background Extinction

By the time that Jefferson died in 1826, most people accepted extinction; as Scottish geologist Hugh Miller stated emphatically in 1857 in his best-selling *The Testimony of the Rocks*, "All geologic history is full of the beginning and the ends of species—of their first and last days." Today we know from overwhelming evidence that extinctions have occurred throughout Earth's history. Some extinct species have become famous (e.g., dodo birds,

Tyranosaurus rex), but most have vanished quietly and remain unknown.[13] Many species have been eliminated by factors such as disease, climate change, loss of habitat, and the competition inherent in natural selection. These ongoing "normal" extinctions account for approximately 80% of all extinct species and are called **background extinction.** Although dodo birds and American passenger pigeons are extinct because they were hunted until none were left, most extinct species disappeared long before humans existed. Currently, increasing numbers of species are threatened with extinction because of human activities, which are destroying or altering habitats at unprecedented rates.

Mass Extinction

Although rates of background extinction have varied, there have been several relatively short geologic periods during which especially large numbers of species disappeared from broad taxonomic groups, diverse ecosystems, and vast geographic ranges. These pulses of extinctions, or **mass extinctions**, account for approximately 20% of all extinctions (Figure 3.10). There have been several mass extinctions in Earth's history, and proposed explanations for mass extinctions have involved factors ranging from drastic declines in sea level to catastrophic bursts of cosmic radiation. Within the last few decades, however, geologists have linked some mass extinctions with environmental changes produced by meteor impacts and volcanic activity.

The K-T Extinction

We'll start with the most famous of these mass extinctions—the **K-T event** 65 million years ago (at the end of the Cretaceous [K, for the German *Kreide*] and the beginning of the Tertiary [T]) that was probably triggered by environmental changes caused by meteor impacts and volcanic activity.[14] This mass extinction is most famous not because it eliminated more species than any other mass extinction; the Permian mass extinction 251 million years ago (discussed in the next section) was much more devastating. The K-T extinction is most famous because of its most celebrated victims: the nonavian dinosaurs.

The K-T mass extinction involved environmental changes produced by volcanic eruptions and meteor impacts. Not all meteors trigger mass

---

13. Dodos became famous after mathematician Charles Dodgson (also known as Lewis Carroll) included the dodo in *Alice in Wonderland,* which is one of the most popular children's stories of all time. Today, the dodo is a symbol of human-caused extinction.

14. *Cretaceous* is Latin for *chalky* and is designated K for its German translation *Kreide* (*chalk*). The K-T boundary is marked by a chalky sediment.

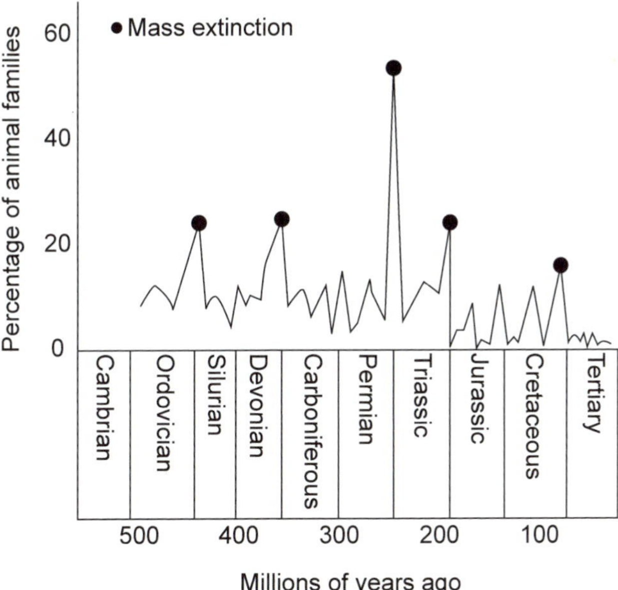

Figure 3.10 Although low rates of extinction have been a common and natural occurrence, the history of life on Earth has been punctuated by several pulses of mass extinctions, during which large numbers of species have disappeared within a few million years. The Cretaceous-Tertiary (K-T) mass extinction eliminated the dinosaurs, but a far more severe die-off ended the Permian. Some of the less-dramatic peaks of extinction may result from distortions in collected data rather than from catastrophic events. (Jeff Dixon)

extinctions when they hit Earth; there have been hundreds of impacts that have *not* caused mass extinctions. However, volcanic eruptions combined with meteor impacts could affect life on earth in many ways. The vaporization of compounds caused by a major meteor impact would alter atmospheric chemistry, and that, together with vast amounts of floating debris, would result in global cooling. There is evidence of widespread fires accompanying the K-T event, which would have produced soot and added to global cooling. The meteor impacts themselves may have also increased volcanic activity, thereby producing more atmospheric disturbance. Note, however, that the meteors themselves did not cause the mass extinction by crushing Earth's animals and plants. Rather, the meteors had an indirect effect—their impact disrupted Earth's climate, geochemical cycles, ocean currents, and ultimately, ecological interactions. These changes, which dramatically changed Earth's environment, had long-term extinction effects that lasted for half a million years.

This linkage of meteor impacts with the demise of dinosaurs is an interesting story, but how do we know it's true? That is, how do scientists know that the K-T mass extinction was associated with a several-kilometers-wide meteorite that struck Earth? The story begins in Gubbio, Italy, where exposed limestone formations bracket the end of the Mesozoic and the beginning of the Cenozoic. There, in 1980, Walter and Luis Alvarez noted that the Cretaceous (K) and Tertiary (T) are separated by a thin band of clay. This clay marks the so-called K-T boundary throughout the world. When the Alvarezes examined this clay, they found that it contained large amounts of iridium, a noble metal rare on Earth's surface but abundant in meteors (the sediments immediately above and below the band of clay contained only trace amounts of iridium). But there was much more evidence. For example, tiny grains of melted and recrystallized minerals called microtektites created by the heat of an impact occur throughout the Caribbean. The shape of these tiny glasslike grains indicates that they cooled as they splattered away from an impact site. There are also sediments in the Caribbean that document tidal waves that occurred at exactly the time of the impact. And, finally, scientists found the impact site itself: a 180-kilometer-wide crater near Chicxulub, Yucatán, which was formed 65 million years ago.

For many years, most scientists accepted that the impact and aftermath of the Chicxulub meteor accounted for the dinosaurs' demise. However, the explanation was expanded by the discovery of an even larger, 65-million-year-old crater just off the coast of Mumbai, India. This crater, which is 500 kilometers in diameter, was formed by a meteor approximately 40 kilometers in diameter. The impact formed the Shiva crater (named after the Hindu deity of destruction and renewal) and may have triggered volcanic eruptions that helped form the Deccan Traps, which are vast fields of basalt in west-central India. The Deccan Traps are one of the largest volcanic formations on Earth; they are more than 2,000 meters thick and cover more than 500,000 square kilometers. These volcanic eruptions could have created a tremendous dust cloud that decreased photosynthesis, lowered Earth's temperature, and ultimately eliminated 55%–65% of the species alive at that time. The combined effects of these huge impacts drove some groups of plants and animals—most notably the nonavian dinosaurs—into extinction, but others (e.g., other reptiles, such as turtles, crocodilians, lizards, and snakes) were relatively unaffected.

The Permian-Triassic Extinction

Although the K-T extinction is the most famous mass extinction, the most devastating mass extinction was the Permian-Triassic mass extinction, which occurred approximately 251 million years ago and marked the transition from the Paleozoic to the Mesozoic. This extinction, which is called "The Great Dying," eliminated more than half of all genera and as many as 90% of all species. Immense shallow-water reefs died off completely.

This extinction, which probably took less than 200,000 years, was associated with continental drift (Chapter 4). At the end of the Permian, the continents had just joined to form a single land mass called Pangaea (prior to that, continental masses had joined and separated several times, but scientists have been unable to reconstruct those arrangements with much precision). This altered oceanic currents which, in turn, changed Earth's climate (as do today's changes in ocean currents such as those caused by El Niño). Supporting this claim about climate change is the fact that glaciers formed and retreated several times during this extinction. If currents did not provide enough oxygen to ocean depths to sustain life, these deep waters would have become stagnant. If surface photosynthesis had persisted, then atmospheric carbon dioxide would have been depleted, and global temperatures would therefore have decreased. Advancing glaciers would have increased ocean circulation, which in turn would have brought the deep ocean waters to the surface, poisoning surface life, releasing carbon dioxide to the atmosphere, and increasing temperatures. This cycle probably continued until Pangaea fragmented.

The Permian-Triassic mass extinction (Figure 3.10) was also associated with vast ozone-depleting volcanic eruptions in Siberia that formed step-like slabs of flood basalts—the so-called Siberian Traps—that covered

SIDEBAR 3.2 Famous Fossils and Fossil Sites

Earth's tropical areas, especially southern and Southeast Asia, are usually covered with plants, and therefore have yielded relatively few fossils. However, arid regions such as deserts have exposed—and often eroded—landscapes in which fossils are abundant.

Dinosaur National Monument

Dinosaur National Monument is along the southeastern side of the Uinta Mountains on the border between Utah and Colorado at the confluence of the Green and Yampa Rivers. These fossil beds, which were discovered in 1909 by Earl Douglass of the Carnegie Museum of Natural History, are the largest known quarry of Jurassic Period dinosaurs ever discovered. More than 2,000 bones are exhibited in the quarry's sandstone wall.

Lyme Regis

Lyme Regis in Dorset, southwest England, is where famed fossil-hunter Mary Anning (1799–1847) found one of the first ichthyosaurs (dolphin-shaped marine reptiles) in 1814, a virtually complete plesiosaur (a long-necked marine reptile) in 1824, and a pterosaur in 1828. The coastline of Lyme Regis is continually eroded by water and wind, and fossil hunters continue to search the so-called Jurassic Coast for fossils every day.

Hell Creek

Hell Creek, Montana, is where the first *Tyrannosaurus rex*—the world's most famous real dinosaur—was unearthed. *T. rex* soon became legendary—it was up to 15 meters long, weighed up to 110 kilograms, had a skull 1.7 meters long, and had a little toe 40 centimeters long; some of its teeth (including their roots) were 30 centimeters long. Chicago's Field Museum paid $8.4 million for *T. rex* Sue, the largest amount ever paid for a fossil. Hell Creek houses one of the finest collections of Mesozoic fossils in North America; these fossils are displayed at museums throughout the world, but many of the best are in the Museum of the Rockies in Bozeman, Montana.

Rancho La Brea Tar Pits

The Rancho La Brea Tar Pits in suburban Los Angeles are a series of approximately 100 tar pits from which scientists have extracted more than 600 species that lived between 40,000 and 10,000 years ago. These species include mammoths, mastodons, and saber-toothed cats.

Burgess Shale

The Burgess Shale in the Rocky Mountains near Field, British Columbia, was discovered in 1909 by American paleontologist Charles Walcott. These shales, which date from the Cambrian (530 million years ago), document the Cambrian Explosion, which is one of the earliest-known bursts of evolution. The Burgess Shale organisms include *Hallucigenia,* a wormlike organism having a blob-like head and narrow tail (or the reverse). This organism had several pairs of spines protruding from the upper and lower sides of its body.

Dinosaur Ridge

Dinosaur Ridge (just outside Denver, Colorado) is where fossils of some of the best-known dinosaurs—for example, *Stegosaurus, Apatosaurus,* and *Allosaurus*—were first discovered. These were among the first collections from the famed Morrison Formation of the late Jurassic.

Neander Valley

Neander Valley is a small valley near Düsseldorf, Germany, where portions of a human skeleton were discovered in a limestone cave (Little Feldhofer Grotto) in 1856. In 1863 these fossils became the type specimen of Neandertal Man, the archetypal "caveman" and an extinct cousin of our own species. Interestingly, recent studies of the Neandertal genome have shown that humans and Neandertals interbred; 1%–4% of DNA in today's Asians and Europeans was inherited from Neandertals (see Chapter 9). The "Lagar Velho Child," a 24,500-year-old child's skeleton discovered in the 1990s in Portugal, has a mix of Neandertal and *Homo sapiens* traits. Neandertal traits include short legs, large front teeth, a broad chest, and a large jaw; human traits include small lower arms and a well-formed chin.

Hadar, Ethiopia, and Olduvai Gorge, Tanzania

Hadar, Ethiopia and Olduvai Gorge, Tanzania, are where several famous hominins were discovered, including "Lucy" (*Australopithecus afarensis*). These fossils are discussed further in Chapter 9 (also see Appendix 3).

Petrified Forest National Park and Yellowstone National Park

In Arizona's Petrified Forest National Park, 1,300 feet of volcanic ash buried and petrified more than 40 genera of plants. Most of these fossils are *Araucaria*-like trees.* At Yellowstone National Park, more than 27 individual forests are preserved in more than 600 meters of volcanic ash.

Gobi Desert

Mongolia's Gobi Desert was first explored for fossils in the 1920s when scientists led by adventurer Roy Chapman Andrews were searching for the "missing link" between humans and apes. The rocks explored by Andrews were 25 to 70 million years old—far too old for human remains—but the expedition did uncover the first known nest of dinosaur eggs (of the dog-sized *Protoceratops*) and the largest-known terrestrial mammal, the 20-ton *Paraceratherium* (formerly known as *Indricotherium* and *Baluchitherium*). Andrews helped inspire the fictional movie hero Indiana Jones.

Dinosaur Valley State Park

Dinosaur Valley State Park, just northwest of Glen Rose, Texas, is where many of the best-preserved dinosaur tracks have been excavated (Figure 3.1c). The original tracks, excavated here from the Paluxy River in 1939 by famed fossil-hunter Roland Bird, are now exhibited in the American Museum of Natural History behind *Apatosaurus,* which in 1905 became the first sauropod dinosaur ever mounted. Visitors to Dinosaur Valley State Park can still see dinosaur tracks in the Paluxy River, as well as on display in the Glen Rose town square.

Liaoning

Liaoning in northeast China produced *Microraptor gui,* a small predatory dinosaur that had feathers on all four limbs. This small, predatory dinosaur is the only known animal having this feature. This site has also produced other dinosaurs (e.g., *Sinornithosaurus*) having featherlike structures, as well as *Repenomamus gigantus,* a dog-sized mammal that overturned the earlier belief that no mammal from the Mesozoic (i.e., the Age of Dinosaurs) was much larger than a rat.

---

* Petrified wood occurs throughout the world, but is exceptionally well preserved at Petrified Forest National Park. Interestingly, Petrified Forest National Park was not a forest, but instead an area where tree trunks were carried by water and buried in sediment and volcanic ash.

an area almost as large as the continental United States. The enormous amount of lava that was released by these eruptions could have covered Earth with a 3-meter-deep blanket of lava (the volume was 2.5-million km^3; for comparison, the eruption of Mount St. Helens in 1980 released 0.75 km^3 of lava). These eruptions poisoned the atmosphere, depleted the oceans of oxygen, and produced a runaway greenhouse effect (CO_2 levels were 1,000–1,500 ppm, far greater than today's level of ~390 ppm). The eruptions may have been triggered by an asteroid that struck Earth at about this time; craters ranging from 200 to 500 kilometers in diameter formed at this time have been found northwest of Australia and beneath the East Antarctic ice sheet. Drilling at these sites has recovered rocks formed at sudden and extreme temperatures and pressures similar to those generated by an impact, as well as unusual isotopes of chromium, suggesting an extraterrestrial impact. (Similarly, a swarm of comets that hit North America 12,900 years ago may have triggered a 1,300-year-long ice age that wiped out "megafauna" species such as mammoths and saber-toothed cats.) However, these claims are controversial, for changes in Earth's crust have destroyed most of the crustal geology of the Permian.

Extinctions do more than just reduce the number of species; they also reorganize ecosystems. In doing so, extinction creates new opportunities for adaptive radiation and species formation. Just as the Permian extinction made room for reptiles, the extinction of nonavian dinosaurs at the K-T extinction opened ecological niches that allowed the adaptive radiation of mammals. Although mammals coexisted with dinosaurs, mammals dominated their environments only after nonavian dinosaurs disappeared.

Human-Caused Extinction

Today, many scientists believe that we are in the midst of another mass extinction. This extinction, unlike previous ones, is not being caused by rogue meteorites or erupting volcanoes, but instead by a potentially more deadly threat: human activities.

Humans' impact on other species was noted as early as 1864, when George Perkins Marsh wrote in *Man and Nature* that "Man is everywhere a disturbing agent. Wherever he plants his foot, the harmonies of nature are turned to discords." Five years later, famed biologist Alfred Russel Wallace—who had a strong aesthetic interest in species—made a similar observation. While pondering a "perfect little organism" he had collected in the Malay Archipelago, Wallace speculated in *The Malay Archipelago* (1869) about the destructiveness of "civilized" societies:

> I thought of the long ages of the past, during which the successive generations of this little creature had run their course—year by year—being born, living and dying amid these dark and gloomy woods,

> with no intelligent eye to gaze upon their loveliness—to all appearances such a wanton waste of beauty. Such ideas excite a feeling of melancholy. It seems sad that on the one hand such exquisite creatures should live out their lives and exhibit their charms only in these wild inhospitable regions, doomed for ages yet to come to hopeless barbarism; while on the other hand, should civilized man ever reach these distant lands, and bring moral, intellectual, and physical light in the recesses of these virgin forests, we may be sure that he will so disturb the nicely balanced relations of organic and inorganic nature as to cause the disappearance, and finally the extinction, of these very beings whose wonderful structure and beauty he alone is fitted to appreciate and enjoy. This consideration must surely tell us that all living things were *not* made for man. Many of them have no relation to him. The cycle of their existence has gone on independently of his, and is disturbed or broken by every advance in man's intellectual development; and their happiness and enjoyments, their loves and hates, their struggles for existence, their vigorous life and early death, would seem to be immediately related to their own well-being and perpetuation alone, limited only by the equal well-being and perpetuation of the numberless other organisms with which each is more or less intimately connected.

Wallace's words have been prophetic, for the arrival of humans to pristine areas has almost always been associated with soaring extinction rates of native organisms. One of the most famous of these extinctions was the dodo (*Raphus cucullatus*), a turkey-sized flightless bird that lived in rainforests on the island of Mauritius in the Indian Ocean. The dodo had no natural predators until humans arrived on the island. However, by 1662—approximately 30 years after humans arrived—it had been hunted to extinction (rats introduced to Mauritius by humans also ate many of the dodos' eggs). Since then, more than 60% of the other endemic birds of Mauritius have also disappeared.

Dodo birds are not an isolated example of human-driven extinctions:

Approximately 45,000 years ago in Australia, 90% of mammalian species were driven to extinction within 5,000 years of human settlement.

Approximately 4,500 years ago in the Caribbean, several species of sloth were driven to extinction within 800 years of human settlement.

Approximately 1,000 years ago in Madagascar, elephant birds (*Aepyornis,* which were up to 3-meters tall and weighed more than 300 kilograms) and other large animals were extinct within 1,000 years of humans' arrival. Elephant birds produced the biggest egg known to be laid by any animal—up to a meter in circumference and containing about 11 liters of liquid.

Approximately 500 years ago in New Zealand, large flightless birds called moas (e.g., *Dinornis,* which was up to 3.5 meters tall and weighed as much as 200 kilograms) were hunted to extinction within 200 years of human settlement (and the simultaneous introduction of rats and dogs).

The extinction of ground sloths and other large mammals followed human colonization, and especially the development and spread of advanced agriculture (including fire) and hunting techniques.

In Hawaii, Polynesians may have caused the extinction of almost half of the native avifauna. Today, Hawaii has more endangered and threatened species than any state in America.

The passenger pigeon (*Ectopistes migratorius*) was once the most abundant bird in North America, and was hunted for its tasty meat. After millions of the birds had been killed, a law was passed to protect them, but it was too late. The last known passenger pigeon living in the wild was shot in early 1899. "Martha," the last passenger pigeon, died in the Cincinnati Zoological Garden in 1914.

The heath hen (*Tympanuchus cupido*) was abundant in the northeastern United States during colonial times. However, overhunting reduced the population to approximately 70 hens by 1900, and to only 12 hens in 1927. The last hen, nicknamed "Booming Ben," died on Martha's Vineyard in 1932.

"Incas," the last Carolina parakeet (*Conuropsis carolinensis*) and the only parrot native to North America, died in captivity in 1918 at the Cincinnati Zoological Garden in the same cage in which "Martha," the last passenger pigeon, had died four years earlier.

The Great Auk (*Pinguinus impennis*), a meter-tall bird that was once abundant along North Atlantic coasts of Canada, Norway, Greenland, and other countries, was hunted to extinction in the mid-19th century. The last pair of these birds was killed in July, 1844.

The last Tasmanian wolf (*Thylacinus cynocephalus*)[15] died of neglect in captivity on September 7, 1936. Today, Australia commemorates that extinction by observing National Threatened Species Day on September 7 every year.

Today, species seldom become extinct because of human hunting, but instead from habitat degradation due to population growth, the effects of introduced species, and our growing consumption of natural resources. Indeed, newspapers are filled with stories of pollution, habitat destruction, atmospheric and oceanic warming, coral reef destruction, and massive

15. The Tasmanian wolf is also referred to as the Tasmanian tiger because of its striped back.

nutrient runoffs that are degrading the environment at rates measurable in tens or hundreds of years instead of thousands or more. Taken together, these impacts have produced extinction rates more than 1,000-times the normal background rate of extinction:

A 2009 report by the International Union for the Conservation of Nature reported more than 11,000 endangered plants and animals. These organisms include more than 4,500 plants, one-fourth of all mammals, one-third of amphibians, and 15% of all birds. The 2009 report listed more than 2,500 more species than the 2008 report.

In 1990 there were more than 100,000 tigers in Asia, but in 2010, tigers were on the verge of extinction, with fewer than 3,300 remaining in the wild.

There are fewer than 60 Javan rhinoceros (Indonesia, Vietnam), fewer than 300 of the crustacean vaquita (Gulf of California), fewer than 1,500 black-footed ferrets (Great Plains of North America), fewer than 2,000 giant pandas (China, Burma, Vietnam), and fewer than 80 golden-headed langurs (Vietnam) remaining on Earth.

All the great apes—our closest relatives—are endangered.

Of about 19,000 species of birds worldwide, more than 120 have disappeared in the last 500 years. Another 1,200 species are threatened—all but just a handful by human activities.

Humans have identified and described fewer than 2 million of the estimated 8–10 million species on Earth today. Yet many species are disappearing before we can even describe them, much less determine if they could be useful sources of food or medicines. Current extinction-rates are exceedingly high, even without the assistance of misplaced ocean currents, volcanic eruptions, or rogue meteors. Extinctions deprive us of natural wonders, the raw materials for industries such as engineering and medicine, and the opportunity to learn more about the history of life on Earth.

Summary

Life on earth today is the product of repeated changes of past life, many of which are documented by fossils. The chronological appearance of life throughout Earth's history, as documented by fossils, makes sense if these organisms arose by evolution, but would be highly improbable otherwise. Extinctions represent biological failures; these organisms have disappeared from the Earth. Extinct organisms are those that could not adapt to competitors, changing climate, or other forms of environmental change. Mass extinctions have opened new niches, thereby allowing the adaptive radiation of other organisms.

4

Biogeography

> When on board H.M.S. "Beagle" as naturalist, I was much struck with certain facts in the distribution of the inhabitants of South America, and in the geological relations of the present to the past inhabitants of that continent. These facts seemed to me to throw some light on the origin of species—that mystery of mysteries, as it has been called by one of our greatest philosophers.
>
> —*Charles Darwin, 1859*

PREDICTIONS

If evolution has occurred, we predict that

1. Life will be extremely diverse, and this diversity will be influenced by geologic history.
2. Species that are most alike will usually be near each other geographically, regardless of differences in environment.
3. Major, long-term changes in the distribution of life's diversity will be influenced by plate tectonics.
4. Different species in similar habitats will evolve similar adaptations.
5. Isolated habitats such as oceanic islands will be populated by descendants of organisms from the nearest mainland.

Darwin's Observations of Life's Diversity

Charles Darwin's ideas about life's diversity were influenced by his observations during and after his five-year voyage aboard the *Beagle.* Darwin was impressed by a variety of things that he saw, including fossils of glyptodonts, which were giant—and now extinct—armadillo-like animals that lived in Argentina. Darwin formulated a hypothesis to account for his discovery, which he described in a letter to his mentor John Stevens Henslow:

"Immediately when I saw [the fossils] I thought they must belong to an enormous Armadillo, living species of which genus are so abundant here." Darwin had good reason for creating such a hypothesis: Modern armadillos are the only modern animals that resemble glyptodonts, and armadillos live in environments similar to those in which glyptodonts lived. But this reasoning also raised a troubling question: If armadillos and glyptodonts were created at the same time (e.g. during a week-long "creation week"), why are only armadillos still alive? One explanation was that they were *not* created at the same time, and that glyptodonts are the ancestors of armadillos.

During his five weeks in Galápagos in 1835 aboard HMS *Beagle,* Darwin noted that the animals there were similar, but not identical, to those living along the western coast of South America (i.e., the nearest mainland). These similarities were especially striking because of the remarkably different environments of these two places: Galápagos are dry, rocky, volcanic islands, whereas the western coast of South America is humid and includes a lush tropical jungle. Why did such similar animals live in such different environments? How could this be? Darwin believed that such questions "must have struck every traveler," but he could not explain what he was seeing.

Darwin also realized that there were several different species of Galápagos tortoises and iguanas, and later learned that each island was inhabited by a different species of tortoise and iguana. However, Darwin's most famous observations involved birds (especially mockingbirds) that he and some of his shipmates collected while in Galápagos (Figure 4.1). Darwin wondered why Galápagos mockingbirds more closely resembled those on the South American mainland than those living on islands having similar environments but that were much further away. For Darwin, the simplest explanation for these observations was not a special creation event on each island; rather, the Galápagos birds, tortoises, and iguanas had evolved from their South American ancestors that had colonized the Galápagos islands long ago, and the geographic isolation of the archipelago had promoted the formation of new species on each island. This conclusion was based on **biogeography**, which is the study of the distribution of life's diversity.

If evolution occurs, we would expect it to be local; that is, each type of organism would be restricted to the part of the world in which it appears. This prediction is strongly supported by biogeography.

Life Is Extremely Diverse, and This Diversity Is Influenced by Geologic History

Earth's Biodiversity

Our census of life on Earth is incomplete, but nevertheless impressive: Earth's biodiversity is staggering. To appreciate this, examine the estimates of the numbers of species of the various groups of organisms listed in Table 4.1.

Figure 4.1 Small birds called finches are adapted to the different islands of Galápagos. Galápagos finches descended from a group of birds called tanagers that now live in South and Central America and the Caribbean Islands. Contrary to popular belief, "Darwin's finches" did not inspire Darwin's earliest ideas about evolution. Instead, Darwin's views of evolution later enabled him to understand the diversity of finches in the archipelago. Shown here is the medium ground finch (*Geospiza fortis*). (Randy Moore)

Table 4.1 Estimates of Number of Species

Group	Number of Known Species	Number of Probable Species	Percentage of Group Known
Insects	950,000	8,000,000	12
Arachnids	75,000	750,000	10
Nematodes	15,000	500,000	3
Bacteria	5,000	500,000	1
Algae	40,000	200,000	20
Fungi	70,000	1,000,000	7
Vascular plants	250,000	300,000	83
Protozoans	40,000	200,000	20
Molluscs	70,000	200,000	35
Crustaceans	40,000	150,000	27
Vertebrates	45,000	50,000	90

This diversity is often hard to appreciate, because much of it goes unnoticed. For example, insects—the most diverse group of animals—first evolved 400 million years ago. Anyone who's visited a warm, humid area knows that there are many different species of insects—but just how many? Probably *millions.* Indeed, there are more than 350,000 known species of beetles alone; this is more than 70-times the number of species of mammals. Some live in remarkably specialized places—for example, one beetle (*Aphaenops cronei*) lives only in some caves in southern France; in turn, this insect is the only home of a fungus (*Laboulbenia*) that lives only on the rear part of the covering wings of this beetle. Similarly, the larvae of *Drosophila carcinophila* can develop only in specialized grooves under the flaps of the third pair of oral appendages of a land crab that lives only on certain Caribbean islands.

Given the abundance of insects, one would expect that insects would live in many different habitats. They do. In an experiment done in Panama, researchers collected more than 1,000 different species of beetles from *Luehea seemannii,* a medium-sized forest evergreen. Moreover, more than 160 of these species live *only* in *Luehea* trees. Some animals are stunningly abundant—for example, there are more than a million billion ants on Earth. However, other species are endangered, and still others are disappearing. Famed American biologist E. O. Wilson estimates that rainforests alone lose 6,000 species per year (approximately 17 species per day), despite the fact that these biomes cover only 6% of Earth's land.

Biogeography

Biogeography, which was recognized as a new discipline in 1866 by Ernst Haeckel, documents and interprets how Earth's organisms are distributed.[1] It addresses a variety of questions:

How many species are there? Why are some species more abundant than others? For example, why are there more species of insects than all other species of animals combined?

Where do these species live? Why do certain plants and animals live in some places and not others? Why are organisms distributed this way? Have organisms always been distributed this way?

Why do large, isolated regions have such distinctive inhabitants compared to those of continents? For example, why are nearly all of Australia's native mammals marsupials, but placental mammals dominate the rest of the world? Why aren't there any polar bears in Antarctica? Why aren't there any giraffes in Hawaii? What accounts for these distributions?

---

1. At the time, Haeckel referred to biogeography as *chorology*.

Some principles of biogeography are obvious to us now. For example, asked why there are no giraffes in Hawaii, most people would answer that giraffes could not get to Hawaii. This simple answer presumes that giraffes originated elsewhere (i.e., on a continent and not on oceanic islands such as Hawaii). This may seem obvious now, but this was not always the case. Indeed, before we had an understanding of evolution, special creation was a commonly accepted idea, and an omnipotent Creator could presumably have created or placed species *anywhere.* Moreover, the prevailing view in those days about Earth and Earth's life was stasis—Earth and its species were immutable, and Earth's organisms were found where they were created, perfectly adapted to their environments. To many people, this perfection seemed obvious; after all, the webbed feet of ducks help the ducks swim, the sticky webs of spiders help the spiders trap prey, and the long, slender beaks of hummingbirds help the hummingbirds sip nectar from flowers. But some people had questions. For example, Leonardo da Vinci, Charles Darwin, and many others had found fossil seashells atop mountains, suggesting that what are now mountains were formerly underwater. How could this happen, and what happened to the organisms that lived there?

There were other questions. What could explain the occurrence of fossils of several of the same species on different continents of the Southern Hemisphere? For example, fossils of the freshwater reptile *Mesosaurus* are found in South America and Africa, as are fossils of the land reptile *Cynognathus.* Fossils of another land reptile (*Lystrosaurus*) occur in Africa, India, and Antarctica. How could this be? Some biologists explained these similarities by claiming that the continents were once connected by vast land bridges, just as Central America now links South and North America. However, there was little evidence for such bridges. Darwin was skeptical of this explanation, scoffing in a letter to his friend Charles Lyell that such bridges were invoked as explanations for biogeography "as easily as a cook does pancakes."

Biogeography was the most convincing evidence that Darwin cited in his monumental *On the Origin of Species* to make his "one long argument" for evolution by natural selection. Darwin understood the importance of biogeography—he referred to it in 1845 as "that grand subject, that almost keystone of the laws of creation." When he published *On the Origin of Species* 14 years later, Darwin dedicated two chapters (both titled "Geographical Distribution") to the subject, and in Chapter XI he used biogeography to bolster his claim that if we allow for the imperfections of the fossil record, the known fossils are distributed as one would expect based on his theory. In this chapter, you'll see that biogeography provides powerful evidence for evolution, and helps to explain many observations that would otherwise be baffling.

SIDEBAR 4.1 The Mystery of the Galápagos Iguanas

Visitors to Galápagos have long been intrigued by the archipelago's iguanas. For example, when Darwin visited Galápagos in 1835, he described the land iguana (*Conolophus subcristatus*; Figure 4.2a) as "ugly" and having a "singularly stupid appearance"; the marine iguana (*Amblyrhynchus cristatus;* Figure 4.2b, c) was described as "clumsy" and "disgusting." More recently, these organisms were also the central figures in an evolutionary mystery. In 1997, a study based on molecular biology reported that the Galápagos land and marine iguanas (which appeared 10–20 million years ago) are older than the islands on which they now live (which are 3–4 million years old). How could this be? How could the iguanas have colonized islands that didn't yet exist? Solving this evolutionary mystery required an understanding of biogeography.

Figure 4.2a, b, and c Iguanas in Galápagos, an isolated group of oceanic islands. (A) Land iguanas (*Conolophus subcristatus*) have a pointed snout and light coloration. (B, C) Marine iguanas (*Amblyrhynchus cristatus*) are the only iguanas in the world adapted to life in the ocean; adaptations include a long, laterally flattened tail for swimming, a blunt nose for biting algae from underwater rocks, nasal glands to expel salt, and dark pigmentation that absorbs light and thereby minimizes its lethargy after emerging from the cold ocean. Interestingly, in areas where the territories of these iguanas overlap (e.g., South Plaza Island), the iguanas sometimes interbreed and produce hybrids. A newly discovered pink species of land iguana (*C. marthae*) lives exclusively around Wolf Volcano on Isabela Island, the highest point in the archipelago (elevation = 1,700 meters). (Randy Moore)

The Galápagos islands, like all volcanic islands, are the tops of high underwater mountains formed by volcanoes. The youngest islands of Galápagos—Isabela and Fernandina—are along the western edge of the archipelago and are still being formed by active volcanoes. The older islands of Galápagos toward the east are now volcanically inactive. However, the oldest islands of Galápagos are no longer visible; they no longer protrude above sea level, but their remains are still present near the western coastline of South America. Although the visible islands of Galápagos are only 3–4 million years old, the older, now-submerged islands of Galápagos are more than 17 million years old. Thus, the Galápagos iguanas colonized and diversified the oldest (i.e., now submerged) islands and later moved to their current locations. A similar explanation accounts for the current distributions of 10 endemic species of flightless weevils in Galápagos.

This story documents the dynamic, ongoing interplay between life and Earth. The current, tourist-laden islands of Galápagos will eventually disappear. The only future survivors in Galápagos will be those species able to colonize newly formed islands of the archipelago.

Early Theories of Biogeography

In the 1700s, Swedish biologist Carolus Linnaeus (1707–1778)—who believed that God speaks through nature—developed the binomial system of nomenclature to classify the burgeoning list of Earth's organisms. Linnaeus was confident that his system, which we still use today, would explain patterns of biodiversity and, in the process, reveal the divinely created order of nature. Linnaeus believed life originated at the equator on the slopes of a "Paradisical Mountain," where each species was adapted to particular habitats along the mountain's slope. To reconcile his ideas with a biblical flood, Linnaeus argued that Noah's Ark had landed on a tall mountain (e.g., Mount Ararat in Turkey), and that all the Ark's passengers migrated down the mountain and eventually colonized the rest of the world when the floodwaters subsided. But Linnaeus also had questions; for example, how could organisms that had originated at a single site (such as Ararat) become adapted to diverse environments such as deserts and rainforests? And how could they have gotten to remote islands?

Georges-Louis Leclerc, Comte de Buffon (1707–1788), was a contemporary of Linnaeus who had different ideas. If Linnaeus was correct, why were similar habitats in different parts of the world inhabited by different kinds of organisms? How could organisms dispersed from a mountain get to their present locations? And if species were immutable, how could they adapt to new environments? Wouldn't new environments such as mountain ranges, deserts, and oceans have blocked the organisms' dispersal? Buffon rejected Linnaeus's claims about a Paradisical Mountain, instead claiming that life had originated in northwestern Europe at a time when climates were relatively mild; when the environment later changed, organisms migrated south to South America and eventually to the rest of Earth. Although Buffon acknowledged that "Man is totally the production of heaven," he also claimed in *Histoire Naturelle* (1761) that Earth and its species were not static and had, in fact, "improved" or "degenerated" as they became isolated in different areas and environments, often to the point that they became "so changed that they are hardly to be recognized." (Alfred Russel Wallace and Charles Darwin later attributed many of these and other changes to evolution by natural selection.) Buffon's recognition that isolated habitats having similar environments are populated by distinctive groups of organisms may seem obvious today, but it was a revolutionary claim in the 18th century. His observation, which became known as **Buffon's Law,** was the first principle of biogeography.[2]

But just how accurate or applicable were these explanations about the distribution of life on Earth? Answers to these and related questions awaited the exploration of new parts of Earth.

Species That Are Most Alike Usually Live Near Each Other Geographically, Regardless of Differences in Environment

Biogeography in the Age of Exploration

The further development of biogeography was linked to the Age of European Exploration, a period during the 18th and 19th centuries in which explorers began traveling the world in search of new land, knowledge, and treasures. Although Darwin's voyage aboard the *Beagle* is one of the most famous of these expeditions, there were many others that also profoundly influenced our knowledge of biogeography and evolution.

The earliest explorers were preoccupied with cataloging the distribution of Earth's plants and animals and, in the process, produced remarkable scientific advances. For example, while on British sailor Captain James Cook's trips around the globe aboard HMS *Endeavor* during the 1760s and 1770s, Joseph Banks collected thousands of unknown species of plants. In 1778, Johann Reinhold Forster, who also accompanied Cook, used the distributions of plants to describe several of Earth's biotic regions. By 1792, German botanist Karl Willdenow (1765–1812) concluded from his studies of biogeography that creation had occurred simultaneously in many places, after which life had dispersed from each site once floodwaters had receded. These and other voyages noted that there are more species near the equator than in areas farther from the equator. How could this be explained?

---

2. Buffon's claim that the New World's organisms were inferior to those of the Old World spurred U.S. President Thomas Jefferson to commission Lewis and Clark to conduct their famous biogeographic survey of the central and northwestern United States (see Chapter 3).

Soon after Willdenow made his claims, one of his students—German explorer Alexander von Humboldt (1769–1859)—conducted a famous expedition to South America that established the discipline of plant biogeography. When von Humboldt scaled mountains such as Ecuador's 6,268-meters-high Mount Chimborazo, he noted that zonation patterns typical of changes in latitude (that had been described earlier by Forster and others) were also created by changes in elevation: tropical plants often inhabited the lower altitudes, temperate plants the intermediate altitudes, and arctic plants the highest altitudes.[3] Willdenow, von Humboldt, and others established that the world can be divided into several natural regions, each of which is characterized by distinctive groups of plants and animals. In the early 1800s, Swiss botanist Augustin P. de Candolle (1778–1841) affirmed the presence of these zones, adding that plants living in each zone have different physiological adaptations. De Candolle also noted that organisms' distributions are influenced by competition, a key ingredient in what would become Darwin's theory of evolution by natural selection (Chapter 1).

The most influential of the early biogeographers was Alfred Russel Wallace. Unlike Darwin, who inherited vast wealth and enjoyed a life of remarkable privilege, Wallace earned a living by collecting and selling animals. In 1848, Wallace moved to Brazil to collect animals, and in 1854 he departed for the Malay Archipelago, where he lived for eight years while collecting more than 100,000 specimens. It was there that Wallace made his most important contributions to biogeography. In 1855, while in Sarawak (a state on the island of Borneo), Wallace published what came to be known as his "Sarawak paper." That paper, titled "On the Law which has Regulated the Introduction of New Species," began modern biogeography by noting that life and Earth evolve together; as Wallace put it, "Every species has come into existence coincident both in time and space with a pre-existing closely allied species." Three of Wallace's subsequent books—*The Malay Archipelago* (1869, dedicated to Darwin), *The Geographic Distribution of Animals* (1876), and *Island Life* (1880, dedicated to botanist Joseph Hooker)—became standards for understanding evolution and biogeography, including how factors such as climate, extinctions, dispersal, and competition affect the distributions of Earth's plants and animals. For example, in Indonesia, Wallace noted that the western islands closest to Asia (e.g., Sumatra and Java) were inhabited by animals that resembled those of Asia (e.g., tigers and pheasants), whereas eastern islands closest to Australia (e.g., New Guinea) were inhabited by animals that resembled those of Australia (e.g., birds of paradise, kangaroos, and cockatoos). The line that

3. Von Humboldt and his co-explorers made it 5,875 meters up Mount Chimborazo, but altitude sickness prevented them from reaching the summit. This was a world-record height that von Humboldt held for 30 years.

Figure 4.3 While in the Malay Archipelago, Alfred Russel Wallace noted the striking differences in the distributions of animals in the islands. Today, the boundary between animals resembling those of Asia and the animals resembling those of Australia is marked by the Wallace Line. (Jeff Dixon)

divided these groups of organisms was often remarkably sharp—for example, Bali and Lombok, which are on opposite sides of the line, are only 40 kilometers apart. This line of demarcation became known as the **Wallace Line,** and it remains the most famous biotic barrier in the world (Figure 4.3). Today, we know that the Wallace Line marks a boundary of the Australian continental plate (see Geologic Evidence of Continental Movement).

Biogeography and Geology

Explorers reported that around the world, species that are most alike usually live near each other geographically, regardless of differences in environments. And in areas having similar climates (e.g., South Africa, Chile, and Australia), the explorers found unrelated plants and animals. These observations raised several perplexing questions. For example, why are distant areas that have similar climates and terrain, such as the deserts of Africa and the Americas, inhabited by species that have similar appearances but are otherwise different? In deserts of the Americas, the succulents are cacti, but the deserts of Africa (and Asia and Australia) have no native cacti; instead, they are populated by euphorbs, a different family of succulent plants. Why do such ecologically similar environments house similar, but different, plants?

These expeditions, as well as subsequent expeditions to the New World, Africa, Asia, Pacific islands, and elsewhere, showed that the world is much bigger than Europe and revealed incredible stories about unimagined plants, animals, and habitats. For example, marsupials are mammals having distinctive pouches in which females transport their young through early infancy.[4] These animals, which include kangaroos, opossums, wombats, and wallabies, live in the Americas, Australia, and New Guinea. Placental mammals dominate the rest of the world. How did kangaroos get to Australia, and why aren't they in Kansas? There do not appear to be any migratory routes linking these populations of marsupials, nor have marsupials been found swimming across the Pacific Ocean. How did marsupials get to their present locations? Why aren't they elsewhere?

As our knowledge of Earth's organisms began to accumulate, biologists began to understand that although biogeography tells us about life's history and the evolution of Earth's biodiversity, it also tells us about the history of the Earth itself. Indeed, the history of life is entwined with geology, and answering many of the questions posed by early explorers required not just knowledge of evolution, but also knowledge of Earth's dynamic past.

Major, Long-term Changes in the Distribution of Life's Diversity Are Influenced by Plate Tectonics

Continental Geography

In Darwin and Wallace's era, fossils (Chapter 3) suggested that life had changed throughout Earth's history, and that new species had repeatedly arisen and disappeared. However, when biologists noted *where* the fossils were found, they were often puzzled. It was not surprising that fossils of the same species were found near each other, but how could fossils of the

4. Almost all marsupials give birth to undeveloped offspring, which mature in pouches. Placental mammals have placentas that enable young to be born at a more developed stage.

same species be found on different continents? A major obstacle to answering this and other questions about biogeography was the assumption that continental geography was stable and unchanging, and that a map of Earth was fixed. If continents have always been in their present positions, how could similar organisms such as marsupials that today are separated by thousands of kilometers have gotten to their present locations?

As long ago as 1596, Flemish mapmaker Abraham Ortelius (1527–1598) had noted a curious fact about our planet that children continue to rediscover to this day: Africa and the Americas—that is, the two landmasses on either side of the Atlantic—seem to fit together as if they were interlocking pieces of a giant jigsaw puzzle. This, Ortelius claimed, was no coincidence, and instead showed that the Americas had been "torn away from Europe and Africa . . . by earthquakes and floods." By 1858, French geographer Antonio Snider-Pelligrini (a contemporary of Charles Lyell; see Chapter 2) speculated that Europe, Africa, and the Americas had once been connected, but had later been separated by Noah's flood. Later, amateur American geologist Frank Taylor also claimed that Africa and the Americas had once been connected, and suggested that collisions between drifting continents could explain the formation of mountain ranges such as the Alps. During times when geologists were questioning such "catastrophic" explanations of Earth's history, these were bold claims that were summarily dismissed. Instead, many geologists accepted the prevailing theory of the Shrinking Earth, which claimed that the once-molten Earth was still cooling, and that this cooling caused Earth to shrink. This shrinkage squeezed Earth's surface into an ever-smaller area, causing "wrinkles" to appear as mountain ranges.

But the idea of moving continents would not go away, and evidence began to suggest it could be valid. For example, many animals of the same species live on the facing coastlines of South America and subtropical Africa. If continents have always been in their present position, how could this be possible? But what if the continents have *not* always been where they are today? Joseph Hooker, a friend of Charles Darwin, believed that analyses of plant distributions suggested that many widely separated populations of plants had once been a "continuous extensive flora. . . that once spread over a larger and more continuous tract of land . . . which has been broken up by geological and climatic causes." Was Hooker right? Such explanations were often dismissed because people could not imagine how different continents could have ever been linked, or how entire continents could have moved at all, much less have moved large distances. Was "terra firma" not so firm after all? What forces could drive such movements? This idea just didn't seem possible.

Continental Drift

The assumption about Earth's stasis was challenged in 1912 by German geophysicist Alfred Wegener (1880–1930). Wegener based his claims on geologic data, the similarity of fossils found on opposite sides of the Atlantic, and on evidence that he gathered on expeditions to tectonically active areas of Greenland. Wegener noted that along the facing coastlines of South America and Africa, geological formations such as specific rock strata and oceanic ridges matched, and that the same species of fossil plants and animals (e.g., the long-snouted reptile *Mesosaurus*) also occurred on the facing coastlines. In his small, famous book *The Origin of Continents and Oceans* (1915),[5] Wegener proposed that the areas housing the matching fossils and geologic formations were once joined in a single supercontinent called Pangaea (from the Greek for "all land") that had covered almost one-third of Earth's surface in the late Paleozoic and early Mesozoic. Wegener also proposed that that this supercontinent later split apart as today's continents floated on Earth's semi-liquid mantle to their present positions. But how could this be? How could entire continents split apart and move such vast distances?

Wegener, like Darwin and Wallace before him, understood that life and Earth evolve together, that oceans and mountains are as much a part of a landscape as plants and animals, and that plants and animals reflect the geographic and climatic changes that have occurred on Earth. Wegener admitted that his evidence was not entirely convincing, but was confident that his claims about moving continents were accurate. However, he could not devise a convincing mechanism to account for continental movement; his proposed explanations based on Earth's magnetic poles and rotation (e.g., Eötvös Force), which were known to be capable of driving air westward and away from the poles, were easy targets for geologists, who quickly determined that these forces were far too weak to move continents. Moreover, Wegener was not a geologist, and he was consequently viewed as an "outsider" by mainstream geologists. As a result, his idea—now known as **continental drift**—was ignored or ridiculed. Many prominent scientists, including famed biologist George Gaylord Simpson, were especially hostile toward Wegener's idea, claiming that its supporters were "without special competence." Other academics dismissed Wegener's claims as everything from "very dangerous" to "utter, damned rot." Despite this opposition, Wegener continued to defend his idea. The fourth, and final, edition of Wegener's book was published in 1929, a year before he died in a blizzard while on an expedition in Greenland to uncover more evidence for his theory.

---

5. Wegener's book became widely known only after the third edition (1922) was translated into English and four other languages. In 1930, Wegener proposed the previous existence of a supercontinent called *Urkontinent*; the term *Pangaea* appeared in a 1928 symposium convened to discuss Wegener's idea.

The first inkling of additional support for Wegener's idea came in 1927, when British geologist Arthur Holmes (1890–1965) outlined how the presence of radioactivity in the Earth could generate heat. Holmes' discovery questioned the Shrinking Earth theory, for it showed that Earth was neither cooling nor shrinking, Moreover, Holmes' claim provided possible support for Wegener's idea, for it accounted for heat that could drive convective currents to move continents.[6]

Geologic Evidence of Continental Movement

Additional evidence supporting Wegener's bold claim began appearing in the 1960s, thanks to two technologies developed during World War II: submarines and sonar. When scientists used submarines and sonar to map the bottoms of oceans, they were surprised to discover underwater mountain ranges spanning thousands of kilometers. These **mid-oceanic ridges** are where new seafloor is formed by the upwelling of magma (molten rock) from Earth's interior; this magma hardens and, in doing so, spreads the seafloor by a process called *seafloor spreading* (some of the magma is also trapped beneath the crust and moves laterally, creating currents that help push the overlying plates; Figure 4.4). Far from the mid-oceanic ridges are V-shaped **oceanic trenches**—some more than 10 kilometers deep—where the seafloor is pulled downward into Earth's core, thereby returning the material to the magma. This ongoing formation and consumption of the ocean floor (i.e., along ridges and trenches, respectively), driven by heat from the Earth's core, move giant continental plates.[7] For example, the ring of mid-oceanic ridges surrounding Antarctica has pushed other plates northward, thereby accounting for why Antarctica is isolated on the South Pole and why water covers more of the Southern Hemisphere than the Northern Hemisphere (81% vs. 61%, respectively). On continents, plates diverge to form rift zones (Figure 4.5), which eventually form oceans; for example, the rift zone created by the Mid-Atlantic Ridge between the eastern edge of the Americas and the western edge of Europe and Africa filled with seawater that ultimately became the Atlantic Ocean. Today's East Africa Rift Valley will eventually split Africa and make way for a new ocean.

Collisions of continental plates can create mountain ranges; for example, the collision of the Indian Plate and the Eurasian Plate 50 million years

---

6. Holmes later produced an absolute timescale for geologic history (Appendix 2). As Holmes noted in 1964, "Earth has grown older much more rapidly than I have—from about 6,000 years when I was 10, to 4 or 5 billion years by the time I reached sixty."

7. Earth's oceanic plates are 50–100 kilometers thick, and its continental plates are 100–250 kilometers thick. Continental plates consist of more-buoyant rock, such as granite.

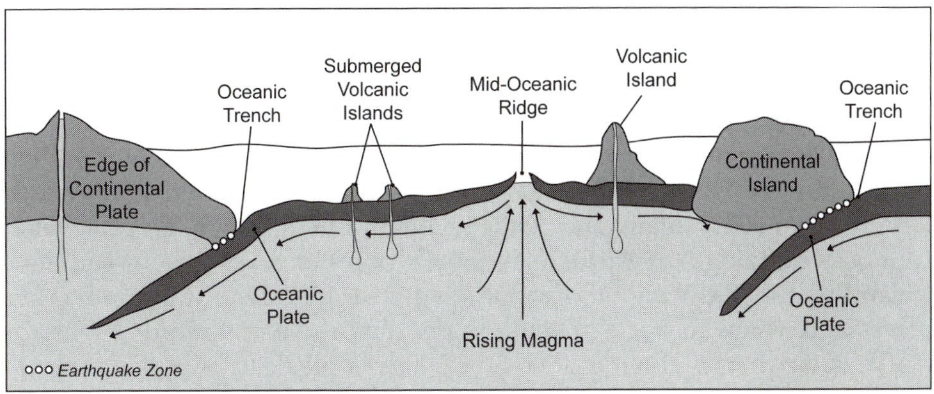

Figure 4.4 In this simplified model of seafloor spreading, oceanic plates are pushed apart by the upwelling of magma along mid-oceanic ridges. Magma may also produce volcanic islands. The oceanic plate and submerged volcanic islands eventually disappear into oceanic trenches. The places where plates contact each other are the sites of epicenters of earthquakes. (Jeff Dixon)

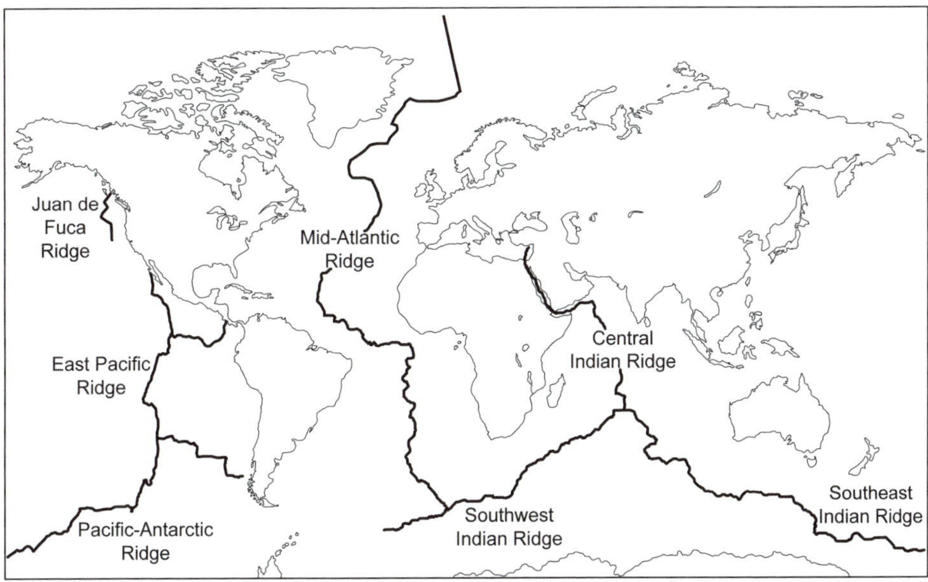

Figure 4.5 There is a global system of mid-oceanic ridges, which mark areas where seafloor is being created by hardening magma. (Jeff Dixon)

ago created a geologic crumple zone known as the Himalayas. Today, those plates continue to move at approximately 5 centimeters per year and, in the process, put the foundations of China under incredible stress. On occasion, those stresses are relieved, such as on May 12, 2008, when an intense (i.e., 8.0 on the Richter scale) earthquake struck the province of Sichuan in southwest China. The energy that was released, which was equivalent to the detonation of a billion tons of high explosive, destroyed hundreds of buildings, killed more than 90,000 people, and left almost 5 million people homeless. Similarly, in February, 2010, a magnitude 8.8 earthquake struck off the coast of Chile at the fault zone where the Nazca plate (the section of Earth's crust under the Eastern Pacific Ocean south of the equator) slides beneath the South American plate. The quake killed hundreds of people and leveled countless buildings. In April 2010, the eruption of a volcano in southern Iceland (in the mid-Atlantic rift zone) halted air traffic in Europe and other places at costs exceeding $200 million per day, and a magnitude 9.0 earthquake in Japan in 2011 left more than 27,000 people dead or missing.

Paleomagnetism

Additional evidence for Wegener's movement of continents came from another source: **paleomagnetism.** Igneous rocks contain tiny, magnetized crystals that, when the rocks solidify, are locked into position by Earth's magnetic field, much like tiny frozen compass-needles. When these crystals were discovered, geologists realized that the crystals could be used to trace the movements of rocks and land masses in which the rocks were positioned. If continents had never moved, all the crystals would point the same way (i.e., to Earth's present magnetic poles), regardless of Earth's age. However, after studying igneous rocks having different ages (Chapter 2) and from different continents, geologists discovered that the crystals did *not* point the same direction, indicating that the continents *had* moved. When the continents' positions were plotted on a globe at various stages throughout Earth's history, scientists realized that the continents had once been joined, just as Wegener had claimed, and that this supercontinent had later split apart, just as Wegener had claimed (Figure 4.6).

Radiometric Dating of the Seafloor

The final pieces of evidence supporting Wegener's claim involved radiometric dating (Chapter 2). Seafloor basalt is youngest adjacent to the ridges, and older farther from the ridges. Although the stable cores of continents are billions of years old, the oldest areas of basalt comprising the seafloor are only about 200 million years old. These results are consistent with new seafloor being constantly produced at mid-ocean ridges, from which it spreads out to both sides. The oldest seafloor is continually disappearing into oceanic trenches, where it is reincorporated into Earth's mantle.

Pangaea and Biogeography

The break-up of Pangaea, which began 175 million years ago, explains many aspects of biogeography that had previously baffled scientists. For

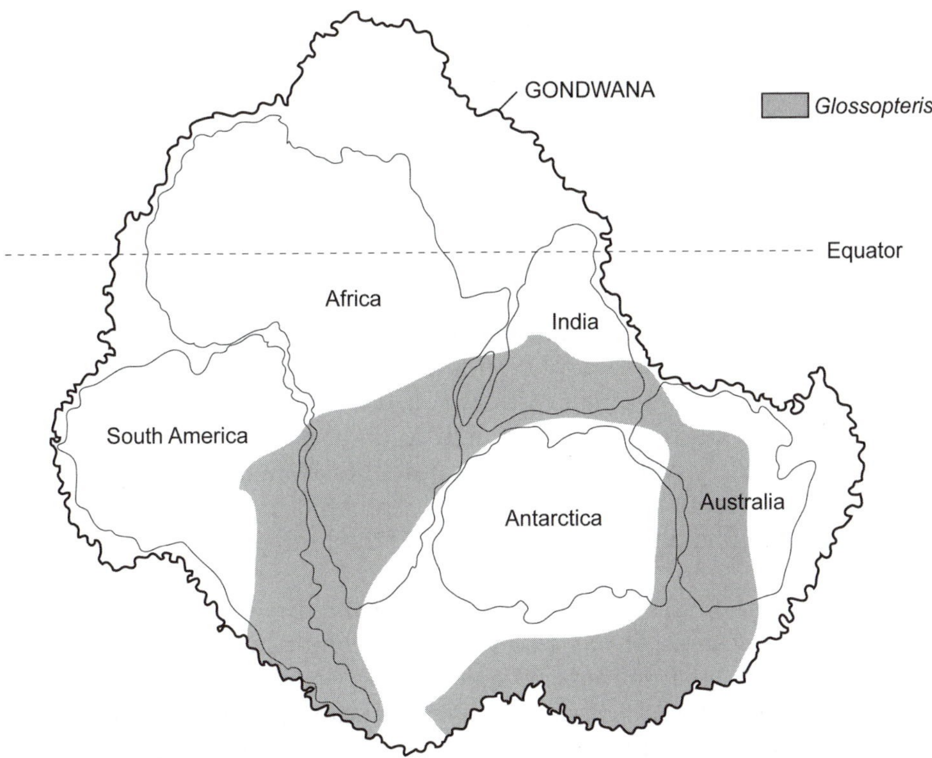

Figure 4.6 Plate tectonics accounts for the evolutionary biogeography of organisms such as *Glossopteris,* an extinct gymnosperm that was widespread (shaded areas) in the Permian on the supercontinent Gondwana. As Gondwana was split apart into South America, Africa, India, Australia, and Antarctica by seafloor spreading along mid-oceanic ridges (see Figure 4.5), the populations of *Glossopteris* and other organisms rode the moving continents to their current positions. (Jeff Dixon)

example, consider a group of organisms that we introduced earlier in this chapter: marsupials. We know from fossil evidence that marsupials originated more than 150 million years ago in China, at which time Asia was linked to North America. This linkage enabled marsupials to spread to North America during the next 50 million years. Although marsupials in North America became extinct 30–25 million years ago, the marsupials that had dispersed to South America did not. This group of marsupials expanded into Antarctica and Australia, both of which were then attached to South America. When South America, Antarctica, and Australia later drifted apart, each continent carried with it a population of marsupials. (One of the South American marsupials, the Virginia opossum, eventually re-colonized North America when the isthmus of Panama emerged approximately 3 million years ago.) Australia's diverse habitats produced differing selection pressures, and the marsupials there evolved into the more than 150 species of bandicoots, koalas, kangaroos, and other marsupials that inhabit Australia today. Only when Australia moved closer to Asia (about 15 million years ago) did placental mammals such as bats and rats colonize Australia.

In light of the above, biogeographers posed the following hypothesis: If South America was once connected to Australia, migrating marsupials must have passed through Antarctica. This was testable by answering the following question: Are fossilized marsupials found in Antarctica? Yes, they are; more than 10 species of fossilized marsupials have been discovered in Antarctica. Just as predicted, these fossils are 30 to 40 million years old. Clearly, then, biogeography provides a convincing explanation for the distribution of marsupials. Marsupials did not need to migrate across land bridges to get from one continent to another, nor did they need to swim—they simply rode the moving continents to their present positions. Although continents move only a few centimeters per year, Earth's vast age (Chapter 2) accounts for continents (and their inhabitants) moving vast distances.

The power of biogeography to account for the distribution of life is not limited to marsupials; it has also produced convincing explanations for the distributions of hundreds of species. For example, *Glossopteris* is an extinct arborescent gymnosperm that was dominant during the Permian. *Glossopteris* fossils are found only in the Southern Hemisphere, but they are found there in South America, Africa, India, Australia, and Antarctica (Figure 4.6). It's highly unlikely that the trees were dispersed to these different continents by seeds, because *Glossopteris* seeds are large and dense, and wouldn't float. *Glossopteris* also was not dispersed by birds because birds did not exist at that time (Appendix 2). The puzzling distribution of *Glossopteris* was understood when geologists realized where the present-day continents in the Southern Hemisphere were during the late Permian. When these continents are arranged in their Permian positions, the current-day patches of *Glossopteris* are united. Thus, the trees did not migrate from one continent to another. Instead, the continents moved, carrying the trees with them. Thanks to evolution and biogeography, the distributions of *Glossopteris* fossils—including those in Antarctica—make sense.

Tectonic Movement and Evolution

We now know that convective forces in Earth's mantle move eight major plates (and over a dozen smaller ones) over Earth's surface and, in the process, have joined and split continents. By bouncing laser beams off of orbiting satellites, we also know that the plates are moving *now*. For example, the North American and European plates are separating at a rate of about 2.5 cm per year, and the Tonga and Pacific plates are moving toward each other 25 cm per year (this is the fastest movement of tectonic plates). In the past, these movements have separated related groups of organisms (such as in Africa and South America), and collisions of plates have united unrelated organisms in the same area. These movements have also affected the distributions and evolution of life in indirect ways by altering climate. For example, the humidity and rainfall of any area are influenced by its distance from the ocean. Just as the central region of modern Eurasia is dry, so too would have been the central region of giant earlier land masses such as Pangaea. The breakup of these supercontinents would have produced wetter, more moderate climates, and would have therefore affected the organisms inhabiting the resulting land masses. The breakup of giant land masses would have also altered oceanic currents and, in doing so, altered the climate. These alterations of climate, and the resulting new habitats and selection pressures, are some of the many ways that geology has fueled the engine of evolution.

Unlike in Darwin's day, we now know that Earth's surface consists of large, rocky, overlapping plates that comprise continents and ocean floors (or, in some instances, only ocean floors), and that these plates move. These movements are collectively referred to as **plate tectonics**. Although the plates move very slowly (5–10 cm yr^{-1}), tracing the movements of the plates throughout geologic history has enabled scientists to analyze the changing distributions of living organisms throughout Earth's history and to discover the historical patterns of biogeography that are visible today. The result is that plate tectonics is the most important factor responsible for major, long-term changes in the distributions of Earth's organisms.

Different Species in Similar Habitats Often Evolve Similar Adaptations

Convergent Evolution

When different species in similar habitats are under similar selective pressures, they evolve similar adaptations. That is, they *converge;* despite being different species, they come to behave similarly and look alike. This phenomenon is *convergent evolution* (introduced in Chapter 1). For example, animals as diverse as hares, foxes, terns, sheep, owls, and bears living in the arctic have all evolved white coloration. Similarly, marsupial mammals such as moles, mice, and anteaters have all evolved adaptations similar to those of placental mammals living in similar habitats (Figure 4.7), and the plants and animals in the Arabian Desert resemble those of the Mojave Desert. To appreciate this, again consider the succulent plants that inhabit deserts. Succulent plants have adaptations for maximizing water storage and minimizing water loss. However, the succulents in American deserts are cacti, while those in African deserts are euphorbs. These plants are a famous example of how convergent evolution has resulted in different species filling similar roles in similar environments.

Figure 4.7 The convergent evolution of mammals. Although the marsupial mammals (e.g., sugar gliders, moles, and anteaters) of Australia evolved independently of (i.e., are not related to) their placental-mammal counterparts in North, Central, and South America, their forms are similar. (Jeff Dixon)

Generations ago, many people believed that the different plants of these deserts were specially created to be best suited to their respective habitats. Succulent plants growing in deserts *do* have traits that minimize water loss (e.g., reduced leaves) and maximize water storage (e.g., fleshy stems). But why would a Creator produce on different continents fundamentally different plants that function and look so much alike? African euphorbs are not inherently superior to American cacti—but when cacti are introduced into African or Australian deserts, they thrive, and often out-compete the native plants. For example, the North American prickly pear cactus was introduced into Australia in the early 1800s as a source of a crimson dye (carmine) that was extracted from cochineal beetles (*Dactylopius cocus*) that feed on the cacti (this dye gives Persian rugs their deep red color). But within a few decades, prickly pear had taken over much of the continent, prompting numerous dramatic attempts to control the introduced plant. (It was finally controlled in the 1920s with the introduction of the cactus moth [*Cactoblastis cactorum*], whose caterpillars eat the cactus.) Similarly, rabbits that were introduced decades ago in Australia continue to replace native marsupials such as the bilby (prompting some to urge replacing the Easter Bunny with the Easter Bilby). Clearly, native species are not perfectly adapted to where they occur (if they were, the introduction of invasive species such as prickly pear and rabbits would have no effect on the native species).

Environments in a particular area are not usually inhabited by the same species that occupy similar environments in other parts of the world, but by *different* species that have evolved independently in those areas. A simple explanation for this distribution of organisms is that convergent evolution has produced similar adaptations in otherwise different species.

Isolated Habitats Such as Oceanic Islands Are Populated by Descendants of Organisms from the Nearest Mainland

Continental and Oceanic Islands

The biogeography of islands also provides powerful evidence for evolution. There are two basic types of islands: continental islands and oceanic islands. *Continental islands* are islands that were once connected to, but later separated from, continents by moving plates or increases in sea level. Examples of continental islands include Madagascar, Newfoundland, and Japan. If continental islands were once connected to the nearby continent, one would predict that continental islands would share species with the continent to which they were once connected. This is what is observed; for example, India and Madagascar (which split 80 million years ago) share

Figure 4.8 The Galápagos islands are intensely studied examples of oceanic islands. These volcanic islands are more than 900 kilometers from the nearest continent (South America) and were colonized by organisms capable of crossing the ocean. Shown here is South Plaza Island, which is located just off the eastern coast of Santa Cruz. (Randy Moore)

many freshwater fish and amphibians, and the plants and animals of the British Isles are almost identical to those of mainland Europe (from which the British Isles separated).

Oceanic islands are islands that were never connected to a continent. There are tens of thousands of oceanic islands, the most famous of which are Hawaii and Galápagos. Unlike continental islands, which have always been inhabited by organisms, these archipelagos arose from underwater volcanoes, and therefore had to be colonized by organisms that could somehow cross oceans to reach the newly formed islands (Figure 4.8). What organisms could reach oceanic islands?

Colonization of Oceanic Islands

If oceanic islands were originally colonized by species from nearby continents, one would predict that oceanic islands would be inhabited by organisms having the best chances of crossing oceans to reach oceanic islands. This is exactly what is observed; oceanic islands are usually inhabited by plants, birds, and small insects—precisely the types of organisms that would have the best chances of reaching and colonizing distant lands.

Birds could fly to the islands, carrying insects and seeds (e.g., in their digestive track and/or in mud on their feet). Lightweight insects, seeds, and spores could be blown to the islands by winds; for example, orchid seeds have been transported more than 120 miles by winds, and plants such as *Metrosideros*—a tree having tiny, windborne seeds—are dispersed throughout the Pacific islands. Coconut palm trees (*Cocos nucifera*) are common on islands because coconuts float and can endure long exposures to seawater. In contrast, many terrestrial and freshwater organisms—which cannot survive long exposures to seawater—are absent from oceanic islands. This is why Hawaii—which is 2,400 miles from the nearest continent—has no native frogs, lizards, snakes, ants, or terrestrial mammals (except bats).

Island Species and Mainland Species

The native inhabitants of oceanic islands are usually closely related to species of the nearest mainland. For example, species native to Galápagos are similar to those of the western coast of South America, and species native to Hawaii are similar to those of the Americas or the nearby mainlands (i.e., Indonesia, Fiji, and New Guinea). These observations suggest that species on the nearby mainland are the ancestors of species now living on oceanic islands. For example, in Galápagos, the mockingbirds that Darwin collected in 1835 were unique to particular islands, but they also resembled each other and, to a lesser extent, the mockingbirds of mainland South America from which they evolved. As Darwin noted, "[A]lmost every product of the land and water [of Galápagos] bears the unmistakable stamp of the American continent," and the organisms have an "affinity to those of the nearest mainland, without being actually the same species." Ancestors of the archipelago's finches colonized Galápagos 2–3 million years ago, after which they evolved into the 14 species that today live nowhere else on Earth except in Galápagos. Through natural selection, each species of Galápagos finch adapted to different niches. For example, heavy-beaked finches crack large seeds and nuts, and small-beaked finches eat smaller, softer seeds and, sometimes, insects; woodpecker finches (*Camarhynchus pallidus*) use tools to dislodge insects from bark, and vampire finches (*Geospiza difficilis septentrionalis*) on Wolf Island peck and drink the blood of birds.

Adaptive Radiation

These observations are consistent with the inhabitants of oceanic islands being descendants of earlier species that colonized the islands from nearby continents. As Darwin noted, "[S]eeing this gradation and diversity of structure in one small, intimately related group of birds, one might really fancy that from an original paucity of birds in this archipelago, one species had been taken and modified for different ends." These are examples

SIDEBAR 4.2 Evolution in Action in the Galápagos

Long-term studies of natural populations are rare because of funding and logistical challenges. However, a remarkable exception to this generalization is the work of Peter and Rosemary Grant. Since the 1970s, the Grants have studied finches on Daphne Major, a small island in Galápagos that is about the size of 80 football fields. The Grants' work has documented evolution by natural selection in action.

The Grants and their co-workers had been studying the finches for several years when, in 1977, there was a severe drought; rainfall decreased from an average of 130 mm per year to only 24 mm in 1977. This drought dramatically reduced the supply of seeds that the birds use for food from 10 g seeds m^{-2} to only 3 g seeds m^{-2}. Once the preferred, smaller seeds had been eaten, birds with larger beaks had a competitive advantage because they could open the larger, tougher seeds that remained (and which are so difficult to crack that they are ignored when food supplies are typical). Death by starvation was most common for finches that ate only the soft seeds; for example, the famine reduced the number of medium ground finches (*Geospiza fortis*) by 84% (from 1,200 to 180 birds). When compared to the finches that died, the survivors were nearly 5% larger and had deeper, longer beaks. That is, larger-beaked birds were more likely to survive and reproduce than birds with smaller beaks, and in 1978 the large-beaked trait was passed to the next generation at a higher frequency. Thus, large, deep beaks were an evolutionary adaptation for cracking large seeds.

In 1983 there was another dramatic change in climate when, in the span of seven months, a total of 1,359 mm of rain fell on Daphne Major. This resulted in 80% of the available seeds being soft. The populations of finches that ate those soft seeds—for example, *G. fortis*—increased by 400%. This reversal of the 1977 changes was dictated by environmental conditions and occurred within six years—a period far too short to be documented in the fossil record, if there had been one.

The Grants' ongoing work shows that beak sizes in the population of finches shift in response to the availability of food, which, in turn, is affected by climate. This is a clear case of differential reproductive success causing a population to evolve. Today, the Grants' work, which has encompassed more than 25 generations and 20,000 birds, focuses on speciation, extinction, and character displacement due to interspecific competition.

of adaptive radiation (introduced in Chapter 1), which is an evolutionary process by which a single ancestor evolves into several species, each having a different niche.

On continents and continental islands, different niches are usually filled by a variety of different species. This is *not* what is observed on oceanic islands. When plants and animals colonized oceanic islands, they would have found empty habitats and niches that were free of competitors and predators. This would have resulted in the radiation of the colonizing plants and animals into the new niches. As a result, one would predict that the different niches on oceanic islands would be filled by similar species. This is precisely what is observed. For example, the niches filled by different species of finches in Galápagos are filled by flycatchers, parrots, toucans, and other birds on the mainland of South America. Similarly, in Hawaii, the 25 or so remaining native species of honeycreepers (a group of small, colorful passerine birds) have niches ranging from nectar drinkers (with long, slender bills) to seed crackers (with shorter, more muscular bills). That is, native Hawaiian honeycreepers have radiated to fill niches that are occupied by different species on continents and continental islands. In Australia, marsupials occupy niches that placental mammals occupy elsewhere. Evolution provides a straightforward explanation for these observations: the ancestors of finches and honeycreepers had colonized the islands before other birds, and there—free from competition and predators—they adapted to diets and lifestyles not normally available to them on the mainland.

Unique Features of Island Biogeography

Clearly, the animals and plants of oceanic islands are *unbalanced*—that is, oceanic islands often lack many types of organisms that are common on continents and continental islands. For example, virtually all continents and continental islands contain a wide variety of plants and animals, but most oceanic islands lack some of the same types of organisms—usually native freshwater fish, land mammals, and amphibians—because, as Charles Darwin noted, "no terrestrial mammal can be transported across a wide space of sea." Aquatic mammals such as sea lions are native on many oceanic islands, but native terrestrial mammals are absent. These patterns help explain why islands have smaller numbers of species than do comparable areas of the nearest mainland. For example, a recent study showed that a Panamanian rainforest housed 56 species of birds, and a nearby shrubland housed 58 species of birds. The nearest island, which had an environment intermediate between a rainforest and shrubland, housed only 20 species of birds.

Life is also more dangerous on islands than on the mainland because organisms cannot readily flee their islands. This helps explain why extinction rates on islands are up to 50-times higher than those on continents. These observations are hard to reconcile with creationists' claims that organisms were created specially for their current locations. Why are extinction rates so high on islands? Why would a Creator leave amphibians, land mammals, and freshwater fish off of oceanic islands but not continental islands?

Oceanic islands are also often characterized by another curious trait: flightless birds and other animals (Figure 4.9). For example, on the island

Figure 4.9 Flightless birds such as this flightless cormorant (*Phalacrocorax harrisi*) often evolve on oceanic islands. Flightless cormorants live on only two islands of Galápagos (Fernandina and Isabela) and are one of the world's rarest birds: as of 2010, there were fewer than 1,500 individuals alive. (Randy Moore)

of Mauritius, dodo birds (which evolved from pigeons that flew there from Southeast Asia) were common until they were hunted to extinction (see Chapter 3). Flightless birds also occur on oceanic islands such as Galápagos. In Hawaii, geese from Canada settled and became flightless, and New Zealand's kiwi and kakapo (as well as the now-extinct moa; see Chapter 3) are also flightless. In each of these locations, the evolution of flightless birds correlated with little competition for resources and the absence of large predators. Instead of investing energy in unneeded flight muscles, the birds that had the greatest fitness reallocated resources to other functions. The variation that ultimately produced flightlessness on these islands also occurs on the mainland, but is quickly extinguished by carnivores.

In 1967, American biologists Robert MacArthur and E. O. Wilson's *The Theory of Island Biogeography* presented a biogeographical framework for understanding and studying the distribution of species on islands. Although some of the ideas were later qualified (e.g., that the distribution of organisms is a balance between immigration and colonization), other ideas had a dramatic influence on biogeography. For example, much evidence now supports the claim that the number of species on an island is strongly correlated with the size of the island (Table 4.2).

Table 4.2 Flora and Fauna of Pacific Islands

Island(s)	Area (km^2)	Genera of Flowering Plants	Genera of Birds
Solomon Islands	40,000	654	126
New Caledonia	22,000	655	64
Fiji Islands	18,500	476	54
Samoa group	3,100	302	33
Cook Islands	250	16	10

The area of islands is related to the number of genera of birds and nonendemic flowering plants in some islands of the Pacific.

Although the major ideas presented in *The Theory of Island Biogeography* were developed for actual islands, the principles of island biogeography were later extended to various analogues of islands (i.e., ecosystems that are surrounded by unlike ecosystems), such as springs surrounded by desert, lakes surrounded by dry land, mountains surrounded by lowlands, and forest fragments surrounded by human-altered ecosystems. Today, the principles of island biogeography are a foundation for studies of habitat fragmentation and conservation biology.

Creationists' Explanations for Biogeography

The earliest "natural theologians" often tried to reconcile the distribution of organisms with a single site of origin—namely, their dispersal from where Noah's Ark landed in the Ararat region of eastern Turkey. For example, Linnaeus claimed that animals from the Ark settled different environments as the flood receded. Today, many creationists continue to try to explain biogeography with similar scriptural claims; for example, John Morris of the Institute for Creation Research claims "present-day animal distributions must be explained on the basis of migrations from the mountains of Ararat." But this religion-based model of biogeography is problematic as well as inconsistent with mountains of evidence. For example, how did kangaroos get to Australia? And did each of the 13 different species of Galápagos finch fly from Ararat to North America, then to South America, and then finally to Galápagos, without any individuals ending up anywhere else? Similarly, what was it about marsupial mammals that made them *all* migrate as a group *only* to Australia?

Creationists in Darwin's era sometimes voiced other explanations for island biogeography. For example, some creationists claimed that mammals, amphibians, and freshwater fish are not native to oceanic islands because they are poorly suited for life on these islands. It didn't take much thought to reject his claim, for these organisms often thrive when they are introduced to oceanic islands. Indeed, on some oceanic islands, introduced species have taken over, out-competing—and often destroying—the native organisms. For example, introduced rats, goats, pigs, and toads have overrun parts of Hawaii, and Galápagos authorities have spent millions of dollars trying to eradicate introduced dogs, cats, goats, donkeys, and pigs that have destroyed populations of many native species on many islands.[8] Elsewhere, similar problems have been created by other invasive (i.e., introduced) species such as kudzu, purple loosestrife, European starlings, and zebra mussels. When cheatgrass (*Bromus tectorum,* a native plant from Eurasia) was accidentally introduced into North America in 1889 in a shipment of crop seeds, it began to spread rapidly and out-competed native plants. Thirty years later, cheatgrass covered tens of thousands of square kilometers of North America. These are not isolated examples; for example, today in Galápagos, introduced species (~700) outnumber native species (~500). As Darwin noted more than a century ago, "He who admits the doctrine of the creation of each separate species, will have to admit, that a sufficient number of the best adapted plants and animals were not created on oceanic islands; for man has stocked them from various sources far more fully and perfectly than nature." These introductions of nonnative species can have catastrophic consequences; for example, in 2007, facing mounting damage inflicted by invasive species and the burgeoning tourist industry, Galápagos was placed on the UNESCO List of World Heritage in Danger (it was removed from the list in 2010).

Creationists continue to voice unusual explanations for biogeography. For example, Answers in Genesis—an organization that rejects science in favor of young-Earth creationism—claims that "forces involved in the [biblical flood] were certainly sufficient" to explain the movement of continents, and that "mountains of Ararat existed for the Ark to land on." Other explanations have invoked a Creator who placed plants, animals, and fossils in their current positions to make it *appear* that species had evolved. Other creationists, however, have rejected such explanations for biogeography because such explanations depict God as a trickster and charlatan. This is why virtually all creationists are silent about the vast amount of biogeographical evidence for evolution.

---

8. As an example, in 1959 fishermen introduced one male and two female goats onto Pinta Island of the Galápagos. By 1973, the population of goats on this island exceeded 30,000.

Summary

Charles Darwin's theory of evolution by natural selection accounts for myriad observations of the distribution of life's diversity. Instead of organisms having been specially created in different environments, species from neighboring areas and continents migrated or were carried to new areas, where natural selection led to the evolution of adaptations to the new environments. Just as plate tectonics has been the engine for the formation of new continents and islands, so too has evolution by natural selection been the engine for the formation of new forms of life. Organisms are found in particular environments because they evolved from ancestors that lived in those environments. The geographic distribution of Earth's organisms fits the predictions of evolution.

5

Molecular Evidence for Evolution

Almost all aspects of life are engineered at the molecular level, and without understanding molecules we can only have a very sketchy understanding of life itself.

—*Francis Crick, 1988*

Fossil bones and footsteps and ruined homes are the solid facts of history, but the surest hints, the most enduring signs, lie in those miniscule genes. For a moment we protect them with our lives, then like relay runners with a baton, we pass them on to be carried by our descendents. There is a poetry in genetics which is more difficult to discern in broken bones, and genes are the only unbroken living thread that weaves back and forth through all those boneyards.

—*Jonathan Kingdon, 1996*

PREDICTIONS

If evolution has occurred, we predict that

1. Common ancestry will be revealed by a common hereditary material (DNA or RNA).
2. Related organisms will reveal their genetic similarity through similar protein-behaviors.
3. Ubiquitous proteins will exist in seemingly disparate taxa.
4. Organisms with more recent common ancestry will have more hereditary material in common than organisms further removed evolutionarily.
5. Comparing the coding regions of ubiquitous proteins will confirm common ancestry and allow a reconstruction of phylogeny.
6. Organisms with common ancestry will share randomly generated transposition elements and homologous genes.
7. Shared ancestry can be inferred through the occurrence of vestigial molecular elements, or pseudogenes.

8. Shared ancestry will be revealed within a lineage showing evidence of a shared viral pathogen, specifically in the form of endogenous retroviruses.

The geological record and the fossils embedded within (Chapters 2 and 3) tell a convincing story of our evolutionary history. Comparative anatomy (Chapter 6) allows us to appreciate fundamental similarities among, and profound differences between, groups of organisms. Although biogeography (Chapter 4) can illuminate the source of a lineage and the pattern of adaptive radiation that followed, molecular biology—in revealing the coding patterns of entire genomes—gives us a historical record like no other. DNA lacks the gaps characteristic of the fossil record, is audible when the rocks are silent, and provides clarity where homology is questionable. In the past few decades, scientists have identified the hereditary material (DNA and RNA), described its structure and operating system, and used this information to establish close relationships between some organisms thought to be distant relatives. This new information confirms what was only inferred a few years ago, and enables us to test many predictions with respect to the molecular evidence for evolution.

Common Ancestry Is Revealed by a Common Hereditary Material (DNA or RNA)

The ultimate universal homologies—that is, features shared by all living organisms—are our cellular architecture (all living things are made of cells), our use of ATP to do work at the cellular level, and the genetic code. These features lend unequivocal support for common ancestry. Moreover, the genetic code provides additional details about key events in evolution (e.g., the origin of photosynthesis, advent of multicellularity, transition to land, and development of bipedality).

Related Organisms Reveal Their Genetic Similarity through Protein Similarity

Antigen Specificity

Early attempts to determine evolutionary relationships using molecular techniques focused on antigen specificity. In most instances, a rabbit was exposed to *antigenic* (or foreign) compounds from two organisms, and its immunological reaction (as an expression of antisera, the compounds rallied by the body to fight invaders) was evaluated. Often, the antigenic compound was the blood protein albumin. If the two organisms in question

SIDEBAR 5.1 Looking at Chromosomes and DNA

The nucleic acids *deoxyribonucleic acid* (DNA) and *ribonucleic acid* (RNA) are large molecules found in the cells of all living organisms. DNA and RNA control all cellular processes through the synthesis of proteins, another group of macromolecules that are specialized for various physiological roles—some proteins are involved in muscle contraction, others conduct nutrients throughout the body, others fight infection, and so on. In addition to their role in protein synthesis, nucleic acids are the material of heredity. We'll discuss the role of DNA in cell division and growth, conveying information across generations, and occasionally incurring the very mistakes that make change possible.

DNA consists of *nucleotides* that are themselves relatively simple: each nucleotide includes a sugar (deoxyribose), a phosphate ion, and a nitrogen-rich base such as guanine (G), adenine (A), cytosine (C), and thymine (T). Nucleotides are joined in a linear sequence that represents one of DNA's strands. One molecule of DNA involves two strands, joined at the bases, and wound in a helix. In many organisms, several molecules of DNA exist as individual *chromosomes.* When we talk about chromosomes, we're talking about how our cells have solved the problem of packing so much information (long strands of DNA) into a small, tightly compacted structure.

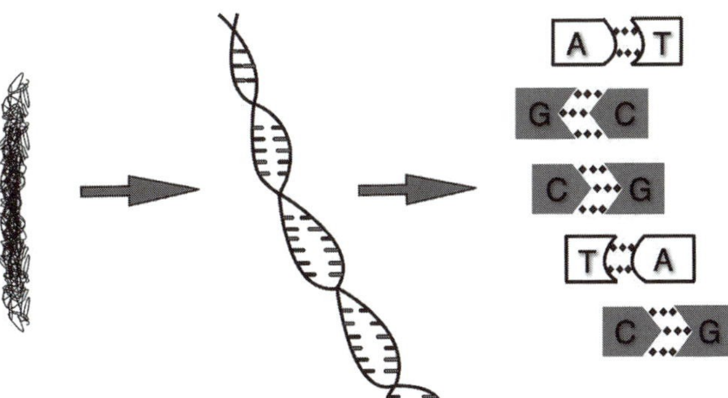

Figure 5.1 Chromosomes and DNA. Deoxyribonucleic acid (DNA) is coiled around proteins and packaged in chromosomes. DNA itself is a helical structure that consists of complementary linear sequences of nucleotides. Nucleotides contain a nitrogen-rich base such as guanine (G), adenine (A), cytosine (C) and thymine (T). (Sehoya Cotner)

SIDEBAR 5.2 Genes and Protein Synthesis

The information encoded in DNA results from how our cells read DNA's four-letter alphabet of G, C, A, and T. Each three-letter sequence of bases can be read as a *codon,* a molecular word that can be translated into an amino acid. The primary structure of a protein is merely a string of amino acids that have been encoded by a string of DNA bases. For example, TCC ACC CGC ACA in DNA codes for the synthesis of a simple protein consisting of the four amino acids serine (from TCC), threonine (from ACC), arginine (from CGC), and threonine (from ACA). The *translation* of DNA into amino acids is accomplished by *transcribing* DNA into an RNA intermediate, specifically *messenger RNA* (or mRNA). RNA nucleotides are complementary to DNA, yet the nitrogenous base *Uracil* occurs in place of Thymine. We can think of a gene as a discrete sequence of DNA bases that encode for a functional sequence of amino acids, which comprise proteins.

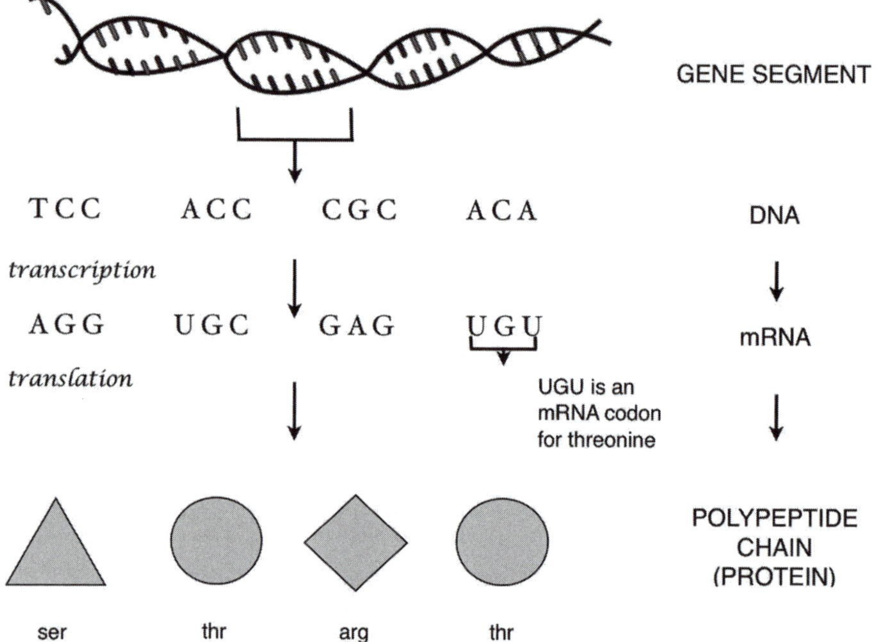

Figure 5.2 DNA and protein synthesis. DNA codes for proteins by using an RNA intermediary. Specifically, DNA is transcribed into RNA, which is then translated into a corresponding sequence of amino acids. This amino acid sequence is the primary structure of a protein. (Sehoya Cotner)

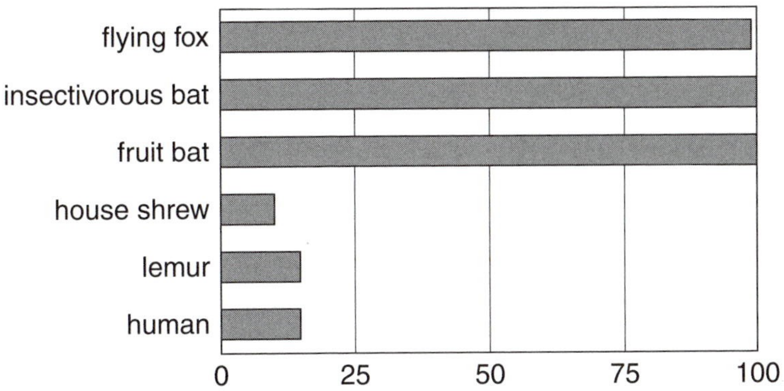

Figure 5.3 Bats, insectivore, and primate immunological cross-reactivity. Bats and flying foxes react similarly to anti-bat immunoglobulin G. Numbers indicate percent reaction to ant-bat IgG, and suggest a common ancestor for the bats and flying foxes. From Omatsu et al. (2003). (Sehoya Cotner)

possessed identical antigenic properties, the rabbit would not be expected to react at all to a second exposure. Such cross-reactivity studies were used to categorize hundreds of organisms, including extinct organisms. For example, in 1980, Ellen Prager and colleagues used albumin extracted from the thigh muscle of a 40,000-year-old frozen mammoth to estimate the degree of relatedness among the extinct mammoths and extant elephants and sea cows (manatees and dugongs). More recently, similar work by Tsutomu Omatsu and colleagues used cross-reactivity studies to establish the monophyly (i.e., shared ancestry) of all bats (Figure 5.3).

Protein Relatedness

Throughout the 1970s, proteins (and eventually nucleic acids themselves) were separated according to their electric potential or molecular weight. Separating proteins on a gel, via gel electrophoresis, enabled investigators to estimate degrees of relatedness based on proxy measures of similarity. Separating proteins also allowed for a within-group estimate of heterozygosity, which is a simple way to assess overall genetic variation. For example, a highly inbred population will have very low within-group heterozygosity and lack the genetic variability on which natural selection operates.

Ubiquitous Proteins Exist in Seemingly Disparate Taxa

Ubiquitous Proteins

Any assemblage of several amino acids is highly unlikely to arise twice by chance. However, some proteins have been so conserved through

SIDEBAR 5.3 Genes and Evolution

When DNA is passed from one generation to the next, what is being transmitted is the genetic code for making proteins. In a group of individuals of the same species, most of their coding regions, or genes, will be the same. However, where there is variation there is the potential for some changes within a lineage. There is also the potential for some individuals to fare better or worse than others; in other words, this variation leads to the differential survival and reproduction we associate with evolution by natural selection (Chapter 1). An organism that inherits two nonfunctioning copies of an otherwise useful gene (one from each parent) may be at a disadvantage, may not survive to reproductive age, and may fail to achieve genetic representation in the next generation. For example, a pocket mouse that inherits two nonfunctioning gene variants, or alleles, for a coat-color gene will not produce fur pigment. This lightly-pigmented pocket mouse may experience increased risk of predation on a dark substrate, and, depending on its environment, could be less likely to survive to reproductive age.

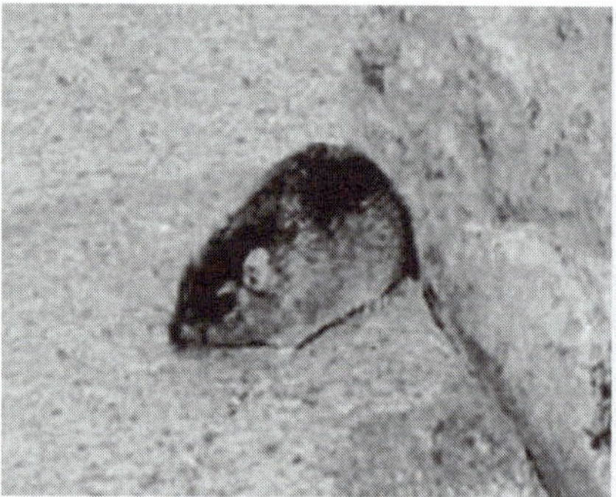

Figure 5.4 Light and dark pocket mice. A point mutation in a single gene alters coat color in rock pocket mice, rendering them more or less vulnerable to predation. A dark mouse will be concealed against a dark background yet will be vulnerable on lighter substrates. (Michael W. Nachman, Hopi E. Hoekstra, and Susan L. D'Agostino, 2003, "The Genetic Basis of Adaptive Melanism in Pocket Mice," *Proceedings of the National Academy of Sciences of the United States of America*, 100 (9): 5268–5273. Copyright © 2003, National Academy of Sciences.)

evolutionary time that they are seemingly ubiquitous. One such group of proteins is the iron-sulfur proteins, small, electron-shuffling proteins involved in diverse, albeit foundational, biochemical processes (e.g., photosynthesis, oxidative phosphorylation, and nitrogen fixation). Universally conserved iron-sulfur proteins occur in all eukaryotes and the eubacteria, an observation that underscores the importance of these molecules and

SIDEBAR 5.4 Mutations Provide the Raw Material for Evolution

DNA accomplishes its tasks with alarming fidelity. However, if it never incurred mistakes we would not have an evolutionary story to tell. *Mutations* are changes that arise in DNA, either through random errors in DNA replication or through any one of a number of mutagenic agents such as UV radiation and toxic compounds. Although mutations can be induced through certain known environmental factors, the alterations to the genome are themselves random: mutation is not directional, nor does it arise on demand to suit the needs of a population.

Point mutation is a random change to one or a few DNA bases. These mutations can be manifest in a number of ways, ranging from a "silent" mutation that has no effect, to mutations that alter a protein's structure without changing the fundamental nature of the protein, to mutations that render that gene nonfunctional, and may thereby have detrimental effects on an organism.

Light	Ser	Thr	Arg	Thr	Gly
Light	TCC	ACC	CGC	ACA	GGC
Dark	TCC	ACC	TGC	ACA	GGC
Dark	Ser	Thr	Cys	Thr	Gly

Figure 5.5 Example of a coding region from the light and dark variants of the pocket mouse (Figure 5.4). A point mutation in the melanocortin-1-receptor (or MC1R) gene in rock pocket mice changes a CGC to TGC; the resulting amino acid change (Arg, or arginine, to Cys, or cysteine) is one of several mutations involved in the interference of pigment production. Mice with this mutation are lightly colored; mice without it are dark (see Figure 5.4). From Nachman et al. (2003). (Sehoya Cotner)

strongly suggests shared ancestry—how else can we explain variability in less essential proteins but fidelity in those critical for basic functions?

Cytochrome *c*

Another highly conserved electron-transport protein is **cytochrome** *c* (*cyt c*). A small protein at 100–104 amino acids long, *cyt c* has been used to study relationships among different taxa, or groups, of organisms. *Cyt c* is involved in electron transport and is therefore part of every aerobic eukaryote. However, the actual structure of *cyt c* is simple and could arise through billions of different permutations in a protein's primary structure; that is, *cyt c* is functionally redundant. If *cyt c* evolved independently in different lineages of organisms, we would expect to see a range of amino acid structures accomplishing the electron-shuffling work we attribute to *cyt c*. But we don't. For example, pigs, cows, and sheep have identical *cyt c* molecules—amino acid for amino acid. Similarly, *cyt c* of chickens and turkeys is identical, and ducks only differ from chickens and turkeys by one amino acid. In fact, researchers have inserted *cyt c* genes from fish, horses, humans, and birds into the genomes of yeasts, and the yeasts have successfully produced *cyt c* protein (Figure 5.6). It is unlikely that this functional equivalence would be seen in molecules that were not commonly inherited.

Hemoglobin

Similarly, the oxygen-transporting protein **hemoglobin** is found in all vertebrates, and has also been used to study evolution from various perspectives. A comparison of amino acid differences in the β-hemoglobin of vertebrates reveals a pattern identical to that revealed by *cyt c*. Just as hemoglobin and the iron-sulfur proteins suggest common descent, so too does *cyt c* and many other essential proteins.

Organisms with More Recent Common Ancestry Have More Hereditary Material in Common Than Organisms Further Removed Evolutionarily

DNA Hybridization

Whole genomes can be compared by denaturing double-stranded DNA from two organisms, allowing the resulting single-stranded fragments to recombine, and then measuring the temperature at which double-stranded hybrids unravel, or denature. In situations in which the two organisms have similar genomes, the denaturing temperature will be almost as high as it is for non-hybrid DNA, reflecting the degree to which the strands are complementary. When genomes differ significantly, the two strands won't attach as strongly, and less heat will be required to separate the strands. These differences in the energy required to separate the strands can be used to assess degrees of relatedness among hybrid pairs. In the 1980s, DNA hybridization studies were used to document the evolution of numerous lineages—notably, of avian, nematode, and rodent groups. In 1991, hybridization studies documented the relationship between the Australasian marsupials

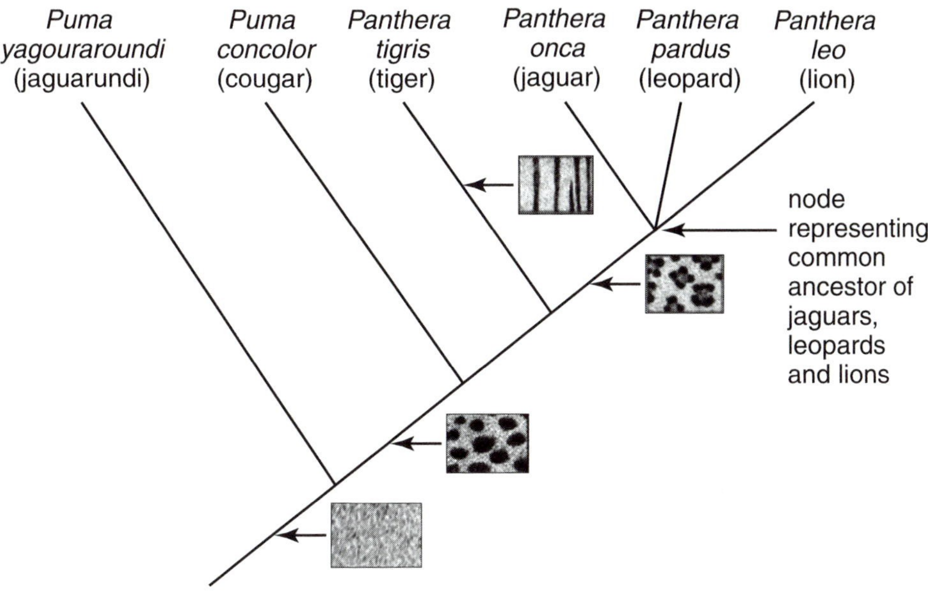

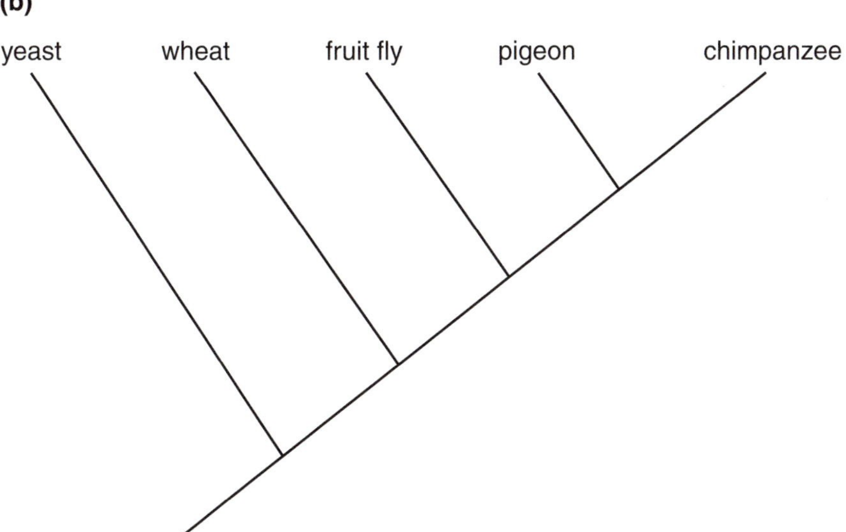

Figure 5.6a, b Using shared and novel traits to construct phylogenies. We can visualize relatedness among taxa, and thus evolutionary histories, with the construction of phylogenetic trees. A phylogenetic tree is organized around branches (representing a known lineage) and nodes (representing common ancestry), and may include information about evolved characteristics. Phylogenetic trees can be constructed with (a) morphological structures (e.g., coat coloration in felid evolution; from Werdelin and Olsson 1997) and (b) molecular information (e.g., *cyt c* nucleotide differences in the reconstruction of animal evolution). (Sehoya Cotner)

and the South American monito del monte (Spanish for "little mountain monkey"); namely, melting temperatures of combined DNA strands suggested that the monito del monte is a close relative of the Australian wallabies and kangaroos, having diverged from the Australian group prior to the breakup of the Southern landmasses. DNA hybridization was an early foray into molecular techniques; in recent years, it has been supplanted by genetic sequence data.

Genetic Sequence Data

Although hybridization techniques tell a useful and consistent story of how living organisms evolved, nothing is as specific, or as unassailable, as the actual sequences of As, Gs, Cs, and Ts that form our genomes. DNA, with its insertions, duplications, and point mutations, gives us not only a play-by-play of molecular evolution, but also a timeline in which key events transpired. Prior to the advent of automatic sequencing devices, restriction fragment length polymorphisms (RFLPs) were used to visualize different-sized fragments of DNA on a gel. However, it is now possible to obtain sequence data for a growing number of whole genomes. With this information, we can compare organisms based on overall genetic similarity, or we can focus on specific coding or non-coding regions of the genome. For example, a 2002 report in the journal *Science* compared mouse chromosome 16 to several corresponding units on human chromosomes. The researchers were unable to find corresponding loci in humans for only 14 (i.e., 1.9%) of the 731 mouse genes that they investigated. Recently, biochemist Douglas Theobold used statistical analyses to test the probability of different evolutionary models—specifically, a single common ancestor versus multiple ancestors for all extant life forms. His work—focused on 23 common proteins from 12 disparate species—dramatically supported the single ancestor model, calculating it to be 10^{2860} times more probable than the multiple-ancestor hypothesis.

Genomes and Genetic Timelines

Not only can sequence data tell us about ancestral patterns, but degrees of difference have also been used to infer the time since taxa diverged. Given a standard mutational rate and known fossil constraints, the number of differences between two lineages gives us data with which to construct a timeline of events. In this way, genomes serve as a **molecular clock** that can be helpful in reconstructing far more than phylogenies, but also the events that may have led to diversification. For example, molecular comparisons have shed some light on the question of whether the mammalian radiation, historically described as a consequence of the end-Cretaceous volcanic activity and asteroid impact (that famously led to the dinosaurs' extinction), was already underway before the dinosaurs went extinct. William Murphy,

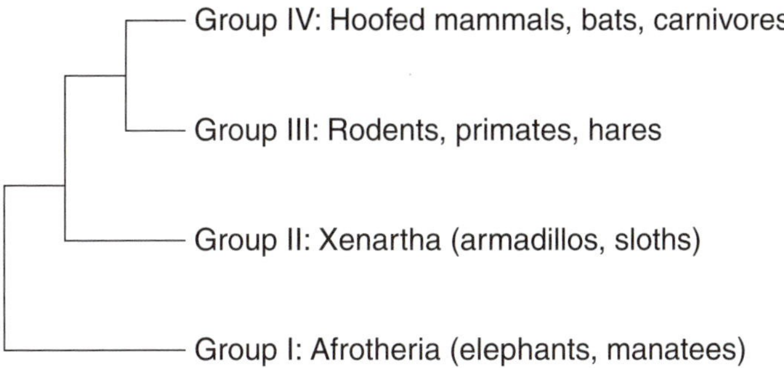

Divergence times of the four major placental mammal groups, as estimated by average sequence divergence:

Group I vs II+III+IV: 76–102 million years ago
Group II vs III+IV: 73–104 million years ago
Group III vs IV: 64–95 million years ago

Figure 5.7 The origin of major placental mammal groups. A timeline for the origin of the major placental mammal groups, applying a molecular clock by using relative degrees of genetic divergence and fossil benchmarks. From Eizirik et al. (2001). (Sehoya Cotner)

Eduardo Eizirik and colleagues at the U.S. National Cancer Institute analyzed 18 genes (totaling almost 10,000 DNA bases) in 64 mammalian species and, in addition to reconstructing their relationships, concluded that the major mammalian groupings had diverged 70 million years ago, and several million years prior to the asteroid impact that ended the Cretaceous period (Figure 5.7; see also discussion in Chapter 3).

Comparing the Coding Regions of Ubiquitous Proteins Confirms Common Ancestry

Genes as Evidence of Common Descent

Genes for ubiquitous proteins provide excellent evidence for common descent. Comparing sequence differences in the *cyt c* coding regions confirms a pattern of descent for the eukaryotes suggested by earlier evidence (such as morphological and physiological similarities). For example, yeasts and great apes differ by 42 (of a possible 104) amino acids, frogs and monkeys differ by 19, and wheat and pigeons differ by 25; the great apes have identical amino acid sequences for *cyt c,* and at most any one mammal differs from another by only 10 amino acids. These degrees of divergence within taxa are predictable (e.g., the *cyt c* genes of all vertebrates differ from those of all yeasts by roughly the same amount) and lend support to the molecular clock discussed earlier.

Heat-Shock Proteins and Adaptation

Heat-shock proteins (Hsp) are a class of proteins found in organisms ranging from bacteria to mammals; these proteins exhibit elevated expression when the organism is exposed to stressors such as elevated or low temperature, food deprivation, dehydration, and hypoxia. The intracellular stress response is considered one of the most highly conserved adaptations in nature, and molecular data echo this claim. Hsp coding-regions can be used to clarify the origins of eukaryotic organelles such as chloroplasts and mitochondria. For example, in 1995 Radhey Gupta documented similar amino acid sequences in eukaryotic chloroplasts and present-day photosynthetic cyanobacteria. This work, which indicates that chloroplasts originated as cyanobacteria, provides additional evidence for the endosymbiotic origins of eukaryotes—namely, that today's eukaryotic cells are the result of long-ago symbioses between prokaryotes (discussed further in Chapter 8).

Organisms with Common Ancestry Share Randomly Generated Transposable Elements and Homologous Genes

Transposable Elements and Shared Ancestry

A large percentage of whole genomes consists of repetitive DNA, which are short sequences that are repeated many—a few hundred to many thousand—times in succession. Many of these repetitive sequences originate as transposable elements that have moved from one part of the genome to another. Transposable elements (TEs), or jumping genes, can arise during an error in DNA replication, or they can move via an RNA intermediate. TEs are common in DNA, especially in regions that do not code for a protein, and their mutation rate is high. Given that TEs are inherited and move randomly, they can be used to identify parentage, suspects in criminal cases, and evolutionary lineages. The sharing by two organisms of the same TE in the same chromosomal location, whether they be cousins or sister taxa, strongly implies shared ancestry. For example, an analysis of the short interspersed elements (SINEs, a type of TE) in mammalian groups reveals that whales and hippos share genetic signatures that they do not share with camels and pigs (see Chapter 3). This finding suggests a hidden history that challenges conventional beliefs about the relationships among cetaceans such as whales, and hoofed mammals such as pigs.

Gene Duplication

During meiosis (the cell division that produces sperm and eggs), chromosomal crossover—whereby paternal DNA is combined with maternal DNA on a single chromosome—is common; this is one of the ways that sexual reproduction generates genetic diversity. The placement of numerous identical TEs along a chromosome makes misaligned crossovers likely. As a consequence, functional genes can be deleted or duplicated during meiosis. Although the deletion of chromosomal segments may produce sperm or eggs that are not viable, the genetic duplications can accumulate over time, and ultimately generate novel genes in the process. Gene duplication leads to the existence of redundant genes in an organism's genome, which can relax the selection pressure on a given gene and allow for a build-up of mutations. In this way, families of homologous genes can arise from one ancestral gene. Within genomes that have been fully sequenced, gene duplications appear to contribute to between 17% (for the bacterium *Helicobacter pylori*) to 65% (for the plant *Arabidopsis thaliana*) of the genome. In vertebrate animals, the olfactory receptor genes are the largest group of duplicated genes; mice have approximately 1,300 olfactory genes at 46 different genomic locations.

Shared Ancestry Is Inferred through the Occurrence of Vestigial Molecular Elements, or Pseudogenes

The Origin of Pseudogenes

A **pseudogene**, a gene that no longer functions, is another potential fate of duplicated genes. A redundant gene may accumulate mutations to the point where it is no longer functional. Pseudogenes are a common part of genomes; for example, the roundworm *Caenorhabditis elegans* has over 2000 pseudogenes, or approximately one pseudogene for every eight functioning genes.

Pseudogenes as Evidence of Common Inheritance

The presence of the same nonfunctional gene in multiple species is evidence of common inheritance. Like the whale's pelvis or the snake's limbs (Chapters 3 and 6), these vestigial elements are key to our understanding of evolutionary history. For example, genes for the egg yolk protein, vitellogenin, exist as remnants in mammals. Of the three vitellogenin genes found in mammals, all three are nonfunctional in the marsupial and placental mammals, but one remains functional in the egg-laying mammal, the platypus. Scientists studying the egg yolk genes also identified the genes coding for the milk protein casein; using molecular dating techniques, they concluded that the emergence of lactation likely reduced the

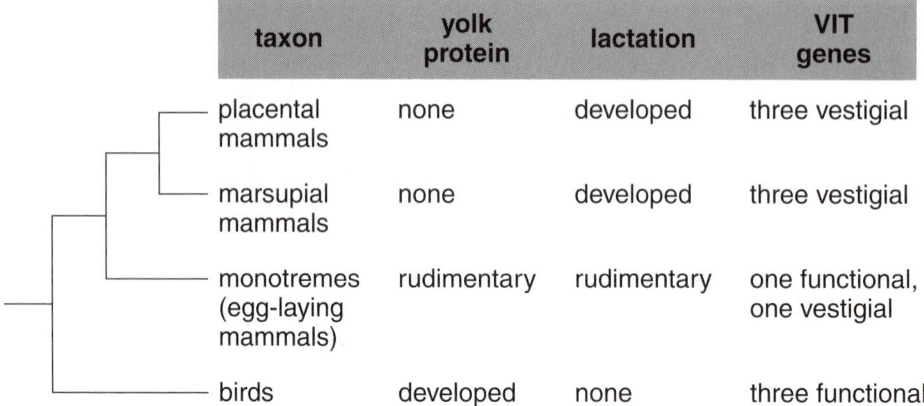

taxon	yolk protein	lactation	VIT genes
placental mammals	none	developed	three vestigial
marsupial mammals	none	developed	three vestigial
monotremes (egg-laying mammals)	rudimentary	rudimentary	one functional, one vestigial
birds	developed	none	three functional

Figure 5.8 VIT (genes for vitellogenin) evolution in tetrapods. Pseudogenes, or vestigial molecular elements, illuminate key events in mammalian evolution. While mammals are known for lactation, their sister taxa—the reptiles and birds—do not produce milk. The evolution of lactation may be key to the loss of function in genes for vitellogenin, or yolk protein, in mammals. A clue resides in the egg-laying mammals such as the platypus: these animals are at a developmental and evolutionary halfway point, producing some milk and some yolk protein, but excelling at neither. From Brawand et al. (2008). (Sehoya Cotner)

selection pressure on the egg yolk proteins. In other words, when young could get their protein from milk, egg protein was less essential and, as the genes mutated to nonfunctional variants, selection did not eliminate individuals possessing these mutant alleles (Figure 5.8). Our current understanding of the evolution of lactation exemplifies the knowledge that can be gained from studying pseudogenes.

Shared Ancestry Is Revealed within a Lineage Showing Evidence of a Shared Viral Pathogen

Endogenous Retroviruses and Common Descent

Viruses operate by commandeering the cellular machinery of their hosts. The *endogenous retroviruses* (ERVs) leave behind some complementary nucleic acid that becomes incorporated into the host cell's DNA. If retroviral DNA infects a host's genome by way of sperm or egg cells, then this remnant of the pathogen can be inherited.

Retroviral DNA can be identified by the presence of virus-specific (and essential) genes: *gag* (which encodes structural proteins that will encapsulate the DNA in future viruses); *env* (surface proteins, that allow for viral entry into host cells); and *pol* (viral enzymes). Since insertion into gamete DNA is a relatively rare and random event, the co-occurrence of ERV sequences is evidence of common descent. Work on one group of ERVs, the human endogenous retrovirus family (HERV-K), has supported a primate phylogeny that groups the great apes into one clade, separate from the monkeys and prosiminians (Chapter 9): some strains of HERV-K are exclusive to the great apes, others exclusive to apes in general, and others characteristic of all primates—apes, monkeys and prosimians.

As with any genetic novelty, an ERV insertion can be manifest in three ways: it can have no effect on the host; it can have a negative effect on the host; or it can benefit the host. Most ERVs appear to have no effect. However, some are associated with increased cancer susceptibility (e.g., a family of ERVs is linked to mouse leukemia), whereas others are associated with resistance to infection by related retroviruses (e.g., an ERV is associated with resistance to feline leukemia). In the latter case, ERVs may contribute resistance by constructing envelope proteins at the host's cell surfaces that interfere with the ability of other viruses to penetrate the cell. As more whole genomes are sequenced and described, we will learn more about the role of ERVs in our evolutionary history.

Summary

The molecular evidence for evolution is powerful, from universal homologies (e.g. DNA, RNA and the genetic code) to vestigial molecular elements. The sequence similarities we observe are extremely improbable without shared ancestry. Also, these similarities mirror similarities observed in, and trajectories implied by, the fossil record and biogeography. The predictions about molecular evidence for evolution have been, and continue to be, confirmed.

6

Anatomical Evidence for Evolution

Nature may be said to have taken pains to reveal, by rudimentary organs and by homologous structures, her scheme of modification, which it seems that we wilfully will not understand.

Embryology will reveal to us the structure, in some degree obscured, of the prototypes of each great class.

—*Charles Darwin, 1859*

What better evidence for Darwin's belief in the commonality of all species than to find the same gene doing the same job in birds and fish, continents apart?

—*Matt Ridley, 2009*

PREDICTIONS

If evolution has occurred, we predict that

1. There will be anatomical similarities among related organisms.
2. Organisms will possess vestigial traits that serve as evolutionary relics and clues to ancestry.
3. There will be developmental similarities among organisms, in terms of gross morphology and molecular machinery.

Hundreds of years of careful observations have helped us understand many things about the relationships among organisms and the evolutionary histories of many divergent lineages. Initial observations were rooted in the most visible clues—an organism's anatomy. Birds have always been linked by feathers, mammals by fur, and spiders by their eight legs. However, anatomical information can be misleading—for example, we know that dolphins, despite their aquatic habitat and streamlined bodies, are not fish, and most biologists agree that crocodiles and birds share a recent

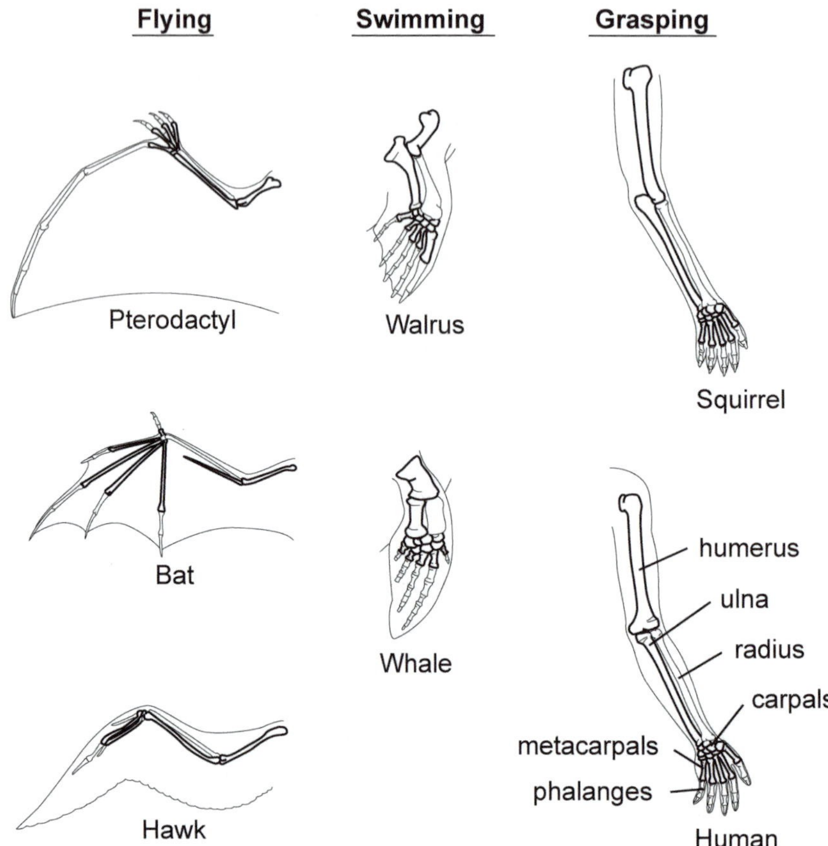

Figure 6.1a Homologous structures. Although the forelimbs of pterodactyls, bats, hawks, walruses, whales, squirrels, and humans have different functions, they all contain the same set and arrangement of bones, suggesting that the animals inherited this feature from a common ancestral vertebrate. (Jeff Dixon)

ancestry that is separate from the snakes and lizards. Also, new work in the emerging field of evolutionary development reveals developmental similarities at the molecular level, and may challenge some of the conventional wisdom established by the pioneers of comparative anatomy.

There Are Anatomical Similarities among Related Organisms

Comparative Anatomy

The inferences about evolution derived from paleontology are reinforced by **comparative anatomy,** which is the comparison of structures across species. Related organisms have anatomical similarities; for example, the

Figure 6.1b These postage stamps from North Korea acknowledge how homologous structures—in this case, the forelimbs of birds, bats, porpoises, horses, and humans—arise during evolution from a common ancestor.

skeletons of humans, rats, and bats are similar, despite their different life histories and different environments—there is a bone-by-bone similarity throughout the bodies of these organisms. This similarity is evident in a comparison of the forelimbs of a pterodactyl, bat, hawk, walrus, whale, squirrel, and human shown in Figure 6.1. These animals live in different types of environments, and their limbs have remarkably different appearances and functions—some are used for flying, others for swimming, and even others for grasping and throwing. Despite their different appearances and functions, these animals' limbs have a similar pentadactyl (i.e., five-digit) structure consisting of a single proximal bone (humerus), two distal bones in the forearm (ulna and radius), several carpals (wrist bones), and five series of metacarpals (palm bones) and phalanges (digits). *All* tetrapods—from amphibians to mammals—have such pentadactyl limbs; this limb structure is also present in ancestral reptiles and amphibians, and even in the first fish that came out of the water onto land hundreds of millions of years ago (Figure 6.1; Chapter 3). Moreover, these structures are derived from the same structures in the embryo. These structures that are derived from a common ancestor and are embryologically similar, but have different functions, are called *homologous structures* (introduced in Chapter 1).

Homologous Structures

Homologous structures suggest that these animals evolved from a common ancestor. Like embryological similarities, homologous structures are difficult to explain if the organisms originated independently of (and were therefore unrelated to) each other, for such an origin would not require pentadactyl limbs or, for that matter, *any* other shared traits. Moreover, there is no obvious reason—be it mechanical, functional, ecological, or otherwise—why *all* vertebrates *should* have 5-digit limbs instead of, say, 3-digit or 21-digit limbs. However, the facts remain: all of these organisms *do* have five-digit limbs that develop in the same way, consist of the same arrangements of the same tissues, and always form in the same position on the animals' bodies, even though such events are not functionally or mechanically necessary. That is, the bones humans use to write with are the same ones used by a bat to fly, by a dog to run, and a squirrel to grasp. This makes sense in the light of evolution because it is evidence of common ancestry—namely, that all tetrapods descended from a common ancestor that had pentadactyl limbs.

Creationists have explained these developmental and structural similarities as the handiwork of a Creator who saved time and work by varying a basic theme. Darwin, however, considered such explanations to be useless, noting that "nothing can be more hopeless than to attempt to explain this similarity. . . by utility or by the doctrine of final causes," and that the similarities are "inexplicable" by traditional views of creationism. Darwin added, "What can be more curious than that the hand of a man, formed for grasping, that of a mole for digging, the leg of a horse, the paddle of a porpoise, and the wing of a bat, should all be constructed on the same pattern, and should include the same bones, in the same relative positions." In fact, not much "new" has happened to these limbs—the large differences in their structure and function result from a simple regulatory shift of a basic shared pattern. In these different animals, the morphological distinctions in homologous traits are due largely to the simple fact that different parts grow at different rates.

Structural Similarities in Plants

Structural similarities are common in plants. Although most plants use leaves for photosynthesis, many plants have leaves that have been modified by evolution for functions other than photosynthesis. For example, a cactus spine and a pea tendril have different appearances and different functions: a spine protects the cactus stem, and a tendril helps support the climbing stem of a pea plant. In insectivorous plants such as Venus's flytrap and pitcher plants, leaves are modified into devices that lure, trap, and digest unsuspecting insects. However, spines, tendrils, and insect traps are all modified leaves; they develop in the same way as leaves and consist of the same parts as leaves. Such modifications of organs used for different functions are an expected outcome of a common evolutionary origin.

Analogous Structures and Convergent Evolution

Not all structural similarities result from common ancestry. As noted in Chapter 4, some structural similarities result from convergent evolution, in which natural selection produces nonhomologous structures having similar functions that resemble one another despite their different origins. To appreciate this, consider flying insects and flying birds. Although birds and insects each have wings, this similarity arose not from the evolutionary modification of a structure that birds and insects inherited from a common ancestor, but instead from the evolutionarily independent modification of two different, nonhomologous structures that eventually gave rise to superficially similar structures (i.e., wings). Stated another way, natural selection favored flight in these two groups of animals. These nonhomologous but superficially similar structures are called *analogous structures.* Analogous structures often differ internally because they are not derived from structures inherited from a common ancestor.

Organisms Possess Vestigial Structures That Serve as Evolutionary Baggage

Vestigial Structures as Remnants of Ancestral Organisms

During Darwin's time, people throughout Victorian England believed that organisms were the perfect creation of a deity. However, this belief was undermined by anatomical studies showing that organisms have many parts that have lost all or most of their original function. Such structures, which may have at one time been useful in an ancestor but now have no function or a different function, are called **vestigial structures**. Vestigial structures, which are common throughout nature, are evolutionary baggage carried by virtually all species. Examples include the following:

Pigs, cattle, deer, and dogs have vestigial toes that do not touch the ground. For example, on the foot of a pig, one digit has been lost completely, two other digits are greatly reduced, and only the two remaining digits support the pig's body.

Many species of flightless beetles have nonfunctioning wings because they are covered by permanently fused wing-covers that never open. These structures are a reminder that flightlessness is an acquired feature in beetles.

The tiny splint-bones in the feet of horses are remnants of when horses had three toes. If these bones break, the horse is permanently crippled.

Toothed whales have full sets of teeth throughout their lives, but baleen whales only have teeth in the early fetal stage, and these teeth are reabsorbed before birth. The teeth in baleen whale embryos are evidence of their common ancestry with toothed whales and other mammals. Similarly, terrestrial salamanders at one stage of development have gills and fins, but lose them before hatching. These and other silent traits are consistent with these animals having evolved from species that possessed and used these traits.

Some snakes (e.g., boa constrictors) have rudiments of a pelvis and tiny legs buried in their sleek bodies, left over from an ancestor that had legs. For example, *Pachyrhachis problematicus* was a Cretaceous snake that *did* have legs.

Wingless birds (e.g., ostriches, emus, kiwis; see Chapter 4) have vestigial wing bones that are remnants of their flying ancestors' wings.

Dandelions—a ubiquitous and irritating pest in lawns and gardens—reproduce asexually, yet possess flowers and produce sterile pollen.

Many blind, cave-dwelling, and burrowing animals have nonfunctional, rudimentary eyes that are remnants of their ancestors' functional eyes. Interestingly, blindness evolved independently in several different populations, and the genes that produced the blindness in these populations are often different.

Modern whales and other cetaceans lack hind limbs, but do have small pelvic bones and thighbones buried in their bodies. These bones have no apparent function and are not anchored to any other bones. Why? Because whales' ancestors lived on land (Chapter 3). Whales' pelvic bones are the evolutionary remnants of their ancestors' life on land.

The great apes—orangutans, gorillas, bonobos, chimpanzees and humans—lose their tails during embryologic development. However, they retain reduced tailbones that may serve a minor function in sitting upright and anchoring muscles.

Vestigial organs are often homologous to organs that are useful in other species; for example, the vestigial tailbone of the great apes is homologous to the functional tail of other primates. Some monkeys use their tails for several functions (e.g., climbing, balance, communication), but the great apes' tails are less essential. Human vestigia are discussed further in Chapter 9.

Vestigial structures are difficult to explain if one denies evolution. For example, why would a deity create an organism that has useless parts? However, vestigial structures are consistent with Darwin's suggestion that they were useful in the species' ancestors. The more sensible explanation for vestigial structures is not that they were specially created, but instead that they are the remnants of structures that were present and functional in ancestral organisms.

Creationists' Criticisms of Vestigial Organs as Evidence of Evolution

Many creationists claim that vestigial organs are not evidence for evolution because such organs do, in fact, sometimes have functions. This criticism is misguided, for *vestigial* does not necessarily mean that an organ is useless. Indeed, a vestige is a "trace or visible sign left by something lost or vanished." Whether these organs have a function is irrelevant; they do not have the function that we would expect from such organs in other animals, for which the creationists claim the parts were "designed."

There Are Developmental Similarities among Organisms

Developmental Similarities

Related organisms develop similarly. In fact, the oddly shaped, colorful aquatic sponges (Phylum Porifera) are classified with the animals on the basis of their initial development. Like shrimp, sea stars, and kangaroos, sponges transition from the zygote (fusion of egg and sperm) stage to the blastula, a hollow ball of cells. An infolding of this ball, *gastrulation,* creates a gastrula, in which tissue types begin to form. Sponges do not undergo gastrulation and do not develop true tissues, but all other animals (including jellies and reef-building corals) do. Organisms with the embryological underpinnings of three distinct tissue-types are called triploblasts, a developmental similarity that unites the flatworms and vertebrates, but excludes the corals and jellies. Triploblasts include bilaterally symmetrical organisms with and without body cavities; those with body cavities can be further divided into the protostomes and deuterostomes, a developmental distinction based on whether the site of gastrulation becomes the animal's mouth or anus.

Embryological Similarities

To appreciate these developmental similarities, consider the embryos of the vertebrates shown in Figure 6.2. These embryos look similar, and it is often difficult to tell them apart. For example, the embryos of all these organisms have tails, and these tails have important functions in many organisms (e.g., in fish, the tail is used for swimming). In human embryos, the tail is approximately 10% the length of the embryo at about six weeks after conception. However, the tail is then reabsorbed, which explains why most humans aren't born with visible tails. Thus, animal embryos are similar early in development; animals' complex adult forms emerge later as embryos develop. As discussed in the next sections, genes active in corresponding

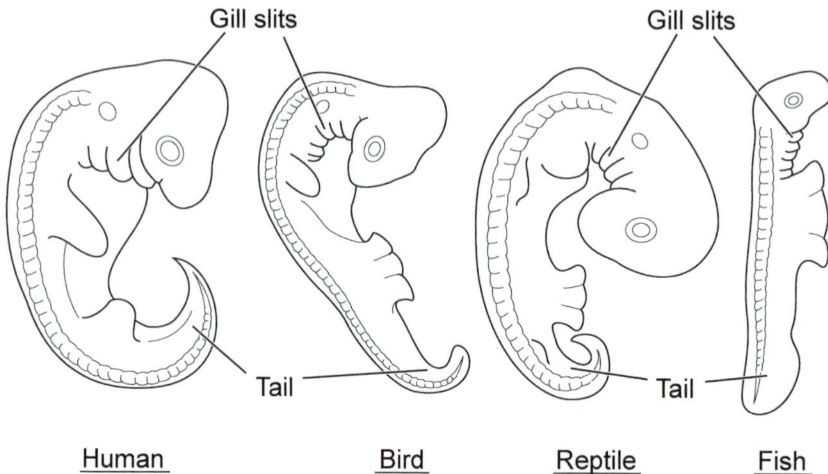

Figure 6.2 Embryos show evolutionary relationships. The embryos of humans, birds, reptiles, and fish have many shared features (including a tail and gill slits), suggesting that these animals share a common ancestor. (Jeff Dixon)

parts of embryos guide the development of different animals, thereby converting the forelimbs of different organisms into fins, wings, arms, and other structures from the same general organization of bones and tissues.

The limbs of all tetrapods (i.e., animals having four limbs such as arms and legs; *tetra* = four, *poda* = foot) develop from limb buds in similar ways, and embryos of all vertebrates also have gill-like branchial arches. The branchial arches of fish become gills, but in primates these structures do not become gills; instead, they become modified for other functions, such as the Eustachian tubes that connect the middle ear with the throat in the great apes.

Development and Embryology as Evidence of Common Ancestry

Embryological and developmental similarities are difficult to explain if one assumes that each was created independently. If all vertebrates were created independently, why do chimpanzee embryos have tails and branchial arches? A simpler explanation for the similar developments of the embryos of all vertebrates is that they share a common ancestor.

Evolution is conservative; instead of starting from scratch, it builds on what has come before. For example, instead of producing new developmental genes, it is more likely that natural selection will favor modifications of preexisting genes. This is why embryology provides such strong evidence for evolution. As Darwin noted, an organism's embryo "reveals the structure of its progenitors" and is "by far the strongest single class of facts in favor of change of forms."

The Study of Evolutionary Development

Anatomical similarities in fully formed and developing organisms are evidence of a universal shared ancestry that mirrors the trajectories inferred by fossil and biogeographical evidence. Yet, questions remain: What is the genetic basis of our similarities? Can we track these similarities and differences from genes to the formed, fully functioning individual? How can we explain our differences? The emerging field of **evolutionary developmental biology (evo-devo)** has great promise for answering these questions. Biologists studying evolutionary development hope to unite the disciplines of molecular biology, developmental biology, and evolution.

Homeotic Genes

Similar structures—head, trunk, appendages—pervade the animals, from ticks to tigers. Studies of the genes controlling body formation reveal that our similarities are not coincidence; rather, we possess shared genetic master-switches, the *homeotic genes,* that regulate the expression of other developmental genes. The coding regions of homeotic genes (for an overview of genetics, see Chapter 5) have been highly conserved evolutionarily; that is, there are few differences in the DNA sequences at these sites. For example, the *homeobox* is a 180-base pair region that codes for a key regulatory protein, the *homeodomain*. The homeobox is 90% identical in fruit flies and humans. Homeobox sequences are characteristic of a group of developmental genes, the *Hox* genes, which control body alignment and the placement of appendages such as legs and antennae. *Hox* gene homologues have even been identified in sea anemones, suggesting a history of at least 600 million years. In some cases, these genes exhibit functional equivalence across taxa; for example, chicken Hox proteins have been used to experimentally manipulate fly development. A fly will develop from an embryo that has had its own *Hox* sequence replaced with a mouse *Hox* sequence. In addition, the *Hox* genes that regulate head-tail body-axis development in fruit flies have homologous genes in mammals, and they exhibit the same linear arrangement across taxa. In fact, the linear arrangement of *Hox* genes mirrors the organism's head-to-tail development, with anterior genes grouped on one end of the series, and posterior genes grouped on the other.

Developmental Variation

The similarities of *Hox* genes tell a convincing tale of shared ancestry and the functional significance of developmental regulation. Yet it is also intriguing to consider the factors underlying our developmental *differences.* If our master switches are so similar, why is there such variation in body shapes? A look at bat wings and finch beaks can address some of this variation.

As the only flying mammals, bats have modified the activity of several of their master genes to make wings from a hand. Other amniotes, such as mammals that are born with distinct digits, initially develop with webbing. However, this webbing typically undergoes massive cell-death prior to birth, in large part due to the expression of bone morphogenetic protein (BMP). Some amniotes, such as ducks, retain webbing in their feet by blocking the expression of BMP. Similarly, bats avoid interdigital cell death partly due to a BMP suppressor, and partly due to enhanced fibroblast growth factor (FGF) that blocks cell death. Additional FGF molecules may control the elongated digits characteristic of bat wings. BMP and FGF are not unique to bats, but they are expressed—at a combination of times during development, and in specific amounts—in a way that has allowed these mammals to take flight.

Variable morphs that have received a lot of scrutiny are the finches of Galápagos—Darwin's finches (see discussion in Chapter 4). The Galápagos archipelago is home to 14 species of these finches, distinguished in large part by their beaks. For example, the large ground finch (*Geospiza magnirostris)* has a deep, wide beak that can effectively crack hard shells to reach the seeds inside. And the long, narrow beak of the common cactus finch (*Geospiza scandens)* can retrieve nectar from cactuses while avoiding the spines. These differences have been discussed by evolutionary biologists in the context of character displacement (Chapter 8) and adaptive radiation (Chapters 1 and 4). But recent work on regulator proteins draws connections between genotype, development, and adaptation. For example, another bone morphogenetic protein—BMP4—explains much of the variation in beak sizes that characterizes these finches. BMP4 is a key signaling molecule that activates other genes, which in turn regulate bone mineralization. Finches with larger, thicker beaks show earlier and more pronounced BMP4 activity during embryonic development. Adding BMP4 to developing chick tissues results in birds with thicker and wider beaks, and blocking BMP4 with an antagonizing protein produces chicks with smaller beaks. Another protein, calmodulin (or calcium-modulated protein, CaM), is a calcium-binding regulatory protein that plays a role in beak variability. Specifically, higher levels of CaM are associated with the longer beaks of the cactus finches. Apparently, fundamental differences between these species is not due to any dramatic genetic alterations, but rather is the result of small changes in the timing and extent of expression of these two proteins. This modification must have granted the finches' progenitors some fitness advantages in their respective habitats. Darwin could not have foreseen how "one species had been taken and modified for different ends," but ongoing work with regulatory proteins such as CaM and the BMPs gives us a clue.

Hox genes (discussed earlier) can also affect developmental variation. Recent work by Portuguese scientists may explain variability in the number of ribs in the vertebrate animals. Humans have 12 pairs of ribs, mice have 13, and snakes can have more than 100 pairs of ribs. This variation appears to be the result of a combination of *Hox*-gene inhibition (i.e., *Hox-10* genes suppress rib development) and *Hox*-gene activation (*Hox-6* genes promote rib growth). When inactive *Hox-6* genes were forcibly activated in typically rib-less regions of developing mice, the mice generated ribs extending down to their tails, just as snakes do.

Eye Development and Animal Evolution

Pax-6, or paired box gene 6, is another master regulator that is highly conserved—*Pax-6* is found in animals ranging from flatworms to vertebrates. *Pax-6* plays a key role in eye development, and its presence across disparate taxonomic groups requires some revision of our understanding of animal evolution (Figure 6.3). The compound eyes of insects, the pigment cups of flatworms, and the simple eyes of mammals and cephalopods (e.g., squid) were previously assumed to be examples of convergence—their eye structures are so different! In fruit flies, *Pax-6* initially went by the name *eyeless* because, in its mutant state, it is found in flies without functional

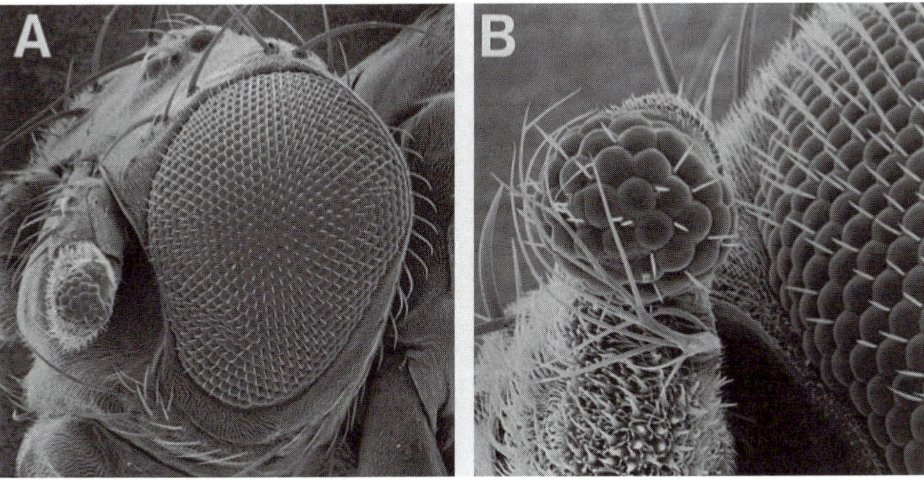

Figure 6.3 Eye structures and *Pax-6* homology. In this example, mouse *Pax-6* induced growth of an eye on the tip of a fruit fly's antenna. The image on the right (B) is magnified to show the structural similarities between the large, normal eye and the small, artificially induced eye. (Reprinted from W. J. Gehring and K. Ikeo, 1999, "Pax 6: Mastering Eye Morphogenesis and Eye Evolution," *Trends in Genetics* 15 (9): 371–377. With permission from Elsevier.)

eyes. However, this gene is now known to be homologous to a gene for the protein "Small" eye in mice. Mice with the mutant gene have much smaller eyes than do those without the mutation.

Further evidence of a shared eye-building apparatus comes from experimental manipulations: similar in outcome to the *Hox* transplants discussed above, mouse *Pax-6* genes induce eye development in flies (Figure 6.3). And our shared eye-building apparatus is not restricted to *Pax-6*: from sea anemones to great apes, organisms possess crystallin proteins for directing light, and opsin proteins for capturing it. These similarities in eye construction (despite our vastly different eyes) and body formation (despite our distinct morphologies) underscore the importance of developmental changes in evolution, and whet the appetite for discoveries that lie ahead.

Summary

Anatomical similarities within lineages, vestigial traits, and developmental histories all confirm what fossils, biogeography, and the molecular evidence have likewise suggested: living organisms share common ancestry and have diverged, over time, through evolutionary processes.

7

Conflict and Cooperation Part I: Behavior

Would I lay down my life to save my brother? No, but I would to save two brothers or eight cousins.

—*John B. S. Haldane*

We are the recorders and reporters of facts—not the judges of the behaviors we describe.

—*Alfred Kinsey*

PREDICTIONS

In general, assuming a heritable and variable nature to behavior, we expect organisms to behave in ways that maximize their own inclusive fitness. Specifically, if evolution has occurred, we predict that

1. Organisms will exhibit behaviors that increase their chances of survival.
2. Behavioral adaptations that enhance survival will be selected over those that do not, even when a cost is incurred by other individuals in the population.
3. Behaviors, like other measurable adaptations, can be artificially selected.
4. Individuals will behave preferentially toward kin.
5. Individuals will exhibit behaviors that acquire mates.
6. There will be an adaptive advantage in behaviors that help ensure parentage.
7. In a conflict between individual survival and reproductive potential, reproductive potential will win.
8. Parents will behave preferentially toward those offspring most likely to augment their inclusive fitness.

Many wasps of the family Ichneumonidae are characterized by larval cannibalism, in which the female wasp deposits her eggs into the larvae of another wasp species. Upon emerging from the eggs, the parasitic larvae feed upon the hosts' tissues, slowly killing the larval hosts in the process. This grisly behavior troubled Charles Darwin, who wrote, in a letter to his friend Asa Gray, "I cannot persuade myself that a beneficent and omnipotent God would have designedly created the Ichneumonidae with the express intention of their feeding within the living bodies of Caterpillars." Darwin wrote this letter one year after the publication of *On the Origin of Species*, when he was having doubts about a benevolent creator, and as he admitted, "I cannot see as plainly as others do, and as I should wish to do, evidence of design and beneficence on all sides of us." Of course, Darwin had already provided a cogent explanation for the ichneumon's behavior: wasps that lay eggs in defenseless hosts are the result of selection acting on this feature of wasp behavior. Behaviors, like other manifestations of an organism's genes, are the product of evolution by natural selection.

Behavior

Behavior is how an organism responds to an internal or external stimulus (or stimuli). A tulip orients its photosynthetic surfaces toward the light, a brittle star releases one of its legs to escape a predator, a dog barks at an intruder—these are behaviors. Behaviors are often termed *innate* (e.g., phototropism in plants) or *learned* (e.g., many common bird-songs), *simple* (e.g., sunning postures in many lizards) or *complex* (e.g., the mating dance of the blue-footed booby). In fact, there is no real dichotomy between innate and learned behavior: all behaviors have some genetic basis, including the capacity for learning. And often, apparently simple behaviors are revealed to have intricate mechanistic underpinnings. Many behaviors consist of a series of steps that occur predictably in response to a particular stimulus—these stereotypical behaviors are called *fixed action patterns* and have been analyzed for causation and heritability in organisms from roundworms to human beings. But regardless of complexity, behaviors can be seen with an evolutionary lens: certain behaviors have been selected for their fitness benefits.

Behavior can be complicated to evaluate. Many behaviors aren't rigid, but rather fall on a presentation continuum (e.g., wheel running in mice). Also, determining the degree of heritability of a behavior is not always straightforward. It is essential to distinguish between proximate and ultimate causation in our analysis of behavior (introduced in Chapter 1). Although proximate explanations for a given behavior may suffice for some, the ultimate (evolutionary) explanations will be our goal in this chapter. In addition, it is often necessary to clarify the nature of the fitness (reproductive success) advantage afforded by a particular behavior. A response that increases an individual's own survival and reproductive potential provides

SIDEBAR 7.1 Is Behavior Inherited?

Evaluating behavior from an evolutionary perspective would be pointless if behaviors weren't at least partially heritable. Studies of various species have revealed some behaviors that are inherited in a simple Mendelian fashion. For example, a single gene in *Drosophila melanogaster* determines whether flies are of the wayward rover or stay-put sitter phenotype: rovers have a single copy of the rover gene, while sitters are homozygous recessives at this locus. This pattern was initially confirmed by crosses in which true-breeding sitters and true-breeding rovers produced thousands of offspring with the roving behavior. Second-generation crosses (or F_1 crosses) produced a 3:1 rover-sitter ratio in the flies that emerged. Similarly, a mutation in the *fosB* gene occurs in female mice that neglect their offspring. And the roundworm *Caenorhabditis elegans* has either a solitary or social nature depending on a single amino-acid substitution in one of its membrane proteins.

Genetic-transfer studies have illustrated the power of a single gene in influencing behavior. In one instance, researchers successfully transferred the *V-1a-receptor* gene from the monogamous prairie vole to the mountain vole, altering the mountain vole's promiscuous nature. Genetic knockout experiments (which render a gene nonfunctional) with this same gene led the faithful male prairie vole to abandon his mate.

Of course, inheritance patterns aren't always this clear-cut, and the inheritance of behavior is no exception. Some behaviors seem to have only a portion of their variation due to inheritance, whereas in others, it is merely the ability to learn a certain behavior that appears to be heritable. Regardless, behavior is heritable, and as a result, can be evaluated evolutionarily.

a *direct* fitness benefit, while one that increases the survival and reproduction of the individual's relatives provides an *indirect* fitness benefit. The combination of fitness-enhancing features, direct and indirect, combines to form an individual's *inclusive* fitness.

Organisms Exhibit Behaviors That Increase Their Chances of Survival

Examples that fulfill this prediction abound! A few that have received much study include the following:

Tropic or Taxic Behavior

Tropic (or taxic) behaviors include movement in response to light (phototaxis), touch (thigmotaxis), and a current (rheotropism). Phototaxis has been documented in essentially all taxa from prokaryotes to chordates. Many zooplankton exhibit diel vertical migration (DVM), whereby they move up or down in the water column to—in most cases—maximize food availability while minimizing predation risk. While movement of the crustacean *Daphnia* is often triggered by light, some strains exhibit positive phototaxis (i.e., movement toward light), some exhibit negative phototaxis (i.e., movement away from light), and some are both positively and negatively influenced by light.

Migration

Migratory feats have long impressed evolutionary biologists. The spiny lobster is noted for single-file processions of dozens of lobsters in search of spawning grounds. Annually, millions of monarch butterflies fly over 3,000 kilometers from their overwintering grounds in Mexico to breeding areas in the temperate United States and Canada (Figure 7.1). The life cycle of the European eel involves spawning in the Sargasso Sea and eventually returning to freshwaters of Europe, a migratory accomplishment that can take up to a year and span thousands of kilometers. And there are the impressive global migrations of birds such as the bar-tailed godwit, which flies nearly 10,000 kilometers nonstop, depleting half its body weight in the process, and the arctic tern, which travels from its arctic breeding grounds to the Antarctic in a journey that can exceed 60,000 kilometers. The huge costs incurred by such energy-demanding, often-perilous migrations appear to be offset by fitness gains associated with reduced competition for resources, greater food availability, reduced predation pressure, warmth, or access to mates. Recent work by biologists at the University of Western Ontario pinpoints tradeoffs involved in migration. W. Alice Boyle and colleagues studied the white-ruffed manakin, a tropical bird for whom migrating is possible, but not essential. Male manakins that migrate are rewarded with superior food availability, yet males that do not migrate ("residents") are compensated with higher social standing and better breeding sites (and possibly more and better females). This work illustrates some of the costs (loss of social standing and decreased access to females) and benefits (better food) of migration in a scenario that mandates male manakins choose between food and sex.

Thermoregulation

Thermoregulation involves keeping the body within a target range of temperatures in spite of environmental fluctuations. Many organisms exhibit behaviors that have a thermoregulatory advantage. *Pieris* butterflies orient their wings for maximum light absorption to increase their body temperature, desert iguanas do pushups to circulate air and cool off, elephants spray themselves with water, dogs pant, and the Cape ground squirrel cools off by shading its body with its own tail. Many organisms, whether they regulate their heat internally (endotherms) or externally

Figure 7.1 Monarchs during migration. Migratory feats such as those achieved by the monarch butterfly (shown here, resting en masse) have intrigued biologists for many years. During the breeding season, monarchs migrate to the site where their parents bred. These sites can be thousands of kilometers away and are entirely unfamiliar to the monarchs. (AP Photo/Marco Ugarte)

(ectotherms), use a combination of physiological, anatomical, and behavioral adaptations to maintain a desired body-temperature. Ultimately, thermoregulatory behaviors confer a fitness advantage by allowing organisms to do things that are strongly temperature dependent. For example, ectothermic snakes can only digest food within a narrow range of temperatures, typically around 30°C.

Storage and Retrieval

Food caching has been documented extensively in birds, mammals, and the social insects (Hymenoptera). Animals that store food typically live in variable habitats that change seasonally or diurnally. For example, many high-latitude animals store summer seeds, nuts, and berries for winter consumption; tropical caching is often regulated by wet and dry seasons; and organisms that forage in the intertidal zone may cache food for consumption during high tides. The behaviors that have evolved in association with food caching include selecting appropriate foods to cache, employing strategies to avoid food spoilage, concealing stockpiles, and relocating them when necessary.

Many behavioral adaptations protect food from microbial contamination and spoilage. Pine squirrels, pikas, and foxes dry their food prior to storage; pikas differentially store unpalatable vegetation for delayed consumption when the toxins (which are also preservatives) have dissipated; and some shrews and moles wound and immobilize their prey so the prey animals stay alive and unspoiled but cannot move from the cache site. In an attempt to conceal the stockpile, hoarders often store their food in many dispersed locations, as is seen in tree squirrels, flying squirrels, canids, many hawks, and most jays. Alternatively, animals may cache in one location, a strategy employed by beavers, pocket gophers, brown bears, and the snowy owl. Some animals, such as Merriam's kangaroo rats, use both of these strategies. Numerous studies have documented the importance of spatial, olfactory, and visual clues in locating previously stored food. Regardless, the adaptive significance is clear: animals that store food for lean times enjoy a distinct fitness benefit.

Behavioral Adaptations That Enhance Survival Are Selected over Those That Do Not, Even When a Cost Is Incurred by Other Individuals in the Population

Egg Eviction by Birds

Viewed sympathetically, some behaviors are particularly difficult to understand outside of an evolutionary context. Egg eviction, whereby incubating eggs are fatally removed from their nests, is well documented in birds. For example, the cuckoo chick emerges from its egg and, while still frail, blind, and featherless, rolls the host eggs out of the nest to their certain death (Figure 7.2). If this behavior could not be interpreted evolutionarily, with a measurable fitness reward, it would just seem cruel. Yet the brood reduction that results from egg eviction reduces competition and presumably allows the cuckoo chick—a parasite—to extract sufficient resources from its host.

Conspecific Infanticide

Likewise, killing infants of the same species, or *conspecific infanticide,* is documented throughout Animalia, from water bugs to damselfish to

Figure 7.2 Egg eviction by the common cuckoo. Brood parasites gain access to resources by evicting the eggs or emerged young of the host bird. Shown here is a blind and naked, newly emerged cuckoo (*Cuculus canorus*) hatchling evicting a host's egg. (Photo courtesy Miklós Bán)

wattled jacanas and many species of primates (Figure 7.3). Explanations for infanticide include the benefits of obtaining nutrients by consuming the kill (as is practiced in some species), reducing competition for immediate resources, reducing competition for future resources, or, as is well documented in lions and many primates, encouraging the infant's mother to become sexually receptive to the infanticidal male. Ultimately, the infanticidal adult is rewarded with fitness, and killing young of the same species can be seen as adaptive.

Behaviors, Like Other Measurable Adaptations, Can Be Artificially Selected

Experimental Manipulation of Behavioral Traits

Excellent evidence for evolution, particularly the power of selection, has been demonstrated by artificial selection for various traits, and behavioral traits are no exception. Experimental manipulation has selected for maternal aggression, nest building, and wheel-running behavior in mice; development time, courtship behavior, and taxis in fruit flies; stress response in zebra finches; light sensitivity in guppies; and activity levels and antipredator behavior in flour beetles. In the late 1960s, Theodosius Dobzhansky

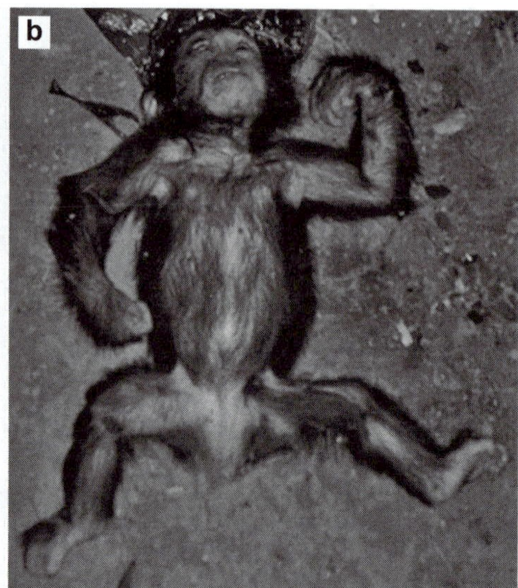

Figure 7.3 Infanticide in chimpanzees. Understanding natural selection makes it easier to envision how infanticide could arise. A male that kills (and may or may not consume) the offspring of rival males is more likely to sire his own offspring with the murdered infant's mother. Shown here is Andromeda the chimp (a) before, and (b) after an infanticidal attack by rival males of a neighboring group. Infanticide can be adaptive if, as is often the case, the mother of the infant becomes sexually receptive soon after her loss. (Left: Courtesy of Michael Wilson. Right: Courtesy of Deus Mjungu)

demonstrated how fruit fly lineages could be divergently selected for both positive and negative geotaxis and phototaxis. After several generations, selective pressure was relaxed, and flies converged on behaviors similar to those exhibited before selection (Figure 7.4).

Heritability of Behaviors

Experiments involving artificial selection allow us to calculate the heritability of behaviors and gauge the costs and benefits associated with the behavior. For example, selection for antiparasitic behaviors in fruit flies has demonstrated that these behaviors exact a fitness cost—specifically, reduced *fecundity* (reproductive ability) is correlated with experimental selection for parasite avoidance behaviors—that may regulate their expression in wild populations.

Recently, in a refinement of conventional artificial selection methods, biologists at the University of Rochester were able, through hybridization of *Nasonia vitripennis* with *N. giraulti,* to introgress a behavioral gene (or genes) from one organism onto the genome of another. *Nasonia* are parasitic wasps that lay their eggs in developing flies. Given a choice, *N. vitripennis* prefer flesh flies (*Sarcophaga*) to blowflies (*Protocalliphora*); *N. giraulti* prefer

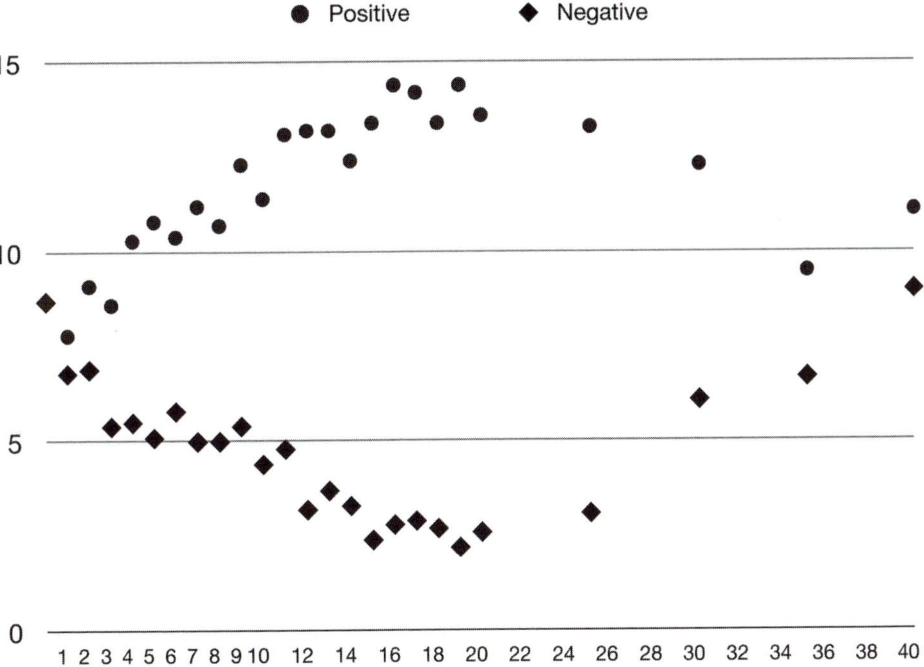

Figure 7.4 Artificial selection in fruit flies. In the 1960s, evolutionary biologist Theodosius Dobzhansky demonstrated how behavioral traits could be directionally selected. Shown here are the results of selection for two *Drosophila* lineages—one selected for greater positive phototaxis, the other for negative phototaxis. Numbers 0–15 are phototactic scores for fly behaviors in an experimental maze; a 0 is completely averse to light, a 15 is entirely light-seeking. The two lines diverged within a few generations, but selection was relaxed at generation 20 (arrow); shortly thereafter, both lines converged on the neutral phototactic behavior exhibited pre-manipulation. (Sehoya Cotner)

blowflies. By reciprocal exchange of a specific region of chromosome 4, the investigators altered the host preferences of these two species of parasitoid wasps. This work emphasized not only the heritability of discrete behaviors, but also our ability to artificially select parasites with specific host preferences.

Individuals Behave Preferentially toward Kin When There Is Likely to Be an Inclusive Fitness Benefit

Altruistic Behavior and Kin Selection

Conceptually, altruism poses an evolutionary "problem" because it is difficult to explain generosity that may lower the fitness of the altruist while benefiting someone else. Examples of behaviors often cited as altruistic include helping (observed in animals—birds, mammals, and social insects—that forego or delay reproduction to help provide for others' offspring), alarm calling (as seen in many birds and mammals), and defensive suicide (extensively documented in social insects).

However, most acts that appear altruistic are anything but, and in fact serve to increase an individual's inclusive fitness by providing for his or her offspring or other close relatives. **Kin selection,** the selection that acts on fitness gains experienced by relatives, provides a model for understanding many examples of altruism based on the recipient's degree of relatedness to the donor. According to kin selection, an individual should engage in risky or selfless behavior when the recipients of his generosity have a degree of relatedness that essentially outweighs the cost of the act, symbolized in *Hamilton's Rule:* An apparently altruistic act can be adaptive when $rB > C$ (where r is the coefficient of relatedness, B is the benefit to the recipient, and C is the cost to the donor).

In determining inclusive fitness gains, we are interested not only in a single individual's survival, but also in that of the individual's relatives (Figure 7.5). For example, a sibling shares, on average, half of your genes, and the coefficient of relatedness, or *r*-value, between two individuals with the same parents, or siblings, is 0.5 (1/2). The *r*-value for parent-offspring is also 0.5 (1/2), for half-siblings it is 0.25 (1/2 x 1/2 = 1/4), and for first cousins it is 0.125 (1/8). Thus Haldane's comment (quoted at the beginning of the chapter) that he would lay down his life for two brothers or eight cousins makes sense: either combination gives an *r*-value of 1.0.

Helping

We can apply Hamiltonian logic to the phenomenon of helping, often observed in avian or mammalian helpers at the nest. In cooperatively breeding Kalahari meerkats, helpers defend the young and provide supplemental feedings pre- and post-weaning. Helpers significantly increase the probability of lifetime reproductive success in those they help, largely by increasing the weight at maturity of the assisted pups (Figure 7.6).

The long-tailed tit is one of many cooperatively breeding birds that demonstrate helping. In this species, birds only become helpers after a failed breeding attempt, but not all failed breeders become helpers. In fact, in a group devoid of kin, essentially none of the failed breeders choose to help, yet helping is extensive when relatives are present. Likewise, a study of helping in the sociable weaver of South Africa confirmed that, when given a choice of whom to help, helpers invariably choose to help their kin. This tendency to help when indirect fitness gains are likely is exactly what one would expect if helping at the nest is an evolved trait.

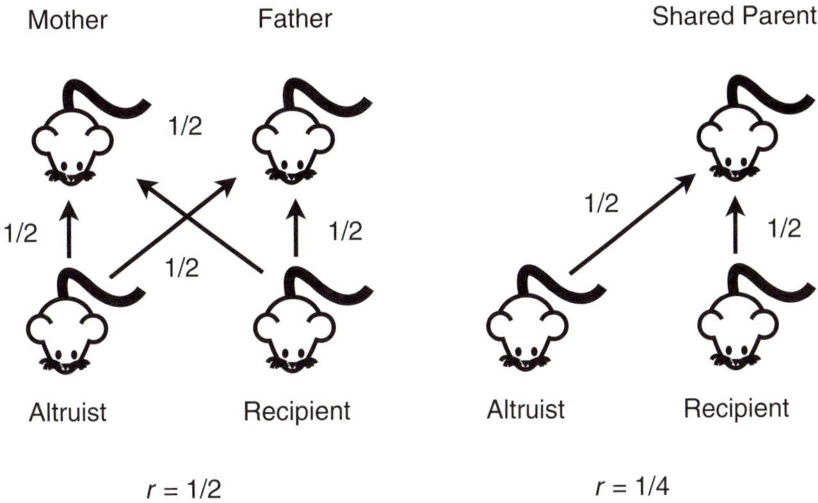

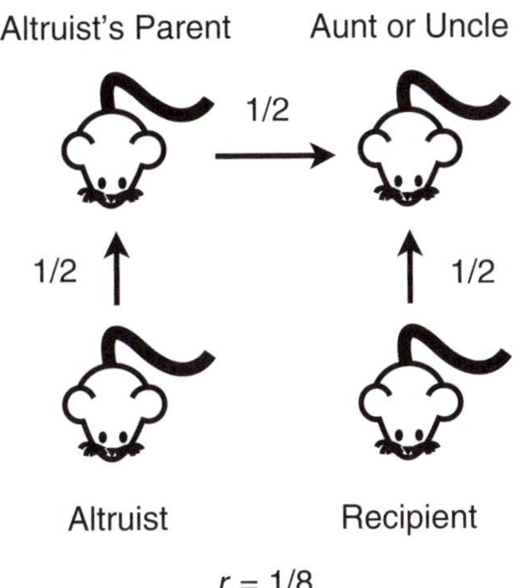

Figure 7.5 Calculating *r*, the coefficient of relatedness. The coefficient of relatedness (*r*) between any two individuals is the probability that the two share identical alleles at a given gene as a result of shared descent. A mouse with a mutant *MC1R* allele has a 50% chance (i.e., 0.5 probability) of sharing that identical allele with his full sibling; the *r*-value for the full-sibling relationship is 0.5. The *r*-value for parent-offspring is also 0.5, for half-siblings and for grandparents-grandoffspring it is 0.25, and for first cousins it is 0.125 (or 1/8). (Sehoya Cotner)

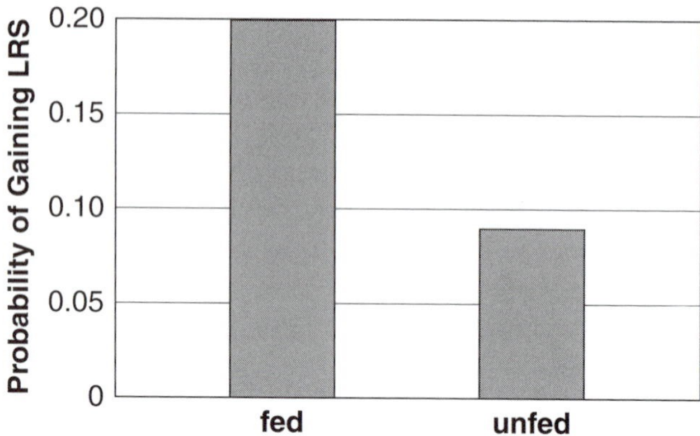

Figure 7.6a The effect of helping behavior on meerkat young. The probability of gaining lifetime reproductive success (LRS) is higher in helped versus unhelped meerkats. Meerkats live communally with relatives, thus gains associated with being helped contribute to the helper's inclusive fitness. (Modified from Russell, A. F., A. J. Young, G. Spong, N. R. Jordan, and T. H. Clutton-Brock. "Helpers Increase the Reproductive Potential of Offspring in Cooperative Meerkats." *Proceedings of the Royal Society B: Biological Sciences* 274, no. 1609 [2007]: 513.)

Figure 7.6b A helper with a young meerkat. (AP Photo/Science, Andrew Radford)

Alarm Calling

Alarm calling has been studied in 20 of the 53 rodent families, including ground squirrels, prairie dogs, and marmots. While alarm calling is an often-cited example of altruism in nature, its suggested costs—energetic expense, opportunity loss, and increased risk of predation—have not been well established. Almost equally elusive is evidence for fitness gains (direct or indirect) from calling behavior. There have been several studies, however, that have documented the tendency for calling to increase when kin are present, or for calling to increase when an individual's offspring are particularly young and vulnerable.

Suicide

Although not exclusive to the social insects, defensive suicide (whereby an individual sacrifices its life to protect other individuals in the population against predators or pathogens) has been well studied in termites, ants, bees, and pea aphids. Soldier termites of the genus *Globitermes sulphureus* risk death by *autothysis* (intentional rupture of an internal body part) when they force a liquid out of a large frontal gland. Upon

SIDEBAR 7.2 The Importance of Kin Recognition

Behaviors that selectively benefit one's relatives cannot evolve without some way of discriminating kin from non-kin. Various kin-recognition mechanisms have been documented, with an emphasis on olfactory detection and a preference for the familiar. When relatives are those individuals with whom you begin life in the nest, being placated by familiar scents is an adaptive strategy. In paper wasps, nest odor mediates wasp behavior: adults that do not smell like the natal nest are not tolerated. Likewise, females are not aggressive toward non-kin wasps that are experimentally exposed to the females' nest materials prior to their interaction.

Interesting work with Belding's ground squirrels has shown that even siblings experimentally reared apart treat each other with a familiar type of passivity, a kin-discrimination phenomenon known as *phenotype matching.* Phenotype matching, whereby an individual can identify relatives based on degree of similarity to the individual's own phenotype, has been suggested for many mammals and social insects. In a cross-fostering experiment with golden hamsters, Jill Mateo noted that females can discriminate kin that they've never met from unfamiliar non-kin, presumably by some form of phenotype matching through scent. Much work centers on odor differences due to variation in the genes of the Major Histocompatibility Complex (MHC), which code for cell-surface glycoproteins that are involved in self-recognition and immune response.

contact with the air, this yellow liquid becomes sticky, trapping potential invaders (typically ants) and giving off a scent that rallies nearby soldier termites.

Pea aphids live in genetically identical groups and are thereby expected to display high levels of self-sacrificing behavior. In fact, pea aphids that are parasitized by a parasitoid wasp often hasten their own demise (and thus protect group-mates from the emerging wasp larvae) by dropping off of their host plant. The cost of such self-sacrifice seems high, but in evolutionary terms, suicide carries no cost for the pea aphid; once parasitized, it dies prior to reproduction.

The ant *Temnothorax unifasciatus* also exhibits a type of social withdrawal (leaving the nest) when death is imminent. This pre-death withdrawal occurs whether or not the ant has been parasitized, refuting the idea that the behavior is mediated by pathogens and suggesting that it is done to protect the colony (and the individual's inclusive fitness).

Worker Castes

In many social insects (as well as the naked mole rat), large numbers of individuals forego reproduction entirely by assignation to a *worker caste*. Not only do they fail to achieve direct fitness, but they constantly toil and self-sacrifice, and thus the term *extreme altruism* is apt.

Applying Hamilton's rule is particularly useful in understanding worker castes in *haplodiploid* (diploid females and haploid males) insects such as the ants, bees, and wasps (the Hymenoptera). With haplodiploidy, worker females have a higher coefficient of relatedness with sisters ($r = 0.75$) than they do with daughters or mothers ($r = 0.5$) or brothers ($r = 0.25$). These differences suggest an adaptive advantage of foregoing reproduction to serve the colony, a hypothesis that has been tested in light of several predictions. The brother-sister inequalities posed by the haplodiploid condition lead to the prediction that females will invest more in caring for sisters than brothers. This prediction has been confirmed by studies taking a ratio of combined weight of female nest-mates to combined weight of males, revealing a 3:1 ratio of investment.

Worker castes are not restricted to the haplodiploid, or even to the social insects. The naked mole rat of East Africa is remarkable as one of only a few mammals known to exhibit the *eusociality*—marked by nonreproductive, worker castes—that is characteristic of the hymenopterans. A naked mole-rat colony is composed of a reproductive queen, a few inseminator males, and predominantly worker males and females that suppress reproduction and perform tasks such as guarding the colony, constructing underground tunnels, and feeding the babies. Early work on the relatedness of these mole rats suggested a colony-wide average r-value of 0.81, supporting kin selection as a mechanism for foregoing or delaying reproduction.

Additional studies have cited dispersal and avoidance of familiar mole rats as mechanisms to avoid even more extreme inbreeding in the colony.

Altruism by Reciprocal Exchange

Altruistic acts are also observed when the donor is likely to be the recipient in future generous acts. Altruism by reciprocal exchange requires the participants (giver and receiver) to be likely to encounter each other again, and to recognize each other when they do. Vampire bats (*Desmodus rotundus*) engage in reciprocal food-sharing with their colony-mates. Bats will share regurgitated blood with non-relatives that they encounter frequently, as well as with close kin. These patterns suggest that for vampire bats, reciprocity potential and inclusive fitness motivate generosity.

Individuals Exhibit Behaviors That Serve Simply to Acquire Mates

Sexual Selection

Sexual selection describes the differential reproductive success that results when individuals differ in their ability to compete with others for mates or to attract members of the opposite sex for mating. The power of sexual selection is evident from an analysis of behaviors that appear to serve little function beyond acquiring mates, either through intra- or intersexual competition. Darwin recognized the existence, if not the full importance, of sexual selection: "And this leads me to say a few words on what I call Sexual Selection. This depends, not on a struggle for existence, but on a struggle between the males for possession of the females; the result is not death to the unsuccessful competitor, but few or no offspring."

Intrasexual Competition

Intrasexual competition occurs when members of one sex compete for reproductive access to the opposite sex. Examples of intrasexual competition are skewed toward male-male rivalry, and range from the bloody brawls of elephant seals to the cryptic sperm-competition of sand lizards. T. S. McCann's work on dominance hierarchies in elephant seals established that social rank, a function of winning or losing violent battles, predicts reproductive success in males. Specifically, the most dominant male in a colony acquires the most copulations; the second male in the dominance hierarchy is a distant second in terms of copulatory success. Similar studies of Cape ground squirrels confirmed this connection between social rank and the number of copulations, demonstrating that the risks associated with physical competition are balanced by the reproductive gains enjoyed by the victors.

Intersexual Selection and Behavior

Intersexual selection involves mate choice in which members of the limiting sex (typically females) choose mates from the nonlimiting sex (typically males) based on certain advertised traits. A large body of work supports the role of intersexual selection in the evolution of behavior. The advertising sex may dance, sing, glow, decorate, or bring gifts. In Darwin's words, "I can see no good reason to doubt that female birds, by selecting, during thousands of generations, the most melodious or beautiful males, according to their standard of beauty, might produce a marked effect."

Some birds, fiddler crabs, and fish manipulate or decorate their environment to attract mates. Bowerbirds are known for their decorating ability, in that males construct stick bowers fronted by displays, at which the male architects will display via dancing and strutting. These displays are often elaborate, sometimes consisting of several thousand stones, bones, shells, and human-made objects arranged by size (from largest to smallest) to create a path leading to the court. This makes the court seem smaller, and the males deceptively large. (If a researcher rearranges the bower, the male bird quickly returns the objects to their original position.) These bowers and their associated decor appear to serve little function beyond soliciting females—they are not nests and are not used for incubating or rearing chicks. Females select for bower adornments, appearing to prefer elaborate decorations with uncommon objects. This female preference has pushed males to amazing feats in the construction of bowers, with instances of decoration theft from, and outright destruction of, rival males' bowers. A male fiddler crab protects a hole in the sand that can serve as a place to retreat when threatened, as well as a love nest for copulating with complicit females (Figure 7.7). Males of some species adorn the exterior of these burrows with sandy pillars, mounds, or overhanging arches.

The pulse of light emitted by fireflies is used to court females. Males flash a distinctive pattern, females choose whether to answer back, and they locate each other by reciprocal flashes. Female preference for flash type is species specific, with some preferring short flashes to long, or long intervals between flashes rather than short. Light emission as a form of sexual display has also been suggested for some fireworms, ostracods, octopuses, and anglerfish.

Nuptial gifts are used by males of many species to gain sexual access to females. This phenomenon has been documented throughout the Arthropoda (especially the insects) and in some birds and mammals. For over a century, scientists have known about the courtship ritual of the spider *Pisaura mirabilis*: males present females with a silk-wrapped prey item, and females select males based on the quality of this gift (Figure 7.8). While consuming the gift, the male transfers his sperm to the female. The nuptial gift does not appear to be essential, but it has its advantages. Males without a gift successfully copulate with females 40% of the time, whereas males with gifts are far more successful—copulating 90% of the time. In addition, females are less likely to terminate copulation with males bearing large gifts, and copulation duration is longest with generous males.

The adaptive significance of dancing to attract mates has been verified in the long-tailed manakin of Central America; specific dance elements (e.g., those called the butterfly and the hop) have been linked to copulatory success in male manakins. Birds' songs also appear to have evolved as a sexually selected behavior. In general, females prefer males with long, complex song repertoires as opposed to those with simpler songs. Female European starlings show a preference for song repertoire that surpasses their preference for territory, a finding that begs the question: what do birds' songs communicate to females?

Theories of Female-Choice Mechanics

Explanations for why females choose certain behavioral traits over others center on the meaning of the display. Research has focused on whether the trait being selected communicates anything about the male's underlying genetic quality.

Figure 7.7 Male fiddler crabs are distinguished by their enlarged cheliped, which serves in both intra- and intersexual competition. In this image, *Uca tomentosa* waves his large claw in front of a sand pillar he constructed outside his burrow. (Courtesy Frank Barnwell)

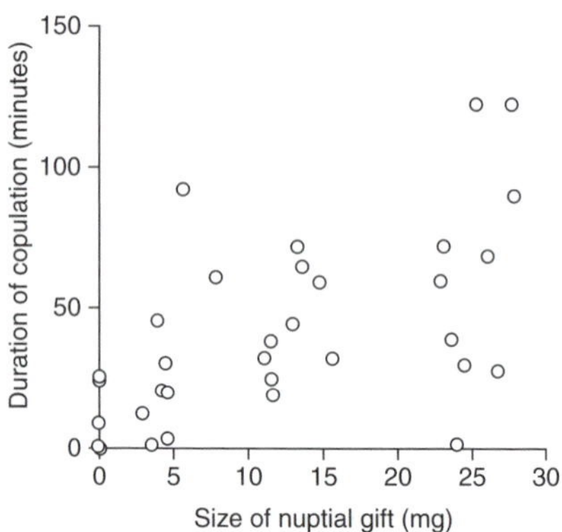

Figure 7.8 Nuptial gifts in insects. A male predatory fly (*Empis* sp.), far left, presents a female, center, with a silk-wrapped nuptial gift, at right. She consumed the carcass during copulation. A generous male enjoys a fitness advantage over his stingier counterparts: females copulate longer while consuming large nuptial gifts. (Photo courtesy of Welwyn; figure modified from Stalhandske, P. "Nuptial Gift in the Spider *Pisaura mirabilis* Maintained by Sexual Selection." *Behavioral Ecology* 12, no. 6 [2001]: 691.)

- According to the *runaway* hypothesis for female choice, females select for certain traits in males, and these traits become subject to extreme development merely due to positive feedback associated with this preference.
- A variant of classic runaway selection is the *sexy son* hypothesis for female choice, whereby females mate with displaying males because their sons will inherit the father's capacity to display; her sons inherit sexiness from the male, and she is rewarded with grandchildren and inclusive fitness.
- According to the *good genes* hypothesis, females choose features that advertise something about the male's fitness potential and that of their future offspring. Some biologists refer to the good-genes hypothesis as "truth in advertising" for conveying an honest message about the male's genetic quality—beyond mere sex appeal.
- Another hypothesis involves *compatible genes,* in which females attempt to mate with a male most likely to complement her genetically.

Several tests of the sexy son hypothesis have focused on facultatively *polygynous* female birds—that is, females that pair with an already-paired male rather than an available one (the term *facultative* implies that females have options besides polygyny). Since females that enter into such an arrangement experience reduced paternal help with the brood, it is logical to question whether they enjoy the indirect fitness benefit associated with a sexy son (who is likely to be polygynous himself). Observations of the pied flycatcher, a facultatively polygynous bird, confirm that a female fledges fewer young when mated to a *bigynous* male (i.e., a male mated to more than one female). However, she is indirectly compensated with grandoffspring equal in number to that of monogamous females.

Tests of the good-genes hypothesis can be fairly straightforward, involving female preference for a trait tightly associated with a measurable male quality. In his work on the Satin Bowerbird, Gerald Borgia observed that bower-constructing males have higher levels of circulating testosterone in their blood; testosterone levels are also higher in males that adorn their bowers with more decorations. In addition, males with higher levels of testosterone achieve more copulations. This work suggests that females may select males with lavish bowers over those without (or over those with drab) bowers because of what bower construction communicates about the male's quality.

The male of the ground cricket, *Allonemobius socius,* transfers nutrients to the female through a special spur on his hind limb. The female selects a male based on his call and then eats his nuptial gift during copulation. Experimental activation of the male cricket's immune system has been associated with an altered call (reduced pulse-duration) and a smaller nuptial gift. In keeping with the good-genes hypothesis, females prefer males with longer pulse calls, intimating that this preference is mediated by the male's quality (specifically, the strength of his immune system, and generally, his overall health).

Work on bird song preferences also supports the good-genes hypothesis. Female song sparrows can discriminate between males based on the male's ability to learn complex repertoires. Specifically, song-learning ability may reflect the male's response to developmental stress, and therefore be an indicator of his quality.

Males may also use deception to secure mates. For example, male splendid fairy wrens sing a special song when they hear the call of a butcherbird, their predator. Although the fairy wrens attract attention to themselves by doing so, they also attract fearful females that have been alerted to the presence of danger. The scared females may exhibit heightened attraction to the males, much like when a scary movie brings a date closer. Similarly, recent work by Jakob Bro-Jorgensen described trickery in topi antelopes; specifically, males respond to a straying female by making a characteristic snort, one typically employed to warn of an approaching predator such as a leopard or lion. Bro-Jorgensen notes, "It's such an obvious lie. Clearly there's no lion." But upon hearing this alarm, a female antelope is less likely to leave the male's side, in turn affording him greater mating access.

SIDEBAR 7.3 Anisogamy and Mating Behavior

Why do males often take the risks and develop the elaborate displays to attract mates? Why do females usually do the choosing? The answer seems to lie in our gametes, notable for being extremely small in males, and large and nutrient-rich in females. In fact, we use this difference in gamete size, **anisogamy,** to distinguish male from female. This fundamental asymmetry of sex really means that females tend to invest more in their offspring, beginning with the development of gametes. In some organisms, this asymmetry escalates when the female takes on most or all of the work of caring for the offspring. As a consequence, females are limited by resources, and males are limited by the availability of mates; females should work hard to ensure sufficient resources for her children, and males should work hard to attract mates—either by competing among themselves or advertising their goods to potential conquests. There are literally *thousands* of examples illustrating how females tend to be choosy in mating, whereas males tend to be less discriminating. The examples in these pages are no exception: male fiddler crabs wield the large claw; male bower birds construct the elaborate bowers and do the sexy dance; and male spiders and crickets offer nuptial gifts for discerning females.

There Is an Adaptive Advantage in Behaviors That Help Ensure Parentage

Copulation with internal fertilization creates situations in which mothers are assured of parentage, but paternity can be questionable. Several male behaviors increase a male's likelihood of paternity. For example, "mate guarding" is documented throughout Animalia and occurs in contexts that suggest its utility: males are more likely to exercise vigilance when females are promiscuous, when other males are present, and during peak periods of female receptiveness and fertility.

Competitive Sperm

If selection results in greater reproductive success among the selected, we would expect the male gametes to exhibit intrasexual selection in competitive situations, a phenomenon referred to as *sperm competition.* When sperm compete to fertilize an egg, competitive adaptations, specifically focused on successful insemination, should arise to increase the success of an individual's sperm. Consequently, sperm competition—both before and after copulation—should be most intense when females typically mate with more than one male. Several studies support this prediction.

Gametes themselves exhibit behaviors that are competitively adaptive. A study of sperm characteristics in the polygynous Iberian red deer revealed that fertility was associated with sperm velocity, suggesting a

SIDEBAR 7.4 What Happens When the Tables Are Turned?

Additional evidence for the predictive power of evolution is evident when we look at what behavioral adaptations arise when males invest more in offspring. Species exhibiting a role reversal include the jacanas, the seahorses, and the phalaropes. Some female seahorses appear to provision males (who carry and protect developing embryos; Figure 7.9) with a substance akin to a nuptial gift. Phalarope females are large and brightly colored, while the males are small and drab. And some male jacanas incubate eggs in the nest while the females pursue additional copulations. In jacanas, the role-reversal conforms to what is predicted when males are limited and constrained by the young of other females: a female jacana, seeking additional copulations, will destroy a male's eggs or kill his nestlings so that he will be available to copulate with her and care for the eggs they make together. In pipefish, a close relative of seahorses, the female injects eggs into the male, which then bears live young. However, the males prefer large females; offspring of larger females have higher rates of survival that those of smaller females. Researchers suspect that males may use "cryptic choice," which involves selecting a mother *after* mating. Males may do this by providing more resources (e.g., nutrients) to broods mothered by large females than to those mothered by small females.

Figure 7.9 Seahorse father releasing a baby. Seahorses exhibit a parenting strategy in which males protect the eggs during incubation. When the young emerge from the eggs, the newly hatched seahorses are released from a special abdominal pouch. (AP Photo/John G. Shedd Aquarium, Edward G. Lines Jr.)

selective advantage for those males with faster sperm. Similar work with turkeys demonstrated that those males with higher-mobility sperm had a higher probability of paternity.

Cooperative Sperm

The converse of competitive sperm is *cooperative sperm*. Sperm that selectively cooperate illustrate an intriguing twist on competitive-sperm interactions. In some mollusks, insects, and mammals, sperm form aggregations that increase the overall mobility of the in-group gametes (sperm from the same individual or from close relatives). A large proportion of the clumping sperm take on a sacrificial role, presumably to the advantage of closely related gametes. Recently, Heidi Fisher and Hopi Hoekstra tested the hypothesis that selection will favor cooperation in closely related sperm. Sperm mixtures were created experimentally with gametes from deer mice having varying degrees of relatedness and tested for aggregating tendencies. Sperm from a monogamous species, *Peromyscus polionotus*, failed to show any discrimination in clumping. However, sperm from the promiscuous *Peromyscus maniculatus* clumped and did so in a discriminatory way: sperm were more likely to clump with close relatives than with sperm from unrelated individuals. In supporting their hypothesis, the authors noted that in the deer mouse, "competition drives cooperation" with clear evolutionary implications.

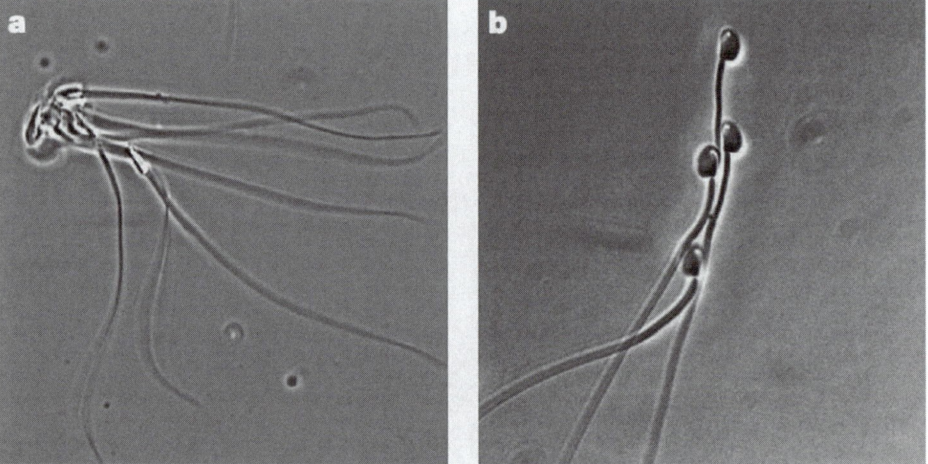

Figure 7.10 Sperm cooperation as a form of sperm competition. Related sperm from the promiscuous deer mouse, *Peromyscus maniculatus*, clump together. This clumping allows individual sperm (in the group) to swim faster and gives them a fertilization advantage. (From Heidi S. Fisher and Hopi E. Hoekstra, 2010, "Competition Drives Cooperation Among Closely Related Sperm of Deer Mice," *Nature* 463 (7282): 801–803. Reprinted by permission from Macmillan Publishers.)

Sperm Competition and Behavior

Examples of behaviors that appear to be driven by sperm competition include the mate guarding mentioned above, evicting the sperm of rival males, prolonged copulation, and deposition of a *copulatory plug* (i.e., any physical barrier—including a body part—deposited by the male in the female's reproductive tract). In the snow crab, a species in which females mate with several males in quick succession and the last male to copulate fertilizes the most eggs, post-copulatory mate guarding can be intense. As male-female sex ratios increase, so does the amount of time males spend guarding females.

Prolonged copulation can result in more successful fertilization. For example, female guppies are more likely to prolong copulation with more colorful (and therefore more carotenoid-rich) males, and this prolonged copulation results in increased paternity for the copulating males—and decreased opportunities for other males. Numerous other organisms increase fertilization success with prolonged copulation, including laboratory rats, golden hamsters, sand lizards, stinkbugs, and damselflies.

Copulatory plugs of various guises are used by many organisms, presumably—in whole or in part—to block sperm from other males. Typically, the copulatory plug, or mating plug, consists of a gelatinous secretion submitted as the final constituent of a male's ejaculate. The secretion can either glue the genitalia shut or harden into a plug; either way, this deposit can block the female from further insemination. The use of copulatory plugs has been documented in many animals, including arthropods (e.g., bumblebees and scorpions), reptiles (e.g., sand lizards and garter snakes), and mammals (e.g., squirrels and primates—including the great apes).

In a Conflict between Individual Survival and Reproductive Potential, Reproductive Potential Wins

Destructive Reproductive Behavior in an Evolutionary Context

If reproductive success is the ultimate currency of evolution, then successful reproduction should be more important than life without reproduction. That is, sex should be more important than life itself. For some organisms, this is played out through copulatory suicide, a phenomenon that has received scrutiny in many insects and spiders that risk cannibalism to achieve fertilization. Male redback spiders characteristically somersault into the waiting jaws of the females, who consume him during copulation. This behavior would be nonsensical outside of its evolutionary context. However, males who die in this manner leave behind more offspring than survivor males because (1) females are less interested in future males after a cannibalizing copulation, and (2) eating the male prolongs copulation and increases insemination success. Similarly, in the cannibalistic praying mantid, females that consumed their mates had larger egg masses than did non-cannibals. Female mantids in poor condition (and presumably more in need of additional resources) were more likely to eat their mates than were their healthier counterparts.

Parents Behave Preferentially toward Those Offspring Most Likely to Augment the Parents' Inclusive Fitness

Differential Treatment of Offspring

Parents play an important role in maximizing their indirect fitness by differential treatment of their offspring. Specifically, parents should divert resources to children most likely to make plentiful, high-quality grandoffspring. A 2009 study of house sparrow nestlings attempted to determine the relationship between fitness potential and parental favoritism. Specifically, house sparrows' mouths are surrounded by fleshy "flanges" having varying degrees of yellow-to-red coloration based on the presence of carotenoids. These flange colors are present in begging nestlings and dissipate as the birds mature. Previous work with many species of birds has demonstrated the correlation between carotenoid-based coloration and general health (e.g., body mass, immune function) as well as the tendency for females to prefer mates displaying such coloration. By experimentally manipulating mouthpart color in house sparrow nestlings, Matthew Dugas showed that parents (especially mothers) gave more food to those nestlings with apparently carotenoid-rich flanges. By extension, work with eastern bluebirds suggests that parental favoritism (in this case, for sons with brighter plumage) is only significant when resources are limited. In both cases, coloration indicated nestling quality and was a cue for parental effort.

Parental Favoritism of the Sexes

In accordance with the parental favoritism prediction, parents seem to divert resources to the sex most likely to serve their fitness goals. Given that males typically are more variable in reproductive success (a loser male is likely to have zero fitness, whereas a winner may sire a huge number of offspring), parents are hypothesized to favor daughters when resources are scarce or when parents are of lower quality, and sons when the converse is true. In a 2009 test of this assumption, biologists at Auburn University observed that mother birds fed sons disproportionately more than daughters when their mates (i.e., the sons' fathers) had brighter plumage coloration.

Parents may even manipulate the sex ratio of their offspring. The Trivers-Willard hypothesis for sex-ratio manipulation states that parents will benefit from having sons when resources are plentiful, and daughters when resources are scarce. This flows logically from the observation that daughters can typically make offspring even if they themselves are of relatively low quality. Sons, given their high-risk, high-reward status (whereby they produce zero-to-many offspring), would therefore be predicted in greater numbers from parents in good overall health. Numerous studies have documented skewed sex-ratios as a result of resource availability or parental quality. For example, female opossums that were given supplemental food were significantly more likely to produce a male-biased sex ratio. In some cases, social dominance influences the sex ratio of offspring, as demonstrated by red deer: the male offspring of dominant females have much greater reproductive success than do the male offspring of submissive females. The mechanisms underlying sex-ratio adjustment are unclear, but recent work with pigeons suggests the adjustment may occur in birds prior to ovulation with the absorption of ovarian follicles of the undesired sex.

Siblicide

Parents also appear to maximize their indirect fitness by failing to intervene when one of their children kills a sibling. Siblicide is common in birds and some mammals, such as the spotted hyena. Parents of cattle egrets actively promote siblicidal interactions by laying their three eggs asynchronously and supplying the first two eggs with more androgens than the third. Thus, the first two offspring are bigger and more aggressive than the third and can typically kill the third chick. Parents may benefit by having the chicks fight it out so that the parents can then invest in only the offspring most likely to survive and reproduce.

Summary

Behaviors, like other heritable components of an individual's phenotype, are under selection pressures to maximize fitness. Organisms exhibit behaviors that maximize inclusive fitness by increasing their chances of surviving, acquiring quality mates, and raising successful offspring.

8

Conflict and Cooperation Part II: Coevolution

If you took away the lichens and mycorrhizae from terrestrial communities, the reef-building corals and their zooanthellae from the oceans, and the sulfur-converting symbioses from deep-sea vents, those ecosystems would collapse.

—*J. N. Thompson, 2009*

If all mankind were to disappear, the world would regenerate back to the rich state of equilibrium that existed ten thousand years ago. If insects were to vanish, the environment would collapse into chaos.

—*E. O. Wilson, 2006*

PREDICTIONS

When two or more species interact, a population should adapt to the presence of its symbionts in ways that maximize individual fitness. Specifically, if evolution has occurred, we predict that

1. Prey species will possess ecologically relevant adaptations against predation.
2. Predators will possess ecologically relevant adaptations that allow them to conquer prey.
3. Organisms will exploit different ecological niches to reduce the negative effects of competition.
4. Parasites will evolve strategies that allow them to exploit host resources.
5. Host species will evolve ways to minimize the costs of being parasitized.
6. Symbiont species that serve a life history function (e.g., pollination, incubation, defense) will be rewarded to continue.
7. Rewarded symbionts will be penalized for cheating (e.g., failing to provide the life history function for which they are being rewarded).

Coevolution

Many adaptations involve intraspecific competition for resources or mating opportunities. However, populations also accommodate the biotic environment, which includes the other living organisms they typically encounter. When individuals of two or more species exert selective pressures on each other, the changes that occur are termed **coevolution.** Organisms that interact are *symbionts* and the general term for their relationship is **symbiosis.** Specifically, a symbiosis can be *mutualistic,* or beneficial to all symbionts, or *antagonistic,* whereby one or more of the symbionts is negatively affected. Antagonism includes *competitive, predatory, herbivorous,* and *parasitic* relationships, which often pit interacting lineages in a coevolutionary escalation colloquially termed an "arms race." In addition, *commensalism* describes an interaction in which one symbiont benefits and the other appears unaffected. In *amensalism,* one symbiont is harmed and the other is unaffected. The importance of these relationships should not be minimized; in fact, most adaptive radiations are attributed to coevolutionary mechanisms.

Ecological Relevance

Coevolution is typically inferred when an adaptation is *ecologically relevant*—that is, when an adaptation can be attributed to the presence of other taxa in an organism's ecosystem. There are many examples of adaptations that are only observed in one species when another species, or group of species, is present. In these cases, the adaptation is ecologically relevant and coevolution is the likely mechanism.

Prey Species Possess Ecologically Relevant Adaptations against Predation

Adaptations that counter predation include cryptic coloration (i.e., camouflage), mimicry, warning calls and signals, toxic emissions and distasteful secretions.

Coloration, Mimicry, and Warning Signals

Selection makes prey either more visible (e.g., the warning, or *aposematic,* coloration of poison arrow frogs) or less visible (e.g., the ability of a cuttlefish to change colors to match its background) to potential predators. Some fascinating work has been done on the evolution of mimicry, a phenomenon that has been studied since no later than 1862 when the naturalist Henry Bates shared his observations of Amazonian butterflies

(Figure 8.1). Particularly, Bates noted that palatable forms of the butterflies appeared to gain protection from predation by mimicking noxious forms, a phenomenon aptly termed *Batesian mimicry*. Seven years later, Johannes Friedrich Müller introduced the concept of *Müllerian mimicry*, whereby noxious forms mimic each other as a way to co-opt the preexisting protection offered by those with similar phenotypes. In 2009, scientists at East Carolina University documented the presence of a mimicry ring (in which several organisms converge on a phenotype for a shared protective benefit) in a group of distasteful millipedes that co-occur in the Appalachian Mountains. Notably, these millipedes are blind, supporting the notion that their distinct color patterns are an adaptation for interspecies, rather than intraspecies, communication (of toxicity, presumably).

Figure 8.1 Batesian mimicry in butterflies. Henry Bates included this drawing in his 1862 book, *Contributions to an Insect Fauna of the Amazon Valley;* Lepidoptera: Heliconiidae. The *Dismorphia* butterflies on the top and third rows are palatable mimics of the toxic *Ithomiini* butterflies on the second and bottom row. (From H. W. Bates, 1862, "Contributions to an Insect Fauna of the Amazon Valley; *Lepidoptera: Heliconidae,*" *Transactions of the Entomological Society* 23 (3): 495–566.)

In a similar fashion, plants use coloration to communicate with would-be herbivores, and this may explain variation in autumnal leaf coloration. In temperate regions, the leaves of many deciduous trees change color in the fall, transforming entire landscapes into a mosaic of reds and yellows. Red coloration is particularly intriguing because it is produced in the fall and is not merely a byproduct of the breakdown of green pigments (as is the case with yellows). Among the proposed explanations for fall's red coloration is the *coevolutionary hypothesis,* which posits that red leaves advertise something—for example, unpalatability or low nutritional quality—to herbivorous insects on their fall migration. Evidence for this hypothesis focuses on aphids and their potential winter host trees. For example, tree species with the strongest autumnal coloration tend to be those trees threatened by the greatest number of migrating aphids. There is reduced herbivory on red (versus green) leaves, and the peak of fall coloration coincides with the most active migratory period for aphids.

A novel twist on antipredator warning calls was recently discovered in the larva of the walnut sphinx moth. Like many insects, the sphinx moth caterpillar breathes through tubes connected to spiracles on its abdomen. It also can use these spiracles to produce a high-frequency whistle that, in laboratory tests, startles predatory birds. This whistling behavior, combined with the caterpillar's leaf-like shape and coloration, provide this slow-moving larva with a full arsenal of antipredator adaptations.

Toxic Emissions and Secretions

Many other prey species produce toxic substances that would-be predators avoid. For example, newts of the genus *Taricha* in the western United States produce a deadly neurotoxin, tetrodotoxin (TTX). Alternatively, some prey species are adapted to sense predator presence through chemical cues, either compounds emitted by the predators themselves or those produced by other prey organisms. Even simple planarian flatworms can learn to associate conspecific alarm cues (from wounded planarians) with a novel predator. By sensing alarm cues of fellow prey organisms, these planarians have adapted to predation in general, rather than the presence of a specific predator. This ability to generalize is adaptive, given the numerous aquatic organisms that prey on planarians.

Influence of Predation Pressure

Some prey behaviors fluctuate depending on the extent of predation pressure, a finding consistent with the occurrence of a coevolutionary history. Diel vertical migration (DVM) in zooplankton (discussed in Chapter 7) is tempered by predation pressure. In some cases, the DVM of *Daphnia* coincides seasonally with high predation pressure, and it wanes when fish predators are less abundant. Likewise, DVM can disappear altogether when predators are experimentally removed. For example, *Enallagma* damselflies have evolved predator-avoidance strategies consistent with predation regime—whether predation from fish, dragonflies, or both. Damselflies from dragonfly-only lakes swim away from danger, whereas those from lakes with fish predators remain motionless and reduce foraging forays to evade detection. The damselflies' increased activity in the *absence* of predators illustrates not only the powerful selective pressure posed by predation, but also the costs incurred by some antipredator adaptations (in this case, the reduced opportunity to feed).

Similarly, pig-tailed lemurs exhibit different behaviors under different predation regimes. Lemurs on the Indonesian mainland are prey to tigers and leopards and flee from the sound of these large cats. Lemurs on the nearby Mentawi islands do not co-occur with predatory cats, nor have they for half-a-million years. Consequently, island lemurs do not react to recordings of tiger or leopard howls. These examples, in animals as disparate as primates and crustaceans, illustrate how selection pressure for antipredatory behavior can become relaxed in the absence of an established threat.

Third-Party Exploitation

There are also many species that borrow—or steal—the cryptic or aposematic (i.e., warning) coloration of another organism to defend themselves against yet a third organism. For example, decorator crabs adorn their shells with algae to deter would-be predators. These crabs selectively decorate with a brown alga that is toxic to many predatory fish. In the presence of predators, decorator crabs will stop feeding but continue adorning their shells, suggesting the relative importance of protection versus nourishment.

A similar type of third-party exploitation occurs in primates that self-anoint with arthropods. Lemurs and capuchin monkeys both rub their bodies with millipedes, spreading millipede exudates into their skin and fur. The monkeys are immediately rewarded with pleasurable sensations (they appear to get "high" on millipedes). However, this behavior is presumed adaptive, because millipede exudates contain mosquito-repelling chemicals, specifically benzoquinones. A monkey that is proximately motivated to get high on millipedes is ultimately rewarded by deterring mosquitos, and thus evading numerous deadly mosquito-borne pathogens.

Predators Possess Ecologically Relevant Adaptations That Allow Them to Conquer Prey

In the newt-TTX example discussed above, *Taricha* newts counter predators by producing a deadly neurotoxin. This has presumably resulted in strong selection for TTX resistance, a phenomenon well documented in

SIDEBAR 8.1 Coevolution and the Red Queen

Coevolution may be responsible for *sexual reproduction,* the intriguing natural phenomenon whereby genes are rearranged, gametes are exchanged, and diverse offspring result. The evolution of sex, with all of its associated costs (energetic, ecological, and genetic), has long intrigued evolutionary biologists. There are several competing and possibly overlapping hypotheses under investigation, including one for which there is considerable empirical evidence—the Red Queen. According to the Red Queen Hypothesis for the evolution and maintenance of sex, the genetic variation afforded by sex gives individuals a competitive edge on their coevolving symbionts—especially their predators and parasites. A key prediction of the Red Queen Hypothesis is that parasites will be more likely to prey upon those individuals in a population expressing the most common phenotype, thus lending an advantage to novelty for novelty's sake. Evidence for this coevolutionary basis of sex includes several studies on mixed—asexual and sexual—populations of organisms such as snails and water fleas. In general, the asexual (or clonal) variants tend to be more susceptible to parasitic infection; the sexuals are, like the Red Queen in *Alice's Adventures in Wonderland,* running (or changing) as fast as they can just to stay in one place.

snakes of the genus *Thamnophis.* The ecological relevance of this adaptation is compelling: snakes that do not co-occur with the TTX newts, or snakes that co-occur with nontoxic newts, do not possess natural resistance.

Coloration in Predators

Prey organisms are not alone in resorting to camouflage for survival. Many organisms use camouflage to attract or approach their prey. Lions possess tan coloration that may allow them to sneak up on prey. Orb-weaving spiders align their bodies in their webs in a fashion that makes them difficult for their prey to detect until it is too late. And ambush bugs blend into their surroundings until it is time to attack unsuspecting prey.

Odor Used to Attract Prey

Coevolutionary mechanisms underlie a variety of interesting behaviors by spiders, such as the affinity of *Nephila edulis* for rotting carcasses (Figure 8.2). This orb-weaving spider incorporates dead organic matter into a band of debris in its web; prey such as sheep blowflies are attracted to the scent of rotting carcasses in the web, an impulse that leads them to their death. Crab spiders also use odors to capture prey. Spiders such as *Thomisus spectabilis* are sit-and-wait predators that position themselves on flowers to ambush their preferred prey—pollinating insects such as honeybees. Spiders have evolved to exploit the odor—rather than size, shape, or color—preferences of honeybees, choosing to wait on flowers with a bee-enticing aroma.

Selection for Predatory Ability

Selection for predatory ability may have rewarded bigger brains in hunting birds. In birds, there is a strong correlation between eye size, brain size, and nocturnal hunting. Selection for good vision in night-hunting birds may have led to the large brains associated with visual acuity. This and many other examples illustrate how powerful interactions between species have been in shaping present-day biodiversity.

Figure 8.2 *Nephila edulis* and debris in the web. The spider *Nephila edulis* collects and stores rotting carcasses in its web, attracting the attention of tasty, corpse-loving blowflies. (Courtesy of Arthur D. Chapman)

Organisms Exploit Different Ecological Niches to Reduce the Negative Effects of Competition

Character Displacement and Niche Partitioning

Competing organisms pay the costs of competition in many ways: they expend time and energy, they may be physically harmed, and they may lose out on resources altogether. Given these potentially high costs, we expect selection to favor those phenotypes that minimize competition in overlapping species. Sympatric species (i.e., species that have overlapping ranges) that reduce conflict by diverging in some similar phenotypic trait are said to exhibit *character displacement.* In contrast, *niche partitioning* occurs when organisms alter some niche component (such as their preferred habitat, food type, or hours of activity) to reduce competition.

Displacement from Antagonistic Coevolution

Where displacement results from antagonistic (i.e., competitive, parasitic, or predatory) coevolution, we expect to see differences magnified where the species coexist, but reduced where their ranges do not overlap. This has been observed with many competing taxa, and was convincingly demonstrated by Joseph Connell in the 1960s. Connell studied two species of barnacles—*Chthamalus stellatus* and *Balanus balanoides*—that colonize intertidal rocks. These barnacles exhibit a stratified distribution, with *Chthamalus* inhabiting rocks further away from the ocean's splash zone, and *Balanus* occurring closer to the water. However, when Connell removed *Balanus,* the *Chthamalus* spread to occupy the entire intertidal zone; *Balanus* did not alter its distribution when its competitor was removed. From these results, Connell inferred that where they co-occur, *Chthamalus* (presumably the weaker competitor) alters its habitat preferences to reduce competition with *Balanus,* which cannot survive farther from the water.

More recently, a similar case was made for niche partitioning in sympatric snakes (Figure 8.3). Studies of the prey preferences of *Masticophis flagellum* (coachwhip) and *Coluber constrictor* (Eastern racer) revealed that, in spite of the snakes' similar sizes and foraging behavior, they eat different prey organisms and do not compete for food, suggesting another form of coevolutionary displacement.

Diversifying Coevolution

In the 1940s, David Lack described character displacement in the finches of the Galápagos islands: where species overlap, beak sizes diverge (also discussed in Chapters 4 and 6). Where the same species occur apart, their beaks are more similar in size. A similar pattern occurs in threespine sticklebacks of British Columbia. Where these fish coexist, we observe magnified

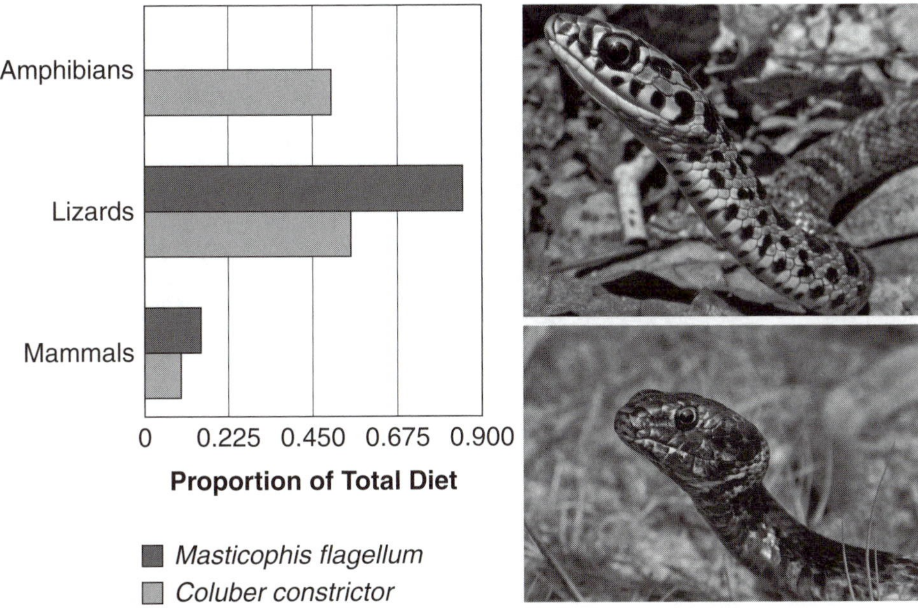

Figure 8.3 Niche partitioning in sympatric snakes. Where they co-occur, the coachwhip (*Masticophis flagellum*) and the Eastern racer (*Coluber constrictor*) diverge in their food choices, a form of coevolutionary displacement. (Images courtesy Alan Cressler (racer) and Andrew Hoffman (coachwhip); graph modified from Halstead et al. 2008.)

differences in food preferences, habitat, and fish morphology. Where they are allopatric (i.e., they do not co-occur), these differences are minimized or nonexistent. Likewise, diversifying coevolution between crossbill birds and conifers appears to be responsible for much of the diversity observed in crossbill beaks.

Parasites Evolve Strategies That Allow Them to Exploit Host Resources

Reciprocal Adaptations

Parasites, which are organisms that extract resources from a living host, exhibit myriad host-specific adaptations that tell stories of conjoined evolutionary paths. Reciprocal adaptations are evident in the Chilean cactus-mistletoe-mockingbird symbiosis. *Tristerix aphyllus* is a leafless mistletoe that parasitizes only cacti. Seeds of *T. aphyllus* are spread by one vector—the Chilean mockingbird, which defecates seeds directly onto cacti. Mistletoe seeds adhere to and germinate among cactus spines, where their primary roots penetrate the epidermis of the cactus. These roots appear to have been selected for their length, while the spines of the cacti have been

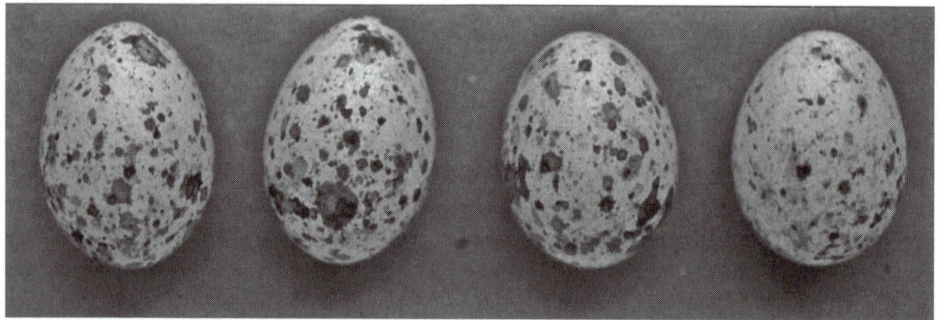

Figure 8.4 Egg mimicry as a strategy for parasitism. Can you tell the difference between the great reed warbler's (*Acrocephalus arundinaceus*) eggs and that of the common cuckoo (*Cuculus canorus*)? Neither can the great reed warbler. The egg on the far right is the cuckoo egg. (Courtesy Csaba Moskat)

counter-evolving to evade this parasitism. Evidence for this coevolutionary arms-race includes the following observations: (1) mockingbirds are less likely to perch on cacti with long spines; (2) mistletoe seedlings with long primary roots are more successful parasites; and (3) there is a relationship between cactus spine length and parasitism status, whereby cacti with the longest spines tend to experience the most intense parasitism.

Avian brood parasites, such as the cuckoos and cowbirds, also appear to be locked in evolutionary arms-races with their hosts. In Britain, cuckoos lay eggs having one of four different color-patterns; these four types correspond well with the eggs of their four primary host species—meadow pipits, dunnocks, reed warblers, and pied wagtails (Figure 8.4). Behavioral adaptations for parasitism include the cuckoo's ability to evict a host egg prior to laying its own; the speed with which the female cuckoo can lay her egg; and the parasitizing chick's (variable) proclivity to kill the host's offspring (see Figure 7.2).

Selection-Driven Virulence in Parasites

Selection has enhanced the ability of parasites to spread among their host organisms. Although conventional wisdom states that it is in the parasite's best interest to minimize virulence to keep its host alive, this may not always be the case. Parasites follow the rules of selection too, and if fitness can be enhanced by increased virulence (and even death), the host's ability to continue supplying resources may be a secondary concern. This tradeoff between fecundity and longevity may underlie the increased virulence documented for pathogens (e.g., the intestinal bacterium *Escherichia coli*) in hospitals. If there are many opportunities for transmitting offspring to novel hosts (in this case, by contact with doctors, nurses, and hospital patients), the parent parasite can afford to kill its host. Degrees of relatedness among parasites inhabiting the same host may also determine virulence; if parasites within the host are closely related, they may gain fitness by reducing virulence and keeping the host alive for their kin to exploit.

Parasite-Mediated Behavior and Coevolutionary Escalation

A strong case for coevolutionary escalation can be made from observations of parasite-mediated behavior, whereby parasites alter ultimate or intermediate (vector) host-behavior in ways that maximize the parasite's fitness. Examples include the following:

- Mosquitoes infected with *Plasmodium* spp., the causative agents of malaria, appear to do their parasite's bidding by increasing their biting rate and human blood consumption, as well as biting more individuals in a single night (Figure 8.5). This behavior should enhance the fitness of the *Plasmodium,* which must ultimately make it into a human host. Also, infected mosquitoes prefer to bite human hosts that carry *Plasmodium* gametocytes, which are necessary for completing the parasite's life cycle. Similarly, the protozoan *Leishmania* (the causative agent of the infectious disease leishmaniasis) manipulates feeding in its sandfly vector, leading the sandfly to increase its biting rate and its number of mammalian targets.
- The flatworm *Microphallus* sp. must be transmitted from its intermediate host, a freshwater snail, to its ultimate host, waterfowl. Accordingly, snails infected with the worm are more likely to feed in locations conspicuous to birds, where they are more likely to be consumed than are uninfected snails. This behavior may result from the flatworm's ability to alter the snail's typical phototactic, geotactic, or photokinetic behavior.
- An acanthocephalan parasitic worm, *Pomphorhynchus laevis,* moves from its intermediate host, the amphipod *Gammarus pulex,* to its definitive (i.e., final) host, the bullhead fish *Cottus gobio. Gammarus* infested with worms exhibit increased drifting behavior (i.e., they move more in the water, and may therefore be easier for predators to detect) and decreased aversion to light, when compared to hosts not infected with the worms. Infected *Gammarus* are more active and more likely to be found in illuminated areas, and are also more common in the stomachs of bullhead. Infected *Gammarus* appear to be under the influence of serotonin, a chemical that can contribute to an organism's affinity for light. Other work on *Gammarus* movement suggests that

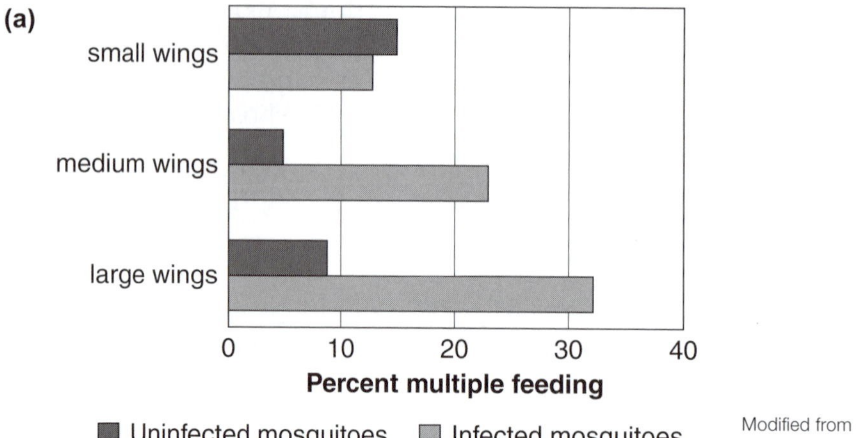

Modified from Koella, et al., 1998.

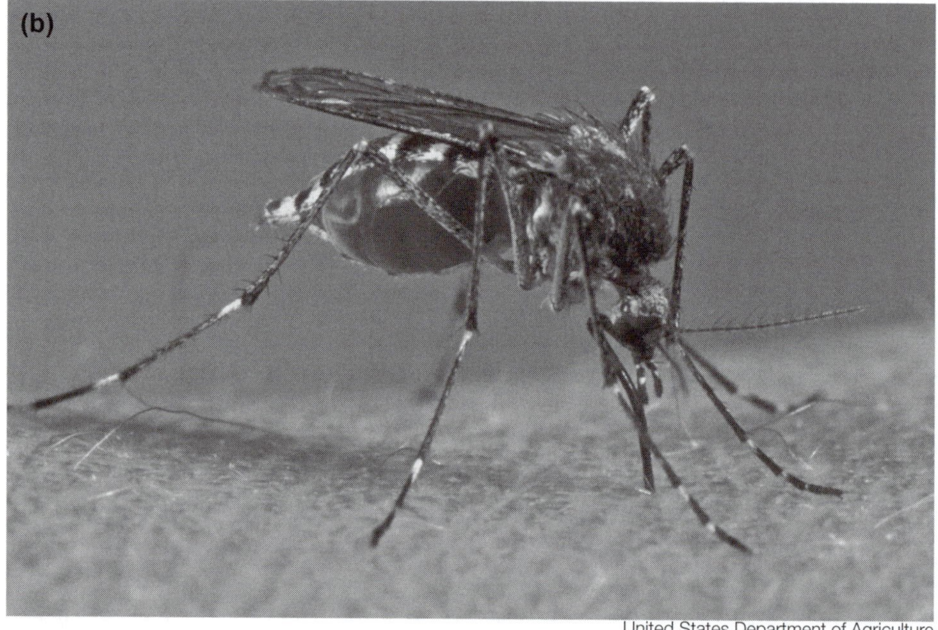

United States Department of Agriculture

Figure 8.5a and b *Plasmodium* affects mosquito feeding. Anopheline mosquitoes infected with *Plasmodium* are more prone to multiple feedings, whereby they suck blood from (and exchange parasites with) more than one human host. This type of host manipulation is adaptive for the *Plasmodium,* as exchanging blood between its human hosts (via mosquitoes) is essential for reproduction.

infected amphipods respond positively to the smell of bullhead, suggesting an olfactory manipulation by the parasite.

- The protozoan parasite, *Toxoplasma gondii,* the causative agent of toxoplasmosis, depends on cats as its definitive hosts, but it can be transmitted by several birds and small mammals. Individual rats infected with *T. gondii* are more active and less feline-averse than are their uninfected counterparts. In fact, infected rats in experimental enclosures are more likely to approach the smell of cat urine than to avoid it, an apparently successful strategy for the cat-dependent parasite (but suicidal, and counteradaptive, for the rat). It would be very difficult to explain the infected rat's behavior outside of its coevolutionary context.

Host Species Evolve Ways to Minimize the Costs of Being Parasitized

Parasite-Host Coevolution in Avian Brood Parasites

Avian brood parasites, such as the cuckoos discussed above, and their avian hosts have provided investigators with a window on parasite-host coevolutionary mechanisms. While brood parasites increase fitness when they successfully trick another bird to raise their young, their hosts are under selection pressure to avoid, or minimize, the costs of being parasitized. Selection appears to have favored hosts that are alert to the dangers of brood parasites and have evolved strategies to counter their effects. For example, experimental egg-manipulations show that many hosts tend to eject eggs that are an unrecognized color, or abandon the nest altogether. In further support of the adaptive significance of egg ejection, pied wagtails in Iceland—which do not co-occur with cuckoos—are less discriminating about egg color and shape than are the pied wagtails in cuckoo-rich Britain. Some host species attempt to puncture foreign eggs, a behavior that presumably has escalated selection for the thicker eggshells characteristic of many parasitic species.

Specificity of Hosts and Intensity of Reciprocal Adaptations

Reciprocal adaptations in hosts and parasites should be most intense where there is high specificity for hosts, and less intense where a variety of hosts are available. This prediction has been confirmed in many symbioses, including the brood parasite–avian host relationships described above. Additional support comes from studies of resistance in wild parsnip to its parasite, the parsnip webworm. Wild parsnip produces a highly specific cocktail of defensive chemicals, which is in turn detoxified by the webworms. Where an alternate host is available, coevolutionary relationships are not as strong, and the parsnip's chemical defenses aren't so well matched to their parasites.

Host Tolerance

In contrast to host resistance, whereby parasitic infection is limited, is host tolerance, in which the occurrence of infection is not limited but the fitness consequences for the parasite *are* restricted. Tolerant hosts have evolved strategies to survive with their parasites, a phenomenon that is well documented, especially in plants. While the mechanisms underlying tolerance are largely unknown, clear patterns have been established whereby areas with a history of intense herbivory (e.g., from grazing animals) are characterized by more tolerant plants.

Host Responses to Parasitism

On the symbiotic continuum, a host can respond in many ways to parasitism: the host can be killed, severely compromised, resistant, or tolerant. And in some cases, parasitic interactions can evolve into **mutualisms,** whereby the host adapts to the parasite in such a way that the host's fitness is increased rather than compromised. Such a case has been documented in the *Wolbachia-Drosophila simulans* interaction. *Wolbachia* are a group of maternally inherited **bacteria** that spread by manipulating their host's reproduction. Laboratory and field observations indicate that, in a ten-year period, *Wolbachia* in California evolved to lend their host females noticeably increased fecundity over uninfected *D. simulans* females, suggesting a switch from parasitism to mutualism in this particular symbiosis.

SIDEBAR 8.2 Endosymbiosis and the Origin of Complexity

A switch from consumption to cohabitation (or from predation to mutualism) is likely responsible for the origin of cells with internal compartmentalization, or organelles. These cells, the *eukaryotes*—plants, animals, protists, and fungi—have resulted from symbiotic associations between two or more *prokaryotes,* those cells that lack organelles such as mitochondria and chloroplasts. According to the *endosymbiotic theory,* large, simple cells engulfed smaller, simple cells that they failed to digest. Some of the engulfed cells likely performed a photosynthetic function, contributing a supply of energy to the larger cell; these smaller cells were the precursors to today's chloroplasts, the organelles that convert light energy into carbohydrates used as food. Mitochondria may have had similar origins. This theory is supported by several lines of evidence, including the facts that mitochondria and chloroplasts possess their own, prokaryote-like circular DNA and they are both bound by double membrane structures, suggesting a primitive background in food vesicles.

Symbiont Species That Serve a Life-History Function Are Rewarded and Continue

Mutualism

Mutualistic associations, whereby all symbionts benefit, are based on provisioning rewards for services rendered. These services, and their associated rewards, are numerous and well documented. Examples include the following:

- Many mutualisms involve cleaners that remove various forms of debris—for example, dead skin, external parasites, and stray food particles—from hosts. Cleaner-host associations are widespread, and include birds that clean crocodile teeth, shrimp that remove parasites from fish, and fish such as wrasses and gobies that reside at *cleaning stations* where other fish (the clients) congregate. Some cleaner fish appear to be nutritionally dependent on client refuse and will die if they cannot perform this custodial service.
- Many plants require assistance with pollination. Pollen that is distributed by wind or water is not at the mercy of a living vector, but many flowering plants rely on animals to transfer sperm to an egg. Darwin discussed this in *On the Origin of Species:* "Let us now suppose a little sweet juice or nectar to be excreted by the inner bases of the petals of a flower. In this case insects in seeking the nectar would get dusted with pollen, and would certainly often transport the pollen from one flower to the stigma of another flower. The flowers of two distinct individuals of the same species would thus get crossed; and the act of crossing, we have good reason to believe (as will hereafter be more fully alluded to), would produce very vigorous seedlings, which consequently would have the best chance of flourishing and surviving." Common animal pollinators, such as bees, are fine-tuned to identify plants that pay well (typically, in nectar) for this service. Plants, in turn, have evolved advertising—usually flowers—featuring scent, shape, or coloration to solicit their pollinators.

Obligate Mutualism

- Numerous examples of specificity between plants and their pollinators suggest coevolutionary escalation, such as the obligate mutualism exhibited by yucca plants and yucca moths, in which neither species can reproduce without the other (Figure 8.6). The female yucca moth arrives at a yucca flower with pollen and eggs that are ready to be deposited. She places her eggs inside the flower's ovary and then transfers her pollen load to the stigma, the site of pollination. This deliberate

Figure 8.6 Yucca moths in a yucca flower. The yucca moth-yucca flower symbiosis is an obligate mutualism. If yucca moths lay too many eggs in a single flower, thus overexploiting the mutualism, they are penalized by selective flower abortion, and none of the eggs survives. (Courtesy Alan Cressler)

transfer is self-interested, in that without yucca seeds to consume, her offspring will emerge from their eggs and starve. Thus, the yucca moth has a vested interest in the yucca's successful fertilization, and the yucca has a vested interest in provisioning future pollinators.

- Acacia ants and acacia trees exhibit another obligate mutualism, whereby the stinging ants occupy the hollow thorns of the acacia branches and defend the tree against potential herbivores. In addition to housing, the ants are rewarded with proteins and lipids produced in specialized structures—*Beltian bodies*—that are found at the tips of acacia leaflets. For the cost of producing this food, the acacia is protected from herbivory. The acacia also absorbs some of the nitrogen secreted in the ants' urine.
- Another obligate mutualism involves the fungal/algal associations called lichens. Typically, the fungal component anchors the lichen and stores water, and the algal component produces food, via photosynthesis, for the lichen. The fungus-algal relationship is so specific that lichens are classified as unique species, rather than as two separate (and taxonomically distinct) entities.

Mutualism in the Oceans and Atmosphere

- In the energy-poor void of the ocean depths there occur energy- and nutrient-rich oases at seemingly uninhabitable hydrothermal vents (see Chapter 3). These vents, which are fissures in the ocean floor that seep hot, acidic water and toxic hydrogen sulfide, support diverse communities that could not exist without the mutualistic interactions at their foundation. The mollusks and tubeworms that constitute much of the biomass at the vents cannot feed themselves because they lack essential nutrients and digestive tracts. Instead, they are supported by endosymbiotic bacteria that use the hydrogen sulfide to make organic compounds that can be consumed by the hosts.
- In a similar fashion, coral reefs are built on mutualism. Photosynthetic dinoflagellate endosymbionts, the zooxanthellae, donate some of the sugars they produce to the corals that house them; in return, the dinoflagellates receive nitrogen-rich waste products from the coral's captured prey organisms. Zooxanthellae contribute up to 95% of the coral's required carbon, and are therefore essential to the growth and maintenance of reef communities.
- To sustain life as we know it, atmospheric nitrogen must be converted to biologically available nitrogen; this conversion occurs naturally in the atmosphere when lightning strikes, as well as in nitrogen-fixing bacteria. These bacteria possess nitrogenase, an enzyme that converts atmospheric nitrogen (N_2) to ammonia (NH_3). Plants such as legumes play a role in the process by housing nitrogen-fixing bacteria, such as *Rhizobia,* in root nodules. This plant-rhizobium symbiosis is a classic mutualism: the plant receives amino acids from the bacteria, and the bacteria receive carbohydrates from the plant.

The list goes on and on, confirming the prediction that mutualisms are sustained by symbionts that are rewarded to provide goods and services. However, symbionts would benefit if they could successfully increase their reward without expending any provisioning effort. How do symbionts avoid the cost of these cheaters?

Rewarded Symbionts Are Penalized for Cheating

Given that significant resources are required to reward a symbiont, those who fail to provide the life history function for which they are being rewarded should be penalized to discourage future cheating. Such penalties for cheating have been suggested in several cases, including the cleaner fish and yucca moth examples introduced previously.

Penalties in the Cleaner-Fish Symbiosis

The cleaner wrasse *Labroides dimidiatus* and its client *Ctenochaetus striatus* engage in a mutualistic symbiosis whereby the cleaner removes ectoparasites from the client at a cleaning station in their shared territory. However, when *C. striatus* are experimentally anesthetized (and thus cannot defend themselves), *L. dimidiatus* selectively consume the client's scales and external mucus instead of its parasitic pests, demonstrating the cleaner's preferences. These cleaners will try to take bites from the client, clearly cheating on the mutualism, a risk that the nonanesthetized client typically punishes with jolting, aggression toward the cleaner, or departure from the cleaning station. *L. dimidiatus* that are penalized in this way apparently get the message, for they are less likely to cheat during subsequent cleaning bouts.

Penalties in the Yucca-Yucca Moth Symbiosis

In the yucca flower-yucca moth symbiosis described previously, the yucca protects and nourishes the moth's emergent young, and the moth pollinates the yucca flower. The inclination for the moth to exploit this situation, mainly by depositing too many eggs, has been demonstrated and is considered a form of cheating. Yucca penalize cheating moths by selectively aborting flowers that either receive poor pollen-loads or too many moth eggs. Thus, moths that overload a flower with their eggs will lose fitness in the end, while those that exhibit restraint fare better overall. A similar case has recently been documented in the mutualism between *Glochidium* trees and *Epicephala* moths.

Summary

Many intricate coevolutionary relationships illustrate the power of biotic interactions in shaping biodiversity. Much of the phenotypic variation we observe, in terms of appearance, behavior, and physiology, is due to symbiotic interactions. Confirmed coevolutionary predictions in the examples discussed in this chapter support ecologist John Thompson's assertion that "[w]ithout these coevolved interactions, highly diverse ecosystems would collapse immediately."

9

Human Evolution

I demand of you, and of the whole world, that you show me a generic character . . . by which to distinguish between Man and Ape. I myself most assuredly know of none.

—*Carl Linnaeus, 1788*

Natural selection, as it has operated in human history, favors not only the clever but the murderous.

—*Barbara Ehrenreich, 1991*

An evolutionary perspective of our place in the history of the earth reminds us that *Homo sapiens* has occupied the planet for the tiniest fraction of that planet's four and a half thousand million years of existence. In many ways we are a biological accident, the product of countless propitious circumstances. As we peer back through the fossil record, through layer upon layer of long-extinct species, many of which thrived far longer than the human species is ever likely to do, we are reminded of our mortality as a species. There is no law that declares the human animal to be different, as seen in this broad biological perspective, from any other animal. There is no law that declares the human species to be immortal.

—*Richard E. Leakey and Roger A. Lewin, 1977*

Light will be thrown on the origin of man and his history.

—*Charles Darwin, 1859*

PREDICTIONS

If humans have evolved, we predict that evidence supporting evolution will include

1. Biogeographical evidence.
2. Fossil evidence.

3. Molecular evidence.
4. Anatomical and developmental evidence.
5. Behavioral evidence.
6. Coevolutionary evidence.

This chapter discusses a subject that has fascinated—and often troubled—people for generations: human evolution. Where did we come from? Who are our ancestors? How long have we been here? And how do we know?

Virtually all cultures and religions have answered these questions in a variety of ways, often invoking deities and supernatural events. Some of these claims are untestable, and others (e.g., young-Earth creationism) have long been rejected by science. In this chapter, we present a science-based discussion for human evolution that, in the words of Charles Darwin's friend Thomas Henry Huxley, "excludes creation and all other kinds of supernatural intervention."

In relegating human evolution to a separate chapter, we do not mean to imply that the evidence for human evolution is somehow different from that of the myriad of other organisms you've encountered throughout this book. On the contrary, humans are the outcome of the same natural forces that have shaped all of life. Consequently, we have ancestral ties with other forms of life, and all of the major topics and predictions you read about throughout this book can be applied directly to humans. The evidence clearly indicates that humans evolved in Africa from a common ancestor that we share with apes.

Human Evolution: Biogeographic Evidence

Sites of Origin

One of Charles Darwin's most influential opponents was famed Swiss biologist Louis Agassiz (1807–1873). Agassiz, who rejected evolution, claimed that organisms had several places of origin, and believed that species remained "at or near their site of creation." Darwin rejected these claims, instead arguing that species had one site of origin from which they could disperse to other areas. If this were true, then one would predict that species living in a particular area would be the descendants of earlier species that had lived in the same area. This hypothesis has been repeatedly tested and confirmed. For example, modern armadillos live only in North America, Central America, and South America. Where are fossil armadillos found? In North America, Central America, and South America. Similarly, modern kangaroos and koalas live in Australia. Where are fossil kangaroos and koalas found? Australia.

Darwin's Prediction: The African Origins of Humans

Darwin barely mentioned human evolution in *On the Origin of Species* (1859). However, he understood humans' similarities with other organisms, later noting that "[w]e must . . . acknowledge, as it seems to me, that man with all his noble qualities . . . still bears in his bodily frame the indelible stamp of his lowly origin." In *The Descent of Man, and Selection in Relation to Sex* (1871), Darwin used biogeography (e.g., the co-occurrences of fossilized ancestors and their descendants; Chapter 4) as a basis for one of the most famous predictions in the history of evolutionary biology—namely, that humans originated in Africa:

> We are naturally led to enquire, where was the birthplace of man at that state of descent when our progenitors diverged from the [Old World monkeys and apes]? The fact that they belonged to this stock clearly shews that they inhabited the Old World; but not Australia or any oceanic island, as we may infer from the laws of geographical distribution. In each great region of the world the living mammals are closely related to the extinct species of the same region. It is therefore probably that Africa was formerly inhabited by extinct apes closely allied to the gorilla and chimpanzee; and as these two species are now man's nearest allies, it is somewhat more probable that our early progenitors lived on the African continent than elsewhere.

Darwin's prediction was not based on hard evidence; in fact, the only hominin fossils that had been discovered at the time of Darwin's prediction were those of Neandertals (*Homo neanderthalensis*). These fossils, which were discovered in Feldhofer Cave in the Neander Valley near Düsseldorf, Germany, in 1856, were from hominins that had lived in Europe fewer than 100,000 years ago (the word *Neanderthal* was derived from *Neander Valley*, and *thal*, which is Old German for *valley*). Nevertheless, Darwin used biogeography to reason that human ancestors (i.e., human "progenitors") came from Africa because (1) Africa had a tropical climate hospitable to apes, and (2) they were more allied to African apes (e.g., chimpanzees and gorillas) than to Asian apes (e.g., orangutans and gibbons). Was Darwin right?

Human Evolution: Fossil Evidence

Dart's Discovery of Hominin Fossils

At first, it appeared that Darwin was wrong, for in 1891 Dutch army surgeon Eugène Dubois—while working in Java—discovered a 500,000-year-old fossilized skullcap and teeth of "Java Man" (*Pithecanthropus erectus*, later classified as *Homo erectus*). This was the first hominin fossil that was different from us (Figure 9.1). Dubois' discovery was followed in 1929 by Canadian anatomist Davidson Black's discovery of "Peking Man" in a cave in Zhoukoudian, China ("Peking Man" was later reclassified as *H. erectus,* which has also been unearthed in Ethiopia, China, Kenya, and elsewhere). However, in 1924, Raymond Dart found what came to be known as "Taung Child" (*Australopithecus africanus*) in rocks collected in a Pleistocene limestone quarry in the Transvaal region near Taung, South Africa. Taung Child, which was 1–2 million years old, was the first early hominin fossil found in Africa; its skull resembles the skull of a young chimpanzee, but many of its parts—notably its teeth—resemble those of humans. (The name *Australopithecus africanus* was derived from *austral,* for *south; pithekos,* for *ape;* and *africanus,* for *from Africa.*) Dart's claims that *Australopithecus africanus* was "an extinct link between man and his simian ancestor," and related to the ancestral stock of humans rather than to the great apes, were controversial, and were rejected by many scientists. However, Dart's claims were strengthened in the 1930s when famed Scottish paleontologist Robert Broom unearthed adult skulls of *Australopithecus* from several caves in South Africa. As Broom noted, "In *Australopithecus* we have a connecting link between the higher apes and one of the lowest human types." Broom's discovery, which confirmed the hominin anatomy seen in Dart's Taung Child, vindicated "the Darwinian claim that Africa would prove to be the cradle of mankind."

Other Hominin Discoveries in Africa

Dart's groundbreaking find was the first of a procession of remarkable discoveries of hominin fossils from Africa (Figure 9.2):

In 1959, Mary and Louis Leakey uncovered bits of a jawbone of the 1.75-million-year-old *Paranthropus boisei* from Tanzania's 40-kilometers-long Olduvai Gorge. This discovery, which was the first ancient hominin found in East Africa, triggered a search of the famed Great Rift Valley, a geologic fault extending through Ethiopia, Tanzania, and Kenya that exposes rocks millions of years old. Although *Paranthropus* means "next to man," *Paranthropus* was probably not a direct ancestor of contemporary humans. After a million or so years, *Paranthropus* became extinct.

In 1961, the Leakeys discovered the jaw and hand of what later became known as *Homo habilis,* the first direct ancestor of humans in Africa. *Homo habilis* lived 1.8 million years ago; its *habilis* name means *handy,* from the inferred ability of this hominin to make stone tools. *Homo habilis* thrived for nearly 500,000 years before it became extinct. For the rest of his life, Louis Leakey claimed that *H. habilis* was on the true evolutionary line to humans, and that all australopithecines were not.

In 1974, Donald Johanson discovered the partial skeleton of *Australopithecus afarensis,* a 3.2-million-year-old female who lived in Hadar, Ethiopia

Figure 9.1 In 1891, Dutch anatomist Eugène Dubois—working along the Bengawan Solo River near the village of Trinil, Java—discovered *Pithecanthropus erectus,* which Dubois named "P.e." Dubois described P.e. as "the first known transitional form linking Man more closely with his next of kin among mammals." Although the fossil became known as Java Man, it was later reclassified as *Homo erectus,* a widespread species that lived in Asia, Africa, and Europe 1–2 million years ago. The skullcap of P.e. that Dubois discovered is the type specimen of *H. erectus.* The monument shown here directs tourists to the site of Dubois' discovery. (Courtesy of Charity Putuhena)

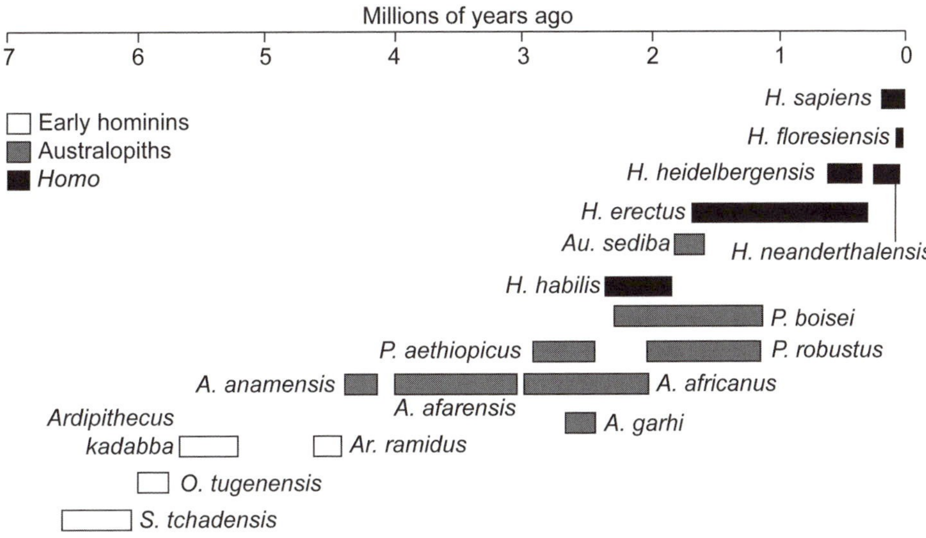

Figure 9.2 The key species of humans. Although evolution has produced many species of hominins (some of which overlapped in time), all but *Homo sapiens*—us—are extinct. This diagram shows approximately when some of the known species of the human family lived. (Jeff Dixon)

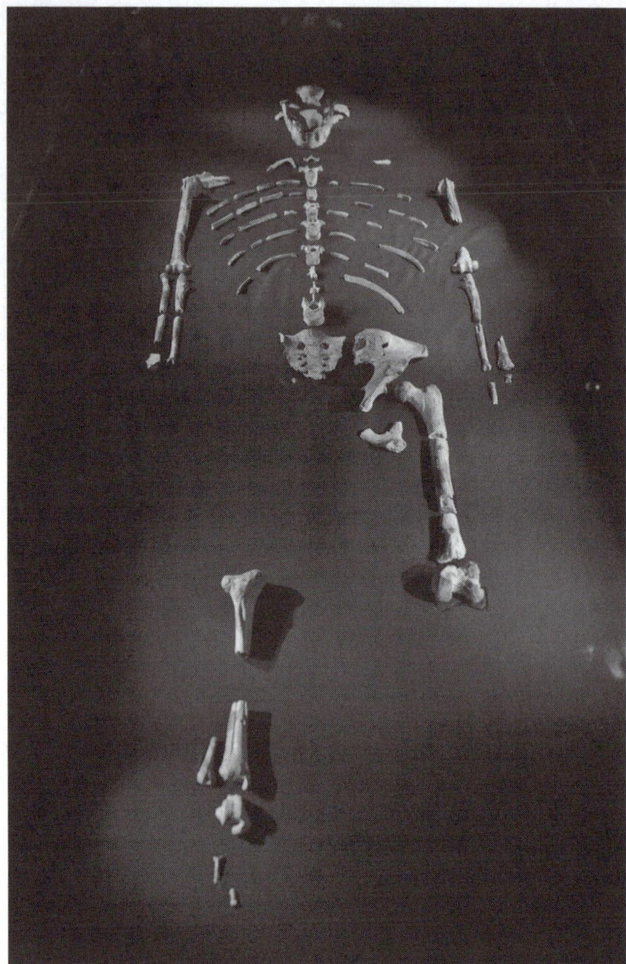

Figure 9.3a Lucy (*Australopithecus afarensis*) was a small woman (less than 1.5 meters tall) who lived 3.2 million years ago in Ethiopia. Lucy is the most famous fossilized hominin. (AP Photo/Michael Stravato)

(Figure 9.3). This fossil, which was named "Lucy" because the Beatles' "Lucy in the Sky with Diamonds" was playing in camp when the discovery was made, walked upright. Indeed, in 2011 paleontologists reported the discovery of a tiny fossil foot bone showing that Lucy's feet supported a two-legged stride—that is, Lucy and her kin walked like modern humans. In addition, the discovery in 2010 of 1.5-meter-tall "Big Man" suggested that *A. afarensis* had an upright, smooth stride. This trait evolved before humans' use of stone tools and an expanded brain. At the time of its discovery, Lucy was the earliest known hominin. Although researchers have discovered older and more complete fossil hominins, Lucy remains a landmark to which others are compared.

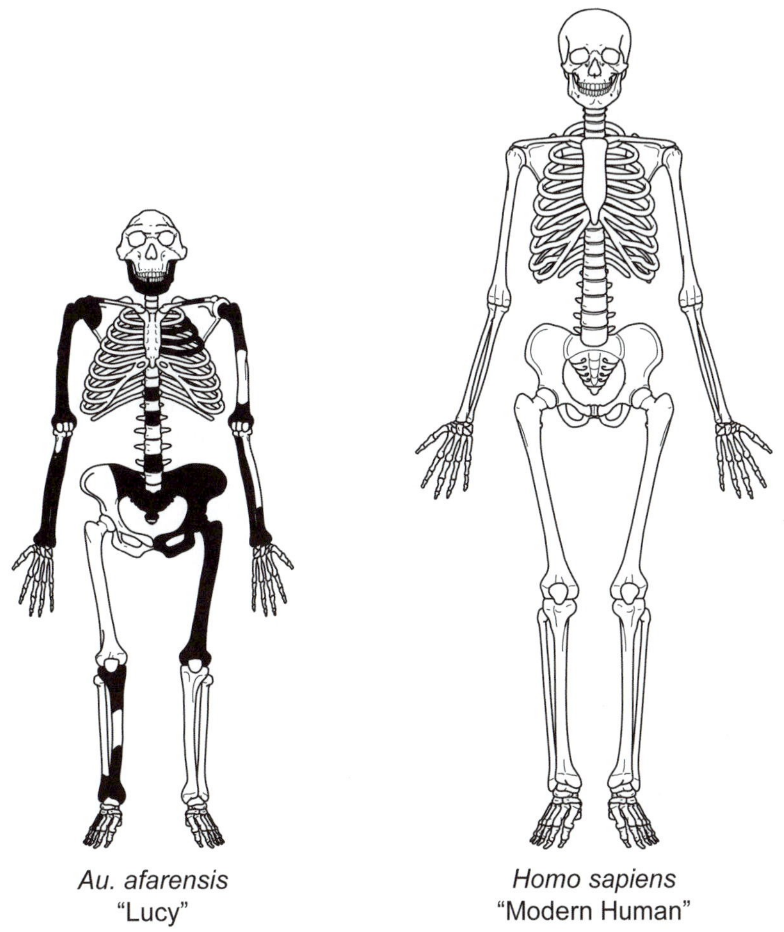

Figure 9.3b These diagrams compare the skeletal features of a modern human with a reconstruction of Lucy. The fossil components are shown in black, and reconstructed parts are based on mirror images of known parts of the skeleton and on other fossils. (Jeff Dixon)

In 1976, Mary Leakey and her co-workers discovered a series of footprints made by at least two upright hominins 3.6 million years ago at Laetoli in northern Tanzania. These footprints are remarkably similar to those of modern humans.

In 1984, Kamoya Kimeu, a member of Richard Leakey's research team, discovered a 1.6-million-year-old skeleton of a young boy in Kenya's Turkana basin. Because "Turkana Boy" had several features that distinguished it from the Asian *Homo erectus,* the hominin was set apart as a separate species, *Homo ergaster.* In 2007, Turkana Boy was the focus of a protest by Kenyan religious leaders denouncing human evolution.

In 1992, *Ardipithecus ramidus*—the oldest known hominin skeleton—was found in Ethiopia, and was described in *Nature* in 1994 as the "long sought potential root species for the Hominidae." This fossil was the first in 20 years to challenge Lucy for her status as the earliest human ancestor. Two years later, American paleontologist Tim White and his co-workers excavating the Middle Awash found a specimen of *A. ramidus* named "Ardi," a 110-pound, 1.25-meter-tall female that was a mosaic of ape and human traits: she walked upright, yet had an opposable toe, which was good for climbing trees in the forest in which she may have lived.[1] Ardi, who was not much like a chimpanzee, gorilla, or human, is by far the most complete of the earliest hominins—most of her teeth and skull, as well as rare bones of her feet, hands, arms, legs, and pelvis, have been found. As of 2010, scientists had found more than 300 specimens from seven different hominin species that lived in the Middle Awash from six million to only 160,000 years ago. Ardi, one of the earliest human relatives, is near the base of the human family tree.

Also in 1992, Meave Leakey's team discovered *Australopithecus anamensis*—the direct ancestor of *Australopithecus afarensis*—in Kenya.

In 1997, graduate student Yohannes Haile-Selassie—who in 1994 had uncovered the first fossils of Ardi—found *Ardipithecus kadabba* in the Middle Awash region of Ethiopia. *A. kadabba* was a precursor of *A. ramidus*, who lived almost six million years ago in an area close to where *A. kadabba* lived (*ramidus* is Afar for *root*, and *kadabba* is Afar for *basal family ancestor*).

In 2000, Martin Pickford and Brigitte Senut's team found 13 fossils of the chimp-sized *Orrorin tugenensis* in the Tugen Hills of Kenya. Although the hominin was popularly known as "Millennium Man," the name *Orrorin tugenensis* was derived from a Tugen legend of the "original man" who settled the Tugen Hills. *Orrorin tugenensis* lived approximately six million years ago.

In 2001, Chadian undergraduate student Ahounta Djimdoumalbaye discovered fossils of *Sahelanthropus tchadensis*—the oldest known hominin—in the Djurab Desert of Chad, 2,500 kilometers west of the eastern Africa's Great Rift Valley, where most of the ancient hominins had been found. *Sahelanthropus tchadensis* lived 6–7 million years ago. Because only the skull is available for study, anthropologists continue to debate whether *Sahelanthropus tchadensis* was bipedal. The nickname of

1. The official announcement of Ardi in *Science* in late 2009 prompted *Time* magazine to include White in its list of the top 100 "people who most affect our world."

Sahelanthropus tchadensis is "Toumaï," which means "hope of life" in the Goran language. Because *Sahelanthropus tchadensis* dates to or before the time when the great apes and the line leading to humans diverged, it is not surprising that *Sahelanthropus tchadensis* has a mixture of ape and human traits. For example, the small skull and large brow ridges resemble those of a chimp, while the flattened face and reduced canine teeth resemble those of a human. *S. tchadensis* is the oldest known human ancestor.

In 2006, the cover of *Nature* reported the discovery of "Selam," an infant *Australopithecus afarensis.* Selam, the most complete hominin infant thus far discovered, was found in the Dikika region of the Afar region, only a few kilometers from where Lucy was discovered.

In 2007, Meave Leakey and her colleagues described the remains of *H. habilis* and *H. erectus* a short distance apart in the same rock layer, suggesting that *H. erectus* probably did not descend from *H. habilis* (as previously thought) and that the two species coexisted in the same area for up to 500,000 years.

In 2010, paleontologist Lee Berger reported the discovery of *Australopithecus sediba* (meaning *southern ape wellspring* in Sotho) from caves just north of Johannesburg, South Africa. These hominins, which were unearthed in 2008 by Berger's son Matthew while chasing his dog, have a mix of early human and ape features; for example, they have long, orangutan-like forearms and a small skull, along with long-striding hips and the flat face of human species. Thus, *Australopithecus sediba* probably strode upright on long legs (with human-shaped hips and pelvis) and climbed through trees with ape-like arms. The fossils date to approximately 1.95 million years ago, which is near the point when *Homo* was beginning to emerge. Berger suggested that *Australopithecus sediba* was a descendant of *A. africanus* and a possible ancestor of *H. erectus,* an immediate precursor to *H. sapiens,* or a close "side branch" that did not lead to humans.

Fossil hominins have been found outside of Africa; for example, in 2003 the bones and artifacts of an 18,000-year-old, "hobbit"-sized *Homo floresiensis* were discovered on the island of Flores in Indonesia. The 1-meter-tall humans had small skulls (their brains were only about one-third as big as ours) that were closer in size to that of a chimp than to any other species of humans. The origins of *H. floresiensis* are controversial, for some of their traits (e.g., robust jawbone, small brain) are more similar to older African hominins (e.g., australopithecines) that to *H. erectus.* Some critics argue that the hobbits are merely human pygmies similar to those still living on Flores. However, the oldest fossil hominins are all from Africa, as would be predicted if humans originated in Africa.

Africa as the Birthplace of Humanity

Darwin was right; Africa is the birthplace of humanity. It is where modern humans first appeared in fossil form about 200,000 years ago, and where numerous hominin ancestors originated during the past six million years. The discovery in 2011 of ancient stone tools near the Persian Gulf produced the controversial claim that humans first left Africa 125,000 years ago. However, genetic analyses of people from around the world today suggest that humans migrated from Africa to Asia around 60,000 years ago, and by 1,500 years ago, people had reached even the Arctic and the most remote islands of the Pacific.

Our predecessor species did not simply succeed one another, with each being more evolved than the last. Some (e.g., *P. boisei* and *H. erectus*) overlapped in time and adapted different survival strategies as they competed for resources (Figure 9.2). Modern humans (*Homo sapiens*) are merely another iteration of our evolutionary family tree, not the inevitable pinnacle of evolution.

Human Evolution: Molecular Evidence

Universal Homology, Ubiquitous Proteins, and Human Evolution

The molecular evidence for the hominin lineage is consistent with the theory that humans represent a recent twig on the ape branch of primate evolution. Humans possess universal homology—that is, we share hereditary material and a genetic code with all other organisms—and express proteins in common with other organisms from seemingly disparate taxa. Our genomes are chock-full of evolutionary relicts—genes that have a history of millions (or billions) of years and code for the same proteins in us that they do in microbes, fruit flies, and sea urchins. For example, cytochrome *c*, the electron-shuffling protein discussed in Chapter 5, is essential in aerobic respiration, and is highly conserved in organisms throughout the biological spectrum. Beyond making the same 104 amino acids, humans and chimps share identical *cyt c* genetic base sequences; the likelihood of this occurring by chance (rather than heredity) is around 1×10^{-94}. Our base sequences for *cyt c* differ from those in rhesus monkeys by only one amino acid, and from mice by only nine. Given all the possible combinations of bases that could account for this type of protein, the probability that we would *coincidentally* share this much with the mouse is approximately 8×10^{-34}. In other words, it makes much more sense that our similarities are due to common ancestry than to chance.

In the last couple of decades, entire genomes have been sequenced for representatives from various taxa (including humans, chimpanzees, mice, roundworms, fruit flies, and hundreds of prokaryotes). Our genetic similarity with other organisms is most striking when viewed alongside our sister species—the chimps and bonobos. For example, the DNA of humans and chimps diverges by less than 2% overall, and histories that are more refined can be gleaned by comparing the DNA of coding regions (Figure 9.4). Studies involving molecular-clock techniques indicate that the human lineage diverged from that of the chimps approximately 4–6 million years ago, a time period that is consistent with the fossil and biogeographical evidence discussed previously in this chapter.

Molecular analyses have also addressed an intriguing question about the relationship between humans and other hominins—namely, did we mate with Neandertals? A 2010 comparison of the Neandertal genome, as gleaned from the thigh bone of a 38,000-year-old-female, suggests that the two species did interbreed, and that 1% to 4% of the DNA in modern humans was inherited from Neandertals. Also, any mixing appears to have occurred after humans left Africa but before the European and Asian lineages diverged—the similarities with Neandertals are exclusive to people with European and Asian backgrounds, and are not shared by those of African descent. As the authors concluded, "The analysis of the Neandertal genome shows that they are likely to have had a role in the genetic ancestry of present-day humans outside of Africa, although this role was relatively minor given that only a few percent of the genomes of present-day people outside Africa are derived from Neandertals."

Fossil Genes in the Human Genome

Comparative studies of whole genomes have also shed light on the presence of vestigial molecular elements, or pseudogenes, in humans. Just as the egg-yolk genes are molecular fossils in marsupial and placental mammals (Chapter 5), so are several mammalian—and even primate—genes vestigial in humans. The ability of a gene to lose function, in an otherwise productive lineage, provides clues about key selection pressures in our evolution. A recent study of the mouse, dog, and human genomes detected 72 genes that had become nonfunctional in humans since these three groups diverged (~75 million years ago). Of these 72 human pseudogenes, as many as 27 may have coded for olfactory receptors, confirming how much more important scent is to dogs and rodents than it is to humans.

An example of one of our non-olfactory pseudogenes is *GULO*, which codes for L-glucono-γ-lactone oxidase (LGGLO), a key enzyme in the synthesis of vitamin C. As a result of our vestigial *GULO* gene, humans cannot synthesize vitamin C and therefore must rely on external sources for this essential nutrient. If our primate ancestors enjoyed a diet of vitamin-C-rich fruits, a loss of function in this gene would not have been penalized too

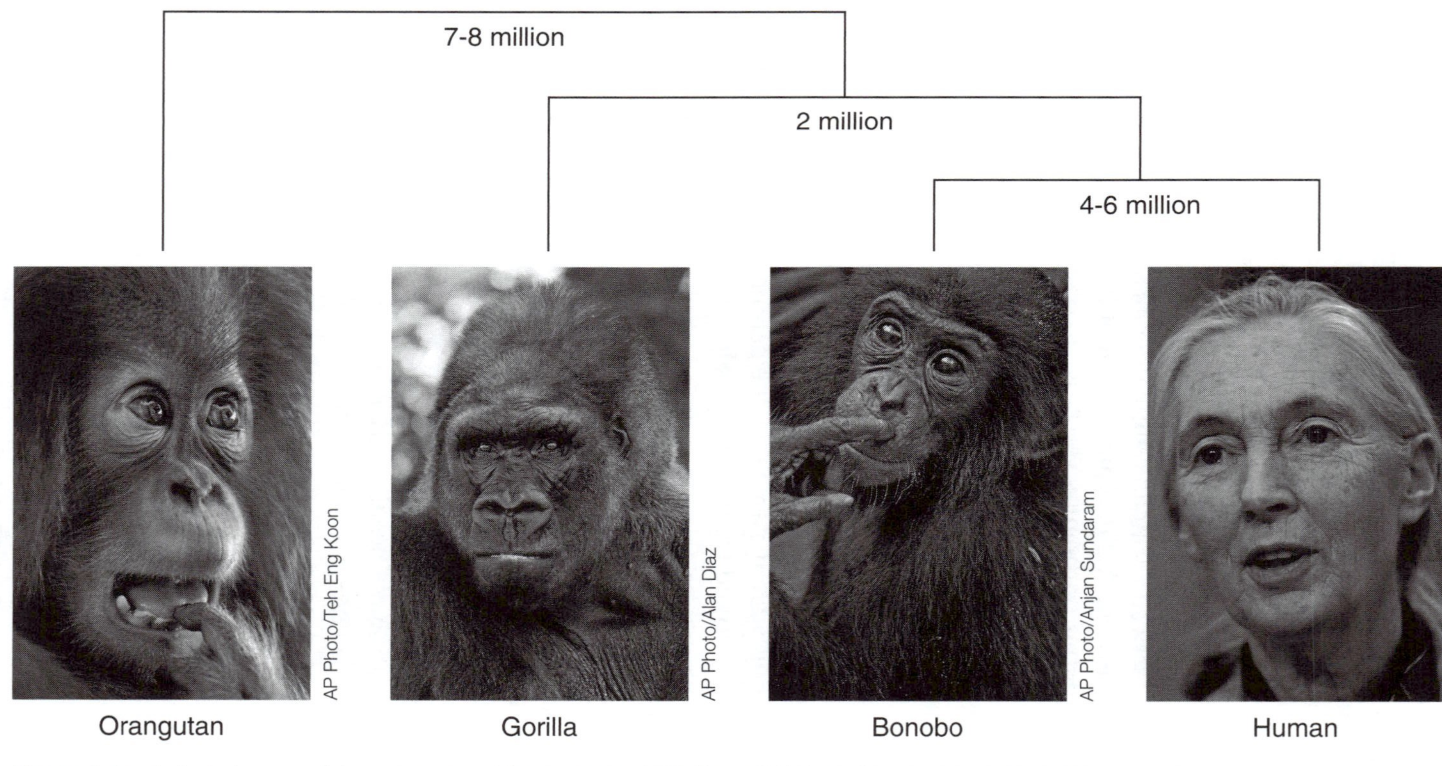

Figure 9.4a–d A phylogeny of the great apes. The greatest similarities of DNA coding regions among the great apes occur between humans and chimps and bonobos, lineages that diverged between 4 and 6 million years ago. Dates since groups diverged are based on molecular-clock estimates.

heavily. In fact, an analysis of four primates—human, chimp, orangutan, and macaque—reveals that we share the same mutation that led to loss of function of this gene. Yet humans have paid a price for this lapse: throughout recent history, many people have died from scurvy, a debilitating disease due to vitamin C deficiency.

The loss of function of some genes could confer a selective advantage, thus leading to a hastening of the gene's demise. This may be the case with the gene for caspase-12 (*CASP-12*), an anti-inflammatory enzyme that cannot function in its mutant state. The functional form of *CASP-12* is gone from Europe and Asia, and is present in only 20%–30% of African populations (Figure 9.5). *In vitro* work with human blood has suggested that people having functional *CASP-12* show a reduced immune response to bacterial pathogens (namely, their immune system doesn't react strongly to the lipopolysaccharides that occur on some bacterial cell membranes). Indeed, a clinical study of a group of several hundred African Americans confirmed that people with the functional *CASP-12* gene are more susceptible to sepsis, an often fatal, whole-body inflammatory response to toxic microbes. Chris Tyler-Smith and colleagues have suggested that the mutant *CASP-12* allele arose over 100,000 years ago in Africa and was initially selectively neutral (meaning it conferred neither a fitness cost nor advantage). However, increased human population densities and the associated spread of microbial diseases would have given an advantage to individuals with the mutation, due to their innate ability to resist sepsis. Thus, with the recent (and possibly ongoing) demise of functional *CASP-12*, we may be witnessing the demise of a functional human gene.

Some vestigial molecular elements can be reactivated. For example, an area of inactive genetic material on chromosome 4 typically initiates transcription and then disintegrates without making a functional product. However, a specific mutation, in which a sequence for the amino acid adenine is inserted into the gene, reactivates the gene and causes the disease facioscapulohumeral muscular dystrophy (FSHD).

Humans and the *MC1R* Gene

Just like our mammalian relatives the rock pocket mice (discussed in Chapter 5), humans have noticeably different pigments. Our skin color is a function of melanin production, whereby people who produce more of the melanin pigment have darker skin. Melanin production in humans is largely due to the action of the *MC1R* gene, the same gene responsible for color polymorphisms throughout the mammals, as well as in fish, reptiles, and amphibians. Human populations vary globally—and predictably—in skin color, with the darkest individuals occurring near the equator and the lightest individuals closer to the poles. Notably,

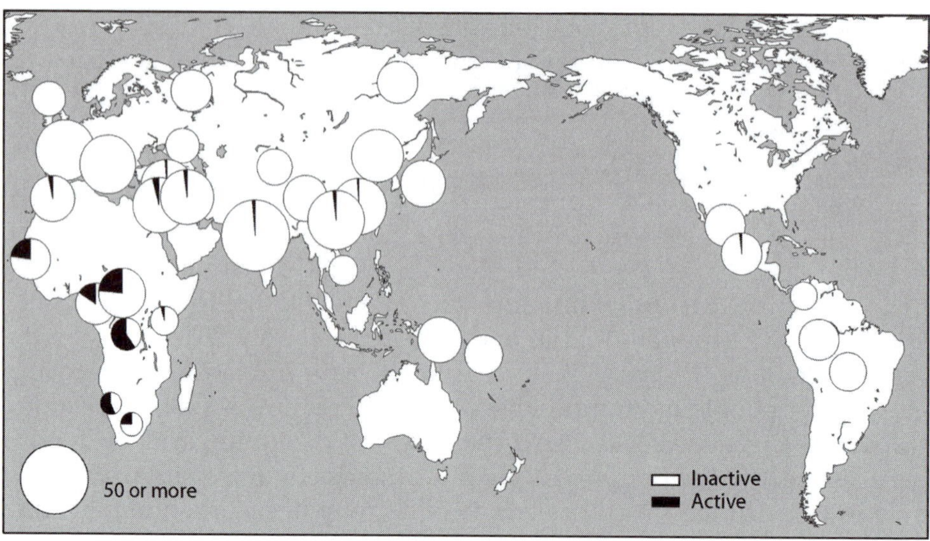

Figure 9.5 Humans and fossil genes. Because the loss of function in the *CASP-12* gene confers resistance to sepsis, the loss of function may itself be the result of selection. A global map with *CASP-12* gene frequencies reveals that functional *CASP-12* is essentially absent from Europe and the Americas, but persists in older, African and Middle-Eastern populations. (Reprinted from Y. Xue, A. Daly, et al., 2006, "Spread of an Inactive Form of Caspase-12 in Humans Is Due to Recent Positive Selection," *American Journal of Human Genetics* 78 (4): 659–670. Copyright © 2006, with permission from Elsevier.)

there is a close correlation between UV-light penetration and skin pigmentation that suggests a connection between increased UV light and enhanced melanin production, and thus a selective advantage of a functional *MC1R* gene. Evidence for the adaptive significance of this gene is seen in a global comparison of mutations at this locus: a high level of variability (due to mutation) persists in populations outside Africa, but virtually none exists in Africans, who appear to have been highly constrained by selection against *MC1R* mutations. One suggestion is that dark skin protects against the detrimental effects of UV—specifically, the breakdown of folate, a nutrient necessary during fetal development, and later, in sperm production. However, UV is necessary for the synthesis of vitamin D, a nutrient essential for calcium absorption. Thus, in northern climes, humans may have responded positively to mutations that crippled the melanin-producing gene, for this would have allowed them to make good use of their limited light levels. Further support for this hypothesis may involve human sex differences: women around the globe tend to be fairer skinned than men. This makes sense, as women would need more UV light to synthesize more vitamin D during gestation and lactation periods.

Humans and Endogenous Retroviruses

The recent sequencing of the human genome revealed that approximately 45% of our genome consists of transposable elements, and another 8% of retroviral sequences. One class of the endogenous retroviruses, the human endogenous retrovirus K grouping, *HERV-K* (discussed in Chapter 5), has been used to test and confirm primate phylogenies. HERVs may be medically relevant as well: many alleles that confer resistance to retroviral infection in the mouse (a model organism for human physiology) include ERV sequences. This suggests that these endoparasites paid their way in our genomes by conferring resistance to other, ecologically relevant, parasites.

Human Evolution: Anatomical and Developmental Evidence

Vestigial Body-Parts

Humans, like all other organisms, are living museums, full of useless parts that are remnants of and lessons about our evolutionary histories (Chapter 6). Humans have more than 100 non-molecular vestigial structures. For example,

Our body-hair has no known function. Tiny muscles at the base of hairs helped our ancestors move their hair and thereby regulate their insulation and look larger to predators. However, in humans these muscles do nothing but give us "goose bumps."

Humans' ears have three extrinsic muscles that enable us to move our ears independently of our heads. In animals such as dogs and rabbits, orienting ears could help the organisms detect potential threats. In humans, however, these muscles do little more than enable us to wiggle our ears.

Like the other great apes, humans don't have tails, but we do have bones similar to those that form the tails in other animals. These bones, which are the terminal 3–5 vertebrae, are fused into a tail-like structure (3–12 cm long) called the coccyx. These bones are what are left of a tail that other mammals still use for balance and communication. We also have remnants of muscles for moving our vestigial tails.

Human Development and Evolutionary History

Human development also provides powerful clues to our past. In fact, differences in the timing and extent of master-gene activation may explain much of the difference between humans and the other great apes.

M. C. King and A. C. Wilson, in their 1975 work on DNA sequence similarities between humans and chimps, noted that "a relatively small number of genetic changes in systems controlling the expression of genes may account for major organisational differences between human and chimpanzees."

For example, the FOXP2 protein regulates the expression of many other genes and appears to play a key role in speech development. Humans, with our sophisticated language, have a FOXP2 protein that differs from that of the chimps by only two amino-acid substitutions. (Neandertals, however, shared our *FOXP2* gene, a fact that suggests they could speak.)

Genetic mistakes can reveal the extent of the genes we share with the other apes, yet no longer express. For example, a repressed gene is occasionally activated and a human baby is born with a fully formed tail.

Human Behavior from an Evolutionary Perspective

Behavior as an Adaptation in Humans

Human behavior provides some of the most persuasive evidence for the importance of natural selection in shaping our evolution. The analysis of the evolution of human behavior is known by many names—human sociobiology, evolutionary psychology, human behavioral ecology—but is nonetheless a relatively new discipline. However, in spite of its nascent status, much evidence confirms that human behaviors are shaped by selection just as are human pigments and tailbones. Humans exhibit behaviors that increase the chances of survival and reproduction: we avoid foods that smell of decay; we cry when we need attention; and we are afraid of heights, large dogs, and snakes—the list goes on and on.

We also exhibit behaviors that were presumably adaptive in an evolutionary context but have apparently lost their utility. One example may be the suite of behaviors referred to as attention deficit/hyperactivity disorder (ADHD), which is found in 5%–10% of people worldwide and is generally characterized by novelty-seeking behaviors, impulsivity, and distractibility. A genetic variant in a dopamine receptor (*DRD4-7R*) is associated with many diagnoses of ADHD, and it may have been selected *for* in our evolutionary past. Individuals may gain fitness from novelty-seeking behavior (e.g., the ability to exploit new habitats and food sources), as well as unpredictability (an advantage in combat) and demonstrated need (e.g., mothers may direct their attention to demanding children more than to their more contented siblings). Thus, a gene associated with behaviors considered maladaptive in many contemporary cultures could still be the result of positive selection. Similarly, some researchers have suggested that schizophrenia could also be a malignant by-product of genes that were positively selected.

Kin Preference in Human Behavior

Like the meerkat helpers, the prairie dogs that sound alarms, and the soldier termites (Chapter 7), humans exhibit behaviors that increase inclusive fitness. We sacrifice for our offspring, our siblings, and our siblings' children, and have evolved kin-recognition mechanisms to facilitate preferential behaviors. A contemporary analysis of inheritance distribution from 1,000 wills of deceased individuals tested whether humans used their wealth to maximize their overall fitness. The data confirmed that, on average, people allocated very little (only 7.7%) of their estate to non-kin; also, the kin that were beneficiaries were mostly close relatives (46.3% of the total went to $r = 0.5$ individuals such as offspring, while only 8.3% went to $r = 0.25$ individuals, and 0.6% went to $r = 0.125$ relatives such as cousins; r values are introduced in Chapter 7). Finally, people bequeathed far more of their resources to those individuals with higher reproductive potential, such as younger offspring instead of older siblings.

The converse of kin preference is non-kin abuse, which is demonstrated in many organisms (e.g., infanticidal lions and chimps), including humans. For example, a study of preschool children found that those living with a stepparent were 40-times more likely to be abused than were those living with both biological parents. Likewise, a similar study of young child (<5 years old) fatalities due to abuse and neglect found that children living with an unrelated adult (most often, a male stepparent) are 8-times more likely to die from maltreatment than are their counterparts in either a single-mother home or a household with both biological parents.

Mating in the Evolved Human

Like elephant seals, sand lizards, and ground squirrels (Chapter 7), humans compete for mating opportunities. This competition can be overt—derogation or violence toward perceived same-sex rivals—or more subtle. For example, males have lower, deeper voices than females, a trait that appears to be the result of male competition for dominance and access to females. Evidence for this includes the fact that men rate as more dominant males with lower voices (both natural and experimentally manipulated). Also, males modulate the pitch of their voices to accommodate the presence of other males: men who perceive themselves as physically superior to a competitor lower their voices in his presence, whereas in the presence of a more dominant male, men raise the pitch of their voices. Females appear to be receptive to these low voices, especially as they reach sexual maturity. Voice pitch may be a reliable indicator of a male's testosterone level, for low-frequency voices show a significant positive correlation to circulating testosterone.

Like the spiders that bear nuptial gifts (Chapter 7), humans entice potential mates with resources. And like fiddler crabs that construct showy sand pillars, or fireworms that glow, we display our potential to our sexual targets. Indeed, males are more likely to display resources (e.g., buying a woman an expensive meal, driving an expensive car, flashing money) or indicate their ability to acquire resources (e.g., discussing financial prospects, bragging about status at work) than are women. Women are more likely to exhibit coyness, chastity, and physical desirability; this finding is consistent with the different mate-choice mechanics proposed for an organism in which females are limited, have reduced opportunities for mating, and are expected to choose males that advertise provisioning ability. These displays have evolved in tandem with sex-specific mate-preferences: in a study of over 10,000 people in over 100 cultures around the globe, males consistently expressed a preference for youth (and reproductive potential), while women preferred males with good financial prospects (Figure 9.6).

Men and women also display fitness-enhancing behaviors that vary with a woman's ovulatory cycle. A human female ovulates approximately every 28 days. Because women are most likely to conceive offspring during ovulation, we would predict that women would also be most selective about their mate choices during ovulation. Many recent studies confirm this prediction:

- During ovulation (i.e., the most fertile period), women show an increased preference for more masculine faces and voices, two qualities that are a reflection of testosterone levels in the male.
- Ovulating women are more attracted to more symmetrical male faces (symmetry is a hypothesized indicator of genetic quality) than are their non-ovulating peers. Fertile women also express an increased preference for the scent of symmetrical men (as measured by their ratings of the smell of previously worn t-shirts).
- When ovulating, women prefer creativity to wealth in a short-term (but not long-term) potential mate. This preference is minimized when women are not fertile.
- Ovulating women show an increased interest in social activity, or "ranging behavior," a preference that would presumably put them in contact with more potential mates.
- A 2011 study of women's shopping-preferences indicates women dress more provocatively when they are ovulating. In an experimental shopping task, fertile women were more likely to choose short skirts over long skirts, plunging necklines over modest necklines, and dramatic accessories over subtle. These tendencies were augmented by viewing images of attractive women, yet were unaffected by viewing images of unattractive women. In addition, when told these attractive women lived far away, shopping preferences did not vary with fertility; viewing images of supposedly local women, however, was associated with making more provocative choices. These behaviors suggest that women are primed by the presence of attractive,

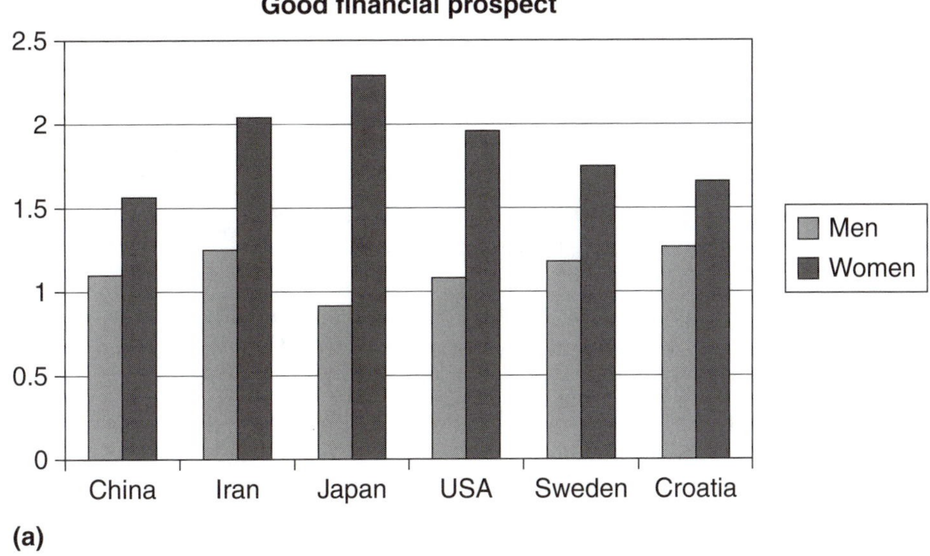

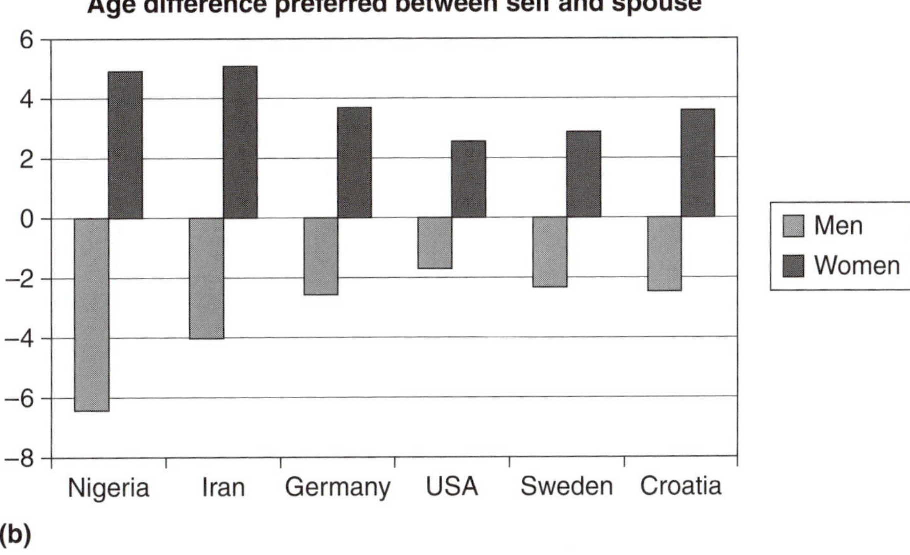

Figure 9.6a and b Human mating preferences. Males and females differ in their mating preferences, in ways that span cultures and suggest biological underpinnings. (a) In six countries sampled, women were more likely to rate as important—on a three-point scale, with "3" being essential—good financial prospects in a potential mate. (b) Cross-culturally, women prefer older men (which may ultimately reflect survival ability, or may be linked to resource acquisition), and men prefer younger women (with higher reproductive potential). Data are relative to the individual's own age; for example, Nigerian women, on average, prefer a spouse who is over four years older than themselves, while Nigerian men prefer a spouse at least six years younger than themselves. (D. M. Buss. 2006, "Strategies of Human Mating," *Psychological Topics* 15 (2): 239–260.)

local women to compete for available males—phenomena that are most pronounced during peak fertility when the costs of competition are greater.
- At peak fertility, women in a relationship show a reduced commitment toward that relationship, and an increased interest in extra-pair copulation (cheating). Cheating exacts a cost (e.g., abandonment by, or injury from, the primary mate) that may only be worth the risk when conception is most likely.

Human males appear to take fertility status into account in their behavior toward women. Men should be most receptive to a woman when she is most fertile, and also most alert to possible rivalry from other males. Again, a growing body of work supports these predictions:

- Women and men report an increase in the men's mate-guarding behavior (e.g. increased attention, vigilance, jealousy, etc.) during a woman's fertility peak.
- Men rate most high the scent of ovulating women, and prefer composite images of fertile women to composite images of non-ovulating women.
- Men act on this ability to distinguish fertile women. For example, one recent study documented that men tipped female sex-performers (e.g., strippers, lap dancers) twice as much when the women were ovulating than when they were not ovulating (Figure 9.7).

Many primates advertise peak fertility with colorful displays (especially red ones) around their genitalia or faces. This reddening is associated with the elevated estrogen-progesterone ratios that accompany ovulation and fertility. Just like the nonhuman primates, human females and males exploit the sex appeal of red. Women are more likely to wear red when they are ovulating, and the use of red cosmetics (lipstick and blush) may mimic an estrogenic signal. Men also rate women in red clothes as more sexually appealing. Men evaluating the attractiveness of women, seen as images on white versus red backgrounds, rated women on red backgrounds as more attractive; however, when queried about the role of color, they indicated that it played little part in their assessment.

Paternity Assurance in the Human Male

Like post-copulatory snow crabs (Chapter 7), human males exhibit behaviors best understood as mate-guarding to assure paternity. Globally, males prevent females from copulating with other males by preventing access (e.g., via enforced sequestration, or suturing of the female's external genitalia), discouraging cheating (e.g., threats of violence, actual violence, removal of the clitoris), or elevated vigilance (e.g., calling a mate frequently,

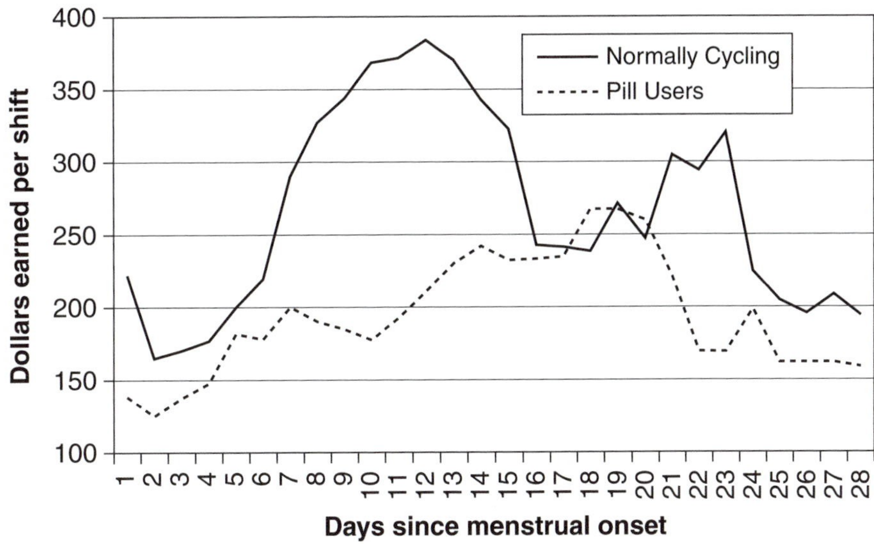

Figure 9.7 Ovulating strippers make more money than their non-ovulating counterparts. On average, female exotic dancers report greater earnings when they are at peak fertility. This may be due to signals the women are sending to their clients, an ability for men to perceive fertility as more appealing, or a combination of factors. (Reprinted from G. Miller, J. M. Tybur, and B. D. Jordan, 2007, "Ovulatory Cycle Effects on Tip Earnings by Lap Dancers: Economic Evidence for Human Estrus?" *Evolution and Human Behavior* 28 (6): 375–381. Copyright (c) 2007, with permission from Elsevier.)

accompanying her in public). For these behaviors to result from selection, we would expect them to be disproportionately directed at women of high reproductive status, an expectation that has been confirmed repeatedly. For example, a study of nearly 4,000 instances of male-perpetrated violence towards a female partner revealed that women of reproductive age were almost 10-times more likely to be victims of aggression than were older women.

Women are aware of the importance of paternity assurance, as demonstrated in studies whereby family members report on their newborn's appearance. Mothers, fathers, and relatives on both sides are significantly more likely to detect a paternal resemblance than a maternal one. This is a proclivity with obvious fitness advantages for both sides if questionable paternity could result in abandonment of the family or death of the infant. For example, a study of New Mexico men showed that males expressing low confidence in paternity were more likely to divorce their spouses than were men who were highly confident of their paternity. A similar study indicated that abortions were more likely for pregnancies with low paternity confidence. Also, studies of men in a domestic-violence program found that the number of times a man could recall hearing from others that his child looked like him was positively correlated with his reported quality of the father-child relationship, and negatively correlated with the severity of injuries the man inflicted on the child's mother.

Parental Favoritism as an Adaptation

Like house sparrows and eastern bluebirds, humans disproportionately allocate resources to those offspring most likely to increase the parents' overall fitness. Much of this differential investment hinges on the sex of the child, and can often be interpreted in light of the Trivers-Willard hypothesis (discussed in Chapter 7). Specifically, parents will invest more in offspring of the sex most likely to be reproductively successful. A male bias is frequently assumed, and often supported—for example, a study of over 25,000 children in India in the early 1990s indicated that males were more likely than females to be fully immunized and well fed. This preference for males may be adaptive, given a male's seemingly limitless ability to procreate. However, males are risky as well—a low-quality male may not reproduce at all, whereas a low-quality female can usually become a mother (a phenomenon introduced in Chapter 7). One prediction—that high-status parents will invest more in males, while low-status parents will invest more in females—has some empirical support. Several studies have documented that low-status individuals bias their investment in daughters (possibly helping the daughters marry higher-ranking males), whereas high-status individuals typically show an extreme bias toward sons.

Dramatic evidence of a sex bias is found in studies of human sex ratios, which, given the assumption of a 50% chance of conceiving a boy or a girl, is expected to hover around 1 male:1 female at birth. Dramatic deviations from this expected ratio have been observed. Some cultures demonstrate an overt bias toward male offspring, and studies have cited millions of "missing girls" in China, India and Afghanistan, among other places. Abortion records from a study of hospitals in Bombay reveal that, of 8,000 abortions performed after sex identification, 7,999 were of female fetuses (see Figure 9.8). In China, there is an established practice of killing baby girls at birth, and of the babies adopted from China, essentially all of them are female.

More subtle are reports of presumably unintentional sex-ratio manipulation, whereby environmental influences appear to affect human sex ratios. Stressful situations seem to favor female births, with slightly (albeit noticeably) lowered male-female sex ratios observed after a serious earthquake in Japan, the ten-day war in Slovenia, and the attacks on the World Trade Center in New York in September 2001. A comparison of sex ratios in East and West Germany in the latter part of the 20th century indicated that, when East Germany's economy collapsed in 1991, the male-female ratio was at its lowest, and notably lower than in West Germany during the same time period. In contrast, power, money, and nutritional status appear

Figure 9.8 In India, where over 99% of aborted fetuses (in which the prenatal sex is known) are female, recent social pressure has led to penalties for sex-specific abortions. This healthcare clinic advertises its inability to conduct prenatal sex-determination tests. (AP Photo/Gurinder Osan)

to skew sex ratios in favor of males, and in some of these cases the results are dramatic. For example, a study of mothers in rural Ethiopia demonstrated that well-fed women had more sons—the male-female sex ratio was nearly 1.6 in the healthiest group (Figure 9.9). Recent observations of the offspring of billionaires emphasize the role of paternal status: male billionaires have children with an elevated male-female sex ratio, while the offspring of female billionaires does not deviate greatly from 1:1. Various suggestions exist for the mechanisms underlying this type of sex-ratio manipulation, including the hypothesis that the stress-related hormone, cortisol, disproportionately induces miscarriages of male fetuses. Regardless of the proximate cause, human sex-ratio adjustment is consistent with patterns observed in other animals and provides strong evidence for humans as evolved beings.

Human Evolution: Coevolutionary Evidence

Given the extreme fitness consequences of avoiding predators and acquiring prey, predators and prey must have been powerful agents of selection in our recent past. Several innate fear-responses have been documented,

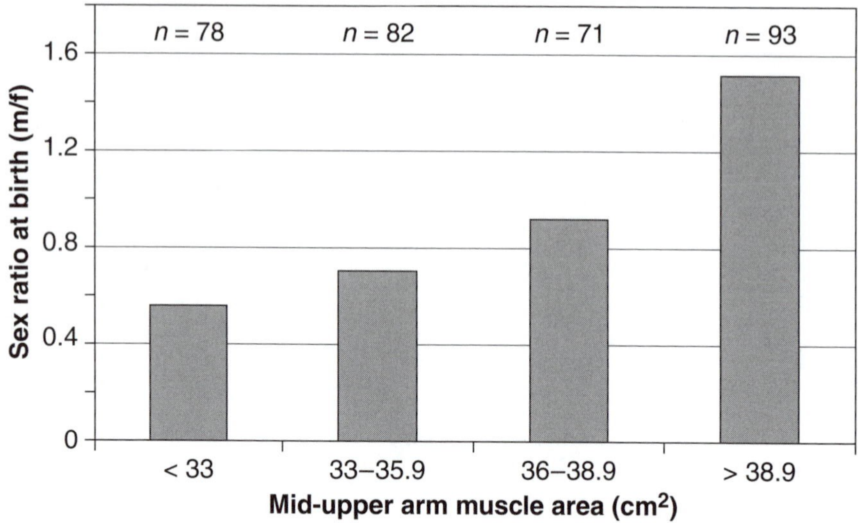

Figure 9.9 Humans in good condition (here reflected as those women with strong arms) have a male-biased sex ratio. This finding is consistent with the Trivers-Willard hypothesis for sex-ratio manipulation (discussed in Chapter 7). (Reprinted from M. A. Gibson and R. Mace. 2003, "Strong Mothers Bear More Sons in Rural Ethiopia." *Proceedings, Royal Society B: Biological Sciences* 270 [Suppl. 1]: S108. With permission from the Royal Society.)

such as an inborn fear of snakes, spiders, and looming objects. Humans appear to have heightened visual sensitivity to potentially dangerous organisms, a phenomenon suggested by studies in which humans locate pictures of snakes and spiders in a matrix of images more quickly than pictures of harmless organisms.

The Shared Evolution of Humans and *Plasmodium*

As host species, we have ecologically relevant adaptations against parasitism. *Plasmodium* species, which cause malaria, are some of our best-studied endosymbionts. *Plasmodium* parasites are highly specific, with one genus of mosquitoes as the only vector, and humans as the only vertebrate host (and the only host in which they can complete their life cycles). This level of parasite-host specificity, especially in such a successful pathogen (e.g., malaria affects hundreds of millions of people worldwide, and kills over one million per year), suggests a long history of reciprocal adaptation. *Plasmodium* species, for their part, engage in some sophisticated host-manipulation (discussed in Chapter 8). In turn, humans possess a suite of molecular mutations that have been positively selected because of their anti-malarial properties. In fact, several of these mutations confer a risk, if not outright harm, to the bearer. For example, a mutation in hemoglobin, a protein in red blood cells responsible for conveying oxygen (introduced in Chapter 5), confers on its bearer a harmful and potentially fatal disease—sickle-cell anemia—when two copies of the allele are inherited. However, an individual who inherits only one copy of the allele has red blood cells that are inhospitable to *Plasmodium,* and thus decreased susceptibility to malaria. In areas where malaria is rampant, selection has maintained high levels of the sickle-cell alleles in the population, thus explaining why sickle-cell anemia still persists at such high frequencies, and why malaria and sickle-cell anemia co-occur at high frequencies.

Similarly, a mutation in the red blood cell enzyme glucose-6-phosphate dehydrogenase confers resistance to malaria in millions of people in Africa and the Middle East. However, this mutation also exacts a cost on its bearer, in that he or she cannot digest a key regional food item, fava beans. Yet another mutation with anti-malarial properties, the nonfunctioning Duffy Antigen Receptor for Chemokines (DARC), blocks *Plasmodium* from entering red blood cells but may make individuals more susceptible to HIV.

Together these three mutant alleles paint a picture of human evolution mechanistically on par with evolution in all other organisms—an immediate and powerful selection pressure (such as a deadly disease) will win out over any future benefits (such as resistance to HIV) or costs (e.g., nutritional) to the organism.

Coevolution with Pathogens and HIV Resistance

Our coevolved interactions with pathogens have also selected for pseudogenes. For example, individuals with a mutation in the *CCR5* gene express a reduced susceptibility to HIV infection (and therefore a reduction in the symptoms associated with AIDS). The *CCR5* membrane protein permits the entry of HIV viral particles into the host's cell, and a deletion-mutation (CCR5-Δ32) at the site disallows entry and infection. This mutation is present in high frequencies (e.g., 10% or greater) in human populations in Europe, but is virtually absent in African, Asian, and American Indian populations, thus suggesting a fairly recent origin. The geographic distribution of the mutation supports an origin in Northern Europe approximately 1,200 years ago, and a southern migration associated with the Viking conquests of the 9th–11th centuries C.E. Also, we have only recently—in the last century—begun our symbiosis with HIV, and given the long generation times in humans, it is unlikely that HIV could have been the agent of selection for this mutation. So, where did this recent mutation, to an even more recent viral pathogen, come from?

Initially, the mutation may have conferred resistance to a pathogen similar to HIV, such as a similar virus. Evidence for smallpox as the selective force includes similarities between the pox viruses and HIV, epidemics of smallpox that just predate the suggested rise of the CCR5-Δ32 mutation, and a geographic distribution that reflects a coevolutionary past between the mutation and the smallpox virus. If the prevalence of the CCR5-Δ32 mutation in some human populations is due to a recent history with a now-eradicated virus, then the resistance to HIV conferred by the mutation is an example of an exaptation (introduced in Chapter 1)—a feature that was originally adaptive in one context (e.g., resistance to smallpox) but has been co-opted for use in another (e.g., resistance to AIDS). The geographic distribution of the mutation—in Europe, rather than in Africa or Southeast Asia where life expectancies are much lower due to AIDS—also emphasizes that mutations do not arise on demand, but rather are generated randomly and then constrained by time and place.

Coevolution with Pathogens and Urban Living

Urban living leads to increased disease transmission, and presumably selection for resistance has operated throughout human history. Recent work by Mark Thomas and colleagues demonstrated a correlation between how long a population has lived in cities and the presence of an allele that confers resistance to tuberculosis. In a study of 17 human groups, the lowest incidences of the resistant allele (specifically, a deletion mutation; see Chapter 4) are seen in Malawian and Saami populations, which have only begun living in established settlements within the past few hundred years. In contrast, the Iranians, Italians, and Turks—settled for several millennia—have the highest incidence of the allele. This, and numerous other examples, illustrate that humans, like so many other coevolved organisms, coevolve in ecologically relevant ways with pathogens.

Human Exploitation of Third-Party Organisms

Just as decorator crabs adorn their shells with algae to deter fish predators (Chapter 8), humans have learned to exploit third-party organisms for our benefit. The use of spices is highly variable in human societies, with some cultures being known for their heavily spiced foods, and others tending toward more bland fare. These differences cannot be explained by the availability of local plants alone, as hundreds of years of global spice trade can attest. Rather, many common spices have antimicrobial properties and may decontaminate foods prior to consumption. Also, many foods we eat are attractive to microbes, and we therefore run the risk of food poisoning—and death—from bacterial toxins. This is especially true with the meat we eat, as animal carcasses are quickly consumed by microbial decomposers. Plants, on the other hand, have cell walls and antimicrobial compounds that afford some postmortem protection against contamination. Given that the use of spices exerts a cost—for example, secondary plant compounds can be toxic in large doses, serving as allergens or mutagens—the coevolved human should not spice foods that are otherwise safe to consume. To test this prediction, a recent study reviewed traditionally prepared foods from 36 countries, comparing the spices used in 4,578 meat dishes to those used in 2,129 vegetable dishes. Meat-based recipes use an average of 3.9 spices per recipe, significantly more than the 2.4 average for vegetable dishes. Also, spice use is highest in warmer (and more microbe-friendly) climates, with places such as India, Ethiopia, and Indonesia averaging over 7 spices per dish; in cooler Ireland, Poland, and Finland, recipes typically call for fewer than 4 spices. We can infer that ancestors who inhabited areas of high risk to food-borne pathogens *and* enjoyed the sensations of consuming certain spices (such as onion, pepper, lime, garlic, and ginger) were rewarded with health and increased fitness.

The Coevolved Human and Mate Choice

Further evidence of our evolutionary history with disease-causing organisms can be seen in our mate-screening tactics. Just as song learning in birds may advertise disease resistance to potential mates, human appearance and odors appear to convey specific information about a suitor's immune system. Several studies have documented how females prefer the scent and appearance of males who differ in their alleles of the major histocompatibility complex (or MHC); these are immune system genes, and strategic mate-choices can boost a female's fitness by conferring an immune benefit to her offspring. In addition, women pair-bonded to men that are MHC incompatible are more likely to seek extra-pair copulations (i.e., cheat on their mate) than are those mated with males that are more MHC complementary.

If pathogen resistance is conveyed as physical attractiveness by various means, then we would predict that people in areas with a high incidence of infectious disease would be more discriminating with respect to a potential mate's appearance. An analysis of the mating preferences of over 7,000 people from 29 globally distributed populations showed just that: there is a correlation between the expressed importance of physical beauty and the prevalence of pathogens in a given locality.

Human Coevolution and Lactose Tolerance

Several lines of evidence suggest reciprocal adaptations between humans and the animals (e.g., dogs, horses) and plants (e.g., corn, potatoes) we've domesticated. For example, our recent history with dairy farming appears to have had a profound effect on our genomes. Most humans lose their ability to produce lactase, an enzyme necessary for digesting lactose (milk sugar), soon after weaning. However, in areas with a history of keeping animals (e.g., cows, goats) for milk, humans are more likely to persist in their production of lactase into adulthood, a trait that is genetically variable and therefore subject to selection.

Genetic analysis of the gene encoding lactase has revealed different genetic signatures for lactase persistence in European versus African milk-drinking populations, suggesting more than one origin for this trait. In other words, humans have converged on this ability to digest milk due to similar coevolutionary pressures associated with dairy farming. Milked animals have experienced parallel evolution with dairy farmers in their short (5,000–10,000-year) history: cows from areas steeped in a milk-drinking history—for example, north central Europe—have greater milk gene diversity than do cows in which dairy farming is relatively new (e.g., the Mediterranean, the Middle East). The story of the human lactose gene is one of humans, biogeography, molecules, behavior (culture), and coevolution—an elegant example emphasizing our ongoing participation in our own evolution. Time will tell how this story unfolds.

Summary

Humans once shared Earth with other hominins. We competed and interbred with them, and—not surprisingly—our DNA now contains their genes. All other species of hominins, like so many species of other genera, are now extinct. Although we often view human evolution to be a linear progression from a crouching ape that culminates with a tall, erect man, modern humans are actually just another evolutionary experiment in a hominin family tree. For now, we are the last hominin standing.

In the final analysis, humans are puzzling and complex creatures. Why do we show such geographic variability, especially with respect to skin color? Why do we get goosebumps when we're scared? Why are deep voices so sexy in men? Why are we more likely to abort baby girls? Why do billionaires have so many sons? Why are we so afraid of snakes and spiders? Why is prepared food so spicy in India, and so bland in Ireland? Why can some of us eat ice cream, while others get sick? The ultimate answer, of course, is evolution. Like the Madagascan orchid and its pollinating moth, like pocket mice, dragonflies, horseshoe crabs, feathered dinosaurs, marine iguanas, stickleback fish, yucca, mistletoe, *Plasmodium,* meerkats, reed warblers, cleaner wrasse and chimpanzees, we have been, and continue to be, evolved.

Appendix 1

Charles Darwin's *On the Origin of Species* (1859, first edition), Chapters IV ("Natural Selection") and XIV ("Recapitulation and Conclusion")

IV. Natural Selection

How will the struggle for existence, discussed too briefly in the last chapter, act in regard to variation? Can the principle of selection, which we have seen is so potent in the hands of man, apply in nature? I think we shall see that it can act most effectually. Let it be borne in mind in what an endless number of strange peculiarities our domestic productions, and, in a lesser degree, those under nature, vary; and how strong the hereditary tendency is. Under domestication, it may be truly said that the whole organisation becomes in some degree plastic. Let it be borne in mind how infinitely complex and close-fitting are the mutual relations of all organic beings to each other and to their physical conditions of life. Can it, then, be thought improbable, seeing that variations useful to man have undoubtedly occurred, that other variations useful in some way to each being in the great and complex battle of life, should sometimes occur in the course of thousands of generations? If such do occur, can we doubt (remembering that many more individuals are born than can possibly survive) that individuals having any advantage, however slight, over others, would have the best chance of surviving and of procreating their kind? On the other hand, we may feel sure that any variation in the least degree injurious would be rigidly destroyed. This preservation of favourable variations and the rejection of injurious variations, I call Natural Selection. Variations neither useful nor injurious would not be affected by natural selection, and would be left a fluctuating element, as perhaps we see in the species called polymorphic.

We shall best understand the probable course of natural selection by taking the case of a country undergoing some physical change, for instance, of climate. The proportional numbers of its inhabitants would almost immediately undergo a change, and some species might become extinct. We may conclude, from what we have seen of the intimate and complex manner in which the inhabitants of each country are bound together, that any change in the numerical proportions of some of the inhabitants, independently of the change of climate itself, would most seriously affect many of the others. If the country were open on its borders, new forms would certainly immigrate, and this also would seriously disturb the relations of some of the former inhabitants. Let it be remembered how powerful the influence of a single introduced tree or mammal has been shown to be. But in the case of an island, or of a country partly surrounded by barriers, into which new and better adapted forms could not freely enter, we should then have places in the economy of nature which would assuredly be better filled up, if some of the original inhabitants were in some manner modified; for, had the area been open to immigration, these same places would have been seized on by intruders. In such case, every slight modification, which in the course of ages chanced to arise, and which in any way favoured the individuals of any of the species, by better adapting them to their altered conditions, would tend to be preserved; and natural selection would thus have free scope for the work of improvement.

We have reason to believe, as stated in the first chapter, that a change in the conditions of life, by specially acting on the reproductive system, causes or increases variability; and in the foregoing case the conditions of life are supposed to have undergone a change, and this would manifestly be favourable to natural selection, by giving a better chance of profitable variations occurring; and unless profitable variations do occur, natural selection can do nothing. Not that, as I believe, any extreme amount of variability is necessary; as man can certainly produce great results by adding up in any given direction mere individual differences, so could Nature, but far more easily, from having incomparably longer time at her disposal. Nor do I believe that any great physical change, as of climate, or any unusual degree of isolation to check immigration, is actually necessary to produce new and unoccupied places for natural selection to fill up by modifying and improving some of the varying inhabitants. For as all the inhabitants of each country are struggling together with nicely balanced forces, extremely slight modifications in the structure or habits of one inhabitant would often give it an advantage over others; and still further modifications of the same kind would often still further increase the advantage. No country can be named in which all the native inhabitants are now so perfectly adapted to each other and to the physical conditions under which they live, that none of them could anyhow be improved; for in all countries, the natives have been so far conquered by naturalised productions, that they have allowed foreigners to take firm possession of the land. And as foreigners have thus everywhere beaten some of the natives, we may safely conclude that the natives might have been modified with advantage, so as to have better resisted such intruders.

As man can produce and certainly has produced a great result by his methodical and unconscious means of selection, what may not nature effect? Man can act only on external and visible characters: nature cares nothing for appearances, except in so far as they may be useful to any being. She can act on every internal organ, on every shade of constitutional difference, on the whole machinery of life. Man selects only for his own good; Nature only for that of the being which she tends. Every selected character is fully exercised by her; and the being is placed under well-suited conditions of life. Man keeps the natives of many climates in the same country; he seldom exercises each selected character in some peculiar and fitting manner; he feeds a long and a short beaked pigeon on the same food; he does not exercise a long-backed or long-legged quadruped in any peculiar manner; he exposes sheep with long and short wool to the same climate. He does not allow the most vigorous males to struggle for the females. He does not rigidly destroy all inferior animals, but protects during each varying season, as far as lies in his power, all his productions. He often begins his selection by some half-monstrous form; or at least by some modification prominent enough to catch his eye, or to be plainly useful to him. Under nature, the slightest difference of structure or constitution may well turn the nicely-balanced scale in the struggle for life, and so be preserved. How fleeting are the wishes and efforts of man! how short his time! and consequently how poor will his products be, compared with those accumulated by nature during whole geological periods. Can we wonder, then, that nature's productions should be far "truer" in character than man's productions; that they should be infinitely better adapted to the most complex conditions of life, and should plainly bear the stamp of far higher workmanship?

It may be said that natural selection is daily and hourly scrutinising, throughout the world, every variation, even the slightest; rejecting that which is bad, preserving and adding up all that is good; silently and insensibly working, whenever and wherever opportunity offers, at the improvement of each organic being in relation to its organic and inorganic conditions of life. We see nothing of these slow changes in progress, until the hand of time has marked the long lapses of ages, and then so imperfect is our view into long past geological ages, that we only see that the forms of life are now different from what they formerly were.

Although natural selection can act only through and for the good of each being, yet characters and structures, which we are apt to consider as of very trifling importance, may thus be acted on. When we see leaf-eating insects green, and bark-feeders mottled-grey; the alpine ptarmigan white in winter, the red-grouse the colour of heather, and the black-grouse that of peaty earth, we must believe that these tints are of service to these birds and insects in preserving them from danger. Grouse, if not destroyed at some period of their lives, would increase in countless numbers; they are known to suffer largely from birds of prey; and hawks are guided by eyesight to their prey, so much so, that on parts of the Continent persons are warned not to keep white pigeons, as being the most liable to destruction. Hence I can see no reason to doubt that natural selection might be most effective in giving the proper colour to each kind of grouse, and in keeping that colour, when once acquired, true and constant. Nor ought we to think that the occasional destruction of an animal of any particular colour would produce little effect: we should remember how essential it is in a flock of white sheep to destroy every lamb with the faintest trace of black. In plants the down on the fruit and the colour of the flesh are considered by botanists as characters of the most trifling importance: yet we hear from an excellent horticulturist, Downing, that in the United States smooth-skinned fruits suffer far more from a beetle, a curculio, than those with down; that purple plums suffer far more from a certain disease than yellow plums; whereas another disease attacks yellow-fleshed peaches far more than those with other coloured flesh. If, with all the aids of art, these slight differences make a great difference in cultivating the several varieties, assuredly, in a state of nature, where the trees would have to struggle with other trees and with a host of enemies, such differences would effectually settle which variety, whether a smooth or downy, a yellow or purple fleshed fruit, should succeed.

In looking at many small points of difference between species, which, as far as our ignorance permits us to judge, seem to be quite unimportant, we must not forget that climate, food, &c., probably produce some slight and direct effect. It is, however, far more necessary to bear in mind that there are many unknown laws of correlation of growth, which, when one part of the organisation is modified through variation, and the modifications are accumulated by natural selection for the good of the being, will cause other modifications, often of the most unexpected nature.

As we see that those variations which under domestication appear at any particular period of life, tend to reappear in the offspring at the same period; for instance, in the seeds of the many varieties of our culinary and agricultural plants; in the caterpillar and cocoon stages of the varieties of the silkworm; in the eggs of poultry, and in the colour of the down of their chickens; in the horns of our sheep and cattle when nearly adult; so in a state of nature, natural selection will be enabled to act on and modify organic beings at any age, by the accumulation of profitable variations at that age, and by their inheritance at a corresponding age. If it profit a plant to have its seeds more and more widely disseminated by the wind, I can see no greater difficulty in this being effected through natural selection, than in the cotton-planter increasing and improving by selection the down in the pods on his cotton-trees. Natural selection may modify and adapt the larva of an insect to a score of contingencies, wholly different from those which concern the mature insect. These modifications will no doubt affect, through the laws of correlation, the structure of the adult; and probably in the case of those insects which live only for a few hours, and which never feed, a large part of their structure is merely the correlated result of successive changes in the structure of their larvae. So, conversely, modifications in the adult will probably often affect the structure of the larva; but in all cases natural selection will ensure that modifications consequent on other modifications at a different period of life, shall not be in the least degree injurious: for if they became so, they would cause the extinction of the species.

Natural selection will modify the structure of the young in relation to the parent, and of the parent in relation to the young. In social animals it will adapt the structure of each individual for the benefit of the community; if each in consequence profits by the selected change. What natural selection cannot do, is to modify the structure of one species, without giving it any advantage, for the good of another species; and though statements to this effect may be found in works of natural history, I cannot find one case which will bear investigation. A structure used only once in an animal's whole life, if of high importance to it, might be modified to any extent by natural selection; for instance, the great jaws possessed by certain insects, and used exclusively for opening the cocoon or the hard tip to the beak of nestling birds, used for breaking the egg. It has been asserted, that of the best short-beaked tumbler-pigeons more perish in the egg than are able to get out of it; so that fanciers assist in the act of hatching. Now, if nature had to make the beak of a full-grown pigeon very short for the bird's own advantage, the process of modification would be very slow, and there would be simultaneously the most rigorous selection of the young birds within the egg, which had the most powerful and hardest beaks, for all with weak beaks would inevitably perish: or, more delicate and more easily broken shells might be selected, the thickness of the shell being known to vary like every other structure.

Sexual Selection

Inasmuch as peculiarities often appear under domestication in one sex and become hereditarily attached to that sex, the same fact probably occurs under nature, and if so, natural selection will be able to modify one sex in its functional relations to the other sex, or in relation to wholly different habits of life in the two sexes, as is sometimes the case with insects. And this leads me to say a few words on what I call Sexual Selection. This depends, not on a struggle for existence, but on a struggle between the males for possession of the females; the result is not death to the unsuccessful competitor, but few or no offspring. Sexual selection is, therefore, less rigorous than natural selection. Generally, the most vigorous males, those which are best fitted for their places in nature, will leave most progeny. But in many cases, victory will depend not on general vigour, but on having special weapons, confined to the male sex. A hornless stag or spurless cock would have a poor chance of leaving offspring. Sexual selection by always allowing the victor to breed might surely give indomitable courage, length to the spur, and strength to the wing to strike in the spurred leg, as well as the brutal cock-fighter, who knows well that he can improve his breed by careful selection of the best cocks. How low in the scale of nature this law of battle descends, I know not; male alligators have been described as fighting, bellowing, and whirling round, like Indians in a war-dance, for the possession of the females; male salmons have been seen fighting all day long; male stag-beetles often bear wounds from the huge mandibles of other males. The war is, perhaps, severest between the males of polygamous animals, and these seem oftenest provided with special weapons. The males of carnivorous animals are already well armed; though to them and to others, special means of defence may be given through means of sexual selection, as the mane to the lion, the shoulder-pad to the boar, and the hooked jaw to the male salmon; for the shield may be as important for victory, as the sword or spear.

Amongst birds, the contest is often of a more peaceful character. All those who have attended to the subject, believe that there is the severest rivalry between the males of many species to attract by singing the females. The rock-thrush of Guiana, birds of paradise, and some others, congregate; and successive males display their gorgeous plumage and perform strange antics before the females, which standing by as spectators, at last choose the most attractive partner. Those who have closely attended to birds in confinement well know that they often take individual preferences and dislikes: thus Sir R. Heron has described how one pied peacock was eminently attractive to all his hen birds. It may appear childish to attribute any effect to such apparently weak means: I cannot here enter on the details necessary to support this view; but if man can in a short time give elegant carriage and beauty to his bantams, according to his standard of beauty, I can see no good reason to doubt that female birds, by selecting, during thousands of generations, the most melodious or beautiful males, according to their standard of beauty, might produce a marked effect. I strongly suspect that some well-known laws with respect to the plumage of male and female birds, in comparison with the plumage of the young, can be explained on the view of plumage having been chiefly modified by sexual selection, acting when the birds have come to the breeding age or during the breeding season; the modifications thus produced being inherited at corresponding ages or seasons, either by the males alone, or by the males and females; but I have not space here to enter on this subject.

Thus it is, as I believe, that when the males and females of any animal have the same general habits of life, but differ in structure, colour, or ornament, such differences have been mainly caused by sexual selection; that is, individual males have had, in successive generations, some slight advantage over other males, in their weapons, means of defence, or charms; and have transmitted these advantages to their male offspring. Yet, I would not wish to attribute all such sexual differences to this agency: for we see peculiarities arising and becoming attached to the male sex in our domestic animals (as the wattle in male carriers, horn-like protuberances in the cocks of certain fowls, &c.), which we cannot believe to be either useful to the males in battle, or attractive to the females. We see analogous cases under nature, for instance, the tuft of hair on the breast of the turkey-cock, which can hardly be either useful or ornamental to this bird; indeed, had the tuft appeared under domestication, it would have been called a monstrosity.

Illustrations of the Action of Natural Selection

In order to make it clear how, as I believe, natural selection acts, I must beg permission to give one or two imaginary illustrations. Let us take the case of a wolf, which preys on various animals, securing some by craft, some by strength, and some by fleetness; and let us suppose that the fleetest prey, a deer for instance, had from any change in the country increased in numbers, or that other prey had decreased in numbers, during that season of the year when the wolf is hardest pressed for food. I can under such circumstances see no reason to doubt that the swiftest and slimmest wolves would have the best chance of surviving, and so be preserved or selected, provided always that they retained strength to master their prey at this or at some other period of the year, when they might be compelled to prey on other animals. I can see no more reason to doubt this, than that man can improve the fleetness of his greyhounds by careful and methodical selection, or by that unconscious selection which results from each man trying to keep the best dogs without any thought of modifying the breed.

Even without any change in the proportional numbers of the animals on which our wolf preyed, a cub might be born with an innate tendency to pursue certain kinds of prey. Nor can this be thought very improbable; for we often observe great differences in the natural tendencies of our domestic animals; one cat, for instance, taking to catch rats, another mice; one cat, according to Mr. St. John, bringing home winged game, another hares or rabbits, and another hunting on marshy ground and almost nightly catching woodcocks or snipes. The tendency to catch rats rather than mice is known to be inherited. Now, if any slight innate change of habit or of structure benefited an individual wolf, it would have the best chance of surviving and of leaving offspring. Some of its young would probably inherit the same habits or structure, and by the repetition of this process, a new variety might be formed which would either supplant or coexist with the parent-form of wolf. Or, again, the wolves inhabiting a mountainous district, and those frequenting the lowlands, would naturally be forced to hunt different prey; and from the continued preservation of the individuals best fitted for the two sites, two varieties might slowly be formed. These varieties would cross and blend where they met; but to this subject of intercrossing we shall soon have to return. I may add, that, according to Mr. Pierce, there are two varieties of the wolf inhabiting the Catskill Mountains in the United States, one with a light greyhound-like form, which pursues deer, and the other more bulky, with shorter legs, which more frequently attacks the shepherd's flocks.

Let us now take a more complex case. Certain plants excrete a sweet juice, apparently for the sake of eliminating something injurious from their sap: this is effected by glands at the base of the stipules in some Leguminosae, and at the back of the leaf of the common laurel. This juice, though small in quantity, is greedily sought by insects. Let us now suppose a little sweet juice or nectar to be excreted by the inner bases of the petals of a flower. In this case insects in seeking the nectar would get dusted with pollen, and would certainly often transport the pollen from one flower to the stigma of another flower. The flowers of two distinct individuals of the same species would thus get crossed; and the act of crossing, we have good reason to believe (as will hereafter be more fully alluded to), would produce very vigorous seedlings, which consequently would have the best chance of flourishing and surviving. Some of these seedlings would probably inherit the nectar-excreting power. Those in individual flowers which had the largest glands or nectaries, and which excreted most nectar, would be oftenest visited by insects, and would be oftenest crossed; and so in the long-run would gain the upper hand. Those flowers, also, which had their stamens and pistils placed, in relation to the size and habits of the particular insects which visited them, so as to favour in any degree the transportal of their pollen from flower to flower, would likewise be favoured or selected. We might have taken the case of insects visiting flowers for the sake of collecting pollen instead of nectar; and as pollen is formed for the sole object of fertilisation, its destruction appears a simple loss to the plant; yet if a little pollen were carried, at first occasionally and then habitually, by the pollen-devouring insects from flower to flower, and a cross thus effected, although nine-tenths of the pollen were destroyed, it might still be a great gain to the plant; and those individuals which produced more and more pollen, and had larger and larger anthers, would be selected.

When our plant, by this process of the continued preservation or natural selection of more and more attractive flowers, had been rendered highly attractive to insects, they would, unintentionally on their part, regularly carry pollen from flower to flower; and that they can most effectually do this, I could easily show by many striking instances. I will give only one not as a very striking case, but as likewise illustrating one step in the separation of the sexes of plants, presently to be alluded to. Some holly-trees bear only male flowers, which have four stamens producing rather a small quantity of pollen, and a rudimentary pistil; other holly-trees bear only female flowers; these have a full-sized pistil, and four stamens with shrivelled anthers, in which not a grain of pollen can be detected. Having found a female tree exactly sixty yards from a male tree, I put the stigmas of twenty flowers, taken from different branches, under the microscope, and on all, without exception, there were pollen-grains, and on some a profusion of pollen. As the wind had set for several days from the female to the male tree, the pollen could not thus have been carried. The weather had been cold and boisterous, and therefore not favourable to bees, nevertheless every female flower which I examined had been effectually fertilised by the bees, accidentally dusted with pollen, having flown from tree to tree in search of nectar. But to return to our imaginary case: as soon as the plant had been rendered so highly attractive to insects that pollen was regularly carried from flower to flower, another process might commence. No naturalist doubts the advantage of what has been called the "physiological division of labour;" hence we may believe that it would be advantageous to a plant to produce stamens alone in one flower or on one whole plant, and pistils alone in another flower or on another plant. In plants under culture and placed under new conditions of life, sometimes the male organs and sometimes the female organs become more or less impotent; now if we suppose this to occur in ever so slight a degree under nature, then as pollen is already carried regularly from flower to flower, and as a more complete separation of the sexes of our plant would be advantageous on the principle of the division of labour, individuals with this tendency more and more increased, would be continually favoured or selected, until at last a complete separation of the sexes would be effected.

Let us now turn to the nectar-feeding insects in our imaginary case: we may suppose the plant of which we have been slowly increasing the nectar by continued selection, to be a common plant; and that certain insects depended in main part on its nectar for food. I could give many facts, showing how anxious bees are to save time; for instance, their habit of cutting holes and sucking the nectar at the bases of certain flowers, which they can, with a very little more trouble, enter by the mouth. Bearing such facts in mind, I can see no reason to doubt that an accidental deviation in the size and form of the body, or in the curvature and length of the proboscis, &c., far too slight to be appreciated by us, might profit a bee or other insect, so that an individual so characterised would be able to obtain its food more quickly, and so have a better chance of living and leaving descendants. Its descendants would probably inherit a tendency to a similar slight deviation of structure. The tubes of the corollas of the common red and incarnate clovers (Trifolium pratense and incarnatum) do not on a hasty glance appear to differ in length; yet the hive-bee can easily suck the nectar out of the incarnate clover, but not out of the common red clover, which is visited by humble-bees alone; so that whole fields of the red clover offer in vain an abundant supply of precious nectar to the hive-bee. Thus it might be a great advantage to the hive-bee to have a slightly longer or differently constructed proboscis. On the other hand, I have found by experiment that the fertility of clover greatly depends on bees visiting and moving parts of the corolla, so as to push the pollen on to the stigmatic surface. Hence, again, if humble-bees were to become rare in any country, it might be a great advantage to the red clover to have a shorter or more deeply divided tube to its corolla, so that the hive-bee could visit its flowers. Thus I can understand how a flower and a bee might slowly become, either simultaneously or one after the other, modified and adapted in the most perfect manner to each other, by the continued preservation of individuals presenting mutual and slightly favourable deviations of structure.

I am well aware that this doctrine of natural selection, exemplified in the above imaginary instances, is open to the same objections which were at first urged against Sir Charles Lyell's noble views on "the modern changes of the earth, as illustrative of geology;" but we now very seldom hear the action, for instance, of the coast-waves, called a trifling and insignificant cause, when applied to the excavation of gigantic valleys or to the formation of the longest lines of inland cliffs. Natural selection can act only by the preservation and accumulation of infinitesimally small inherited modifications, each profitable to the preserved being; and as modern geology has almost banished such views as the excavation of a great valley by a single diluvial wave, so will natural selection, if it be a true principle, banish the belief of the continued creation of new organic beings, or of any great and sudden modification in their structure.

On the Intercrossing of Individuals

I must here introduce a short digression. In the case of animals and plants with separated sexes, it is of course obvious that two individuals must always unite for each birth; but in the case of hermaphrodites this is far from obvious. Nevertheless I am strongly inclined to believe that with all hermaphrodites two individuals, either occasionally or habitually, concur for the reproduction of their kind. This view, I may add, was first suggested by Andrew Knight. We shall presently see its importance; but I must here treat the subject with extreme brevity, though I have the materials prepared for an ample discussion. All vertebrate animals, all insects, and some other large groups of animals, pair for each birth. Modern research has much diminished the number of supposed hermaphrodites, and of real hermaphrodites a large number pair; that is, two individuals regularly unite for reproduction, which is all that concerns us. But still there are many hermaphrodite animals which certainly do not habitually pair, and a vast majority of plants are hermaphrodites. What reason, it may be asked, is there for supposing in these cases that two individuals ever concur in reproduction? As it is impossible here to enter on details, I must trust to some general considerations alone.

In the first place, I have collected so large a body of facts, showing, in accordance with the almost universal belief of breeders, that with animals and plants a cross between different varieties, or between individuals of the same variety but of another strain, gives vigour and fertility to the offspring; and on the other hand, that *close* interbreeding diminishes vigour and fertility; that these facts alone incline me to believe that it is a general law of nature (utterly ignorant though we be of the meaning of the law) that no organic being self-fertilises itself for an eternity of generations; but that a cross with another individual is occasionally perhaps at very long intervals—indispensable.

On the belief that this is a law of nature, we can, I think, understand several large classes of facts, such as the following, which on any other view are inexplicable. Every hybridizer knows how unfavourable exposure to wet is to the fertilisation of a flower, yet what a multitude of flowers have their anthers and stigmas fully exposed to the weather! but if an occasional cross be indispensable, the fullest freedom for the entrance of pollen from another individual will explain this state of exposure, more especially as the plant's own anthers and pistil generally stand so close together that self-fertilisation seems almost inevitable. Many flowers, on the other hand, have their organs of fructification closely enclosed, as in the great papilionaceous or pea-family; but in several, perhaps in all, such flowers, there is a very curious adaptation between the structure of the flower and the manner in which bees suck the nectar; for, in doing this, they either push the flower's own pollen on the stigma, or bring pollen from another flower. So necessary are the visits of bees to papilionaceous flowers, that I have found, by experiments published elsewhere, that their fertility is greatly diminished if these visits be prevented. Now, it is scarcely possible that bees should fly from flower to flower, and not carry pollen from one to the other, to the great good, as I believe, of the plant. Bees will act like a camel-hair pencil, and it is quite sufficient just to touch the anthers of one flower and then the stigma of another with the same brush to ensure fertilisation; but it must not be supposed that bees would thus produce a multitude of hybrids between distinct species; for if you bring on the same brush a plant's own pollen and pollen from another species, the former will have such a prepotent effect, that it will invariably and completely destroy, as has been shown by Gärtner, any influence from the foreign pollen.

When the stamens of a flower suddenly spring towards the pistil, or slowly move one after the other towards it, the contrivance seems adapted solely to ensure self-fertilisation; and no doubt it is useful for this end: but, the agency of insects is often required to cause the stamens to spring forward, as Kölreuter has shown to be the case with the barberry; and curiously in this very genus, which seems to have a special contrivance for self-fertilisation, it is well known that if very closely-allied forms or varieties are planted near each other, it is hardly possible to raise pure seedlings, so largely do they naturally cross. In many other cases, far from there being any aids for self-fertilisation, there are special contrivances, as I could show from the writings of C. C. Sprengel and from my own observations, which effectually prevent the stigma receiving pollen from its own flower: for instance, in Lobelia fulgens, there is a really beautiful and elaborate contrivance by which every one of the infinitely numerous pollen-granules are swept out of the conjoined anthers of each flower, before the stigma of that individual flower is ready to receive them; and as this flower is never visited, at least in my garden, by insects, it never sets a seed, though by placing pollen from one flower on the stigma of another, I raised plenty of seedlings; and whilst another species of Lobelia growing close by, which is visited by bees, seeds freely. In very many other cases, though there be no special mechanical contrivance to prevent the stigma of a flower receiving its own pollen, yet, as C. C. Sprengel has shown, and as I can confirm, either the anthers burst before the stigma is ready for fertilisation, or the stigma is ready before the pollen of that flower is ready, so that these plants have in fact separated sexes, and must habitually be crossed. How strange are these facts! How strange that the pollen and stigmatic surface of the same flower, though placed so close together, as if for the very purpose of self-fertilisation, should in so many cases be mutually useless to each other! How simply are these facts explained on the view of an occasional cross with a distinct individual being advantageous or indispensable!

If several varieties of the cabbage, radish, onion, and of some other plants, be allowed to seed near each other, a large majority, as I have found, of the seedlings thus raised will turn out mongrels: for instance, I raised 233 seedling cabbages from some plants of different varieties growing near each other, and of these only 78 were true to their kind, and some even of these were not perfectly true. Yet the pistil of each cabbage-flower is surrounded not only by its own six stamens, but by those of the many other flowers on the same plant. How, then, comes it that such a vast number of the seedlings are mongrelised? I suspect that it must arise from the pollen of a distinct *variety* having a prepotent effect over a flower's own pollen; and that this is part of the general law of good being derived from the intercrossing of distinct individuals of the same species. When distinct *species* are crossed the case is directly the reverse, for a plant's own pollen is always prepotent over foreign pollen; but to this subject we shall return in a future chapter.

In the case of a gigantic tree covered with innumerable flowers, it may be objected that pollen could seldom be carried from tree to tree, and at most only from flower to flower on the same tree, and that flowers on the same tree can be considered as distinct individuals only in a limited sense. I believe this objection to be valid, but that nature has largely provided against it by giving to trees a strong tendency to bear flowers with separated sexes. When the sexes are separated, although the male and female flowers may be produced on the same tree, we can see that pollen must be regularly carried from flower to flower; and this will give a better chance of pollen being occasionally carried from tree to tree. That trees belonging to all Orders have their sexes more often separated than other plants, I find to be the case in this country; and at my request Dr Hooker tabulated the trees of New Zealand, and Dr Asa Gray those of the United States, and the result was as I anticipated. On the other hand, Dr Hooker has recently informed me that he finds that the rule does not hold in Australia; and I have made these few remarks on the sexes of trees simply to call attention to the subject.

Turning for a very brief space to animals: on the land there are some hermaphrodites, as land-mollusca and earth-worms; but these all pair. As yet I have not found a single case of a terrestrial animal which fertilises itself. We can understand this remarkable fact, which offers so strong a contrast with terrestrial plants, on the view of an occasional cross being indispensable, by considering the medium in which terrestrial animals live, and the nature of the fertilising element; for we know of no means, analogous to the action of insects and of the wind in the case of plants, by which an occasional cross could be effected with terrestrial animals without the concurrence of two individuals. Of aquatic animals, there are many self-fertilising hermaphrodites; but here currents in the water offer an obvious means for an occasional cross. And, as in the case of flowers, I have as yet failed, after consultation with one of the highest authorities, namely, Professor Huxley, to discover a single case of an hermaphrodite animal with the organs of reproduction so perfectly enclosed within the body, that access from without and the occasional influence of a distinct individual can be shown to be physically impossible. Cirripedes long appeared to me to present a case of very great difficulty under this point of view; but I have been enabled, by a fortunate chance, elsewhere to prove that two individuals, though both are self-fertilising hermaphrodites, do sometimes cross.

It must have struck most naturalists as a strange anomaly that, in the case of both animals and plants, species of the same family and even of the same genus, though agreeing closely with each other in almost their whole organisation, yet are not rarely, some of them hermaphrodites, and some of them unisexual. But if, in fact, all hermaphrodites do occasionally intercross with other individuals, the difference between hermaphrodites and unisexual species, as far as function is concerned, becomes very small.

From these several considerations and from the many special facts which I have collected, but which I am not here able to give, I am strongly inclined to suspect that, both in the vegetable and animal kingdoms, an occasional intercross with a distinct individual is a law of nature. I am well aware that there are, on this view, many cases of difficulty, some of which I am trying to investigate. Finally then, we may conclude that in many organic beings, a cross between two individuals is an obvious necessity for each birth; in many others it occurs perhaps only at long intervals; but in none, as I suspect, can self-fertilisation go on for perpetuity.

Circumstances Favourable to Natural Selection

This is an extremely intricate subject. A large amount of inheritable and diversified variability is favourable, but I believe mere individual differences suffice for the work. A large number of individuals, by giving a better chance for the appearance within any given period of profitable variations, will compensate for a lesser amount of variability in each individual, and is, I believe, an extremely important element of success. Though nature grants vast periods of time for the work of natural selection, she does not grant an indefinite period; for as all organic beings are striving, it may be said, to seize on each place in the economy of nature, if any one species does not become modified and improved in a corresponding degree with its competitors, it will soon be exterminated.

In man's methodical selection, a breeder selects for some definite object, and free intercrossing will wholly stop his work. But when many men, without intending to alter the breed, have a nearly common standard of perfection, and all try to get and breed from the best animals, much improvement and modification surely but slowly follow from this unconscious process of selection, notwithstanding a large amount of crossing with inferior animals. Thus it will be in nature; for within a confined area, with some place in its polity not so perfectly occupied as might be, natural selection will always tend to preserve all the individuals varying in the right direction, though in different degrees, so as better to fill up the unoccupied place. But if the area be large, its several districts will almost certainly present different conditions of life; and then if natural selection be modifying and improving a species in the several districts, there will be intercrossing with the other individuals of the same species on the confines of each. And in this case the effects of intercrossing can hardly be counterbalanced by natural selection always tending to modify all the individuals in each district in exactly the same manner to the conditions of each; for in a continuous area, the conditions will generally graduate away insensibly from one district to another. The intercrossing will most affect those animals which unite for each birth, which wander much, and which do not breed at a very quick rate. Hence in animals of this nature, for instance in birds, varieties will generally be confined to separated countries; and this I believe to be the case. In hermaphrodite organisms which cross only occasionally, and likewise in animals which unite for each birth, but which wander little and which can increase at a very rapid rate, a new and improved variety might be quickly formed on any one spot, and might there maintain itself in a body, so that whatever intercrossing took place would be chiefly between the individuals of the same new variety. A local variety when once thus formed might subsequently slowly spread to other districts. On the above principle, nurserymen always prefer getting seed from a large body of plants of the same variety, as the chance of intercrossing with other varieties is thus lessened.

Even in the case of slow-breeding animals, which unite for each birth, we must not overrate the effects of intercrosses in retarding natural selection; for I can bring a considerable catalogue of facts, showing that within the same area, varieties of the same animal can long remain distinct, from haunting different stations, from breeding at slightly different seasons, or from varieties of the same kind preferring to pair together.

Intercrossing plays a very important part in nature in keeping the individuals of the same species, or of the same variety, true and uniform in character. It will obviously thus act far more efficiently with those animals which unite for each birth; but I have already attempted to show that we have reason to believe that occasional intercrosses take place with all animals and with all plants. Even if these take place only at long intervals, I am convinced that the young thus produced will gain so much in vigour and fertility over the offspring from long-continued self-fertilisation, that they will have a better chance of surviving and propagating their kind; and thus, in the long run, the influence of intercrosses, even at rare intervals, will be great. If there exist organic beings which never intercross, uniformity of character can be retained amongst them, as long as their conditions of life remain the same, only through the principle of inheritance, and through natural selection destroying any which depart from the proper type; but if their conditions of life change and they undergo modification, uniformity of character can be given to their modified offspring, solely by natural selection preserving the same favourable variations.

Isolation, also, is an important element in the process of natural selection. In a confined or isolated area, if not very large, the organic and inorganic conditions of life will generally be in a great degree uniform; so that natural selection will tend to modify all the individuals of a varying species throughout the area in the same manner in relation to the same conditions. Intercrosses, also, with the individuals of the same species, which otherwise would have inhabited the surrounding and differently circumstanced districts, will be prevented. But isolation probably acts more efficiently in checking the immigration of better adapted organisms, after any physical change, such as of climate or elevation of the land, &c.; and thus new places in the natural economy of the country are left open for the old inhabitants to struggle for, and become adapted to, through modifications in their structure and constitution. Lastly, isolation, by checking immigration and consequently competition, will give time for any new variety to be slowly improved; and this may sometimes be of importance in the production of new species. If, however, an isolated area be very small, either from being surrounded by barriers, or from having very peculiar physical conditions, the total number of the individuals supported on it will necessarily be very small; and fewness of individuals will greatly retard the production of new species through natural selection, by decreasing the chance of the appearance of favourable variations.

If we turn to nature to test the truth of these remarks, and look at any small isolated area, such as an oceanic island, although the total number of the species inhabiting it, will be found to be small, as we shall see in our chapter on geographical distribution; yet of these species a very large proportion are endemic, that is, have been produced there, and nowhere else. Hence an oceanic island at first sight seems to have been highly favourable for the production of new species. But we may thus greatly deceive ourselves, for to ascertain whether a small isolated area, or a large open area like a continent, has been most favourable for the production of new organic forms, we ought to make the comparison within equal times; and this we are incapable of doing.

Although I do not doubt that isolation is of considerable importance in the production of new species, on the whole I am inclined to believe that largeness of area is of more importance, more especially in the production of species, which will prove capable of enduring for a long period, and of spreading widely. Throughout a great and open area, not only will there be a better chance of favourable variations arising from the large number of individuals of the same species there supported, but the conditions of life are infinitely complex from the large number of already existing species; and if some of these many species become modified and improved, others will have to be improved in a corresponding degree or they will be exterminated. Each new form, also, as soon as it has been much improved, will be able to spread over the open and continuous area, and will thus come into competition with many others. Hence more new places will be formed, and the competition to fill them will be more severe, on a large than on a small and isolated area. Moreover, great areas, though now continuous, owing to oscillations of level, will often have recently existed in a broken condition, so that the good effects of isolation will generally, to a certain extent, have concurred. Finally, I conclude that, although small isolated areas probably have been in some respects highly favourable for the production of new species, yet that the course of modification will generally have been more rapid on large areas; and what is more important, that the new forms produced on large areas, which already have been victorious over many competitors, will be those that will spread most widely, will give rise to most new varieties and species, and will thus play an important part in the changing history of the organic world.

We can, perhaps, on these views, understand some facts which will be again alluded to in our chapter on geographical distribution; for instance, that the productions of the smaller continent of Australia have formerly yielded, and apparently are now yielding, before those of the larger

Europaeo-Asiatic area. Thus, also, it is that continental productions have everywhere become so largely naturalised on islands. On a small island, the race for life will have been less severe, and there will have been less modification and less extermination. Hence, perhaps, it comes that the flora of Madeira, according to Oswald Heer, resembles the extinct tertiary flora of Europe. All fresh-water basins, taken together, make a small area compared with that of the sea or of the land; and, consequently, the competition between fresh-water productions will have been less severe than elsewhere; new forms will have been more slowly formed, and old forms more slowly exterminated. And it is in fresh water that we find seven genera of Ganoid fishes, remnants of a once preponderant order: and in fresh water we find some of the most anomalous forms now known in the world, as the Ornithorhynchus and Lepidosiren, which, like fossils, connect to a certain extent orders now widely separated in the natural scale. These anomalous forms may almost be called living fossils; they have endured to the present day, from having inhabited a confined area, and from having thus been exposed to less severe competition.

To sum up the circumstances favourable and unfavourable to natural selection, as far as the extreme intricacy of the subject permits[,] I conclude, looking to the future, that for terrestrial productions a large continental area, which will probably undergo many oscillations of level, and which consequently will exist for long periods in a broken condition, will be the most favourable for the production of many new forms of life, likely to endure long and to spread widely. For the area will first have existed as a continent, and the inhabitants, at this period numerous in individuals and kinds, will have been subjected to very severe competition. When converted by subsidence into large separate islands, there will still exist many individuals of the same species on each island: intercrossing on the confines of the range of each species will thus be checked: after physical changes of any kind, immigration will be prevented, so that new places in the polity of each island will have to be filled up by modifications of the old inhabitants; and time will be allowed for the varieties in each to become well modified and perfected. When, by renewed elevation, the islands shall be re-converted into a continental area, there will again be severe competition: the most favoured or improved varieties will be enabled to spread: there will be much extinction of the less improved forms, and the relative proportional numbers of the various inhabitants of the renewed continent will again be changed; and again there will be a fair field for natural selection to improve still further the inhabitants, and thus produce new species.

That natural selection will always act with extreme slowness, I fully admit. Its action depends on there being places in the polity of nature, which can be better occupied by some of the inhabitants of the country undergoing modification of some kind. The existence of such places will often depend on physical changes, which are generally very slow, and on the immigration of better adapted forms having been checked. But the action of natural selection will probably still oftener depend on some of the inhabitants becoming slowly modified; the mutual relations of many of the other inhabitants being thus disturbed. Nothing can be effected, unless favourable variations occur, and variation itself is apparently always a very slow process. The process will often be greatly retarded by free intercrossing. Many will exclaim that these several causes are amply sufficient wholly to stop the action of natural selection. I do not believe so. On the other hand, I do believe that natural selection will always act very slowly, often only at long intervals of time, and generally on only a very few of the inhabitants of the same region at the same time. I further believe, that this very slow, intermittent action of natural selection accords perfectly well with what geology tells us of the rate and manner at which the inhabitants of this world have changed.

Slow though the process of selection may be, if feeble man can do much by his powers of artificial selection, I can see no limit to the amount of change, to the beauty and infinite complexity of the coadaptations between all organic beings, one with another and with their physical conditions of life, which may be effected in the long course of time by nature's power of selection.

Extinction

This subject will be more fully discussed in our chapter on Geology; but it must be here alluded to from being intimately connected with natural selection. Natural selection acts solely through the preservation of variations in some way advantageous, which consequently endure. But as from the high geometrical powers of increase of all organic beings, each area is already fully stocked with inhabitants, it follows that as each selected and favoured form increases in number, so will the less favoured forms decrease and become rare. Rarity, as geology tells us, is the precursor to extinction. We can, also, see that any form represented by few individuals will, during fluctuations in the seasons or in the number of its enemies, run a good chance of utter extinction. But we may go further than this; for as new forms are continually and slowly being produced, unless we believe that the number of specific forms goes on perpetually and almost indefinitely increasing, numbers inevitably must become extinct. That the number of specific forms has not indefinitely increased, geology shows us plainly; and indeed we can see reason why they should not have thus increased, for the number of places in the polity of nature is not indefinitely great, not that we have any means of knowing that any one region has as yet got its maximum of species. probably no region is as yet fully stocked, for at the Cape of Good Hope, where more species of plants are crowded together than in any other quarter of the world, some foreign plants have become naturalised, without causing, as far as we know, the extinction of any natives.

Furthermore, the species which are most numerous in individuals will have the best chance of producing within any given period favourable variations. We have evidence of this, in the facts given in the second chapter, showing that it is the common species which afford the greatest number of recorded varieties, or incipient species. Hence, rare species will be less quickly modified or improved within any given period, and they will consequently be beaten in the race for life by the modified descendants of the commoner species.

From these several considerations I think it inevitably follows, that as new species in the course of time are formed through natural selection, others will become rarer and rarer, and finally extinct. The forms which stand in closest competition with those undergoing modification and improvement, will naturally suffer most. And we have seen in the chapter on the Struggle for Existence that it is the most closely-allied forms, varieties of the same species, and species of the same genus or of related genera, which, from having nearly the same structure, constitution, and habits, generally come into the severest competition with each other. Consequently, each new variety or species, during the progress of its formation, will generally press hardest on its nearest kindred, and tend to exterminate them. We see the same process of extermination amongst our domesticated productions, through the selection of improved forms by man. Many curious instances could be given showing how quickly new breeds of cattle, sheep, and other animals, and varieties of flowers, take the place of older and inferior kinds. In Yorkshire, it is historically known that the ancient black cattle were displaced by the long-horns, and that these "were swept away by the short-horns" (I quote the words of an agricultural writer) "as if by some murderous pestilence."

Divergence of Character

The principle, which I have designated by this term, is of high importance on my theory, and explains, as I believe, several important facts. In the first place, varieties, even strongly-marked ones, though having somewhat of the character of species as is shown by the hopeless doubts in many cases how to rank them yet certainly differ from each other far less than do good and distinct species. Nevertheless, according to my view, varieties are species in the process of formation, or are, as I have called them, incipient species. How, then, does the lesser difference between varieties become augmented into the greater difference between species? That this does habitually happen, we must infer from most of the innumerable species throughout nature presenting well-marked differences; whereas varieties, the supposed prototypes and parents of future well-marked species, present slight and ill-defined differences. Mere chance, as we may call it, might cause one variety to differ in some character from its parents, and the offspring of this variety again to differ from its parent in the very same character and in a greater degree; but this alone would never account for so habitual and large an amount of difference as that between varieties of the same species and species of the same genus.

As has always been my practice, let us seek light on this head from our domestic productions. We shall here find something analogous. A fancier is struck by a pigeon having a slightly shorter beak; another fancier is struck by a pigeon having a rather longer beak; and on the acknowledged principle that "fanciers do not and will not admire a medium standard, but like extremes," they both go on (as has actually occurred with tumbler-pigeons) choosing and breeding from birds with longer and longer beaks, or with shorter and shorter beaks. Again, we may suppose that at an early period one man preferred swifter horses; another stronger and more bulky horses. The early differences would be very slight; in the course of time, from the continued selection of swifter horses by some breeders, and of stronger ones by others, the differences would become greater, and would be noted as forming two sub-breeds; finally, after the lapse of centuries, the sub-breeds would become converted into two well-established and distinct breeds. As the differences slowly become greater, the inferior animals with intermediate characters, being neither very swift nor very strong, will have been neglected, and will have tended to disappear. Here, then, we see in man's productions the action of what may be called the principle of divergence, causing differences, at first barely appreciable, steadily to increase, and the breeds to diverge in character both from each other and from their common parent.

But how, it may be asked, can any analogous principle apply in nature? I believe it can and does apply most efficiently, from the simple circumstance that the more diversified the descendants from any one species become in structure, constitution, and habits, by so much will they be better enabled to seize on many and widely diversified places in the polity of nature, and so be enabled to increase in numbers.

We can clearly see this in the case of animals with simple habits. Take the case of a carnivorous quadruped, of which the number that can be supported in any country has long ago arrived at its full average. If its natural powers of increase be allowed to act, it can succeed in increasing (the country not undergoing any change in its conditions) only by its varying descendants seizing on places at present occupied by other animals: some of them, for instance, being enabled to feed on new kinds of prey, either dead or alive; some inhabiting new stations, climbing trees, frequenting water, and some perhaps becoming less carnivorous. The more diversified in habits and structure the descendants of our carnivorous animal became, the more places they would be enabled to occupy. What applies to one animal will apply throughout all time to all animals that is, if they vary for otherwise natural selection can do nothing. So it will be with plants. It has been experimentally proved, that if a plot of ground be sown with several distinct genera of grasses, a greater number of plants and a greater weight of dry herbage can thus be raised. The same has been found to hold good when first one variety and then several mixed varieties of wheat have been sown on equal spaces of ground. Hence, if any one species of grass were to go on varying, and those varieties were continually selected which differed from each other in at all the same manner as distinct species and genera of grasses differ from each other, a greater number of individual plants of this species of grass, including its modified descendants, would succeed in living on the same piece of ground. And we well know that each species and each variety of grass is annually sowing almost countless seeds; and thus, as it may be said, is striving its utmost to increase its numbers. Consequently, I cannot doubt that in the course of many thousands of generations, the most distinct varieties of any one species of grass would always have the best chance of succeeding and of increasing in numbers, and thus of supplanting the less distinct varieties; and varieties, when rendered very distinct from each other, take the rank of species.

The truth of the principle, that the greatest amount of life can be supported by great diversification of structure, is seen under many natural circumstances. In an extremely small area, especially if freely open to immigration, and where the contest between individual and individual must be severe, we always find great diversity in its inhabitants. For instance, I found that a piece of turf, three feet by four in size, which had been exposed for many years to exactly the same conditions, supported twenty species of plants, and these belonged to eighteen genera and to eight orders, which shows how much these plants differed from each other. So it is with the plants and insects on small and uniform islets; and so in small ponds of fresh water. Farmers find that they can raise most food by a rotation of plants belonging to the most different orders: nature follows what may be called a simultaneous rotation. Most of the animals and plants which live close round any small piece of ground, could live on it (supposing it not to be in any way peculiar in its nature), and may be said to be striving to the utmost to live there; but, it is seen, that where they come into the closest competition with each other, the advantages of diversification of structure, with the accompanying differences of habit and constitution, determine that the inhabitants, which thus jostle each other most closely, shall, as a general rule, belong to what we call different genera and orders.

The same principle is seen in the naturalisation of plants through man's agency in foreign lands. It might have been expected that the plants which have succeeded in becoming naturalised in any land would generally have been closely allied to the indigenes; for these are commonly looked at as specially created and adapted for their own country. It might, also, perhaps have been expected that naturalised plants would have belonged to a few groups more especially adapted to certain stations in their new homes. But the case is very different; and Alph. De Candolle has well remarked in his great and admirable work, that floras gain by naturalisation, proportionally with the number of the native genera and species, far more in new genera than in new species. To give a single instance: in the last edition of Dr Asa Gray's "Manual of the Flora of the Northern United States," 260 naturalised plants are enumerated, and these belong to 162 genera. We thus see that these naturalised plants are of a highly diversified nature. They differ, moreover, to a large extent from the indigenes, for out of the 162 genera, no less than 100 genera are not there indigenous, and thus a large proportional addition is made to the genera of these States.

By considering the nature of the plants or animals which have struggled successfully with the indigenes of any country, and have there become naturalised, we can gain some crude idea in what manner some of the natives would have had to be modified, in order to have gained an advantage over the other natives; and we may, I think, at least safely infer that diversification of structure, amounting to new generic differences, would have been profitable to them.

The advantage of diversification in the inhabitants of the same region is, in fact, the same as that of the physiological division of labour in the organs of the same individual body a subject so well elucidated by Milne Edwards. No physiologist doubts that a stomach by being adapted to digest vegetable matter alone, or flesh alone, draws most nutriment from these substances. So in the general economy of any land, the more widely and perfectly the animals and plants are diversified for different habits of life, so will a greater number of individuals be capable of there [*sic*] supporting themselves. A set of animals, with their organisation but little diversified, could hardly compete with a set more perfectly diversified in structure. It may be doubted, for instance, whether the Australian marsupials, which are divided into groups differing but little from each other, and feebly representing, as Mr Waterhouse and others have remarked, our carnivorous, ruminant, and rodent mammals, could successfully compete with these well-pronounced orders. In the Australian mammals, we see the process of diversification in an early and incomplete stage of development.

After the foregoing discussion, which ought to have been much amplified, we may, I think, assume that the modified descendants of any one species will succeed by so much the better as they become more diversified in structure, and are thus enabled to encroach on places occupied by other beings. Now let us see how this principle of great benefit being derived from divergence of character, combined with the principles of natural selection and of extinction, will tend to act.

The accompanying diagram will aid us in understanding this rather perplexing subject. Let A to L represent the species of a genus large in its own country; these species are supposed to resemble each other in unequal degrees, as is so generally the case in nature, and as is represented in the diagram by the letters standing at unequal distances. I have said a large genus, because we have seen in the second chapter, that on an average more of the species of large genera vary than of small genera; and the varying species of the large genera present a greater number of varieties. We have, also, seen that the species, which are the commonest and the most widely-diffused, vary more than rare species with restricted ranges. Let (A) be a common, widely-diffused, and varying species, belonging to a genus large in its own country. The little fan of diverging dotted lines of unequal lengths proceeding from (A) may represent its varying offspring. The variations are supposed to be extremely slight, but of the most diversified nature; they are not supposed all to appear simultaneously, but often after long intervals of time; nor are they all supposed to endure for equal periods. Only those variations which are in some way profitable will be preserved or naturally selected. And here the importance of the principle of benefit being derived from divergence of character comes in; for this will generally lead to the most different or divergent variations (represented by the outer dotted lines) being preserved and accumulated by natural selection. When a dotted line reaches one of the horizontal lines, and is there marked by a small numbered letter, a sufficient amount of variation is supposed to have been accumulated to have formed a fairly well-marked variety, such as would be thought worthy of record in a systematic work.

The intervals between the horizontal lines in the diagram, may represent each a thousand generations; but it would have been better if each had represented ten thousand generations. After a thousand generations, species (A) is supposed to have produced two fairly well-marked varieties, namely *a*1 and *m*1. These two varieties will generally continue to be exposed to the same conditions which made their parents variable, and the tendency to variability is in itself hereditary, consequently they will tend to vary, and generally to vary in nearly the same manner as their parents varied. Moreover, these two varieties, being only slightly modified forms, will tend to inherit those advantages which made their common parent (A) more numerous than most of the other inhabitants of the same country; they will likewise partake of those more general advantages which made the genus to which the parent-species belonged, a large genus in its own country. And these circumstances we know to be favourable to the production of new varieties.

If, then, these two varieties be variable, the most divergent of their variations will generally be preserved during the next thousand generations. And after this interval, variety *a*1 is supposed in the diagram to have produced variety *a*2, which will, owing to the principle of divergence, differ more from (A) than did variety *a*1. Variety *m*1 is supposed to have produced two varieties, namely *m*2 and *s*2, differing from each other, and more considerably from their common parent (A). We may continue the process by similar steps for any length of time; some of the varieties, after each thousand generations, producing only a single variety, but in a more and more modified condition, some producing two or three varieties, and some failing to produce any. Thus the varieties or modified descendants, proceeding from the common parent (A), will generally go on increasing in number and diverging in character. In the diagram the process is represented up to the ten-thousandth generation, and under a condensed and simplified form up to the fourteen-thousandth generation.

But I must here remark that I do not suppose that the process ever goes on so regularly as is represented in the diagram, though in itself made somewhat irregular. I am far from thinking that the most divergent varieties will invariably prevail and multiply: a medium form may often long endure, and may or may not produce more than one modified descendant; for natural selection will always act according to the nature of the places which are either unoccupied or not perfectly occupied by other beings; and this will depend on infinitely complex relations. But as a general rule, the more diversified in structure the descendants from any one species can be rendered, the more places they will be enabled to seize on, and the more their modified progeny will be increased. In our diagram the line of succession is broken at regular intervals by small numbered letters marking the successive forms which have become sufficiently distinct to be recorded as varieties. But these breaks are imaginary, and might have been inserted anywhere, after intervals long enough to have allowed the accumulation of a considerable amount of divergent variation.

As all the modified descendants from a common and widely-diffused species, belonging to a large genus, will tend to partake of the same advantages which made their parent successful in life, they will generally go on multiplying in number as well as diverging in character: this is represented in the diagram by the several divergent branches proceeding from (A). The modified offspring from the later and more highly improved branches in the lines of descent, will, it is probable, often take the place of, and so destroy, the earlier and less improved branches: this is represented in the diagram by some of the lower branches not reaching to the upper horizontal lines. In some cases I do not doubt that the process of modification will be confined to a single line of descent, and the number of the descendants will not be increased; although the amount of divergent modification may have been increased in the successive generations. This case would be represented in the diagram, if all the lines proceeding from (A) were removed, excepting that from *a1* to *a10*[.] In the same way, for instance, the English race-horse and English pointer have apparently both gone on slowly diverging in character from their original stocks, without either having given off any fresh branches or races.

After ten thousand generations, species (A) is supposed to have produced three forms, *a*10, *f*10, and *m10*, which, from having diverged in character during the successive generations, will have come to differ largely, but perhaps unequally, from each other and from their common parent. If we suppose the amount of change between each horizontal line in our diagram to be excessively small, these three forms may still be only well-marked varieties; or they may have arrived at the doubtful category of sub-species; but we have only to suppose the steps in the process of modification to be more numerous or greater in amount, to convert these three forms into well-defined species: thus the diagram illustrates the steps by which the small differences distinguishing varieties are increased into the larger differences distinguishing species. By continuing the same process for a greater number of generations (as shown in the diagram in a condensed and simplified manner), we get eight species, marked by the letters between *a*14 and *m*14, all descended from (A). Thus, as I believe, species are multiplied and genera are formed.

In a large genus it is probable that more than one species would vary. In the diagram I have assumed that a second species (I) has produced, by analogous steps, after ten thousand generations, either two well-marked varieties (*w*10 and *z*10) or two species, according to the amount of change supposed to be represented between the horizontal lines. After fourteen thousand generations, six new species, marked by the letters *n*14 to *z*14, are supposed to have been produced. In each genus, the species, which are already extremely different in character, will generally tend to produce the greatest number of modified descendants; for these will have the best chance of filling new and widely different places in the polity of nature: hence in the diagram I have chosen the extreme species (A), and the nearly extreme species (I), as those which have largely varied, and have given rise to new varieties and species. The other nine species (marked by capital letters) of our original genus, may for a long period continue transmitting unaltered descendants; and this is shown in the diagram by the dotted lines not prolonged far upwards from want of space.

But during the process of modification, represented in the diagram, another of our principles, namely that of extinction, will have played an important part. As in each fully stocked country natural selection necessarily acts by the selected form having some advantage in the struggle for life over other forms, there will be a constant tendency in the improved descendants of any one species to supplant and exterminate in each stage of descent their predecessors and their original parent. For it should be remembered that the competition will generally be most severe between those forms which are most nearly related to each other in habits, constitution, and structure. Hence all the intermediate forms between the earlier and later states, that is between the less and more improved state of a species, as well as the original parent-species itself, will generally tend to become extinct. So it probably will be with many whole collateral lines of descent, which will be conquered by later and improved lines of descent. If, however, the modified offspring of a species get into some distinct country, or become quickly adapted to some quite new station, in which child and parent do not come into competition, both may continue to exist.

If then our diagram be assumed to represent a considerable amount of modification, species (A) and all the earlier varieties will have become extinct, having been replaced by eight new species (*a*14 to *m*14); and (I) will have been replaced by six (*n*14 to *z*14) new species.

But we may go further than this. The original species of our genus were supposed to resemble each other in unequal degrees, as is so generally the case in nature; species (A) being more nearly related to B, C, and D, than to the other species; and species (I) more to G, H, K, L, than to the others. These two species (A) and (I), were also supposed to be very common and widely diffused species, so that they must originally have had some advantage over most of the other species of the genus. Their modified descendants, fourteen in number at the fourteen-thousandth generation, will probably have inherited some of the same advantages: they have also been modified and improved in a diversified manner at each stage of descent, so as to have become adapted to many related places in the natural economy of their country. It seems, therefore, to me extremely probable that they will have taken the places of, and thus exterminated, not only their parents (A) and (I), but likewise some of the original species which were most nearly related to their parents. Hence very few of the original species will have transmitted offspring to the fourteen-thousandth generation. We may suppose that only one (F), of the two species which were least closely related to the other nine original species, has transmitted descendants to this late stage of descent.

The new species in our diagram descended from the original eleven species, will now be fifteen in number. Owing to the divergent tendency of natural selection, the extreme amount of difference in character between species *a*14 and *z*14 will be much greater than that between the most different of the original eleven species. The new species, moreover, will be allied to each other in a widely different manner. Of the eight descendants from (A) the three marked *a*14, *q*14, *p*14, will be nearly related from having recently branched off from *a*14; *b*14 and *f*14, from having diverged at an earlier period from *a*5, will be in some degree distinct from the three first-named species; and lastly, *o*14, *e*14, and *m*14, will be nearly related one to the other, but from having diverged at the first commencement of the process of modification, will be widely different from the other five species, and may constitute a sub-genus or even a distinct genus. The six descendants from *(I)* will form two sub-genera or even genera. But as the original species (I) differed largely from (A), standing nearly at the extreme points of the original genus, the six descendants from (I) will, owing to inheritance, differ considerably from the eight descendants from (A); the two groups, moreover, are supposed to have gone on diverging in different directions. The intermediate species, also (and this is a very important consideration), which connected the original species (A) and (I), have all become, excepting (F), extinct, and have left no descendants. Hence the six new species descended from (I), and the eight descended from (A), will have to be ranked as very distinct genera, or even as distinct sub-families.

Thus it is, as I believe, that two or more genera are produced by descent, with modification, from two or more species of the same genus. And the two or more parent-species are supposed to have descended from some one species of an earlier genus. In our diagram, this is indicated by the broken lines, beneath the capital letters, converging in sub-branches downwards towards a single point; this point representing a single species, the supposed single parent of our several new sub-genera and genera.

It is worth while to reflect for a moment on the character of the new species F14, which is supposed not to have diverged much in character, but to have retained the form of (F), either unaltered or altered only in a slight degree. In this case, its affinities to the other fourteen new species will be of a curious and circuitous nature. Having descended from a form which stood between the two parent-species (A) and (I), now supposed to be extinct and unknown, it will be in some degree intermediate in character between the two groups descended from these species. But as these two groups have gone on diverging in character from the type of their parents, the new species (F14) will not be directly intermediate between them, but rather between types of the two groups; and every naturalist will be able to bring some such case before his mind.

In the diagram, each horizontal line has hitherto been supposed to represent a thousand generations, but each may represent a million or hundred million generations, and likewise a section of the successive strata of the earth's crust including extinct remains. We shall, when we come to our chapter on Geology, have to refer again to this subject, and I think we shall then see that the diagram throws light on the affinities of extinct beings, which, though generally belonging to the same orders, or families, or genera, with those now living, yet are often, in some degree, intermediate in character between existing groups; and we can understand this fact, for the extinct species lived at very ancient epochs when the branching lines of descent had diverged less.

I see no reason to limit the process of modification, as now explained, to the formation of genera alone. If, in our diagram, we suppose the amount of change represented by each successive group of diverging dotted lines to be very great, the forms marked *a214 to p*14, those marked *b*14 and *f*14, and those marked *o*14 to *m*14, will form three very distinct genera. We shall also have two very distinct genera descended from (I) and as these latter two genera, both from continued divergence of character and from inheritance from a different parent, will differ widely from the three genera descended from (A), the two little groups of genera will form two distinct families, or even orders, according to the amount of divergent modification supposed to be represented in the diagram. And the two new families, or orders, will have descended from two species of the original genus; and these two species are supposed to have descended from one species of a still more ancient and unknown genus.

We have seen that in each country it is the species of the larger genera which oftenest present varieties or incipient species. This, indeed, might have been expected; for as natural selection acts through one form having some advantage over other forms in the struggle for existence, it will chiefly act on those which already have some advantage; and the largeness of any group shows that its species have inherited from a common ancestor some advantage in common. Hence, the struggle for the production of new and modified descendants, will mainly lie between the larger groups, which are all trying to increase in number. One large group will slowly conquer another large group, reduce its numbers, and thus lessen its chance of further variation and improvement. Within the same large group, the later and more highly perfected sub-groups, from branching out and seizing on many new places in the polity of Nature, will constantly tend to supplant and destroy the earlier and less improved sub-groups. Small and broken groups and sub-groups will finally tend to disappear. Looking to the future, we can predict that the groups of organic beings which are now large and triumphant, and which are least broken up, that is, which as yet have suffered least extinction, will for a long period continue to increase. But which groups will ultimately prevail, no man can predict; for we well know that many groups, formerly most extensively developed, have now become extinct. Looking still more remotely to the future, we may predict that, owing to the continued and steady increase of the larger groups, a multitude of smaller groups will become utterly extinct, and leave no modified descendants; and consequently that of the species living at any one period, extremely few will transmit descendants to a remote futurity. I shall have to return to this subject in the chapter on Classification, but I may add that on this view of extremely few of the more ancient species having transmitted descendants, and on the view of all the descendants of the same species making a class, we can understand how it is that there exist but very few classes in each main division of the animal and vegetable kingdoms. Although extremely few of the most ancient species may now have living and modified descendants, yet at the most remote geological period, the earth may have been as well peopled with many species of many genera, families, orders, and classes, as at the present day.

Summary of Chapter

If during the long course of ages and under varying conditions of life, organic beings vary at all in the several parts of their organisation, and I think this cannot be disputed; if there be, owing to the high geometrical powers of increase of each species, at some age, season, or year, a severe struggle for life, and this certainly cannot be disputed; then, considering the infinite complexity of the relations of all organic beings to each other and to their conditions of existence, causing an infinite diversity in structure, constitution, and habits, to be advantageous to them, I think it would be a most extraordinary fact if no variation ever had occurred useful to each being's own welfare, in the same way as so many variations have occurred useful to man. But if variations useful to any organic being do occur, assuredly individuals thus characterised will have the best chance of being preserved in the struggle for life; and from the strong principle of inheritance they will tend to produce offspring similarly characterised. This principle of preservation, I have called, for the sake of brevity, Natural Selection. Natural selection, on the principle of qualities being inherited at corresponding ages, can modify the egg, seed, or young, as easily as the adult. Amongst many animals, sexual selection will give its aid to ordinary selection, by assuring to the most vigorous and best adapted males the greatest number of offspring. Sexual selection will also give characters useful to the males alone, in their struggles with other males.

Whether natural selection has really thus acted in nature, in modifying and adapting the various forms of life to their several conditions and stations, must be judged by the general tenour and balance of evidence given in the following chapters. But we already see how it entails extinction; and how largely extinction has acted in the world's history, geology plainly declares. Natural selection, also, leads to divergence of character; for more living beings can be supported on the same area the more they diverge in structure, habits, and constitution, of which we see proof by looking at the inhabitants of any small spot or at naturalised productions. Therefore during the modification of the descendants of any one species, and during the incessant struggle of all species to increase in numbers, the more diversified these descendants become, the better will be their chance of succeeding in the battle of life. Thus the small differences distinguishing varieties of the same species, will steadily tend to increase till they come to equal the greater differences between species of the same genus, or even of distinct genera.

We have seen that it is the common, the widely-diffused, and widely-ranging species, belonging to the larger genera, which vary most; and these will tend to transmit to their modified offspring that superiority which now makes them dominant in their own countries. Natural selection, as has just been remarked, leads to divergence of character and to much extinction of the less improved and intermediate forms of life. On these principles, I believe, the nature of the affinities of all organic beings may be explained. It is a truly wonderful fact the wonder of which we are apt to overlook from familiarity that all animals and all plants throughout all time and space should be related to each other in group subordinate to group, in the manner which we everywhere behold namely, varieties of the same species most closely related together, species of the same genus less closely and unequally related together, forming sections and sub-genera, species of distinct genera much less closely related, and genera related in different degrees, forming sub-families, families, orders, sub-classes, and classes. The several subordinate groups in any class cannot be ranked in a single file, but seem rather to be clustered round points, and these round other points, and so on in almost endless cycles. On the view that each species has been independently created, I can see no explanation of this great fact in the classification of all organic beings; but, to the best of my judgment, it is explained through inheritance and the complex action of natural selection, entailing extinction and divergence of character, as we have seen illustrated in the diagram.

The affinities of all the beings of the same class have sometimes been represented by a great tree. I believe this simile largely speaks the truth. The green and budding twigs may represent existing species; and those produced during each former year may represent the long succession of extinct species. At each period of growth all the growing twigs have tried to branch out on all sides, and to overtop and kill the surrounding twigs and branches, in the same manner as species and groups of species have tried to overmaster other species in the great battle for life. The limbs divided into great branches, and these into lesser and lesser branches, were themselves once, when the tree was small, budding twigs; and this connexion of the former and present buds by ramifying branches may well represent the classification of all extinct and living species in groups subordinate to groups. Of the many twigs which flourished when the tree was a mere bush, only two or three, now grown into great branches, yet survive and bear all the other branches; so with the species which lived during long-past geological periods, very few now have living and modified descendants. From the first growth of the tree, many a limb and branch has decayed and dropped off; and these lost branches of various sizes may represent those whole orders, families, and genera which have now no living representatives, and which are known to us only from having been found in a fossil state. As we here and there see a thin straggling branch springing from a fork low down in a tree, and which by some chance has been favoured and is still alive on its summit, so we occasionally see an animal like the Ornithorhynchus or Lepidosiren, which in some small degree connects by its affinities two large branches of life, and which has apparently been saved from fatal competition by having inhabited a protected station. As buds give rise by growth to fresh buds, and these, if vigorous, branch out and overtop on all sides many a feebler branch, so by generation I believe it has been with the great Tree of Life, which fills with its dead and broken branches the crust of the earth, and covers the surface with its ever branching and beautiful ramifications.

XIV. Recapitulation and Conclusion

As this whole volume is one long argument, it may be convenient to the reader to have the leading facts and inferences briefly recapitulated.

That many and grave objections may be advanced against the theory of descent with modification through natural selection, I do not deny. I have endeavoured to give to them their full force. Nothing at first can appear more difficult to believe than that the more complex organs and instincts should have been perfected not by means superior to, though analogous with, human reason, but by the accumulation of innumerable slight variations, each good for the individual possessor. Nevertheless, this difficulty, though appearing to our imagination insuperably great, cannot be considered real if we admit the following propositions, namely,—that gradations in the perfection of any organ or instinct, which we may consider, either do now exist or could have existed, each good of its kind,—that all organs and instincts are, in ever so slight a degree, variable,—and, lastly, that there is a struggle for existence leading to the preservation of each profitable deviation of structure or instinct. The truth of these propositions cannot, I think, be disputed.

It is, no doubt, extremely difficult even to conjecture by what gradations many structures have been perfected, more especially amongst broken and failing groups of organic beings; but we see so many strange gradations in nature, as is proclaimed by the canon, "Natura non facit saltum," that we ought to be extremely cautious in saying that any organ or instinct, or any whole being, could not have arrived at its present state by many graduated steps. There are, it must be admitted, cases of special difficulty on the theory of natural selection; and one of the most curious of these is the existence of two or three defined castes of workers or sterile females in the same community of ants but I have attempted to show how this difficulty can be mastered. With respect to the almost universal sterility of species when first crossed, which forms so remarkable a contrast with the almost universal fertility of varieties when crossed, I must refer the reader to the recapitulation of the facts given at the end of the eighth chapter, which seem to me conclusively to show that this sterility is no more a special endowment than is the incapacity of two trees to be grafted together, but that it is incidental on constitutional differences in the reproductive systems of the intercrossed species. We see the truth of this conclusion in the vast difference in the result, when the same two species are crossed reciprocally; that is, when one species is first used as the father and then as the mother.

The fertility of varieties when intercrossed and of their mongrel offspring cannot be considered as universal; nor is their very general fertility surprising when we remember that it is not likely that either their constitutions or their reproductive systems should have been profoundly modified. Moreover, most of the varieties which have been experimentised on have been produced under domestication; and as domestication apparently tends to eliminate sterility, we ought not to expect it also to produce sterility.

The sterility of hybrids is a very different case from that of first crosses, for their reproductive organs are more or less functionally impotent; whereas in first crosses the organs on both sides are in a perfect condition. As we continually see that organisms of all kinds are rendered in some degree sterile from their constitutions having been disturbed by slightly different and new conditions of life, we need not feel surprise at hybrids being in some degree sterile, for their constitutions can hardly fail to have been disturbed from being compounded of two distinct organisations. This parallelism is supported by another parallel, but directly opposite, class of facts; namely, that the vigour and fertility of all organic beings are increased by slight changes in their conditions of life, and that the offspring of slightly modified forms or varieties acquire from being crossed increased vigour and fertility. So that, on the one hand, considerable changes in the conditions of life and crosses between greatly modified forms, lessen fertility; and on the other hand, lesser changes in the conditions of life and crosses between less modified forms, increase fertility.

Turning to geographical distribution, the difficulties encountered on the theory of descent with modification are grave enough. All the individuals of the same species, and all the species of the same genus, or even higher group, must have descended from common parents; and therefore, in however distant and isolated parts of the world they are now found, they must in the course of successive generations have passed from some one part to the others. We are often wholly unable even to conjecture how this could have been effected. Yet, as we have reason to believe that some species have retained the same specific form for very long periods, enormously long as measured by years, too much stress ought not to be laid on the occasional wide diffusion of the same species; for during very long periods of time there will always be a good chance for wide migration by many means. A broken or interrupted range may often be accounted for by the extinction of the species in the intermediate regions. It cannot be denied that we are as yet very ignorant of the full extent of the various climatal and geographical changes which have affected the earth during modern periods; and such changes will obviously have greatly facilitated migration. As an example, I have attempted to show how potent has been the influence of the Glacial period on the distribution both of the same and of representative species throughout the world. We are as yet profoundly ignorant of the many occasional means of transport. With respect to distinct species of the same genus inhabiting very distant and isolated regions, as the process of modification has necessarily been slow, all the means of migration will have been possible during a very long period; and consequently the difficulty of the wide diffusion of species of the same genus is in some degree lessened.

As on the theory of natural selection an interminable number of intermediate forms must have existed, linking together all the species in each group by gradations as fine as our present varieties, it may be asked, Why do we not see these linking forms all around us? Why are not all organic beings blended together in an inextricable chaos? With respect to existing forms, we should remember that we have no right to expect (excepting in rare cases) to discover *directly* connecting links between them, but only between each and some extinct and supplanted form. Even on a wide area, which has during a long period remained continuous, and of which the climate and other conditions of life change insensibly in going from a district occupied by one species into another district occupied by a closely allied species, we have no just right to expect often to find intermediate varieties in the intermediate zone. For we have reason to believe that only a few species are undergoing change at any one period; and all changes are slowly effected. I have also shown that the intermediate varieties which will at first probably exist in the intermediate zones, will be liable to be supplanted by the allied forms on either hand; and the latter, from existing in greater numbers, will generally be modified and improved at a quicker rate than the intermediate varieties, which exist in lesser numbers; so that the intermediate varieties will, in the long run, be supplanted and exterminated.

On this doctrine of the extermination of an infinitude of connecting links, between the living and extinct inhabitants of the world, and at each successive period between the extinct and still older species, why is not every geological formation charged with such links? Why does not every collection of fossil remains afford plain evidence of the gradation and mutation of the forms of life? We meet with no such evidence, and this is the most obvious and forcible of the many objections which may be urged against my theory. Why, again, do whole groups of allied species appear, though certainly they often falsely appear, to have come in suddenly on the several geological stages? Why do we not find great piles of strata beneath the Silurian system, stored with the remains of the progenitors of the Silurian groups of fossils? For certainly on my theory such strata must somewhere have been deposited at these ancient and utterly unknown epochs in the world's history.

I can answer these questions and grave objections only on the supposition that the geological record is far more imperfect than most geologists believe. It cannot be objected that there has not been time sufficient for any amount of organic change; for the lapse of time has been so great as to be utterly inappreciable by the human intellect. The number of specimens in all our museums is absolutely as nothing compared with the countless generations of countless species which certainly have existed. We should not be able to recognise a species as the parent of any one or more species if we were to examine them ever so closely, unless we likewise possessed many of the intermediate links between their past or parent and present states; and these many links we could hardly ever expect to discover, owing to the imperfection of the geological record. Numerous existing doubtful forms could be named which are probably varieties; but who will pretend that in future ages so many fossil links will be discovered, that naturalists will be able to decide, on the common view, whether or not these doubtful forms are varieties? As long as most of the links between any two species are unknown, if any one link or intermediate variety be discovered, it will simply be classed as another and distinct species. Only a small portion of the world has been geologically explored. Only organic beings of certain classes can be preserved in a fossil condition, at least in any great number. Widely ranging species vary most, and varieties are often at first local,—both causes rendering the discovery of intermediate links less likely. Local varieties will not spread into other and distant regions until they are considerably modified and improved; and when they do spread, if discovered in a geological formation, they will appear as if suddenly created there, and will be simply classed as new species. Most formations have been intermittent in their accumulation; and their duration, I am inclined to believe, has been shorter than the average duration of specific forms. Successive formations are separated from each other by enormous blank intervals of time; for fossiliferous formations, thick enough to resist future degradation, can be accumulated only where much sediment is deposited on the subsiding bed of the sea. During the alternate periods of elevation and of stationary level the record will be blank. During these latter periods there will probably be more variability in the forms of life; during periods of subsidence, more extinction.

With respect to the absence of fossiliferous formations beneath the lowest Silurian strata, I can only recur to the hypothesis given in the ninth chapter. That the geological record is imperfect all will admit; but that it is imperfect to the degree which I require, few will be inclined to admit. If we look to long enough intervals of time, geology plainly declares that all species have changed; and they have changed in the manner which my theory requires, for they have changed slowly and in a graduated manner. We clearly see this in the fossil remains from consecutive formations invariably being much more closely related to each other, than are the fossils from formations distant from each other in time.

Such is the sum of the several chief objections and difficulties which may justly be urged against my theory; and I have now briefly recapitulated the answers and explanations which can be given to them. I have felt these difficulties far too heavily during many years to doubt their weight. But it deserves especial notice that the more important objections relate to questions on which we are confessedly ignorant; nor do we know how ignorant we are. We do not know all the possible transitional gradations between the simplest and the most perfect organs; it cannot be pretended that we know all the varied means of Distribution during the long lapse of years, or that we know how imperfect the Geological Record is. Grave as these several difficulties are, in my judgment they do not overthrow the theory of descent with modification.

Now let us turn to the other side of the argument. Under domestication we see much variability. This seems to be mainly due to the reproductive system being eminently susceptible to changes in the conditions of life so that this system, when not rendered impotent, fails to reproduce offspring exactly like the parent-form. Variability is governed by many complex laws,—by correlation of growth, by use and disuse, and by the direct action of the physical conditions of life. There is much difficulty in ascertaining how much modification our domestic productions have undergone; but we may safely infer that the amount has been large, and that modifications can be inherited for long periods. As long as the conditions of life remain the same, we have reason to believe that a modification, which has already been inherited for many generations, may continue to be inherited for an almost infinite number of generations. On the other hand we have evidence that variability, when it has once come into play, does not wholly cease; for new varieties are still occasionally produced by our most anciently domesticated productions.

Man does not actually produce variability; he only unintentionally exposes organic beings to new conditions of life, and then nature acts on the organisation, and causes variability. But man can and does select the variations given to him by nature, and thus accumulate them in any desired manner. He thus adapts animals and plants for his own benefit or pleasure. He may do this methodically, or he may do it unconsciously by preserving the individuals most useful to him at the time, without any thought of altering the breed. It is certain that he can largely influence the character of a breed by selecting, in each successive generation, individual differences so slight as to be quite inappreciable by an uneducated eye. This process of selection has been the great agency in the production of the most distinct and useful domestic breeds. That many of the breeds produced by man have to a large extent the character of natural species, is shown by the inextricable doubts whether very many of them are varieties or aboriginal species.

There is no obvious reason why the principles which have acted so efficiently under domestication should not have acted under nature. In the preservation of favoured individuals and races, during the constantly-recurrent Struggle for Existence, we see the most powerful and ever-acting means of selection. The struggle for existence inevitably follows from the high geometrical ratio of increase which is common to all organic beings. This high rate of increase is proved by calculation, by the effects of a succession of peculiar seasons, and by the results of naturalisation, as explained in the third chapter. More individuals are born than can possibly survive. A grain in the balance will determine which individual shall live and which shall die,—which variety or species shall increase in number, and which shall decrease, or finally become extinct. As the individuals of the same species come in all respects into the closest competition with each other, the struggle will generally be most severe between them; it will be almost equally severe between the varieties of the same species, and next in severity between the species of the same genus. But the struggle will often be very severe between beings most remote in the scale of nature. The slightest advantage in one being, at any age or during any season, over those with which it comes into competition, or better adaptation in however slight a degree to the surrounding physical conditions, will turn the balance.

With animals having separated sexes there will in most cases be a struggle between the males for possession of the females. The most vigorous individuals, or those which have most successfully struggled with their conditions of life, will generally leave most progeny. But success will often depend on having special weapons or means of defence, or on the charms of the males; and the slightest advantage will lead to victory.

As geology plainly proclaims that each land has undergone great physical changes, we might have expected that organic beings would have varied under nature, in the same way as they generally have varied under the changed conditions of domestication. And if there be any variability under nature, it would be an unaccountable fact if natural selection had not come into play. It has often been asserted, but the assertion is quite incapable of proof, that the amount of variation under nature is a strictly limited quantity. Man, though acting on external characters alone and often capriciously, can produce within a short period a great result by adding up mere individual differences in his domestic productions; and every one admits that there are at least individual differences in species under nature. But, besides such differences, all naturalists have admitted the existence of varieties, which they think sufficiently distinct to be worthy of record in systematic works. No one can draw any clear distinction between individual differences and slight varieties; or between more plainly marked varieties and subspecies, and species. Let it be observed how naturalists differ in the rank which they assign to the many representative forms in Europe and North America.

If then we have under nature variability and a powerful agent always ready to act and select, why should we doubt that variations in any way useful to beings, under their excessively complex relations of life, would be preserved, accumulated, and inherited? Why, if man can by patience select variations most useful to himself, should nature fail in selecting variations useful, under changing conditions of life, to her living products? What limit can be put to this power, acting during long ages and rigidly scrutinising the whole constitution, structure, and habits of each creature, favouring the good and rejecting the bad? I can see no limit to this power, in slowly and beautifully adapting each form to the most complex relations of life. The theory of natural selection, even if we looked no further than this, seems to me to be in itself probable. I have already recapitulated, as fairly as I could, the opposed difficulties and objections: now let us turn to the special facts and arguments in favour of the theory.

On the view that species are only strongly marked and permanent varieties, and that each species first existed as a variety, we can see why it is that no line of demarcation can be drawn between species, commonly supposed to have been produced by special acts of creation, and varieties which are acknowledged to have been produced by secondary laws. On this same view we can understand how it is that in each region where many species of a genus have been produced, and where they now flourish, these same species should present many varieties; for where the manufactory of species has been active, we might expect, as a general rule, to find it still in action; and this is the case if varieties be incipient species. Moreover, the species of the large genera, which afford the greater number of varieties or incipient species, retain to a certain degree the character of varieties; for they differ from each other by a less amount of difference than do the species of smaller genera. The closely allied species also of the larger genera apparently have restricted ranges, and they are clustered in little groups round other species—in which respects they resemble varieties. These are strange relations on the view of each species having been independently created, but are intelligible if all species first existed as varieties.

As each species tends by its geometrical ratio of reproduction to increase inordinately in number; and as the modified descendants of each species will be enabled to increase by so much the more as they become more diversified in habits and structure, so as to be enabled to seize on many and widely different places in the economy of nature, there will be a constant tendency in natural selection to preserve the most divergent offspring of any one species. Hence during a long-continued course of modification, the slight differences, characteristic of varieties of the same species, tend to be augmented into the greater differences characteristic of species of the same genus. New and improved varieties will inevitably supplant and exterminate the older, less improved and intermediate varieties; and thus species are rendered to a large extent defined and distinct objects. Dominant species belonging to the larger groups tend to give birth to new and dominant forms; so that each large group tends to become still larger, and at the same time more divergent in character. But as all groups cannot thus succeed in increasing in size, for the world would not hold them, the more dominant groups beat the less dominant. This tendency in the large groups to go on increasing in size and diverging in character, together with the almost inevitable contingency of much extinction, explains the arrangement of all the forms of life, in groups subordinate to groups, all within a few great classes, which we now see everywhere around us, and which has prevailed throughout all time. This grand fact of the grouping of all organic beings seems to me utterly inexplicable on the theory of creation.

As natural selection acts solely by accumulating slight, successive, favourable variations, it can produce no great or sudden modification; it can act only by very short and slow steps. Hence the canon of "Natura non facit saltum," which every fresh addition to our knowledge tends to make more strictly correct, is on this theory simply intelligible. We can plainly see why nature is prodigal in variety, though niggard in innovation. But why this should be a law of nature if each species has been independently created, no man can explain.

Many other facts are, as it seems to me, explicable on this theory. How strange it is that a bird, under the form of woodpecker, should have been created to prey on insects on the ground; that upland geese, which never or rarely swim, should have been created with webbed feet; that a thrush should have been created to dive and feed on sub-aquatic insects; and that a petrel should have been created with habits and structure fitting it for the life of an auk or grebe! and so on in endless other cases. But on the view of each species constantly trying to increase in number, with natural selection always ready to adapt the slowly varying descendants of each to any unoccupied or ill-occupied place in nature, these facts cease to be strange, or perhaps might even have been anticipated.

As natural selection acts by competition, it adapts the inhabitants of each country only in relation to the degree of perfection of their associates; so that we need feel no surprise at the inhabitants of any one country, although on the ordinary view supposed to have been specially created and adapted for that country, being beaten and supplanted by the naturalised productions from another land. Nor ought we to marvel if all the contrivances in nature be not, as far as we can judge, absolutely perfect; and if some of them be abhorrent to our ideas of fitness. We need not marvel at the sting of the bee causing the bee's own death; at drones being produced in such vast numbers for one single act, and being then slaughtered by their sterile sisters; at the astonishing waste of pollen by our fir-trees; at the instinctive hatred of the queen bee for her own fertile daughters; at ichneumonidae feeding within the live bodies of caterpillars; and at other such cases. The wonder indeed is, on the theory of natural selection, that more cases of the want of absolute perfection have not been observed.

The complex and little known laws governing variation are the same, as far as we can see, with the laws which have governed the production of so-called specific forms. In both cases physical conditions seem to have produced but little direct effect; yet when varieties enter any zone, they occasionally assume some of the characters of the species proper to that zone. In both varieties and species, use and disuse seem to have produced some effect; for it is difficult to resist this conclusion when we look, for instance, at the logger-headed duck, which has wings incapable of flight, in nearly the same condition as in the domestic duck; or when we look at the burrowing tucutucu, which is occasionally blind, and then at certain moles, which are habitually blind and have their eyes covered with skin; or when we look at the blind animals inhabiting the dark caves of America and Europe. In both varieties and species correction of growth seems to have played a most important part, so that when one part has been modified other parts are necessarily modified. In both varieties and species reversions to long-lost characters occur. How inexplicable on the theory of creation is the occasional appearance of stripes on the shoulder and legs of the several species of the horse-genus and in their hybrids! How simply is this fact explained if we believe that these species have descended from a striped progenitor, in the same manner as the several domestic breeds of pigeon have descended from the blue and barred rock-pigeon!

On the ordinary view of each species having been independently created, why should the specific characters, or those by which the species of the same genus differ from each other, be more variable than the generic characters in which they all agree? Why, for instance, should the colour of a flower be more likely to vary in any one species of a genus, if the other species, supposed to have been created independently, have differently coloured flowers, than if all the species of the genus have the same coloured flowers? If species are only well-marked varieties, of which the characters have become in a high degree permanent, we can understand this fact; for they have already varied since they branched off from a common progenitor in certain characters, by which they have come to be specifically distinct from each other; and therefore these same characters would be more likely still to be variable than the generic characters which have been inherited without change for an enormous period. It is inexplicable on the theory of creation why a part developed in a very unusual manner in any one species of a genus, and therefore, as we may naturally infer, of great importance to the species, should be eminently liable to variation; but, on my view, this part has undergone, since the several species branched off from a common progenitor, an unusual amount of variability and modification, and therefore we might expect this part generally to be still variable. But a part may be developed in the most unusual manner, like the wing of a bat, and yet not be more variable than any other structure, if the part be common to many subordinate forms, that is, if it has been inherited for a very long period; for in this case it will have been rendered constant by long-continued natural selection.

Glancing at instincts, marvellous as some are, they offer no greater difficulty than does corporeal structure on the theory of the natural selection of successive, slight, but profitable modifications. We can thus understand why nature moves by graduated steps in endowing different animals of the same class with their several instincts. I have attempted to show how much light the principle of gradation throws on the admirable architectural powers of the hive-bee. Habit no doubt sometimes comes into play in modifying instincts; but it certainly is not indispensable, as we see, in the case of neuter insects, which leave no progeny to inherit the effects of long-continued habit. On the view of all the species of the same genus having descended from a common parent, and having inherited much in common, we can understand how it is that allied species, when placed under considerably different conditions of life, yet should follow nearly the same instincts; why the thrush of South America, for instance, lines her nest with mud like our British species. On the view of instincts having been slowly acquired through natural selection we need not marvel at some instincts being apparently not perfect and liable to mistakes, and at many instincts causing other animals to suffer.

If species be only well-marked and permanent varieties, we can at once see why their crossed offspring should follow the same complex laws in their degrees and kinds of resemblance to their parents,—in being absorbed into each other by successive crosses, and in other such points,—as do the crossed offspring of acknowledged varieties. On the other hand, these would be strange facts if species have been independently created, and varieties have been produced by secondary laws.

If we admit that the geological record is imperfect in an extreme degree, then such facts as the record gives, support the theory of descent with modification. New species have come on the stage slowly and at successive intervals; and the amount of change, after equal intervals of time, is widely different in different groups. The extinction of species and of whole groups of species, which has played so conspicuous a part in the history of the organic world, almost inevitably follows on the principle of natural selection; for old forms will be supplanted by new and improved forms. Neither single species nor groups of species reappear when the chain of ordinary generation has once been broken. The gradual diffusion of dominant forms, with the slow modification of their descendants, causes the forms of life, after long intervals of time, to appear as if they had changed simultaneously throughout the world. The fact of the fossil remains of each formation being in some degree intermediate in character between the fossils in the formations above and below, is simply explained by their intermediate position in the chain of descent. The grand fact that all extinct organic beings belong to the same system with recent beings, falling either into the same or into intermediate groups, follows from the living and the extinct being the offspring of common parents. As the groups which have descended from an ancient progenitor have generally diverged in character, the progenitor with its early descendants will often be intermediate in character in comparison with its later descendants; and thus we can see why the more ancient a fossil is, the oftener it stands in some degree intermediate between existing and allied groups. Recent forms are generally looked at as being, in some vague sense, higher than ancient and extinct forms; and they are in so far higher as the later and more improved forms have conquered the older and less improved organic beings in the struggle for life. Lastly, the law of the long endurance of allied forms on the same continent, of marsupials in Australia, of edentata in America, and other such cases,—is intelligible, for within a confined country, the recent and the extinct will naturally be allied by descent.

Looking to geographical distribution, if we admit that there has been during the long course of ages much migration from one part of the world to another, owing to former climatal and geographical changes and to the many occasional and unknown means of dispersal, then we can understand, on the theory of descent with modification, most of the great leading facts in Distribution. We can see why there should be so striking a parallelism in the distribution of organic beings throughout space, and in their geological succession throughout time; for in both cases the beings have been connected by the bond of ordinary generation, and the means of modification have been the same. We see the full meaning of the wonderful fact, which must have struck every traveller, namely, that on the same continent, under the most diverse conditions, under heat and cold, on mountain and lowland, on deserts and marshes, most of the inhabitants within each great class are plainly related; for they will generally be descendants of the same progenitors and early colonists. On this same principle of former migration, combined in most cases with modification, we can understand, by the aid of the Glacial period, the identity of some few plants, and the close alliance of many others, on the most distant mountains, under the most different climates; and likewise the close alliance of some of the inhabitants of the sea in the northern and southern temperate zones, though separated by the whole intertropical ocean. Although two areas may present the same physical conditions of life, we need feel no surprise at their inhabitants being widely different, if they have been for a long period completely separated from each other; for as the relation of organism to organism is the most important of all relations, and as the two areas will have received colonists from some third source or from each other, at various periods and in different proportions, the course of modification in the two areas will inevitably be different.

On this view of migration, with subsequent modification, we can see why oceanic islands should be inhabited by few species, but of these, that many should be peculiar. We can clearly see why those animals which cannot cross wide spaces of ocean, as frogs and terrestrial mammals, should not inhabit oceanic islands; and why, on the other hand, new and peculiar species of bats, which can traverse the ocean, should so often be found on islands far distant from any continent. Such facts as the presence of peculiar species of bats, and the absence of all other mammals, on oceanic islands, are utterly inexplicable on the theory of independent acts of creation.

The existence of closely allied or representative species in any two areas, implies, on the theory of descent with modification, that the same parents formerly inhabited both areas; and we almost invariably find that wherever many closely allied species inhabit two areas, some identical species common to both still exist. Wherever many closely allied yet distinct species occur, many doubtful forms and varieties of the same species likewise occur. It is a rule of high generality that the inhabitants of each area are related to the inhabitants of the nearest source whence immigrants might have been derived. We see this in nearly all the plants and animals of the Galapagos archipelago, of Juan Fernandez, and of the other American islands being related in the most striking manner to the plants and animals of the neighbouring American mainland; and those of the Cape de Verde archipelago and other African islands to the African mainland. It must be admitted that these facts receive no explanation on the theory of creation.

The fact, as we have seen, that all past and present organic beings constitute one grand natural system, with group subordinate to group, and with extinct groups often falling in between recent groups, is intelligible on the theory of natural selection with its contingencies of extinction and divergence of character. On these same principles we see how it is, that the mutual affinities of the species and genera within each class are so complex and circuitous. We see why certain characters are far more serviceable than others for classification;—why adaptive characters, though of paramount importance to the being, are of hardly any importance in classification; why characters derived from rudimentary parts, though of no service to the being, are often of high classificatory value; and why embryological characters are the most valuable of all. The real affinities of all organic beings are due to inheritance or community of descent. The natural system is a genealogical arrangement, in which we have to discover the lines of descent by the most permanent characters, however slight their vital importance may be.

The framework of bones being the same in the hand of a man, wing of a bat, fin of the porpoise, and leg of the horse,—the same number of vertebrae forming the neck of the giraffe and of the elephant,—and innumerable other such facts, at once explain themselves on the theory of descent with slow and slight successive modifications. The similarity of pattern in the wing and leg of a bat, though used for such different purposes,—in the jaws and legs of a crab,—in the petals, stamens, and pistils of a flower, is likewise intelligible on the view of the gradual modification of parts or organs, which were alike in the early progenitor of each class. On the principle of successive variations not always supervening at an early age, and being inherited at a corresponding not early period of life, we can clearly see why the embryos of mammals, birds, reptiles, and fishes should be so closely alike, and should be so unlike the adult forms. We may cease marvelling at the embryo of an air-breathing mammal or bird having branchial slits and arteries running in loops, like those in a fish which has to breathe the air dissolved in water, by the aid of well-developed branchiae.

Disuse, aided sometimes by natural selection, will often tend to reduce an organ, when it has become useless by changed habits or under changed conditions of life; and we can clearly understand on this view the meaning of rudimentary organs. But disuse and selection will generally act on each creature, when it has come to maturity and has to play its full part in the struggle for existence, and will thus have little power of acting on an organ during early life; hence the organ will not be much reduced or rendered rudimentary at this early age. The calf, for instance, has inherited teeth, which never cut through the gums of the upper jaw, from an early progenitor having well-developed teeth; and we may believe, that the teeth in the mature animal were reduced, during successive generations, by disuse or by the tongue and palate having been fitted by natural selection to browse without their aid; whereas in the calf, the teeth have been left untouched by selection or disuse, and on the principle of inheritance at corresponding ages have been inherited from a remote period to the present day. On the view of each organic being and each separate organ having been specially created, how utterly inexplicable it is that parts, like the teeth in the embryonic calf or like the shrivelled wings under the soldered wing-covers of some beetles, should thus so frequently bear the plain stamp of inutility! Nature may be said to have taken pains to reveal, by rudimentary organs and by homologous structures, her scheme of modification, which it seems that we wilfully will not understand.

I have now recapitulated the chief facts and considerations which have thoroughly convinced me that species have changed, and are still slowly changing by the preservation and accumulation of successive slight favourable variations. Why, it may be asked, have all the most eminent living naturalists and geologists rejected this view of the mutability of species? It cannot be asserted that organic beings in a state of nature are subject to no variation; it cannot be proved that the amount of variation in the course of long ages is a limited quantity; no clear distinction has been, or can be, drawn between species and well-marked varieties. It cannot be maintained that species when intercrossed are invariably sterile, and varieties invariably fertile; or that sterility is a special endowment and sign of creation. The belief that species were immutable productions was almost unavoidable as long as the history of the world was thought to be of short duration; and now that we have acquired some idea of the lapse of time, we are too apt to assume, without proof, that the geological record is so perfect that it would have afforded us plain evidence of the mutation of species, if they had undergone mutation.

But the chief cause of our natural unwillingness to admit that one species has given birth to other and distinct species, is that we are always slow in admitting any great change of which we do not see the intermediate steps. The difficulty is the same as that felt by so many geologists, when Lyell first insisted that long lines of inland cliffs had been formed, and great valleys excavated, by the slow action of the coast-waves. The mind cannot possibly grasp the full meaning of the term of a hundred million years; it cannot add up and perceive the full effects of many slight variations, accumulated during an almost infinite number of generations.

Although I am fully convinced of the truth of the views given in this volume under the form of an abstract, I by no means expect to convince experienced naturalists whose minds are stocked with a multitude of facts all viewed, during a long course of years, from a point of view directly opposite to mine. It is so easy to hide our ignorance under such expressions as the "plan of creation," "unity of design," &c., and to think that we give an explanation when we only restate a fact. Any one whose disposition leads him to attach more weight to unexplained difficulties than to the explanation of a certain number of facts will certainly reject my theory. A few naturalists, endowed with much flexibility of mind, and who have already begun to doubt on the immutability of species, may be influenced by this volume; but I look with confidence to the future, to young and rising naturalists, who will be able to view both sides of the question with impartiality. Whoever is led to believe that species are mutable will do good service by conscientiously expressing his conviction; for only thus can the load of prejudice by which this subject is overwhelmed be removed.

Several eminent naturalists have of late published their belief that a multitude of reputed species in each genus are not real species; but that other species are real, that is, have been independently created. This seems to me a strange conclusion to arrive at. They admit that a multitude of forms, which till lately they themselves thought were special creations, and which are still thus looked at by the majority of naturalists, and which consequently have every external characteristic feature of true species,—they admit that these have been produced by variation, but they refuse to extend the same view to other and very slightly different forms. Nevertheless they do not pretend that they can define, or even conjecture, which are the created forms of life, and which are those produced by secondary laws. They admit variation as a *vera causa* in one case, they arbitrarily reject it in another, without assigning any distinction in the two cases. The day will come when this will be given as a curious illustration of the blindness of preconceived opinion. These authors seem no more startled at a miraculous act of creation than at an ordinary birth. But do they really believe that at innumerable periods in the earth's history certain elemental atoms have been commanded suddenly to flash into living tissues? Do they believe that at each supposed act of creation one individual or many were produced? Were all the infinitely numerous kinds of animals and plants created as eggs or seed, or as full grown? and in the case of mammals, were they created bearing the false marks of nourishment from the mother's womb? Although naturalists very properly demand a full explanation of every difficulty from those who believe in the mutability of species, on their own side they ignore the whole subject of the first appearance of species in what they consider reverent silence.

It may be asked how far I extend the doctrine of the modification of species. The question is difficult to answer, because the more distinct the forms are which we may consider, by so much the arguments fall away in force. But some arguments of the greatest weight extend very far. All the members of whole classes can be connected together by chains of affinities, and all can be classified on the same principle, in groups subordinate to groups. Fossil remains sometimes tend to fill up very wide intervals between existing orders. Organs in a rudimentary condition plainly show that an early progenitor had the organ in a fully developed state; and this in some instances necessarily implies an enormous amount of modification in the descendants. Throughout whole classes various structures are formed on the same pattern, and at an embryonic age the species closely resemble each other. Therefore I cannot doubt that the theory of descent with modification embraces all the members of the same class. I believe that animals have descended from at most only four or five progenitors, and plants from an equal or lesser number.

Analogy would lead me one step further, namely, to the belief that all animals and plants have descended from some one prototype. But analogy may be a deceitful guide. Nevertheless all living things have much in common, in their chemical composition, their germinal vesicles, their cellular structure, and their laws of growth and reproduction. We see this even in so trifling a circumstance as that the same poison often similarly affects plants and animals; or that the poison secreted by the gall-fly produces monstrous growths on the wild rose or oak-tree. Therefore I should infer from analogy that probably all the organic beings which have ever lived on this earth have descended from some one primordial form, into which life was first breathed.

When the views entertained in this volume on the origin of species, or when analogous views are generally admitted, we can dimly foresee that there will be a considerable revolution in natural history. Systematists will be able to pursue their labours as at present; but they will not be incessantly haunted by the shadowy doubt whether this or that form be in essence a species. This I feel sure, and I speak after experience, will be no slight relief. The endless disputes whether or not some fifty species of British brambles are true species will cease. Systematists will have only to decide (not that this will be easy) whether any form be sufficiently constant and distinct from other forms, to be capable of definition; and if definable, whether the differences be sufficiently important to deserve a specific name. This latter point will become a far more essential consideration than it is at present; for differences, however slight, between any two forms, if not blended by intermediate gradations, are looked at by most naturalists as sufficient to raise both forms to the rank of species. Hereafter we shall be compelled to acknowledge that the only distinction between species and well-marked varieties is, that the latter are known, or believed, to be connected at the present day by intermediate gradations, whereas species were formerly thus connected. Hence, without quite rejecting the consideration of the present existence of intermediate gradations between any two forms, we shall be led to weigh more carefully and to value higher the actual amount of difference between them. It is quite possible that forms now generally acknowledged to be merely varieties may hereafter be thought worthy of specific names, as with the primrose and cowslip; and in this case scientific and common language will come into accordance. In short, we shall have to treat species in the same manner as those naturalists treat genera, who admit that genera are merely artificial combinations made for convenience. This may not be a cheering prospect; but we shall at least be freed from the vain search for the undiscovered and undiscoverable essence of the term species.

The other and more general departments of natural history will rise greatly in interest. The terms used by naturalists of affinity, relationship, community of type, paternity, morphology, adaptive characters, rudimentary and aborted organs, &c., will cease to be metaphorical, and will have a plain signification. When we no longer look at an organic being as a savage looks at a ship, as at something wholly beyond his comprehension; when we regard every production of nature as one which has had a history; when we contemplate every complex structure and instinct as the summing up of many contrivances, each useful to the possessor, nearly in the same way as when we look at any great mechanical invention as the summing up of the labour, the experience, the reason, and even the blunders of numerous workmen; when we thus view each organic being, how far more interesting, I speak from experience, will the study of natural history become!

A grand and almost untrodden field of inquiry will be opened, on the causes and laws of variation, on correlation of growth, on the effects of use and disuse, on the direct action of external conditions, and so forth. The study of domestic productions will rise immensely in value. A new variety raised by man will be a far more important and interesting subject for study than one more species added to the infinitude of already recorded species. Our classifications will come to be, as far as they can be so made, genealogies; and will then truly give what may be called the plan of creation. The rules for classifying will no doubt become simpler when we have a definite object in view. We possess no pedigrees or armorial bearings; and we have to discover and trace the many diverging lines of descent in our natural genealogies, by characters of any kind which have long been inherited. Rudimentary organs will speak infallibly with respect to the nature of long-lost structures. Species and groups of species, which are called aberrant, and which may fancifully be called living fossils, will aid us in forming a picture of the ancient forms of life. Embryology will reveal to us the structure, in some degree obscured, of the prototypes of each great class.

When we can feel assured that all the individuals of the same species, and all the closely allied species of most genera, have within a not very remote period descended from one parent, and have migrated from some one birthplace; and when we better know the many means of migration, then, by the light which geology now throws, and will continue to throw, on former changes of climate and of the level of the land, we shall surely be enabled to trace in an admirable manner the former migrations of the inhabitants of the whole world. Even at present, by comparing the differences of the inhabitants of the sea on the opposite sides of a continent, and the nature of the various inhabitants of that continent in relation to their apparent means of immigration, some light can be thrown on ancient geography.

The noble science of Geology loses glory from the extreme imperfection of the record. The crust of the earth with its embedded remains must not be looked at as a well-filled museum, but as a poor collection made at hazard and at rare intervals. The accumulation of each great fossiliferous formation will be recognised as having depended on an unusual concurrence of circumstances, and the blank intervals between the successive stages as having been of vast duration. But we shall be able to gauge with some security the duration of these intervals by a comparison of the preceding and succeeding organic forms. We must be cautious in attempting to correlate as strictly contemporaneous two formations, which include few identical species, by the general succession of their forms of life. As species are produced and exterminated by slowly acting and still existing causes, and not by miraculous acts of creation and by catastrophes; and as the most important of all causes of organic change is one which is almost independent of altered and perhaps suddenly altered physical conditions, namely, the mutual relation of organism to organism,—the improvement of one being entailing the improvement or the extermination of others; it follows, that the amount of organic change in the fossils of consecutive formations probably serves as a fair measure of the lapse of actual time. A number of species, however, keeping in a body might remain for a long period unchanged, whilst within this same period, several of these species, by migrating into new countries and coming into competition with foreign associates, might become modified; so that we must not overrate the accuracy of organic change as a measure of time. During early periods of the earth's history, when the forms of life were probably fewer and simpler, the rate of change was probably slower; and at the first dawn of life, when very few forms of the simplest structure existed, the rate of change may have been slow in an extreme degree. The whole history of the world, as at present known, although of a length quite incomprehensible by us, will hereafter be recognised as a mere fragment of time, compared with the ages which have elapsed since the first creature, the progenitor of innumerable extinct and living descendants, was created.

In the distant future I see open fields for far more important researches. Psychology will be based on a new foundation, that of the necessary acquirement of each mental power and capacity by gradation. Light will be thrown on the origin of man and his history.

Authors of the highest eminence seem to be fully satisfied with the view that each species has been independently created. To my mind it accords better with what we know of the laws impressed on matter by the Creator, that the production and extinction of the past and present inhabitants of the world should have been due to secondary causes, like those determining the birth and death of the individual. When I view all beings not as special creations, but as the lineal descendants of some few beings which lived long before the first bed of the Silurian system was deposited, they seem to me to become ennobled. Judging from the past, we may safely infer that not one living species will transmit its unaltered likeness to a distant futurity. And of the species now living very few will transmit progeny of any kind to a far distant futurity; for the manner in which all organic beings are grouped, shows that the greater number of species of each genus, and all the species of many genera, have left no descendants, but have become utterly extinct. We can so far take a prophetic glance into futurity as to foretel that it will be the common and widely-spread species, belonging to the larger and dominant groups, which will ultimately prevail and procreate new and dominant species. As all the living forms of life are the lineal descendants of those which lived long before the Silurian epoch, we may feel certain that the ordinary succession by generation has never once been broken, and that no cataclysm has desolated the whole world. Hence we may look with some confidence to a secure future of equally inappreciable length. And as natural selection works solely by and for the good of each being, all corporeal and mental endowments will tend to progress towards perfection.

It is interesting to contemplate an entangled bank, clothed with many plants of many kinds, with birds singing on the bushes, with various insects flitting about, and with worms crawling through the damp earth, and to reflect that these elaborately constructed forms, so different from each other, and dependent on each other in so complex a manner, have all been produced by laws acting around us. These laws, taken in the largest sense, being Growth with Reproduction; inheritance which is almost implied by reproduction; Variability from the indirect and direct action of the external conditions of life, and from use and disuse; a Ratio of Increase so high as to lead to a Struggle for Life, and as a consequence to Natural Selection, entailing Divergence of Character and the Extinction of less-improved forms. Thus, from the war of nature, from famine and death, the most exalted object which we are capable of conceiving, namely, the production of the higher animals, directly follows. There is grandeur in this view of life, with its several powers, having been originally breathed into a few forms or into one; and that, whilst this planet has gone cycling on according to the fixed law of gravity, from so simple a beginning endless forms most beautiful and most wonderful have been, and are being, evolved.

Appendix 2

The Geologic Timescale

Most of the major geological time-spans were named in the 1800s by geologists who were either indifferent or hostile to evolution. Some of these names commemorate areas where rocks of a particular age—and containing distinctive **fossils**—were first collected and studied. For example, the Cambrian period was named in 1835 by Adam Sedgwick while he was studying rocks from Northern Wales (*Cambria* is a Roman-Latin term for *Welsh*). Similarly, the Permian was named after the town of Perm on the western edge of the Ural Mountains, and in 1799 Alexander von Humboldt applied the name Jurassic to the predominantly limestone rocks exposed in the Jura Mountains of France and Switzerland.

The names of some other geological time-spans commemorate the main types of substances deposited at a particular time. For example, the Carboniferous was named in 1822 for the coal beds and associated strata of north central England, and the Cretaceous was named for the chalk deposited as the Cretaceous progressed (*Kreta* is Latin for *chalk*). As first noted by William Smith in 1796, different geological ages can be distinguished and recognized by their distinctive fossils.

Finally, other periods are named in a numerical scheme (Table A.1). For example, north of the Jura Mountains, the strata below the Jurassic consist of limestone sandwiched between sandstone and clay. In 1834, Friedrich von Alberti named this three-layered zone the Triassic.

Table A.1 The International Stratigraphic Chart

Era	Period and Symbol* (Year Named)	Epoch (Year Named)	Millions of Years (Ma) from Start to Present**	Major Events
Cenozoic (Cz) "Recent Life"	Quaternary (Q)*** (1829)	Holocene (1833)	0.0117	Human activities reduce biological diversity to the lowest levels since the Mesozoic Flowering plants dominate land Rise of agriculture and civilizations
		Pleistocene (1839)	1.806	Modern humans appear Continents in modern positions Repeated glaciations and lowering of sea level Apelike ancestors of humans appear in Africa
	Neogene (N) (1856))	Pliocene (1833)	5.332	Increasingly dry, cool climate
		Miocene (1833)	23.03 ± 0.5	First hominins appear
	Paleogene (PE) (1856)	Oligocene (1854)	33.9 ± 0.1	Radiation of mammals, birds, snakes, angiosperms, teleost fishes, and pollinating insects
		Eocene (1833)	55.8 ± 0.2	
		Paleocene (1874)	66.5 ± 0.3	Continents near modern positions; climate increasingly cool
The Cretaceous-Tertiary (K-T) mass extinction eliminates more than 60% of all species. This is the most famous mass extinction, not because of its magnitude (the Permian mass extinction wiped out many more species), but because of its most famous victims: dinosaurs. Crocodiles, lizards, birds, and mammals are relatively unaffected.				
Mesozoic (Mz)	Cretaceous (K) (1822)		145.5 ± 4.0	Increasing diversity of mammals, birds, social insects, and angiosperms Continued radiation of dinosaurs Primitive birds replace pterosaurs Breakup of Gondwana Earth's climate cools
	Jurassic (J) (1799)		199.6 ± 0.6	First birds, lizards, and angiosperms

Era	Period	Millions of years ago	Events
			Gymnosperms dominate land Diversification of dinosaurs Mammals common but small Breakup of Pangaea into Gondwana and Laurasia Radiation of dinosaurs
The Triassic mass extinction eradicates about 20% of all species, thereby opening niches that allow the diversification of dinosaurs.			
	Triassic (TR) (1834)	251.0 ± 0.4	Increasing diversity of marine life Continents begin to separate Gymnosperms rise to dominance Diversification of reptiles, including archosaurs First mammals and crocodilia First dinosaurs
Paleozoic (Pz) "Ancient Life"	Permian (P) (1841)	299.0 ± 0.8	Glaciations and low sea level Diversification of insects, including beetles and flies Vertebrate land animals disperse Decline of amphibians Continents aggregate into Pangaea, creating the Appalachians Appearance of gymnosperms
The Permian mass extinction kills more than 90% of all species. The Permian mass extinction is the most devastating mass extinction.			
	Carboniferous (C) (1822)	359.2 + 2.5	Extensive forests of early vascular plants dominate land Early orders of winged insects Appearance of reptiles Diversification of amphibians Gondwana and small northern continents form

Continued

Table A.1 The International Stratigraphic Chart *(Continued)*

Era	Period and Symbol* (Year Named)	Epoch (Year Named)	Millions of Years (Ma) from Start to Present**	Major Events
	Devonian (D) (1839)		416.0 + 2.8	First amphibians, seed plants, insects, ferns, trees, and ammonoids Radiation of land plants Diversification of bony fishes Early tetrapods First wingless insects
The Devonian mass extinction eliminates as many as two-thirds of all species. Most of these extinctions are marine species; there are far fewer losses on land.				
	Silurian (S) (1835)		443.7 ± 1.5	Origin of jawed fishes Earliest vascular plants First millipedes and arthropleurids on land
	Ordovician (O) (1879)		488.3 ± 1.7	Diversification of invertebrates; arthropods and mollusks dominate the sea Early fish, corals, and echinoderms First green plants and fungi on land Ice Age at end of period
The Ordovician-Cambrian mass extinction eliminates most marine species; many groups lose more than half their species. Biologists know relatively little about the cause and impact of this mass extinction because most animals were soft bodied and therefore were unlikely to become fossils.				
	Cambrian (€) (1835)		542.0 ± 1.0	First appearance of many animal phyla in the Cambrian Explosion Diversification of marine animals and algae

		Abundant marine invertebrates First shelled organisms
Proterozoic “First Life”	2500	Earliest eukaryotes Trace fossils of animals Multicellular animals First invertebrates Atmosphere becomes oxygenic Possible “Snowball Earth” period
Archaean “Old”	4000	Origin of life First fossils Diversification of bacteria Photosynthetic bacteria produce O_2 Evolution of aerobic respiration
Hadean (informal) “Hell”	Lower limit not defined	Oldest known rocks and minerals Formation of Earth (c. 4550 Ma)

*Paleontologists often refer to faunal *stages* rather than *geological periods*.

**These dates are accurate ± 1%, but the boundary dates continue to change as geologists learn more about rocks and refine their methods of radiometric dating. The dates listed here are from the *International Stratigraphic Chart of the International Commission on Stratigraphy 2004.*

***The definitions of the Quaternary and the Pleistocene are under revision. The historic “Tertiary” includes the Paleogene and Neogene, but now has no official rank.

Appendix 3

Major Species of Known Hominins

Much evidence documents the evolution of modern humans (*Homo sapiens*). Although some dates and relationships between species remain unresolved (e.g., whether "Turkana Boy" is *Homo erectus* or *H. ergaster*), most species are well documented and firmly established. For example, there are more than 120 species of *Astralopithecus africanus* (of which "Taung Child" is but one), 160 individuals of *A. afarensis* (of which "Lucy" is but one), 150 individuals of *H. erectus,* 90 individuals of *A. robustus,* and 500 individuals of *H. neanderthalensis.*

Table A.2

Age of Hominin (Ma)*	Hominin ("Nickname")	Date of Discovery	Location of Discovery	Type Specimen**
7.0-6.0	*Sahelanthropus tchadensis* ("Toumaï")	2001	Toros-Menalla Chad,	Adult cranium
6	*Orrorin tugenensis* ("Millenium ancestor")	2001	Tugen Hills, Kenya	Adult mandible
5.8-5.2	*Ardipithecus kadabba*	1997	Middle Awash, Ethiopia	Jawbone
4.5-4.3	*Ardipithecus ramidus* ("Ardi")	1992	Aramis, Ethiopia	Adult teeth
3.5-3.3	*Kenyanthropus platyops* ("Flat-faced Man")	1998	Lake Turkana, Kenya	Adult cranium
4.1-3.5	*Australopithecus anamensis*	1965	Kanapoi, Kenya	Adult cranium
2.5-?	*Australopithecus garhi*	1997	Middle Awash, Ethiopia	Adult cranium
3.9-3.0	*Australopithecus afarensis* ("Lucy," "A.L. 288-1")	1974	Hadar, Ethiopia	Adult mandible
3.0-2.6	*Australopithecus africanus* ("Taung Child," "Mrs. Pless")	1924	Taung, South Africa	Immature Skull
2.7-2.2	*Australopithecus aethiopicus* (Paranthropus boisei) ("Black Skull")	1985	West Turkana, Kenya	Adult mandible
2.9-1.4	*Australopithicus boisei* ("Nutcracker Man," "Zinj")	1959	Olduvai, Tanzania	Adult Cranium
1.95	*Australopithecus sediba*	2010	Johannesburg, South Africa	Partial skeleton
1.9-1.4	*Australopithecus robustus* (Paranthropus robustus)	1938	South Africa	Adult Cranium
2.3-1.6	*Homo habilis* ("Handyman," "Lucy's Child")	1960	Olduvai, Tanzania	Adult mandible
2.0-1.6	*Homo rudolfensis*	1972	Lake Turkana, Kenya	Adult Cranium
1.8	*Homo georgicus*	2000	Republic of Georgia	Adult mandible

1.9-1.8	*Australopithicus sediba*	2010	South Africa	Skull, clavicle
1.9-1.5	*Homo ergaster* ("Turkana Boy")	1984	Lake Turkana, South Africa	Adult mandible
1.9-0.3	*Homo erectus* ("Java Man," "Peking Man")	1891	Trinil, Java	Adult skullcap
0.5-0.2	*Homo heidelbergensis* ("Mauer Jaw")	1907	Mauer, Germany	Adult mandible
0.3-0.03	*Homo neanderthalensis* ("Neandertal Man")***	1856	Neander Valley, Germany	Partial adult skeleton
0.2-0.01	*Homo floresiensis* ("Hobbit")	2003	Flores, Indonesia	Partial adult skeleton
0.3-0	*Homo sapiens*			

*Geologic time is typically measured in millions of years, or mega-annums (Ma), where the word "ago" is taken as read (e.g., 65 Ma can mean 65 million years ago).

**The type specimen of a species is a particular specimen to which the species name was first properly applied. The rules governing the naming of a species are described by the *International Code of Zoological Nomenclature*.

***Around 1900, German orthography changed and the silent *h* in some words was eliminated. Thus, "Neanderthal Man" and "Neandertal Man" are both appropriate, but this change did not apply to the spelling of the specific epithet, *H. neanderthalensis*.

Glossary

adaptation An inherited *trait* that confers a *fitness* benefit by improving an organism's chances of survival and reproduction.

adaptive radiation The development of many different forms from an originally homogeneous group of organisms as they fill different ecological niches; the diversification of *species* into separate forms that each adapt to occupy a specific environmental niche.

allele A variant of a *gene*. *Evolution* is measured as changes in allelic frequencies in a population over time.

allopatric Occupying separate, nonoverlapping geographic ranges. This term is often used in ecology to describe two or more *species* or populations that do not co-occur. Compare with *sympatric.*

altruism Self-sacrificing behavior that benefits others at a cost to the altruist. In biology, altruistic behavior decreases the actual or potential reproductive success of the altruist while increasing the actual or potential reproductive success of others.

Alvarez event See *K-T event.*

amino acid An organic molecule that consists of an amino group (NH_2), a carboxylic acid group (COOH), and a characteristic functional group. Amino acids form the primary structure of *proteins,* assembling in a sequence determined by the *DNA* bases in a gene.

amniote Vertebrates whose embryos are encased in the amnion, a protective membranous sac. Birds, reptiles, and mammals are amniotes.

analogous trait A *trait* shared by a set of species but not present in their common ancestor. An example of an analogous trait is the streamlined shape of fish and whales; this trait is shared not from common ancestry, but instead from convergent selective pressures. Also called convergently evolved trait. Compare with *homologous trait.*

anisogamy A condition in which gametes are distinguished based on size. In organisms with two sexes—male and female—these different-sized gametes define the sex (males have small, often-flagellated sperm; females have large eggs).

antagonism An adversarial, contentious, or spiteful relationship that includes competition as well as predator/prey and host/parasite interactions.

antediluvial The period before the biblical flood described in Genesis. Compare with diluvial.

antibiotic resistance The process by which bacteria evolve, via *natural selection,* immunity to antibiotics.

antigen A foreign substance that elicits an immune response.

aposematism Conspicuous warning coloration or patterning that advertises the toxicity or potency of its bearer.

appearance of age The claim that God created the world in six literal days relatively recently, and that this world was mature from its birth (i.e., it did not have to grow or develop from a simple beginning). That is, Adam and Eve were never infants, trees were created with tree rings, and geologic sediments were created with *fossils* already in them. Advocates of this idea have included Phillip Gosse, Henry Morris, and John Whitcomb, Jr. Also known as ideal-time creationism.

Ararat A mountainous region of extreme eastern Turkey near the Iranian border. The region's two highest peaks are Little Ararat (elevation = 3,914 m) and Great Ararat (elevation = 5,155 m), the latter of which is often claimed to be the resting place of Noah's Ark after the Genesis Flood (Genesis 8:4).

Archaea One of the two prokaryotic domains of life, the other being *Bacteria.*

Archaeopteryx The oldest and most primitive fossil bird yet discovered. *Archaeopteryx,* one of the most famous fossils in the world, lived 150 million years ago; it was about 6 inches high and 12–18 inches long, and weighed 1.5–3 kilograms. The first *Archaeopteryx* was discovered in 1861 in Upper Jurassic limestone in Germany. *Archaeopteryx* ("ancient feather"), which was named in 1863 by Richard Owen, had feathers, wings, and hollow bones like a bird, and teeth, legs, and a bony tail like a small coelurosaur (a subgroup of theropods that includes *Tyrannosaurus, Velociraptor,* and *Microraptor*).

archetype An abstract concept of a primitive form or body plan from which a group of organisms presumably developed. Anatomist Richard Owen described an archetype as a "divine idea" and "primal pattern."

argument from design The argument that order in nature is evidence that nature was created by a divine power. This argument was stated most famously by William Paley in his book *Natural Theology* (1802). Intelligent design creationism (ID) is a version of the argument from design.

artificial selection The deliberate selection of organisms by humans, performed to emphasize desirable or useful traits.

atavism The reappearance in an organism of a *trait* last seen in a remote ancestor but not seen more recently. Atavisms are similar to *vestigial traits,* but occur only in rare individuals instead of entire species. Examples of atavisms include extra limbs on dolphins, teeth in birds, and tails on humans.

australopithecines A subfamily (Australopithecinae) consisting of a single genus (*Australopithecus*) of early African bipedal *hominins* who lived 4.4–2.0 million years ago. Australopithecines walked upright, had relatively small brains (less than 500 cm^3), and did not use stone tools. The most famous australopithecine is *Lucy.* Also see Appendix 3.

background extinction The ongoing extinction of individual species due to environmental or ecological factors such as climate change, disease, loss of habitat, or competitive disadvantage in relation to other species. Background extinction occurs at a fairly steady rate over geological time and is the result of normal evolutionary processes, with only a limited number of species in an ecosystem being affected at any one time. Compare with *mass extinction.*

Bacteria One of two prokarytic domains of life, the other being *Archaea.*

basalt The most common type of solidified lava; a dense, dark grey, fine-grained, igneous rock made primarily of plagioclase feldspar and pyroxene.

Batesian mimicry A type of mimicry in which the individuals of a harmless species gain a fitness advantage through resembling individuals of a toxic species.

Beagle The British gunship on which Charles Darwin spent the years 1831–1836. During this time, Darwin sailed more than 64,000 kilometers, most of which were focused on South America. Without his experiences aboard the *Beagle,* it is unlikely that Darwin would have discovered *evolution* by *natural selection.*

behavior An action in response to an external or internal stimulus.

binomial A two-part, Latinized name of a species. For example, modern humans are members of *Homo sapiens* ("wise man"). In this binomial, *Homo* is the genus, and *sapiens* is our species epithet. Although humans are the only extant species of *Homo,* some genera are more diverse: *Canis* includes the gray wolf (*Canis lupus*), the coyote (*Canis latrans*), and the dog (*Canis familiaris).* Today's classification schemes, which continue to use binomials, try to reflect evolutionary relatedness, and are said to be *phylogenetic.* The foundation of binomial nomenclature was established in the 1753 edition of Linnaeus's *Species Plantarum.*

biogeography The study of the geographic distribution of organisms, and the changes in those distributions over time. Biogeography comprises a major source of evidence for *evolution.*

biological species concept The definition of a species as a group of organisms that can interbreed. See *species.*

biostratigraphy A branch of stratigraphy that correlates and assigns relative ages to rock strata based on fossils present in the strata.

bipedal Walking on hind limbs, especially in an upright, human manner.

blastula A single, spherical layer of cells that characterizes all animals during early embryonic development.

Buffon's law The biological principle stating that different geographic regions, although perhaps having similar environments, have their own distinctive plants and animals.

Burgess Shale A rock formation in the western Canadian Rockies containing a variety of fossilized invertebrates of the early Cambrian period that were buried by an underwater avalanche of fine silt, preserving many details of their soft parts. Fossils of the Burgess Shale have provided valuable information about the *evolution* of early life.

Cambrian Explosion An event at the beginning of the Cambrian period lasting 40–50 million years in which there was an abrupt (on a geological scale) increase in the diversity of animal *species.* Many of the major extant animal phyla appear in the fossil record during the Cambrian Explosion, which is often called the "big bang of animal evolution." Also see Appendix 2.

catastrophism The geological theory espoused in the 18th and 19th centuries by Georges Cuvier, Louis Agassiz, and others that Earth's geological features have resulted primarily from sudden, widespread (worldwide), violent, or unusual catastrophes (presumably caused by capricious natural forces) that are beyond our current experiences (e.g., a worldwide flood). These catastrophes were followed by the creation of new forms of life. Compare with *uniformitarianism.*

chromosome A threadlike structure made of DNA and proteins that carries genetic information.

coevolution Reciprocal adaptation occurring between *species.* Examples of coevolution have occurred between flowers and their pollinators, predators and their prey, and parasites and their hosts.

common ancestor The most recent ancestral form or *species* from which two species evolved.

comparative anatomy The science of testing predictions, and making deductions, based on structural and developmental similarities among, and differences between, different types of organisms.

competition A contest between individuals for control of a resource. Competition is a type of *symbiosis.*

continental drift The movement of continents in geologic time over Earth's surface due to *plate tectonics.*

convergently evolved trait See *analogous trait.*

co-option See *exaptation.*

creation Creationists' term for the beginning when God created life and the universe. For young-Earth creationists, creation occurred 6,000–10,000 years ago as is literally described in Genesis; for old-Earth creationists, creation occurred billions of years ago, and Genesis is interpreted allegorically.

creationism The religious belief in a supernatural deity or force that intervenes, or has intervened, directly in nature. Some forms of creationism (e.g., *young-Earth creationism*) stipulate that each species (or perhaps higher *taxon*) was created separately and in essentially its present form rather than by natural processes such as *evolution.* Although Western civilizations are largely dominated by Christian stories of creation, there are many interpretations of the Genesis story of creation (e.g., theistic evolution, young-Earth creationism, intelligent design creationism, and gap creationism).

Cretaceous system The group of stratified rocks normally below the oldest Tertiary deposits and above the Jurassic system. The Cretaceous system closed the Mesozoic era. The system, which is famous for its abundant chalk, is usually represented by a K (for *Kreide,* which is German for *chalk*). The Cretaceous ended with the *K-T event.*

crypsis antipredator adaptations that serve to conceal prey.

cytochrome *c* A pigment *protein,* used in cellular metabolism, that has been evolutionarily conserved across disparate taxa.

Darwinism The concept proposed by Charles Darwin that biological *evolution* has produced many highly adapted species through natural selection acting on hereditary variation in populations. See *natural selection.*

Darwin's finches The 14 species of finches unique to the Galápagos Islands. Charles Darwin collected some of these finches in 1835. Each species is adapted to exploit a different food source. These finches are not found on the mainland because *competition* there for these food sources from other birds is fiercer. All of the Galápagos finches are descendants of a few that strayed from the mainland; this provides important evidence for Darwin's theory of *evolution.* See also *adaptive radiation.*

deep time The extreme antiquity of Earth, as reflected by the geological record, but in contrast to the young age of Earth based on biblical literalism.

descent with modification The phrase used by Charles Darwin to refer to the process by which *natural selection* favors some variations, resulting in their becoming more common in the next generation. Descent with modification was Darwin's term for *evolution.*

deuterostome A taxonomic category referring to an organism in which the anus originates from the blastopore of the gastrula during embryonic development. Chordates and Echinoderms are deuterostomes. Compare with *protostome.*

DNA Deoxyribonucleic acid. DNA stores genetic information in all cellular organisms. DNA and the three-letter genetic code are examples of a *universal homology,* and evidence of a common ancestry for all living organisms.

DNA hybridization A molecular technique in which complementary strands of DNA from different organisms are joined and then separated with heat. The lower the temperature at which the strands denature, the less complementary the strands are.

drift See *genetic drift.*

electrophoresis A molecular technique in which components are separated by an electric current.

endosymbiotic hypothesis The proposal that eukaryotic cells, including organelles such as mitochondria and chloroplasts, arose from symbiotic associations of once free-living prokaryotic cells.

environment An organism's biotic and abiotic surroundings.

Eukarya The domain of life that includes all eukaryotic organisms.

evolution Change in the genetic makeup of a population over time. Evolution occurs over successive generations, not in the lifetimes of individual organisms. Charles Darwin referred to evolution as *descent with modification.*

evolutionary developmental biology (evo-devo) A discipline of biology that compares the development of different organisms to determine ancestral relationships and the developmental mechanisms that produce evolutionary change.

exaptation A term coined in 1992 by Stephen Gould and Elisabeth Vrba to refer to a *trait* evolved for one function and later co-opted (and perhaps further evolved) for another. A trait that may be beneficial, but was favored by *natural selection* for a different function. Also called co-option.

extinction The permanent disappearance of a *species* or a group of species.

fission-track dating A type of radiometric dating based on the number of tracks made in volcanic rock as uranium-238 spontaneously decays into lead. The number of tracks is proportional to the age of the sample.

fitness The relative success of an organism measured by its ability to contribute *genes* to future generations.

flood geology The claim originated by George McCready Price that a biblical, worldwide flood (as described in Genesis 6–9) accounts for geological formations and the *fossil* record (e.g., that helpless invertebrates were buried first, while larger animals fled or floated to what became higher strata in the geological record). Price's idea was developed by John Whitcomb, Jr., and Henry Morris in *The Genesis Flood* (1961), which began the modern *creationism* movement. Flood geology, which is also referred to as diluvial geology and creation geology, is based on biblical literalism and is a hallmark of *young-Earth creationism.*

fossil A preserved remnant, impression, or trace of an organism that lived in the past. Today, there are more than 250,000 known species of fossil animals, most of which are shelled creatures that lived in shallow seas. Although Charles Darwin's ideas for *On the Origin of Species* were based almost entirely on living organisms, today fossils are a primary source of evidence for Darwin's theory. Few organisms become fossils.

Galápagos archipelago The volcanic islands in the Pacific Ocean 1,000 kilometers west of Ecuador (the country to which they belong) visited by Charles Darwin in September, 1835. Observations at these islands by Darwin later helped him formulate the theory of *evolution* by *natural selection.*

gastrula An invaginated ball of cells that characterizes most animals during early embryonic development.

gene A unit of heredity. Most genes are specific sequences of *DNA* that contain information needed to make a *protein.*

gene flow The movement of genetic information from one population to another; migration.

gene pool All of the *genes* in a population.

genetic code The sequence of nucleotides in *DNA* or RNA that specifies the *amino acids* involved in making a *protein.*

genetic drift A change in the allelic frequencies of a small population due to chance rather than *natural selection.*

genetics The science of heredity; the study of *genes* and their relationship to the traits of organisms.

geologic time The 4.6-billion-year timescale used by scientists to describe events in Earth's history. See Appendix 2.

gradualism The proposition that large phenotypic differences have evolved through many slightly different intermediate steps. Also called phyletic gradualism. Compare with *punctuated equilibrium.*

Grand Canyon A gorge of the Colorado River in northwest Arizona. Grand Canyon, which is up to 1.6 kilometers deep, from 6.4 to 29 kilometers wide, and more than 321.8 kilometers long, is cited by young-Earth creationists as evidence of the Genesis Flood.

group selection The idea that the frequency of *alleles* can increase in a population because of the benefits those alleles bestow on groups, regardless of the alleles' effect on the *fitness* of individuals within that group.

half-life The time required for half the nuclei in a sample of a specific isotopic species to undergo radioactive decay.

hemoglobin An oxygen-binding pigment protein that is highly conserved across disparate taxa.

herbivore An organism that consumes primarily plant matter for food.

homeobox A *DNA* sequence that characterizes several *genes* involved in body segmentation during embryonic development. Homeobox sequences characterize *Hox* genes, which code for *homeodomains.*

homeodomain One of several highly conserved transcription-factor proteins that control body segmentation during embryonic development. Homeodomains are the product of *homeobox*-containing *Hox* genes.

hominid A member of the taxonomic family Hominidae that includes humans, gorillas, chimpanzees, and orangutans. Compare with *hominin.* Also see Appendix 3.

hominin A member of the taxonomic tribe Hominini that includes the human lineage; any human relative whose last common ancestor postdates the divergence of humans and chimpanzee lineages. Aside from modern humans, all hominins (e.g., *Sahelanthropus, Australopithecus*) are now extinct. Compare with *hominid.* Also see Appendix 3.

homologous trait A similarity observed in related species that results from common ancestry.

host In a parasitic relationship, the individual who provides resources—often unwillingly—and, in so doing, experiences a fitness cost.

Hox genes Highly conserved *genes* that specify body-axis development (front-to-back) and segmentation during embryonic development. Hox genes all house *homeobox* sequences and code for *homeodomain* proteins.

hydraulic sorting A model invoked by many young-Earth creationists to explain the *fossil* record. According to the model, the biblical flood buried the heavy shells of immobile marine invertebrates and fish in the lower sediments, and motile animals such as amphibians—who were fleeing the floodwaters—in higher, intermediate sediments. Smart animals such as humans, who were also fleeing the rising floodwaters, allegedly climbed to the highest levels before drowning and being buried.

hydrothermal vent A fissure in the planet's surface, from which volcanically heated water spews forth. Hydrothermal vent communities are metabolically unique and can be biologically diverse.

igneous rock Any of various crystalline or glassy, noncrystalline rocks formed by the cooling and solidification of molten earth material (magma).

inclusive fitness A measure of reproductive success that includes not only your direct descendents (e.g., your offspring) but also your indirect descendents (e.g., nieces and nephews).

inheritance of acquired traits The claim that *traits* acquired during a parent's lifetime can be passed to offspring. This was the basis for the first (albeit inaccurate) recorded scientific theory of *evolution.* The inheritance of acquired traits, which is associated with Jean-Baptiste Lamarck, was popular until the establishment of modern genetics. Also known as the use-disuse theory, soft inheritance, and *Lamarckism.*

innate A feature that is genetically programmed or congenital, and not learned or acquired through experience.

interspecific Between individuals of two or more *species.*

intraspecific Among individuals of the same *species.*

isochron A line on a chart linking rocks of the same age, especially as measured using the ratios of lead isotopes.

isotope One of two or more atoms with the same atomic number but with different numbers of neutrons.

K-T event An ancient cataclysm involving at least one large meteorite that hit Earth near the present-day Yucatán Peninsula approximately 65 million years ago at the end of the Cretaceous (the third and final period of the Mesozoic era). This event triggered a *mass extinction,* the most famous victims of which were the nonavian dinosaurs. Also called the Alvarez event for its early proponents, Luis and Walter Alvarez. See Appendix 2.

kin selection A type of selection that acts on indirect *fitness* (i.e., fitness gains from the increased reproduction of relatives).

Lamarckism A discredited but historically influential proposal of Jean-Baptiste Lamarck claiming that traits developed or lost during an organism's life can be passed to the organism's offspring. Also see *inheritance of acquired traits.*

lichen An obligate mutualism between an alga and a fungus.

LUCA Last universal common ancestor. If the *evolution* of all life is shown in a tree of life, the organism represented by the ancestral node (i.e., the putative forebearer of all life) is LUCA. Just as humans and chimps share a common ancestry, all modern forms of life share a common history back to the divergence that produced the three domains of life (i.e., as far back as LUCA, estimated to be 3.6–4.1 billion years ago). LUCA was not necessarily the first living organism or the only common ancestor of life, but rather the most recent one (or group of organisms).

Lucy Nickname for a partial (47 of 206 bones) female skeleton of a *fossil hominin* discovered by David Johanson in 1974 in Ethiopia. Lucy was 1.2 meters tall, lived approximately 3.2 million years ago, and was a member of *Australopithecus afarensis.* Lucy, who is formally known as "A. L. 288–1," is regarded by some paleontologists as an ancestor of all subsequent *Australopithecus* and *Homo* species. Lucy is the most famous human ancestor. Also see Appendix 3.

macroevolution A vague term used to describe major evolutionary change involving groups above the level of *species.*

mass extinction The collective extinction of large numbers of *species* in a relatively short geological period of time. There have been several mass extinctions during Earth's history, the largest of which occurred 250 million years ago, at the end of the Permian; this mass extinction (The Great Dying) eliminated more than 80% of the species on Earth. However, the most famous mass extinction occurred 65 million years ago at the end of the Cretaceous. That extinction, which killed two-thirds of existing species, is best known for its most famous victims: the nonavian dinosaurs. Also see Appendix 2.

metamorphic rock Rock that has been transformed by pressure, heat, or other natural causes.

microevolution Changes in *gene* frequencies within *species* or populations.

mid-oceanic ridge An underwater mountain range formed by *plate tectonics* and typically having a valley known as a rift running along its spine. A mid-ocean ridge marks the boundary between two tectonic plates. The mid-ocean ridges of the world are connected and form a single global mid-oceanic ridge system that is part of every ocean. The mid-oceanic ridge system is the longest mountain range in the world.

mimicry A natural phenomenon in which one species gains a fitness advantage through resembling another species.

missing link A popular term used during Charles Darwin's era to denote hypothetical organisms that linked different groups of organisms, and especially humans with anthropoid apes. Most biologists no longer use the term "missing link" because it implies that organisms are linked by a hierarchal chain or ladder, when, in fact, organisms share common ancestors. The use of "missing link" also inaccurately implies that if a certain fossil has not yet been found, then *evolution* cannot be valid. See *transitional form.*

Modern Synthesis The integration of Charles Darwin's theory of *evolution* by *natural selection* with other scientific disciplines—especially those involving paleontology and contemporary genetics—to produce a comprehensive theory of evolution. The Modern Synthesis was devised largely between 1937–1947.

molecular clock The clocklike regularity of the change of a molecule (e.g., a gene) or a whole genotype over geologic time.

Mullerian mimicry A type of mimicry in which two or more toxic species gain protection through their shared appearance.

mutation A random change in genetic information. Mutations, which can be caused by mistakes during replication or by damage from external agents such as chemicals or radiation, produce the genetic changes that provide new genetic information. Mutations create new *alleles.* Heritable mutations provide the template of variation upon which *natural selection* acts.

mutualism An interaction between organisms in which each participant experiences a gain in *fitness.*

natural selection An evolutionary mechanism that produces differences in survival and reproduction among organisms with different heritable traits. Natural selection is the mechanism for adaptive evolutionary change proposed by Charles Darwin in *On the Origin of Species* (1859).

natural theology Theology or knowledge of God based on observations of nature rather than on divine revelation.

Neandertal A *hominin* that was similar to, but distinct from, modern humans. Neandertals lived in Western Asia and Europe from 150,000 to 30,000 years ago. Neandertals were discovered in 1856 in Germany's Neander Valley (*Neander Thal* in Old German). In 1901, when German spelling was made more consistent with pronunciation, *Thal* was changed to *Tal;* this is why Neandertals are sometimes referred to as Neanderthals.

Noachian Flood Genesis 7 reports that "all the fountains of the great deep broken up, and the windows of heaven were opened" and "the rain was upon the earth forty days and forty nights," which ultimately meant that "every living substance was destroyed which was upon the face of the ground." Before the Flood, Noah and his family herded "two and two of all flesh" of "every beast after his kind" into the Ark; when the rain ended, organisms aboard the Ark were released and repopulated Earth. The reality of the Noachian Flood is a critical component of most young-Earth creationists' perspective of Earth's history, and flood geologists interpret geological formations (e.g., the *Grand Canyon*) as proof of the action of the biblical Flood. Also known as the biblical Deluge.

oceanic trench A long, narrow, and deep depression in the ocean floor where, at the junction of two tectonic plates, one plate dives steeply beneath another. Oceanic trenches are the deepest parts of oceans.

ontogeny The development of an organism over its lifetime, from zygote to death. Compare with *phylogeny.*

ontogeny recapitulates phylogeny Development repeats *evolution.* The idea that one can see the assemblage of evolutionary stages in embryonic development (i.e., that embryological development is a flashback to past ancestral events). Also known as the biogenic law.

paleomagnetism The permanent magnetism in rocks, resulting from the orientation of Earth's magnetic field at the time of rock formation in a past geologic age. Paleomagnetism is a source of information for the paleomagnetic studies of *plate tectonics.*

paleontology The scientific study of *fossils* to reconstruct the history of life.

Pangaea A supercontinent that existed 250 million years ago consisting of most or all of today's continents. The breakup of Pangaea during the late Jurassic caused the isolation (and therefore the separate *evolution*) of different groups of organisms from each other.

pangenesis A hereditary proposal of Charles Darwin and others in which small "pangenes" or "gemmules" are produced by all of an organism's tissues and are sent to reproductive structures, where they are incorporated into gametes. A change in the amount of a specific pangene resulting from the use or disuse of a particular structure was proposed to explain the *inheritance of acquired traits.*

parasitism An interaction between organisms in which one, the host, loses *fitness* while the other, the parasite, extracts resources and gains fitness.

phylogenetics The study of evolutionary relationships between groups of organisms, especially of the branching patterns of different lineages.

phylogenetic tree A branching diagram depicting the evolutionary relationships among groups of organisms.

phylogeny The evolutionary history of a *species* or higher taxonomic group of organisms. Compare with *ontogeny.*

plate tectonics The theory that Earth's crust consists of movable plates that can join or separate over geologic time. The movements of plates explains *continental drift,* earthquakes, volcanoes, mountain building, and some aspects of *biogeography.*

pollination In plants, the transfer of sperm (as pollen) from male reproductive structures (anthers) to female reproductive structures (stigmas). Pollination can be conducted abiotically—by wind or water—or biotically by the movement of animals such as insects, birds, and mammals.

polymorphism Genetic variation within a population. Polymorphisms are essential for *natural selection* because they provide the variability that contributes to differential survival and reproduction.

population A geographically localized group of interbreeding organisms that share a *gene pool.*

predation The act of one animal killing another for food.

principle of superposition The claim by Nicolas Steno in 1669 that in undisturbed *sedimentary rock,* older strata are lower in the geologic column than younger strata.

protein An organic molecule specified by a *DNA* sequence, and consisting of *amino acids*. Proteins serve many functions in an organism, including structural (e.g., collagen), respiratory (e.g., hemoglobin), and protective (e.g., melanin, antibodies) roles.

protostome A taxonomic category referring to an organism in which the mouth originates from the blastopore of the *gastrula* during embryonic development. Annelids, arthropods and mollusks are protostomes. Compare with *deuterostome.*

pseudogene A *gene* that is no longer functioning at full or even partial capacity. Pseudogenes are vestigial molecular elements that exist as evolutionary relics.

punctuated equilibrium A concept developed by Niles Eldredge in 1971 (and popularized by Eldredge and Stephen Gould in 1972) suggesting that the tempo of *evolution* is more sporadic than gradual. Populations evolve rapidly into new *species,* after which there are long periods of equilibrium with little evolutionary change. Compare with *gradualism.*

radiometric dating A common way of estimating the age of a *fossil* or rock by analyzing the elemental isotopes and the products of their decay within the accompanying rock.

RNA world The name given by Nobel laureate Walter Gilbert (b. 1932) to the concept that pieces of RNA having catalytic and self-replicating abilities predated protein synthesis before life appeared on Earth.

saltation Large-scale evolutionary change between successive generations generally leading to speciation. Prior to an adequate understanding of the underlying genetics of *evolution,* it was widely held that evolutionary novelties (including *species*) could arise only through major mutational changes occurring in a single generation. Integration of an understanding of inheritance and the nature of the underlying genetic variation of *traits* has allowed understanding of how speciation can occur through gradual genetic change across many generations.

sedimentary rock Rock formed by consolidated sediments that are deposited in layers.

selection pressure Environmental forces (e.g., the scarcity of food) that result in the differential survival and reproduction of some organisms having *traits* that provide resistance.

sexual selection Differences in *fitness* as a result of differences in the ability to obtain mates.

social Darwinism A trend in social theory holding that Darwin's theory of *evolution* by *natural selection* can substantiate a political ideology (e.g., that ruthless egoism is the most successful policy) and provide a critique human social institutions. Although biological evolution in its pure form is descriptive (i.e., it tells us, without judgment, what has happened to life on Earth), Social Darwinism is prescriptive.

sociobiology The biological study of social behavior (including that of humans), with particular reference to the adaptive features of those behaviors.

special creation The belief that all forms of life were created by God as separate, distinct *species.* This belief implie that species do not change through time, and that there are no evolutionary relationships between different species.

speciation The *evolution* of new species. Evolution usually proceeds by speciation—the splitting of one lineage from a parental stock—and not by a slow transformation of large parental stocks. Repeated episodes of speciation (i.e., evolutionary sequences) are not rungs of a ladder, but instead are retrospective reconstructions of evolution.

species The fundamental unit of biological classification commonly defined as a group of organisms that can interbreed to produce fertile, viable offspring (i.e., the biological species concept). Species are considered by most biologists to be real biological entities—unlike higher levels of taxonomic organization (e.g., genera, orders, families) that are human devices to catalog biological diversity—that have arisen due to specific evolutionary histories. Although the biological species concept is used most widely, it has limited applicability to many types of organisms, especially those that are extinct or that do not reproduce sexually; as a result, there have been dozens of species concepts devised. This has led, for example, to recent proposals that the identification of unbranching lineages—that is, groups of organisms that have a unique, shared evolutionary history—should be used to define species, although this information may be difficult to collect for many organisms.

strata Layers of rock, typically horizontal.

superposition The principle formulated by Danish geologist Nicholas Steno that layers of rock are arranged in a time sequence, with the oldest on the bottom and the youngest on the top, unless later processes disturb this arrangement. This principle is Steno's most famous contribution to geology.

survival of the fittest A phrase coined by philosopher Herbert Spencer in 1852 to describe the competition for survival and preeminence in a population. Although Charles Darwin used survival of the fittest as a synonym for *natural selection,* it is a metaphor seldom used by modern biologists.

symbiosis A relationship between two or more interacting organisms. Symbioses occur on a continuum from antagonisms (e.g., parasitism, predation, competition) to obligate mutualisms, in which the symbionts cannot function without each other.

sympatric Occupying overlapping geographic ranges. This term is often used in ecology to describe two or more *species* or populations that co-occur. Compare with *allopatric.*

synteny The relationship between *genes* occurring on the same *chromosome* in different species.

taxon A grouping of organisms, such as a *species,* genus, or family.

taxonomy The theory and practice of naming and classifying groups of organisms.

tetrapod A vertebrate with four limbs, such as an amphibian, reptile, or mammal.

theory In science, as opposed to common usage, a theory is a well-substantiated explanation of some aspect of the natural world that usually incorporates many confirmed observational and experimental facts. A scientific theory makes predictions consistent with what we see. A scientific theory is not a guess; on the contrary, a scientific theory is widely accepted within the scientific community (e.g., the germ theory claims that certain infectious diseases are caused by microorganisms). Scientific theories do not become facts; scientific theories *explain* facts.

trait A characteristic, condition, or property of an organism or population.

transitional form An organism having anatomical features intermediate between those of two major groups of organisms in an evolutionary sequence. Transitional forms show evolutionary sequences between lineages by having characteristics of ancestral and newer lineages. Since all populations are in evolutionary transition, a transitional form represents a particular evolutionary stage that is recognized in hindsight. *Archaeopteryx* is a transitional form between birds and dinosaurs, and *Tiktaalik* is a transitional form between land-dwelling tetrapods evolved from fish ancestors. Compare with *missing link.*

transmutation The concept of evolution—especially when leading to the production of new species—was generally referred to as transmutation for hundreds of years. For example, Jean-Baptiste Lamarck referred to his proposal of evolution via inheritance of acquired characteristics as his "transmutation hypothesis," and Charles Darwin (who did not use the word *evolution* in *On the Origin of Species*) developed many of his ideas about evolution in his "transmutation notebooks."

triploblast An organism with three embryonic tissue types, originating with the *gastrula.*

unconformity The surface between two successive *strata* representing a missing interval in the geologic record. Scotland's Siccar Point, which helped James Hutton appreciate Earth's vast age, is an unconformity.

uniformitarianism A theory suggested by James Hutton and developed by Charles Lyell summarized by the phrase "the present is the key to the past"—that Earth's geological features have developed over long periods of time through a variety of slow geologic processes involving common events such as rain, volcanic activity, and wind. Uniformitarianism does not imply that change occurred at a uniform rate or deny the occurrence of localized catastrophes. Indeed, much evidence attests to occasional catastrophic events in Earth's history, most famously the impact of a meteor or comet off the Yucatán Peninsula approximately 65 million years ago that led to the extinction of much Mesozoic life (including nonavian dinosaurs). Such events that depart from gradual geologic change do not necessarily support *young-Earth creationism* or biblical catastrophes such as a worldwide flood. Compare with *catastrophism.*

universal homology A similar *trait* found in all organisms. Gene replication, transcription, and translation—the basis of molecular biology—are examples of universal homologies.

vestigial traits Traits of organisms that are identifiable and often characteristic of that species but that either have no apparent function or have a function different from that for which they evolved. Vestigial traits reflect the evolutionary history of a lineage where new selective forces favored the change or loss of once-useful characteristics (adaptations). Examples of vestigial traits in humans include the coccyx, appendix, and muscles that move our ears.

virus A very small infectious agent that relies completely on the cellular machinery of its hosts.

Wallace line A boundary that separates the flora of Asia and Australia. Organisms west of the line are related to Asiatic species, whereas those east of the line are most closely related to Australian species. See Figure 4.3.

young-Earth creationism The claim that Earth is 6,000 to 10,000 years old, that the six days of creation described in Genesis each lasted 24 hours, and that catastrophic events such as Noah's Flood produced the Grand Canyon and other geological features. Young-Earth creationists, who reject many aspects of modern biology, geology, physics, and other sciences, are biblical literalists who believe that modern organisms were created by God, and that life did not evolve. Young-Earth creationism, which involves divine intervention that suspends the laws of nature, also contradicts theistic evolution, which claims that God works through natural laws. Today, when most people hear the word *creationism,* they think of young-Earth creationism.

References

Age of Earth

Baadsgaard, H., J. F. Lerbekmo, and J. R. Wijbrans. 1993. "Multimethod Radiometric Ages of a Bentonite Near the Top of the *Baculites reesidei* Zone of Southwest Saskatchewan (Campanian-Maastrichtian Stage Boundary?)." *Canadian Journal of Earth Science* 30: 769–775.

Barrell, J. 1917. "Rhythms and the Measurement of Geologic Time." *Bulletin of the Geological Society of America* 28: 745–904.

Brice, W. R. 1982. "Bishop Ussher, John Lightfoot and the Age of Creation." *Journal of Geological Education* 30: 18–24.

Dalrymple, G. B. 1991. *The Age of the Earth.* Stanford, CA: Stanford University Press.

Dalrymple, G. B. 1994. *Ancient Earth, Ancient Skies.* Stanford, CA: Stanford University Press.

Darwin, G. H. 1879. "On the Precession of a Viscous Spheroid, and on the Remote History of the Earth. *Philosophical Transactions of the Royal Society of London* 170: 447–538.

de Maillet, Benoit. 1748. *Telliamed: or Conversations between an Indian Philosopher and a Frence Missionary on the Diminution of the Sea*. Amsterdam: L'Honoré et Fils.

Eldredge, Niles. 1982. *The Monkey Business: A Scientist Looks at Creationism.* New York: Pocket Books.

Faul, H. 1966. *Ages of Rocks, Planets, and Stars.* New York: McGraw-Hill.

Fleck, John. 2009. *The Tree Rings' Tale: Understanding Our Changing Climate.* Albuquerque, NM: University of New Mexico Press.

Haber, F. C. 1959. *The Age of the Earth: Moses to Darwin.* Baltimore: Johns Hopkins Press.

Hazen, R. M. 2010. "How Old is Earth, and How Do We Know?" *Evolution Education and Outreach* 3: 198–205.

Holmes, A. 1927. *The Age of the Earth: An Introduction to Geological Ideas.* London: Ernest Benn.

Huxley, T. H. 1869. "The Anniversary Address of the President." *Quarterly Journal of the Geological Society of London* 25: xxviii–liii.

Joly, J. 1899. "An Estimate of the Geological Age of the Earth." *Annual Report of the Smithsonian Institution for 1899:* 247–288.

Kerr, R. A. 2008. "Two Geologic Clocks Finally Keeping Same Time." *Science* 320: 434–435.

Lamarck, J. B. 1802[1964]. *Hydrogéologie.* Translated as *Hydrogeology* by Albert V. Carozzi. Urbana: University of Illinois Press.

Miller, Hugh. 1857. *The Testimony of the Rocks.* Boston: Gould and Lincoln.

Numbers, R. 1992. *The Creationists: The Evolution of Scientific Creationism.* New York: Knopf.

Patterson, C. 1956. "Age of Meteorites and the Earth." *Geochimica et Cosmochimica Acta* 10: 230–237.

Reese, R. L., S. M. Everett, and E. D. Craun. 1981. "The Chronology of Archbishop James Ussher." *Sky and Telescope* 62: 404–405.

Schuchert, C. 1931. "Geochronology or Age of the Earth on the Basis of Sediments and Life." In *Physics of the Earth, Part 4: The Age of the Earth,* 10–64. Washington, DC: National Research Council of the National Academy of Sciences, Bulletin 80.

Snelling, Andrew. 2010. "Radiometric Dating." *Answers* (April–June): 51.

Thomson, W. 1862a. "On the Age of the Sun's Heat." *Macmillan's Magazine* (March): 388–393.

Thomson, W. 1862b. "On the Secular Cooling of the Earth." *Royal Society of Edinburgh Transactions* 23: 157–159.

Ussher, J. [1650] 2003. *Annals of the World: James Ussher's Classic Survey of World History.* Modern English republication, ed. Larry Pierce and Marion Pierce. Reprint, Green Forest, AR: Master Books.

Fossils

Ash, Sidney. 2005. *Petrified Forest: A Story in Stone.* Petrified Forest, AZ: Petrified Forest Museum Association.

Barrow, Mark V., Jr. 2009. *Nature's Ghosts: Confronting Extinction from the Age of Jefferson to the Age of Ecology.* Chicago: University of Chicago Press.

Clack, Jennifer A. 2009. "The Fish-Tetrapod Transition: New Fossils and Interpretations." *Evolution Education and Outreach* 2:doi:10.1007/s12052-009-0119-2.

Daeschler, E. B., and Shubin, Neil. 1998. "Fish with Fingers?" *Nature* 391:133.

Daniels, F. J. and R. D. Dayvault. 2006. *Ancient Forests: A Closer Look at Fossil Wood.* Grand Junction, CO: Western Colorado Publishing.

Edwards, L. E., and Pojeta, J., Jr. 1994. *Fossils, Rocks, and Time.* Washington, DC: U.S. Government Printing Office.

Emling, Shelley. 2009. *The Fossil Hunter: Dinosaurs, Evolution, and the Woman Whose Discoveries Changed the World.* New York: Palgrave Macmillan.

Gaz, Stan. 2009. *Sites of Impact: Craters around the World.* New York: Princeton Architectural Press.

Parker, S. 2007. *The Complete Guide to Fossils and Fossil-Collecting.* London: Anness Publishing.

Pojeta, J., Jr., and D. A. Springer. 2001. *Evolution and the Fossil Record.* Alexandria, VA: American Geological Institute.

Prothero, D. 2004. *Bringing Fossils to Life: An Introduction to Paleobiology,* 2nd ed. New York: W. H. Freeman.

Prothero, D., and F. Schwab. 2004. *Sedimentary Geology,* 2nd ed. New York: W. H. Freeman.

Shubin, N. H., E. B. Daeschler, and F. A. Jenkins. 2006. "The Pectoral Fin of *Tiktaalik roseae* and the Origin of the Tetrapod Limb." *Nature* 440:764–771.

Shubin, Neil. 2008. *Your Inner Fish: A Journey into the 3.5-Billion-Year History of the Human Body.* New York: Pantheon Books.

Winchester, Simon. 2002. *The Map That Changed the World: William Smith and the Birth of Modern Geology.* New York: HarperCollins.

Biogeography

Cowen, R. 2000. *History of Life.* 3rd ed. Boston: Blackwell Science.

Cox, A., and R. B. Hart. 1986. *Plate Tectonics: How It Works.* Oxford: Blackwell.

Cox, C. Barry, and Peter D. Moore. 2005. *Biogeography: An Ecological and Evolutionary Approach.* 7th ed. Malden, MA: Blackwell Publishing.

Crisci, J. V., L. Katinas, and P. Posadas. 2003. *Historical Biogeography: An Introduction.* Cambridge, MA: Harvard University Press.

Darwin, Charles. 1845. *Journal of Researches in the National History of Geology of the Countries Visited During the Voyage of H.M.S. Beagle Round the World.* London: John Murray.

Darwin, Charles. 1859. *On the Origin of Species by Natural Selection, or the Preservation of Favoured Races in the Struggle for Life.* London: John Murray.

Ebach, M. C., and R. S. Tangney, eds. 2006. *Biogeography in a Changing World.* Boca Raton, FL: Taylor and Francis, CRC Press.

Grant, K. Thalia, and Gregory B. Estes. 2009. *Darwin in Galápagos: Footsteps to a New World.* Princeton, NJ: Princeton University Press.

Lavers, C. 2000. *Why Elephants Have Big Ears: Understanding Patterns of Life and Earth.* New York: St. Martin's Press.

Lomolino, Mark V., Brett R. Riddle, and James H. Brown. 2006. *Biogeography.* 3rd ed. Sunderland, MA: Sinauer Associates.

MacArthur, R. H., and E. O. Wilson. 1967. *The Theory of Island Biogeography.* Princeton, NJ: Princeton University Press.

McCalman, Iain. 2009. *Darwin's Armada: Four Voyages and the Battle for the Theory of Evolution.* New York: Simon and Schuster.

McCarthy, Dennis. 2009. *Here Be Dragons: How the Study of Animal and Plant Distributions Revolutionized Our Views of Life and Earth.* Oxford: Oxford University Press.

Quammen, David. 1996. *The Song of the Dodo: Island Biogeography in an Age of Extinction.* London: Hutchinson.

Simpson, George G. 1953. *Evolution and Geography.* Eugene, OR: Oregon State System of Higher Education.

Stanley, S. 1986. *Earth and Life through Time.* New York: W. H. Freeman.

Tallis, J. H. 1991. *Plant Community History: Long-Term Changes in Plant Distribution and Diversity.* London: Chapman & Hall.

Vakhrameev, V. A. 1991. *Jurassic and Cretaceous Floras and Climates of the Earth.* Cambridge, UK: Cambridge University Press.

Van Oosterzee, Penny. *Where Worlds Collide: The Wallace Line.* Ithaca, NY: Cornell University, 1997.

Wallace, Alfred R. 1869. *The Malay Archipelago.* New York: Harper.

Wallace, Alfred R. 1876. *The Geographical Distribution of Animals.* London: Macmillan.

Wallace, Alfred R. 1880. *Island Life.* London: Macmillan.

Wegener, Alfred. 1915. *The Origin of Continents and Oceans.* New York: Dover.

Weiner, Johanthan. 1994. *The Beak of the Finch: A Story of Evolution in Our Own Time.* New York: Knopf.

Whittaker, R. J. 1998. *Island Biogeography: Ecology, Evolution and Conservation.* Oxford: Oxford University Press.

Wilson, E. O. 1988. *Biodiversity.* New York: National Academic Press.

Molecules

Bleiweiss, R., J. A. Kirsch, and J. C. Matheus.1997. "DNA Hybridization Evidence for the Principal Lineages of Hummingbirds (Aves:Trochilidae)." *Molecular Biology and Evolution* 14 (3): 325–343.

Blomme, Tine, Klaas Vandepoele, Stefanie De Bodt, Cedric Simillion, Steven Maere, and Yves Van de Peer. 2006. "The Gain and Loss of Genes during 600 Million Years of Vertebrate Evolution." *Genome biology* 7 (5): R43.

Brawand, David, Walter Wahli, and Henrik Kaessmann. 2008. "Loss of Egg Yolk Genes in Mammals and the Origin of Lactation and Placentation." *PLoS Biology* 6 (3): E63.

Bromham, L., M. J. Phillips, and D. Penny. 1999. "Growing Up with Dinosaurs: Molecular Dates and the Mammalian Radiation." *Trends in Ecology & Evolution (Personal ed.)* 14 (3): 113–118.

Brown, J. R., H. Ye, R. T. Bronson, P. Dikkes, and M. E. Greenberg. 1996. "A Defect in Nurturing in Mice Lacking the Immediate Early Gene Fosb." *Cell* 86 (2): 297–309.

Dobzhansky, T., and B. Spassky. 1969."Artificial and Natural Selection for Two Behavioral Traits in *Drosophila Pseudoobscura.*" *Proceedings of the National Academy of Sciences of the United States of America* 62 (1): 75–80.

Eizirik, E., W. J. Murphy, and S. J. O'Brien. 2001. "Molecular Dating and Biogeography of the Early Placental Mammal Radiation." *Journal of Heredity* 92 (2): 212–219.

Griffiths, D. J. 2001. "Endogenous Retroviruses in the Human Genome Sequence." *Genome Biology* 2 (6): 1017.

Gupta, R. S. 2006. "Evolution of the Chaperonin Families (HSP60, HSP 10 and TCP-1) of Proteins and the Origin of Eukaryotic Cells." *Molecular Microbiology* 15 (1): 1–11.

Hall, D. O., R. Cammack, and K. K. Rao. 1974. "The Iron-Sulphur Proteins: Evolution of a Ubiquitous Protein from Model Systems to Higher Organisms." *Origins of Life and Evolution of Biospheres* 5 (3): 363–386.

Hardison, R. 1998. "Hemoglobins from Bacteria to Man: Evolution of Different Patterns of Gene Expression." *Journal of Experimental Biology* 201 (8): 1099.

Hardison, R. 1999. "The Evolution of Hemoglobin Studies of a Very Ancient Protein Suggest That Changes in Gene Regulation Are an Important Part of the Evolutionary Story." *American Scientist* 87:126–137.

Hoekstra, H. E., and M. W. Nachman. 2006. "Coat Color Variation in Rock Pocket Mice (*Chaetodipus intermedius*): From Genotype to Phenotype." *Mammalian Diversification: From Chromosomes to Phylogeography.*

Hoekstra, H. E., J. G. Krenz, and M. W. Nachman. 2005. "Local Adaptation in the Rock Pocket Mouse (*Chaetodipus intermedius*): Natural Selection and Phylogenetic History of Populations." *Heredity* 94 (2): 217–228.

Imes, D. L., L. A. Geary, R. A. Grahn, and L A .Lyons. 2006. "Albinism in the Domestic Cat (*Felis catus*) Is Associated with a Tyrosinase (TYR) Mutation." *Animal Genetics* 37 (2): 175–178.

de Jong, Wilfried W., Jack A. M. Leunissen, and Christina E. M. Voorter. 1993. "Evolution of the Alpha-Crystallin/Small Heat-Shock Protein Family." *Molecular Biology and Evolution.* 10 (1): 103–126.

Kijas, J. M., R. Wales, A. Törnsten, P. Chardon, M. Moller, and L. Andersson. 1998. "Melanocortin Receptor 1 (MC1R) Mutations and Coat Color in Pigs." *Genetics* 150 (3): 1177–1185.

Kirsch, J. A. W., A. W. Dickerman, O. A. Reig, and M. S. Springer. 1991. "DNA Hybridization Evidence for the Australasian Affinity of the American Marsupial *Dromiciops australis.*" *Proceedings of the National Academy of Science USA* 88: 10465–69.

Koonin, E. V., L. Aravind, and A. S. Kondrashov. 2000. "The Impact of Comparative Genomics on Our Understanding of Evolution." *Cell* 101 (6): 573–576.

Lebedev, Y. B., O. S. Belonovitch, N. V. Zybrova, P. P. Khil, S. G. Kurdyukov, T. V. Vinogradova, G. Hunsmann, and E. D. Sverdlov. 2000. "Differences in HERV-K LTR Insertions in Orthologous Loci of Humans and Great Apes." *Gene* 247 (1–2): 265–277.

Lindquist, S., and E. A. Craig. 1988. "The Heat-Shock Proteins." *Annual Review of Genetics* 22 (1): 631–677.

Mundy, Nicholas I. 2005. "A Window on the Genetics of Evolution: MC1R and Plumage Colouration in Birds." *Proceedings. Biological Sciences/The Royal Society* 272 (1573): 1633–1640.

Mural, Richard J., Mark D. Adams, Eugene W. Myers, Hamilton O. Smith, George L. Gabor Miklos, Ron Wides, Aaron Halpern, et al. 2002. "A Comparison of Whole-Genome Shotgun-Derived Mouse Chromosome 16 and the Human Genome." *Science* 296 (5573): 1661–1671.

Murphy, W. J., E. Eizirik, W. E. Johnson, Y. P. Zhang, O. A. Ryder, and S. J. O'Brien. 2001. "Molecular Phylogenetics and the Origins of Placental Mammals." *Nature* 409 (6820): 189–196.

Nachman, Michael W, Hopi E. Hoekstra, and Susan L. D'Agostino. 2003. "The Genetic Basis of Adaptive Melanism in Pocket Mice." *Proceedings of the National Academy of Sciences of the United States of America* 100 (9): 5268–5273.

Omatsu, Tsutomu, Yoshiyuki Ishii, Shigeru Kyuwa, Elizabeth G. Milanda, Keiji Terao, and Yasuhiro Yoshikawa. 2003. "Molecular Evolution Inferred from Immunological Cross-Reactivity of Immunoglobulin G among Chiroptera and Closely Related Species." *Experimental Animals* 52 (5): 425–428.

Prager, E. M., A. C. Wilson, J. M. Lowenstein, and V. M. Sarich. 1980. "Mammoth Albumin." *Science* 209 (4453): 287–289.

Sarich, V. M., and A. C. Wilson. 1967. "Rates of Albumin Evolution in Primates." *Proceedings of the National Academy of Sciences of the United States of America* 58 (1): 142–148.

Sémon, Marie, and Kenneth H. Wolfe. 2007. "Consequences of Genome Duplication." *Current Opinion in Genetics and Development* 17 (6): 505–512.

Sibley, C. G., and J. E. Ahlquist. 1984. "The Phylogeny of the Hominoid Primates, as Indicated by DNA-DNA Hybridization." *Journal of Molecular Evolution* 20 (1): 2–15.

Smit, A. 1999. "Interspersed Repeats and Other Mementos of Transposable Elements in Mammalian Genomes." *Current Opinion in Genetics and Development* 9:657–663.

Wu, Miao-Lun, Tsan-Piao Lin, Min-Yi Lin, Yu-Pin Cheng, and Shih-Ying Hwang. 2007. "Divergent Evolution of the Chloroplast Small Heat Shock Protein Gene in the Genera Rhododendron (Ericaceae) and Machilus (Lauraceae)." *Annals of Botany* 99 (3): 1–15.

Comparative Anatomy

Abzhanov, A., W. P. Kuo, C. Hartmann, B. R. Grant, P. R. Grant, and C. J. Tabin. 2006. "The Calmodulin Pathway and Evolution of Elongated Beak Morphology in Darwin's Finches." *Nature* 442 (7102): 563–567.

Adams, R. A. 2008. "Morphogenesis in Bat Wings: Linking Development, Evolution and Ecology." *Cells Tissues Organs* 187 (1): 13–23.

Bejder, L., and B. K. Hall. 2002. "Limbs in Whales and Limblessness in Other Vertebrates: Mechanisms of Evolutionary and Developmental Transformation and Loss." *Evolution and Development* 4 (6): 445–458.

Callaerts, P., G. Halder, and W. J. Gehring. 1997. "PAX-6 in Development and Evolution." *Annual Review of Neuroscience* 20 (1): 483–532.

Carroll, Sean. 2005. *Endless Forms Most Beautiful: The New Science of Evo/Devo.* New York: W. W. Norton.

Cohn, M. J., and C. Tickle. 1999. "Developmental Basis of Limblessness and Axial Patterning in Snakes." *Nature* 399 (6735): 474–478.

Gehring, W. J., and K. Ikeo. 1999. "Pax 6: Mastering Eye Morphogenesis and Eye Evolution." *Trends in Genetics* 15 (9): 371–377.

Lamb, T. D., E. N. Pugh, and S. P. Collin. 2008. "The Origin of the Vertebrate Eye." *Evolution: Education and Outreach* 1 (4): 415–426.

Lamb, T. D., S. P. Collin, and E. N. Pugh. 2007. "Evolution of the Vertebrate Eye: Opsins, Photoreceptors, Retina and Eye Cup." *Nature Reviews Neuroscience* 8 (12): 960–976.

Luo, G., C. Hofmann, A. L .Bronckers, M. Sohocki, A. Bradley, and G. Karsenty. 1995. "BMP-7 Is an Inducer of Nephrogenesis, and Is Also Required for Eye Development and Skeletal Patterning." *Genes and development* 9 (22): 2808.

Merino, R., J. Rodriguez-Leon, D. Macias, Y. Ganan, A. N. Economides, and J. M. Hurle. 1999. "The BMP Antagonist Gremlin Regulates Outgrowth, Chondrogenesis and Programmed Cell Death in the Developing Limb." *Development* 126 (23): 5515.

Nielsen, Claus. 2008. "Six Major Steps in Animal Evolution: Are We Derived Sponge Larvae?" *Evolution and development* 10 (2): 241–257.

Niven, Jeremy E. 2008. "Evolution: Convergent Eye Losses in Fishy Circumstances." *Current Biology* 18 (1).

Oakley, T. H., and M. S. Pankey. 2008. "Opening the "Black Box": The Genetic and Biochemical Basis of Eye Evolution." *Evolution: Education and Outreach* 1 (4): 390–402.

Quiring, R., U. Walldorf, U. Kloter, and W. J .Gehring. 1994. "Homology of the Eyeless Gene of Drosophila to the Small Eye Gene in Mice and Aniridia in Humans." *Science* 265 (5173): 785.

Richardson, M. K., S. M. H. Gobes, A. C. van Leeuwen, J. A. E. Polman, C. Pieau, and M. R. Sánchez-Villagra. 2009. "Heterochrony in Limb Evolution: Developmental Mechanisms and Natural Selection." *Journal of Experimental Zoology Part B: Molecular and Developmental Evolution* 312 (6): 639–664.

Roberts, D. J., and C. Tabin. 1994. "The Genetics of Human Limb Development." *American Journal of Human Genetics* 55 (1): 1.

Ryan, Joseph F., Patrick M. Burton, Maureen E. Mazza, Grace K. Kwong, James C. Mullikin, and John R. Finnerty. 2006. "The Cnidarian-Bilaterian Ancestor Possessed at Least 56 Homeoboxes: Evidence from the Starlet Sea Anemone, *Nematostella vectensis*." *Genome Biology* 7 (7): doi:10.1186/gb-2006-7-7-R64.

Sears, K. E. 2007. "Molecular Determinants of Bat Wing Development." *Cells Tissues Organs* 187 (1): 6–12.

Selim, J. 2004. "Useless Body Parts." *Discover-New-York* 25 (6): 42–47.

Tchernov, E., O. Rieppel, H. Zaher, M. J. Polcyn, and L. L. Jacobs. 2000. "A Fossil Snake with Limbs." *Science* 287 (5460): 2010.

Vinagre, T., N. Moncaut, M. Carapuco, A. Novoa, J. Bom, and M. Mallo. 2010. "Evidence for a Myotomal Hox/Myf Cascade Governing Nanautonomous Control of Rib Specification within Global Vertebral Domains." *Developmental Cell* 18 (4): 655.

Weatherbee, S. D., R. R. Behringer, J. J. Rasweiler, and L. A. Niswander. 2006. "Interdigital Webbing Retention in Bat Wings Illustrates Genetic Changes Underlying Amniote Limb Diversification." *Proceedings of the National Academy of Sciences of the United States of America* 103 (41): 15103.

Behavior

Anderson, David J. 1995. "The Role of Parents in Siblicidal Brood Reduction of Two Booby Species." *Auk* 112 (4): 860–869.

Anderson, Michael G, Csaba Moskát, Miklós Bán, Tomás Grim, Phillip Cassey, and Mark E Hauber. 2009. "Egg Eviction Imposes a Recoverable Cost of Virulence in Chicks of a Brood Parasite." *PLoS ONE* 4 (11): doi:10.1371/journal.pone.0007725.

Andrade, M. C. B. 1996. "Sexual Selection for Male Sacrifice in the Australian Redback Spider." *Science* 271 (5245): 70.

Angilletta, M. J, et al. 2002. "The Evolution of Thermal Physiology in Ectotherms." *Journal of Thermal Biology* 27 (4): 249–268.

Austad, S. N., and M, E, Sunquist. 1986. "Sex-Ratio Manipulation in the Common Opossum." *Nature* 324 (6092): 58–60.

Barry, K. L., G. I. Holwell, and M. E. Herberstein. 2008. "Female Praying Mantids Use Sexual Cannibalism as a Foraging Strategy to Increase Fecundity." *Behavioral Ecology*: 710–715.

Berthouly, Anne, Fabrice Helfenstein, Marion Tanner, and Heinz Richner. 2008. "Sex-Related Effects of Maternal Egg Investment on Offspring in Relation to Carotenoid Availability in the Great Tit." *Journal of Animal Ecology* 77 (1): 74–82..

Birkhead, Timothy R., and Tommaso Pizzari. 2002. "Postcopulatory Sexual Selection." *Nature Reviews Genetics* 3 (4): 262–273.

Blumstein, D. 2007. "The Evolution of Alarm Communication in Rodents: Structure, Function, and the Puzzle of Apparently Altruistic Calling in Rodents." In *Rodent Societies*, ed. J. O. Wolff and P.W. Sherman, 317–327. University of Chicago Press, Chicago, Illinois.

de Bono, M., and C. I. Bargmann. 1998. "Natural Variation in a Neuropeptide Y Receptor Homolog Modifies Social Behavior and Food Response in *C. Elegans.*" *Cell* 94 (5): 679–89.

Borgia, G. 1986. "Sexual Selection in Bowerbirds." *Scientific American* 254 (6): 92–100.

Boyle, W. A., C. G. Guglielmo, K. A. Hobson, and D. R. Noris. 2011. "Lekking Birds in a Tropical Forest Forego Sex for Migration." *Biology Letters.*

Brown, S. P. 1999. "Cooperation and Conflict in Host-Manipulating Parasites." *Proceedings of the Royal Society B: Biological Sciences* 266 (1431): 1899.

Cade, William H. 1981. "Alternative Male Strategies: Genetic Differences in Crickets." *Science* 212 (4494): doi:10.1126/science.212.4494.563.

Cariello, Mariana O., Regina H. F. Macedo, and Hubert G. Schwabl. 2006. "Maternal Androgens in Eggs of Communally Breeding Guira Cuckoos (*Guira guira*)." *Hormones and Behavior* 49 (5): 654–662.

Chapuisat, Michel. 2010. "Social Evolution: Sick Ants Face Death Alone." *Current Biology* 20 (3): R104–105.

Clutton-Brock, T. H., S. D. Albon, and F. E. Guinness. 1984. "Maternal Dominance, Breeding Success and Birth Sex Ratios in Red Deer." *Nature* 308 (5957): 358–360.

Covas, R., A. Dalecky, A. Caizergues, and C. Doutrelant.2006. "Kin Associations and Direct Vs Indirect Fitness Benefits in Colonial Cooperatively Breeding Sociable Weavers *Philetairus socius*." *Behavioral Ecology and Sociobiology* 60 (3): 323–331.

Dobzhansky, T., and B. Spassky. 1969. "Artificial and Natural Selection for Two Behavioral Traits in *Drosophila pseudoobscura*." *Proceedings of the National Academy of Sciences of the United States of America* 62 (1): 75–80.

Donoghue, A. M., T. S. Sonstegard, L. M. King, E. J. Smith, and D. W. Burt. 1999. "Turkey Sperm Mobility Influences Paternity in the Context of Competitive Fertilization." *Biology of Reproduction* 61 (2): 422–427.

Dugas, M. B. 2009. "House Sparrow, *Passer domesticus*, Parents Preferentially Feed Nestlings with Mouth Colours That Appear Carotenoid-Rich." *Animal Behaviour* 78 (3): 767–772.

Faulkes, C. G., and N. C. Bennett. 2001. "Family Values: Group Dynamics and Social Control of Reproduction in African Mole-Rats." *Trends in Ecology and Evolution (Personal ed.)* 16 (4): 184–190.

Fisher, Heidi S, and Hopi E Hoekstra. 2010. "Competition Drives Cooperation among Closely Related Sperm of Deer Mice." *Nature* 463 (7282): 801–803.

Fitzpatrick, J. L., R. Montgomerie, J. K. Desjardins, K. A. Stiver, N. Kolm, and S. Balshine. 2009. "Female Promiscuity Promotes the Evolution of Faster Sperm in Cichlid Fishes." *Proceedings of the National Academy of Sciences* 106 (4).

Foster, Kevin R., Tom Wenseleers, and Francis L. W. Ratnieks. 2006. "Kin Selection Is the Key to Altruism." *Trends in Ecology and Evolution (Personal ed.)* 21 (2): 57–60.

Fusani, L., Beani, L., Lupo, C., and Dessì-fulgheri, F. 1997. "Sexually Selected Vigilance Behaviour of the Grey Partridge Is Affected by Plasma Androgen Levels." *Animal Behavior* 54 (4): 1013–8.

Gadagkar, R. (2003). "Is the Peacock Merely Beautiful or Also Honest?" *Current Science* 85 (7): 1012.

Gammie, S. C., A. P. Auger, H. M. Jessen, R. J. Vanzo, T. A. Awad, and S. A. Stevenson. 2007. "Altered Gene Expression in Mice Selected for High Maternal Aggression." *Genes, Brain, and Behavior* 6 (5): 432–443.

Gammie, Stephen C., Theodore Garland, and Sharon A Stevenson. 2006. "Artificial Selection for Increased Maternal Defense Behavior in Mice." *Behavioral Genetics* 36 (5): 713–22..

Griffin, A. S., J. M. Pemberton, P. N. M. Brotherton, G. McIlrath, D. Gaynor, R. Kansky, J. O'Riain, and T. H. Clutton-Brock. 2003. "A Genetic Analysis of Breeding Success in the Cooperative Meerkat (*Suricata suricata*)." *Behavioral Ecology* 14 (4): 472.

Heinze, Jürgen, and Bartosz Walter. 2010. "Moribund Ants Leave Their Nests to Die in Social Isolation." *Current Biology* 20 (3): 249–252.

Helbig, A. J. 1991. "Inheritance of Migratory Direction in a Bird Species: A Cross-Breeding Experiment with SE-and SW-Migrating Blackcaps (*Sylvia atricapilla*)." *Behavioral Ecology and Sociobiology* 28 (1): 9–12.

Jansen, Patrick A., Frans Bongers, and Lia Hemerick. 2004. "Seed Mass and Mast Seeding Enhance Dispersal by a Neotropical Scatter-Hoarding Rodent." *Ecological Monographs* 74 (4): 569–589.

Jenkins, S. H., A. Rothstein, and W. C. H. Green. 1995. "Food Hoarding by Merriam's Kangaroo Rats: A Test of Alternative Hypotheses." *Ecology* 76 (8): 2470–2481.

Kamil, A. C., and K. L. Gould. "Memory in Food Caching Animals." 2008. *Papers in Behavior and Biological Sciences:* 61.

Kilner, R. M. 2003. "How Selfish Is a Cowbird Nestling?" *Animal Behaviour* 66 (3): doi:10.1006/anbe.2003.2204.

Kilner, R. M., and H. Drummond. 2007. "Parent-Offspring Conflict in Avian Families." *Journal of Ornithology* 148:doi:10.1007/s10336-007-0224-3.

LaDage, Lara D., Timothy C. Roth, Rebecca A. Fox, and Vladimir V. Pravosudov. 2009. "Flexible Cue Use in Food-Caching Birds." *Animal Cognition* 12 (3): doi:10.1007/s10071-008-0201-0.

Lewis, S. M., C. K. Cratsley, and J. A. Rooney. 2004. "Nuptial Gifts and Sexual Selection in Photinus Fireflies." *Integrative and Comparative Biology* 44 (3): 234.

Ligon, R. A., and G. E. Hill. 2010. "Sex-Biased Parental Investment Is Correlated with Mate Ornamentation in Eastern Bluebirds." *Animal Behaviour* 79 (3): 727–734.

Locatello, L., M. B. Rasotto, J. P. Evans, and A. Pilastro. 2006. "Colourful Male Guppies Produce Faster and More Viable Sperm." *Journal of Evolutionary Biology* 19 (5): 1595–1602.

Lohmann, K. J., N. D. Pentcheff, G. A. Nevitt, G. D. Stetten, R. K. Zimmer-Faust, H. E. Jarrard, L. C. Boles. 1995. "Magnetic Orientation of Spiny Lobsters in the Ocean: Experiments with Undersea Coil Systems." *Journal of Experimental Biology* 198 (Pt. 10): 2041–2048.

Lougheed, L. W., and D. J. Anderson. 1999. "Parent Blue-Footed Boobies Suppress Siblicidal Behavior of Offspring." *Behavioral Ecology and Sociobiology* 45 (1): 11–18.

Lynch, C. B. 1980. "Response to Divergent Selection for Nesting Behavior in *Mus musculus*." *Genetics* 96(3): 757–765.

Malo, Aurelio F., J. Julián Garde, Ana J. Soler, Andrés J. García, Montserrat Gomendio, and Eduardo R. S. Roldan. 2005. "Male Fertility in Natural Populations of Red Deer Is Determined by Sperm Velocity and the Proportion of Normal Spermatozoa." *Biology of Reproduction* 72 (4).

Mateo, J. M. 2003. "Kin Recognition in Ground Squirrels and Other Rodents." *Journal Information* 84 (4).

Mateo, J. M. "The Nature and Representation of Individual Recognition Odours in Belding's Ground Squirrels." *Animal Behaviour* 71 (1).

Mateo, J. M., and R. E. Johnston. 2000. "Kin Recognition and the 'Armpit Effect': Evidence of Self-Referent Phenotype Matching." *Proceedings. Biological Sciences/ The Royal Society* 267 (1444).

McCann, T.S. 1981. "Aggression and Sexual Activity of Male Southern Elephant Seals, *Mirounga leonine*." *Journal of Zoology* 195 (3): 295–310.

Meester, L., and H. J. Dumont. 1988. "The Genetics of Phototaxis in Daphnia Magna: Existence of Three Phenotypes for Vertical Migration Among Parthenogenetic Females." *Hydrobiologia* 162 (1): 47–55.

Miller, Jeremy A. 2007. "Repeated Evolution of Male Sacrifice Behavior in Spiders Correlated with Genital Mutilation." *Evolution: International Journal of Organic Evolution* 61 (6).

Moore, Harry, Katerina Dvoráková, Nicholas Jenkins, and William Breed. 2002. "Exceptional Sperm Cooperation in the Wood Mouse." *Nature* 418 (6894).

Nakayama, S., Y. Nishi, and T. Miyatake. 2010. "Genetic Correlation between Behavioural Traits in Relation to Death-Feigning Behaviour." *Population Ecology* 52 (2).

Nelson, D. A. 1999. "Ecological Influences on Vocal Development in the White-Crowned Sparrow." *Animal Behavior* 58 (1).

Nelson, D. A. 2000. "A Preference for Own-Subspecies' Song Guides Vocal Learning in a Song Bird." *Proceedings of the National Academy of Sciences of the United States of America* 97 (24).

Nowak, M. A., and K. Sigmund. 2005. "Evolution of Indirect Reciprocity." *Nature* 437 (7063): 1291–1298.

Nowicki, S., W. A. Searcy, M. Hughes, and J. Podos. 2001. "The Evolution of Bird Song: Male and Female Response to Song Innovation in Swamp Sparrows." *Animal Behaviour* 62 (6).

Nowicki, Stephen, William A Searcy, and Susan Peters. "Quality of Song Learning Affects Female Response to Male Bird Song." *Proceedings. Biological Sciences/The Royal Society* 269 (1503).

Olsén, K. H., M. Grahn, J. Lohm., and A. Langefors. 1998. "MHC and Kin Discrimination in Juvenile Arctic Charr, *Salvelinus alpinus* (L.)." *Animal Behavior* 56 (2): 319–327.

Olsson, M, T. Madsen, B. Ujvari, and E. Wapstra. 2004. "Fecundity and MHC Affects Ejaculation Tactics and Paternity Bias in Sand Lizards." *Evolution: International Journal of Organic Evolution* 58 (4): 906–909.

Olsson, M., R. Shine, T. Madsen, A. Gulberg, and H. Tegelstrom. 1996. "Sperm Selection by Females." *Nature* 383:585.

Osorno, J. L., and H. Drummond. 2003. "Is Obligate Siblicidal Aggression Food Sensitive?" *Behavioral Ecology and Sociobiology* 54 (6).

Perrigo, G., W. C. Bryant, and F. S. vom Saal. 1989. "Fetal, Hormonal and Experiential Factors Influencing the Mating-Induced Regulation of Infanticide in Male House Mice." *Physiology and Behavior* 46 (2): 121–128.

Petrie, M, et al. 1994. "Improved Growth and Survival of Offspring of Peacocks with More Elaborate Trains." *Nature* 371 (6498): 598–599.

Pfennig, D. W., G. J. Gamboa, H. K. Reeve, J. S. Reeve, and I. D. Ferguson.1983. "The Mechanism of Nestmate Discrimination in Social Wasps (*Polistes*, Hymenoptera: Vespidae)." *Behavioral Ecology and Sociobiology* 13 (4): 299–305.

Pike, T. W. 2005. "Sex Ratio Manipulation in Response to Maternal Condition in Pigeons: Evidence for Pre-Ovulatory Follicle Selection." *Behavioral Ecology and Sociobiology* 58 (4): doi:10.1007/s00265-005-0931-9.

Pilastro, A., M. Mandelli, C. Gasparini, M. Dadda, and A. Bisazza. 2007. "Copulation Duration, Insemination Efficiency and Male Attractiveness in Guppies." *Animal Behaviour* 74 (2).

Pilastro, A., M. Simonato, A. Bisazza, and J. P. Evans. 2004. "Cryptic Female Preference for Colorful Males in Guppies." *Evolution: International Journal of Organic Evolution* 58 (3): 665–669.

Pitcher, T. E., F. H. Rodd, and L. Rowe. 2007. "Sexual Colouration and Sperm Traits in Guppies." *Journal of Fish Biology* 70 (1).

Pravosudov, Vladimir V. 2008. "Mountain Chickadees Discriminate between Potential Cache Pilferers and Non-Pilferers." *Proceedings. Biological Sciences/The Royal Society* 275 (1630). (2008).

Ratnieks, Francis L. W., and Heikki Helanterä. 2009. "The Evolution of Extreme Altruism and Inequality in Insect Societies." *Philosophical Transactions of the Royal Society of London. Series B, Biological Sciences* 364 (1533).

Reeve, H. K., D. F. Westneat, W. A. Noon, P. W. Sherman, and C. F. Aquadro. 1990. "DNA 'Fingerprinting' Reveals High Levels of Inbreeding in Colonies of the Eusocial Naked Mole-Rat." *Proceedings of the National Academy of Sciences of the United States of America* 87 (7): 2496–500.

Rondeau, A., and B. Sainte-Marie. 2001. "Variable Mate-Guarding Time and Sperm Allocation by Male Snow Crabs (*Chionoecetes opilio*) in Response to Sexual Competition, and Their Impact on the Mating Success of Females." *Biological Bulletin* 201 (2): 204–217.

Russell, A. F., A. J. Young, G. Spong, N. R. Jordan, and T. H. Clutton-Brock. 2007. "Helpers Increase the Reproductive Potential of Offspring in Cooperative Meerkats." *Proceedings of the Royal Society B: Biological Sciences* 274 (1609).

Russell, A. F., and B. J. Hatchwell. 2001. "Experimental Evidence for Kin-Biased Helping in a Cooperatively Breeding Vertebrate." *Proceedings. Biological Sciences/The Royal Society* 268 (1481).

Shine, R., T. Langkilde, and R. T. Mason. 2003. "Cryptic Forcible Insemination: Male Snakes Exploit Female Physiology, Anatomy, and Behavior to Obtain Coercive Matings." *American Naturalist* 162 (5): 653–667.

Smith, C. C., and O. J. Reichman. 1984. "The Evolution of Food Caching by Birds and Mammals." *Annual Review of Ecology and Systematics* 15 (1): 329–351.

Snow, L. S. E., A. Abdel-Mesih, and M. C. B. Andrade. 2006. "Broken Copulatory Organs Are Low-Cost Adaptations to Sperm Competition in Redback Spiders." *Ethology* 112 (4).

Swallow, J. G., P. A. Carter, and T. Garland. 1998. "Artificial Selection for Increased Wheel-Running Behavior in House Mice." *Behavior Genetics* 28 (3): 227–237.

Vander Wall, S. B., and S. H. Jenkins. 2003. "Reciprocal Pilferage and the Evolution of Food-Hoarding Behavior." *Behavioral Ecology* 14 (5).

Vander Wall, S. B. 2000. "The Influence of Environmental Conditions on Cache Recovery and Cache Pilferage by Yellow Pine Chipmunks (*Tamias amoenus*) and Deer Mice (*Peromyscus maniculatus*)." *Behavioral Ecology* 11:544–549.

Veiga, J. P. 2004. "Replacement Female House Sparrows Regularly Commit Infanticide: Gaining Time or Signaling Status?" *Behavioral Ecology* 15 (2).

Waterman, J. M. 1998. "Mating Tactics of Male Cape Ground Squirrels, *Xerus inauris:* Consequences of Year-Round Breeding." *Animal Behavior* 56 (2): 459–466.

Wilkinson, G. S. (1984) "Reciprocal Food Sharing in the Vampire Bat." *Nature* 308 (5955): 181–184.

Wilson, M. L., and R. W. Wrangham. 2003. "Intergroup Relations in Chimpanzees." *Annual Review of Anthropology* 32 (1): 363–392.

Woolfenden, G. E. 1975. "Florida Scrub Jay Helpers at the Nest." *Auk* 92 (1):1–15.

Coevolution

Altshuler, Douglas L., and Christopher James Clark. 2003. "Darwin's Hummingbirds." *Science* 300 (5619): 588–589.

Archetti, M. "Classification of Hypotheses on the Evolution of Autumn Colours." *Oikos* 118, no. 3 (2009).

Archetti, Marco, Thomas F. Döring, Snorre B. Hagen, Nicole M. Hughes, Simon R. Leather, David W. Lee, Simcha Lev-Yadun, et al. 2009. "Unravelling the Evolution of Autumn Colours: An Interdisciplinary Approach." *Trends in Ecology and Evolution (Personal ed.)*. 24 (3).

Baldauf, Sebastian A., Timo Thünken, Joachim G. Frommen, Theo C. M. Bakker, Oliver Heupel, and Harald Kullmann. 2007. "Infection with an Acanthocephalan Manipulates An Amphipod's Reaction to a Fish Predator's Odours." *International Journal for Parasitology* 37 (1).

Benkman, C. W. 2010. "Diversifying Coevolution between Crossbills and Conifers." *Evolution: Education and Outreach* 3: 47–53.

Berdoy, M., J. P. Webster, and D. W. Macdonald. 2000. "Fatal Attraction in Rats Infected with *Toxoplasma gondii*." *Proceedings. Biological Sciences/The Royal Society* 267 (1452).

Bernot, R. J. "Trematode Infection Alters the Antipredator Behavior of a Pulmonate Snail." *Journal of the North American Benthological Society* 22 (2): 241–248.

Bjorkman-Chiswell, Bojun T., Melissa M. Kulinski, Robert L. Muscat, Kim A. Nguyen, Briony A. Norton, Matthew R. E. Symonds, Gina E. Westhorpe, and Mark A. Elgar. 2004. "Web-Building Spiders Attract Prey by Storing Decaying Matter." *Die Naturwissenschaften* 91 (5).

Boag, P. T., and P. R. Grant. 1981. "Intense Natural Selection in a Population of Darwin's Finches (Geospizinae) in the Galapagos." *Science* 214 (4516): 82.

Bonfante, Paola, and Iulia-Andra Anca. 2009. "Plants, Mycorrhizal Fungi, and Bacteria: A Network of Interactions." *Annual Review of Microbiology* 63.

Bshary, R., and A. S. Grutter. 2002. "Asymmetric Cheating Opportunities and Partner Control in a Cleaner Fish Mutualism." *Animal Behaviour* 63 (2).

Bshary, R., and D. Schäffer. 2002. "Choosy Reef Fish Select Cleaner Fish That Provide High-Quality Service." *Animal Behaviour* 63 (3): 557–564.

Bura, V. L., V. G. Rohwer, P. R. Martin, and J. E. Yack. 2010. "Whistling in Caterpillars (*Amorpha juglandis*, Bombycoidea): Sound-producing Mechanism and Function." *Journal of Experimental Biology* 214: 30–37.

Chao, L., K. A. Hanley, C. L. Burch, C. Dahlberg, and P. E. Turner. 2000. "Kin Selection and Parasite Evolution: Higher and Lower Virulence with Hard and Soft Selection." *Quarterly Review of Biology* 75 (3).

Cheney, Karen L., and Isabelle M Côté. 2005. "Frequency-Dependent Success of Aggressive Mimics in a Cleaning Symbiosis." *Proceedings. Biological Sciences/The Royal Society* 272 (1581).

Connell, J. H. 1961."The Influence of Interspecific Competition and Other Factors on the Distribution of the Barnacle Chthamalus Stellatus." *Ecology* 42 (4).

Davies, N. B., and M. De L. Brooke. 1989. "An Experimental Study of Co-Evolution between the Cuckoo, *Cuculus canorus*, and Its Hosts." *Journal of Animal Ecology* 58 (1): 225–236.

Dufaÿ, Mathilde, and Marie-Charlotte Anstett. 2003. "Conflicts between Plants and Pollinators That Reproduce within Inflorescences: Evolutionary Variations on a Theme." *Oikos* 100 (1).

Edwards, D. P., and D. W. Yu. "The Roles of Sensory Traps in the Origin, Maintenance, and Breakdown of Mutualism." *Behavioral Ecology and Sociobiology* 61 (9).

Fisher, L .M. 2000. "The Effect of a Trematode Parasite (*Microphallus* sp.) on the Response of the Freshwater Snail Potamopyrgus Antipodarum to Light and Gravity." *Behaviour:* 1141–1151.

Frank, S. A. 1996. "Models of Parasite Virulence." *Quarterly Review of Biology* 71 (1).

Fredensborg, B. L., and R. Poulin. 2006. "Parasitism Shaping Host Life-History Evolution: Adaptive Responses in a Marine Gastropod to Infection by Trematodes." *Journal of Animal Ecology* 75 (1).

Futuyma, D. J. "How Species Affect Each Other's Evolution." *Evolution: Education and Outreach.*

Garamszegi, László Zsolt, Anders Pape Møller, and Johannes Erritzøe. 2002. "Coevolving Avian Eye Size and Brain Size in Relation to Prey Capture and Nocturnality." *Proceedings. Biological Sciences/The Royal Society* 269 (1494).

Grim, T., J. Rutila, P. Cassey, and M. E. Hauber. 2000. "The Cost of Virulence: An Experimental Study of Egg Eviction by Brood Parasitic Chicks." *Behavioral Ecology* 20 (5).

Hafner, M. S., and S. A. Nadler. 1988. "Phylogenetic Trees Support the Coevolution of Parasites and Their Hosts." *Nature* 332 (6161).

Halstead, B. J., H. R. Mushinsky, and E. D. McCoy. 2008. "Sympatric *Masticophis flagellum* and *Coluber constrictor* Select Vertebrate Prey at Different Levels of Taxonomy." *Journal Information* (4).

Hamilton, W. D., and S. P. Brown. 2001. "Autumn Tree Colours as a Handicap Signal." *Proceedings of the Royal Society of London. Series B: Biological Sciences* 268 (1475): 1489.

Hanifin, Charles T., and Edmund D. Brodie. 2008. "Phenotypic Mismatches Reveal Escape from Arms-Race Coevolution." *PLoS Biology* 6 (3).

Harper, George R., and David W. Pfennig. 2007. "Mimicry on the Edge: Why Do Mimics Vary in Resemblance to Their Model in Different Parts of Their Geographical Range?" *Proceedings. Biological Sciences/The Royal Society* 274 (1621).

Hay, M. E., J. D. Parker, D. E. Burkepile, C. C. Caudill, A. E. Wilson, Z. P. Hallinan, and A. D. Chequer. 2004. "Mutualisms and Aquatic Community Structure: The Enemy of My Enemy Is My Friend." *Annual Review of Ecology, Evolution, and Systematics* 35: 175–197.

Heiling, A. M., K. Cheng, and M. E. Herberstein. 2004. "Exploitation of Floral Signals by Crab Spiders (*Thomisus spectabilis,* Thomisidae)." *Behavioral Ecology* 15 (2): 321–326.

Hoekstra, R. F. 2005. "Why Sex Is Good." *Nature (London)* 434 (7033): 571–573.

Hoese, F. J., E. A. J. Law, D. Rao, and M. E. Herberstein. 2006. "Distinctive Yellow Bands on a Sit-And-Wait Predator: Prey Attractant or Camouflage?" *Behaviour* 143 (6): 763–781.

Howard, R. S., and C. M. Lively. 1994. "Parasitism, Mutation Accumulation and the Maintenance of Sex." *Nature* 367 (6463).

Hughes, D. P., D. J. C. Kronauer, and J. J. Boomsma. "Extended Phenotype: Nematodes Turn Ants into Bird-Dispersed Fruits." *Current Biology* 18 (7): R294–R295.

Huth, C. J., and O. Pellmyr. 2008. "Pollen-Mediated Selective Abortion in Yuccas and Its Consequences for the Plant-Pollinator Mutualism." *Ecology* 81(4): 1100–1107.

Huyse, Tine, Robert Poulin, and André Théron. 2005. "Speciation in Parasites: A Population Genetics Approach." *Trends in Parasitology* 21 (10).

Johnson, S. D., and B. Anderson. 2010. "Coevolution between Food-Rewarding Flowers and Their Pollinators." *Evolution: Education and Outreach.*

Jokela, J., C. M. Lively, M. F. Dybdahl, and J. A. Fox. 2003. "Genetic Variation in Sexual and Clonal Lineages of a Freshwater Snail." *Biological Journal of the Linnean Society* 79 (1): 165–181.

Kelman, E. J., R. J. Baddeley, A. J. Shohet, and D. Osorio. 2007. "Perception of Visual Texture and the Expression of Disruptive Camouflage by the Cuttlefish, *Sepia officinalis*." *Proceedings. Biological Sciences/The Royal Society* 274 (1616).

Kelman, Emma J, Daniel Osorio, and Roland J Baddeley. 2008. "A Review of Cuttlefish Camouflage and Object Recognition and Evidence for Depth Perception." *Journal of Experimental Biology* 211 (Pt. 11).

Kilner, R. M. 2003. "How Selfish Is a Cowbird Nestling?" *Animal Behaviour* 66 (3).

Lagrue, C., N. Kaldonski, M. J. Perrot-Minnot, S. Motreuil, and L. Bollache. 2007. "Modification of Hosts' Behavior by a Parasite: Field Evidence for Adaptive Manipulation." *Ecology* 88 (11): 2839–2847.

Letters, E. 2009. "Selective Flower Abortion Maintains Moth Cooperation in a Newly Discovered Pollination Mutualism." *Ecology Letters* 13:321–329.

Lively, C. M. 1987. "Evidence from a New Zealand Snail for the Maintenance of Sex by Parasitism." *Nature* 328 (6130): 519–521.

Lively, C. M. 2010. "Antagonistic Coevolution and Sex." *Evolution: Education and Outreach* 3: 19–25.

Lively, C. M., C. Craddock, and R. C. Vrijenhoek. 1990. "Red Queen Hypothesis Supported by Parasitism in Sexual and Clonal Fish." *Nature* 344 (6269): 864–866.

Maitland, D. P. 1994. "A Parasitic Fungus Infecting Yellow Dungflies Manipulates Host Perching Behaviour." *Proceedings: Biological Sciences* 258 (1352): 187–193.

Marek, Paul E., and Jason E. Bond. 2009. "A Mullerian Mimicry Ring in Appalachian Millipedes." *Proceedings of the National Academy of Sciences* 106 (24): 9755–9760.

Mauricio, R. 2000. "Natural Selection and the Joint Evolution of Tolerance and Resistance as Plant Defenses." *Evolutionary Ecology* 14 (4): 491–507.

McCurdy, D. G. 2001. "Asexual Reproduction in *Pygospio elegans* Claparede (Annelida, Polychaeta) in Relation to Parasitism by *Lepocreadium setiferoides* (Miller and Northup) (Platyhelminthes, Trematoda)." *Biological Bulletin* 201 (1): 45.

Medel, R., M. A. Mendez, C. G. Ossa, and C. Botto-Mahan. "Arms Race Coevolution: The Local and Geographical Structure of a Host–Parasite Interaction." *Evolution: Education and Outreach* 3: 26–31.

Merrill, Richard M., and Chris D. Jiggins. 2009. "Müllerian Mimicry: Sharing the Load Reduces the Legwork." *Current Biology* 19 (16).

Moland, Even and Geoffrey P. Jones. 2004. "Experimental Confirmation of Aggressive Mimicry by a Coral Reef Fish." *Oecologia* 140 (4): 676–683.

Moore, Sarah L., and Kenneth Wilson. 2002. "Parasites as a Viability Cost of Sexual Selection in Natural Populations of Mammals." *Science* 297 (5589).

Moran, Nancy A., John P. McCutcheon, and Atsushi Nakabachi. 2008. "Genomics and Evolution of Heritable Bacterial Symbionts." *Annual Review of Genetics* 42: 165–190.

Mouritsen, K. N., and R. Poulin. 2004. "Parasites Boost Biodiversity and Change Animal Community Structure by Trait-Mediated Indirect Effects." *Oikos* 108 (2): 344–350.

Neiman, Maurine, Gery Hehman, Joseph T. Miller, John M. Logsdon, and Douglas R. Taylor. 2010. "Accelerated Mutation Accumulation in Asexual Lineages of a Freshwater Snail." *Molecular Biology and Evolution* 27 (4): 954–963.

Neuhauser, C., and J. E. Fargione. "A Mutualism-Parasitism Continuum Model and Its Application to Plant-Mycorrhizae Interactions." *Ecological Modelling* 177 (3–4): 337–352.

Nowak, M. A. 2006. "Five Rules for the Evolution of Cooperation." *Science* 314 (5805): 1560.

Pellmyr, O., and C. J. Huth. 1994. "Evolutionary Stability of Mutualism between Yuccas and Yucca Moths." *Nature* 372 (6503): 257–260.

Penn, D., and W. K. Potts. 1998. "Chemical Signals and Parasite-Mediated Sexual Selection." *Trends in Ecology and Evolution* 13 (10): 391–396.

Pfennig, Karin S., David W. Pfennig. 2009. "Character Displacement: Ecological and Reproductive Responses to a Common Evolutionary Problem." *Quarterly Review of Biology* 84 (3): 253–276.

Piculell, Bridget J., Jason D. Hoeksema, and John N. Thompson. 2008. "Interactions of Biotic and Abiotic Environmental Factors in an Ectomycorrhizal Symbiosis, and the Potential for Selection Mosaics." *BMC Biology* 6: 23.

Polak, M, L. T. Luong, and W. T. Starmer. 2007. "Parasites Physically Block Host Copulation: A Potent Mechanism of Parasite-Mediated Sexual Selection." *Behavioral Ecology* 18 (5).

Read, A. F. 1994. "The Evolution of Virulence." *Trends in Microbiology* 2 (3): 73–76.

Rogers, M. E., and P. A. Bates. 2007. "Leishmania Manipulation of Sand Fly Feeding Behavior Results in Enhanced Transmission." *PLoS Pathogens* 3:e91.

Rolshausen, G., and H. M. Schaefer. 2007. "Do Aphids Paint the Tree Red (or Yellow)—Can Herbivore Resistance or Photoprotection Explain Colourful Leaves in Autumn?" *Plant Ecology* 191 (1).

Roy, B. A., and J. W. Kirchner. 2000. "Evolutionary Dynamics of Pathogen Resistance and Tolerance." *Evolution: International Journal of Organic Evolution* 54 (1): 51–63.

Sachs, Joel L., Ulrich G. Mueller, Thomas P. Wilcox and James J. Bull. 2004. "The Evolution of Cooperation." *Quarterly Review of Biology* 79 (2): 135–160.

Schaefer, H. M., and G. Rolshausen. 2007. "Aphids Do Not Attend to Leaf Colour As Visual Signal, but to the Handicap of Reproductive Investment." *Biology Letters* 3 (1).

Schluter, D., and P. R. Grant. 1982. "The Distribution of *Geospiza difficilis* in Relation to *G. fuliginosa* in the Galapagos Islands: Tests of Three Hypotheses." *Evolution: International Journal of Organic Evolution* 36 (6): 1213–1226.

Schluter, Dolph, Trevor D. Price, and Peter R. Grant. 1985. "Ecological Character Displacement in Darwin's Finches." *Science* 227 (4690).

Serbus, Laura R., Catharina Casper-Lindley, Frédéric Landmann, and William Sullivan. 2008. "The Genetics and Cell Biology of *Wolbachia*-Host Interactions." *Annual Review of Genetics* 42.

Soler, J. J. and M. Soler. 2000. "Brood-Parasite Interactions between Great Spotted Cuckoos and Magpies: A Model System for Studying Coevolutionary Relationships." *Oecologia* 125: 309–320.

Stachowicz, John J. and Mark E. Hay. 1999. "Reducing Predation through Chemically Mediated Camouflage: Indirect Effects of Plant Defenses on Herbivores." *Ecology* 80 (2): 495–509.

Stevens, Martin, and Sami Merilaita. 2009. "Animal Camouflage: Current Issues and New Perspectives." *Philosophical Transactions of the Royal Society of London. Series B. Biological Sciences* 364 (1516).

Stoks, R., M. A. McPeek and J. L. Mitchell. 2003. "Evolution of Prey Behavior in Response to Changes in Predation Regime: Damselflies in Fish and Dragonfly Lakes." *Evolution: International Journal of Organic Evolution* 57 (3): 574–585.

Strohm, Erhard, Johannes Kroiss, Gudrun Herzner, Claudia Laurien-Kehnen, Wilhelm Boland, Peter Schreier, and Thomas Schmitt. 2008. "A Cuckoo in Wolves' Clothing? Chemical Mimicry in a Specialized Cuckoo Wasp of the European Beewolf (Hymenoptera, Chrysididae and Crabronidae)." *Frontiers in Zoology* 5 (2).

Temeles, Ethan J. and W. John Kress. 2003. "Adaptation in a Plant-Hummingbird Association." *Science* 300 (5619): 630–633.

Thomas, F., R. Poulin, J.-F. Guégan, Y. Michalakis, and F. Renaud. 2000. "Are There Pros as Well as Cons to Being Parasitized?" *Parasitology Today* 16 (12).

Thomas, F., S. Adamo, and J. Moore. 2005. "Parasitic Manipulation: Where Are We and Where Should We Go?" *Behavioural Processes* 68 (3): 185–199.

Thompson, J. N. 2010. "Four Central Points about Coevolution." *Evolution: Education and Outreach* 3: 7–13.

Thompson, John N. 2009. "The Coevolving Web of Life." *American Naturalist* 173 (2): 125–140.

Vaclav, Radovan, and Pavol Prokop. 2006. "Does the Appearance of Orbweaving Spiders Attract Prey?" *Annales Zoologici Fennici* 43:65–71.

Webster, Joanne P. 2005. "Parasitic Manipulation: Where Else Should We Go?" *Behavioural Processes* 68 (3).

Weeks, A. R., M. Turelli, W. R. Harcombe, K. T. Reynolds, and A. A. Hoffmann. 2007. "From Parasite to Mutualist: Rapid Evolution of *Wolbachia* in Natural Populations of *Drosophila*." *PLoS Biology* 5:e114.

Weldon, P. J., J. R. Aldrich, J. A. Klun, J. E. Oliver, M. Debboun, and Anonymous. 2003. "Benzoquinones from Millipedes Deter Mosquitoes and Elicit Self-Anointing in Capuchin Monkeys (*Cebus* spp.)." *Die Naturwissenschaften* 90 (7): 301–304.

Winfree, R. 1999. "Cuckoos, Cowbirds and the Persistence of Brood Parasitism." *Trends in Ecology and Evolution (Personal ed.)* 14 (9): 338–343.

Woolhouse, M. E. J., J. P. Webster, E. Domingo, B. Charlesworth, and B. R. Levin. 2002. "Biological and Biomedical Implications of the Co-Evolution of Pathogens and Their Hosts." *Nature Genetics* 32 (4): 569–577.

Yamazaki, Kazuo. 2008. "Autumn Leaf Colouration: A New Hypothesis Involving Plant-Ant Mutualism Via Aphids." *Die Naturwissenschaften* 95 (7): 671–676.

Human Evolution

Abrejo, F. G., B. T. Shaikh, and N. Rizvi. 2009. "'And They Kill Me, Only Because I Am a Girl'. . . : A Review of Sex-Selective Abortions in South Asia." *The European Journal of Contraception and Reproductive Health Care* 14 (1): 10–16.

Adamo, S. A., and R. J. Spiteri. 2009. "He's Healthy, but Will He Survive the Plague? Possible Constraints on Mate Choice for Disease Resistance." *Animal Behaviour* 77 (1): 67–78.

Anderson, K. G. 2006. "How Well Does Paternity Confidence Match Actual Paternity?" *Current Anthropology* 47 (3): 513–520.

Anderson, K. G., H. Kaplan, and J. B. Lancaster. 2007. "Confidence of Paternity, Divorce, and Investment in Children by Albuquerque Men." *Evolution and Human Behavior* 28 (1): doi:10.1016/j.evolhumbehav.2006.06.004.

Anderson, K. G., H. Kaplan, and J. B. Lancaster. 2006. "Demographic Correlates of Paternity Confidence and Pregnancy Outcomes Among Albuquerque Men." *American Journal of Physical Anthropology* 131 (4): 560.

Ansell, J., K. A. Hamilton, M. Pinder, G. E. L .Walraven, and S. W. Lindsay. 2002. "Short-Range Attractiveness of Pregnant Women to Anopheles Gambiae Mosquitoes." *Transactions of the Royal Society of Tropical Medicine and Hygiene* 96 (2): 113–116.

Apicella, C. L., and F. W. Marlowe. 2004. "Perceived Mate Fidelity and Paternal Resemblance Predict Men's Investment in Children." *Evolution and Human behavior* 25 (6): 371–378.

Arnold, F., M. K. Choe, and T. K. Roy. 1998. "Son Preference, the Family-Building Process and Child Mortality in India." *Population Studies* 52 (3): 301–315.

Baker, R. R., and M. A. Bellis. 1989. "Number of Sperm in Human Ejaculates Varies in Accordance with Sperm Competition Theory." *Animal Behavior* 37 (5): 867–869.

Baker, R. R., and M. A. Bellis. 1993a. "Human Sperm Competition: Ejaculate Adjustment by Males and the Function of Masturbation." *Animal Behaviour* 46:861–885.

Baker, R. R., and M. A. Bellis. 1993b. "Human Sperm Competition: Ejaculate Manipulation by Females and a Function for the Female Orgasm." *Animal Behaviour* 46:887–909.

Barnes, I., A. Duda, O. G. Pybus, M. G. Thomas. 2011. "Ancient Urbanization Predicts Genetic Resistance to Tuberculosis." *Evolution* 65 (3): 842–848.

Barrett, H. C. 2005. "Adaptations to Predators and Prey." *The Handbook of Evolutionary Psychology*: 200–223.

Bedard, K., and O .Deschenes. 2005. "Sex Preferences, Marital Dissolution, and the Economic Status of Women." *Journal of Human Resources* 40 (2): 411.

Bersaglieri, T., P. C. Sabeti, N. Patterson, T. Vanderploeg, S. F. Schaffner, J. A. Drake, M. Rhodes, D. E. Reich, and J. N. Hirschhorn. 2004. "Genetic Signatures of Strong Recent Positive Selection at the Lactase Gene." *American Journal of Human Genetics* 74 (6): 1111–1120.

Bowles, Samuel. 2006. "Group Competition, Reproductive Leveling, and the Evolution of Human Altruism." *Science* 314 (5805).

Burch, R. L., et al. 2000. "Perceptions of Paternal Resemblance Predict Family Violence." *Evolution and Human Behavior* 21 (6): 429–435.

Burger, J., M. Kirchner, B. Bramanti, W. Haak, and M. G. Thomas. 2007. "Absence of the Lactase-Persistence-Associated Allele in Early Neolithic Europeans." *Proceedings of the National Academy of Sciences of the United States of America* 104 (10): 3736.

Buss, D. M. 1988. "The Evolution of Human Intrasexual Competition: Tactics of Mate Attraction." *Journal of Personality and Social Psychology* 54 (4): 616–628.

Buss, D. M. 2006. "Strategies of Human Mating." *Psychological Topics* 15 (2): 239–260.

Buss, D. M. 2010. "Sex Differences in Human Mate Preferences: Evolutionary Hypotheses Tested in 37 Cultures." *Behavioral and Brain Sciences* 12 (1): 1–14.

Buss, D. M., and T. K. Shackelford. 1997. "Human Aggression in Evolutionary Psychological Perspective." *Clinical Psychology Review* 17 (6): 605–619.

Cagnacci, A., A. Renzi, S. Arangino, C. Alessandrini, and A. Volpe. 2004. "Influences of Maternal Weight on the Secondary Sex Ratio of Human Offspring." *Human Reproduction* 19 (2).

Cameron, Elissa Z., and Fredrik Dalerum. 2009. "A Trivers-Willard Effect in Contemporary Humans: Male-Biased Sex Ratios among Billionaires." *PLoS ONE* 4 (1).

Catalano, R. A. "Sex Ratios in the Two Germanies: A Test of the Economic Stress Hypothesis." *Human Reproduction* 18 (9): 1972–1975.

Catalano, R., T. Bruckner, A. R. Marks, and B. Eskenazi. 2006. "Exogenous Shocks to the Human Sex Ratio: The Case of September 11, 2001 in New York City." *Human Reproduction* 21 (12): 3127–3131.

Cavalli-Sforza, L. L., and M. W. Feldman. 2003. "The Application of Molecular Genetic Approaches to the Study of Human Evolution." *Nature Genetics* 33: 266–275.

Chavanne, T. J., and G. G. Gallup. 1998. "Variation in Risk Taking Behavior among Female College Students as a Function of the Menstrual Cycle." *Evolution and Human Behavior* 19 (1): 27–32.

Coreil, J. 1983. "Allocation of Family Resources for Health Care in Rural Haiti." *Social Science and Medicine* 17 (11): 709–719.

Cronk, L. 2007. "Boy or Girl: Gender Preferences from a Darwinian Point of View." *Reproductive Biomedicine Online* 15 (Supp. 2): 23–32.

Dadley-Moore, Davina. 2004. "Caspase-12: The Long and the Short of It." *Nature Reviews Immunology* 4 (6): doi:10.1038/nri1384.

Dahl, Edgar, Ruchi S. Gupta, Manfred Beutel, Yve Stoebel-Richter, Burkhard Brosig, Hans-Rudolf Tinneberg, and Tarun Jain. 2006. "Preconception Sex Selection Demand and Preferences in the United States." *Fertility and Sterility* 85 (2): 468–473.

Dahl, G. B., and E. Moretti. 2004. "The Demand for Sons: Evidence from Divorce, Fertility, and Shotgun Marriage." *NBER Working Paper Series.*

Daly, M., and M. I. Wilson. 1982. "Whom Are Newborn Babies Said to Resemble?" *Ethology and Sociobiology* 3 (2): 69–78.

Daly, M, and M. I. Wilson. 1985. "Child Abuse and Other Risks of Not Living with Both Parents." *Ethology and Sociobiology* 6 (4): 197–210.

Daly, M., and M. I. Wilson. 1999. "Human Evolutionary Psychology and Animal Behaviour." *Animal Behaviour* 57 (3): 509–519.

Dart, Raymond. 1959. *Adventures with the Missing Link.* New York: Viking.

Darwin, Charles. 1871. *The Descent of Man and Selection in Relation to Sex.* London: John Murray.

DeBruine, L. M. 2004. "Resemblance to Self Increases the Appeal of Child Faces to Both Men and Women." *Evolution and Human Behavior* 25 (3): 142–154.

DeBruine, L. M., B. C. Jones, A. C. Little, and D. I. Perrett. 2008. "Social Perception of Facial Resemblance in Humans." *Archives of Sexual Behavior* 37 (1): 64–77.

Dolan, R. J. 2002. "Emotion, Cognition, and Behavior." *Science* 298 (5596): 1191.

Durante, Kristina M., and Norman P. Li. 2009. "Oestradiol Level and Opportunistic Mating in Women." *Biology letters* 5 (2).

Durante, Kristina M., V. Griskevicius, S. E. Hill, C. Perilloux, and N. P. Li. 2011. "Ovulation, Female Competition, and Product Choice: Hormonal Influences on Consumer Behavior." *Journal of Consumer Research* 37 (6): 921–934.

Elliot, Andrew J., and Daniela Niesta. 2008. "Romantic Red: Red Enhances Men's Attraction to Women." *Journal of Personality and Social Psychology* 95 (5): 1150–1164.

Etkin, N. L. 2007. "The Co-Evolution of People, Plants, and Parasites: Biological and Cultural Adaptations to Malaria." *Proceedings of the Nutrition Society* 62 (2): 311–317.

Evans, S., N. Neave, D. Wakelin, and C. Hamilton. 2008. "The Relationship between Testosterone and Vocal Frequencies in Human Males." *Physiology and Behavior* 93 (4–5): 783–788.

Ferwerda, B., M. B. B. McCall, M. C. De Vries, J. Hopman, B. Maiga, A. Dolo, O. Doumbo, et al. 2009. "Caspase-12 and the Inflammatory Response to Yersinia Pestis." *PLoSOne* 4 (9): e6870.

Fukuda, M., K. Fukuda, T. Shimizu, and H. Møller. 1998. "Decline in Sex Ratio at Birth after Kobe Earthquake." *Human Reproduction* 13 (8): 2321–2.

Gangestad, Steven, and David M. Buss. 1993. "Pathogen Prevalence and Human Mate Preferences." *Ethology and Sociobiology* 14 (2): 89–96.

Gangestad, Steven W., Jeffry A. Simpson, Alita J. Cousins, Christine E. Garver-Apgar, and P. Niels Christensen. 2004. "Women's Preferences for Male Behavioral Displays Change across the Menstrual Cycle." *Psychological Science* 15 (3): 203–207.

Garver-Apgar, Christine E., Steven W. Gangestad, Randy Thornhill, Robert D. Miller, and Jon J. Olp. 2006. "Major Histocompatibility Complex Alleles, Sexual

Responsivity, and Unfaithfulness in Romantic Couples." *Psychological Science* 17 (10): 830–835.

Gibbons, Ann. 2010. "The Human Family's Earliest Ancestors." *Smithsonian* 40 (12): 34–41.

Gibson, Mhairi A., and Ruth Mace. 2003. "Strong Mothers Bear More Sons in Rural Ethiopia." *Proceedings. Biological Sciences/The Royal Society* 270 (Supp. 1).

Goetz, A. T., T. K. Shackelford, G. A. Romero, F. Kaighobadi, and E. J. Miner. 2008. "Punishment, Proprietariness, and Paternity: Men's Violence against Women from an Evolutionary Perspective." *Aggression and Violent Behavior* 13(6): 481–489.

Griffiths, D. J. 2001. "Endogenous Retroviruses in the Human Genome Sequence." *Genome Biology* 2 (6): 1017.

Hansen, D., H. Moller, and J. Olsen. 1999. "Severe Periconceptional Life Events and the Sex Ratio in Offspring: Follow Up Study Based on Five National Registers." *BMJ* 319:548–549.

Havlicek, J., R. Dvorakova, L. Bartos, and J. Flegr. 2006. "Non-Advertized Does Not Mean Concealed: Body Odour Changes across the Human Menstrual Cycle." *Ethology* 112 (1): 81–90.

Hesketh, T., and Z. W. Xing. 2006. "Abnormal Sex Ratios in Human Populations: Causes and Consequences." *Proceedings of the National Academy of Sciences of the United States of America* 103 (36): 13271.

Hill, A. V., A. Jepson, M. Plebanski, and S. C. Gilbert. 1997. "Genetic Analysis of Host-Parasite Coevolution in Human Malaria." *Philosophical Transactions of the Royal Society B: Biological Sciences* 352 (1359): 1317.

Hopcroft, R. L. 2004. "Parental Status and Differential Investment in Sons and Daughters: Trivers-Willard Revisited." *Social Forces* 83:1111.

Hughes, J. F., and J. M. Coffin. 2001. "Evidence for Genomic Rearrangements Mediated by Human Endogenous Retroviruses During Primate Evolution." *Nature Genetics* 29 (4): 487–489.

Jablonski, N. G., and G. Chaplin. 2000. "The Evolution of Human Skin Coloration." *Journal of Human Evolution* 39 (1): 57–106.

Jablonski, N., and G. Chaplin. 2003. "Skin Deep." *Scientific American Special Editions*, 72–79.

Johanson, D., B. Edgar, and D. Brill. 2006. *From Lucy to Language: Revised, Updated, and Expanded.* New York: Simon and Schuster.

Jones, B. C., L. M. DeBruine, D. I. Perrett, A. C. Little, D. R. Feinberg, and M. J. Law Smith. 2008. "Effects of Menstrual Cycle Phase on Face Preferences." *Archives of Sexual Behavior* 37(1): 78–84.

King, M.C., and A.C. Wilson. 1975. "Evolution at Two Levels Human and Chimpanzee." *Science* 188: 107–116.

Kusum. 1993. "The Use of Pre-Natal Diagnostic Techniques for Sex Selection: The Indian Scene." *Bioethics* 7 (2–3): 149–165.

Kwiatkowski, Dominic P. 2005. "How Malaria Has Affected the Human Genome and What Human Genetics Can Teach Us About Malaria." *American Journal of Human Genetics* 77 (2): doi:10.1086/432519.

Lightcap, J. L., J. A. Kurland, and R. L. Burgess. 1982. "Child Abuse: A Test of Some Predictions from Evolutionary Theory." *Ethology and Sociobiology* 3 (2): 61–67.

Lummaa, V., T. Vuorisalo, R. G. Barr, and L. Lehtonen. 1998. "Why Cry? Adaptive Significance of Intensive Crying in Human Infants." *Evolution and Human Behavior* 19 (3): 193–202.

Macintyre, K., J. Keating, S. Sosler, L. Kibe, C. M. Mbogo, A. K. Githeko, and J. C. Beier. 2002. "Examining the Determinants of Mosquito-Avoidance Practices in Two Kenyan Cities." *Malaria Journal* 1 (1): 14.

Manning, J. T., D. Scutt, G. H. Whitehouse, S. J. Leinster, and J. M. Walton. 1996. "Asymmetry and the Menstrual Cycle in Women." *Ethology and Sociobiology* 17 (2): 129–143.

Manouvrier-Hanu, S., M. Holder-Espinasse, and S. Lyonnet. 1999. "Genetics of Limb Anomalies in Humans." *Trends in Genetics* 15 (10): 409–417.

McLain, D. K., D. Setters, M. P. Moulton, and A. E. Pratt. 2000. "Ascription of Resemblance of Newborns by Parents and Nonrelatives." *Evolution and Human Behavior* 21 (1): 11–23.

Meston, Cindy M., and David M. Buss. 2007. "Why Humans Have Sex." *Archives of Sexual Behavior* 36 (4): 477–507.

Miller, G., J. M. Tyber, B. D. Jordan. "Ovulatory Cycle Effects on Tip Earnings by Lap Dancers: Economic Evidence for Human Estrus?" *Evolution and Human Behavior* 28 (6): 375–381.

Morell, Virginia. 1995. *Ancestral Passions: The Leakey Family and the Quest for Humankind's Beginnings.* New York: Simon & Schuster.

Nielsen, R., I. Hellmann, M. Hubisz, C. Bustamante, and A. G. Clark. 2007. "Recent and Ongoing Selection in the Human Genome." *Nature Reviews Genetics* 8 (11): 857–868.

Ober, C. 1999. "Studies of HLA, Fertility and Mate Choice in a Human Isolate." *Human Reproduction Update* 5 (2): 103–7.

Ober, C., T. Steck, K. van der Ven, C. Billstrand, L. Messer, J. Kwak, K. Beaman, and A. Beer. 1993. "MHC Class II Compatibility in Aborted Fetuses and Term Infants of Couples with Recurrent Spontaneous Abortion." *Journal of Reproductive Immunology* 25 (3): 195–207.

Ohman, A., A. Flykt, and F. Esteves. 2001. "Emotion Drives Attention: Detecting the Snake in the Grass." *Journal of Experimental Psychology General* 130 (3): 466–478.

Pande, Rohini P. 2003. "Selective Gender Differences in Childhood Nutrition and Immunization in Rural India: The Role of Siblings." *Demography* 40 (3): 395–418.

Patel, R. "The Practice of Sex Selective Abortion in India: May You Be the Mother of a Hundred Sons." *UCIS paper* 7.

Pääbo, S. 1999. "Human Evolution." *Trends in Cell Biology* 9 (12): M13–M16.

Peritz, E., and P. F. Rust. 1972. "On the Estimation of the Nonpaternity Rate Using More Than One Blood-Group System." *American Journal of Human Genetics* 24 (1): 46–53.

Pound, N. 2002. "Male Interest in Visual Cues of Sperm Competition Risk." *Evolution and Human Behavior* 23 (6): 443–466.

Puts, D. A., C. R. Hodges, R. A. Cárdenas, and S. J. C. Gaulin. 2007. "Men's Voices as Dominance Signals: Vocal Fundamental and Formant Frequencies Influence Dominance Attributions Among Men." *Evolution and Human Behavior* 28 (5): 340–344.

Puts, D. A., S. J. C. Gaulin, and K. Verdolini. 2006. "Dominance and the Evolution of Sexual Dimorphism in Human Voice Pitch." *Evolution and Human Behavior* 27 (4): 283–296.

Raley, S., and S. Bianchi. 2006. "Sons, Daughters, and Family Processes: Does Gender of Children Matter?" *Annual Review of Sociology* 32: 401–421.

Rana, B. K., D. Hewett-Emmett, L. Jin, B. H. J. Chang, N. Sambuughin, M. Lin, S. Watkins, et al. 1999. "High Polymorphism at the Human Melanocortin 1 Receptor Locus." *Genetics* 151 (4): 1547.

Retherford, R. D., and T. K. Roy. 2002. "Factors Affecting Sex-Selective Abortion in Punjab, India." *20th Population Census Conference, Ulaanbaatar, Mongolia.*

Roberts, S. Craig, Jan Havlicek, Jaroslav Flegr, Martina Hruskova, Anthony C. Little, Benedict C. Jones, David I. Perrett, and Marion Petrie. 2004. "Female Facial Attractiveness Increases During the Fertile Phase of the Menstrual Cycle." *Proceedings. Biological Sciences/The Royal Society* 271 (Supp. 5).

Roberts, S. Craig, L. Morris Gosling, Vaughan Carter, and Marion Petrie. 2008. "Mhc-Correlated Odour Preferences in Humans and the Use of Oral Contraceptives." *Proceedings. Biological Sciences/The Royal Society* 275 (1652).

Saleh, Maya, John P. Vaillancourt, Rona K. Graham, Matthew Huyck, Srinivasa M. Srinivasula, Emad S. Alnemri, Martin H. Steinberg, et al. 2004. "Differential Modulation of Endotoxin Responsiveness by Human Caspase-12 Polymorphisms." *Nature* 429 (6987): 75–78.

Santos, Pablo Sandro Carvalho, Juliano Augusto Schinemann, Juarez Gabardo, and Maria da Graça Bicalho. 2005. "New Evidence That the MHC Influences Odor Perception in Humans: A Study with 58 Southern Brazilian Students." *Hormones and Behavior* 47 (4): 384–388.

Sawyer, G. J., and V. Deak. 2007. *The Last Human: A Guide to 22 Species of Extinct Humans.* New Haven: Yale University Press.

Shackelford, T. K., D. P. Schmitt, and D. M. Buss. 2005. "Universal Dimensions of Human Mate Preferences." *Personality and Individual Differences* 39 (2): 447–458.

Shipman, Pat. 2002. *The Man Who Found the Missing Link: Eugene Dubois and His Lifelong Quest to Prove Darwin Right.* Cambridge, MA: Harvard University.

Shreeve, Jamie. 2010. "The Evolutionary Road." *National Geographic* 218 (1): 34–61.

Smit, A. 1999. "Interspersed Repeats and Other Mementos of Transposable Elements in Mammalian Genomes." *Current Opinion in Genetics and Development* 9:657–663.

Smith, M. S., B. J. Kish, and C. B. Crawford. 1978. "Inheritance of Wealth as Human Kin Investment." *Ethology and Sociobiology* 8 (3): 171–182.

Stiffman, M. N., P. G. Schnitzer, P. Adam, R. L. Kruse, and B. G. Ewigman. 2002. "Household Composition and Risk of Fatal Child Maltreatment." *Pediatrics* 109 (4): 615–621.

Tattersall, Ian. 2009. "Charles Darwin and Human Evolution." *Evolution Education and Outreach* 2: 28–34.

Tattersall, Ian. 2009. *The Fossil Trail: How We Know What We Think We Know about Human Evolution.* 2nd ed. New York: Oxford University Press.

Tishkoff, Sarah A., Floyd A. Reed, Alessia Ranciaro, Benjamin F. Voight, Courtney C. Babbitt, Jesse S. Silverman, Kweli Powell, et al. 2007. "Convergent Adaptation of Human Lactase Persistence in Africa and Europe." *Nature Genetics* 39 (1): 31–40.

Torrents, D., M. Suyama, E. Zdobnov, and P. Bork. 2003. "A Genome-Wide Survey of Human Pseudogenes." *Genome Research* 13 (12): 2559.

Tregenza, T., and N. Wedell. 2000. "Genetic Compatibility, Mate Choice and Patterns of Parentage: Invited Review." *Molecular Ecology* 9 (8): 1013–1027.

Webb, J. L. A., Jr. 2005. "Malaria and the Peopling of Early Tropical Africa." *Journal of World History:* 269–291.

Wedekind, C., T. Seebeck, F. Bettens, and A. J. Paepke. 1995. "Mhc-Dependent Mate Preferences in Humans." *Proceedings. Biological Sciences/The Royal Society* 260 (1359): 245–249.

Widdig, Anja. 2007. "Paternal Kin Discrimination: The Evidence and Likely Mechanisms." *Biological Reviews of the Cambridge Philosophical Society* 82 (2): 319–334.

Williams, J., and E. Taylor. 2006. "The Evolution of Hyperactivity, Impulsivity and Cognitive Diversity." *Journal of the Royal Society Interface* 3 (8): 399.

Wilson, A. C., and V. M. Sarich. 1969. "A Molecular Time Scale for Human Evolution." *Proceedings of the National Academy of Sciences of the United States of America* 63 (4): 1088–1093.

Wilson, M. I., M. Daly, and S. J. Weghorst. 2008. "Household Composition and the Risk of Child Abuse and Neglect." *Journal of Biosocial Science* 12 (3): 333–340.

Xue, Yali, Allan Daly, Bryndis Yngvadottir, Mengning Liu, Graham Coop, Yuseob Kim, Pardis Sabeti, et al. 2006. "Spread of an Inactive Form of Caspase-12 in Humans Is Due to Recent Positive Selection." *American Journal of Human Genetics* 78 (4). 659–670.

Zhang, Z. L., and M. Gerstein. 2004. "Large-Scale Analysis of Pseudogenes in the Human Genome." *Current Opinion in Genetics and Development* 14 (4): 328–335.

Zorn, Branko, Veselin Sucur, Janez Stare, and Helena Meden-Vrtovec. 2002. "Decline in Sex Ratio at Birth after 10-Day War in Slovenia: Brief Communication." *Human Reproduction* 17 (12): 3173–3177.

Index

About the Authors

RANDY MOORE earned a PhD in biology from UCLA, after which he worked as a biology professor at several large universities. He edited *The American Biology Teacher* for 20 years, teaches courses about evolution and creationism, and has authored and co-authored several books about the evolution-creationism controversy, including *Evolution 101* (2006), *More Than Darwin: The People and Places of the Evolution-Creationism Controversy* (2008), and *Chronology of the Evolution-Creationism Controversy* (2009). Randy is H.T. Morse-Alumni Distinguished Teaching Professor of Biology at the University of Minnesota.

SEHOYA COTNER earned a PhD in conservation biology from the University of Minnesota. Currently a member of the Biology Program at the University of Minnesota, Sehoya's research is focused on science education, specifically, effective strategies for teaching about evolution and the nature of science. She is a co-author of *Chronology of the Evolution-Creationism Controversy* (2009).

Part B of *Planetary astronomy from the Renaissance to the rise of astrophysics* is the sequel to Part A (Tycho Brahe to Newton), and continues the history of celestial mechanics and observational discovery through the eighteenth and nineteenth centuries. Twelve different authors (astronomers, historians of astronomy, celestial mechanists and a statistician) have contributed their expertise in some 17 chapters, each of them intended to be accessible to the interested layman. An initial section (six chapters) deals with stages in the reception of Newton's inverse-square law as exact. In the remainder of the book a large place is given to the development of the mathematical theory of celestial mechanics from Clairaut and Euler to LeVerrier, Newcomb, Hill, and Poincaré – a topic rarely discussed – treated at once synoptically and in some detail. This emphasis is balanced by other chapters on observational discoveries and the rapprochement of observation and theory (for instance, the discovery of Uranus and the asteroids, use of Venus transits to refine solar parallax, introduction of the method of least squares, and the development of planetary and satellite ephemerides). Lists of 'further reading' provide entrée to the literature of the several topics.

THE GENERAL HISTORY OF ASTRONOMY

Volume 2

Planetary astronomy from the Renaissance to the rise of astrophysics

Part B: The eighteenth and nineteenth centuries

Published under the auspices of the International Astronomical Union
and the
International Union for the History and Philosophy of Science

THE GENERAL HISTORY OF ASTRONOMY

General Editor: Michael Hoskin, University of Cambridge

Volume 2

Planetary astronomy from the Renaissance to the rise of astrophysics
Part B: The eighteenth and nineteenth centuries

EDITED BY

RENÉ TATON

Centre Alexandre Koyré, Paris

and

CURTIS WILSON

St John's College, Annapolis, Maryland

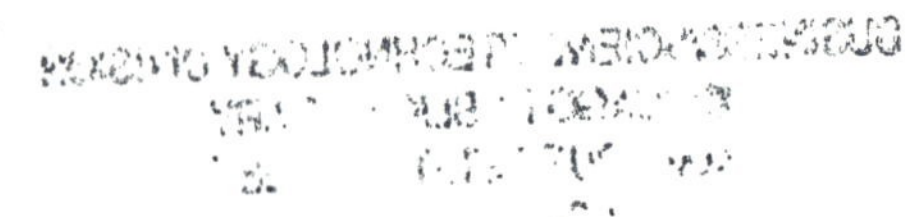

Published by the Press Syndicate of the University of Cambridge
The Pitt Building, Trumpington Street, Cambridge CB2 1RP
40 West 20th Street, New York, NY 10011-4211, USA
10 Stamford Road, Oakleigh, Melbourne 3166, Australia

First published 1995

Printed in Great Britain at the University Press, Cambridge

A catalogue record for this book is available from the British Library

Library of Congress cataloguing in publication data

Planetary astronomy from the Renaissance to the rise
of astrophysics.
(The general history of astronomy; v. 2)
"Published under the auspices of the International
Astronomical Union and the International Union for the
History and Philosophy of Science."
Includes bibliographies and index.
Contents: pt. A. Tycho Brahe to Newton – pt. B. The
eighteenth and nineteenth centuries.
1. Astronomy – History. 2. Astrophysics – History.
I. Taton, René. II. Wilson, Curtis. III. Series.
QB15.G38 1984 vol. 2 520′.9s 88-25817
ISBN 0-521-24265-1 (pt. A) [520′.9′03]

ISBN 0 521 35168 5 hardback

SE

CONTENTS

CONTENTS OF VOLUME 2A

PREFACE

Volume 2 of *The General History of Astronomy* deals with the history of the descriptive and theoretical astronomy of the solar system, from the late sixteenth century to the end of the nineteenth century.

In the European tradition from the time of Plato to the sixteenth century, theoretical astronomy viewed its task as the reduction of the apparent celestial movements to combinations of uniform circular motions. This formulation was still axiomatic for Copernicus; indeed, Ptolemy's violation of the axiom in his *Almagest* was an important stimulus leading Copernicus to undertake his renovation of astronomy. To many astronomers working in the later sixteenth century, after the publication of Copernicus's *De revolutionibus* (1543), the major achievement of this work was that it had freed astronomy from such violations of principle; in contrast, the associated rearrangement of circles that put the Sun at the centre and attributed motion to the Earth was, in the view of many, an absurd error. Copernicus's work, both in its adherence to the long-held axiom of uniform circular motion and in its organization, was thoroughly traditional; and it is thus fitting that Volume 1 of *The General History of Astronomy*, which is devoted to the history of ancient and medieval astronomy, should conclude with it. But, as all the world knows, the Copernican rearrangement contained the seeds of a further transformation.

The story of that transformation begins with the work of the Danish astronomer Tycho Brahe, which is also the starting-point for Volume 2. In 1563 young Tycho, then a student at Leipzig, was shocked to discover that the accepted astronomical tables of the day (Alfonsine based on Ptolemaic theory, and the Prutenic based on Copernican theory) were, both of them, days out in predicting the conjunction of Jupiter and Saturn that took place on 24 August of that year. Tycho's response, over the remaining years of the century, was to give to the accumulation of accurate astronomical observations a priority and importance it had never had before. The store of observations he accumulated in the process became the empirical base for Kepler's justly titled *Astronomia nova* (*New Astronomy*), which made possible a new level of accuracy in the prediction of planetary motions.

Inextricably bound up in Kepler's theory was a new celestial physics, founded on the Copernican vision of the solar system. It was an attempt to account for the observed motions by means of hypothesized quasi-magnetic forces and structures. Kepler did not and could not derive his 'laws' of planetary motion – elliptical orbit and areal rule – from observations alone; on the contrary, his analysis of Tycho's observations was directed by his own highly conjectural celestial physics. Consequently, his contemporaries and successors could not but find the hypothetical foundations dubious. Yet the new predicative accuracy that Kepler achieved was too remarkable to be ignored. Astronomers were ineluctably faced with a challenge: to achieve results as successful as Kepler's, but with more convincing physical foundations. It is not easy to imagine what the history would have been if Kepler's *Astronomia nova* had never appeared to pose this problem. At least we can say: exceedingly different.

The seventeenth century was a period of complex transition in thought, belief, and knowledge of the world; delineation of the aspects of the transition that bear on astronomy is a major concern in Part A of Volume 2. From 1610 onward, telescopic observations of the heavens revealed new facts that had to be incorporated in the system of the world, whether this was conceived to be geocentric or heliocentric. Studies of refraction and parallax showed the Earth to be some 18 times further from the Sun than had previously been believed, and so nearly 6000 times smaller in relation to the volume of the Sun. New studies of motion and force, of falling bodies and the impacts of bodies, were proposed by Galileo, Descartes, and others, a major motive being to show that the heliocentric system was not incompatible with physical principle. The Keplerian planetary tables were tested repeatedly, and continued to prove superior to earlier tables.

An impressively persuasive replacement for Kepler's celestial physics was at length provided in 1687 by Isaac Newton's *Principia*. Part A of Volume 2 concludes with the story of its emergence, and a summary of the astronomical results that Newton succeeded in deriving from his principle of universal gravitation.

Part B, after a section devoted to the gradual acceptance of the Newtonian doctrine during the first half of the eighteenth century, takes up the history of the efforts, from the 1740s onward, to deduce detailed mathematical consequences from universal gravitation. It is the story of what the eighteenth century called 'physical astronomy', but what Laplace in 1799 renamed '*mécanique céleste*', celestial mechanics. The challenge of deducing the consequences triggered the development of new forms of mathematics. Of notable importance here were the trigonometric series, first introduced by Euler, but with later contributions from d'Alembert, Clairaut, Lagrange, and Laplace; a new mechanics of the rotation of rigid bodies, to which d'Alembert and Euler were the chief initial contributors; and the method of variation of the constants of integration in the solution of differential equations, developed by Euler, Lagrange, and Laplace. The story also involves the introduction of statistical calculations into astronomy: at first, with a less than satisfactory outcome, by Euler; then more successfully by Tobias Mayer, whose example was followed by Laplace. With Laplace and Gauss the use of statistical procedures in bringing multiple data to bear on the determination of astronomical constants became *de rigeur*.

Through the nineteenth century the grand theoretical questions remained those that Laplace had forcefully posed and prematurely claimed to answer: the question of the stability of the solar system, and the question of the adequacy of universal gravitation to account for the observed motions of its constituent bodies. By the end of the century Simon Newcomb and his collaborators would achieve a precision in their prediction of planetary motions measured in seconds or fractions of a second of arc. Yet some inconsistencies remained, among them an anomalous motion of the perihelion of Mercury. The first of these was a statistical artifact; the second would become evidence for a new theory of gravitation, which would supersede that of Newtonian physics.

The scope of Volume 2 does not include general relativity. Nor does it embrace astrophysics, which came into being in the 1850s with the development by Kirchhoff and Bunsen of spectral analysis. Both topics belong to volume 4, *Astrophysics and Twentieth-century Astronomy to 1950*. Similarly excluded from the scope of Volume 2 are the topics of stellar astronomy and cosmology, and astronomical instruments, institutions and education, from the Renaissance to the beginnings of astrophysics: these topics constitute the subject-matter of Volume 3. At certain points the concerns of Volumes 2 and 3 overlap. Tycho's sighting instruments were highly relevant to his observational achievement; and from the 1660s and 1670s onward, the pendulum clock, telescopic sights, and the filar micrometer were similarly relevant to the programmes for the improvement of lunar and planetary tables that were adopted by the newly founded Paris Observatory and Greenwich University. Again, Bradley's discoveries of the aberration of light and nutation were prerequisites for the attainment of seconds-of-arc accuracy in the prediction of planetary and lunar positions. For these and similar topics concerned with the stars and with observational instruments and their institutional context the reader must be referred to Volume 3.

Our aim in Volume 2, as in the other volumes of the series, has been to throw light on the development of astronomy as an inventive human activity. We have sought to view the questions and problems of the astronomers of a given time in the very way in which those astronomers saw them, without regard for what has later come to be accepted as 'correct'. We have not attempted an encyclopaedic completeness of coverage; rather, our goal has been to provide an intelligible account of the major endeavours through which astronomy has evolved.

A word is in order with respect to the division of tasks between the two editors of Volume 2, René Taton and myself. Professor Taton resigned his editorship in 1983. By this time he had drawn up the general plan of the volume, and had engaged authors to write rather more than half of the sections envisaged. The engaging of authors for the remaining sections has been my responsibility, and I have also undertaken some reorganization of materials, reducing the original tripartite scheme to a two-part plan, and introducing into Part B a number of new sections which focus on the application of theory to observation.

Finally, I have the grateful duty of acknowledging the extensive guidance and generous assistance provided a neophyte editor by the General Editor of the series, Michael Hoskin. He has again and again given generously of his time, knowledge, and thought to the solution of the problems and difficulties, whether major or minor, encountered in the assembling and editing of this volume. Our undertaking has been, in every aspect, a joint endeavour.

Annapolis, Maryland Curtis Wilson

PART V

Early phases in the reception of Newton's theory

14

The vortex theory in competition with Newtonian celestial dynamics

ERIC J. AITON

Evidence of the intimate acquaintance that Isaac Newton (1642–1727) had with René Descartes's *Principia philosophiae* (1644) is to be found in some of his earliest manuscripts. Thus when Newton in his *Waste Book* formulates the principle of inertia and the "endeavour from the centre" of a body moved circularly, both the ideas and the wording clearly echo Descartes's *Principia.*

In a manuscript beginning "De gravitatione et aequipondio fluidorum . . .", variously dated to the late 1660s and to the early 1680s, Newton raised fundamental objections to certain Cartesian principles. He rejected Descartes's identification of space and body (which, he claimed, offered a path to atheism) and he demonstrated the inconsistency of Descartes's relativistic concept of motion. Moreover, he observed that the rotation of vortices, from which Descartes deduced the force of the aether to recede from the centre (and thus the whole of his mechanical philosophy) implied a space distinct from bodies as a reference frame. Against the existence of Descartes's non-resisting aether or celestial matter, he argued that, if the resistance were set aside, so would be the corporeal nature; for the faculties of stimulating perception and moving other bodies were essential to matter.

Whenever it was that these objections were formulated, it must be allowed that many of Newton's statements during the 1660s and 1670s argue his acceptance during these years of the hypothesis of aethereal vortices. In 1675 he sent to Henry Oldenburg for presentation to the Royal Society a manuscript entitled "An Hypothesis explaining the Properties of Light discoursed of in my severall Papers", in which he developed the hypothesis of a universal aether as agent for electrical, magnetic and gravitational forces. He supposed this aethereal medium to be "much of the same constitution with air, but far rarer, subtiler and more strongly elastic". In this manuscript he also mentioned "the aethers in the vortices of the Sun and planets". Again he alluded to the Cartesian theory of vortices in 1680 in a letter to Thomas Burnet.

But when he took up anew the problem of the planetary motions following the correspondence with Robert Hooke in 1679 and 1680, Newton evidently abandoned the hypothesis of aethereal mechanisms. His objections to the Cartesian principles were brought before the public in a forceful and well-reasoned attack in the first edition of the *Principia* (1687). In particular, Newton adduced experimental evidence against the existence of the aether and claimed to demonstrate that the Cartesian vortices could not explain the motions of the planets in accordance with Kepler's laws.

Newton's experiment, which he remarked had been performed some time before, concerned a comparison of the resistance experienced by a pendulum consisting of a suspended wooden box, first when empty and then when filled with metal. The points of return of the empty box after the first, second and third oscillations having been marked, it was found that the full box required 77 oscillations to return to the first mark, and then the same number to return to each of the others in succession. As the weight of the full box was 78 times that of the empty box, Newton concluded that, if the resistances had been equal, the full box (by virtue of its *vis insita*) should have taken 78 oscillations to return to each of the marks. Consequently, the ratio of the resistance in the case of the empty box to that in the case of the full box was 77 to 78. Taking x to represent the resistance on the external surface of the box and y to represent the resistance on its internal superficies (that is, the internal superficies of the wood, so that the resistance on the internal superficies of the wood and metal of the full box will

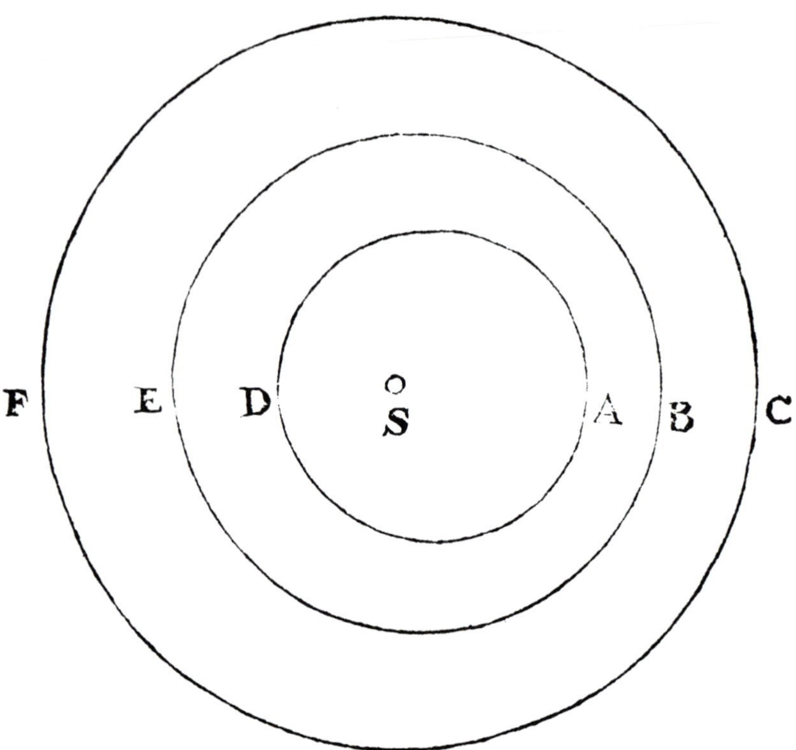

14.1. Newton's diagram in *Principia* (1687) to illustrate his critique of Descartes's explanation of planetary orbits.

be $78y$), the ratio of the total resistance in the case of the empty box to that in the case of the full box becomes $x+y$ to $x+78y$. From the equation $(x+y):(x+78y)=77:78$, the ratio $x:y$ is found to be 5928:1. The resistance on the internal superficies was therefore less than $\frac{1}{5000}$th part of the resistance on the external surface. This reasoning, Newton remarked, depended on the supposition that the greater resistance in the case of the full box "arises from the action of some subtle fluid upon the included metal". But as the period of the full box was less than that of the empty box, the resistance on the external surface was greater because of the greater speed, and greater again because of the longer distances traversed in oscillating. These considerations showed that "the resistance of the internal parts of the box will be either nil or wholly insensible". Newton directed his demonstration against "the opinion of some, that there is a certain aethereal medium extremely rare and subtle, which freely pervades the pores of all bodies".

Descartes had supposed that the planets floated in the aether moving with the same speeds as the layers of the rotating vortex in which they were situated. Newton claimed to demonstrate that a solid planet (subject to no forces except that of the aether) could not remain continually in the same orbit unless it was of the same density as the fluid. For if the planet had a greater density, it would have a greater endeavour to recede from the centre (that is, it would have more centrifugal force) and would move further away. Descartes's aether would thus have had to be as dense as the planets themselves.

In a scholium at the end of the section on the circular motion of fluids, Newton concluded that the planets were not carried around in corporeal vortices. For, according to the "Copernican hypothesis", the planets revolve around the Sun in ellipses having the Sun in their common focus, and by radii drawn to the Sun describe areas proportional to the times, whereas "the parts of a vortex can never revolve with such a motion". Suppose that *AD*, *BE*, *CF* (Figure 14.1) represent three orbits, the outermost (for convenience) being taken concentric with the Sun *S*, while *A* and *B* are the aphelia of the others. Newton argued that,

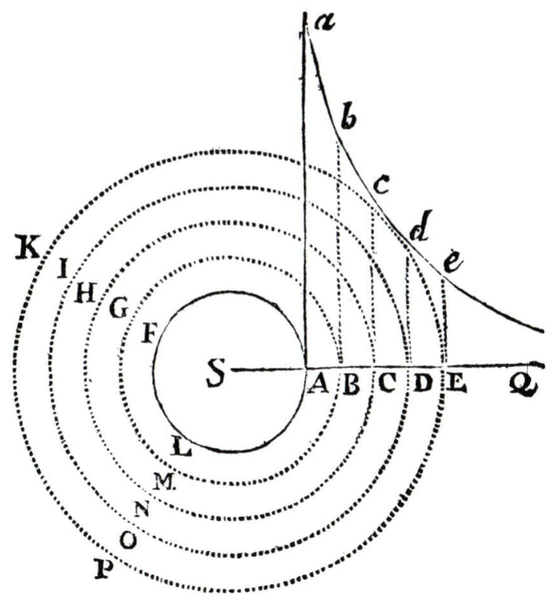

14.2. Newton's diagram in *Principia* (1687) to illustrate his demonstration that planetary motions in a vortex are incompatible with Kepler's third law.

because the space *AC* is narrower than the space *DF*, the fluid of the vortex (and hence the planet) moving in the orbit *BE* would have a greater speed at *B* than at *E*, whereas observations revealed the opposite to be the case.

In order to demonstrate the inconsistency of the explanation of planetary motions in terms of a fluid vortex with Kepler's third law, Newton investigated vortices in an infinite fluid maintained by a rotating cylinder and sphere respectively. He adopted the hypothesis that the resistance or friction between adjacent layers of fluid is proportional to the relative speed and to the area in contact. As Newton followed the same line of reasoning in the two cases, his method may be illustrated by a description of his demonstration for the spherical vortex. First, he supposed the vortex to be divided into solid layers of equal thickness. Let *S* (Figure 14.2) represent the central body, x the radius, dx the thickness and v the speed of a point on the equator of a layer. As each layer moves uniformly, the impressions made by the adjacent superior and inferior layers must be equal and opposite, so that the force must be the same for all layers. According to the hypothesis, the force is proportional to the surface area in contact and the relative speed, here referred to by Newton as the translations of the layers from one another. Perhaps this terminology was intended to indicate the fact that the relative speed varies with the latitude. Although Newton's demonstration lacks a mathematical treatment of the fluid in different latitudes (perhaps he intended this to be supplied by the reader), such an analysis is quite simple and leads to a result consistent with Newton's conclusions. Taking the relative speed at the equator as a measure of the 'translation', the force on a surface of a layer is proportional to the area (consequently to x^2) and to $-x\mathrm{d}(v/x)$. Since the force is the same for all layers, it follows that $-x^3\mathrm{d}(v/x)=c\mathrm{d}x$, where c is a constant, so that $v=bx+c/x$, where b is another constant. Newton in fact only achieved the particular solution $v=c/x$, from which he deduced that the periodic times of the layers are proportional to x^2.

In the case of the cylindrical vortex, the general solution is $v=bx+c$, but Newton gave only the particular solution $v=c$, and deduced that the periodic times are proportional to x.

In both cases Newton had first supposed that the layers rotated as if they were solid. Then, in the spherical vortex, he reasoned that, if the layers were divided into zones and then the matter in these zones made fluid, the vortex would still rotate in exactly the way he had described. Recognizing the weakness of his argument, however, he added:

> *But now, as the circular motion, and the centrifugal force thence arising, is greater at the ecliptic than at the poles, there must be some cause operating to retain the several particles in their circles; otherwise the matter that is at the ecliptic will always recede from the centre, and come round about to the poles by the outside of the vortex, and from thence return by the axis to the ecliptic with a perpetual circulation.*

Although he thus had to admit that his analysis of vortex motion was at least incomplete, Newton nevertheless claimed that the periodic times could not be brought into line with Kepler's law without hypotheses relating density and friction to distance which he considered to be unreasonable. Again, a number of contiguous vortices, such as Descartes described, would in Newton's view run into each other and so gradually destroy themselves. The formation of a vortex wake behind an obstacle had been observed by Leonardo da Vinci but evidently not by Newton. Having demonstrated to his own satisfaction that the "hypothesis of vortices is utterly irreconcileable with astronomical phenomena", Newton referred the reader to Bks I and III of the *Principia* for an explanation of how "these motions are performed in free space without vortices".

Vortices between the first and third editions of Newton's *Principia*

The first edition of Newton's *Principia* appeared one year after the anonymous publication of Bernard le Bouyer de Fontenelle's *Entretiens sur la pluralité des mondes*. In numerous editions, this popular work introduced the general ideas of the vortex theory to a wide readership. The second edition of Newton's work was published in 1713, one year after the sixth (definitive) edition of Nicolas Malebranche's *Recherche de la vérité*. Malebranche, the leading Cartesian among Newton's contemporaries, completely ignored Newton's *Principia*, even after taking up the study of the differential calculus with the Marquis de l'Hospital, though he described Newton's *Opticks* as an excellent work. When the third and definitive edition of the *Principia* appeared one year before Newton's death, the problem of the Moon's motion (which might be regarded as the Achilles's heel of his system) had not yet been solved.

Four reviews of the first edition of Newton's *Principia* had appeared within a year of its publication. The first was in fact a pre-publication review by Edmond Halley in the *Philosophical Transactions*. This gave a description of the contents, including the remark: "the Cartesian doctrine of the vortices of the celestial matter carrying with them the planets about the Sun, is proved to be altogether impossible". Halley praised Newton throughout but there was perhaps a hint of criticism for Newton's failure to give Johannes Kepler adequate credit in his pointed attribution of the phenomena of celestial motions to Kepler.

Although the other reviews are anonymous, there are some indications that the one in the *Bibliothèque universelle* may have been written by John Locke. This review just listed the contents, though Newton's criticism of vortices was emphasized. Again, the review in the *Acta eruditorum*, now known to have been written by Christoph Pfautz, provided an epitome rather than a critical analysis. In particular, Newton's explanation that resistance was wholly on the surface of a body and not on the internal superficies, his criticism that vortices were inconsistent with Kepler's laws and his explanation of the planetary motions in free space, without vortices, were all described. Only the review in the *Journal des sçavans* offered critical comment. Its author described the *Principia* as "the most perfect mechanics that we can imagine" but based on "hypotheses which are generally arbitrary" and which can therefore serve only as the foundation of a treatise on mechanics. To complete the work, he advised, "Newton has only to give us a physics as exact as the mechanics". While the reviewer failed to understand the reasoning by which Newton sought to establish universal gravitation in accordance with the inverse-square law (so that he interpreted this as an arbitrary hypothesis), it would be unfair to blame him for regarding the work only as one of mechanics and not physics; for Newton himself had remarked of the words "attraction, impulse or propensity of any sort towards a centre", that he considered those forces not physically but mathematically. This brief review gave little indication of the contents of the *Principia* but it identified the focus to which the criticisms of Newton's opponents were to be directed; namely, his failure to describe the physical cause of the attraction.

None of the reviewers gave the slightest hint of having recognized the weakness of Newton's treatment of fluid vortices. Although his ideas of solidification and interaction of the layers depended on arbitrary assumptions, so that his propositions did not follow from the axioms and laws of motion set out in Bk I, it was only in 1744 that Jean le Rond d'Alembert demonstrated for the first time that the cylindrical and spherical vortices described by Newton could not exist in a permanent state.

The earliest documented criticism of Newton's section on the motion of fluids is probably the marginal note by Gottfried Wilhelm Leibniz (1646–1716) in his copy of the *Principia*, which casts doubt on Newton's hypothesis that the force of resistance between the quasi-solid layers of fluid is proportional to the relative speed of separation. Leibniz's objection stemmed from his distinction between absolute and respective resistance. The first, which he supposed to be the type envisaged in Newton's hypothesis, arose from the rubbing of the particles of fluid against the solid body and was independent of the speed. On the other hand, the respective resistance (not involved in Newton's hypothesis) Leibniz conceived to arise from the impact of fluid on the body and consequently to be proportional to the speed. Although Leibniz offered no further criticism of Newton's treatment of vortices in his marginal notes (which are available to historians in the modern edition of his *Marginalia*), he had in 1689 published his own hypothesis concerning the explanation of the planetary motions by the motion of fluid vortices. This hypothesis, representing the first attempt to reconcile a vortex explanation of the planetary motions with Kepler's laws, will be described in a later section.

Christiaan Huygens (1629–95), who discovered the satellite Titan and the rings of Saturn, accepted the Newtonian system, though with important reservations. In his *Discours sur la cause de la pésanteur*, published in Leiden in 1690, Huygens declared that he had nothing against the *vis centripeta* or gravity of the planets towards the Sun, not only because it was established by experience but also because it could be explained by mechanical principles. The cause of this *vis centripeta*, he suggested, might be similar to that he had proposed for terrestrial gravity; namely the rapid circulation around the central body of a fluid aether in all directions on spherical surfaces, which displaced solid bodies downwards by virtue of its greater centrifugal force. Huygens admitted that he could not offer any explanation of the inverse-square law. But he rejected Newton's universal attraction extending to the smallest particles as an occult quality not explicable by any principle of mechanics. Moreover, he rejected Newton's void on the ground that light could not be transmitted across it.

According to Huygens's interpretation, Newton demonstrated the elliptical orbit by a counterbalancing of the gravity and the centrifugal force. He therefore regarded Newton's theory as a development of that of Giovanni Alfonso Borelli and also, he added, some of the ancients, according to a report of Plutarch. These ideas were set out by Huygens in his *Cosmotheoros*, published posthumously in 1698, where he also described his own modification of the Cartesian theory. Huygens's vortices (surrounding every star) were neither dense nor contiguous but dispersed in space so as not to hinder each other's free rotations. He offered no reason why such vortices, lacking external constraint, should hold together.

When Newton's *Principia* was published in 1687, the influential Cartesian textbook of Jacques Rohault, *Traité de physique* (1671), was well known in England, as on the Continent, for Théophile Bonet's Latin translation (first published in Geneva in 1674) had been published again in Amsterdam and London in 1682. Samuel Clarke, a follower of Newton, in 1697 prepared a new annotated Latin translation of Rohault's work for the use of Cambridge students. At this time he made no criticism of vortices and did not even mention Newton's attraction. In the second edition of 1702, he began to support Newton against Descartes, but the famous refutation of the Cartesian text in the Newtonian footnotes was not fully realized until the 1710 edition.

A new Cartesian work, Abbé Philippe Villemot's *Nouveau système ou nouvelle explication du mouvement des planètes*, published in Lyons in 1707, was praised for its originality by Fontenelle. Malebranche described it as the embryo of an excellent work. He advised Villemot to abandon the opening chapters (which presented a fantastic hypothesis of the formation of the world) and to begin instead with Kepler's third law, which had been received by all astronomers, and then to seek a demonstration of this law from the properties of centrifugal force. Johann I Bernoulli (1667–1748) was less enthusiastic. For he hoped to see in a new edition of Malebranche's *Recherche de la vérité* "the gravity of bodies towards the centre of the earth and the planets towards the sun (which Mr Newton supposes) soundly explained and the phenomena better demonstrated than with Mr Villemot". To Leibniz and Pierre Varignon (1654–1722), the work seemed to have little merit. Leibniz could not find a shadow of a demonstration in it and Varignon commented adversely on Villemot's knowledge of mathematics.

Villemot knew nothing of the calculus and when he came across a copy of Newton's *Principia* shortly before the publication of his own work, he could find nothing in it to cause him to revise his theory. Villemot offered an explanation of Kepler's third law, though he did not mention either the first or second laws. Like all other Cartesians before him, he used the term 'ellipse' only in reference to the lunar variation, while simply remarking of the planetary orbits in general that they were not perfect circles. In his view, the equilibrium of the vortex required the total centrifugal force of each spherical layer to be the same, so that the speed v was proportional to $1/r^{\frac{1}{2}}$, r being the radius of the layer. Consequently, the periodic time t was proportional to $r^{\frac{3}{2}}$. Villemot claimed this to be a "demonstration of the famous problem of Kepler ... of which up to now it has not been possible to render account in any system". Although he recognized that "Ptolemy's theory" required the times needed by a planet to traverse equal arcs in the neighbourhood of the apsides to be proportional to the squares of the distances, he believed that this was only approximately true and that rigorously they "obey Kepler's [third] law exactly".

Villemot also attempted explanations of the precession of the equinoxes, the motions of the planetary and lunar nodes, rotation of the apse-line and the rotations of the planets on their axes. The causes of all these phenomena he found in the effects of the celestial matter of the solar and terrestrial vortices. For example, having followed Copernicus in attributing the precession of the equinoxes to the near equality of two rotations resulting in a slow rotation of the Earth's axis relative to the stars, he found the cause of the slight inequality in the difference of the speeds of the aether at the distances of the Earth's centre and its lower extremity from the centre of the solar vortex. The numerical results for precession, rotation of the lunar nodes and the Earth's diurnal motion would have been impressive if his theory had been soundly based, but as we have remarked, its superficiality was clearly evident to Varignon and Leibniz.

Gravity

Gravity was the force, Villemot supposed, "which keeps the heavens in equilibrium, which sustains the planets in their spheres, and in a word which arranges and disposes all the parts of the universe". The acceptance, first by Huygens, then by Villemot and Cartesians generally, of the gravity of the planets towards the Sun demonstrated by Newton, introduced new difficulties for Cartesian explanations of the planetary motions. According to Descartes's explanation, the planets were indeed pushed towards the centre of the vortex but only in the same way as the aether itself; that is, by the external pressure which constrained the whole vortex to circulate. As the planet and surrounding aether circulated with the same speed, neither had more centrifugal force than the other to resist the external pressure, so that both continued to circulate at the same distance from the centre. On the other hand, heavy bodies near the surface of a planet were displaced downwards by the more rapidly moving aether on account of its greater centrifugal force. Thus in Descartes's theory, the causes of deflection of a planet from a straight path and the vertical fall of terrestrial bodies were essentially different. For Newton, of course, they were essentially the same.

Villemot sought to identify the gravity of the planets towards the Sun with terrestrial gravity. But as he supposed, following Descartes, that the planets moved with the same speed as the aether, he deduced that the centrifugal force of the aether could not be the cause of gravity. For this purpose he introduced an outward impulse arising from the hot matter in the centre, so that the aether, impelled upwards, displaced solid bodies downwards. By analogy with rays of light, he deduced the inverse-square law for gravity. While the existence of hot matter at the centre of the solar vortex was obvious, the heat experienced in mines, he declared, provided evidence for the existence of a source of heat in the interior of the Earth. Yet the planets did not fall towards the Sun like heavy bodies towards the Earth. To solve this problem, Villemot introduced a vague force of impact of the planetary vortex on the solar vortex, which served to balance the gravity.

In the sixth edition of the *Recherche de la vérité* (1712), Malebranche modified Villemot's theory of terrestrial gravity while retaining Descartes's explanation of the circulation of the planets. Thus he explained that the gravity of the planets did not come from the centre but from the external compression of the solar vortex. Malebranche gave Villemot's theory of terrestrial gravity some semblance of plausibility by reasoning that the effect of the outward impulse was greater on the small vortices, of which he supposed the aether to be composed, than on solid bodies. Nevertheless the main weakness remained; namely that any theory depending on the action of a fluid on a small body immersed in it without relative motion contradicted the known principles of hydrostatics. For Blaise Pascal had shown that, in a fluid at rest, the pressure was the same in all directions. In the case of the terrestrial vortex, Malebranche supposed, the outward impulse did not arise from the action of hot matter in the centre (as Villemot thought) but from the reaction of the solid Earth on the small elastic vortices. This explanation enabled him to account for the absence of an outward impulse in the solar vortex.

Both Villemot and Malebranche had been led to reject Descartes's theory of terrestrial gravity by the insuperable difficulty they had found in the extreme speed of the aether that Huygens had shown to be necessary. In particular, Malebranche could not see how bodies could escape being carried horizontally with the rapidly moving aether.

The Cartesian theory of terrestrial gravity as the effect of a rapid circulation of the aether was defended by Malebranche's friend Joseph Saurin (1659–1737). First, by a misapplication of mechanical principles, he claimed to show that, contrary to the view of Huygens, the simple circulation about an axis supposed by Descartes could account for the fall of heavy bodies towards the centre of the Earth. Moreover he inferred that a plumb-line on a spherical Earth would not be deflected by the Earth's rotation and he thus rejected Huygens's deduction of the spheroidal form of the Earth from the absence of such a deflection. After this unpromising start, Saurin's second contribution, presented to the Royal Academy of Sciences in Paris in 1709, had considerable merit. Taking Mariotte's experiments on the force of impact of a liquid on a solid surface as a basis, he estimated the feebleness of the force of the aether moving with the speed calculated by Huygens that would be needed to render its effect imperceptible when masked by the air, and so explain why heavy bodies were not carried horizontally by the circulation. Having found the force to be at least three or four million times more feeble than that of air, he speculated that such a non-resisting aether would only be possible if the particles of ordinary matter were small compared with the spaces between, say in the ratio of 1 to 100 000, so that the aether could flow freely between them. Experimental confirmation of this bold idea of the constitution of matter was first provided by Daniel Bernoulli, who deduced from the Boyle–Mariotte gas law that the particles were infinitesimal compared with the spaces between them.

Saurin had not, of course, rendered the aether incapable of causing gravity. For the mechanism of gravity, in Cartesian physics, was not one of impulsion but of hydrostatic displacement. Solid bodies were not impelled downwards by the force of impact of the aether but were displaced downwards because they had less centrifugal force than the rapidly circulating aether.

This idea of a non-resisting aether that could nevertheless be the cause of gravity might have led to an early reconciliation of the Newtonian and Cartesian systems if it had been generally accepted, though it should be remarked that Saurin himself was firmly attached to the Cartesian system. The idea of a non-resisting aether causing gravity seemed contradictory to Jean Bouillet, who was awarded a prize for an essay on gravity in 1720 by the Royal Academy of Bordeaux. Bouillet rejected Saurin's theory in favour of the ideas of Villemot and Malebranche, which he sought to combine into the explanation of a gravity common to the solar and planetary vortices. First, he supposed the aether to consist of Malebranche's small elastic vortices. Like the celestial matter of Villemot, these could penetrate to the centre of the Earth. In place of Villemot's vague impact of the planetary vortex on the solar vortex, introduced to balance the gravity of the planet and thus prevent it from falling into the Sun, he substituted the centrifugal force of the planetary vortex arising from its motion around the Sun. Then he deduced the inverse-square law from the elasticity of the small vortices. Supposing that the small vortices must have equal quantities of motion, he reasoned that those further from the centre, having a greater speed of translation, must have a smaller speed of rotation about their own centres and hence a smaller elasticity. There was also another reason, he explained, why the elasticity decreased with distance from the centre of the vortex:

The proper motion of the small vortices, or their elasticity, must weaken or diminish to the extent that the layer which contains them finds less resistance from the superior layers, or is less pressed towards the centre of the circulation. But the greater the circle described by the aether, the less it experiences the resistance of the superior layers, the latter having more liberty for filling their motion and consequently less force for pushing those which are inferior to them.

Since each of the two causes weakened the elasticity in proportion to the distance, Bouillet concluded that the elasticity, and hence the effect of gravity, "must diminish according to the squares of the distances"

Leibniz's explanation of planetary motion

According to his own account, Leibniz's "Tentamen de motuum coelestium causis", published in the *Acta eruditorum* in February 1689, was a hasty extemporization of his ideas concerning planetary motion, stimulated by the review of Newton's work in the same journal but written before he had the opportunity to read the *Principia* itself. It is now known that Leibniz's claim to have composed his essay independently of the *Principia* is untrue. For some recently discovered manuscripts show preparatory calculations for the essay that clearly imply a knowledge of Newton's *Principia*. Leibniz attributed to Kepler the idea of the fluid vortex and also (mistakenly) the idea that circular motion engendered a centrifugal force. Descartes, he believed, made use of these ideas without acknowledgement but did not attempt to explain the physical causes of the laws discovered by Kepler, either because he could not reconcile them with his own principles or because he did not believe them true. Evidently Leibniz did not doubt that Descartes had known Kepler's laws.

Leibniz supposed that the vortex carrying the planets rotated in spherical layers. To account for the elliptical orbit he postulated two motions of the planet; a radial motion from layer to layer and a trans-radial motion in which the planet moved with the same speed as the fluid. Kepler's second law, which Leibniz stated explicitly in the exact area form, required the trans-radial speed to vary inversely as the distance, as Kepler had shown in the *Epitome astronomiae Copernicanae*. Consequently, Leibniz described his harmonic vortex or circulation as one in which the speeds of circulation of the layers were inversely proportional to the radii or distances from the centre of circulation. Having supposed the planet M (Figure 14.3) to move in the curve $M_3M_2M_1$ and in equal times to

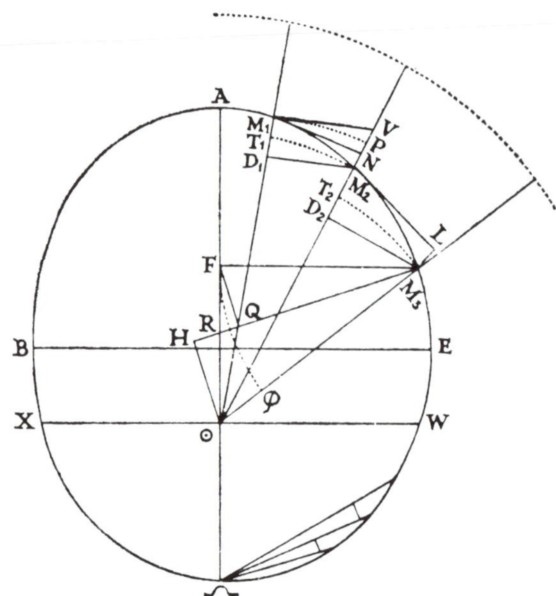

14.3. Diagram by Leibniz to illustrate his theory of planetary motion, published in *Acta eruditorum* in 1689.

describe the elements M_3M_2, M_2M_1 of this curve, he remarked that the motion could be regarded as composed of a circular motion around the Sun ⊙, whose elements were M_3T_2, M_2T_1, and a rectilinear motion whose elements were T_2M_2, T_1M_1.

Leibniz's work contains at least one major error; namely a calculation of the centrifugal force which gives only half the true value. Following correspondence with Varignon, however, he eliminated this error and published some corrections in 1706. His definitive calculation of centrifugal force was based on the approximation of the circle by a polygon with first-order infinitesimal sides, so that the 'tangent' at a vertex was taken to be the prolongation of the preceding infinitesimal chord and the centrifugal force was expressed by a uniform motion from the circle to the tangent. As Leibniz remarked to Varignon, the method was simpler that did not put the acceleration into the elements. Varignon tried in vain to convince Leibniz that, although the motion in the infinitesimal chord gave correct results, in reality the motion was in the infinitesimal arc.

The original calculation of the gravity did not require correction, probably because (consciously or unconsciously) it had been based on the same principles as the corrected calculation of the centrifugal force. Although he did not make an explicit

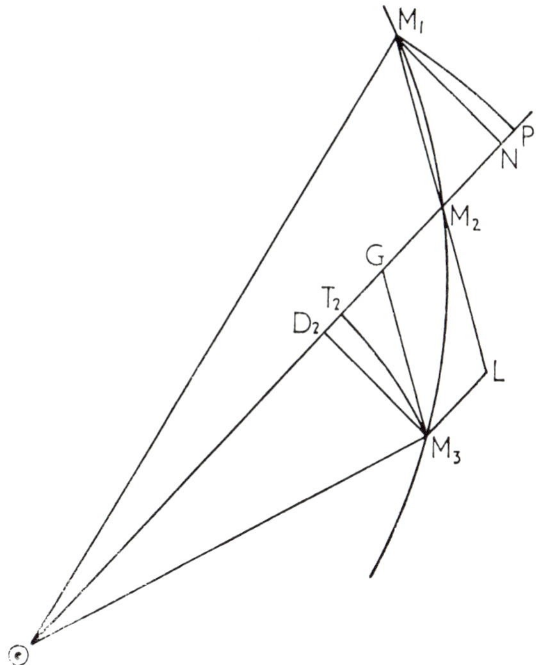

14.4. Interpretation of Leibniz's diagram in accordance with his definition of tangent.

statement that this was how his calculation was to be understood, Leibniz's demonstrations are consistent with this interpretation. For he expressed the gravity by a uniform motion from the 'tangent' to the curve and the interpretation removes a number of apparent errors, all trivial but otherwise seemingly inexplicable. In the description that follows, the corrections are incorporated and it is assumed that the interpretation according to which the curve is approximated by a polygon with first-order infinitesimal sides is the one that Leibniz held. Such an interpretation is open to the same logical objections as Newton's demonstration of Kepler's second law, in which the curve is similarly approximated.

Leibniz first demonstrated that, in the case of a body moving with harmonic circulation, the centrifugal force varied inversely as the cube of the distance. Let $h\,dt$ be the constant area swept out in equal times dt, equal to twice the area of triangle $M_2M_3\odot$ (Figures 14.3 and 14.4); that is, $D_2M_3 \times \odot M_2$ or $D_2M_3 \times r$, taking $\odot M_2 = r$. Then $D_2M_3 = h\,dt/r$, so that the centrifugal force $2D_2T_2$ (Leibniz's corrected measure) $= (D_2M_3)^2/r = h^2dt^2/r^3$.

The radial motion, called by Leibniz *motus paracentricus*, was explained by the combined action of the gravity of the planet towards the Sun and the centrifugal force arising from its circulation with the vortex. Leibniz envisaged two ways of resolving the motion along an infinitesimal element of the curve into components. These are clearly described in a letter to Huygens, where he writes:

the circulation D_1M_2 or D_2M_3 being harmonic, and M_3L parallel to $\odot M_2$, meeting the preceding direction M_1M_2 prolonged in L, then M_1M_2 is equal to M_2L (or to GM_3; the printer has omitted the letter G between T_2 and M_2 marked in my description) and consequently the new direction M_2M_3 is composed as much of the preceding direction M_2L added to the new impression of gravity, that is to say, to LM_3, as of the speed of circulation D_2M_3 of the ambient aether in harmonic progression added to the paracentric speed M_2D_2 already acquired in some progression.

In the case of the first resolution, the planet was supposed to be deflected from its instantaneous inertial path by the action of gravity alone; this would happen in the absence of a vortex.

Consequently Leibniz expressed the effect of the "paracentric solicitation of gravity" by a uniform motion along LM_3. In the case of the second resolution (that of physical reality), the planet was constrained by the vortex to circulate in accordance with Kepler's second law, while the paracentric motion (that is, the motion along the rotating radius vector) was caused by the combined action of gravity and the centrifugal force engendered by the circulation. From the geometry of the figure Leibniz demonstrated that the radial acceleration ddr = centrifugal force – solicitation of gravity.

From M_1 and M_3, let M_1N and M_3D_2 be normal to $\odot M_2$; as, on account of the harmonic circulation, the triangles $M_1M_2\odot$ and $M_2M_3\odot$ are of equal area, the altitudes M_1N and M_3D_2 also will be equal (because of the common base $\odot M_2$). Taking M_2G equal to LM_3, let M_3G be drawn parallel to M_2L; then the triangles M_1NM_2 and M_3D_2G will be congruent, and M_1M_2 will be equal to GM_3 and NM_2 will be equal to GD_2. Again in line $\odot M_2$ (produced if necessary, as is always understood) let $\odot P$ be taken equal to $\odot M_1$ and $\odot T_2$ equal to $\odot M_3$, then PM_2 will be the difference between the radii $\odot M_1$ and $\odot M_2$, and T_2M_2 the difference between the radii $\odot M_2$ and $\odot M_3$. Now PM_2 equals NM_2 or

GD_2+NP, and T_2M_2 equals $M_2G+GD_2-D_2T_2$; consequently $PM_2-T_2M_2$ (the difference of the differences) will be $NP+D_2T_2-M_2G$, that is (because NP and D_2T_2, the versed sines of two angles and radii having differences less than any assignable magnitude, coincide) twice $D_2T_2-M_2G$. Now the difference of the radii expresses the radial speed [and] the difference of the differences expresses the element of the radial speed [or the acceleration].

The centrifugal force in the harmonic vortex, $2D_2T_2$, has already been shown to be $h^2\mathrm{d}t^2/r^3$, so that $\mathrm{dd}r=h^2\mathrm{d}t^2/r^3-$solicitation of gravity. For a body moving in an ellipse in a harmonic vortex, Leibniz demonstrated that $\mathrm{dd}r=h^2\mathrm{d}t^2/r^3-k\mathrm{d}t^2/r^2$, where $k=h^2/$(semilatus rectum), from which he deduced the solicitation of gravity to vary inversely as the square of the distance.

The hypothesis of a fluid vortex seemed to Leibniz to be the only possibility remaining after the ideas of planetary intelligences, solid spheres, sympathies and abstruse qualities had been eliminated. Having introduced the harmonic vortex to explain the motion of the planets in accordance with Kepler's second law, he made no attempt to explain the motion of the vortex itself in terms of basic physical principles. Neither did he offer any explanation of gravity in the printed paper – simply alluding to the Cartesian theory (which he attributed to Kepler) – but in another paper published in 1690, he explained gravity as the effect of an outward impulse transmitted through the aether which pushed the planets towards the centre by circumpulsion. By analogy with rays of light, the outward impulse decreases in intensity in proportion to the square of the distance. The outward impulse, he supposed, arose from the circulation of the aether in great circles; this was, in effect, the theory of Huygens. Leibniz distinguished the vortex causing gravity from the harmonic vortex carrying the planets. The two vortices were held to be independent, the first consisting of very tenuous matter revolving in all directions in great circles, the second consisting of coarser matter revolving about an axis with harmonic circulation.

When Huygens pointed out that the harmonic vortex was superfluous, since Newton had demonstrated that gravity alone provided a sufficient physical basis, Leibniz agreed that the hypothesis of Newton explained the elliptical orbit, adding however that he had retained his own (which was consistent with that of Newton) because it could also explain the circulation of all the planets (as well as the satellites of Jupiter and Saturn) in the same sense and very nearly in the same plane. Thus Leibniz held that Newton's hypothesis and his own could be reconciled but that his own was preferable on account of its greater explanatory power.

David Gregory (1659–1708) was exceptional among Newtonians in recognizing the merits of Leibniz's theory. If Kepler's laws could be reconciled with vortex motion, he remarked, Leibniz was the one to do it. Nevertheless he criticized the theory on two counts; first, comets obey Kepler's second law and should therefore be drawn round with the harmonic vortex, whereas their orbits are often inclined at large angles to the plane of circulation of the vortex, and second, the different planets move in accordance with Kepler's third law, which is inconsistent with the harmonic circulation. He even suggested a reply to the second objection which he believed Leibniz would make; namely, that the harmonic circulation was restricted to narrow bands containing the planets, the fluid between circulating in accordance with Kepler's third law. Leibniz indeed accepted this suggestion, although Gregory had warned that it would hardly carry conviction.

In addition to the objection concerning the comets, Newton (as we know from his manuscript notes) made two basic criticisms of Leibniz's theory; first, he claimed that 'centrifugal force', by his third law, was always equal and opposite to 'centripetal force', and second, that Leibniz's mathematical reasoning was unsound, owing to errors in the handling of second-order infinitesimals. The objection concerning 'centrifugal force' assumed that the term could only be interpreted in Newton's definitive sense. Leibniz in fact used it, as Newton had earlier, to denote a force engendered by circular motion. Concerning the other objection, Newton claimed that second-order errors were involved in the assumptions that $NP=D_2T_2$ and $NM_2=GD_2$. In the case of the second assumption, Newton's assertion would be true if Leibniz's term 'tangent' were interpreted in the Euclidean sense of a tangent at the point M_2 of the curve, but his own understanding of second-order infinitesimals is brought into question by the fact that his statement concerning NP and D_2T_2 is false.

Newton's criticisms formed the basis of the objections of John Keill, published in the *Journal litéraire de la Haye* in 1714. Keill pointed out that NM_2 and GD_2 could be taken equal if the tangent were interpreted as the chord M_1M_2 but that in his view, such an interpretation was not permissible. Newton himself published his objections anonymously in a review of the *Commercium epistolicum*, while Roger Cotes, in his preface to the second edition of the *Principia*, repeated the objection concerning the comets.

Influence of Newtonianism

On the basis of a mistaken inference from a memoir of William Whiston, it has generally been supposed that David Gregory, who was professor of mathematics at Edinburgh between 1683 and 1690, was the first to introduce Newtonianism into courses for students, but a study of manuscripts of Gregory's lectures and a number of extant copies taken by students shows that Newton's theories never formed an integral part of these lectures. The reason was not to be found in any lack of interest on Gregory's part but rather in the immaturity of Scottish students, who entered the universities at about the age of fourteen years, and the utilitarian aims of the curriculum. Gregory began to make a detailed commentary on Newton's *Principia* as soon as he received it in 1687 and in 1690 at least three of his students were being encouraged to "keep Acts" (that is, read out essays) on aspects of Newtonian philosophy. Lectures at Edinburgh began to be based fully on Newton's ideas only towards the turn of the century.

It was on Newton's recommendation that Gregory was appointed Savilian professor of astronomy at Oxford in 1691, where he published his *Astronomiae physicae et geometricae elementa* (1702), which included an exposition of Newton's theories in a form suitable for students.

The first Continental mathematician to interpret Newton's theory in terms of the differential calculus was Varignon. This he did in a series of memoirs contributed to the Royal Academy of Sciences between 1700 and 1710. Having received his first instruction in the differential calculus from Johann Bernoulli in 1692, he was ready by 1698 to apply this calculus to problems of mechanics and planetary motion. Fontenelle brought Varignon's work to the attention of Leibniz, who encouraged him to undertake further investigations. In his letter to Fontenelle of 3 September 1700, accepting appointment as a foreign member of the Academy, Leibniz raised a number of queries; among them was the question whether G. D. Cassini, doyen of the Paris astronomers, and Philippe de La Hire had abandoned Kepler's ellipse. After confirming that Cassini had replaced the ellipse of Kepler by an oval in which the product of the distances from the foci was constant and explaining La Hire's finding that the planets do not describe any perfectly regular curve, Fontenelle remarked that Varignon, "one of our ablest geometers", had developed a general method for finding the "central forces" which push the planets. He added that, while Newton and Leibniz himself had established these forces by geometry, Varignon used only the differential calculus and never failed to render homage to its inventor, Leibniz.

Although Varignon evidently regarded his work simply as mathematics, Fontenelle, in his reviews for the annual *Histoire* of the Academy, attempted to fit it into a Cartesian framework by finding the cause of the "weights of the planets" in the action of the celestial matter of the solar vortex. Varignon began his investigation by considering rectilinear motion, establishing two general rules defined by the equations $v = \mathrm{d}x/\mathrm{d}t$ and $f = \mathrm{d}v/\mathrm{d}t = \mathrm{dd}x/\mathrm{d}t^2$, where x is the distance from the starting point, v the speed, f the force (acting in the direction of the line) and $\mathrm{dd}x$ the distance traversed owing to the increment of speed $\mathrm{d}v$, assumed to be acquired instantaneously at the beginning of the infinitesimal element of time $\mathrm{d}t$. To prove the second rule, Varignon appealed to a generalization of Galileo's law of fall. Taking the distance traversed by a body subjected to a constant force to be proportional to the force and the square of the time, he concluded that $\mathrm{dd}x = f\,\mathrm{d}t^2$. Eliminating the time from the two equations, he obtained the relation $f\,\mathrm{d}x = v\,\mathrm{d}v$ or $\int f\,\mathrm{d}x = \frac{1}{2}v^2$, which he recognized as Newton's Prop. 39 of Bk I. For this result he claimed independent discovery.

In his second memoir Varignon established a formula for the centripetal force of a body moving in a curve. Let PK (Figure 14.5) be $\mathrm{d}s$, r the distance of P from the centre of force, f the centripetal force and f_1 the component of this force in the direction PK. Then $f/f_1 = PK/PN = \mathrm{d}s/(-\mathrm{d}r)$, from which it

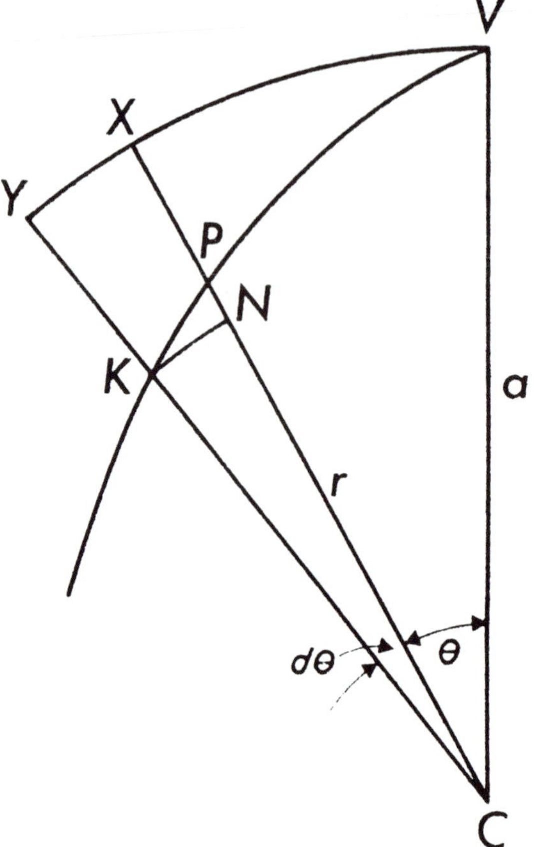

14.5. Varignon's theory of motion in a curve.

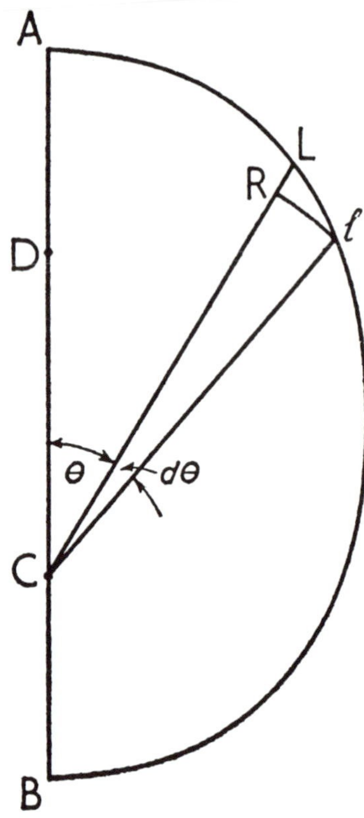

14.6. Varignon's theory of planetary orbits.

follows that $f = -f_1 \mathrm{d}s/\mathrm{d}r$. By Varignon's second rule of motion, $f_1 = \mathrm{dd}s/\mathrm{d}t^2 = \mathrm{d}v/\mathrm{d}t$, so that $f = (-\mathrm{d}v/\mathrm{d}t)(\mathrm{d}s/\mathrm{d}r) = -v\mathrm{d}v/\mathrm{d}r$. Varignon failed to recognize that this result was equivalent to Newton's Prop. 40 of Bk I. He described it as "a very simple formula of *central forces*, centrifugal as well as centripetal, which are the principal foundation of the excellent work of Mr Newton".

Although Varignon in 1700 described Kepler's second law as the "most physical", he recognized that some astronomers preferred other laws. At this time he may even have regarded it as an independent hypothesis, for he made no attempt to deduce it from his rules of motion. It could not in fact have been deduced from his formula for centripetal force, for this embodied what was later called the principle of conservation of energy, whereas something equivalent to the principle of conservation of angular momentum would have been needed to establish Kepler's second law. By 1710, however, Varignon had no doubts about the law, "which Mr Newton has demonstrated ... to be the true one in a space without resistance".

In his third memoir of 1700, Varignon completed the first phase of his investigation of planetary motions, using his formula for centripetal force to demonstrate that, for a planet moving in an ellipse in accordance with Kepler's second law, the centripetal force directed towards a focus must vary inversely as the square of the distance. Suppose that the ellipse (Figure 14.6) with foci C and D represents a planetary orbit, C being the centre of force. Let $CL = r$, $Rl = \mathrm{d}z = r\mathrm{d}\theta$, where $\theta = \angle ACL$, $Ll = \mathrm{d}s$, $AB = a$ and $CD = c$. Also let $b^2 = a^2 - c^2$. Then the equation of the ellipse is $-b\,\mathrm{d}r = \mathrm{d}z\sqrt{(4ar - 4r^2 - b^2)}$. It follows that $(4ar - 4r^2)\mathrm{d}z^2 = b^2(\mathrm{d}r^2 + \mathrm{d}z^2) = b^2\mathrm{d}s^2$ or $(4a - 4r)/r = b^2\mathrm{d}s^2/(r\,\mathrm{d}z)^2 = b^2v^2$, putting $r\,\mathrm{d}z = \mathrm{d}t$ in accordance with Kepler's second law. On differentiation, this equation becomes $-4a\mathrm{d}r/r^2 = 2b^2v\mathrm{d}v$, so that $f = -v\mathrm{d}v/\mathrm{d}r = 2a/(br)^2$. Thus the centripetal force varies inversely as the square of the distance. As corollaries, Varignon extended his result to the cases of the hyperbola and parabola.

These results, Varignon remarked, were in agreement with the principal propositions concerning planetary motion demonstrated by Newton and Leibniz.

Turning his attention next to the theories of Seth Ward and G. D. Cassini, using firstly their own hypotheses of times and secondly the law of Kepler, Varignon obtained some very complicated expressions for the centripetal forces that would be needed. Evidently there was no hypothesis among those that had been proposed by astronomers to which his general formula for central forces could not be applied.

Having established the principal propositions for a planetary orbit, Varignon set about developing another approach, expressing the centripetal force in terms of the radius of curvature. It was on the basis of this work, communicated in a letter, that Leibniz in 1702 encouraged Varignon to continue his researches, for the fundamental memoirs of 1700 were not published until 1703. Leibniz asked Varignon to consider more than one centre of attraction, mentioning that David Gregory had considered the action of the Sun on a planet and a satellite but not the actions of the principal planets on each other. Although he had not seen Gregory's book, Varignon responded immediately, but his examples were limited to hypothetical cases of little relevance to astronomy. Moreover, they concerned the direct problem; given a curve and a number of fixed centres, to find the forces needed to move a planet with constant speed or speed varying according to a given rule. Leibniz, of course, had in mind the general inverse problem; that is, to determine the orbit, given several forces directed towards moving centres.

Newton had considered the motion of the apse-line from the standpoint of the direct problem, assuming the orbit to be a rotating ellipse and finding that the addition of a centripetal force proportional to the cube of the distance would be needed. In a memoir of 1705, Varignon applied the differential calculus to this problem, confirming Newton's result, though without resolving the difficulty of accounting fully for the motion of the lunar apogee. For Varignon, the rotating ellipse was just another "system of astronomy" to which his general theory could be applied. He did not suggest a physical cause for the extra force, though Fontenelle related the moving apogee to the effect of neighbouring bodies on the basis of the vortex theory.

For the case of a single centripetal force varying inversely as the square of the distance, the inverse problem was solved by Newton in the *Principia*. Having solved the direct problem for ellipse, hyperbola and parabola, he concluded that the inverse-square orbit must be a conic section, because a conic could always be constructed to satisfy the initial conditions of speed and direction of a projected body and two different orbits touching at a point were not possible. A more elegant solution was given in *Principia*, Bk I, Prop. 41, which lacks only the evaluation of the integrals. Both Jakob Hermann and Johann Bernoulli independently translated this proposition into terms of the differential calculus and evaluated the integrals to demonstrate that an inverse-square orbit must be a conic section. Their memoirs were presented to the Paris Academy on the same day in 1710. There followed a controversy in which Bernoulli criticized Hermann's solution for lacking a constant and being somewhat contrived. Hermann was defended by Jacopo Francesco Riccati, who commented on the elegance of the solution and pointed out that the lacking constant only changed the origin, so that no loss of generality was involved. When Varignon saw these solutions, he easily deduced the result of Newton's Prop. 41 from Kepler's second law and his own formula for the centripetal force but he lacked the facility in the integral calculus needed to evaluate the integrals. He had sought instruction in this calculus from Bernoulli but this had been withheld in accordance with a secret agreement between Bernoulli and the Marquis de l'Hospital. The following year Varignon wrote to Leibniz that, although he had considered the problem of fixed multiple centres both in a medium and the void, he had not found anything on moving centres. Neither Varignon nor his contemporaries made any further progress towards the solution of this intractable problem.

Following the publication in 1713 of the second edition of Newton's *Principia*, his system was taught in Leiden by W. J. 'sGravesande, who became professor of mathematics and astronomy in 1717, on his return from a visit to England, where he had met Newton and John Theophilus

Desaguliers. Although he made no original contribution to knowledge, 'sGravesande greatly furthered the cause of Newtonianism, especially through his *Physices elementa mathematica experimentis confirmata sive introductio ad philosophiam Newtonianam* (1720–21), which became one of the most influential popular accounts of Newton's system. This work, however, was ignored by the editors of the *Journal des sçavans*, while the review in the *Mémoires à Trevoux* raised objections against the proscription of hypotheses and the acceptance of the void and attraction in the Newtonian system. These principles of Newtonian philosophy had already become the focus of renewed controversy in the Leibniz–Clarke correspondence, first published in 1717 followed by two further editions in 1720. They continued to be the chief obstacle to the acceptance of the Newtonian system by Cartesians and others who saw in them a return to the occult qualities of the Scholastics. Newton's claim to have deduced the attraction from the phenomena did not convince the critics, for as Fontenelle asked rhetorically in his "Eloge de Newton", were not the Scholastic qualities also causes whose effects could be seen?

Vortices after Newton

In his "Eloge de Newton", Fontenelle contrasted the methods of Descartes and Newton. Descartes, he wrote, "proceeds from what he clearly understands to find the cause of what he sees", whereas "Newton proceeds from what he sees to find the cause, whether it be clear or obscure". A year later, reviewing the first of a series of papers presented to the Royal Academy of Sciences by Joseph Privat de Molières (1677–1742), Fontenelle remarked that, although the Newtonian system had "some very advantageous aspects", the Cartesian system was more "agreeable to the intellect".

Privat de Molières based his defence of vortices on the ideas of Villemot and Malebranche. In his first paper, he sought to establish that the gravity of the planets, varying inversely as the square of the distance, was an effect of the external compression of the vortex. First he divided the vortex into a set of imaginary concentric spherical layers and supposed that the speeds of all the particles in a particular layer were equal. The centrifugal force f of a particle in latitude λ was proportional to $v^2/(r \cos \lambda)$, v being the speed and r the radius of the layer. Then Privat de Molières supposed that the reaction between the imaginary layers was normal to the surface. Thus each particle was pushed by the layer immediately above with a centripetal force ϕ, which balanced the normal component $f \cos \lambda$ of its centrifugal force. It followed that $\phi = v^2/r$. Moreover, for equilibrium of the whole vortex, the total centripetal force, according to Privat de Molières, had to be the same for each layer. This implied that, if S was the area of the surface of a layer, ϕS was constant and consequently ϕ was proportional to $1/r^2$.

Having established the inverse-square law for gravity, he proceeded to demonstrate Kepler's third law. In the plane of the ecliptic, $\phi = f$, so that $1/r^2 = v^2/r$. Also v is proportional to r/t, where t is the periodic time, so that, eliminating v, it follows that r^3 is proportional to t^2. By this demonstration, Privat de Molières declares, Kepler's law becomes a "principle of mechanics, from which all the celestial motions can be deduced geometrically, as M. Villemot has already attempted to do, and that sustains and confirms the system of Descartes, far from overthrowing it, as has been claimed in our day".

In his second paper, presented in 1729, Privat de Molières showed how an aether composed of Malebranche's elastic globules (expanding and contracting as a result of the inequalities in the external pressure) could revolve in elliptical layers and consequently carry the planets in elliptical orbits. His third paper, presented in 1733, concerned the reconciliation of Kepler's second and third laws in application to the layers of a vortex. At this stage, however, Privat de Molières attempted to assimilate the mathematical results of Newton into his system. As gravity in a spherical vortex had been shown to vary inversely as the square of the distance, he supposed that the same relation would hold very nearly in an elliptical vortex that was nearly spherical. He then concluded that, as Newton had demonstrated the consistency of Kepler's second and third laws with an inverse-square law for gravity, Kepler's laws were consistent also in his elliptical vortex.

Another problem taken up by Privat de Molières was that of the Earth's diurnal motion, which Descartes had attributed ambiguously to the action of the terrestrial vortex and to the persistence of the rotation impressed on the Earth at the time of its creation. The need for a solution had become urgent, for La Hire had remarked, criticizing Villemot's explanation in terms of the action of the terrestrial vortex, that as the inferior layer of the aether moved faster (from west to east) than the superior, the vortex should cause the planet to rotate in the opposite direction (from east to west) to that observed. A more elaborate solution to the problem was given by Jean Jacques d'Ortous de Mairan. On the basis of plausible assumptions (apart from the confusion of mass and weight), Mairan attempted, by a confused argument involving the differences of the impulses of the aether on the superior and inferior hemispheres of the planet and the supposed differences of the weights of these hemispheres in virtue of their different distances from the centre of the vortex, to determine the periods of rotation of the planets on their axes, arriving at remarkably accurate results. For Jupiter, his theory predicted a period of $10\frac{1}{4}$ hours and some minutes, "very nearly as the observations reveal". Even the case of the Moon could be explained by a hypothetical difference in density of the two hemispheres, and he was able to point out that the Newtonian attraction was quite incapable of accounting for the axial rotation of the planets.

In his *Leçons de physique*, published in four volumes between 1734 and 1739 and consisting of lectures delivered to students at the Collège Royal de France, Privat de Molières completed his reconciliation of the systems of Newton and Descartes, assimilating "the void of Newton, or that *non-resisting* space, of which this philosopher has so invincibly established the existence", as well as the "attraction ... from which Newton, without however having been able to discover the mechanical cause, has taken so many good conclusions". Confusion of mass and weight had led him to the conclusion that the aether offered no resistance owing to its lack of gravity towards the Sun. In effect Privat de Molières had adopted the Newtonian system in his treatise, which was the last Cartesian textbook covering the whole field of terrestrial and astronomical physics. His only reservation was that he could never agree that matter was essentially heavy and hard, "absurd hypotheses" which he claimed "Newton does not frame ... but which he adopts".

Privat de Molières had set out in 1728 to defend the Cartesian system and had come to accept the Newtonian system, apart from the attraction considered as a cause. The same trend towards acceptance of the Newtonian system even by Cartesian sympathizers is evident in a number of essays awarded prizes by the Royal Academy of Sciences between 1728 and 1740. In 1728 the prize was awarded to Georg Bernard Bulffinger for an uncompromisingly Cartesian essay in which he offered a modification of Huygens's theory of gravity and a defence of Leibniz's theory. In particular, he showed that Leibniz's reconciliation of Kepler's second and third laws would require the density to be uniform in the harmonic layers of the solar vortex (the layers in which the planets moved) and to vary inversely as the square root of the distance in the intermediate layers (the layers moving in accordance with Kepler's third law).

Another Cartesian essay, on the causes of the elliptical orbit and the rotation of the line of apsides, was adjudged most worthy of the prize in 1730. This was Johann I Bernoulli's "Nouvelles pensées sur le système de M. Descartes". Bernoulli begins his essay with the remark that the reader may be surprised to find him defending the vortices at a time when many philosophers, particularly the English, regard them as pure fancy. Some unpublished correspondence seems to indicate that he defended the vortices simply to improve his chances of winning the prize. His description of the Cartesian system at the beginning, where he seems to suggest that a miraculous arrangement of the vortices by God would be needed to produce the regular effects of the planets, might well lead the reader to wonder whether he is defending it. Yet he declared the Newtonian principles of attraction and void to be unacceptable and there can be little doubt that his criticisms of Newton's propositions were intended seriously. By these propositions Newton had claimed to demonstrate that the hypothesis of vortices was inconsistent with Kepler's third law. In Bernoulli's view, however, Newton's reasoning was "a manifest sophism, being based on two propositions equally false".

Newton's first error, according to Bernoulli, was to suppose the friction between adjacent layers (which he had first regarded as solid) to be dependent on the relative speed and the area. For Guillaume Amontons had shown experimentally that the friction between solid surfaces depended on the normal force between them but was independent of the area in contact. Newton's second error, as seen by Bernoulli, was the neglect of the lever effect of the frictional force. Bernoulli repeated Newton's calculations for the cylindrical and spherical vortices, in his view correctly, estimating the normal force by an integral taking into account the action of all the inferior layers. He deduced periodic times proportional to $x^{\frac{4}{3}}$ and $x^{\frac{5}{3}}$ (where x is the distance from the centre) for the cylindrical and spherical vortices respectively. These results were nearer the $x^{\frac{3}{2}}$ required by Kepler's third law than were those of Newton. From this he surmised that there might be some shape between the cylindrical and spherical that would give periodic times in accordance with Kepler's law. Just a little flattening in the Sun, similar to that observed in Jupiter, he suggested, would suffice.

Bernoulli then considered the explanation of Kepler's third law in terms of a variation in density with distance from the centre of the vortex, finding that the density would need to be proportional to $x^{-\frac{1}{2}}$. This reconciliation of vortex motion and Kepler's law, he remarked, would have had little effect in silencing his opponents if he had not first demonstrated the falsity of Newton's propositions. Evidently Bernoulli did not know that Bulffinger had arrived at exactly the same relation of density with distance using the assumptions of Newton's propositions.

Turning his attention to the problems proposed for the prize, Bernoulli attributed the elliptical orbit to the combination of a circular motion of the planet about the Sun and a radial oscillation about an equilibrium position, taking the planet in turn into layers of greater and smaller density, while the line of apsides rotated, he thought, owing to a lack of synchronism between the circulation and radial oscillation. His treatment of these problems was entirely qualitative and his solutions would have needed the same miraculous arranging of the vortices mentioned at the beginning. Moreover, he did not even attempt to explain why a planet, immersed in fluid layers circulating in accordance with Kepler's third law, should itself rotate about the Sun in accordance with the second.

The prize having been withheld in 1732, a double prize was offered in 1734 for an essay on the physical cause of the inclinations of the planetary orbits to the plane of the solar equator. On this occasion prizes were awarded to Johann I Bernoulli and his son Daniel, while Pierre Bouguer and Jean Baptiste Duclos received honourable mention. Johann Bernoulli, encouraged perhaps by the appearance of P. L. M. de Maupertuis's *Discours sur les différentes figures des astres* in 1732, undertook to form a new system by combining elements of the Newtonian and Cartesian systems. First, he explained gravity by a central stream of matter from the circumference. Then he postulated an aether consisting mainly of Descartes's first element; that is, he remarked, an aether like Newton's void. Moreover, he abandoned the idea that the planets were carried by the vortex. For if the vortex were to be regarded as an appendage of the Sun, as Bernoulli supposed, the layer in contact with the Sun should have the same speed as the solar rotation. Application of Kepler's third law to the layers of the vortex then gave them speeds of about $\frac{1}{230}$th part of those of the planets. However, the weak circulation enabled him to explain the direction over a period of time of the planets into orbits in the plane of the ecliptic. The small inclinations of the orbits to this plane he attributed to small drifts arising from the spheroidal form of the planets. The comets spent most of their periodic time where the circulation was extremely weak and were therefore hardly affected by it.

Daniel Bernoulli's replacement of the aethereal vortex by a solar atmosphere itself subject to gravity indicates his preference for the Newtonian system, though he speculated (perhaps as a concession to the judges) that the cause of gravity might be found in a complicated system of independent vortices moving through each other without mutual interference. Having concluded on the basis of a calculation of probability that the concentration of the planetary orbits in a narrow zone could not have occurred by chance, he suggested the effect of the circulating solar atmosphere as the cause. A mathematical analysis of the solar atmosphere in accordance with the gas laws relating density, pressure and temperature, showed a falling off of density in the outer layers almost to zero.

Taking the radius of the Sun to be a and the density, temperature and pressure at a distance x from the centre to be D, T and p respectively, let the density, temperature and pressure at the surface of the Sun be 1. Then by the gas law, $D=p/T$. Supposing the temperature to be inversely proportional to the square of the distance, Bernoulli

Table 14.1

Position in the solar atmosphere	Density
Surface of the Sun	1
Mercury	2200
Venus	3000
Earth	2600
Mars	1300
Jupiter	0.4
Saturn	0.000006

deduced that $D=(x^2/a^2)p$. As the difference in pressure between points of two adjacent layers balances the weight, $-dp=c(a^2/x^2)D\,dx$, where c is a constant. Substituting the value of D in this equation gives $-dp=cp\,dx$ and consequently $p=e^{-c(x-a)}$. It follows that $D=(x^2/a^2)e^{-c(x-a)}$ and by differentiation the distance of the layer of maximum density is found to be $x=2/c$.

For a trial calculation, Bernoulli took Venus to be at the distance of maximum density. Then the distribution of densities was found to be as in Table 14.1. This seemed to suggest that the layer of maximum density was in fact further out, say in the region of Jupiter, so that all the planets would be subject to the action of the solar atmosphere and in consequence of this action over a long period of time reduced "within the narrow limits where they are at present". After an infinite time, Bernoulli believed, the orbits would become perfect circles and united exactly in the plane of the solar equator. As the comets spent most of their time in the region of negligible density, he concluded that they had been almost unaffected by the circulation, so that their orbits should be randomly inclined. In fact, the mean inclination of the orbits of 24 observed comets was found to be 43° 39′.

Duclos's astronomical observations won him membership of the Academy of Sciences in Lyons, where the manuscript of his unpublished essay may be found. His solution of the problem of the inclinations of the planetary orbits was based on the idea that, as the particles of fluid tended to move in great circles but were constrained to move in small circles parallel to the equator, there arose in them an impeded endeavour to move along the meridians. By virtue of this "polar" force, Duclos believed, the aether caused the planet to oscillate about the plane of the equator, like a pendulum, while circulating with the vortex. Duclos remained firmly attached to the Cartesian theory, but in an unpublished history of the vortex theory written in 1737 or 1738, he expressed the view that the best defence of the system of vortices would be the reconciliation of this system with that of Newton along the lines suggested by Privat de Molières. By employing "the good theory of Newton on the central forces without having need of the attraction", this philosopher, he remarked, "profits, without any inconvenience, from the advantages of the English physics and loses nothing of the incontestable superiority of the French physics".

Bouguer's essay took the form of a dialogue between a Newtonian, a Cartesian and a character representing himself. While a modified Cartesianism was allowed to prevail, the Newtonian case was stated quite strongly. To explain the inclination of the planetary orbits, he supposed that originally the various spherical layers of the vortex rotated about different axes, whose inclinations have been reduced in course of time by imperceptible degrees to their present values as revealed by the planets.

The Newtonian system was openly defended by Maupertuis in his *Discours sur les différentes figures des astres*. Although he may have derived some ideas from Bouguer, the main source of inspiration of Maupertuis's work was evidently the inconsistency of the conclusions of Newton and Jacques Cassini concerning the figure of the Earth. This, together with the fact that Cassini wrote papers defending the Cartesian system, may have given rise to the erroneous view that the Cartesian system predicts the Earth to be a prolate spheroid. Cassini's figure of the Earth was, in fact, derived entirely from faulty geodetic measurements. Despite the outcome of the controversy, Fontenelle was not unreasonable when he remarked that the actual measures were to be preferred to those resulting from theories. In his advocacy of the Newtonian system, Maupertuis emphasized its predictive power – in determining the position of comets, for example – and also the fact that impulsion was no more intelligible than attraction.

Jacques Cassini in 1735 and 1736 presented papers to the Royal Academy of Sciences offering Cartesian solutions to two outstanding problems. These papers may have been inspired by dissatisfaction with the explanations of Johann Bernoulli and Privat de Molières, resulting from their combination of Cartesian and Newtonian elements. The first problem concerned the inconsistency of the slow rotation of the Sun with the circulation of the layers of the solar vortex carrying the planets in accordance with Kepler's third law. Cassini solved the problem by surrounding the Sun and planets with atmospheres rotating with the same angular speed as the central bodies. The speed of the atmosphere at its outer limit, he supposed, was equal to that of the layer of aether in contact with it. For the terrestrial vortex, the atmosphere would need to extend to a height of six times the Earth's radius. This was considerably greater than the height formerly attributed to the atmosphere. No difficulty would arise in the case of the solar atmosphere, which would need to extend to about half the distance of Mercury. To test the theory, Cassini proposed a comparison of the theoretical period of revolution of Saturn, assuming the atmosphere to extend to the outside of the ring, with the observational evidence that some favourable occasion might provide.

The second problem to which Cassini gave his attention was the compatibility of Kepler's second and third laws. Privat de Molières had not really attempted a physical solution, having been content simply to accept Newton's mathematics. Cassini considered whether the two laws might result from a lack of precision in the observations. As they predicted a difference of 15′ or 16′ in the position of the Sun in the course of a month round the Earth's aphelion or perihelion, he regarded the reality of the problem as established. Cassini based his solution on the continuity of the aether in the plane of the equator of the vortex.

Suppose that the aether at the mean distance *R* (Figure 14.7) of the planet turns circularly about the Sun *S* with the speed required by Kepler's third law. Then Cassini maintains that the aether contained in the space *RST* moves into the space *RSV*, while an equal volume enters the space *RST*. Now suppose that the matter is retained in the ellipse *ARPI* by some force which prevents it from describing the circle. Then, Cassini explains, the aether that would have filled the small space *RVM* must extend in the ellipse beyond *SM* and occupy the small sector *MSO*, so that the planet, approaching perihelion, traverses a greater distance in the ellipse than it would have traversed in the circle.

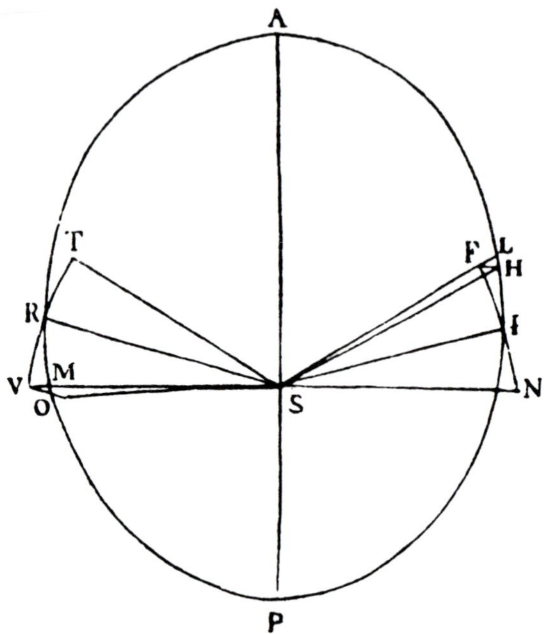

14.7. Jacques Cassini's diagram illustrating his Cartesian theory of planetary orbits.

The area *SRO* being equal to the area *RSV* that would be swept out by the line joining planet and Sun in the same time in a circular orbit, it follows that equal areas are described in equal times, as required by Kepler's second law. Cassini's explanation is basically sound, though incomplete. For the aether contained in one sector does not in fact move as a whole into the next sector, though equal volumes of aether leave one and enter the other. Fontenelle was satisfied that Cassini had shown the agreement of the two laws of Kepler in the hypothesis of vortices.

In 1740 the prize for an essay on the tides was divided between four competitors. One of these, the Jesuit Antoine Cavalleri, gave a Cartesian justification of Newton's results. The others, namely Colin Maclaurin, Leonhard Euler and Daniel Bernoulli, based their explanations unequivocally on the Newtonian system, although Euler insisted that the inverse-square law must be the effect of the action of aethereal vortices. Daniel Bernoulli again emphasized the predictive power of universal gravitation, "this incomprehensible and uncontestable principle, that the great Newton has so well established and that we can no longer hold in doubt".

It was their recognition of the explanatory power of the Newtonian system that led the ablest Cartesians to combine Newton's mathematical theory with physical vortices. Then when they lost faith in vortices, remarks such as those of Maupertuis concerning the equal unintelligibility of impulsion and attraction (of which there are hints in Bouguer's essay) could help them to abandon the vortices with a clear conscience and accept universal gravitation as a physical axiom. In the second edition of his prize essay (published in 1748), Bouguer himself, having in the meantime observed the deflection of a plumb-line in the neighbourhood of a mountain in Peru, announced his conversion to Newtonianism, though with a modified attraction law. To cover all cases, including the phenomena explored by the chemists, he suggested a law consisting of two terms, either of which might be dominant, one inversely proportional to the square and the other to the cube of the distance. Alexis-Claude Clairaut, he remarked, had also needed a modification of Newton's attraction to explain the motion of the Moon's apogee; in fact, the addition of a term inversely proportional to the fourth power of the distance. Later in the year 1748, however, Clairaut succeeded in explaining this motion on the basis of the inverse-square law, thus advancing beyond Newton to achieve a confirmatory triumph for the theory of universal gravitation with an inverse-square law in application to the solar system.

Bouguer clearly indicated the common path to progress when he remarked that, for peaceful coexistence between Newtonians and Cartesians, it was sufficient for them to declare,

the ones perhaps without much belief, and the others without much hope, that the word attraction, *the same as that of weight, simply describes a fact, while awaiting the discovery of the cause.*

A new generation of Continental Newtonian mathematicians, beginning with Clairaut (who helped the Marquise du Châtelet to translate Newton's *Principia* into French), had no difficulty in following Bouguer's advice to confine themselves to the truths of induction or admit only the immediate consequences. When Fontenelle's *Théorie des tourbillons* appeared with thinly veiled anonymity in 1752, this work, like its author (born in 1657), must have seemed to its readers as a survival from a former age. Euler did not even mention vortices in his letters to the Princess of Anhalt-Dessau but expressed the view prevalent among scientists in 1760 when he asserted that the system of universal attraction was established by the most solid reasons. Yet Euler himself maintained the view that gravity had a physical cause, not yet known in detail but certainly arising from the action of a fluid matter filling space.

Further reading

E. J. Aiton, *The Vortex Theory of Planetary Motions* (London and New York, 1972)

P. Brunet, *L'Introduction des théories de Newton en France au XVIIIe siècle* (Paris, 1931)

S. Delorme *et al*, *Fontenelle: sa vie et son oeuvre* (Paris, 1961).

M. A. Hoskin, Mining all within: Clarke's notes to Rohault's *Traité de physique*, *The Thomist*, vol. 24 (1962), 353–63

A. Koyré, La gravitation universelle de Képler à Newton, *Archives internationales d'histoire des sciences*, vol. 4 (1951), 638–53

L. M. Marsak, Cartesianism in Fontenelle and French science, 1686–1752, *Isis*, vol. 50 (1959), 51–60

H. Parenty, Les tourbillons de Descartes et la science moderne, *Mémoires de l'Académie des Sciences, Belles-Lettres et Arts de Clermont–Ferrand*, deuxième série, fasc. 16 (Clermond–Ferrand, 1903)

15

The shape of the Earth

SEYMOUR L. CHAPIN

The great voyages of discovery and exploration in the fifteenth and sixteenth centuries enormously extended man's knowledge of the globe on which he lives. Although all educated men since Antiquity had known that the Earth was spherical, these voyages – and especially the spectacular circumnavigations – confirmed for the ordinary people that man's abode was indeed round, and, in fact, that it was a sphere of very considerable extent. Further, mathematicians and astronomers of the first half of the sixteenth century undertook to determine its size with some precision. Thus, the Frenchman Jean François Fernel, having astronomically established the northern extremity of 1° of the meridian which began in Paris, determined the length of that distance by counting the number of revolutions of a coach's wheel, while a few years later the Netherlander Reiner Gemma Frisius described the method of triangulation for measuring such arcs with greater certainty by means of the terrestrial observation of angles between established sites followed by appropriate trigonometric procedures. It remained for the seventeenth and eighteenth centuries, however, to develop and apply the technique of triangulation as well as to raise the question of the Earth's precise shape and to provide a solution to that question.

An early assault on the first of these tasks was undertaken by Willebrord Snell (1580–1626), a professor at the University of Leiden, who developed the idea of his fellow countryman, Gemma Frisius, to such an extent as to deserve to be called the father of triangulation. Starting his base line with his house in Leiden, he used three spires of two churches as reference sites – solving in the process the so-called recession problem for three points, subsequently named after him – and then established a network of triangles that enabled him to compute the distance between two Dutch towns about 130 kilometres apart. Although he employed large instruments made by the famous cartographer Willem Janszoon Blaeu – a former assistant to Tycho Brahe who had himself in that capacity undertaken a very accurate arc measure by triangulation – Snell was dissatisfied with the results that he published in 1617. He therefore continued to work toward their perfection, but, because of his early death, these efforts did not become publicly available until more than a century later.

Snell's work was probably inspired in part by an understandable desire to establish precise cartographical details about a newly independent country. It must have owed something also to the fact that the knowledge of a length of a degree was essential to the perfection of sea charts. Certainly Blaeu, for example, had extensive contacts with both merchants and navigators, as evidenced by the fact that he was to become the official cartographer of the Dutch East India Company.

The needs of navigation seem clearly to have been the major incentive for an equivalent English interest in the question of the length of a degree of the Earth's spherical surface. Thus, the English cartographer Edward Wright (1561–1615), whose fame rests largely upon his mathematical explanation of the famous Mercator projection (which yielded maps particularly well suited to navigational needs), in 1599 called for an improvement in the traditional value of 60 miles per degree then employed by English seamen in his work *Certaine Errors in Navigation....* His suggestion was acted upon by Richard Norwood (1590–1665) who, after having voyaged himself as a young man and then written several books on the mathematics of navigation, decided in 1635 to measure the distance between London and York. He first determined the latitude of each city and then walked from one to the other furnished with a measuring chain and a theodolite, employing the first – along with the measurement of his own steps – to determine the distance covered, and the second to correct for the fact that the route he followed was neither a straight line nor on a flat surface. His value of about 69.5 miles for a degree compares reasonably with what was to become the classic measure of a degree in the second half of the seventeenth century.

That new standard was established in the France of Louis XIV, whose great finance minister, Jean Baptiste Colbert, was sympathetic to proposals made in the early 1660s for the establishment of a state-sponsored centre for scientific studies. Colbert saw that such an institution could be an efficient consultative body for the government, another arm, so to speak, of his whole mercantilistic economic programme. Thus, it would be able to assist in the increase of the country's productivity, not only by passing judgement upon the usefulness of new inventions but by suggesting various ways in which science could serve the nation – as by improving craft practices through scientific theory or playing a role in another of Colbert's broad goals, the exact mapping of his monarch's kingdom.

The latter aim fitted in well with the extensive astronomical programme proposed for the Academy of Sciences that was established in 1666. The proposed programme was put forward by two of the new Academy's original members. One of these was the great Dutch scientist Christiaan Huygens (1629–95), who in 1656 had made Galileo's discovery of the isochronism of the pendulum into the principle underlying an accurate timepiece and who had subsequently developed a marine variant featuring a short pendulum; the latter had been responsible for Colbert's luring him to Paris in 1665 with the offer of an enormous salary. The other was the native Frenchman, Adrien Auzout, who had only recently perfected – if not independently invented – the filar micrometer, and who called for the creation of a royal observatory in Paris and the dispatch of scientific expeditions for the investigation of significant problems or phenomena. Auzout's active support was soon responsible for bringing into the Academy one Jean Picard (1620–82), the closest approximation to a 'professional' astronomer then to be found in France.

It was largely Picard who carried through the revolution in observational astronomy made possible by the filar micrometer, the astronomical pendulum clock, and the application of telescopes to large-scale graduated instruments appropriate for the measurement of small angles. It was with this equipment that he undertook to measure the distance between two locations approximately on the meridian of Paris, to determine the differences in their latitudes, to measure their separation by triangulation, and to deduce from these results the length of a degree of meridian; from this one could easily calculate the size of the Earth, as had been desired by both Huygens and Auzout. This eminently successful arc measure was executed during the summer and autumn academic vacations in 1668, 1669 and 1670. Picard's result provided the basis on which the desired rectification of French cartography could be carried out, and, equally important, a model for all later undertakings of this kind of 'geometrical' geodesy.

The Academy also implemented Auzout's call for scientific expeditions. This occurred first in the search for a solution to a problem that had become crucial with the great oceanic voyages mentioned at the outset, namely the determination of longitude at sea. Since that problem would be solved if the difference in time between a ship's location and its port of departure or some other agreed-upon reference point could be measured, it is not surprising that one of the solutions proposed early on was to have the ship carry along a timepiece keeping the local time of the chosen reference site. This idea was first put forward by Gemma Frisius in 1553, but sufficiently accurate clocks were not available before the appearance of the pendulum-driven clock. We have already mentioned Huygens's development in 1662 of the marine version of this device. The Academy undertook several expeditions between 1668 and 1670 from the Mediterranean to France's possessions on the Atlantic seaboard of North America for the purpose of testing such clocks. Their unhappy outcome was to demonstrate that the utilization of the timekeeper approach would have to await the perfection of spring-driven mechanisms, something that was to require almost another century of development.

Another of Galileo's ideas was destined to bear fruit in the same area. In 1610 he had discovered the first four satellites of Jupiter, and in 1612 he had observed the eclipse of a satellite. He was struck by the possibility of reading the positions of the satellites as the hands of a celestial clock in the determination of longitudinal differences. An expedition to Marseilles, Malta, Cyprus and Tripoli, dispatched under the aegis of Nicolas Claude Fabri de Peiresc (see Chapter 9 of Volume 2A), gave but dubious results: the available tables of the satellites were too inaccurate. But in 1668 the Italian astronomer, Gian Domenico Cassini (Cassini I, 1625–1712), brought out new tables of the motions of Jupiter's satellites which, because they were accurate enough to predict the eclipses of those bodies, finally rendered Galileo's idea practicable for the determination of longitude on *terra firma*. At sea, because of the difficulty of making precise observations through a long telescope on the deck of a swaying ship, the idea proved unusable. But for the cartographical rectification of France, Cassini's ephemerides provided an important tool. Small wonder, then, that Cassini, like Huygens, was invited to enjoy a large salary from the Sun King on the condition of taking up residence in Paris, where he assumed not only membership of the Academy but a leading role in the affairs of the observatory then being constructed.

In July 1671, Picard himself went on an expedition in execution of one part of a programme of astronomical researches that he had presented to the Academy in late 1669. Its purpose was to establish the exact location of Tycho Brahe's observatory at Uraniborg on the island of Hven and its longitudinal separation from Paris, in order to be able effectively to utilize Tycho's star catalogue as well as his planetary and lunar observations, a massive edition of which Picard had begun to prepare for publication in France. Armed with Cassini's predictions of the eclipses of Jupiter's satellites that would be observable in the autumn of 1671, Picard achieved his goal with notable success.

Another aspect of this voyage is more closely related to our chapter's topic: its association with the question of the Earth's possibly non-spherical shape, a question that arose out of concern about the length of a seconds-pendulum (a pendulum adjusted so that its period of swing in one direction was one second of time). In his early work on the pendulum, Huygens had found that the period of a simple pendulum oscillating through small angles is proportional to the square root of its length, and had established a relation between this period and the motion of free fall, so that we can derive from his proportions the modern formula $T = 2\pi\sqrt{l/g}$, where T is the time for a full back-and-forth swing, l is the pendulum's length, and g is the acceleration of gravity. But g does not appear explicitly in his formulas, and he did not imagine it to be variable. Consequently, he proposed to the Royal Society of London at the end of 1661 that the length of the seconds-pendulum – which he stated as 38 Rhineland inches – be accepted as a universal standard of length. In 1666 he made the same proposal to the new Academy in Paris; but by the summer of 1669 he was indicating that the length of the seconds-pendulum, in Parisian units, ought to be 36 inches $8\frac{1}{2}$ lines (there are 12 lines to the Paris inch, as well as to the Rhineland and English inches).

The latter figure appears to be that arrived at by Picard, who had undertaken careful experimentation with pendulums and enthusiastically embraced the suggestion that a seconds device be universally adopted as a standard of length. Near the end of 1670, in fact, he asked a Parisian correspondent of the Royal Society to dispatch the value that he, Picard, had established to that body's secretary with a view to having this adopted as the standard. By that time, however, Huygens had already noted an apparent discrepancy, for when he employed the accepted ratio of 720/695 for the measure of a length in Rhineland inches and its measure in Paris inches, he found that Picard's value corresponded not to the 38 inches he had found, but to 38 inches and $\frac{3}{10}$ of a line (the difference is about 0.6 millimetre). Alerted to this discrepancy, and having learned that the English were finding a longer length for the seconds-pendulum, Picard took advantage of the voyage to Uraniborg to make accurate pendulum observations. His experiments both in Leyden and in Uraniborg appeared to confirm his conviction that the seconds-pendulum was viable as a universal standard of length. But in achieving that result he altered the Rhineland–Paris ratio to 720/696. (A seconds-pendulum in Uraniborg in fact differs in length from a seconds-pendulum in Paris by about 0.63 millimetres, and from a seconds-pendulum in Leyden by about 0.32 millimetres.)

While Picard was in Uraniborg, the Academy

dispatched one of its young "student" members, Jean Richer, to Cayenne (just off the Atlantic coast of South America a little north of the equator) to conduct a series of astronomical observations. Especially important were those designed to take advantage of the close approach of Mars to Earth in 1672 in order to deduce, with the help of corresponding observations made by Cassini in Paris, a new and improved figure for solar parallax (see Chapter 7 of Volume 2A). But what is significant for our present topic is that Richer took with him a pendulum that had been regulated to beat seconds in Paris. In Cayenne he found it necessary to shorten its length by no less than 1.25 lines (2.8 millimetres) in order to make it continue to do so.

This piece of evidence could not immediately be explained. Cassini was inclined to think that Richer had been mistaken or careless. Picard apparently agreed, for he continued to insist that the seconds-pendulum was invariable in length. Indeed, he persevered in that belief in his own subsequent measures in France even when his own assistant suspected a variation, especially in those taken at Bayonne in the extreme south of the country. In 1682, however, the year of Picard's death and a decade after Richer's measurement at Cayenne, another expedition dispatched by the Academy confirmed his results. These new observations were made at Gorée, a small island off Cape Verde on the west coast of Africa, and at Guadaloupe in the West Indies. They were the work of two of the Sun King's engineers for hydrography, Jean Deshayes and a M. Varin, joined by a M. de Glos, a young man trained – as were they – by Cassini at the Royal Observatory.

Today we recognize that the variation with latitude in the duration of the swing of a pendulum of given length must be due to a change in the acceleration of gravity. How was this possible? The answer was soon forthcoming in the *Principia* of Isaac Newton (1642–1727), which appeared in 1687. In Prop. 19 of Bk III Newton set out from the idea that the Earth could originally have been a homogeneous fluid mass, subject to the law of universal gravitation and rotating about an axis; he postulated without demonstration that it would form an ellipsoid of revolution, flattened at the poles. Fluid in two canals leading from centre to surface, one in the equatorial plane and the other coinciding with the polar axis, would be in balance. By calculation Newton found the centrifugal force at the equator to be to the mean acceleration of gravity as 1:289 (he used 1:290 in the first edition). By a laboriously derived approximation, he argued that the canal along an equatorial radius would exceed the canal along the polar axis by $\frac{1}{289} \times \frac{505}{400} = \frac{1}{229}$th of the polar radius. He claimed (without demonstration and in fact mistakenly) that the ellipticity would necessarily be greater if the Earth were denser toward the centre. In the third edition (1726) he offered the explanation that "the planets are more heated by the Sun's rays toward their equators, and therefore are a little more condensed by that heat than towards their poles".

In Prop. 20 of Bk III Newton undertook to compare effective weights at different latitudes on the surface of the Earth. By the principle of balanced canals, these weights vary inversely as the distances to the centre. It followed that the increase in effective weight varied as the square of the sine of the latitude. From this result he deduced (on the assumption of a homogeneous Earth) that a pendulum beating seconds would have to be shortened by about $\frac{1}{12}$ of an inch when taken from Paris to the Equator (the figure becomes 1.087 twelfths in the second edition of 1713). He likewise supposed that if the Earth were denser toward the centre, the shortening of the seconds pendulum would be greater. And the accumulating evidence suggested that the required shortening was greater.

Newton's treatment of the problem of the shape of the Earth in his *Principia* was the effective beginning of *dynamic* or *physical* geodesy: the attempt to derive the shape of the Earth from assumptions about the forces involved as a rotating, fluid Earth solidified.

Newton's first successor in this endeavour was Huygens, in his *Discours de la cause de la pesanteur* (published in 1690). Huygens did not admit the reciprocal attraction of all particles of matter, but adopted instead a modified Cartesian theory in which gravity was the result of aethereal pressure. He supposed that each particle of a homogeneous fluid mass was impelled only toward the centre of gravity of the mass; the problem of the equilibrium form of the mass under rotation was thus greatly simplified. Huygens, too, arrived at the shape of an oblate spheroid. But, owing to his initial assumption, he found the polar radius to be shorter than the equatorial radius by only $\frac{1}{578}$th of the latter.

Thus towards the end of the seventeenth century we find Huygens with a quasi-Cartesian theory of gravity, and Newton with a theory of universal attraction, in agreement as to the oblateness of the Earth but not as to the extent thereof. Operations were soon to be under way, however, which would contradict the idea of oblateness itself. As early as 1682, Picard, having completed his work on Tycho's observations and sent it to the royal printing house, was making plans for a voyage to Alexandria in order to establish the bases necessary for a similar study and verification of Ptolemy's observations. Meanwhile Cassini had received backing for a plan that called for a southward extension within France of Picard's earlier arc measure. Unfortunately, neither of these projects came to fulfilment, the first succumbing to Picard's death, the second being cancelled after the demise of Colbert the following year. Colbert's replacement by the Marquis de Louvois as Louis XIV's most influential minister brought changes in the Academy's programme: the publication of Tycho's observations was halted, and the astronomers of the Academy who had by then left on the expedition to extend the meridian measure were called back. Their labours were turned instead to more practical projects nearer to the king's heart.

The work of extending the arc measured by Picard was resumed by Cassini only in 1700, but then in both northerly and southerly directions. These efforts were carried to completion in 1718 by Cassini's son Jacques (Cassini II, 1677–1756). The report of their findings was published in 1720 in his *De la grandeur et de la figure de la terre*.

If the Earth has the shape of an oblate spheroid, the length of one degree of latitude – corresponding to a change of 1° in the altitude of the celestial pole – ought to increase as one moves from the equator to the poles. In the measure undertaken in France, therefore, a northern extension of the same latitudinal amplitude as Picard's arc ought to have been slightly longer than the original arc, which, in turn, ought to have exceeded in length a southern extension. The figures published by Jacques Cassini, however, indicated the opposite result. Having found that the southerly portion of the total arc had the greater length, he and a group of followers maintained that the Earth had the shape of a prolate spheroid – that it was elongated rather than flattened at the poles.

In France at this time opposition to Newton's theory of gravitation was nearly universal. Cassini's result nevertheless posed a problem: how to reconcile it with the flattening effect – undeniable, thanks to Huygens – of the centrifugal forces due to the Earth's rotation? Bernard le Bouyer de Fontenelle, secretary of the Academy of Sciences, charged J. J. d'Ortous de Mairan, a member of the Academy and a defender of Cartesianism, with attempting a theoretical reconciliation of these opposed findings. Mairan's solution, presented in 1720, was arrived at by denying the primitive sphericity of the Earth, which Huygens had assumed as a postulate. If the Earth had originally been a prolate spheroid, then rotation, while reducing its prolateness, could leave it still prolate.

During the 1730s the champion of Newtonianism in France was P. L. Moreau de Maupertuis (1698–1759). In 1732 and early 1733 he undertook to present Newton's law of attraction to the Paris Academy. His *Discours sur les différentes figures des astres* (1732) assumed and defended Newtonian principles. But on the question of the Earth's shape, Maupertuis, like other Continental mathematicians, found Newton's argumentation elliptical and ambiguous. Attempting a derivation of his own based on the same principles, he arrived at a value for the Earth's flattening markedly different from Newton's. The resolution of the question, he concluded, called for a turn from dynamic to geometrical geodesy.

Meanwhile, questions had been raised about the inferences that Jacques Cassini had drawn from his measurements. Joseph-Nicolas Delisle (1688–1768), a professor of mathematics at the Royal College, was the first to raise doubts: the excess in length of the southern degree of latitude over the northern could be nothing more than observational error. Delisle thought the Earth's shape could be derived at least as accurately from variations in lengths of degrees of longitude; in measurements of the latter he believed there were ways of reducing the margin of error that were not available in the measurement of latitudinal variations.

In 1720 Delisle, having earlier tried and failed to interest his fellow Academicians in Newton's lunar theory, had apparently little hope of interesting them in Newton's and his own geodetical ideas; he merely confided them to the king's representative in the Academy. Then, in 1725, he accepted an offer from Peter the Great to found an observatory and an associated school of astronomy in Russia. His stay there, which was to have been for four years, stretched into twenty-two. During this period, he and his students engaged in geodetic and geographical expeditions throughout the country, intended – in a transfer of an established French tradition to the opposite end of the European continent – as a basis for a projected large-scale map of Russia, which remained unrealized.

Meanwhile, the thoughts on geodesy that had occurred to Delisle occurred also to Giovanni Poleni, professor of mathematics at the University of Padua, who in 1724 published a pamphlet challenging the conclusiveness of Cassini's measurements of degrees of latitude, and, like Delisle, advocating measurements of lengths of degrees of longitude. A second edition of the pamphlet appeared in 1729, and gave rise to a lengthy review in 1733 in the *Journal historique de la république des lettres*, published in the Netherlands.

It was a memoir presented to the Academy by Maupertuis on 3 June 1733 that touched off the revival of geometrical geodesy in Paris as a means of determining the Earth's shape. Maupertuis had read Poleni's pamphlet, and made use of it in his memoir. It was now his hope that the Earth's shape could be determined simply as a question of observational fact.

Maupertuis's memoir of June 1733 prompted two of the Academy astronomers, Louis Godin and Charles-Marie de La Condamine (1701–74), to present papers discussing the difficulties in measuring degrees of latitude accurately. The papers of Maupertuis, Godin, and La Condamine led directly to a proposal to send teams of academicians to Peru and Lapland; the large latitudinal difference between these places should give rise to differences in lengths of a degree of latitude or longitude that would settle the question of the Earth's shape.

Jacques Cassini first learned of the Dutch journal review in late 1734, on his return from a year and a half spent measuring the length of a perpendicular to the Paris meridian across France. In a rejoinder published in the Academy's *Mémoires* he argued that, unless errors in timing the eclipses of Jupiter's satellites could be reduced, Poleni's method would lack the accuracy claimed for it. This may explain why both expeditionary teams proceeded to measure degrees of latitude rather than degrees of longitude.

The Peruvian expedition set out in 1735, and the expedition to Lapland almost a year later. But the Lapland expedition returned in 1737 – seven years before any members of the Peruvian expedition reappeared in France. Maupertuis, who had headed the northern undertaking, published an account of it in 1738. The degree of latitude measured in Lapland was considerably longer than that measured by Picard in France. No wonder that Maupertuis had himself painted bedecked in furs and holding a globe of the Earth that he was squeezing flat at the poles (Figure 15.1).

The result of the Lapland measurement (Figure 15.2) invalidated the results announced after the prolongation of Picard's arc, and thus led to a call for a remeasurement of the meridian of Paris. This task was undertaken in 1739 and 1740 under the auspices of the Academy and the team was led by the Abbé Nicolas-Louis de Lacaille and Cassini III, César-François Cassini de Thury, as he liked to fashion himself. The report of their operations, entitled *Méridienne vérifiée*, appeared under Cassini de Thury's name without mention of Lacaille, although the latter was in fact its author. It reversed the earlier findings and erased any doubts that might have remained. Thus, in 1740, the question was definitely decided in favour of the oblateness of the Earth. The results to be brought back from Peru would corroborate this conclusion.

The return of members of the Peruvian expedition was delayed, partly because of the difficulty of the operations themselves, which took a long time to execute, and partly because of friction that developed among members of the expedition. Godin, who, as the first proponent of the equatorial venture, had been designated leader, did not return to France after completion of the operations, but stayed on until 1751 as a professor of mathematics at the University of San Marcos. He claimed to be writing an account of the voyage, but it never appeared.

La Condamine, another of the astronomers of the expedition, returned to France by the longest and most dangerous route, the Amazon. After reaching the Atlantic, he went to Cayenne where he repeated the observations of Richer that had first raised the problem of the Earth's shape more

15.1. Maupertuis in Lap costume compressing the Earth, with a poem praising his work in geodesy.

15.2. The triangulation in Lapland used by Maupertuis and his colleagues.

15.3. The triangulation in Peru used by Godin, La Condamine and Bouguer.

than seventy years earlier. He returned to Paris in 1745, and in 1751 published a day-by-day account of the actions of the expedition which served as a kind of introduction to the detailed treatment of the geodetic operations that he brought out later that same year (Figure 15.3). Thanks to his pleasant personality and his talent as a writer, La Condamine has received much of the credit for the success of the expedition; but in fact he was a less gifted astronomer than Godin and a less reliable mathematician than the venture's third chief member, Pierre Bouguer (1698–1758) – the most important of the three.

The son of a royal professor of hydrography, Bouguer was a prodigy who, on the death of his father, was appointed to succeed him at the age of fifteen. He quickly became "the leading French theoretical authority on all things nautical", and was named an associate geometrician in the Academy in 1731; to his fellow Academicians he seemed a natural choice to accompany Godin and La Condamine. In Peru he engaged in a number of investigations beyond the geodetical work itself, including determination of the deviation of a plumb bob from the vertical owing to the gravitational attraction of a nearby mountain. The first of the three to return to France (in 1744), he was also the first to publish an account of the expedition, which he did five years later.

Although the expedition to Peru corroborated the results of the Lapland expedition and the *Méridienne vérifiée* in confirming the oblateness of the Earth, it cannot be said to have fixed with certainty the value of the ellipticity. Different calculators obtained different values: Jorge Juan, a Spanish observer who had worked with Godin, found $\frac{1}{266}$; Bouguer decided for $\frac{1}{179}$. Meanwhile, the question of the relation of the empirically determined shape and variation of gravity to Newton's proposed explanation had been given prominence by new theoretical investigations.

The researches of the Scottish mathematician Colin Maclaurin (1698–1746) were presented in a work entitled *A Treatise of Fluxions*, published in 1742. Maclaurin claimed to have demonstrated that if the Earth were a homogeneous fluid mass it would assume the shape of an oblate spheroid in consequence of its diurnal rotation. Actually, he had not proved that the planet *would* assume that form, but had shown for the first time that the oblate spheroid was *a* form of equilibrium. His value for the ratio of the axes, on the assumption of a homogeneous Earth, was practically the same as Newton's, namely 230 to 229. Since this value disagreed with the ellipticity derived from the measurements in Lapland and in France, Maclaurin proposed to treat the Earth as non-uniform in density, with either less or greater density at the centre. But his investigation of these hypotheses was unsatisfactory in a number of respects, both physical and mathematical; he failed, for instance, to show that any of the hypothesized distributions would be in equilibrium under rotation. One of those aware of the difficulties was the French mathematician, Alexis-Claude Clairaut (1713–65).

Clairaut's epochal *Théorie de la figure de la terre* appeared in 1743. It was the culmination of several years of intensive study of the subject, by a very talented man who had entered the Academy at the extraordinarily young age of eighteen. He had published several papers in the Academy's *Mémoires* on the problem of the Earth's shape, and had been a member of the Academy's Lapland expedition. In his major study of 1743, Clairaut first treated the Earth as a homogeneous fluid; his methods and results were essentially the same as Maclaurin's, his value for the ratio of the axes being 231:230. He then went on to show that, contrary to Newton's claim, the Earth if denser toward the centre need not be more flattened than it would be if homogeneous.

For a heterogeneous stratified ellipsoid of revolution, consisting of ellipsoidal strata of revolution with the denser strata towards the centre, and surrounded by a layer of fluid with its surface everywhere at right angles to the direction of gravity (a "level surface", later called a "geoid"), he showed that the ellipticity would be less than in the homogeneous case, provided that the ellipticity of each sub-surface shell multiplied into the square of its distance from the centre was never greater than the corresponding product at the surface. Generalizing a principle of hydrostatics put forward by Maclaurin, and deriving from it what would later be called the theory of potential, Clairaut argued that the density of the Earth should diminish from centre to surface, with the ellipticity of the strata increasing from centre to surface. The case of the stratified ellipsoid with density decreasing from centre to surface falls between that of the homogeneous ellipsoid, of which the ellipticity is $\frac{1}{230}$, and the case of a spheroid acted upon by a single force directed to the centre (the Huygenian case), in which the ellipticity is about $\frac{1}{578}$. Thus the ellipticity of the stratified ellipsoid – the most plausible model, in Clairaut's view – should fall between these two values, and hence be less than $\frac{1}{230}$.

Newton's principle of balanced canals, Clairaut found, did not hold for the heterogeneous stratified ellipsoid of revolution: the effective gravities at different points on the surface of the ellipsoid did not vary inversely as their distances from the centre. The increase in effective gravity as one moved from equator to pole still proved to be proportional to the square of the sine of the latitude: $g_L - g_E = (g_P - g_E)\sin^2 L$, where g_E, g_P, and g_L are the effective gravities at the equator, pole, and latitude L, respectively. But the total increase compared with the effective gravity at the equator, Clairaut showed, was given by $(g_P - g_E)/g_E = 5\phi/2 - \delta$, where ϕ is the ratio of the centrifugal force at the equator to the effective gravity there, and δ is the ellipticity of the heterogeneous ellipsoid. This is often referred to as "Clairaut's theorem". Since ϕ had been shown to be $\frac{1}{288}$, so that $\frac{5}{2}\phi = \frac{2}{230}$, and measurement of g_L at different latitudes was indicating that $(g_P - g_E)/g_E > \frac{1}{230}$, it followed that, if the Earth were the heterogeneous ellipsoid of revolution of the kind Clairaut favoured, then $\delta < \frac{1}{230}$.

The trouble with this conclusion in 1743 was that the value then current for the ellipticity was the one obtained from comparison of the measurements of the meridian in Lapland and in France, namely $\frac{1}{177}$, which was greater than $\frac{1}{230}$. The measurement of more distant degrees, Clairaut urged, was required to obtain an accurate ratio of the axes; he looked forward to the results of the Peruvian measurement.

Unfortunately, the Peruvian measurement did not lead to general acceptance of an ellipticity less than $\frac{1}{230}$, and so failed to confirm Clairaut's theory. Jean le Rond d'Alembert, for instance, in his *Recherches sur la précession* (1749), obtained from the measurements in Lapland and Peru an ellipticity of $\frac{1}{174}$. By the time he was writing Pt III of his *Recherches sur différens points importans du système du monde* (1756), two further meridian arcs had been measured, one by Lacaille at the Cape of Good Hope, the other by Rudjer J. Bošković and Christopher Maire in the Papal States of Italy. Comparing these with the earlier measurements, d'Alembert concluded that no value of the ellipticity could reconcile them. And Bošković in his account of the Italian measure ended by remarking that the more one measured degrees, the more uncertain the shape of the Earth became.

Eventually, in the late 1740s, Clairaut did reconcile a value of δ greater than $\frac{1}{230}$ with a value of the fraction $(g_P - g_E)/g_E$ greater than $\frac{1}{230}$, but in doing so he had to give up the hypothesis he had favoured as to the internal constitution of the Earth, namely the assumption that the interior strata were in the form of *ellipsoidal* surfaces of revolution. This led him to abandon as well the supposition that the Earth had originally been in a fluid state, for he believed he had shown that the strata in a heterogeneous, stratified, self-gravitating fluid figure were level surfaces, and further that, in a slowly rotating figure, such level surfaces were, to a close approximation, ellipsoidal. D'Alembert later criticized Clairaut's proofs of these propositions as inconclusive; but the propositions in fact hold if the forces are conservative – which is the case with gravitation.

D'Alembert repeatedly criticized various aspects of Clairaut's work. He objected to Clairaut's belief that the laws of hydrostatics required the denser strata to be nearer the centre, and all the strata to be level surfaces. His own treatment of the problem of the Earth's figure is more general than Clairaut's, in that he did not limit consideration to ellipsoids of revolution. The method that he introduced for estimating the attraction of a spheroid by resolving the body into a sphere and a thin additional shell has proved of value. But as Pierre-Simon Laplace (1749–1827) remarks, d'Alembert's investigations "lack the clarity so necessary in complicated calculations". By contrast, in characterizing Clairaut's *Théorie de la figure de la terre*, Laplace stated that the importance of Clairaut's results and the elegance with which he presents them "puts this work in the class of the most beautiful mathematical productions".

Clairaut's original theory was in fact less in error than the value of the ellipticity Clairaut and d'Alembert had thought it necessary to accept would have implied. This conclusion emerged from analyses undertaken by A.-M. Legendre (1752–1833) and Laplace. In these studies it has been said that "Laplace played leapfrog with Legendre".

The game began in January 1783 when Legendre presented to the Academy a memoir on the attraction of homogeneous spheroids. Here he proved that if the attraction of a solid of revolution is known for every external point on the prolongation of its axis, it is known for every external point. The attraction is given as an integral over the spheroid, and in expanding the integrand as a series, Legendre introduced the 'Legendre polynomials', which serve as coefficients in the series. With the aid of these and on the basis of a suggestion from Laplace, he obtained what was later to be called the potential, from which the attraction was then derivable.

As Legendre was not yet a member of the Academy, his memoir underwent an official review before its eventual publication (in 1785) in the Academy's *Savants étrangers*; the reviewer was Laplace. In March 1783 Legendre was elected a member of the Academy, and so in May 1783 heard Laplace read to the Academy a memoir on the attraction of elliptical spheroids, deepening, as he admitted, his own treatment of the subject.

Laplace's memoir, revised and amplified, became Pt II of his *Théorie du mouvement et de la figure elliptique des planètes*, brought out in 1784 with financial underwriting from Bochart de Saron, honorary member of the Academy. During the 1770s Laplace had written three memoirs on the attraction of solids of revolution, but without using the concept of potential. This concept had been adumbrated in Clairaut's *Théorie* of 1743 and clearly deployed in several of Lagrange's memoirs of the 1770s (see Chapter 21); it was probably from Lagrange's memoirs that Laplace gained his awareness of the concept's value. In his treatise of 1784 he used Legendre's polynomials to calculate the potential of an arbitrary spheroid, not necessarily a solid of revolution, but differing little from a sphere. And here he showed how the departure from sphericity could be developed as a series in what were later to be called spherical functions or spherical harmonics.

Legendre continued his investigations of the attractions of spheroids during the 1780s, publishing three further memoirs in which he demonstrated various properties of the Legendre polynomials and of spherical harmonics.

In the meantime Laplace produced two more memoirs on our subject. The results of these and of his *Théorie* of 1784 were then incorporated, with some revision, in the second volume of his *Traité de mécanique céleste* (1799). From both pendulum experiments and the phenomena of the precession of the equinoxes and nutation of the Earth's axis, Laplace concluded that the flattening of the Earth must be less than $\frac{1}{230}$. The precession and nutation set limits between which this fraction must fall; Laplace in 1799 gave them as $\frac{1}{304}$ and $\frac{1}{578}$ (Nathaniel Bowditch, translator of Laplace's *Mécanique céleste* into English, in taking account of Laplace's later revision of constants, determined that the first of these should be $\frac{1}{279}$).

On the basis of pendulum experiments and the assumption of ellipticity, Laplace found the most probable ellipsoid to have an ellipticity of $\frac{1}{336}$ (Bowditch, correcting errors in the calculation, obtained $\frac{1}{315}$). By "most probable" Laplace here means that the positive and negative errors add to zero, while the sum of their absolute values is a minimum. (The method of least squares was still in the future.)

The total variation in the length of the seconds-pendulum from equator to pole in fact amounts only to 5.24 millimetres. To obtain a reliable value of the Earth's ellipticity from so minute a variation requires that a large number of carefully performed measurements be brought to bear on the formation of a statistical mean. Laplace had only fifteen such measurements at his disposal. Bowditch, who brought out his translation of the second volume of Laplace's *Mécanique céleste* in 1832, had by this date assembled 52 such measurements, and to these he applied the method of least squares; neglecting eight of the measurements as greatly discrepant from the rest, he obtained an ellipticity of $\frac{1}{297}$, precisely the amount adopted internationally some 92 years later.

Laplace also examined the implications of the degrees of the meridian measured at different latitudes; seven of these were available to him in 1799. Determining the ellipticity of an ellipsoid that would reduce the largest error to a minimum, he found this minimum largest error to be 189 metres, and to occur in the degrees of Pennsylvania, the Cape of Good Hope, and Lapland; the ellipticity, on this assumption, was $\frac{1}{277}$. These measures, Laplace believed, could not be so greatly in error; the Earth's shape, therefore, must differ sensibly from the ellipsoidal.

When he calculated the most probable ellipsoid, based on the same seven measures, he found an ellipticity of $\frac{1}{312}$, and the error in the degree of Lapland to be 336 metres. Sure that the error could not be nearly so large – as in fact the remeasurement undertaken by Jons Svanberg in 1801–1803 was to show it to be – he again concluded that the departure from ellipticity must be sizeable. But Bowditch in 1832 found later measurements fitting the elliptic hypothesis more closely. From five measurements in Peru, India, France, England, and Sweden, he obtained an ellipticity of $\frac{1}{310}$.

One of the seven measurements used by Laplace in Vol. 2 of the *Mécanique céleste* was a very recent one, the result of an investigation begun in 1792 and completed in 1799. The establishment of a uniform system of weights and measures had been a goal of the mercantilists from the time of Colbert; and this goal was adopted in 1790 by the revolutionary National Assembly, which directed the Academy of Sciences to prepare a report. An early proposal considered by the Academy's commission on weights and measures was to take the length of the seconds-pendulum at 45° of latitude as the universal standard of length – the idea of Picard, modified as the discovery of latitudinal variation required. It was hoped that this idea would be accepted by the English. It was not, however, and the idea was then set aside in favour of another 'natural' standard: the ten-millionth part of the quarter of the terrestrial meridian. This choice, accompanied by a proposal for the long and expensive operation of measuring an arc of the meridian from Dunkirk to Barcelona, was defended on various grounds, but the chief motivation behind it was to shed new light on the question of the shape of the Earth.

The measurement of the meridian was begun under the direction of two young Academicians, Jean-Baptiste Joseph Delambre and Pierre-François-André Méchain, in 1792. Work was temporarily suspended at the end of the following year, then resumed in 1795 and completed by 1799. In the latter year, the French government invited other states to send representatives to Paris for the purpose of examining the finished work, and the resulting conference may be considered the first international congress of scientists. Its final action was to adopt an ellipticity of $\frac{1}{334}$ based upon the arc measured in Peru and that newly measured in France and Spain. Delambre and Méchain, who had carried the arduous work of measurement through to completion, published an account of their operations in the three-volume *Base du système métrique*, of which the first volume appeared in 1806. In reviewing the calculations for the meridian while preparing this volume for publication, Delambre concluded that the Earth's ellipticity was given more accurately as $\frac{1}{309}$.

The arc measured by Delambre and Méchain was soon extended southward to the Balearic Islands so that the whole measured arc would centre around a latitude of 45°. This extension was begun by Méchain, and on his death was continued by a new pair of young Academicians, Dominique François Jean Arago and Jean-Baptiste Biot. In 1821 they published an account of their work that in effect constitutes a fourth volume of the *Base du système métrique*.

The geodetic operations involved in the establishment of the metric system did not finally settle the question of the exact shape of the Earth. But by introducing new instruments and techniques, they supplied a model for further measurements of the same kind. The example thus set gave rise to a series of new arc measurements, such as Svanberg's revision of the Lapland measurement.

In Vol. 2 of the *Mécanique céleste* (1799), Laplace could claim with assurance that the Earth's flattening was less than $\frac{1}{230}$, the value implied by the supposition of homogeneity; the result was confirmed by pendulum experiments, the degree measurements, and the precession and nutation. In Vol.3, which appeared in 1802, he drew support for the same conclusion from the lunar theory. In the lunar tables of Charles Mason (published in 1780) and those of Johann Tobias Bürg (in preparation in 1802, published in 1806), there appeared in the Moon's longitude a term proportional to the sine of the longitude of the lunar node; it had been determined empirically, and Mason put the coefficient at 7″.7, while Bürg gave it as 6″.8. By gravitational theory, Laplace showed the coefficient to be proportional to the Earth's oblateness; the derivation was free of the uncertainty in derivations of other lunar inequalities, where the convergence of the approximations was slow. By the theory, an oblateness of $\frac{1}{230}$ gave a coefficient of 11″.499, and $\frac{1}{334}$ gave 5″.552; thus Bürg's value implied an oblateness of $\frac{1}{305}$.

From the more accurate measurements of latitudinal degrees available in the 1830s, Bowditch, as previously mentioned, concluded a fairly close agreement with Clairaut's original hypothesis of elliptical strata. The primal fluidity of the Earth, proposed by Newton, was once more likely. The prolonged discussion had ended by showing Newton's and Clairaut's ideas to be the apt ones.

Meanwhile, increasing precision was bringing a new problem into focus, the "deflection of the vertical", or departure of the plumb line from perpendicularity to the reference ellipsoid of revolution, owing to anomalies in the distribution of the Earth's mass. This fact would set the problem for the next generation of investigators of the figure of the Earth – the endpoint in one phase of the investigation becoming a point of departure for the next.

Further reading

John Greenberg, Geodesy in Paris in the 1730s and the Paduan connection, *Historical Studies in the Physical Sciences*, vol. 13 (1983), 239–60

Rob Iliffe, Aplatisseur du monde et de Cassini: Maupertuis, Precision Measurement, and the Shape of the Earth in the 1730s, *History of Science*, vol. 31 (1993), 335–73

Tom B. Jones, *The Figure of the Earth* (Lawrence, Kansas, 1967)

Henri Lacombe and Pierre Costabel (eds.), *La Figure de la terre du XIIIe siècle à l'ère spatiale* (Paris, 1988)

Georges Perrier, *Petite histoire de la géodesie* (Paris, 1939)

Isaac Todhunter, *A History of the Mathematical Theories of Attraction and of the Figure of the Earth* (London, 1873; reprinted New York, 1962)

Joella Yoder, *Unrolling Time: Christiaan Huygens and the Mathematization of Nature* (Cambridge, 1988)

16

Clairaut and the motion of the lunar apse: the inverse-square law undergoes a test

CRAIG B. WAFF

In Bk III of his *Principia* (all editions), Isaac Newton (1642–1727) put forward two different arguments for establishing the inverse-square formulation of the law of gravitation. The first was based on the recognized accuracy or near-accuracy of Kepler's third law as applied to the Jovian and (in the second and third editions) Saturnian satellites orbiting about their primary planets, and to the primary planets orbiting about the Sun, together with Cor. 6 of Prop. 4 of Bk I: "If the periodic times are as the three-halves powers of the radii, the centripetal forces will be inversely as the squares of the radii. . . ." But Prop. 4, including this corollary, was demonstrated only for concentric circular orbits – a condition not strictly satisfied by the orbits of the planets. Newton therefore developed his second argument, in which he established a mathematical relation between the motion of the apsides of an orbit and the centripetal force that would produce it. According to his formula for this relation (given in Cor. 1 of Prop. 45 of Bk I), the virtually immobile aphelia of the planets implied a nearly exact inverse-square centripetal force between each of them and the Sun.

The motion of the Moon's apsides, amounting to about 3° per revolution, was more sizeable. Newton had hoped to account for it as a perturbative effect, but a second formula, which he derived in Cor. 2 of Prop. 45, and which mathematically related apsidal motion to a combination of an inverse-square centripetal force and a radial component of a perturbative force, yielded only half the observed motion.

Newton's statements about this calculation in the *Principia* are confusing if not contradictory. As an illustration of Cor. 2 of Prop. 45 (Figure 16.1), he proposed a radial perturbing force that was subtractive (i.e., directed away from the central body), and equal to $\frac{1}{357.45}$ of the inverse-square force directed toward the central body. (By the latter force alone the planet or satellite would revolve in a stationary ellipse.) He attributed this numerical example to no specific astronomical case in the first two editions of the *Principia*, but in the third edition he added the remark: "The apse of the Moon is about twice as swift." In the second edition, however, he inserted in Prop. 3 of Bk III the additional statement that

the action of the Sun, attracting the Moon from the Earth, is nearly as the Moon's distance from the Earth; and therefore (by what we have shown in Cor. II, Proposition 45, Book I) is to the centripetal force of the Moon as 2 to 357.45, or nearly so; that is, as 1 to $178\frac{29}{40}$.

In Prop. 4 of Bk III (second and third editions), in determining the force that the Earth would exert on the Moon if the latter were brought down to the Earth's surface, Newton proceeded as though there were a *subtractive* radial perturbative force due to the Sun, having to the Earth's force a ratio of 1 to $178\frac{29}{40}$.

In Props. 25 and 26 of Bk III, however, Newton identified a perturbative radial force of this value as *additive*, the remaining solar perturbative force being parallel to the line connecting the Sun and the Earth. The latter force had a *subtractive* radial component, and from the analysis of Prop. 26 one can deduce that the *average* radial perturbative force, when both radial components are taken into account, is subtractive, and bears to the centripetal force exercised by the Earth on the Moon the ratio of 1 to 357.45, precisely the value used in the illustration of Cor. 2 of Prop. 45. A careful reading thus indicates that the analysis Newton provides within the *Principia* leads to only half the Moon's apsidal motion. It is no wonder, however, that readers reached differing conclusions as to what

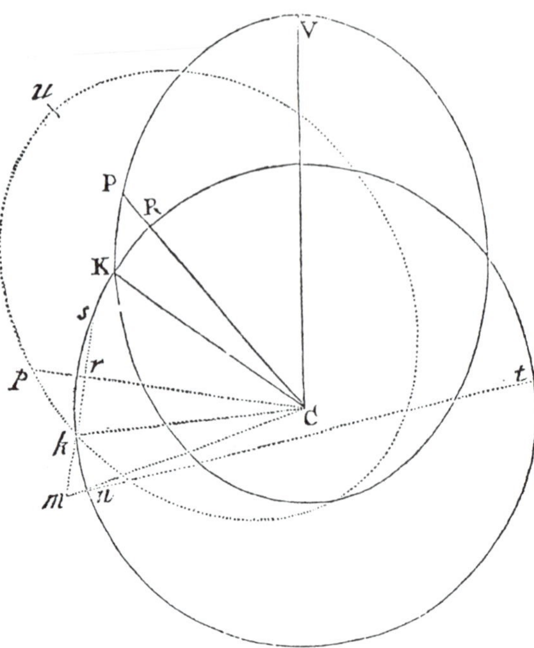

16.1. Newton's figure for Prop. 44 of Bk I of the *Principia*. With the aid of this diagram Newton shows that a body, which by an inverse-square law of force tending to *C* orbits in the stationary ellipse *VPK* about *C*, will orbit in the moving ellipse *upk* if there is added to the inverse-square force an inverse-cube force towards *C*.

In Cor. 2 of Prop. 45, Newton then shows that, in a nearly circular orbit, a force directed away from *C* that is 1/357.45 times the average inverse-square centripetal force will cause the ellipse to advance 1° 31′ 28″ per revolution. (In fact, the radial component of the Sun's perturbing force on the Moon has just this average value.) In the third edition of the *Principia* Newton added the remark: "The apse of the Moon is about twice as swift". In Props. 3 and 4 of Bk III of the *Principia* (all editions), he surreptitiously assumed the average radial component of the Sun's perturbing force to be twice as great, or − 2/357.45 times the Earth's force on the Moon, thus appearing to imply that the radial perturbing force accounted for the observed motion of the lunar apse.

Newton had here accomplished or failed to accomplish.

In the 1740s three of the foremost mathematicians of Europe, Leonhard Euler (1707–83), Alexis-Claude Clairaut (1713–65), and Jean le Rond d'Alembert (1717–83), undertook new, analytical derivations of the Moon's motions from the inverse-square law; and all three initially found, for the motion of the lunar apsides, an anomalous result similar to that which can be inferred from the analysis of Prop. 45 of Bk I of the *Principia*: the inverse-square law appeared to yield only about half the observed motion. As a consequence, a considerable discussion and controversy arose, much of it centred round Clairaut, concerning various aspects of the problem: (1) the interpretation of Newton's statements in the *Principia*; (2) the general applicability of the various "remedies" (in particular Clairaut's proposed emendation of the gravitational law) that were proposed in 1747 and 1748 to account for the rest of the apsidal motion; and (3) the accuracy of the calculations that had produced the apparent anomaly.

Responses to the anomaly

Euler

The first of the eighteenth-century mathematicians publicly to call into question the accuracy of the inverse-square law was Euler, in a paper entitled "Recherches sur le mouvement des corps célestes en général", which he presented to the Berlin Academy of Sciences on 8 June 1747. Since Newton's time, Euler pointed out, astronomers had come to the conclusion that the motions of the planetary apsides and nodes were indeed detectable (their detectability had earlier been questioned by Thomas Streete and Nicholaus Mercator); and such motions implied that the planets were subject not only to an inverse-square force directed toward the Sun, but to additional forces as well. Such forces, Euler pointed out, could arise from (1) an irregular distribution of mass in the Sun; (2) sources of force other than the Sun, such as the mutual actions of the planets on one another; or (3) a gravitational law that did not follow exactly the inverse-square ratio.

In his paper, Euler confessed that he was doubtful that the mutual forces among the planets followed the inverse-square law exactly. His doubts were partly due to his commitment to the idea that the action of all forces is by contact; thus in his thinking on the cause of gravity he took up a suggestion of Daniel Bernoulli, which equated gravitation to a pressure in a hydrodynamic aether, the pressure being greater where the aether's velocity was less. In such a model, it seemed unlikely that the law of gravitation, whatever its formula, would extend in mathematical exactness to all distances.

Euler's *a priori* doubts regarding the accuracy of the inverse-square law were strengthened by empirical evidence. Part of the evidence came from his having been unable to account theoretically (on the basis of the law) for all of the anomalies in the motions of the planets Jupiter and Saturn. In addition, Euler pointed out that in a careful study of the inequalities in the Moon's motion, in which he had been engaged since at least 1744, he had found a number of disagreements between observation and the results derived from the inverse-square law. One of these was particularly striking:

Having first supposed that the forces acting on the Moon from both the Earth and the Sun are perfectly proportional reciprocally to the squares of the distances, I have always found the motion of the apogee to be almost two times slower than the observations make it; and although several small terms that I have been obliged to neglect in the calculation may be able to accelerate the motion of the apogee, I have ascertained after several investigations that they would be far from sufficient to make up for this lack, and that it is absolutely necessary that the forces by which the Moon is at present solicited are a little different from the ones I supposed....

After stating several other discrepancies between observation and theoretical calculation, Euler concluded that "all these reasons joined together appear therefore to prove invincibly that the centripetal forces one conceives in the Heavens do not follow exactly the law established by Newton."

Euler expressed a similar opinion in an essay completed at about the same time as the memoir just cited, namely his "Recherches sur la question des inégalités du mouvement de Saturne et de Jupiter", which was his entry in the prize contest of 1748 held by the Paris Academy of Sciences. The problem set for the contest was to formulate a "theory of Saturn and of Jupiter by which the inequalities [of motion] that these planets appear to cause mutually, principally near the time of conjunction, can be explained". Euler's essay, which would eventually win the contest award, arrived at the Paris Academy on 27 August 1747, and at the final pre-vacation meeting of the Academy on 6 September the judges of the contest were selected and were given the submitted essays, including Euler's. Two of the judges – Clairaut and d'Alembert – had a particular interest in the subject of the contest. As the Marquise du Châtelet (hard at work during this period on a translation of Newton's *Principia*) observed to a correspondent, "Messieurs Clairaut and d'Alembert are after the system of the world, [because] they understandably do not wish to be anticipated by the prize [contest] pieces". In seeking to "crack" what came to be known as the three-body problem, both d'Alembert and Clairaut in the preceding year had undertaken to develop general methods for determining the perturbative effects caused by the mutual actions of the celestial bodies on one another.

Clairaut

Clairaut (Figure 16.2) completed his general solution (by approximation) of the three-body problem toward the end of 1746; this took the form of a memoir placed in a sealed envelope at the Academy on 7 January 1747. In two more sealed envelopes, deposited in March and June, he inserted papers dealing with applications of the theory, in particular to the Moon. He read these three papers to the Academy at meetings in June, July, and August. In a fourth sealed envelope, deposited on 6 September 1747 (the very day on which he received the essays for the prize contest, including Euler's), Clairaut gave the numerical determination of the equations for the Moon's varying radius vector and longitude, and (for the first time) stated his discovery of a discrepancy between the theory and the observed motion of the lunar apogee.

Upon subsequently reading Euler's contest essay, Clairaut was delighted (so he wrote Euler) to find himself in agreement with the doubts that Euler had expressed as to the accuracy of the inverse-square law. Even before reading this essay, however, Clairaut had decided that his discovery of the discrepancy between the observed motion of the lunar apsides and that calculated on the basis of the inverse-square law was sufficiently important to merit discussion in the Paris Academy, whose twice-weekly sessions would resume in November. At the public meeting of 15 November he gave a lecture on the subject, and at two further meetings, on 2 December 1747 and 20 January 1748, he gave readings of the sealed paper he had deposited in September.

Clairaut evidently suspected that his announce-

16.2. Alexis-Claude Clairaut.

ment of the discrepancy, and of the conclusions he had drawn from it, would provoke controversy. He began his public lecture in a good debating style:

In order to justify what I advance here, I am going to expose in all their force the reasons that ordinarily determine [one] in favour of the system of M. Newton; I will then render [an] account of the motives that have impelled me to seek some new proof of this system, of the work that this research required, and of what resulted from it.

An impressive number of phenomena, Clairaut reminded his audience, had been found to agree closely with the results predicted by Newton's law. (1) The general motions of the planets and their satellites were observed to be in near agreement with Kepler's laws, which Newton had shown to be implied by an inverse-square attraction law in the absence of disturbing forces. (2) The motion of the lunar nodes, which Clairaut had himself calculated by a method different from Newton's, appeared "rather consistent" with observation. (3) The theory of the tides "has been verified by the most skilful Mathematicians" (Clairaut no doubt had in mind the "Newtonian" essays on the tides by Daniel Bernoulli, Colin Maclaurin, and Euler, which shared the prize of the Paris Academy in 1740). (4) And there were "several other questions equally favourable to attraction", which Clairaut left unspecified.

Clairaut had two motives for making a particular study of the motion of the Moon's apogee derivable from gravitational theory. One was the "obscurity" of Newton's research and remarks on the subject. The other was the importance of the apsidal motion in determining the Moon's place. The position of the Moon relative to the apogee of its orbit determined the size of the "equation of centre", which is either added to or subtracted from the mean longitude to obtain the true longitude. Because the equation of centre amounted at maximum to more than 6°, errors due to theoretical misplacement of the apogee could be sizeable.

Like Euler, Clairaut stressed the extreme care he had taken in deriving the apsidal motion:

Seeing therefore the great importance of the determination of the motion of the apogee, I have sought to derive it from the solution of the general problem . . .; this operation was more difficult than the solution of the problem itself, because in determining the orbit of a planet, some small quantities, which cannot cause any considerable error during a single revolution, can be neglected; but these quantities can become of infinite consequence in the large number of revolutions necessary for knowing the motion of the apogee.

Clairaut then related his astonishment on discovering that his theoretical derivation, despite such care, generated only half the observed motion, "that is to say that the period of the apogee [i.e., the time it takes for the lunar apogee to return to the same point in the heavens] that follows from the attraction reciprocally proportional to the squares of the distances, would be about 18 years, instead of a little less than 9 which it is in fact."

This result, Clairaut told his audience, swayed him at first toward abandoning attraction entirely. Consideration of the impressive quantitative confirmations of Newton's law, however, tempered his reaction. Some means, he felt, needed to be found of "reconciling the reasons that seem ... contrary and favourable to [gravitational] attraction". The expedient that occurred to him was to suppose that gravitational attraction does indeed take place in nature, but by some law other than the inverse-square. What was needed was a law of gravitational attraction "that will differ very sensibly from the law of the square at small distances, and diverge from it so little at large ones that the difference is undetectable in the observations". As an example, he suggested a law consisting of two terms, one having the square of the distance, and the other the fourth power of the distance, as divisor. Clairaut argued that the difference in the forces assigned by the modified law and by Newton's law at the distance of the Moon might be able to generate the "missing" half of the Moon's apsidal motion. On the other hand, at the greater distance of the planet Mercury from the Sun, any difference in apsidal motion due to the difference in forces assigned by the two laws would take centuries to make itself observationally detectable.

The idea of a non-inverse-square attractive force was not new. Newton himself in Query 31 of the later editions of the *Opticks* had suggested that an attractive force acting sensibly over only rather small distances might account for the phenomena of cohesion and chemical reactions. The Oxford mathematician John Keill, in a paper published in the *Philosophical Transactions of the Royal Society* in 1708, had argued that attractive forces acting only over very small distances would decrease with a triplicate, quadruplicate, or some higher ratio of the increase of the distances. Clairaut himself in 1739 and 1742 had presented papers to the Paris Academy of Sciences in which he proposed to account for the refraction of light and the ascension of liquids in capillary tubes by attractive forces proportional to some unknown function of the distance.

Neither was the suggestion of a multi-term mathematical expression, combining an inverse-square term with one or more terms involving higher inverse powers of the distance, unprecedented. In a passage noticed by Clairaut in the Marquise du Châtelet's *Institutions de physique* (1740), the Marquise reported that

Some Newtonians . . . have had the idea of explaining all the Phenomena, as well the celestial as the terrestrial, by a single and same attractive force, which acts as an algebraic quantity

$$\frac{a}{xx}+\frac{b}{x^3}+\ldots,$$

x marking the distance . . .; at some remote distances, as for example, at those of the Planets, the part of the attractive force that acts as the cube, is almost nil, and disturbs only infinitely little the other part of the attractive force which acts as the square, and on which depends the ellipticity of the orbits.

The same idea was also discussed in another book of which Clairaut had recently made the official examination for the "Approbation": the Abbé Pierre Sigorgne's *Institutions Newtoniennes*, which was published in June 1747. What was novel in Clairaut's proposal of a multi-term law was only the suggestion that a higher-order term could have a sensible effect at the distance of the Moon from the Earth.

D'Alembert

D'Alembert, like Clairaut, completed his general solution of the three-body problem toward the end of 1746. To the Berlin Academy, in December 1746 and January 1747, he sent memoirs on the determination of planetary perturbations and on the theory of the Moon, which he subsequently read (possibly in revised form) before the Paris Academy of Sciences in June 1747. In a paper read before the latter academy on 28 February and 6 March 1748, d'Alembert announced publicly that he too had found only half the observed motion of the lunar apsides to follow from the inverse-square law.

His speculations regarding the cause of this discrepancy, unlike those of Euler and Clairaut, remained private. In letters written to Euler and to the Swiss mathematician Gabriel Cramer in 1748 and 1749, d'Alembert expressed his disapproval of Clairaut's attempt to account for both terrestrial and celestial phenomena with a single modified law of gravitation. Newton, on the basis of the inverse-square law, had successfully derived the lunar *variation* and the inequalities of motion to which the lunar nodes and lunar inclination were subject. D'Alembert had confirmed these results, and so concluded that there was no need to question the mathematical form of the gravitation law.

D'Alembert viewed the discrepancy in the lunar apsidal motion as indicating only that the *particular* force acting between the Earth and the centre of gravity of the Moon followed a different law. To suppose that this force followed a law other than the inverse-square, he realized, would weaken the strong correlation between the Moon's centripetal acceleration and terrestrial gravity that Newton had obtained (in Prop. 4 of Bk III of the *Principia*) by assuming this law; nevertheless, he suggested two ways by which a different force relationship between the Earth and the Moon might arise. One way was to suppose, besides the universal inverse-square gravitational force, an additional non-gravitational force acting solely between the Earth and the Moon. In letters written during 1748, d'Alembert inclined to the view that this force, if it existed, was magnetic in origin.

A second way was to suppose that a radially asymmetric distribution of mass in the Moon – an arrangement that could escape observational detection from the Earth owing to the fact that the Moon always keeps virtually the same face towards its parent body – could affect the apsidal motion. To investigate this possibility, d'Alembert imagined the matter of the Moon to be concentrated into two globes connected by a thin rod and rotating around the rod's midpoint. A calculation for this extreme case showed that in order to produce the "missing" half of the apsidal motion the distance between either globe and the rod's midpoint would have to be about twice the Earth's radius, a result that even d'Alembert conceded to be unrealistic. He soon learned that Euler had undertaken a similar calculation and had reached a similar conclusion.

Interpreting the *Principia*

Meanwhile Clairaut, following his public lecture of 15 November 1747, had been faced with the charge (by his colleague and fellow contest-judge Pierre-Charles Le Monnier) that his announcement of a discrepancy between the inverse-square law and the lunar apsidal motion was based on the research in Euler's prize essay. To rebut this charge, Clairaut on 2 December read the paper he had deposited under seal with the Academy's secretary just prior to receiving the contest essays. In this paper he had made brief mention of Newton's general conclusions in Props. 44 and 45 of Bk I regarding the relation of central force to apsidal motion. This reference led someone to call attention to Newton's "lunar" illustration in Cor.2 of Prop. 45, and as a result Clairaut found himself once more being charged with plagiary, this time in respect of Newton's work.

To answer this new charge, Clairaut prepared yet another paper, which he read to the Academy on 20 and 23 December, and in which he examined, proposition by proposition, all of Newton's work in the *Principia* relevant to the motion of the lunar apogee. Here he sought to show not only *that* Newton had failed to recognize the defect of his law of gravitation, but *why*. To obtain by the calculation of Cor.2 of Prop. 45 the true motion of the lunar apse, it was necessary that the coefficient of the term to be added to the inverse-square force be $1/178\frac{29}{40}$. In Prop. 25 of Bk III, on resolving the Sun's perturbing force into two components, one parallel to the line from Earth to Sun, and the other directed toward the Earth, Newton had demonstrated that the latter component was indeed in its mean quantity $1/178\frac{29}{40}$ of the centripetal force between the Earth and the Moon. Clairaut claimed that Newton had taken this radial component of the Sun's perturbing force, directed *towards* the Earth, as the sole radial component, and had failed to take into account the fact that the component parallel to the line from Earth to Sun could itself be resolved into radial and transverse components, the radial component being *subtractive*. (An additive radial component, we remind the reader, would cause the apsides to *regress*; a subtractive radial component would cause them to *advance*.) For Newton's use of the coefficient $\frac{1}{357.45}$ in Cor. 2 of Prop. 45, Clairaut had an explanation that presupposed that Newton had not known how to evaluate the integral

$$\int_0^{2\pi} \cos\theta \, d\theta,$$

where θ is the elongation between the Moon and the Sun.

Clairaut also had to defend himself against the charge that the deficiency in Newton's calculation of the motion of the apsides had been announced earlier, in the third volume of an edition of the *Principia* edited by the Minim priests Thomas Le Seur and François Jacquier, and published in Geneva in 1742. Here, in an extensive footnote to the scholium following Prop. 35 of Bk III, two different solutions of the problem of determining the motion of the Moon's apogee were supplied, and the second of them, which proceeded by Newton's method of Prop. 45 of Bk I, yielded only half the observed apsidal motion.

The anonymous author of these solutions was the Swiss mathematician Giovanni Ludovico Calandrini (1703–58), who had been professor of mathematics and later of philosophy at the Geneva Academy since 1724. As it happened, Calandrini's long-time colleague Cramer was visiting in Paris in late 1747 and early 1748, and attended the Academy's public meeting on 15 November at which Clairaut announced his discovery of the discrepancy in the motion of the lunar apogee. After being informed by Cramer about this announcement, Calandrini corresponded with both Cramer and Clairaut concerning it. Calandrini did not question the inverse-square law, but considered the method of Newton's Prop. 45 to be defective, in that it failed to take into account the eccentricity of the lunar orbit. Calandrini had been able by a rough calculation using a different method, and taking the Moon's variable orbital eccentricity into account, to obtain a much larger value for the motion of the lunar apse, and he believed that refinement of the calculation (by computing an accurate *mean* value for the eccentricity) would lead to the observed motion.

Clairaut, who exchanged letters with Calandrini in February and March, did not agree. The orbital eccentricity of the Moon, he insisted, could have only a negligible effect on the motion of the lunar apse. He believed Calandrini's method to be uncertain if not erroneous, and in particular objected to Calandrini's having omitted from consideration the component of the Sun's perturbing force at right angles to the Earth–Moon radius vector. Calandrini, on the other hand, was unable to accept Clairaut's contention that Newton had somehow been able to convince himself that the method of Prop. 45 could account for the entire mean motion of the lunar apogee.

Meanwhile, another debater had entered the fray: Clairaut's colleague in the Paris Academy, Georges-Louis Leclerc, Comte de Buffon (1707–88). On 20 and 24 January Buffon read a paper that, among other things, questioned Calandrini's interpretation of Newton's remarks regarding the motion of the lunar apogee.

It was Buffon's mistaken belief that the proper ratio of perturbing force to centripetal force to be substituted into the formula of Cor.2 was $1/178\,\frac{29}{40}$, as calculated by Newton in Prop. 25 of Bk III. He thus shared with Clairaut (although for different reasons) the belief that Newton was convinced of the adequacy of his theory to yield, at least approximately, the entire mean motion of the apogee. Buffon was therefore astonished to find Calandrini claiming that Newton had been well aware that the calculation yielded only half the observational value. Such a conclusion, in Buffon's view, would conflict with the honesty and good faith that had been evident in all of Newton's other publications.

The Buffon–Clairaut polemic

Buffon's main concern in his January paper, however, was not to criticize Calandrini's interpretation – he recognized his own lack of proficiency in mathematical calculations, and proposed to leave the calculational question to others. Rather, he particularly desired to register sharp opposition to Clairaut's conclusion that the inverse-square law of gravitation established by Newton ought to be modified. To base such a conclusion on a discrepancy between theory and observation in but a single phenomenon, as he believed Clairaut had done, was in Buffon's view unreasonable. The correct course would be to "seek the particular reason of this singular phenomenon". Buffon suggested five different possible explanations as to how the missing half of the apogeal motion might be explained, four of them invoking irregular distributions of mass in the Moon, the fifth depending on the Earth's magnetic force. The latter explanation appears to be the one Buffon most favoured.

Was it necessary, Buffon asked, when theoretical calculations of the apogeal motion failed to agree with observation, that Newton's theory be "ruined" at the foundation by changing the law?

M. Clairaut proposes a difficulty contrary to the system of Newton; but it is not a difficulty that should or can become a Principle. It is necessary to seek to resolve it, and not to make from it a Theory all of whose consequences are supported only on a single fact.

According to Buffon, whenever a law is proposed for the variation of a physical quantity, the use of a single term to express this law is mandatory. A physical law must meet two essential requirements: the quantity or quality being measured must vary in a "simple" fashion, and it must be expressed by a single term. This *simplicité* and *unité* is immediately destroyed if a two-term expression is interpreted as implying a complex variation of a single quantity or quality. Two terms had to represent two "qualities" (i.e., two distinct types of force) rather than one.

Buffon was also convinced that Newton's gravitational force could not vary in any ratio other than the inverse-square; this ratio was the only one permitted for a quality or quantity like light or odour propagated in straight lines from a centre.

In the final portion of his paper Buffon posed three questions to Clairaut. (1) If the law of gravitation did not follow the inverse-square ratio, would Newton's "rigorous demonstration" of the identity of terrestrial gravity with the force retaining the Moon in its orbit (in Prop. 4 of Bk III) remain valid? (2) Given Clairaut's newly proposed law, what would be the consequences for comets (such as that of 1680) that came very close to the Sun? Would not the orbital apse of such a cometary orbit undergo a large motion, in contradiction to the empirical verification obtained by Newton and Edmond Halley? (3) Would not the motion of the apsides of the inner planets and of the satellites of Jupiter and Saturn be affected, contrary to what had been observed?

Clairaut responded in a paper read to the Academy on 17 February 1748. He had given detailed arguments against Calandrini's method, and he resented Buffon's failure to examine them. He denied that Newton had "rigorously demonstrated" the identity of terrestrial gravity with the centripetal force of the Moon. In Prop. 4 of Bk III Newton had considered the lunar orbit as a circle, and had not taken into account the radial component of the Sun's perturbing force. (Clairaut at this point appears unaware of Newton's incorrect use in this proposition, as in the more refined argument of Cor.7 of Prop. 37 of Bk III, of a subtractive radial component equal to $1/178\frac{29}{40}$ of the Earth's mean force on the Moon.) As for the other questions that Buffon raised, Clairaut argued that the effect of the inverse-fourth-power term on the comet of 1680 would be observationally undetectable, and that the effects on the apsidal motions of the inner planets would be ascertainable only after a long period of time. The satellites of Jupiter and Saturn moved so nearly in concentric circles that their apsidal motions could not be reliably determined.

Turning to Buffon's metaphysical arguments for the correctness of the inverse-square law, Clairaut granted that metaphysics could be helpful in leading through analogy to tentative hypotheses, but urged that "if we allow ourselves to be led by its torch alone, we can lead ourselves astray at any moment". To suppose that, because an inverse-square law holds for the intensity of light, an entirely analogous law must hold for gravitation, presupposes that the mechanisms are similar, and that a body attracts gravitationally by the emission of particles. Such a supposition, Clairaut observed, is quite conjectural; the only reliable way to determine the law of gravitation is by means of empirical facts.

As for Buffon's contention that a physical law must be "simple" and be expressed by a single algebraic term, Clairaut put it down to a resistance against using complex quantities known as *functions* in the study of physico-mathematical problems. "If we are unable to render [these functions] very simple in expressing them, it is the fault of Algebra which, like language, has its imperfections." To claim as Buffon did that each term of Clairaut's two-term law represented a distinct force was presumptuous: "Is it appropriate for us," Clairaut inquired, "to decide whether the Creator has given the attractive virtue to matter by two different decrees, or has endowed it with two forces at the same time by the sole action of his will?"

At this point we must note that, at the meeting of the Academy of 27 January, Pierre Bouguer (1698–1758), a physicist, geodesist, and astronomer, had read a paper "On the institution of the laws of attraction", which he was planning to add to a new edition of his *Entretiens sur la cause d'inclinaison des orbites des planètes*, first published in 1734. Like Buffon, Bouguer claimed that gravitation could be considered as one of several sensible qualities (e.g. light) that are "exercised" in the direction of straight lines emanating from a single point, and that therefore decrease with the square of the distance. But unlike Buffon he believed that laws of attractive forces could differ from the inverse-square, and that the only way of determining these laws was from the phenomena. Some molecules might attract in accordance with the inverse-square, and others in accordance with the inverse-cube, so that the total effect of, say, the Earth on the Moon would be given by

$$\frac{m}{x^2}+\frac{n}{x^3},$$

where m designates the multitude and intensity of the corpuscles acting according to the one law, and n the multitude and intensity of the corpuscles acting according to the other.

In his response to Buffon of 17 February, Clairaut compared Bouguer's hypothesis with his own:

[M. Bouguer's] idea has nothing contrary to my researches, and I am very far from rejecting it: however I do not prefer it to mine, because I find in my hypothesis the advantage of making only a single law for all the phenomena attributed commonly to attraction, and this advantage appears to me superior to the one of the simplicity of analytical expressions.

Clairaut also replied in detail to Buffon's proposal of various asymmetrical distributions of the Moon's mass in order to account for the apsidal motion. Like d'Alembert and Euler, he concluded that the lack of symmetry necessary to yield a sizeable motion of the apsides would be extreme. "If ever I learn that the Moon has been seen otherwise than round", a sceptical Clairaut guaranteed, "I shall yield myself to this explanation."

Buffon made no immediate public response to Clairaut's paper. Clairaut, consequently, must have assumed that their debate was at an end. He undertook now to unite his several papers on the motion of the lunar apsides into a single composite paper, which he had previously obtained permission to publish in the Academy's *Mémoires* for 1745, then being readied for the press (the collected memoirs for any given year were generally not published till several years later). In preparing the composite paper he ran up against the difficulty of not being able to specify the precise formulation of the proposed additional term in the gravitational law, in the absence of some phenomenon other than the motion of the lunar apsides in which the effect of this term could be determined quantitatively. Among the phenomena in which the effect might reveal itself, capillary action had not yet been sufficiently studied; the figure of the Earth and variation of gravity with latitude depended on the arrangement of the Earth's interior parts, still a matter of considerable controversy; the motion of the apsides of the Jovian and Saturnian satellites were as yet unknown, and those of Mercury and Venus were not precisely enough determined. Clairaut was thus unable to satisfy the desire he had expressed in his sealed paper of 6 September 1747 on the apsidal motion, of demonstrating how a modified law could account for the entire motion of the lunar apsides, while not affecting the agreement that Newton had shown between the inverse-square law and other phenomena attributed to gravitation.

Despite these difficulties, Clairaut still felt that it was simpler to hypothesize a single universal law than to assume two or more different types of attractive force, each following different laws, and perhaps deriving from different physical principles. In Clairaut's opinion, the idea of a modified universal law of gravitation was the most conservative hypothesis that could be made to account for the apsidal discrepancy.

Clairaut's arranging to have his composite paper printed in the *Mémoires* for 1745, together with his omission from this paper of certain paragraphs in the original papers read in 1747, aroused the ire of Buffon, who proceeded to make deletions from his own paper and to arrange to have it published in the same volume of the *Mémoires*. As Clairaut put it in a letter of July 1749 to Cramer, Buffon "subtracted all the things on which I had straightened him up, like the objections derived from comets, the figure of the Moon, etc., and he . . . conserved from all that he had objected to me only some metaphysical arguments that could be disputed eternally."

Meanwhile, in the latter part of 1748 Clairaut undertook a review of his theoretical calculation of the Moon's motions. This review led to his discovery in December of that year that a more complete calculation on the basis of the inverse-square law would account for practically all the observed motion of the lunar apsides. The full calculation would require considerable time; to safeguard his priority, Clairaut in late January 1749 arranged for sealed envelopes, containing a description of the method of his new investigation and of the new result for the apsidal motion, to be deposited at the Paris Academy of Sciences and at the Royal Society of London. By the time of the Paris Academy's meeting of 17 May, he had sufficient confidence in his new calculations to make a public announcement of the new discovery that the inverse-square law alone could account for the full motion of the lunar apse. In doing so he asked to be excused from describing his method in detail until he could finish the recalculation, and specifically denied that Buffon's "vague reasons" had had anything to do with the new discovery.

Buffon, however, was convinced that it had been his objections that had persuaded Clairaut to retract the proposal of a modified law, and he insisted, in response to Clairaut's announcement, that the Academy, in printing his own earlier critique, mention the date on which it had been read. He also publicly reproached Clairaut for making changes in the published versions of his memoirs.

Clairaut, on going to the Academy's secretary to arrange for the printing of his retraction, discovered that Buffon not only had inserted in his critique a new metaphysical argument that he had not read at any meeting of the Academy, but also had surreptiously inserted a separate short note containing two additional arguments of the same character. This last step provoked Clairaut to attack his colleague's behaviour and arguments in a new memoir read to the Academy on 21 May and 4 and 11 June. At the last of these sessions Buffon presented a further addition to his original critique, to which Clairaut responded on 21 June. Buffon continued to insist that a multi-term law involved absurdity; Clairaut believed Buffon's objections to be founded on mistaken algebra and a mistaken view of the relation of algebra to nature.

Clairaut terminated his dispute with Buffon by again asserting that his hypothesis of a modified law had been simply an expedient that he felt could admirably explain a wide variety of both celestial and terrestrial phenomena; he had defended it solely because Buffon had considered it absurd on the basis of metaphysical, mathematical, and physical grounds that he himself considered to have no validity. Buffon had involved him in an unwanted dispute that contributed nothing to the solution of the real problem, which was the use of phenomena to know the true laws of nature.

The resolution of the problem

Clairaut's memoir of January 1749, in which he first described the procedure by which he obtained his new derivation of the apsidal motion, was eventually read to the Academy of Sciences in March 1752, and appeared in the Academy *Mémoires* for 1748, published in 1752. The delay in publication was due to Clairaut's incorporation of the memoir in a larger treatise on the Moon's motion; the latter treatise being Clairaut's award-winning entry in the St Petersburg Academy's prize contest of 1750.

Clairaut's general method is described in Chapter 20 below, and it is there pointed out that, in resolving the equations of motion by approximation, Clairaut substituted for the radius vector r an expression giving the Earth–Moon distance in a rotating ellipse:

$$\frac{k}{r} = 1 - e \cos mv, \qquad (1)$$

where k is the parameter of the rotating ellipse and e its eccentricity, v is the true anomaly (the angle that the radius vector makes with the line of apsides), and m is an undetermined constant differing from 1, so as to allow for motion of the lunar apse. This substitution would find its justification if in the final solution the larger terms could be identified with the terms in (1), and the remaining terms were very small by comparison.

The outcome appeared to justify the substitution: Clairaut obtained for the radius vector the expression

$$\frac{k}{r} = 1 - e\cos mv + 0.007\,098\,8\cos\left(\frac{2v}{n}\right) - 0.009\,497\,05\cos\left(\frac{2}{n} - m\right)v + 0.000\,183\,61\cos\left(\frac{2}{n} + m\right)v, \qquad (2)$$

where n is the ratio of the Moon's mean motion to the difference between the mean motions of the Moon and the Sun. The last three terms in (2) are indeed but small oscillations superimposed on the other terms giving the value of k/r. The value of m, on the other hand, proved to be 0.995 803 6, implying that in one sidereal revolution of the Moon the apse moved $(1 - 0.995\,803\,6) \times 360° = 1°\,30'\,38''$ – only about half the observed motion.

Even before he had undertaken to evaluate the coefficients of his equation for the Moon's orbit, Clairaut had indicated a means whereby more exact values of m, n, k, etc., could be obtained. Instead of substituting for k/r the quantity given by (1) above, it was only necessary to substitute

$$\frac{k}{r} = 1 - e\cos mv + \beta\cos\left(\frac{2v}{n}\right) + \gamma\cos\left(\frac{2}{n} - m\right)v + \delta\cos\left(\frac{2}{n} + m\right)v, \qquad (3)$$

which is derived from (2) by putting the indeterminate quantities β, γ, δ in place of the numerical coefficients in (2). Clairaut obtained his anomalous result for the apsidal motion prior to September 1747, but it was not until the latter half of 1748 that he undertook the longer calculation required if the expression (3) were used for k/r. He appears to have been initially convinced that the new calculation would not yield a significantly different value for the motion of the apse. Two events may be singled out as having possibly led him to question this belief.

In a lost letter to Calandrini of early 1748, Clairaut pointed out, in criticism of the Swiss mathematician's procedures for determining the motion of the lunar apse, that Calandrini had failed to take account of the component of the solar perturbing force at right angles to the radius vector. Clairaut made this critique despite his belief at this time that the effect of this component on the motion of the apse could be proved to be negligible. Calandrini responded on 19–20 February, arguing that the criticism did not reflect on his method, but only on his failure to prove the omission legitimate, and pointing out that in a circular orbit the effects of the perturbing force at right angles to the radius vector would, in successive quadrants, cancel each other. Clairaut in his answer of 6 March rejected the supposition of a circular or nearly circular orbit as irrelevant, and seemed to imply that he was still seeking a demonstration that the effect of the perpendicular component was negligible.

Of what would such a demonstration consist? In Clairaut's first determination of m, only the cos mv term of (2) played a role. If the more refined substitution (3) were used, the last three terms of this expression (of which the last two derived from the perpendicular component of the Sun's perturbing force) would produce some further cos mv terms. Because of the smallness of the coefficients of the last three terms in (2), Clairaut apparently assumed at first that the new cos mv terms would be negligible. By the time he penned his second letter to Calandrini on 6 March, he may have begun to wonder whether the sum total of all the coefficients of these new cos mv terms might affect the value of m more than he had originally suspected. At any rate, in late April he wrote to Euler that he was once more at work on the problem of the Moon, and was making new discoveries regarding gravitational attraction, of which he would inform Euler when he had completed the calculation.

Another stimulus for carrying out the more refined derivation may have come from d'Alembert. In Clairaut's first calculation of both k/r and of the time t as a function of true anomaly v, he had ignored powers of the eccentricity e higher than the first. The square of the eccentricity in fact influenced the equation for k/r very little, but it had a more pronounced effect on the expression for t, owing to the large augmentation that certain terms received by integration. D'Alembert, by his own later account, informed Clairaut of this result in June 1748; and he claimed that the information had been important in persuading Clairaut to carry out the more refined calculation.

In any case, the new calculation was performed.

In the new equations for the orbit and the time, the coefficients of the terms were not very different from the ones determined previously. But, as Clairaut pointed out in his later presentation of the new theory to the Paris Academy,

an important respect in which my new solution differs essentially from the first is the determination of the letter m which gives the motion of the apogee. The term ... by which m is determined is found nearly doubled by the addition that is made of the terms

$$\beta \cos\left(\frac{2v}{n}\right) - \gamma \cos\left(\frac{2}{n} - m\right) v + \text{etc.}$$

to the value of $1 - e \cos mv, \ldots,$ *and by this means the motion of the apogee is found nearly conforming to the observations, without supposing the Moon impelled toward the Earth by any force other than the one which acts inversely as the square of the distance; and there are grounds for believing that [with further refinement of the calculation] the slight difference that is found between the theory and the observations will vanish entirely.*

Thus Clairaut had shown that the inverse-square law of gravitation proposed by Newton could be used to account, in a precise mathematical way, not only for the planets' generally elliptical orbits, but also for the deviation of the Moon's motion from this ideal pattern. The title (here Englished) that he chose for his St Petersburg Academy prize entry reflected his achievement: *Theory of the Moon Deduced from the Single Principle of Attraction Reciprocally Proportional to the Squares of the Distances* (1752).

Euler's praise for Clairaut's achievement was unstinting. In a letter to Clairaut of 29 June 1751 he wrote:

... the more I consider this happy discovery, the more important it seems to me .... For it is very certain that it is only since this discovery that one can regard the law of attraction reciprocally proportional to the squares of the distances as solidly established; and on this depends the entire theory of astronomy.

Consideration of the apsidal motion, according to Euler, offered the safest means of deciding on the sufficiency of the Newtonian theory. And so, in the introduction to the memoir that brought him the prize of 1752 of the Paris Academy, he affirmed that

... because M. Clairaut has made the important discovery that the movement of the apogee of the Moon is perfectly in accord with the Newtonian hypothesis ..., there no longer remains the least doubt about this proportion .... If the calculations that one claims to have drawn from this theory are not found to be in good agreement with the observations, one will always be justified in doubting the correctness of the calculations, rather than the truth of the theory.

Further reading

Craig B. Waff, *Universal Gravitation and the Motion of the Moon's Apogee: The Establishment and Reception of Newton's Inverse-square Law, 1687–1749* (Johns Hopkins University Ph.D. dissertation, 1975; University Microfilms, Ann Arbor, Michigan)

Craig B. Waff, Alexis Clairaut and his proposed modification of Newton's inverse-square law of gravitation, pp. 281–8 in Centré International de Synthèse (ed.), *Avant, avec, après Copernic: La représentation de l'Univers et ses conséquences épistémologiques* (Paris, 1975)

Craig B. Waff, Isaac Newton, the motion of the lunar apogee, and the establishment of the inverse-square law, *Vistas in Astronomy*, vol. 20 (1976), 99–103

17

The precession of the equinoxes from Newton to d'Alembert and Euler

CURTIS WILSON

Isaac Newton was the first to propose a quantifiable cause for the precession of the equinoxes. This slow westward motion of the equinoctial points – the projections from the Earth's centre onto the celestial sphere of the intersections of the terrestrial equator with the ecliptic – had been known since Antiquity, and by Newton's time had come to be evaluated at a steady 50″ of arc per year. From a heliocentric standpoint, it could be viewed as a slow rotation of the Earth's axis about the ecliptic's poles. It was caused, Newton urged, by the gravitational action of the Sun and Moon on the equatorial bulge of the Earth.

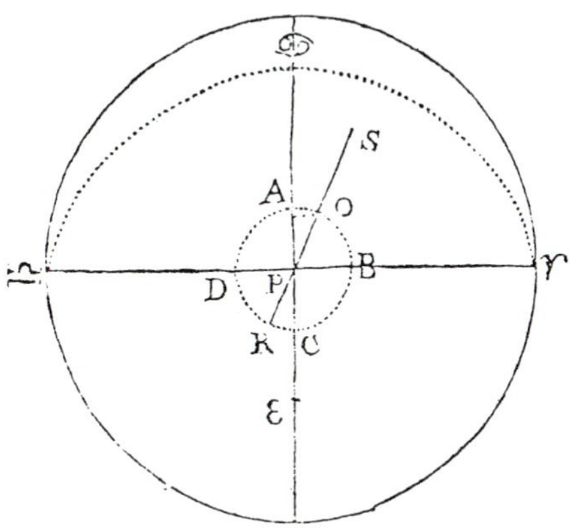

17.1. John Machin's diagram to explain the geometry of nutation (from Bradley's letter announcing nutation published in *Philosophical Transcations* in 1748).

The mean place of the pole of the Earth, as projected onto the celestial sphere, is *P*, which according to Bradley was in the eighteenth century some 29° 29′ from *E*, the pole of the ecliptic's circle. But the true pole moves counterclockwise around *P* in circle *ABCD* with radius 9″ (the size of the circle *ABCD* is greatly exaggerated in the diagram). Later in his letter, to improve the accuracy of the hypothesis, Bradley proposes replacing the circle *ABCD* by an ellipse with major axis $AC = 18''$ and minor axis $DB = 16''$.

The true pole is at *A* when the Moon's ascending node is in the beginning of Aries, and at *C* when the node is in the beginning of Libra; the circle *ABCD* is completed in 18 years, 7 months, which is the period of the revolution of the Moon's nodes.

In relation to the much slower motion of precession – the slow retrogradation of the mean pole *P* in a circle about *E* with a period of nearly 26000 years – the nutation is a small superimposed wobble.

Newton's attempt to derive the precession quantitatively, however, was seriously flawed. The first to achieve an essentially correct derivation was Jean le Rond d'Alembert (1717–83), a mathematician and prominent member of the Paris Academy of Sciences from the early 1740s, during the 1750s co-editor with Denis Diderot of the *Encyclopédie*, and in his last thirty years a powerful and somewhat sinister figure in the politics of the Paris and Berlin academies.

D'Alembert's interest in deriving the precession was sparked by the discovery by James Bradley, Astronomer Royal of England, of the nutation. This is a small wobble in the Earth's precessing axis due to the changing orientation of the lunar orbit, which while maintaining a nearly constant inclination to the ecliptic, rotates westward, completing a circle in 18.6 years. Bradley had detected the nutation in the early 1730s, but delayed announcing his discovery till he had traced the phenomenon through a full cycle of the Moon's nodes. His announcement was published in the *Philosophical Transactions of the Royal Society* in January 1748 (Figure 17.1). Copies reached the Continent only in the summer, at which time the Abbé Nicolas-Louis de Lacaille, an astronomer who was in correspondence with Bradley, read an extract at a meeting of the Academy of Sciences. D'Alembert immediately attempted a calculation of the nutation, but his result did not agree with Bradley's observations.

From December onward d'Alembert devoted himself full-time to the problem. He appears to have informed no-one of this endeavour save Gabriel Cramer in Geneva; he was afraid (unnecessarily, as the event proved) of being anticipated by the English. For some time he was delayed by his failure to see that the Earth's diurnal rotation was of crucial importance in determining the effect. Finally, on 17 May 1749, he was able to present to the Academy the completed manuscript of his *Recherches sur la précession des equinoxes et sur la nutation de l'axe de la terre, dans le système Newtonien*. It was published in July.

The problem of deriving the precession was set for d'Alembert within a context in which the exactitude of the inverse-square law was under vigorous discussion. Clairaut, having earlier claimed that the inverse-square law could not account for the full motion of the lunar apse, reversed his position in December 1748; but he had not yet revealed how the true motion of the apse was to be derived (see Chapter 16). Through most of 1748 d'Alembert had been at work on the lunar theory, seeking to extract from the inverse-square law the observed anomalies. When he turned in December to the problem of the precession, it was with the same object in view. The solution he finally achieved was a striking confirmation of the theory. As he stated in the preface to his memoir,

The nutation of the terrestrial axis, confirmed by both the observations and the theory, furnishes, it seems to me, the most complete demonstration of the gravitation of the Earth toward the Moon, and hence of the principal planets toward their satellites. Previously this tendency had not appeared manifest except in the ocean tides, a phenomenon perhaps too complicated and too little susceptible to a rigorous calculation to reduce to silence the adversaries of reciprocal gravitation.

In the course of his investigation. d'Alembert carried out a thorough critique of Newton's treatment of the precession; it appears to have been the first such critique to be made.

D'Alembert's critique of Newton's derivation of the precession

Newton's qualitative understanding of the cause of the precession emerged out of his study of the perturbations of the Moon, from the analogy between the retrogression of the Moon's nodes and the retrogression of the equinoctial points. That the nodes of the Moon's orbit must retrogress on the ecliptic is caused, he showed in Cor. 11 of Prop. 66 of Bk I of the *Principia*, by the action of the Sun. In Cor. 18 of the same proposition he went on to propose that, if the Moon were multiplied so as to become many moons, and these were liquified so as to form a fluid ring, the nodes of the ring would retrogress in the same way as the nodes of the lunar orbit. He hypothesized further (in Hypothesis II of Bk III in the second and third editions) that the result would be the same if the ring became solid. These claims, according to d'Alembert, were correct not precisely but as giving the *average* result.

Newton's next step was to consider a smaller ring, subject to the same perturbative action of the Sun, but with radius equal to the Earth's radius, and period of rotation equal to the Earth's diurnal period. The retrogression of the nodes of this ring, Newton asserted, would be to that of the lunar nodes as the period of the ring to that of the Moon. D'Alembert confirmed this conclusion.

If the imaginary ring be then attached to the underlying Earth, it must share its motion with the latter. But according to what rule? Newton assumed that the total quantity of motion (mv) previously in the ring in virtue of the retrogression of its nodes would be distributed between the ring and underlying sphere in just the way required for the two bodies to rotate together. What Newton should have done, d'Alembert pointed out, was to set the *moment* of the motion lost by the ring equal to the *moment* of the motion gained by the sphere. In other words, he should have assumed conservation of angular – not of linear – momentum. Because of this error, Newton's result for the angular velocity transferred to the Earth was too large by about 16%.

A spherical Earth with an attached equatorial ring is not the same as a spheroidal Earth, and Newton recognized that the Sun's attraction would have a greater moment in the former case than in the latter. In the first edition of the *Principia* he took the ratio of the two moments to be 4:1, in the second edition he showed by a correct deduction that it is 5:2. But he then assumed that the motions engendered would vary as the moments. This, as d'Alembert pointed out, is wrong: the effect of the torques depends not just on the mass to be moved but on its figure or – to use the Eulerian term – its

17.2. Jean le Rond d'Alembert.

moment of inertia.

Another flaw in Newton's derivation was quantitatively more important: the failure to take into account the diurnal rotation of the Earth. "The globe", Newton claimed in Cor. 20 of Prop. 66, "is perfectly indifferent to the receiving of all impressions"; and he evidently meant by this that the finite motion he had deduced for the precession was simply to be added to the diurnal motion. But only infinitesimal rotations add vectorially. D'Alembert in his initial calculation fell into the same error, but then found it implied a precession due to the solar force that is less than half the correct value.

Finally, Newton attempted to derive the ratio of the solar and lunar forces, acting on the equatorial bulge to wheel the Earth about its centre, from data on the heights of the tides; and his result is disastrously wrong. In the first edition of the *Principia* he took the ratio to be $6\frac{1}{3}$ to 1; in the second edition he fiddled with the same data to obtain 4.4815:1. Daniel Bernoulli, in a memoir that won a prize in the Paris Academy's contest of 1740, sought to determine the ratio not from the heights but from the periods of the tides, and arrived at the value 2.5:1. Bradley, in his announcement of the discovery of the nutation in 1748, suggested determining the ratio from the observed values of the nutation and precession; for the first of these appeared to be due to the action of the Moon alone, while the precession was caused by joint action of the Sun and Moon. D'Alembert was to follow this suggestion, obtaining the ratio 2.35:1. The value accepted today is 2.17:1.

The chief conclusion to be drawn from the foregoing list of errors is that Newton lacked an articulated kinematics and dynamics for extended, rigid bodies. A fully developed mechanics of rigid bodies first appeared in 1765, in Leonhard Euler's great treatise on the subject (*Theoria motus corporum solidorum*). During the 1750s, a number of mathematicians attempted to derive the precession and nutation without relying on d'Alembert's derivation; among them a Benedictine monk, Charles Walmsley (in the *Philosophical Transactions* for 1756), the mathematician Thomas Simpson (in his *Miscellaneous Tracts*, 1757), and Patrick d'Arcy (in the *Mémoires* of the Paris Academy for 1759). These attempts were seriously flawed. Simpson's derivation appeared in slightly modified form in the first edition of J.-J. L. de Lalande's *Astronomie* (1764); d'Alembert, in an essay published in 1768, made a detailed analysis of its errors. All these putative derivations utilized integrations, but none of them formulated differential equations representing accurately the mechanics of a spinning, spheroidal Earth, subject to oblique forces. Such equations first appeared in d'Alembert's *Recherches* of 1749, and then (and in dependence on d'Alembert's work) in Euler's "Découverte d'un nouveau principe de Mécanique", published in the Berlin *Mémoires* for 1750.

D'Alembert's derivation of the precession and nutation

An adequate dynamics of extended, solid bodies must somehow take into account what is happening to all the 'elements' or 'particles' of the body – the forces and accelerations to which they are subject. The mechanics of systems of bodies or mass-elements in rigid or flexible connection was developed quite independently of Newton, starting from the problem of the 'centre of oscillation' of a compound pendulum. The discussion of this and related problems, particularly by Jakob I and Daniel Bernoulli, forms the background for d'Alembert's announcement, in his *Traité de dynamique* of 1743, of the famous "d'Alembert's principle". On this principle turns d'Alembert's derivation of the precession and nutation.

D'Alembert's statement of the principle, both in the *Traité de dynamique* and in his later *Recherches sur la précession*, is less revealing than it could be. It says in effect that the product of each mass-element by its reversed acceleration is to be considered a force, and that the sum of all such forces is in equilibrium with the applied forces. But d'Alembert does not give the true reason why this is so, namely that the forces acting to connect the mass-elements into a rigid body are in equilibrium with each other and therefore equal 0. The actual acceleration $\vec{a}$ of each mass-element is the vector-sum of the acceleration due to the connections, $\vec{a}_c$, and the acceleration imposed by the external forces, $\vec{a}_f$; hence $\vec{a}_c = \vec{a} - \vec{a}_f$. Since the accelerations $\vec{a}_c$ multiplied by their respective masses form a system in equilibrium, so do the accelerations $\vec{a} - \vec{a}_f$ or $\vec{a}_f - \vec{a}$.

According to Pierre-Simon Laplace, d'Alembert was the first to lay down the conditions for translational and rotational equilibrium in the three coordinate directions. In applying these conditions in the *Recherches* of 1749, however, d'Alembert fell into a curious error. Instead of setting the sum of the components of all the forces in each coordinate direction equal to zero, he set the components of the external forces in each direction equal to the corresponding components of the 'forces' represented by the product of each mass-element by its reversed acceleration, summed over the whole body. The three resulting equations for translational equilibrium are wrong, but since d'Alembert never used them, the error did no harm. In expressing the rotational equilibrium d'Alembert similarly set the moments of the external forces about each coordinate axis equal to the corresponding moments of the 'forces' represented by the reversed accelerations of the mass-elements. Because he made a further error in the direction of the externally applied torques, the resulting equations are correct.

Years later, in the first volume of his *Opuscules mathématiques*, d'Alembert stated the conditions of equilibrium correctly, setting the total forces in a given direction, and the total torques about a given axis, equal to zero. If he recognized his original error, he did not acknowledge it. The fact remains that the correctness of his results in the *Recherches* of 1749 depended on compensating errors of sign.

D'Alembert's next task was to derive, from his three equations for rotational equilibrium, differential equations enabling him to determine the motions of the Earth's axis. For this he needed (1) expressions for the external torques in terms of known quantities pertaining to the periods, distances, longitudes and latitudes of the Sun and Moon, and (2) expressions for the moments about three perpendicular axes of the product 'mass × reversed acceleration' for an arbitrary mass-element of the Earth, in terms of the position of the Earth's axis. The latter expressions were to be integrated over all mass-elements, to yield the total moment about each axis.

In deriving expressions for the external torques, d'Alembert first assumed the Earth to be a uniformly dense spheroid with polar radius a and equatorial radius $a(1+\alpha)$, where α is a very small number. Later he showed that the results still held, with minor modification, if the density varied in such a way that the layers of equal density were bounded by surfaces of revolution nearly spherical in shape. Under these assumptions the only forces contributing to unbalanced torque were those on the equatiorial bulge, bounded externally by the spheroidal surface and internally by the inscribed sphere.

D'Alembert resolved the accelerative force due to the Sun's or Moon's attraction of any mass-element of the equatorial bulge into two components, one directed to the Earth's centre, the other parallel to the line joining the Earth's centre to the perturbing body, Sun or Moon. Only the second of these components, or rather its difference from the parallel accelerative force on a mass-element at the Earth's centre, produced torque on the axis. D'Alembert first summed the moments of these differences for an annulus of the equatorial bulge; then he summed the moments for all the annuli.

In determining the moments of the 'internal forces', as we may call them, d'Alembert used three planes of reference: the ecliptic, in which the Earth's centre of gravity was assumed to lie; a plane passing through the Earth's axis perpendicular to the ecliptic; and a third plane passing through the Earth's centre of gravity and perpendicular to the other two. He specified the position of the Earth's axis by two variables: the angle (ϵ) that the projection of the axis onto the ecliptic made with a fixed line, say the line from the Earth's centre of gravity to the first star of Aries; and the angle (Π) that the axis itself made with the ecliptic plane. To specify the position of an arbitrary mass-element of the Earth, he used the distance b along the Earth's axis from one pole, the radius f at right angles to the axis, and the angle X which this radius formed with the plane through the axis and perpendicular to the ecliptic.

The reversed accelerations of an arbitrary mass-element, in directions perpendicular to each of the three reference planes, were expressed in terms of $d^2\epsilon$, $d^2(\cos \Pi)$, and d^2P ($=d^2X$), where P is the angle through which the Earth rotates in its diurnal motion during the time-interval considered. The sum of the moments about each axis of the products 'mass-element × reversed acceleration' was then obtained by integration.

Once all the required substitutions had been made into the equations for rotational equilibrium, d'Alembert was faced with three differential equations. By combining these equations he obtained an integrable equation for d^2P, and was then left with two differential equations in the variables ϵ and Π. These he resolved using approximations suggested by the known observational facts. As the second-order differentials occurred in relatively small terms, d'Alembert left them out of account, so that only first-order equations remained. Integrating these, he obtained

$$\Pi = \Pi_m - \frac{3Am'(1+\beta)\sin \Pi_m}{2Kk(n'-M)} \cos(n'-M), \quad (1)$$

$$\epsilon = -\frac{3A(2+\beta)\sin \Pi_m}{2Kk} z - \frac{3Am'(1+\beta)\cos 2\Pi_m}{2Kk(n'-M)\cos \Pi_m} \sin(n'-M)z. \quad (2)$$

Here Π_m is the mean inclination of the Earth's axis to the ecliptic; $(1+\beta)$ is the ratio of the Moon's force to the Sun's force in moving the equatorial bulge; z is the mean longitude of the Sun; n' is the rate of motion of the nodes of the lunar orbit, compared to the Earth's motion about the Sun; M is the mean rate of precession per year; m' is the tangent of the inclination of the lunar orbit to the ecliptic; and A, K, k are constants.

Equation (1) is in agreement with Bradley's observational result if we equate $3Am'(1+\beta)\sin \Pi_m/2Kk(n'-M)$ with 9″. Equation (2) is in agreement with Bradley's observational determinations provided $(3A/2Kk)(2+\beta)\sin \Pi_m$ is identified with M, the mean rate of precession or 50″ per year. The second term of (2) represents the inequality of the precession, with a maximum value of 16″.8 (but d'Alembert gave it as 15″).

Bradley's empirical determinations allowed d'Alembert to obtain a value for the ratio $(2+\beta)/(1+\beta)$, and thus for $(1+\beta)$; he found it to be 2.35. This result in turn permitted a calculation of the mass-ratio of the Moon to the Earth. D'Alembert put it at $\frac{1}{80}$; the present-day value is $\frac{1}{81.3}$. (But this ratio, we should note, is very sensitive to small changes in the values assigned to the mean precession and maximum nutation.)

D'Alembert's equations (1) and (2) have to do with the motions of the Earth's axis of figure. Since this axis is continuously in motion, it can never be identical with the instantaneous axis of rotation, that is, the line within the Earth's body that is instantaneously at rest with respect to the Earth's centre of gravity. From his differential equations d'Alembert could determine the relation between these two axes; he found the tangent of the angle between them to be

$$\frac{[d\Pi^2 + \cos^2\Pi_m \, d\epsilon^2]}{kdz},$$

where the numerator is the angular distance through which the Earth's axis of figure moves, while the Earth is rotating through the angle kdz about its axis of rotation. The numerator is so small in comparison with the denominator that the angle between the two axes can be regarded as negligible.

In Chap. 11 of his *Recherches* d'Alembert undertook a second derivation, assuming from the start the identity of the axes of figure and rotation. The basic strategy was to combine vectorially the infinitesimal displacements due to the Earth's diurnal rotation with those due to the attractive forces of the Sun and Moon. D'Alembert thus obtained differential equations identical with those he had previously extracted by approximation from the more exact equations of his first derivation. The second derivation thus corroborated the first.

D'Alembert's exposition of his course of reasoning leaves much to be desired. His friend Cramer wrote to him that the work was disorderly and full of typographical errors, and that it contained totally unintelligible diagrams. But what above all mattered to d'Alembert in composing this work was to show that the precession and nutation indeed followed from Newton's law of gravitation. And this, if we set aside two inadvertent and compensating errors, he succeeded in doing.

Euler's work on the precession and mechanics of rigid bodies

D'Alembert's *Recherches sur la précession* appeared in print early in July 1749, and d'Alembert sent off a copy to Euler on 20 July. Not until 3 January 1750 did Euler write to d'Alembert to acknowledge that he had received it, and, albeit with difficulty, had learned enough from it to carry out a derivation of the precession and nutation of his own. A more detailed acknowledgement appears in Euler's letter of 7 March, written just two days before he presented his own derivation to the Berlin Academy:

I applied myself repeatedly and for a long time to the problem of the precession, but I always encountered an obstacle, ... and above all this problem: given a body turning about any axis freely, and acted upon by an oblique force, to find the change caused both in the axis of rotation and in the movement .... With respect to this problem all my investigations had been hitherto unavailing, and I would not have applied myself to it further, if I had not seen that the solution must necessarily be encompassed in your treatise, although I was not able to find it there .... I must confess that I could not follow you in the preliminary propositions you employed, for your way of carrying out the calculation was not yet familiar to me .... But now that I have succeeded better in the investigation of this same subject, having been assisted by some insights in your work by which I was gradually enlightened, I have come to be able to judge your excellent conclusions.

Euler's derivation of the precession was published in 1751 in the Berlin *Mémoires* for 1749, and d'Alembert was astonished and hurt to find that it contained no acknowledgement of his own priority in achieving such a derivation. His complaints constrained Euler to insert, in the Berlin *Mémoires* for 1750 (published in 1752), an *Avertissement* declaring that Euler had written his memoir "only after having read the excellent work of M. d'Alembert on the subject; and that he has not the least pretention to the glory due to him who first resolved this important question."

Euler's derivation of the precession and nutation was a model of clarity. First he determined the moment of inertia of the Earth considered either as a homogeneous sphere or as a sphere with a denser, spherical core. Then, using equatorial coordinates, he computed the torque exerted by a distant attracting body out of the plane of the equator on an Earth considered as spheroidal in shape and either homogeneous or with a denser spheroidal core. To shift from equatorial to ecliptic coordinates, he utilized the propositions of spherical trigonometry, thereby avoiding the geometrical interpretation of differentials by which d'Alembert's two derivations had been bedevilled. Finally, to calculate the angle through which the Earth's axis is shifted by a torque S during the time dt, Euler had recourse to the formula

$$\text{angle} = Sdt^2/2Ids,$$

where I is the Earth's moment of inertia (taken to be the same about all axes), ds is the angle described in the Earth's diurnal rotation during dt, and the factor '2' is required because of Euler's way of defining force. This formula Euler promised to explain elsewhere.

In several respects Euler's derivation was less exacting than d'Alembert's first derivation. Euler started from the assumption that the Earth's axis of figure and axis of rotation are identical, claiming without proof that their angle of separation was negligible. He also assumed that at each instant the Earth's axis of rotation is such that the centrifugal forces generated in rotation about it exactly balance, so that the shift in the axis is solely due to the external torques. Finally, he assumed the formula of the preceding paragraph, to be explained "elsewhere". D'Alembert, followed later by Laplace, pointed out that Euler's derivation was identical in its basic strategy with d'Alembert's second derivation; but it is unlikely that Euler was aware of this identity.

Euler's final numerical equations for precession and nutation differ little from d'Alembert's; but they include small terms that d'Alembert neglected, those in the nutation and inequality of the precession due to the Sun's action.

Euler's "elsewhere" proves to be his "Découverte d'un nouveau principe de mécanique", presented to the Berlin Academy in September 1750. Here Euler mounted a frontal attack on the general problem of the free rotation of a rigid body subject to oblique forces. It was a problem he had already envisaged at the time he wrote his *Mechanica* (published in 1736). He had hazarded a tentative hypothesis as to how it might be dealt with in the first volume of his *Scientia navalis*, completed in 1740 but not published till 1749. But what finally set him on the path to a successful solution was a procedure he took over from d'Alembert's *Recherches sur la précession* of 1749.

The main course of argumentation in the "Découverte" begins with a "determination of the movement in general of which a solid body is capable, while its centre of gravity remains at rest". Euler here derives expressions for the components of acceleration of an arbitrary mass-element in the directions of the x, y, and z axes, in terms of the components λ, μ, ν of the instantaneous angular velocity, and the differentials $d\lambda$, $d\mu$, $d\nu$. Paralleling his steps to d'Alembert's, he then finds expressions for the moments of the products 'mass-element × acceleration' with respect to each axis, and integrates over all mass-elements of the body. The latter step is effected for a body of arbitrary mass-distribution by introducing special symbols for the products and moments of inertia. Finally, the moments thus found are set equal to the moments of the external forces. The formula introduced from "elsewhere" in Euler's *Recherches sur la précession* emerges as a corollary.

With the formulas of the "Découverte", Euler was not altogether satisfied. In general, they were difficult to apply. The single coordinate system used was fixed in space; the coordinates of each mass-element dM changed from one interval dt to the next as the body rotated about its centre of gravity. In general, therefore, integrations over time were impossible.

Euler tackled this difficulty in a study entitled "Du mouvement d'un corps solide quelconque lorsqu'il tourne autour d'un axe mobile", presented to the Berlin Academy in October 1751, but published only in 1767. In this work Euler introduced, besides the original coordinate system fixed in space, a second orthogonal system with the same origin but fixed in the body. To express the relation between the two systems, he introduced the now well-known 'Eulerian angles'. The resulting kinematics makes possible the computation of the angle between the axis of figure of a rotating spheroid and the instantaneous axis of rotation, and Euler accordingly calculated the angle between these two axes in the case of the spinning, precessing Earth.

He also investigated the conditions under which a body would continue to rotate about the same axis without application of any externally applied torque. He derived an equation relating the moments and products of inertia about the axes fixed in the body to the ratio of two components of the angular velocity, when the externally applied torques vanish. The equation is cubic; hence there is at least one real root, and therefore at least one axis about which the body can continue to rotate freely.

In 1755 a protege of Euler's, János-András Segner, professor at Halle, proved geometrically that there must be three such axes. Then Euler, in his "Recherches sur la connaissance mécanique des corps", submitted to the Berlin Academy in July 1758, gave a different proof: he showed that the axes through the centre of inertia of a body with respect to which the moments of inertia are either a maximum or a minimum are such that the products of inertia with respect to them vanish. But there must be a maximum and a minimum, unless all the moments of inertia are equal. Hence there are at least two axes about which the body can rotate freely, and all three roots of the cubic equation have to be real, since imaginary roots come in pairs. To these three axes, which are mutually perpendicular, Euler gave the name *principal axes*. Henceforth he would use these axes for the coordinate system fixed in the body.

The triumphant application of these results came in Euler's "Du mouvement de rotation des corps solides autour d'un axe variable", presented to the Berlin Academy in November 1758. Euler found ("with surprise", he says) that he was able to obtain integral solutions to problems he had previously believed "to surpass the powers of the calculus". These included determining the continuation of the motion of a body on which a motion of rotation about an axis other than a principal axis has been imposed. As a result of such successes, Euler then undertook the writing of his extended treatise, *Theoria motus corporum solidorum seu rigidorum*, finished by 1761 but published in 1765.

Epilogue

The fourth volume of d'Alembert's *Opuscules mathématiques*, published in 1768, contains two essays on the mechanics of rigid bodies. D'Alembert's aim here was to show that the results obtained by Euler were obtainable by his own procedures. The Eulerian method "being [he wrote] very complicated, I thought people would not be displeased to see how one can arrive at the same result by a much simpler analysis".

In 1773 Laplace, in the first treatise he submitted to the Paris Academy, in introducing "the general equations of motion of a body of any figure, acted upon by any forces", stated:

> *M. d'Alembert was the first to give the general solution of this problem, along with the most direct method of arriving at it, in his excellent treatise* Sur la précession des équinoxes, *an original work that shines throughout with the genius of invention, and that one can regard as containing the germ of all that has been done since on the mechanics of solid bodies.*

As Laplace proceeded to outline his approach to the mechanics of rigid bodies, however, it emerged as essentially Eulerian. A half century later he was to write: "the equations at which Euler arrives appear to me to be the most simple it is possible to obtain".

Further reading

For a more extended account of the developments here traced, see Curtis Wilson, D'Alembert *versus* Euler on the precession of the equinoxes and the mechanics of rigid bodies, *Archive for History of Exact Sciences*, vol. 37 (1987), 233–73.

18

The solar tables of Lacaille and the lunar tables of Mayer

ERIC G. FORBES and CURTIS WILSON

It is one thing to deduce mathematically the perturbations of the Moon and planets implied by Newton's law of gravitation, another to develop tables yielding precise predictions. The three mathematicians who first derived perturbations analytically (see Chapter 20) – namely Leonhard Euler (1707–83), Alexis-Claude Clairaut (1713–65), and Jean le Rond d'Alembert (1717–83) – all constructed lunar tables; but the errors of these tables ran as high as 5 arc-minutes, roughly half the maximal error of Isaac Newton's modified Horrocksian theory of the Moon (see Chapter 13 of Volume 2A). To do better, skills were required other than those of the mathematical theorist – skills in the making of observations, and in the reduction of observations both new and old. Also required was an understanding of the analytical theory, and a sensitivity to problems involved in fitting it to observational data.

The earliest published astronomical tables incorporating perturbations deduced analytically from the inverse-square law of gravitation appear to have been Euler's lunar tables of 1746 (in his *Opuscula varii argumenti*). But the first such tables to come into general use were the solar tables of Nicolas-Louis de Lacaille (1713–62) and the lunar tables of Johann Tobias Mayer (1723–62). Lacaille's solar tables were published in 1758, and republished in 1764 and 1771 in the first two editions of the *Astronomie* of Joseph-Jérôme Lefrançais de Lalande (1732–1807). From 1761 into the 1790s (but in modified form after 1779), they were the basis for the solar ephemerides of the annual *Connaissance des temps*. Mayer's lunar tables were published in a first version in Göttingen in 1753 and in Lalande's *Astronomie* of 1764. In a revised form they were published in London in 1770, and were the basis of the lunar ephemerides of the *Nautical Almanac* from 1767 to 1813 (but with successive modifications after 1776).

The achievement of Lacaille's and Mayer's tables was not merely a matter of adding theoretically computed perturbations to the purely Keplerian-style tables used earlier. Keplerian orbital elements needed to be redetermined in the light of the perturbations. Also, owing to uncertainties in the planetary and lunar masses and in the solar and lunar parallaxes, the perturbations deduced from theory had themselves to be assessed on an empirical basis. In addition, the discovery by James Bradley of the aberration of light and the nutation of the Earth's axis (officially announced in 1729 and 1748, respectively; see Volume 3) implied that all astronomical observations must be 'corrected' for these effects; and Lacaille and Mayer were the first astronomers to found their tables solely on observations thus reduced.

In pursuing a rapprochement of observation and theory, Lacaille and Mayer were indefatigable. J.-B. J. Delambre in the 1810s, looking back on their achievements from the vantage-point of the Laplacian era, saw them, along with Bradley on whose discoveries their work was premissed, as the giants of eighteenth-century astronomy.

The solar tables of Lacaille

Supported by a scholarship, Lacaille took the three-year course in theology at the Collège de Navarre. In the midst of his theological and literary studies, he discovered Euclid, and developed a secret love of astronomy, which he proceeded to study entirely on his own. A *contre-temps* between Lacaille and the vice-chancellor of the college at the ceremony for the awarding of degrees fortified Lacaille's resolve to devote himself to the mathematical sciences rather than to literature or theology. At some point he had received the title of *abbé*, but he seems never to have served as a priest.

In 1737 Lacaille became an assistant to Jacques Cassini (Cassini II) at the Paris Observatory. In 1739 and 1740 he worked with César-François Cassini de Thury (Cassini III) on the verification of the meridian of the Observatory from Perpignan to Dunkirk. A result of these new measurements was to show that degrees of terrestrial latitude increased in length toward the poles, implying a flattened Earth, in contradiction to the elongated shape that Jacques Cassini had claimed to deduce from earlier measurements. For this work Lacaille was admitted to the Academy of Sciences on 8 May 1741. An account of the meridian measurements was published in 1744 at the expense of Cassini de Thury, whose name alone appeared on the title page, although according to a letter from Lacaille to Bradley of 28 July 1744, the sole author was Lacaille. Earlier, in 1739, Lacaille had been named to the chair of mathematics at the Collège Mazarin, where an observatory was installed for him.

On 25 April 1742 Lacaille read to the Academy a project for a new catalogue of the fixed stars. Improvements in astronomy, he urged, must begin with a star catalogue, since it is from their relation to the fixed stars that the positions of planets and comets are determined. But all previous star catalogues were faulty because previous solar theories, on which the determination of right ascensions of stars must depend, were accurate at best only to about 1 arc-minute, and because star positions had not previously been corrected for aberration and the newly discovered variations in the obliquity of the ecliptic. Lacaille's aim was to develop a star catalogue free of these faults.

To determine right ascensions (RA), Lacaille proposed dividing the sky into a series of zones between parallels of declination, and fixing a telescope in the meridian in each zone in succession, to clock the transits of stars. To obtain the absolute right ascension of some one bright star in the zone, to which the other stars might be referred, he proposed, in the case of zones entered by the Sun, to use the method of John Flamsteed: the differences between the RA of the Sun and that of a bright star near the solstitial colure were determined on two occasions when the Sun was at the same declination; these differences were then added (after account was taken of the star's precession in the intervening time) to give the total difference in RA between the Sun's two positions. Since the solstitial colure is at 6 hours of RA, the RA of the Sun on each of the two occasions could be calculated; and a simple addition or subtraction would then yield the RA of the star used for comparison.

Where the zone is not entered by the Sun, Lacaille proposed determining the time of meridian transit of the reference star by the method of 'corresponding altitudes': measure the times of passage of the star through a number of different altitudes as it rises in the east, and through the same altitudes as it descends in the west, then average the results to give the time of transit.

Finally, the altitudes of the principal stars were to be determined and corrected for refraction. Lacaille does not enter into the details of how this was to be done; the table of refractions that he later worked out proved to be infected with the errors of division of his instrument.

Lacaille's programme was not taken up by the other astronomers of the Academy, and Lacaille was left to accomplish what he could on his own. The skies in Paris were so frequently overcast as to make singlehanded completion of the programme impracticable; he was able to determine only the positions of the brighter stars of the northern hemisphere. In 1750, however, he won the Academy's endorsement for an expedition to the Cape of Good Hope, and there, between March 1751 and March 1753, he measured the positions of nearly 10 000 stars of the southern hemisphere, while performing other astronomical and geodetical tasks as well, including a determination of solar parallax carried out jointly with Lalande in Berlin, and the first measurement of an arc of a meridian in the southern hemisphere.

How to achieve an accurate solar theory? Most of the known methods for determining both apogee and equation of centre, Lacaille urged, demanded observations more accurate than it is possible to obtain. Methods less direct and less elegant could be surer in practice. Such is the method for the greatest equation of centre proposed by Flamsteed: determine the longitudes of the Sun in spring and autumn when it is approximately 90° from apogee, then from the angle traversed in the interim subtract the angle of mean motion; the difference is just twice the greatest equation of centre. The method assumes an approximate knowledge of the place of the apogee, and an exact knowledge of the year's length, whence the Sun's mean motion is derived.

By this method Lacaille found for the maximum

equation of centre in 1745 two values averaging to 1° 55′ 44″.9, too large by about 7″.8. In Jacques Cassini's *Tables astronomiques* of 1740 the corresponding error was +13″.9; the errors in other tables were larger.

For the solar apogee, Lacaille proposed a method assuming an exact knowledge of the anomalistic year, the time between two successive passages of the Sun through apogee. The apogee and perigee are the only two points of the orbit such that the Sun requires precisely half the anomalistic period to move from one to the other. Hence, given observed longitudes of the Sun very near apogee and very near perigee, and given also the daily motion of the Sun in longitude when at apogee and at perigee, it is possible by trial and error and repeated applications of 'the rule of three' to calculate the position of the apogee. By this method Lacaille found it to be, in mid-1744, at longitude 98° 31′ 35″. According to the tables of Simon Newcomb ("Tables of the Sun", *Astronomical Papers of the American Ephemeris*, Vol. 6), the correct value at this date was 98° 32′ 54″.96, greater by 1′ 20″. No other existing tables came so close (the error in Cassini's tables was −11′ 1″).

Bradley's formulas for the nutation reached Paris in the summer of 1748, and at once Lacaille set out to revise the solar elements, taking the new corrections into account and using a large number of longitudes he had determined from observed differences in RA between the Sun and Procyon, Arcturus, and Vega. At the same time, in order to determine the annual motion of the apogee, and whatever secular changes there might be in the obliquity of the ecliptic and the length of the tropical year, he undertook a study of the solar observations made by Bernhard Walther in Nuremberg around 1500. Astronomers differed as to whether the obliquity of the ecliptic was subject to a secular diminution. Also, Euler in 1746, accepting Ptolemy's equinoxes of AD 139 and 140, and assuming that the Earth's motion was subject to aethereal resistance, had proposed that the length of the year was gradually decreasing. Again, seventeenth-century astronomers such as Thomas Streete had claimed that the apogee was fixed with respect to the stars; others disputed the claim. The construction of solar tables required adjudication of these issues.

In the case of the obliquity, Lacaille found definite evidence of a secular decrease, although his value for this was only about half the correct value. The tropical year he found to be constant to within a few seconds of time (his mean value agrees, to within a fraction of a second, with Newcomb's value for 1750). As for the motion of the apogee, he found it to be 64″.82 per year with respect to the equinox, about 3″ larger than the correct value.

Walther's observations also enabled Lacaille to determine the epoch of the mean longitude of the Sun on 1 January 1500; he found it to be 289° 25′ 36″. Newcomb's value is 1″.69 greater. (The value for the same constant in Cassini's tables errs by +48″.3.) For the equation of centre in Walther's time, Lacaille obtained a value more than 1′ too small, owing primarily to a faulty assumption about the latitude of Walther's observatory; as a result he concluded that the equation of centre was not subject to secular diminution. The establishment of a 17″ to 18″ decrease per century had to wait upon the theoretical work of J. L. Lagrange in the 1780s.

In June 1750 Lacaille presented to the Paris Academy two memoirs on solar theory, reporting his re-determinations of the solar apogee and eccentricity, his investigation of the lunar inequality in the Sun's apparent motion, and a revision in his value for the tropical year. To begin with, Lacaille discusses the problem of precision. He makes it a rule, he states, never to neglect tenths of arc-seconds, whether in the observations or in the calculations. The resulting solar tables, he estimates, are accurate to 12″ or 15″.

Newton had asserted the existence of a detectable lunar inequality in the Sun's apparent motion in Prop. 13 of Bk III of the *Principia* (second and third editions), but did not supply a value for the inequality. Euler in 1744 was the first to incorporate this inequality in solar tables, putting its value at 15″ sin θ, where θ is the elongation of the Moon from the Sun.

Lacaille undertook to establish the existence of the lunar inequality empirically, using eight pairs of observations of the Sun's longitude taken when the Moon was at first quarter, then at last quarter, or vice versa. The difference between the two longitudes, minus the amount of motion predicted by good solar tables that did not take the lunar inequality into account, should be double the maximum lunar inequality, and positive or negative according as the motion is from last to first quarter or vice versa. Lacaille found the differences to be always in the expected direction, but varying over a range from 3″.7 to 35″.9. He concluded that the existence of the lunar inequality, though not its magnitude, was established.

D'Alembert challenged this conclusion in the third part of his *Recherches sur différens points importans du système du monde* (1756), asserting that the scatter in the results was too wide to permit conclusions to be reliably drawn. Lacaille responded in 1757, correcting a calculational error in his earlier memoir and thus somewhat reducing the scatter. D'Alembert returned to the attack in a memoir read on 14 January 1758; Lacaille replied in a section added to his memoir of 1757; d'Alembert replied to the reply in 1762, after Lacaille's death. One of the points at issue was a supposed 40″ variation in the Sun's equation of centre, which the astronomer Pierre-Charles Le Monnier claimed to have observed between September 1746 and April 1747; d'Alembert accepted Le Monnier's observations, while Lacaille challenged the second one as flawed. With hindsight we can conclude that Lacaille was on the right track. The value for the maximum lunar inequality that he initially adopted in 1750 was 12″; the value that he finally adopted in his tables was 8″.5, still too large by about 2″.

In his second memoir of 1750 Lacaille introduced a new method of determining the apogee and equation of centre simultaneously from a triplet of observations, two taken near equinoxes and one near a solstice. The data yield differences in true anomaly and in mean anomaly. By making different assumptions for the eccentricity and place of the apogee, then applying the rule of three to pairs of these assumptions, Lacaille was able to select values for both constants such that they would imply the given differences in true and mean anomaly. He repeated this procedure for eleven triplets formed from nine observations, then took the averages for his final result.

On 29 January 1755 Lacaille presented a supplement to his two memoirs of 1750, in which he once more determined the solar apogee, eccentricity, and epoch of mean longitude, this time from observations of the Sun made at the Cape of Good Hope – observations superior in accuracy, he believed, to those made in Paris.

The entire procedure had to be gone through still one more time, when it became evident that planetary perturbations needed to be taken into account.

It was Euler who first suggested that planetary perturbations were likely to be important in solar theory. Originally he had thought otherwise, claiming in a memoir of 1744 that the astronomical tables of the primary planets, including the Earth, can be constructed on the basis of the exact solution of the two-body problem, just as if they were attracted to the Sun alone. But in studying the inequalities of Jupiter and Saturn in 1747 he came to question this assumption. As he wrote in 1749, when proposing questions for the first prize contest to be sponsored by the St Petersburg Academy of Sciences:

> *Since it is now established beyond doubt that the motion of Saturn, when it approaches Jupiter, is not a little perturbed, it is asked whether the motion of the Earth is not subject to a similar perturbation from the same cause ....*

Although the St Petersburg Academy chose the lunar rather than the solar theory for its first contest, Euler continued to urge the question about solar theory, and through his influence it was made the subject of the contests of the Paris Academy for 1754 and 1756. Euler himself entered the contest of 1756; his essay, which won the prize, was the first systematic derivation of the planetary perturbations in the Earth's motion.

To calculate the perturbations, one must have values for the masses of the perturbing planets. The masses of Jupiter and Saturn can be computed from the periods and distances of satellites – Newton had done this, and Euler accepted his results – but Mars, Venus, and Mercury were without known satellites. Euler had recourse to a conjecture. Newton's values for the masses of Saturn, Jupiter, and the Earth (the latter value being in error by about +97%, owing to a faulty value for solar parallax) suggested to him that the densities of the planets vary reciprocally as the square roots of their periods. This relation, together with the solar distances of the planets and their apparent diameters as determined by Le Monnier, yielded values for the masses. From the masses so determined Euler was able to derive a value for the annual motion of the Earth's aphelion, namely 12″.75 with respect to the stars, which agreed very nearly with Cassini's empirically determined value of the same constant; he therefore considered that his conjecture had been satisfactorily corroborated. In the case of Venus his value differs from the modern value by only +0.7%; in the cases of Mars and Mercury his results are much more in error, but in both these cases he found the perturbations to be very small, as they are in fact.

The perturbations in longitude that Euler derived are of the form $A \sin \theta + B \sin 2\theta$, where θ is the difference in mean longitude between the perturbing planet and the Sun. For all the perturbing planets, Euler's values for A and B are surprisingly good; only in one case does the error, as determined from a comparison with Newcomb's tables, rise as high as 1″.4.

Lacaille, however, did not use Euler's values, but delayed revision of his solar elements until Clairaut had completed his own derivation of these same perturbations. Clairaut, in undertaking the derivation anew, desired to show that his method was of easier application than Euler's; he was also eager to compare the theory with the large number of accurate solar observations that Lacaille had accumulated. Of the planetary perturbations, he dealt only with those due to Jupiter and Venus. Unlike Euler, he took account of the eccentricity of the Earth's orbit, so arriving at formulas with three or four terms each. Unfortunately, the later terms of his formula for the lunar inequality were mistaken, and so was one of the terms in his formula for the Jovian inequality. In the case of both the lunar and Venusian inequalities Clairaut suggested determining the absolute values of the coefficients by a comparison with observations (the formulas were thus proposed as giving only the relative values of the coefficients).

With Clairaut's results in hand, Lacaille proceeded to refine the elements of his solar theory. His calculations were presented in a memoir of 1757. First, he selected observations of the Sun's longitude when the lunar inequality was zero, and comparing these with his provisional solar theory, concluded that the coefficients for the Venusian inequality should be reduced to 82.4% of the values Clairaut had given them (the correct percentage would have been 48.4%). Also, he revised Clairaut's coefficients for the lunar inequality on the basis of the value of the Moon:Earth mass-ratio derived by d'Alembert in his treatise on the precession and nutation, and so arrived at the formula $7''.7 \sin \theta + 3''.5 \sin \theta \sin M$, where θ represents the difference in mean longitude between the Moon and the Sun, and M the mean anomaly of the Sun.

Table 18.1

	Lacaille (L)	Newcomb (N)	L − N
Equation of centre	1°55′ 31″.6	1°55′ 36″.3	−4″.7
Apogee, 1 Jan. 1750	98°38′ 4″	98°38′ 35″.2	−31″.2
Epoch, 1 Jan. 1750	280° 0′ 43″.4	280° 0′ 37″.3	+6″.1

Next, from 22 empirically determined longitudes of the Sun (each based on a series of observations taken on successive days), Lacaille subtracted out the lunar, Jovian, and Venusian perturbations, so as to obtain the longitudes the Sun would have had if these inequalities were absent. From the 22 longitudes he then formed 34 triplets, and for each triplet determined the equation of centre, apogee, and epoch of mean longitude by the method of his memoir of 1750. His final averages, compared with Newcomb's values for the same constants, are shown in Table 18.1.

Finally, Lacaille compared the resulting theory with 144 observed longitudes of the Sun. Aside from six observations that differed from the theory by more than 15″ (five of them by more than 27″), and which Lacaille discarded as defective, he found the average deviation between theory and observation to be 5″.4. A comparison between Lacaille's tables and Newcomb's for the 22 dates of Lacaille's fundamental observations yields a standard error of 6″.6; for the tables of Edmond Halley and Jacques Cassini, by contrast, the corresponding figures are 56″.1 and 32″.2. Such was the increment in accuracy that Lacaille achieved.

The observational basis of Lacaille's solar theory was published in his *Astronomiae fundamenta* of 1757 (Figure 18.1), "a work of many vigils", as he described it, giving the positions of 400 bright stars and the previously mentioned 144 places of the Sun in the ecliptic. This, and the *Tabulae solares* of 1758 (Figure 18.2) as well, were printed in limited editions at the author's own expense, and distributed to the chief astronomers and academies of science in Europe.

The reception of Lacaille's solar tables, at least by his colleagues in Paris, was mixed. Cassini's son complained that Lacaille should have confined

ASTRONOMIÆ
FUNDAMENTA
NOVISSIMIS
SOLIS ET STELLARUM
OBSERVATIONIBUS
STABILITA
LUTETIÆ IN COLLEGIO MAZARINÆO
ET IN AFRICA AD CAPUT BONÆ SPEI
PERACTIS

A Nicolao-Ludovico DE LA CAILLE, in almâ Studiorum Univerſitate Pariſienſi Matheſeon Profeſſore, Regiæ Scientiarum Academiæ Aſtronomo, & earum quæ Petropoli, Berolini, Holmiæ & Bononiæ florent, Academiarum Socio.

PARISIIS;
E Typographiâ J. J. Stephani COLLOMBAT, Typographi ordinarii Regis.

M. DCC. LVII.

18.1. The title-page of Lacaille's *Astronomiae fundamenta* of 1757.

himself to correcting the Cassinian solar tables. D'Alembert maintained an attitude of superior scepticism, and Le Monnier, attached to the project of correcting Flamsteed's solar and lunar tables by means of observations, continued to insist that "neither the epochs nor the several other elements of the theory of the Sun are as yet well known to us". Lalande, however, defended the new tables, and when he became editor of the *Connaissance des temps* in 1760, adopted them as the basis for the annual ephemerides.

Mayer, on examining Lacaille's solar tables "found that they very nearly agreed with the many and careful observations made by myself from the year 1756 forward with an excellent mural quadrant. Wherefore it did not seem necessary ... to

TABULÆ SOLARES

Quas è noviſſimis ſuis Obſervationibus deduxit N. L. DE LA CAILLE, *in almâ Studiorum Univerſitate Pariſienſi Matheſeon Profeſſor, Regiæ Scientiarum Academiæ Aſtronomus, & earum quæ Petropoli, Berolini, Holmiæ, Bononiæ & Gottingæ florent, Academiarum Socio.*

PARISIIS,

E Typographia H.L. GUERIN & L.F. DELATOUR.

M. DCC. LVIII.

18.2. The title-page of Lacaille's *Tabulae solares* of 1758.

construct solar tables entirely new, but only ... to correct his tables as far as my observations seemed to require." Mayer left Lacaille's equation of centre unaltered, revised the epoch of mean longitude and the apogee (removing 6″ of error from the former and adding 30″ of error to the latter), and introduced his own tables for the perturbations, which were a good deal less in error than those that Lacaille had derived from Clairaut's formulas. Mayer's version of Lacaille's tables became the basis for the ephemerides of the British *Nautical Almanac* from 1767, and for those of the *Connaissance des temps* from 1779.

The solar theory would not be revised again until the 1780s, when Delambre, inspired by the example of Lacaille, and with the observations of the British Astronomer Royal Nevil Maskelyne at his disposal, undertook to re-do the determination of elements once more from the beginning.

The lunar tables of Mayer

Mayer's investigation of the Moon's motion arose out of his work in cartography, for which more accurate determinations of terrestrial longitudes were greatly in demand.

Mayer, who grew up in Esslingen, showed early talent for drawing and a fascination with practical geometry and mathematical analysis. His first book, published when he turned 18, dealt with the analytic solution of geometrical problems. A second work, entitled *Mathematischer Atlas*, appeared in Augsburg in 1745; in sixty coloured plates it surveyed the range of elementary mathematics, with astronomical, geographical, and other applications.

In 1746 Mayer joined the Homann Cartographic Bureau in Nuremberg. In constructing a map to illustrate the course of a partial lunar eclipse due to occur in August 1748, he discovered that the method of orthographic projection hitherto used in such constructions was inaccurate, because of its neglect of lunar parallax. Also, he found that the best of the existing lunar maps – those of the seventeenth-century observers Giambattista Riccioli and Johann Hevelius – were unreliable.

At the time, lunar eclipses were a principal means for determining differences in terrestrial longitude: one had only to compare the local times at which the same topographical features on the Moon's surface were obscured by (or emerged from) the Earth's shadow, as estimated by observers situated on different meridians of longitude.

One outcome was that Mayer undertook to lay the foundations for an accurate lunar map. Between April 1748 and August 1749, using a telescope fitted with a glass micrometer of his own design, he made a series of measurements of both the Moon's diameter and the positions of various lunar markings, in terms of their relative distances from the centre of the disk and the angle made at the centre with respect to the north–south line.

The task of obtaining an accurate lunar map was complicated by the lunar libration: some 18% of the lunar surface is alternately visible and invisible, owing to (a) the inclination of the Moon's equator to the plane of its orbit, (b) the non-uniformity of its orbital motion contrasted with the uniformity of its axial rotation, and (c) the displacement of the terrestrial observer from the Earth's centre. It was therefore essential to fix a coordinate system in the body of the Moon itself, independent of the position of the terrestrial observer.

For this purpose Mayer used 27 observations of the distance and angle of the crater Manilius from the apparent centre of the lunar disk. From these he derived 27 'equations of condition' involving three unknowns: the inclination of the Moon's equator to the ecliptic, the longitude of the intersection of these two circles, and the 'geographic latitude' of Manilius with respect to the Moon's equator. He next separated the equations into three groups of nine each, and added the equations of each group together to form a single equation; the three resulting equations were then solved for the three unknowns. In separating the original equations into three groups, Mayer was guided by the idea of maximizing the coefficient of a given unknown in each group.

Essentially the same method had been described by Euler in his 1748 prize essay *Recherches sur la question des inégalités du mouvement de Saturne et de Jupiter* (which was published in Paris as a book the following year), as a way of determining empirical corrections to orbital elements and the coefficients of perturbational terms; and since Mayer, in a letter to Euler of 4 July 1751, acknowledged his familiarity with this work, it is likely that he derived the method from Euler's memoir. Euler's application of the method had not been particularly successful, because of the inadequacy of his theory of Saturn – its failure to include certain large perturbational terms first discovered in 1785 by P. S. Laplace. Thus it was Mayer who made the first clearly successful use of the method of equations of condition. His results were published in a memoir of 1750, "Abhandlung über die Umwälzung des Monds um seine Axe".

These results substantiated and refined the theory of G. D. Cassini (Cassini I) on the Moon's libration. Lagrange made use of them in his prize-winning memoir of 1764 on the lunar libration. Lalande in the second edition of his *Astronomie* (1771) reviewed Mayer's procedure as a method generally applicable to the combination of large numbers of observational equations, and Laplace made use of it in his epoch-making "Théorie de Jupiter et de Saturne" (1788). Thereafter this 'method of equations of condition' was widely applied by astronomers to the fitting of planetary and lunar theories to observations, until the superiority of the method of least squares came to be generally recognized (see Chapter 27).

As a means to the determination of differences in geographical longitude, lunar eclipses left much to be desired. They were infrequent, and the edge of the Earth's umbra did not have the sharpness that would permit a very precise fixing of the instant at which a given feature of the lunar surface was eclipsed. The occultation of a star by the Moon, in contrast, could be timed to within a second with the aid of a good telescope, and several such occultations occurred each month. During 1747 and 1748 Mayer observed occultations of the Pleiades and a number of other zodiacal stars. For the timing of immersion and emersion, he utilized not only the directly observed times of these events, but also a number of measured distances between the star and the Moon's limb at times before immersion and after emersion, and applied these multiple data to the determination of an average result. The suddenness with which stars were occulted convinced him that the Moon had no atmosphere.

Mayer's memoirs describing his micrometer, his determination of the lunar libration, his observations of lunar occultations, and his arguments for denying an atmosphere to the Moon, appeared in 1750 in the *Kosmographische Nachrichten und Sammlungen auf das Jahr 1748*. These publications led to his being called in November 1750 to a professorship at the Georg-August Academy in Göttingen. Here he was to have the use of the Göttingen Observatory, then under construction, and after 1755 to be its sole director until his death in 1762.

On 4 July 1751 Mayer initiated a correspondence with Euler. He had tried several times, but in vain, he told Euler, to deduce the lunar inequalities from Newton's theory, until he met with Euler's memoir on the motions of Saturn and Jupiter, which pointed the way. However, unlike Euler, he retained the mean anomalies of the Moon and Sun as the variables from which the inequalities were to be calculated (Euler had used for this purpose the eccentric anomaly, which falls about midway between the mean anomaly and the true anomaly, and is related to the mean anomaly by Kepler's equation, $E+e \sin E=M$). Mean anomalies are derivable immediately from the time; eccentric anomalies must be obtained from a series expansion or tables constructed for the appropriate eccentricity. Euler in his response pointed out that the perturbations depend on the *true* angular distance between the Moon and Sun, and the attempt to express them in terms of mean anomalies and the differences of mean longitudes inevitably leads to complicated formulas. The point was not lost on Mayer, as we shall see.

In his letter of July 1751, Mayer allowed that the agreement of his theory with observation was less than satisfactory, but expressed disapproval of Clairaut's proposal to emend the Newtonian law of attraction: "there [are] so many circumstances which make the calculation and approximation doubtful, that the error can rather be on the latter side." To obtain more accurate predictions, he proposed using the Saros cycle or period of 223 lunations ($18^y\ 11^d\ 7^h\ 43^m\ 30^s$), by which Halley had earlier sought to provide observational corrections to Newton's lunar theory. Mayer, however, employed his own lunar theory, expressible in ten groups of terms incorporated in the same number of tables. By means of these tables he compared observations separated by the Saros period, and always found his predictions accurate to within 10″ or 20″. "Consequently, if one had a continuing series of observed positions of the Moon throughout 18 years, one could determine the Moon's position very accurately for any future time by means of these tables."

Euler, in replying on 27 July, gave high praise to Mayer's project for predicting the Moon's longitude. He had now succeeded, he informed Mayer, in deducing the whole motion of the lunar apogee from the Newtonian law of attraction. (Clairaut had been the first to succeed in this deduction, as Mayer learned at about this time from newspapers.) Mayer, responding on 15 November, stated that he had not yet determined accurately enough the coefficients of a number of equations, but hoped in repeating the calculation to obtain everything more precisely. However, he urged publication of Euler's and Clairaut's theories, which would make this labour superfluous. In passing he noted that, since the nutation of the Earth's axis is caused by the attraction of the Moon on the spheroidal Earth, there must be reciprocal inequalities in the Moon's motion due to the Earth's shape.

During the second half of 1751 Mayer continued to work on the lunar theory. In a letter to Euler of 6 January 1752 he gave a list of 23 terms for the Moon's inequalities in longitude. "Among these many inequalities there are some which I can completely determine, and some which I must correct, through observations. I have completely discarded still more on account of their smallness." Euler, in his reply of 18 March 1752, compared the terms of Mayer's theory with the corresponding terms in Clairaut's theory (Euler's own theory was not directly comparable, because of his use of the eccentric anomaly as independent variable). Both theories contained considerable error at this stage.

Mayer continued to work on his derivation. In January 1753 he was able to announce to Euler that, although he had still not been able to derive the total observed motion of the Moon's apogee,

I now have the equations of longitude as accurately as I have wished. There are no more than 13 of them, which I have arranged in tables, and through which I shall be able to calculate the longitude of the Moon so correctly, that the error never amounts to 2 minutes.

The equations that Mayer went on to list are, except for a few slight changes in coefficients, the same as those given in his "Novae tabulae motuum solis et lunae", published in Vol. 2 of the *Commentarii* of the Royal Society of Sciences of Göttingen in the spring of 1753. A number of the tables incorporate more than one sinusoidal term. But what chiefly enabled Mayer to reduce the number of tables to thirteen, aside from the neglect of small terms, was a shift in strategy. This was first explained and justified in Mayer's *Theoria lunae juxta systema Newtonianum*, sent (as we shall see) to the British Admiralty in late 1754 but not printed until 1767.

The *Theoria lunae* is a systematic derivation of the lunar inequalities from the inverse-square law. Like Euler, Mayer develops formulas relating mean and true anomaly to eccentric anomaly; unlike Euler, he then eliminates the eccentric anomaly so as to yield true anomaly in terms of mean anomaly. The resulting formula contained 46 sinusoidal terms. Mayer commented:

Although this formula ... can easily be reduced to tables entirely similar in form to those of the celebrated Clairaut; yet the huge number of inequalities and tables thence derived would make the calculation of the Moon's place so laborious and troublesome that even the most patient calculator would give in to weariness .... I therefore sought a way by which the inequalities comprised in this formula could be reduced in number.

The procedure Mayer now adopted entailed progressive correction of the independent arguments.

(1) First, the Moon's mean longitude was corrected for the secular acceleration discovered by Halley. In his tables of 1753 Mayer gave the resulting increase in longitude per century as 7″; later he revised this figure to 9″.

(2) Next, the Moon's mean longitude and the mean longitude of its apogee and node were corrected for the annual inequalities to which they were subject, depending on the changing distance of the Sun from the Earth during the course of the year. According to Mayer's tables of 1753, their maxima were 11′ 20″, 20′ 36″, and 10′ 18″ respectively. The corrected mean longitude of the Moon less its corrected mean apogee yields the corrected mean anomaly.

(3) The Moon's mean longitude was then corrected for seven inequalities dependent on certain linear combinations of the corrected mean anomaly of the Moon, the mean anomaly of the Sun, and twice the difference in mean longitude of the Moon and Sun. The largest of these in Mayer's tables of 1753 came at maximum to 1′ 48″.

(4) The Moon's mean longitude was then further corrected for two inequalities, dependent on the distances of the Moon's apogee and node, respectively, from the *true* longitude of the Sun. In the tables of 1753 the maximum of the first of these was 3′ 45″; of the second, 44″.

(5) With the mean anomaly of the Moon resulting from all the preceding corrections, Mayer now calculated the Moon's equation of centre, amounting at maximum to about $6\frac{1}{3}$ degrees. The Moon's corrected apogee (step 1) plus the equation of centre (step 4) yielded the "equated longitude of the Moon".

(6) In his tables of 1753 Mayer next computed the 'evection', dependent on twice the corrected distance between the Moon and Sun minus the corrected mean anomaly of the Moon, and amounting at maximum to 1° 20′ 43″. In his later tables he found it possible to combine the calculation of this inequality with that of a similar inequality at stage 2.

(7) Finally, Mayer calculated the Tychonic 'variation', depending on the equated longitude of the Moon minus the true longitude of the Sun. This is a further correction to the Moon's longitude, amounting at maximum to about 37′.

In his *Theoria lunae* Mayer showed that if the terms derived from theory, and depending solely on mean arguments, were transformed so as to depend on the progressively modified arguments employed in the foregoing sequence of steps, then many of the resulting terms – those that he left out of account in his tables – were so small as to be safely neglected. This theoretical justification, however, was not made public till long after the tables of 1753 were published.

In the preface to his tables of 1753, Mayer made a large claim for their accuracy. In comparisons with more than 200 observations of his own and the preceding century, he claimed to have found scarcely ten observations that differed from the predictions of his tables by as much as $\frac{1}{2}'$, and none differing by as much as 2′. By contrast, he noted, the tables of Euler and Clairaut were subject to errors (as their authors conceded) of as much as 4′ or 5′.

D'Alembert, in remarks appended to his *Théorie de la lune* (1754), doubted that Mayer's tables were as accurate as claimed. The manner in which Mayer had derived the formulas underlying his tables was left unclear, and the tables failed to take account of terms that d'Alembert believed to be sizeable.

In nothing Mayer published did he provide a description of his manner of fitting the lunar theory to observations. In the introduction to the tables of 1753 he claims to have spared no labour in his investigation of the mean motions and the secular acceleration of the Moon. Values for these constants must be so chosen as to fit both the less reliable ancient observations and the more accurate modern ones, each with an appropriate degree of precision. In determining the secular acceleration Mayer made use of the Saros cycle, and he also employed two solar eclipses observed by Ibn Yūnus on 13 December AD 977 and 8 June AD 978. These eclipses were especially precious because Ibn Yūnus recorded the altitudes of the Sun on the two occasions, so that the precise times of the eclipses could be calculated. Independently of Mayer, Richard Dunthorne in the *Philosophical Transactions* for 1749 had used these same observations in the same calculation, obtaining 10″ as the change in mean longitude per century, as compared with Mayer's initial value of 7″.

The tables that Mayer published in 1753 turned out to be only a first version. In December 1754, with Euler's encouragement, Mayer entered the competition for one of the prizes offered by the British Parliament: £20 000 for a method giving the longitude at sea to within 30 nautical miles, £10 000 for a method good to within 60 nautical miles. The lunar tables that Mayer submitted to the British Admiralty in late 1754 differed little in form from the published tables of 1753, but contained revised numerical coefficients. Mayer's last revision of the tables was sent to the Admiralty in 1763 by his widow.

We do not know how Mayer computed these revisions. In the preface to his *Theoria lunae*, he states:

In exhibiting here the way in which I investigated the inequalities in the motion of the Moon on the basis of theory, I am not aiming to demonstrate the reliability and truth of my lunar tables. For the theory has this inconvenience, that many of the inequalities cannot be deduced from it accurately, unless one should pursue the calculation – in which I have now exhausted nearly all my patience – much further. My aim is rather to show that at least no argument against the goodness of my tables can be drawn from the theory. This is most evidently gathered from the fact that the inequalities found in the tables, which have been corrected by comparison with many observations, never differ from those that the theory alone supplies by more than $\frac{1}{2}'$. Therefore . . . it is evident that these small errors derive rather from the side of the theoretical calculus than from the tables. This is put beyond all doubt if the numerical results deduced by others from the same *theory – and especially those of the celebrated Euler, Clairaut, and d'Alembert – are compared either with my tables or my theoretical calculation. They sometimes differ by 3' or more; nor do they agree better with each other, except in a few places ....*

Later in the same treatise Mayer states that, besides the terms that can be evaluated only on the basis of observation, such as those depending on the orbital eccentricity or on solar parallax, there are terms whose derivation from the theory is so arduous that they are best evaluated observationally; and he adds that he has corrected all terms of the theory by adding or subtracting a few seconds of arc, to bring about better agreement with the heavens. For so massive an operation of correction, it is plausible to assume that Mayer made use of equations of condition – the only then known statistical procedure for determining a large number of unknowns simultaneously, and one which Mayer alone had previously used with success.

In Vol.3 of the *Commentarii* of the Royal Society of Sciences of Göttingen Mayer published a comparison between his tables of 1753 and some 55 lunar eclipses observed telescopically since 1612, and 139 lunar longitudes observed by Bradley from 1743 to 1745 (Figure 18.3). In the latter group, the average deviation between the tables and the observations was 27", the maximum being 1' 37".

When the Göttingen observatory became ready for use in the summer of 1753, Mayer began his own series of observations, determining the Moon's position through occultations of fixed stars – first through occultations of a single fixed star, so that any error in the position assumed for this star would be reflected only in the epoch of mean motion of the Moon, which could later be corrected through observation of a solar eclipse. In this manner Mayer found considerable errors in the then accepted positions for a number of zodiacal stars, and at the same time obtained precise corrections for the numerical coefficients of his lunar theory.

By 1760 Bradley had completed a laborious comparison of Mayer's tables with some 1100 observations made at Greenwich, and so confirmed that the tables were generally reliable to within 1'.25 – an accuracy not quite high enough to justify the maximum Parliamentary award, but certainly sufficient to merit the lesser bounty of £10000 offered for a method both "useful and practicable" for determining longitude at sea to within 60 nautical miles. A test of the method at sea was postponed, however, because of the British navy's involvement in the Seven Years War, and in the interim John Harrison completed his fourth marine chronometer, which thus came into rivalry with Mayer's lunar method for the longitude.

Although the revised tables that Mayer's widow submitted to the Board of Longitude in August 1763 were tested on a voyage to Barbados, they were not considered in the final recommendation of the Board of Longitude, made on 9 February 1765. Harrison was to receive £10000 for a method that gave the longitude to within 30 nautical miles; Mayer's heirs were to receive a prize "not exceeding £5000". In May 1765, before Parliament made its final allocation of the prizes, *Gentleman's Magazine* published a letter by Clairaut, in which he protested that he and Euler had developed lunar theories more rigorous than Mayer's, and that Mayer's tables derived their accuracy primarily from his skilful discussion of observational data. Parliament, a few days later, awarded only £3000 to Mayer's heirs, along with £300 to a surprised Euler, as the supposed author of the theory on which Mayer's tables were based.

No doubt the accuracy of Mayer's tables *was* primarily due to the skilful fitting of theory to observational data, but Mayer's artfulness in such empirical determinations of numerical coefficients was new and unmatched among his contemporaries. The lunar tables of Johann Tobias Bürg and Johan Karl Burckhardt, which appeared in 1806 and 1812 respectively, would still depend on the empirical determination of perturbational coefficients by the same method of equations of condition that Mayer probably used. Only after 1820 was the theoretical derivation carried out in sufficient detail to begin to yield observationally satisfactory values of these numbers.

Another consequence of Mayer's careful attention to the accord of theory and observation may be mentioned: his refutation of Euler's hypothesis that the Earth's motion about the Sun was detectably accelerating. Euler had first argued for this hypothesis in his *Opuscula* of 1746, and had repeated the same views in a letter to Caspar Wetstein, published in the *Philosophical Transactions* for 1752.

139. LOCA LUNAE EX OBSERVATIONIBUS CEL. D. BRADLEYI. 393

1743.	temp med. Grenovic. ftyl. vet.	Longitudo Lunae vera obfervata.		Different. calculi ab obferv.
M. d.	h. ′ ″	S ° ′ ″		′ ″
Sept. 11.	3. 20. 39	7. 24. 10. 57		+ 0. 40
12	4. 8. 26	8. 6. 43. 28		+ 0. 56
14	5. 48. 40	9. 1. 22. 43		+ 0. 14
16	7. 30. 36	9. 26. 1. 38		— 0. 29
17	8. 20. 5	10. 8. 33. 15		— 0. 36
18	9. 7. 52	10. 21. 18. 12		— 0. 36
20	10. 39. 10	11. 17. 39. 19		— 0. 15
22	12. 11. 36	0. 15. 15. 0		+ 0. 24
24	13. 49. 19	1. 13. 54. 50		— 0. 2
26	15. 41. 50	2. 13. 18. 12		— 0. 51
28	17. 46. 1	3. 12. 57. 28		— 0. 27
Octob. 14	6. 11. 30	10. 3. 12. 28		+ 0. 13
15	6. 59. 24	10. 15. 41. 29		— 0. 22
19	9. 59. 38	0. 8. 34. 14		— 0. 10
21	11. 35. 57	1. 7. 21. 57		0. 0
24	14. 33. 17	2. 22. 45. 57		— 0. 7
26	16. 41. 9	3. 23. 8. 7		— 0. 11
27	17. 40. 52	4. 7. 59. 7		+ 0. 20
28	18. 35. 48	4. 22. 31. 41		+ 0. 49
Nov. 9	3. 14. 33	9. 16. 6. 0		+ 0. 2
13	6. 22. 48	11. 5. 38. 34		— 0. 11
17	9. 21. 24	0. 29. 42. 59		+ 0. 31
18	10. 12. 50	1. 14. 27. 13		+ 0. 21
20	12. 14. 8	2. 15. 15. 59		+ 0. 49
21	13. 20. 15	3. 1. 3. 15		+ 0. 30
22	14. 27. 2	3. 16. 49. 17		+ 0. 15
27	18. 56. 42	6. 0. 12. 18		+ 0. 54
Dec. 8	2. 48. 6	10. 6. 16. 52		+ 0. 7
10	4. 18. 41	11. 0. 56. 32		— 0. 5
11	5. 1. 20	11. 13. 28. 13		— 0. 2
12	5. 43. 18	11. 26. 14. 1		+ 0. 3
13	6. 25. 48	0. 9. 20. 15		+ 0. 1
19	12. 1. 32	3. 8. 18. 51		+ 0. 24
20	13. 10. 21	3. 24. 24. 10		+ 0. 32
23	16. 4. 27	5. 10. 48. 4		0. 0
26	18. 21. 40	6. 22. 30. 40		+ 0. 13
27	19. 5. 30	7. 5. 31. 27		+ 0. 10

1744.

Ddd 2

18.3. The first of four pages in which, for various dates between September 1743 and April 1745, Mayer gives the longitude of the Moon as observed by Bradley at Greenwich, and the difference between this observed longitude and that calculated by himself. Note that on this page the error never reaches 1′ (and on the other pages exceeds 1′ only rarely). (From Vol. 3 (1754) of the *Commentarii* of the Royal Society of Sciences of Göttingen.)

Writing to Euler on 22 August 1753 Mayer cited empirical evidence for assuming the tropical year to be essentially constant. Euler was misled by the equinoxes that Ptolemy claimed to have observed in AD 139 and 140: these were surely in error. Ptolemy had constructed his planetary tables before 139, and such construction presupposed solar tables. He presumably borrowed the solar tables of Hipparchus, thus putting the length of the year at $365^d\ 5^h\ 55^m$ – so large, Mayer thought, that it was probably deduced from the luni-solar cycle of 76 years, rather than from actual observations.

This year is now too large by approximately $6\frac{1}{2}$ minutes, which in the 300 years which intervened between Hipparchus and Ptolemy, amount to approximately $1\frac{1}{4}$ day .... And the Ptolemaic equinoxes are found to be too late by just about this much. It can be that Ptolemy perceived this error of his solar tables in his observations of the equinoxes, ... but preferred to *discard his observations rather than attempt to revise his system from the ground up.*

Reluctantly Euler gave up his belief in the Earth's secular acceleration. Light, in his view, was a wave-motion, requiring an aether for its propagation; and the aether must offer resistance to planetary motion, causing planets to fall toward the Sun and so to increase their angular motion. (Even if, as the Newtonians claimed, light consisted of corpuscular projectiles, these corpuscles, Euler argued, having inertia would cause resistance to planetary motion.) At least, he wrote to Mayer, the well-confirmed secular acceleration of the Moon could be attributed to aethereal resistance.

Mayer replied (on 25 November 1753) that it might also be due to a long-term inequality. In addition, he informed Euler that ancient observations of equinoxes and eclipses indicated a secular decrease in the Earth's orbital eccentricity, which likewise could be due to a long-term inequality. The correctness of both suggestions would be confirmed in the work of Lagrange and Laplace (see Chapters 21 and 22).

Further reading

Eric G. Forbes, *Tobias Mayer (1723–1762): Pioneer of Enlightened Science in Germany* (Gottingen, 1980)

Eric G. Forbes, *The Euler–Mayer Correspondence (1751–1755)* (New York, 1971)

E. Doublet, *Le Bicentenaire de l'Abbé de la Caille* (Bordeaux, 1914)

Curtis Wilson, Perturbations and solar tables from Lacaille to Delambre: the rapprochement of observation and theory, Part I, *Archive for History of Exact Sciences*, vol. 22 (1980), 53–188

19

Predicting the mid-eighteenth-century return of Halley's Comet

CRAIG B. WAFF

with an appendix on Clairaut's calculation by Curtis Wilson

In his lecture on 25 April 1759 before the public assembly of the Paris Academy of Sciences, the astronomer Joseph-Jérôme Lefrançais de Lalande (1732–1807) opened with the declaration: "The Universe sees this year the most satisfying phenomenon that Astronomy has ever offered us; unique event up to this day, it changes our doubts into certainty, & our hypotheses into demonstrations." He was speaking of the year's outstanding astronomical event – the apparition of a comet fulfilling the bold prediction of Edmond Halley (1656–1742) that one would appear in late 1758 or early 1759 with orbital elements similar to those of comets observed in 1682, 1607, 1531, and possibly before. Halley, who had laboriously calculated the elements of these comets and had recognized their similarity, had speculated that the comets seen in those earlier years were one and the same and that, if so, this celestial object travelled in a highly elliptical retrograde orbit around the Sun with a period of some 75 or 76 years.

Halley's prediction of a return appearance of a comet was without precedent. Moreover, his suggestion that at least some comets might travel in closed elliptical orbits highlighted the possibility that the motions of these celestial objects, like those of the planets, were governed by the gravitational force posited by Isaac Newton. Although by the late 1750s various eighteenth-century mathematicians had already confirmed that Newton's gravitational theory could explain a wide variety of phenomena (see Chapters 16 and 17), Lalande and other contemporary commentators argued that establishing the periodicity of repeated returns of comets, which are not visible to earthbound observers throughout most of their orbits and whose exact nature was still mysterious, would provide a particularly striking vindication of the theory. Locating the expected comet and verifying that it had orbital elements similar to those of the comets of 1682, 1607, and 1531 thus became a major objective of astronomers and mathematicians in the mid-eighteenth century.

Such a task was no easy matter. Astronomers could most clearly show that a particular comet appearing in the late 1750s was probably identical to those that had appeared in the aforementioned years if they could observe it over a sufficiently long time to permit computation of its orbital elements. They therefore wished to recover the expected comet at the earliest possible moment. The problem for astronomers was deciding when and where in the sky the comet would most likely appear. The theoretical first visibility of the comet would depend on its intrinsic brightness, the distance between it and the Earth, and the magnifying power of an astronomer's telescope. The comet's location in the sky at the time of theoretical first visibility and thereafter would depend on the relative positions of it and the Earth from moment to moment in their respective orbits. Searching the entire sky was out of the question. To select a region in which the comet might possibly appear at any particular time, an astronomer had to make an assumption as to where it was in its orbit at that particular time.

The period of revolution of this supposedly recurring comet was variable by as much as a year, however, as Halley himself recognized in making his prediction. (By contrast, the periods of revolution for each of the planets never varied by more than a few days.) Such a situation obviously rendered uncertain when the comet would reach its closest approach to the Sun (perihelion) or any other specific location in its orbit (including the point of theoretical first visibility). Halley attributed the unevenness of the periods to the possibility that the motion of the comet had been perturbed when it passed near the giant outer planet Jupiter during its repeated journeys to and from the outer regions of the solar system. If so, any accurate prediction of the comet's return to perihelion would require a detailed calculation, based on gravitational theory, of the specific perturbational influences of Jupiter (and possibly other planets) on the comet's motion. Halley himself made no such calculation, but the French mathematician Alexis-Claude Clairaut (1713–65), who had already made significant contributions to the application of gravitational theory to the figure of the Earth and the motion of the Moon (Chapters 15 and 16), would undertake this challenging task together with several colleagues.

Even before Clairaut publicly announced a predicted date of perihelion, however, astronomers and others had devised a variety of means for guiding the search for the expected comet. Some investigators, like Clairaut, would predict a specific date of perihelion. Others confined themselves to suggesting specific times of the year and areas of the sky in which the comet would most likely appear.

Halley's *Synopsis astronomiae cometicae*

In the first edition of his *Principia* (1687), Newton had demonstrated (Prop. 40, Bk III) that comets, like planets, must move in conic sections with foci in the middle of the Sun. He thus foresaw the possibility of a closed elliptical orbit for some comets, but he offered no examples. It was Halley who first undertook the tasks of calculating, by geometrical construction, orbital elements for twenty comets that he considered the best observed, and of examining these elements for similarities that might indicate that a comet had appeared more than once to earthbound observers and that it consequently travelled in a closed elliptical orbit around the Sun.

Although Halley himself had observed several comets in the 1680s, including the one that appeared in 1682, his first intense period of investigation of cometary motions appears to have begun around August 1685, when he visited Newton at Cambridge to determine for the latter the orbits of several comets, including those seen in 1664–65, 1680–81, 1682, and 1683. Even before he undertook detailed calculations of the comet of 1682, Halley perceived features in the motion of this comet that it shared with those seen in 1531 and 1607. Perhaps the most prominent similarity was the retrograde direction of motion, that is, it moved opposite to the direction (counterclockwise to a hypothetical observer stationed 'north' of the solar system) of the motions of Earth and the other planets around the Sun. On 28 September 1695, Halley informed Newton that he was "more and more confirmed that we have seen that Comett now three times, since ye yeare 1531". Further calculations revealed additional similarities in the locations of perihelia, the inclinations of the orbital planes to the plane of the Earth's orbit, the locations of the nodes (points of intersection of the cometary orbits with the plane of the Earth's orbit), and the distances of the comets from the Sun at the times of perihelion. These similarities, plus the fact that nearly equal periods of time had elapsed between the apparitions of 1531 and 1607, and between those of 1607 and 1682, convinced Halley that one and the same comet had made successive appearances in the inner solar system at a regular interval (about 75 years) – a result that he announced to the Royal Society of London on 3 June 1696.

His appointment as Deputy Comptroller of the Chester Mint (1695–98) and several sea voyages (1698–1701) interrupted Halley's further work on comets over the next years, so that it was not until 1705 that he published a summary of his investigations of cometary motions. The principal feature of Halley's *Synopsis astronomiae cometicae* was a table (Figure 19.1) listing orbital elements that he had calculated for twenty-four well observed comets seen since 1337. Near the end of his essay Halley remarked that he had considered the orbits of all these comets as parabolic. The reason for doing so, he acknowledged, was that such orbits were easier to calculate than hyperbolic or elliptical ones. If all cometary orbits were parabolic in reality, however, astronomers could have no hope of ever seeing a comet return. Halley justified his approach by arguing that the frequency of cometary appearances, plus the fact that none had ever been observed to move with a hyperbolic motion, made it probable that they traversed highly eccentric elliptical orbits, whose orbital elements would differ little from the parabolic ones that he had calculated.

In his *Synopsis*, Halley declared himself con-

(1886)

Cometarum Omnium hactenus rite Observatorum, Motuum in Orbe Parabolico Elementa Astronomica.

Cometæ Anni.	Nodus Ascend. gr. ′ ″	Inclin. Orbitæ. gr. ′ ″	Perihelion in Orbe. gr. ′ ″	Perihelion in Ecliptica gr. ′ ″	Latitudo Perihelii gr. ′ ″	Distantia Perihelia à Sole.	Log. dist. Periheliæ à Sole.	Temp. æquat. Perihelii Londini. die. h. ′
1337	♊ 24. 21. 0	32. 11. 0	♉ 7. 59. 0	♉ 12. 45. 15	22. 40. 30 B	40666	9. 609236	Junii 2. 6. 25
1472	♑ 11. 46. 20	5. 20. 0	♉ 15. 33. 30	♉ 15. 40. 20	4. 25. 50 A	54273	9. 734584	Feb. 28. 22. 23
1531	♉ 19. 25. 0	17. 56. 0	♒ 1. 39. 0	♒ 0. 48. 15	17. 3. 05 B	56700	9. 753583	Aug. 24. 21. 18½
1532	♊ 20. 27. 0	32. 36. 0	♋ 21. 7. 0	♋ 16. 59. 40	15. 57. 00 B	50910	9. 706803	Oct. 19. 22. 12
1556	♍ 25. 42. 0	32. 6. 30	♑ 8. 50. 0	♑ 11. 0. 00	31. 10. 20 B	46390	9. 666424	Apr. 21. 20. 3
1577	♈ 25. 52. 0	74. 32. 45	♌ 9. 22. 0	♍ 7. 53. 00	69. 35. 20 A	18342	9. 263447	Octo. 26. 18. 45
1580	♈ 18. 57. 20	64. 40. 0	♋ 19. 5. 50	♋ 19. 17. 10	64. 40. 0 B	59628	9. 775450	Nov. 28. 15. 00
1585	♉ 7. 42. 30	6. 4. 0	♈ 8. 51. 0	♈ 8. 59. 10	2. 55. 25 A	109358	0. 038850	Sept 27. 19. 20
1590	♍ 15. 30. 40	29. 40. 40	♏ 6. 54. 30	♏ 2. 55. 50	22. 45. 50 A	57661	9. 760882	Jan. 29. 3. 45
1596	♒ 12. 12. 30	55. 12. 0	♏ 18. 16. 0	♏ 22. 44. 35	54. 44. 30 B	51293	9. 710058	Julii 31. 19. 55
1607	♉ 20. 21. 0	17. 2. 0	♒ 2. 16. 0	♒ 1. 29. 40	16. 10. 5 B	58680	9. 768490	Oct. 16. 3. 50
1618	♊ 16. 1. 0	37. 34. 0	♈ 2. 14. 0	♈ 6. 10. 00	35. 50. 0 A	37975	9. 579498	Oct. 29. 12. 23
1652	♊ 28. 10. 0	79. 28. 0	♈ 28. 18. 40	♊ 10. 41. 35	58. 14. 0 A	84750	9. 928140	Nov. 2. 15. 40
1661	♊ 22. 30. 30	32. 35. 50	♋ 25. 58. 40	♋ 21. 37. 30	17. 17. 0 B	44851	9. 651772	Jan. 16. 23. 41
1664	♊ 21. 14. 0	21. 18. 30	♌ 10. 41. 25	♌ 8. 40. 35	16. 1. 50 A	102575½	0. 011044	Nov. 24. 11. 52
1665	♏ 18. 02. 0	76. 05. 0	♊ 11. 54. 30	♉ 24. 6. 35	23. 8. 0 B	10649	9. 027309	Apr. 14. 5. 15½
1672	♑ 27. 30. 30	83. 22. 10	♉ 16. 59. 30	♋ 9. 26. 00	69. 27. 40 B	69739	9. 843476	Feb. 20. 8. 37
1677	♏ 26. 49. 10	79. 03. 15	♌ 17. 37. 5	♋ 16. 21. 05	75. 44. 10 B	28059	9. 448072	Apr. 26. 00. 37½
1680	♑ 2. 2. 0	60. 56. 0	♐ 22. 39. 30	♐ 27. 26. 50	8. 11. 10 A	00612½	7. 787106	Dec. 8. 00. 6
1682	♉ 21. 16. 30	17. 56. 0	♒ 2. 52. 45	♒ 2. 0. 30	16. 59. 20 B	58328	9. 765877	Sept 4. 07. 39
1683	♍ 23. 23. 0	83. 11. 0	♊ 25. 29. 30	♋ 10. 36. 55	82. 52. 00 B	56020	9. 748343	Julii 3. 2. 50
1684	♐ 28. 15. 0	65. 48. 40	♏ 28. 52. 0	♐ 15. 15. 25	26. 35. 20 A	96015	9. 982339	Maii 29. 10. 16
1686	♓ 20. 34. 40	31. 21. 40	♊ 17. 00. 30	♊ 16. 24. 00	31. 17. 35 B	32500	9. 511883	Sept. 6. 14. 33
1698	♐ 27. 44. 15	11. 46. 0	♑ 00. 51. 15	♑ 0. 47. 20	0. 38. 10 A	69129	9. 839660	Oct. 8. 16. 57

Hæc Tabula vix indiget explicatione, cum ex titulis satis pateat quid sibi velint **Numeri.** *Distantiæ autem periheliæ æstimantur in ejusmodi partibus quales media distantia* **Terræ à** *Sole habet centies millenas.*

Tabula

19.1. Halley's 1705 table of the orbital elements of twenty-four comets, from *Philosophical Transactions*. Note the similarity of the orbits of the comets of 1531, 1607 and 1682 (which are in fact successive appearances of Halley's Comet).

vinced that the comets seen in 1531, 1607, and 1682 were one and the same. The only seemingly contradictory piece of evidence was the inequality of period (varying between 75 and 76 years), but he argued that the magnitude of this inequality could be explained by gravitational perturbation:

For the Motion of Saturn *is so disturbed by the rest of the Planets, especially* Jupiter, *that the Periodick Time of that Planet is uncertain for some whole Days together. How much more therefore will a Comet be subject to such like Errors, which rises almost Four times higher than* Saturn, *and whose velocity, tho' encreased but a very little, would be sufficient to change its orbit, from an Elliptical to a Parabolical one.*

His belief in the common identity of the three comets was now reinforced by one further piece of evidence: "In the Year 1456, in the Summer time, a comet was seen passing Retrograde between the Earth and the Sun, much after the same Manner: Which, tho' no Body made Observations upon it, yet from its period [75 years before 1531], and the Manner of its Transit, I cannot think differently from those I have just now mentioned." Halley therefore declared that he would "dare venture to foretell" that the comet apparently seen on four previous occasions would return again in 1758, that is, about 76 years after 1682.

Halley promised to discuss the subject of cometary motions in a larger volume "if it shall please God to continue my Life and Health", and he in fact did so in a revised and much expanded version of the *Synopsis*, written sometime before 1719, that eventually appeared in his posthumously published *Tabulae astronomicae* of 1749, soon to be translated into English and French. Halley recalled that after the original version of his essay had appeared, he had discovered in some catalogues of ancient comets new or better records of three earlier retrograde comets – those seen in 1305 around Easter, in 1380 in an unknown month, and in June 1456 – that had preceded at like intervals (75 or 76 years) the three that he had already suspected were the same comet. (The two perihelia of Halley's Comet preceding the perihelia of 1456 were in fact on 25 October 1301 and 10 November 1378.) He also announced that he had subsequently developed a method, which he had applied to the comet of 1682, by which the parameters of a highly eccentric elliptical orbit could be calculated. Calculating the perihelial dates from the best observed positions of the comets seen in 1531 (by Peter Apian), 1607 (by Johannes Kepler and Longomontanus), and 1682 (by John Flamsteed), Halley claimed that "it is manifest that the two Periods of this Comet are finished in 151 years nearly, and that each alternately, the greater and the less, are completed in about 76 and 75 years". Arguing that the comet's close encounter with Jupiter in 1681 had increased both the comet's period and its inclination during the current revolution, he predicted that its next return would take 76 years or more (after 1682), and thus it ought to appear "about the end of the year 1758, or the beginning of the next".

Halley provided no detailed mathematical analysis to back this prediction, however, and so the prediction remained open to question. Until such an analysis was undertaken – one that would examine in detail particularly the perturbational influences of Jupiter and Saturn on the comet's motion and determine a more precise date for its return to perihelion – mid-eighteenth-century astronomers could make only uncertain guesses when, and where in the sky, the expected comet might first be seen.

One comet or two?

The unevenness of the periods of the supposedly identical comet that had appeared in 1305, 1380, 1456, 1531, 1607, and 1682 – appearances separated by alternating intervals of 75 and 76 years – led at least two individuals to suggest that two comets, rather than one, were involved. In his treatise on the spectacular appearance of the comet of 1744, the Swiss astronomer Jean Philippe Löys de Chéseaux (1718–51) argued that it was "very possible & even rather probable" that the apparently alternating periods could be better explained by the presence of two comets, with identical and constant periods, moving in the same orbit, with one comet near but not quite at its aphelion at the same time that the other was at its perihelion. He considered this hypothesis probable because of doubts that gravitation alone could produce substantial changes in a comet's orbit. Calculating that 151 years 10 days had elapsed between the perihelion passages of the supposedly identical comet seen in 1531 and 1682, Chéseaux suggested that the comet seen in 1607 would return after the same period of time and would consequently reach perihelion on 7 November 1758. If so, the comet would appear very brightly in autumn and would pass near the Earth in mid-September.

Fourteen years later the two-comet theory was proposed again by Thomas Stevenson (d. 1764), a plantation owner on the West Indies island of Barbados and a former Surveyor General of that British colony. An amateur astronomer who had observed a comet that appeared in 1757, Stevenson first proposed a two-comet theory, quite possibly without knowledge of Chéseaux's earlier speculation, in a letter to his friend John Rotheram in England that was later published in several London newspapers and magazines late in October 1758. He also discussed his theory in June 1759 in a letter to James Bradley, the English Astronomer Royal. Like Chéseaux, Stevenson resorted to a two-comet theory because of his inability to understand how the force of gravitation could cause major changes in a comet's orbit. He told Bradley that

I cannot apprehend if once Accelerated, how [the comet] was retarded, & after that Accelerated and retarded Alternately, according to the times of his Appearances recorded in History ... Again nor can I conceive, how the attraction of Jupiter or Saturn whose plan[e]s differ little from the plan[e] of the Ecliptic; could increase the Angle of Inclination of the Comits Orbit, which is upwards of 17°.

In his comment on the inclination, Stevenson was referring to the fact that the angle had been 17° 56′ for the comets seen in 1531 and 1682 and 17° 2′ for the one seen in 1607.

Unlike Chéseaux, Stevenson relied only on the supposed 1305, 1456, and 1607 apparitions of the comet, in order to determine its period of revolution and so predict a date for its next return to perihelion. Uncertainty regarding the exact dates of perihelion in 1305 and 1456, however, complicated the problem of determining the period. In his letter to Rotheram, Stevenson argued that the comet would not be observed until at least ten days after it had passed perihelion; it would then be at an angular distance of 40° from the Sun and a linear distance from the Earth of $\frac{825}{1000}$ AU. On the basis of this reasoning, he established 1 April as the date of perihelion in 1305, because Easter had occurred on 11 April of that year. For unknown reasons, he made no attempt to pinpoint a perihelial date in 1456. Calculating that 302 years 198 days had passed between 1 April 1305 and 16 October 1607 (Halley's calculated perihelial date for that year's apparition), Stevenson took half this period (151 years 99 days) and added it to the latter date. By this means he derived a perihelial date of 23 January 1758/59 (OS) or 3 February 1759 (NS) for the forthcoming expected apparition.

After determining the actual perihelial date (13 March) of a comet appearing in 1759 that he identified as the expected one, Stevenson, in an apparent effort to fit the facts, changed the perihelion-to-first-appearance interval in 1305 from 10 to 71 days, and suggested to Bradley that the period of the 1305–1456–1607–1759 comet was increasing by 4 days with each revolution around the Sun (beginning with 151 years 128 days for the revolution between 1305 and 1456). Stevenson considered this regularly occurring small increase per revolution plausible. What he could not comprehend was the variability of a year in period that Halley had postulated in asserting that the comet made its returns to perihelion twice as frequently as Stevenson supposed.

The inequality of periods that concerned Chéseaux and Stevenson was interpreted quite differently by Nils Schenmark, a member of the Swedish Academy of Sciences who apparently had access only to Halley's 1705 publication on comets. Noting that about 76 years had passed between the 1531 and 1607 apparitions, and about 75 years between the 1607 and 1682 apparitions, he suggested in 1755 that the period of the comet might be decreasing in an arithmetic proportion. On the basis of this hypothesis he derived a period of 73 years 228 days for the current revolution and hence a perihelial date of 19 April 1756.

Comet ephemerides

Chéseaux and Stevenson proposed their two-comet theories because of their strongly held belief that a comet's period of revolution around the Sun could not be subject to as large a variation (as much as a year) as Halley had assumed in positing the identity of the comets of 1531, 1607, and 1682. Their assumption of at most only a small amount of variation (a few days rather than a year) of course permitted them (by different methods) to make fairly precise predictions of the date in the late

1750s when the expected comet (last seen in 1607 according to them) ought to return once again to perihelion.

Until a sophisticated theoretical analysis of perturbational effects was performed, those who accepted Halley's view that a comet could have a period with significant variation were unable to make such precise predictions of a perihelial date. Eager nevertheless to observe the expected comet at the earliest possible moment, they took a more cautious and practical approach by constructing (for the benefit of themselves, other astronomers, and the general public) a kind of comet ephemeris – guides to where in the heavens the comet would most likely first appear during a given month (or near a specific date) of the year. The constructors of such ephemerides based them on the known orbit of the Earth (and hence the known pattern of visible background stars at any given time), the orbital elements of the comet that had been determined from observations made at its posited previous apparitions, and various suppositions regarding the perihelial date of the comet at its forthcoming expected apparition.

The first person to construct an ephemeris for the expected comet was Thomas Barker (1722–1809), a grandson of William Whiston living in the English village of Lyndon, near Uppingham, Rutland, and an author mainly of religious works. In a letter to Bradley dated 17 December 1754 and promptly published in the 1755 *Philosophical Transactions*, Barker initially constructed a series of twelve tables that gave the apparent path of the comet with each table supposing the perihelion to occur in a different month of the year, and under those suppositions naming the points in the sky where the comet might then be expected first to appear. From the twelve separate tables he produced a further table (Figure 19.2), summarizing where in the heavens he believed the comet would first be visible for any month of the year. This table proved popular. It was subsequently printed anonymously in 1756 in issues of the *Gentleman's Magazine* and the *Scots Magazine* and in full or abbreviated form in at least two English almanacs for the year 1758.

Contemporary with Barker's effort, another attempt to construct a comet ephemeris was made in 1755 by Dirk Klinkenberg, a Dutch astronomer who had discovered comets in 1743 and 1748. He too constructed twelve tables charting the expected path of the comet, each table (Figure 19.3) assuming a different mid-month perihelial return date.

The next person to construct a comet ephemeris was T. Jamard, an Augustinian of the Abbaye Royale de Sainte-Géneviève in Paris. In a memoir that was favourably reviewed by the Academy of Sciences and published in July 1758, Jamard, like Barker and Klinkenberg, made a range of assumptions regarding the perihelial date. For each he gave such information as the times at and constellations in which the comet would be expected to appear and disappear, its expected route, its probable brightness, how long its tail might be, and the likelihood of discovery.

When a comet was indeed independently discovered by several observers in mid-September 1757 – the first confirmed comet to be seen in seven years – there was naturally speculation as to whether or not it was the one whose return was expected. In England that speculation was mixed with some apprehension among the public, thanks to publications in the preceding two years by the Methodist evangelist John Wesley (1703–91) and the science popularizer Benjamin Martin (1705–82). In November 1755, shortly after a devastating earthquake in Lisbon, Wesley wrote a sermon, later published in London, Bristol, Newcastle, and Dublin, that warned of a forthcoming comet, supposedly on a collision course with Earth, that would cause much havoc unless the populace admitted a belief in the existence of God.

Martin was less inflammatory, but he did suggest in the margins of a popular 1757 broadside (Figure 19.4) that if the expected comet passed through its descending node on 12 May, "we would be in a dangerous Situation, as the denser Part of its blazing Tail would then envelope the Earth, which God forbid". The fear engendered by Wesley and Martin seems to have peaked on 7 October 1757, when, according to a newspaper report, an "uncommon darkness [possibly some freakish weather] occasioned terrible Apprehensions in many People from the Effects the present Comet will have upon the Earth, they being enthusiastically persuaded, that the General Conflagration of the World is near at Hand".

The speculation that the comet appearing in 1757 was the one expected did at first have some

A Table ſhewing where the Comet may be expected to begin to appear any Month

			Lat.	
January		Scarce to be ſeen		
February	end	Retr. between 30° & 15° ♐	Small increaſing S.	7 Weeks after Perihelion
March	begin	——— 30 & 15 ♑	Small N. or S.	a Month after Perihelion
	end	——— 30 & 0 ♒	Small N. decreaſing	
April	begin	——— 15 & 0 ♓	Small N. decreaſing	2 or 3 Weeks after
	end	Stat. 10 ♈ & 20 ♓	Small N.	
May	begin	——— middle ♈	N.	about Perihelion
	end	Dir. begin. ♉		1, 2, or 3 Weeks
June	begin	——— begin. ♉		
	end	——— end ♉	N. increaſing	
July	begin	——— begin. ♊		2 to 5 Weeks before
	end	——— middle ♊		
Auguſt		——— end ♊		5 to 8 Weeks before
September		Stat. 25 & 30 ♊	Small increaſing N.	2 Months before Perihelion
October		Retr. end ♊	Small S. or N.	2 or 3 Months
	begin	begin. ♊	Small S.	
Novem.	mid.	5 ♊ & 20 ♉		3 Months before Perihelion
	end.	begin. ♉		
Decem.	begin	begin. ♉ end. ♈	Small S. or N.	11 to 14 Weeks
	end	begin. ♈	very faint	

19.2. Thomas Barker's table in *Philosophical Transactions* for 1755, showing where he expected the returning comet might first become visible for any month of the year.

Als de Comeet den 14den February in het *Perihelium* komt.

Op de Plaat den Weg Bb.

Dagen		Langte gr.	Langte min.	Breedte gr.	Breedte min.	Distantie à ⊖	Distant. à ☉
Octob.	27 in ♊	20	0	4	35 Z	1 24	2 06
Nov.	6 —	10	10	3	30	1 00	1 92
—	16 in ♉	23	45	1	50	0 79	1 78
—	21 —	13	0	0	50	0 73	1 71
—	26 —	1	50	0	8 N	0 71	1 63
Dec.	1 in ♈	20	0	1	35	0 72	1 57
—	6 —	9	20	2	45	0 77	1 48
—	11 —	1	0	4	5	0 84	1 41
—	16 in ♓	23	30	4	25	0 92	1 33
—	26 —	13	25	5	35	1 09	1 17
Januar.	5 —	7	0	6	15	1 27	1 02
—	15 —	1	30	6	35	1 42	0 87
—	30 in ♒	24	30	6	55	1 57	0 67
Febr.	19 —	14	40	6	0	1 46	0 59
Maart	6 —	7	5	3	15	1 15	0 71
—	21 in ♑	27	30	2	0 Z	0 75	0 94
—	31 —	11	40	11	5	0 49	1 10
April	5 in ♐	26	25	18	35	0 39	1 17
—	10 in ♏	29	15	27	20	0 33	1 25
—	15 in ♎	26	45	30	20	0 37	1 33
—	20 —	5	20	27	10	0 47	1 41
—	25 in ♍	23	15	24	5	0 60	1 48
Mey	5 —	12	25	19	25	0 92	1 63
—	15 —	8	0	16	25	1 25	1 78
—	3[illegible] —	6	45	14	20	1 75	1 99

19.3. Dirk Klinkenberg's table of expected positions of the comet, assuming a mid-February perihelion date. This is one of a dozen tables drawn up by Klinkenberg for hypothetical perihelion dates spread evenly throughout the year.

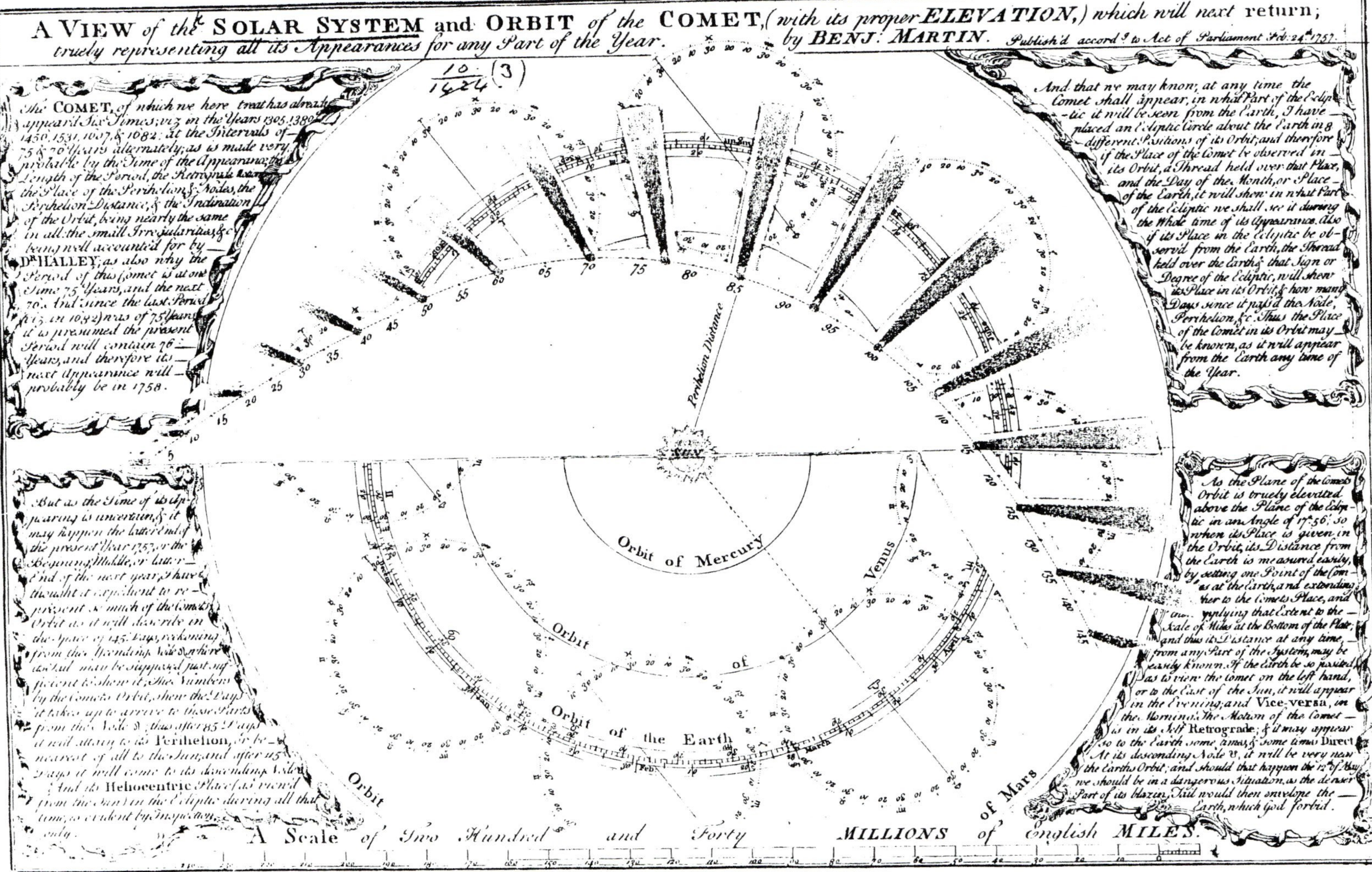

19.4. Benjamin Martin's broadside of 24 February 1757 illustrating the orbit of the comet overlying the Earth's orbit.

supporting evidence. Lalande later noted that this comet, when first seen, had a position conforming to the orbit of the comet of 1682 if one assumed it was then 43 days from reaching its perihelion point. Its subsequent course, however, indicated that it was moving in an entirely different orbit.

The appearance of the comet of 1757 did nevertheless renew public interest in the expected comet and probably inspired the publication shortly thereafter of two new ephemerides. One of these, a twelve-part anonymously written "Account of the COMET's Orbit", appeared serially between October 1757 and October 1758 in the *Boston-Gazette* newspaper, published in the British colony of Massachusetts in North America. For each month, the positions of the ascending and descending nodes, as well as of an intermediate point of the comet's orbit, were given. A circle passing through these three points, the author of the ephemeris noted, "would be, near enough for the present Purpose, the Projection of the Orbit in the Heavens". The author clearly had a good knowledge of astronomy – he explained why he believed that February and September would be, respectively, the most difficult and most favourable times of the year to discover the comet – and was most likely John Winthrop, the Hollis professor of mathematics and natural philosophy at Harvard University and a later observer of the comet appearing in 1759 that was confirmed as Halley's.

Perhaps to give hope to those who were disappointed that the comet of 1757 was not the expected one, Lalande, in the November 1757 issue of the *Mémoires pour l'Histoire des Sciences & Beaux-Arts*, pointed out that in November the Earth would be at the most favourable position for observers to detect the expected comet. It was during this month that the Earth approached the comet's orbit most closely, and if the comet was then in the part of its orbit near the Earth, observers ought easily to see it. The conditions for recovering the comet would become less favourable in the subsequent few months, Lalande explained, both because the distance between the orbits would increase and because around Paris rain often occurred in these months.

An essential factor in knowing where to look for the comet, according to Lalande, was a determination of how far from the Earth the comet could be detected. He pointed out that in 1531 it was perceived at a distance of $\frac{1}{2}$ AU; in 1607, at $\frac{1}{3}$ AU; and in 1682, again at $\frac{1}{2}$ AU. Lalande concluded that those who had the best view and were the most alert ought to be able to perceive the comet when it came within $\frac{2}{5}$ AU of the Earth. On this supposition he then gave the most likely places the comet would be located if it was first seen, respectively, on 1, 5, 10, 15, 20, and 25 November 1757.

As events happened, the various comet ephemerides that were published in the latter half of the 1750s played no direct role in the recovery of the expected comet. Their value to astronomers and the public, however, was not negligible. By considering the many variables that influenced the potential visibility of the comet, the makers of the ephemerides were able to present a reasonable idea as to where the comet might be first spotted for a given time of year. This service prevented needless, and potentially discouraging, searches at times when, and in areas of the heavens where, the expected comet could not possibly be seen.

Clairaut's perturbational analysis

In his article of November 1757 Lalande recalled that Halley had based his prediction of a late-1758 or early-1759 return to perihelion on the argument that the comet had been sensibly delayed by a close pre-perihelion encounter with Jupiter in June 1681. (Jupiter, Halley had argued, would have accelerated the comet, and so lifted it into a higher orbit with a longer period.) Halley had made no mention of a subsequent post-perihelion encounter with Jupiter, at about the same distance, in November 1683, the result of which, in Lalande's opinion, would have tended to cancel the effect of the pre-perihelion encounter. Lalande noted also that Halley had not remarked on a close encounter of the comet with Venus on 22 September 1682 at a distance of $\frac{1}{7}$ AU. He concluded that nothing definitive could be said about the complicated effects on the comet's motion caused by these close encounters until a rigorous perturbational analysis was performed – an analysis, he announced, that his fellow academician Alexis-Claude Clairaut had undertaken, beginning in June 1757, despite the "frightful" length of the calculations involved. In that calculational effort, Clairaut was aided not only by Lalande but also by Mme Nicole-Reine Étable de Labrière Lepaute, whom Clairaut later called *la savante calculatrice*.

In late autumn of 1758, although his calculation was still incomplete, Clairaut decided he must make an announcement so as not to be forestalled by the comet. At the public meeting of the Academy on 14 November, he gave his preliminary results. The action of Jupiter was such that the period from 1607 to 1682 would be 432 days shorter than the period from 1531 to 1607. Without having fully calculated Saturn's action during the same two periods, Clairaut believed that this action would increase the difference by at least 4 days, making the second period 436 days shorter than the first. The difference as determined by observation, he reported, was 469 days, leaving a discrepancy of 33 days. (As was later pointed out, Clairaut here failed to take into account the shift from Julian to Gregorian calendar, and so the observational difference was 459 days and the discrepancy between observation and calculation 23 days.) The near agreement, Clairaut urged, was as striking a confirmation of the Newtonian system as any of those previously given.

As for the effect of perturbation on the periods from 1607 to 1682 and from 1682 to 1759, Clairaut found that Jupiter's action would lengthen the second of these periods so as to make it 518 days longer than the first. His calculation of Saturn's action was still incomplete, but he believed this would cause the second period to be 100 days longer than the first. The period from 1682 to 1759 would thus be 618 days longer than the preceding period, and the perihelion in 1759 should occur in mid-April. On the basis of the calculation of the perihelion of 1682, Clairaut added: give or take a month.

A brief account of Clairaut's procedures, of the completion and later refinement of his calculation, and of the controversy it gave rise to among his contemporaries, is provided in the appendix to this chapter.

The recovery of the comet

At about the same time that Clairaut began his perturbational analysis of the comet's motion, his academic colleague Joseph-Nicolas Delisle (1688–1768) initiated his own approach to finding the expected comet at the earliest possible moment. Rather than predicting the entire expected course of the comet under various hypotheses of perihelial date, as earlier ephemeris-makers had done, this veteran astronomer limited his calculations to the moment when it ought to become first visible, "because having once found it, one would be able to follow it by observations and calculation during the remainder of the apparition". First visibility, coming earlier for telescopic and later for naked-eye observation, would occur when sunlight reflected from the comet was strong enough to make the latter perceptible from Earth, with the aid of the instrument used.

In order to make the best estimate of how soon before perihelion the comet could be seen, Delisle examined the observations of its apparitions in 1531, 1607, and 1682. Unfortunately, for the 1531 and the 1607 apparitions, no information was available either on the size and figure of the comet at the time of discovery or on how it was first perceived to be a comet. Delisle did learn from Apian's *Astronomicum Caesareum* (1540), however, that the comet of 1531 had been seen 18 days before perihelion and that only six days later it had a tail longer than 15°. The comet of 1607 had been first seen 33 days before perihelion. Three days later, according to Johannes Hevelius's *Cometographia* (1668), it had an exceedingly short tail, its head was not quite round, and though brighter than stars of the first magnitude, it had a pale and weak colour. As for the comet of 1682, Delisle was aware that it had first appeared, whitish and without a tail, 24 days before reaching perihelion. On the basis of this information, he conjectured that at each apparition the comet might have been seen telescopically about a month before perihelion if a search had been undertaken in the place where it was at that time. (The 1531 and 1607 apparitions of course preceded the invention of the telescope.)

On the basis of the Earth's continually changing position throughout the year and assuming a fixed orbit for the comet, Delisle then calculated at 10-day intervals where the comet would be on the supposition that it would be first seen either 35 or 25 days before reaching its perihelion point. He gathered these data points in a table (Figure 19.5) and also plotted them on a celestial map (Figure 19.6). When the points were connected on the latter, they formed two ovals. It was along arcs connecting identical days on these two ovals that Delisle instructed his assistant Charles Messier (1730–1817) to look for the comet.

		Pour trente-cinq jours.		Pour vingt-cinq jours.	
		LONGITUDE.	LATITUDE boréale.	LONGITUDE.	LATITUDE boréale.
Novembre..	1	♑ 15^d 45′	24^d 14′	♐ 28^d 5′	17^d 23′
	10	23. 50	17. 31	♑ 7. 25	14. 27
	20	♒ 1. 5	13. 27	15. 20	12. 13
Décembre...	1	8. 5	10. 34	23. 50	10. 21
	10	13. 25	9. 9	♒ 0. 5	9. 18
	20	19. 25	7. 55	7. 5	8. 23
Janvier.....	1	26. 5	6. 56	14. 35	7. 34
	10	♓ 0. 50	6. 21	20. 15	7. 7
	20	6. 20	5. 52	26. 20	6. 40
Février....	1	12. 55	5. 26	♓ 3. 25	6. 17
	10	x 17^d 45′	5^d 10′	x 8^d 35′	6^d 4′
	20	23. 5	4. 57	14. 25	5. 53
Mars......	1	27. 35	4. 48	19. 35	5. 47
	10	♈ 2. 15	4. 41	24. 25	5. 42
	20	7. 30	4. 36	♈ 0. 20	5. 39
Avril.....	1	13. 50	4. 34	7. 25	5. 39
	10	18. 5	4. 32	12. 25	5. 41
	20	23. 15	4. 34	18. 5	5. 47
Mai......	1	28. 55	4. 36	24. 5	5. 55
	10	♉ 3. 50	4. 42	28. 55	6. 6
	20	8. 25	4. 51	♉ 4. 50	6. 20
Juin......	1	14. 27	5. 3	11. 35	6. 41
	10	18. 50	5. 14	16. 35	7. 2
	20	24. 5	5. 32	22. 25	7. 30
Juillet.....	1	29. 25	5. 57	28. 50	8. 12
	10	♊ 3. 55	6. 21	♊ 4. 25	8. 54
	20	9. 5	6. 56	10. 50	9. 54
Août.....	1	15. 5	7. 53	19. 25	11. 31
	10	20. 25	8. 47	26. 5	13. 9
	20	26. 5	10. 12	♋ 4. 50	15. 54
Septembre..	1	♋ 2. 35	12. 52	18. 35	20. 39
	10	8. 25	15. 58	♌ 3. 35	25. 58
	20	16. 25	22. 28	♍ 0. 25	33. 59
Octobre...	1	♌ 1. 45	37. 25	♎ 16. 5	37. 19
	10	♍ 14. 35	62. 24	♏ 18. 25	31. 36
	20	♐ 25. 5	44. 38	♐ 11. 35	23. 40

19.5. Joseph Delisle's table of positions of the comet calculated at ten-day intervals, based on the assumptions that it would be seen either thirty-five days (first pair of columns) or twenty-five days (second pair of columns) before perihelion.

After a search lasting more than a year, Messier, guided to some degree by Delisle's map, spotted an object on 21 January 1759 that he and Delisle, during the course of a series of observations that continued through 14 February, began to suspect was the expected comet. Although he claimed that this object was approximately at the position indicated on Delisle's map for a first-appearance of 21 January, Messier clearly extended his search along the arc for this date beyond the confines of the oval. He in fact later speculated that if Delisle had taken looser limits on how soon the comet might be seen with a telescope before perihelion, he (Messier) might have detected it much earlier. Apparently because he did not wish to make a public announcement regarding the discovery of this object and its probable identification with the expected comet until he knew more about its orbit, Delisle forbade Messier to inform other Parisian astronomers of its existence until after it had rounded the Sun and been recovered again. Delisle apparently felt that the identification of this comet with those seen in 1682, 1607, 1531, and earlier could be definitively established only after it was recovered, in its post-perihelion phase, at a position that clearly indicated its orbit was similar to those determined for its posited earlier apparitions.

Unbeknownst to Delisle and Messier, the same object had already been discovered, 27 days before Messier's first observation, by Johann Georg Palitzsch (1723–88), a German amateur astronomer and prosperous farmer living in Prohlis, a small town near Dresden in Saxony. Whether he made use of any ephemeris to facilitate his discovery remains unclear, but he was certainly aware that a comet was expected. In an account of the discovery published in a Dresden newspaper by his friend Christian Gotthold Hoffman, Palitzsch remarked:

Following my fatiguing habit of wanting to observe as much as possible all that occurred in Nature, and especially the remarkable celestial events, I examined the stars on the 25th of December about six o'clock in the evening with my 8-foot Telescope. The constellation of the Whale [Cetus] presented itself well, and it was also the epoch when the announced Comet ought to approach and show itself. Thus came to pass for me the indescribable pleasure of discovering in the Fish [Pisces], not far from the marvellous star of the Whale [the variable star Mira Ceti], a nebulous star never perceived there before. To tell the truth, it is found between the two stars marked ϵ and δ on Bayer's Uranometria, *or* O *and* N *on Doppelmayer's map.*

The same observation, renewed on the 26th and the 27th, confirmed my supposition that it was a comet; because from the 25th to the 27th, it had effectively moved from the star O *toward the star* N.

Palitzsch did not specifically identify this newly discovered comet as the one expected. Hoffman, who observed the object on 27 December, in fact suggested that it might be the same as the one

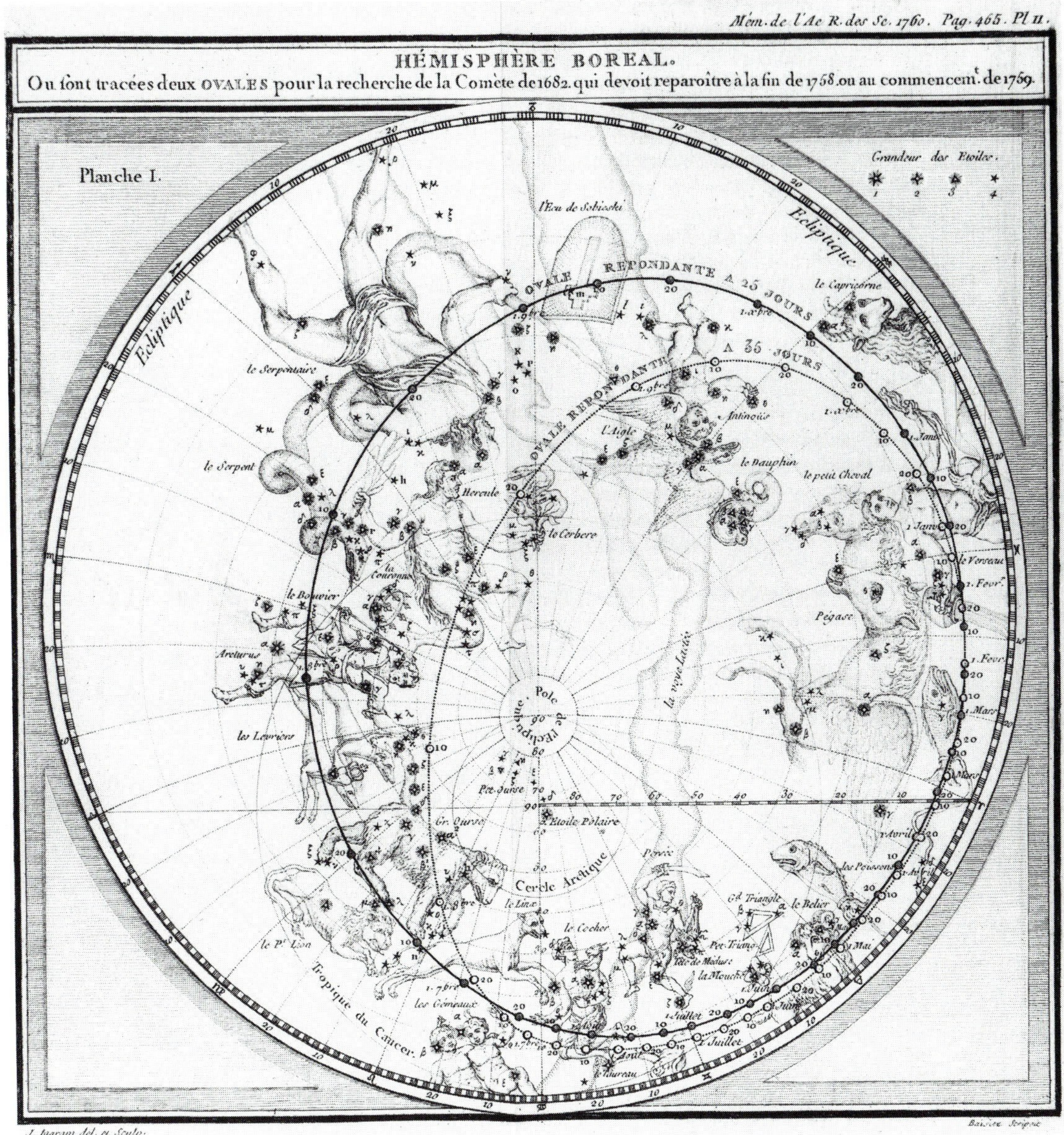

19.6. Delisle's celestial map on which separate lines connecting the thirty-five day and the twenty-five day data points form two oval curves. Charles Messier looked for the comet along arcs connecting identical days on these two ovals.

observed by Thaddeus Hagecius in 1580.

The first person to claim publicly that the object discovered by Palitzsch was the expected comet was the anonymous author of a pamphlet published in Leipzig on 24 January 1759. The author (most likely Gottfried Heinsius, a professor of mathematics at the University of Leipzig), upon learning of the observations of Palitzsch and Hoffman, immediately suspected that the object was the expected comet. After performing a theoretical analysis (probably based on the orbit of the 1682 comet) of where the comet would be likely to appear, he spotted the object himself on 18 and 19 January at positions predicted by his theory. He reported these new observations, as well as those of Paltizsch and Hoffman, in his pamphlet, together with a fairly accurate description of the comet's expected positions and appearances over the next few months.

The Seven Years War then raging in Europe slowed communications between the countries of that continent. As a result, word of Palitzsch's discovery did not reach Paris until 1 April. By coincidence, early that morning Messier recovered the object that he had observed in January and February at a location that convinced Delisle it was the expected comet. With this confirmation he at last permitted Messier, that same day, to inform other Parisian astronomers regarding the appearance of the long-awaited object. Quite understandably, these other astronomers strongly criticized Delisle for having deprived them of an opportunity to observe the comet earlier.

By 25 April, when Lalande gave his lecture at the public assembly of the Academy of Sciences, he and other Parisian astronomers had determined that the object discovered by Palitzsch and Messier had orbital elements very similar to those of the comets seen in 1531, 1607, and 1682, and thus it was virtually certain that all of these comets, as Halley had hypothesized, were one and the same. They interpreted the periodic return of this comet as indicating that its motion, like that of the planets, was affected by the force of gravitation posited by Newton. That interpretation was reinforced after they determined that the comet had passed its perihelion point on 13 March 1759. That date was at the outer limit of the margin of error set by Clairaut in predicting the comet's return to perihelion on 15 April. Its reappearance at a time that seemed in accordance with the predictions of Newton, Halley, and Clairaut, transformed the comet, known forever after as "Halley's Comet", from a mysterious interloper to a long-time member of the solar system.

Further reading

Peter Broughton, The first predicted return of Comet Halley, *Journal for the History of Astronomy*, vol. 16 (1985), 123–33

D. W. Hughes, Edmond Halley: why was he interested in comets?, *Journal of the British Interplanetary Society*, vol. 37 (1984), 32–44

S. P. Rigaud, *Some Account of Halley's Astronomiae Cometicae Synopsis* (Oxford, 1835)

Norman J. W. Thrower (ed.), *Standing on the Shoulders of Giants: A Longer View of Newton and Halley* (Berkeley, 1990), especially Halley, Delisle, and the making of the comet by Simon Schaffer (pp. 254–98); Edmond Halley: his interest in comets by David W. Hughes (pp. 324–72); and The first International Halley Watch: guiding the worldwide search for Comet Halley, 1755–1759 by Craig B. Waff (pp. 373–411)

Craig B. Waff, Comet Halley's first expected return: English apprehensions, 1755–58, *Journal for the History of Astronomy*, vol. 17 (1986), 1–37

Craig B. Waff, Tales from the first International Halley Watch (1755–59): 1. Boston waits for and watches the comet, *International Halley Watch Newsletter*, no. 9 (1986), 16–26

Craig B. Waff and Stephen Skinner, Tales from the first International Halley Watch (1755–59): 2. Thomas Stevenson of Barbados and his two-comet theory, *International Halley Watch Newsletter*, no. 10 (1987), 3–9

Appendix: Clairaut's calculation of the comet's return

CURTIS WILSON

Alexis-Claude Clairaut's calculation of a perihelion date for the eighteenth-century return of Halley's Comet was the first large-scale numerical integration ever performed. For the sheer magnitude of the undertaking and the accuracy of the final result, it has been much praised. The calculation itself has received little attention; for a description of it we must turn to Clairaut's own *Théorie du mouvement des comètes* of 1760.

Initially Clairaut believed that it would be necessary only to compute the perturbations of the comet on its near approaches to Jupiter. This hope soon evaporated. Very slight changes in the comet's speed near perihelion produced large changes in the comet's period. Even when the comet was far from both Jupiter and the Sun, Jupiter's action was changing the Sun's place, and as a consequence influencing the time required for the comet to return to perihelion. Moreover, Saturn with about a third of Jupiter's mass could not be ignored. The perturbations induced by both planets must be calculated for entire periods, to determine how they were affecting the times of return.

The main basis of the calculation would be Clairaut's solution of the three-body problem, which he had applied earlier in determining the perturbations of the Moon and the Earth (see Chapters 16, 18, and 20). It consisted of two fundamental equations, one for the radius vector in a perturbed elliptical orbit and the other for the time in the same orbit. The right-hand member of each equation was in two parts, one valid for the unperturbed ellipse, the other giving the added effect of the perturbations. Clairaut used the second part of the second equation in determining the perturbationally induced changes in the times of return of Halley's Comet.

Thus, given the known dates of the perihelia in 1607 and 1682, Clairaut could compute the perturbational increases or decreases in time for the period 1607–82 and again for the following period, and take the difference; the latter, added to the known length of the period 1607–82, would give the length of the following period and hence the date of perihelion toward the end of the 1750s.

To gauge the accuracy of this computation, Clairaut decided to calculate in an entirely parallel way the date of perihelion in 1682, starting from the dates of the perihelia in 1531 and 1607. The difference between the calculated and actual dates of the perihelion in 1682 would give a measure of the accuracy to be expected in the computation of the perihelion date in the 1750s.

The expression giving the perturbational alteration in the comet's period was an integral; its integrand involved the perturbing forces. To evaluate these forces, it was necessary to know, for any given moment, the positions of both the perturbing planet and the comet. Yet the position of the comet was subject to perturbation. The logical circle was unavoidable: to compute perturbational effects, it was necessary to start from an approximate orbit and motion for the comet.

For this approximate orbit and motion, Clairaut used a single reference ellipse throughout his calculation, namely the ellipse that Halley had determined by assuming a mean period of 75.5 years. Halley had found it to have a semimajor axis of 17.86 AU and an eccentricity of 0.96739. It was this Halleian ellipse, in the slightly altered orientations that Halley found for it in the three successive perihelia of 1531, 1607, and 1682, that Clairaut used when he undertook to approximate the position of the comet in evaluating the integrand of the integral that expressed the perturbationally induced changes in the comet's period.

But Clairaut's integral when applied to the comet could not, in general, be integrated formally.

To be thus integrable, the integrand would have had to be expressed as a function of a single variable, say the eccentric anomaly of the comet. In the case of planets perturbed by planets it had been possible to express the integrand in this way, by using trigonometric series to approximate its value. The goodness of the approximation depended on the rapidity of convergence of the series, which in turn depended on the nearness with which the orbits approximated circles. Such a procedure was inapplicable to the comet because of the large eccentricity of its orbit. Clairaut therefore resorted to 'mechanical quadratures', now called numerical integration. R. J. Bošković, in an essay for the Paris prize contest of 1752, had proposed using mechanical quadratures to compute planetary perturbations (see Chapter 20), but Clairaut was the first actually to launch a large-scale numerical integration.

It was a daunting task. At intervals of 2° of eccentric anomaly, starting at perihelion, it was necessary to compute the distance between perturbing planet and comet – the latter being taken as unperturbed in the orbit Edmond Halley had assigned to it – and then to determine the resulting inverse-square force, with its radial and transverse components. The successive values of the integrand could then be calculated. After that the quadratures could be carried out, essentially by adding up the equally spaced ordinates, but with some use of a correction formula, that had been derived independently by James Gregory and Isaac Newton, to take account of the curvature of the curve atop the ordinates.

With a view to avoiding numerical integration where possible, Clairaut employed two additional processes. When the comet was in the superior half of its orbit, and so farthest from the Sun and from Jupiter and Saturn (the only perturbing planets considered), the main perturbative effect was the action of the perturbing planet on the Sun; and this action could be determined geometrically, to a good approximation, by supposing the comet to be moving in an ellipse about the common centre of gravity of the Sun and the perturbing planet. In some circumstances numerical integration could be avoided by using a 'parabolic line' (a power series in the eccentric anomaly) to approximate the perturbing forces. Clairaut sought shortcuts because he was short of time, and feared the comet would return before the calculation had been completed (it in fact did so).

Given the pressure of time and the magnitude of the task, Clairaut needed collaborators. The astronomer Joseph-Jérôme Lefrançais de Lalande offered to calculate the distances between the comet and the perturbing planets, the resulting forces, their components, and the products into which they entered. Clairaut was to check all these results for consistency (mainly by constructing graphs to determine whether the ordinates varied in a smooth fashion), and then perform the quadratures. Lalande was assisted by Mme Nicole-Reine Étable de Labrière Lepaute, wife of a noted clockmaker. Clairaut wrote an appreciation of her assistance for inclusion in his initial report, but a jealous ladyfriend insisted that it be suppressed, Lalande commenting years afterward that Clairaut was a *savant judicieux mais faible*.

In his reminiscences Lalande also stated that for six months the three of them calculated from morning till night, sometimes even at table, and from this strenuous work he acquired a malady that changed his temperament for the rest of his life.

For the calculations Clairaut divided the cometary orbit into three parts: from 0° to 90° of eccentric anomaly, the first quadrant of the ellipse from perihelion, requiring more than 7 years to be traversed; from 90° to 270° of eccentric anomaly, the superior half of the orbit, taking more than 60 years to be traversed; and from 270° to 360°, the final quadrant of the ellipse, taking once again about 7 years. The finer subdivision for the quadratures proceeded generally in 2° steps, but in the close approaches to Jupiter before and after the perihelion of 1682 Clairaut used 1° steps.

In calculating the Saturnian perturbations, Clairaut separated the action of Saturn on the Sun from its action on the comet throughout the cometary orbit, and left out of account the eccentricity of Saturn's orbit and its inclination to the cometary orbit through all three revolutions of the comet, except for a short interval near the perihelion of 1682, when the comet made a near approach to Saturn.

As reported in the body of this chapter, Clairaut announced preliminary results, before his calculation was complete, at the public meeting of the Academy of Sciences on 14 November 1758. His prediction was for a perihelion in mid-April 1759, give or take a month. After the recovery of the comet on 31 March, Lalande, N.-L. de Lacaille, and others fitted a parabolic orbit to observations from before and after the conjunction with the Sun, and found that the perihelion had occurred on 13 March. Clairaut's prediction was therefore a month off, at the limit of the range of error he had allowed for it.

A controversy now erupted within the Paris Academy of Sciences as to the accuracy of this prediction. The dispute was bitter, and during the summer led to a spate of letters in the various journals. Its chief instigators, the astronomer Pierre-Charles Le Monnier (1715–99) and the mathematician Jean le Rond d'Alembert (1717–83), maintained that Clairaut's error was large, while the partisans of Clairaut maintained that it was small. Clairaut's followers wanted to compare the error with whole periods of the comet, 75 or 151 years; Le Monnier and d'Alembert insisted that it be compared with the *difference* between two successive periods, since this alone, among the results Clairaut had derived purely from theory, admitted of empirical verification.

In the case of the periods from 1607 to 1682 and from 1682 to 1759, the difference in length had proved to be 585 days, and so 33 days would constitute an error of $(33/585) \times 100\% = 5.6\%$. Such an estimate of error might have been acceptable to Clairaut, but d'Alembert was bent on arguing that in some of Clairaut's calculations the error had been much larger. We shall not enter into the dubious paths d'Alembert followed to extract such a conclusion; the only honest way of checking Clairaut's calculations would have been to repeat them. D'Alembert's posture in this controversy arose from jealousy and a contempt for "number-crunching". He called Clairaut's calculation "more laborious than deep", and he had fundamental doubts about the capacity of theory (whether in Newtonian gravitation, hydrodynamics, or the physics of the vibrating string) to apply with precise accuracy to the world.

Clairaut's attitude was very different, and he constantly sought to improve the precision of predictions from Newtonian theory. Following his preliminary report of November 1758, he completed and revised the cometary calculation for his *Théorie du mouvement des comètes*, which was submitted to the Academy in August 1759 and published the following April. In 1761, in order to compete in the contest sponsored by the St Petersburg Academy of Science for 1762, he further refined his calculation; he won one-half of the prize (the other half was won by Johann Albrecht Euler, with an essay on the near-approach of Halley's Comet to the Earth in 1760), and his prize essay was published in 1762. In these two revisions Clairaut claimed to reduce the error of his prediction, first to 23 days, and then to 19 days. From a careful summing up of the component numbers entering into the two sums, they are found to be 26.1 and 23.4 days, respectively.

To evaluate the error of the component computations, it would be necessary to re-do them in an entirely parallel way; no such computation has been or is likely to be made. An approximate evaluation can be attempted, however, on the basis of an integration by computer completed by Tao Kiang in 1971; the general result may be summarized as follows (the evaluations of error are to be taken as accurate only to within a day or two).

In Clairaut's calculation as a whole, there were ten main parts, two pertaining to the period 1531–1607, and four to each of the periods 1607–82 and 1682–1759. In five of the parts Clairaut's errors did not exceed 4 days each; in two of the Saturnian parts the discrepancies were much larger, but here what counted was the difference between the two results, and this was in error by only 4 days. Another of the Saturnian parts was off by 8 days. Finally, in two of the Jovian parts of the calculation for the period 1682–1759, the errors were −15 and +33 days, fortunately opposite in sign. The goodness of Clairaut's final result, we have to conclude, rested in part on a fortuitous cancellation of opposing errors.

Clairaut was well aware of possible and probable sources of error in his calculation. He nevertheless believed his result to be a verification of the Newtonian system "as striking as any of those previously given". In his final response to d'Alembert (in the *Journal des sçavans* for June 1762) he dismissed the question of a precise measurement of the error of his calculations as unimportant. The important point, he insisted, was that he had silenced the critics of the Newtonian theory of comets. What mattered, it seems, was the general "feel" of the verification, the sense that it augured well for the future of the calculation of cometary perturbations. And since d'Alembert did not have the patience for the calculation of cometary perturbations, Clairaut urged him to leave in peace those who did.

Later developments have vindicated Clairaut's attitude, but not the complete sufficiency of the Newtonian theory to account for cometary returns. Perturbations due to planets other than Jupiter and Saturn caused the return in 1759 to be 8 days earlier than it would otherwise have been. In addition there was a non-gravitational acceleration. Clairaut had considered the possibility of an acceleration due to aethereal resistance, but had dismissed it, marshalling the evidence against it in the last part of his prize essay of 1762.

A second periodic comet was discovered by J.-L. Pons in 1818. It proved to have a period of 3.3 years, and when J. F. Encke computed the perturbations he found a non-gravitational decrease of about 2.5 hours per period, which he attributed to aethereal resistance. As a result celestial mechanicians computing the return of Halley's Comet in 1835 were looking for a similar effect; what they found, however, was a perihelial date later than that calculated from the perturbations, and thus an effect not explicable by resistance. In the 1980s, direct numerical integration from the differential equations has shown that the non-gravitational acceleration of Halley's Comet is constant in each return, and amounts to an increase of about 4.1 days per period. This (like the non-gravitational acceleration of Comet Encke) has been explained by the hypothesis that comets are rotating icy conglomerates. On their near approaches to the Sun they are strongly irradiated, with the result that there is a jet effect due to out-gassing; a time-delay between maximum reception of solar radiation and maximum out-gassing, given the angular rotation of the comet, accounts for the fact that the comet is either accelerated or decelerated in the direction of its orbital motion, and thus pushed into a higher or lower orbit.

Further reading

Alexis-Claude Clairaut, *Théorie du mouvement des comètes* (Paris, 1760)

Alexis-Claude Clairaut, *Recherches sur la comète des années 1531, 1607, et 1759* (St Petersburg, 1762)

T. Kiang, The past orbit of Halley's Comet, *Memoirs of the Royal Astronomical Society*, vol. 76 (1972), 27–66

C. Wilson, Clairaut's calculation of the eighteenth-century return of Halley's Comet, *Journal for the History of Astronomy*, vol. 24 (1993), 1–15

D. K. Yeomans, Investigating the motion of Comet Halley over several millennia, in *Comet Halley: Investigations, Results, Interpretations*, ed. J. W. Mason (New York, 1990) vol. 2, pp. 225–31

D. K. Yeomans, and T. Kiang, The long-term motion of comet Halley, *Monthly Notices of the Royal Astronomical Society*, vol. 197 (1981), 633–46

PART VI

Celestial mechanics during the eighteenth century

20

The problem of perturbation analytically treated: Euler, Clairaut, d'Alembert

CURTIS WILSON

In 1840 William Whewell in his *Philosophy of the Inductive Sciences* proclaimed that astronomy

is not only the queen of the sciences, but, in a stricter sense of the term, the only perfect science; – the only branch of human knowledge in which particulars are completely subjugated to generals, effects to causes ... and we have in this case an example of a science in that elevated state of flourishing maturity, in which all that remains is to determine with the extreme of accuracy the consequences of its rules by the profoundest combinations of mathematics, the magnitude of its data by the minutest scrupulousness of observation ....

This praise of astronomy's state in the 1830s sounds excessive today – the work of Henri Poincaré and Albert Einstein has shown it to be simplistic – but it harbours an historical truth. During the century preceding the appearance of Whewell's book, there had emerged by stages, in a series of innovations and refinements, an extensive body of analytical theory concerning the three-body problem – the problem of the perturbations produced in the motions of one body about another, when a third body is present and all three attract in accordance with Newton's law. With Laplace the honing of the theory reached such a point as to make possible the accurate prediction of lunar and most planetary positions to within a few arc-seconds. In intricacy and sophistication, the mathematical machinery marshalled and brought to bear in this development surpassed that in any earlier mathematical theory applying to natural phenomena. *Astronomie physique*, or as Laplace rechristened it in 1796, *mécanique céleste*, was indeed, in the early decades of the nineteenth century, the undisputed queen of the sciences.

The theoretical development was almost exclusively the work of five mathematicians: Leonhard Euler (1707–83), Alexis-Claude Clairaut (1713–65), Jean le Rond d'Alembert (1717–83), Joseph Louis Lagrange (1736–1813), and Pierre-Simon Laplace (1749–1827). Laplace's achievement would depend crucially on Lagrange's accomplishment; Lagrange's accomplishment would depend in turn on earlier advances made by Euler, Clairaut, and d'Alembert, and these are the subject of the present chapter.

Why, it may be asked, did the development of analytical celestial mechanics not get under way till the 1740s? For it would appear that, much earlier, there was motive enough for the development of such a theory. A means of accurately determining longitude at sea was urgently needed; the British Parliament in 1714 had instituted a handsome prize for the invention of such a procedure – £20 000 if it gave the longitude to within $\frac{1}{2}°$, half as much if to within 1°. A lunar theory accurate to 2′ of arc would suffice for the second prize.

That no lunar theory of this accuracy was constructed during the first half of the eighteenth century must be put down to ignorance of how it was to be done. Isaac Newton, as we have seen (Chapter 13 in Volume 2A), for a time hoped to obtain a lunar theory good to 2′ by adding on several inequalities to the Horrocksian theory, some of them (such as the annual equations in the motions of the apse and nodes) inferred from the gravitational law, others from observation alone. The result was disappointing, the errors being much higher than 2′. Edmond Halley then proposed comparing the theory with observations throughout a Saros cycle (a period of 18 years after which solar and lunar eclipses return in the same sequence), on the supposition that the discrepancies so found would be the same in each subsequent cycle. He carried out the comparison during his years as Astronomer Royal (1720–42), and the results were published posthumously in 1749. The attempt to employ them in correcting predictions drawn from the faulty Newtonian lunar theory (a cumbrous procedure!) appears to have been short-lived.

As for attempts to develop the lunar theory mathematically, in England these followed the Newtonian pattern, combining geometry with limiting ratios of infinitesimals (as in John Machin's propositions on the motions of the Moon's nodes, incorporated in the third edition of the *Principia*). How much might have been accomplished by such methods must remain a matter of speculation.

A limitation in the Newtonian procedures, as compared with the Leibnizian calculus which was being developed by mathematicians on the Continent (for instance Pierre Varignon, Jakob I and Johann I Bernoulli, Euler, Alexis Fontaine, Clairaut), was the absence of an explicit representation of the independent variable. Newton's "pricked letters", introduced in the 1690s to rival the Leibnizian fractional notation for the derivative, implied a single independent variable, usually time. The Leibnizian notation, by incorporating an explicit symbol for the independent variable, facilitated changing one such for another, the fractional form for the derivative suggesting the rules to be followed in this transformation. (Such shifts of variable would have been possible in the Newtonian notation, but required a little more doing.) Changes of variable were used from the beginning of the analytic development of perturbation theory, for instance in eliminating time as a variable from the differential equations so as to obtain an equation for the orbit in polar coordinates.

Yet, whatever the advantages of the Leibnizian calculus, it did not "lead one by the hand" to any obvious procedure for treating the three-body problem. As we have seen (Chapter 14), when G. W. Leibniz proposed the problem to Varignon, the latter succeeded only in proposing a way of coping with perturbations caused by a *fixed* centre of force.

The Continental mathematicians were largely committed to aethereal theories of gravity, and their attempts to reconcile such theories with Newton's inverse-square law no doubt absorbed time and effort that could have been expended on the perturbational problem. But the chief obstacle

20.1. Leonhard Euler.

to analytic development of perturbation theory appears to have been a technical one.

This theory as worked out by Euler, Clairaut, d'Alembert, Lagrange and Laplace depends on integration of trigonometrical functions. The development of the calculus of these functions was much slower than that of the other well-known transcendental functions, the logarithm and exponential. Trigonometrical tables were widely employed, but sines, cosines, and tangents tended to be regarded as lines in diagrams rather than as functions of an independent variable. Formulas for the derivatives of the sine, tangent, and secant were given by Roger Cotes in his *On the Estimation of Errors*, published posthumously in 1722, but this promising start was not followed up.

It was Euler (Figure 20.1) in 1739 who effectively introduced the calculus of trigonometrical functions to the community of Continental mathematicians, in papers in which he developed a general procedure for solving linear differential equations with constant coefficients. He later claimed this calculus to be his own invention, "in so far as the special signs and rules are [there] comprised". Already by the early 1740s, presumably from the reading of Euler's papers, the rules had become well-known to such practitioners of analysis as Daniel Bernoulli, Clairaut, and d'Alembert.

It was again Euler who, in the early 1740s, first attempted systematically to derive the lunar inequalities by means of the Leibnizian calculus. In his *Mechanica* of 1736, he had applied this calculus to many of the problems and theorems of Newton's *Principia*, but treatment of the three-body problem was conspicuous by its absence. By 1744 he had constructed lunar tables incorporating the inequalities caused in the Moon's motion by the Sun, as derived from the inverse-square law. What made the difference in the latter year, we hypothesize, was Euler's command of the calculus of trigonometrical functions.

Euler's lunar tables of 1744 were first published (probably in revised form) in his *Opuscula varii argumenti* of 1746, and then without an exposition of the calculations by which they were derived. In his preface Euler characterized these tables as follows.

I come now to my lunar tables. Their nature and basis of construction would require too much space to explain. I therefore only point out that they are derived from the theory of attraction which Newton with such happy success introduced into astronomy. Although it is claimed of several lunar tables that they are based on this theory, I dare to assert that the calculations to which this theory leads are so intricate, that such tables must be considered to differ greatly from the theory. Nor do I claim that I have included in these tables all the inequalities of motion which the theory implies. But I give all those equations which are detectable in observations and are above $\frac{1}{2}$ arc-minute. For I was able to carry the calculation to the point of identifying the arguments of the individual inequalities, and I determined the true quantity of many of the equations by the theory alone, but some I was forced to determine by observations.

Along with his lunar tables, Euler constructed solar tables, but these embodied only one perturbation deriving from gravitational attraction, namely the lunar inequality. Newton had mentioned this perturbation in Prop. 13 of Bk III of the *Principia* (second and third editions). It was the centre of gravity of the Earth and Moon, he stated, rather than the Earth itself, that moves in an ellipse about the Sun according to the area law. The rule thus enunciated is not quite exact on the assumption of the inverse-square law, but it is sufficiently so, and Euler followed it in his solar tables. He also employed Newton's erroneous value for the ratio of the masses of the Moon and Earth, namely 1:39.788, too large by a factor of more than 2, and so arrived at a lunar inequality in the Sun's apparent motion of about 15″ (the correct value is about 6″.5).

Besides the lunar perturbation, Euler incorporated in his solar tables the action of a non-gravitational force: the resistance of an aether. The existence of a subtle material filling the space between gross bodies was for Euler a matter of conviction. He regarded action-at-a-distance as contrary to reason, and in an essay of 1740, he asserted the only possible cause of gravitational attraction to be the centrifugal force developed in vortices. Daniel Bernoulli, writing to Euler in January 1742, objected that vortices in which the centrifugal force varied inversely as the square of the distance from the centre would be unstable. Euler bowed to this argument, and went on to accept an explanation of gravity that Bernoulli had proposed in his *Hydrodynamica* of 1738, according to which gravitational force is the result of variation in aethereal pressure, arising from differences in velocity. Euler's commitment to the aether was reinforced by his theory of light, for he believed that the transmission of light could be plausibly explained only by means of waves in an aether. The aether thus filled interplanetary space. As it did not move with the planets it must necessarily resist their motions.

In an essay of 1746, Euler derived the effects of such aethereal resistance. The planet's velocity, his differential equations showed, would decay exponentially; hence the orbit would gradually shrink and the period diminish. The line of apsides would remain stationary, but the orbital eccentricity would decrease. That astronomers had reported no evidence for such effects, Euler suggested, was due to the low density of the aether. In comparing ancient and modern solar observations, he believed he had found evidence that the Earth's motion was subject to the secular changes that aethereal resistance would cause: the orbital eccentricity was decreasing, and the tropical year diminishing, the latter at a rate of five seconds per century. In 1753 Tobias Mayer would persuade him that he was wrong, and that within the limits of observational error, the tropical year had remained constant.

In his solar tables of 1746, Euler took the solar apogee to be fixed with respect to the stars; the evidence that astronomers had previously adduced for assigning a secular motion to the apogee, he urged, was insufficiently firm. This meant that he was assuming perturbations of the Earth's motion by other planets to be negligible. If there were any motion of the apogee, he remarked, he would prefer to attribute it to the action of some comet passing close to the Earth. Thus at the time of drawing up his solar tables of 1746 Euler believed the Earth's motion to be determined almost exactly by the inverse-square force due to the Sun, the sole modifications being due to the lunar inequality and the resistance of the aether.

He was soon to encounter facts that would cause him to change his mind, and to conclude that perturbations of the planets by one another could not be assumed to be negligible. At the same time, he was entering a phase in which the exactitude of the inverse-square law of gravitation no longer appeared to him a reliable assumption.

The Paris Academy's contest of 1748

The topic for the Paris Academy's contest of 1748 was chosen at a meeting of the prize commission in March 1746. The astronomer Pierre Charles Le Monnier had presented evidence that Jupiter and Saturn were subject to observationally detectable inequalities, attributable, he believed, to the mutual attractions of the two planets. The prize, it was decided, should be offered for a "theory of Jupiter and Saturn explicating the inequalities that these planets appear to cause in each other's motions, especially about the time of their conjunction".

That the mutual action would be especially detectable around the time of conjunction was suggested by Newton's theory of gravitation. As Newton put it in Prop. 13 of Bk III of the *Principia* (second and third editions), from the action of Jupiter upon Saturn "... arises a perturbation of the orbit of Saturn in every conjunction of this planet with Jupiter, so sensible, that astronomers are puzzled with it". The error, Newton stated, could be largely avoided by placing the focus of Saturn's ellipse in the common centre of gravity of Jupiter and the Sun; the anomaly in the mean motion, he claimed, would not exceed 2′ per year.

Astronomers had long been puzzled by the anomalies in the motions of Jupiter and Saturn, but these did not appear to be confined to the times of conjunction. In a letter to Michael Mästlin in 1625 Johannes Kepler had reported finding, by a comparison of ancient observations reported by Ptolemy and the sixteenth-century Tychonic observations with those of Bernhard Walther in the late fifteenth century, that "the mean motions are no longer mean", but subject to sizeable long-term variations. He assumed these inequalities to be periodic, but concluded that their periods and sizes could be determined only after good observations had been accumulated over a period of several centuries. In his *Rudolphine Tables*, he calculated the mean motions of the two planets on the basis of the Ptolemaic and Tychonic observations alone.

That the values so obtained were unsatisfactory was recognized by such seventeenth-century astronomers as Jeremiah Horrocks, Noël Durret, Maria Cunitia, and Ismael Boulliau. Jupiter was moving more rapidly and Saturn more slowly than Kepler's numbers implied. No mere correction of the mean motions, Horrocks found, would accommodate the ancient observations.

John Flamsteed struggled with the problem for decades. During the 1670s and 1680s he tried to solve it by correcting Kepler's numbers. For Jupiter he obtained a theory fitting the observations for a number of years, but then it ceased to be accurate. In the early 1690s Newton was proposing "to conciliate the orbits of Jupiter and Saturn to the heavens by a consideration of physical causes", and Flamsteed furnished him with observations for this purpose. The apsides of the two planets, Newton told him, were "agitated by a libratory motion". This suggests that Newton at this time was attempting to cope with the anomalies by following the analogy of Jeremiah Horrocks's lunar theory, with its oscillating apse and eccentricity.

By the late 1690s Newton had apparently abandoned his efforts, and Flamsteed now set out, with the aid of calculators, to solve it on his own. The anomalies, he believed, should be cyclical, with the period of 59 years in which Jupiter and Saturn return very nearly to the same configuration with respect to each other and the Sun. Repeated attempts to apply this or other periods, however, led nowhere. By 1717 Flamsteed had given up in despair. As he wrote to Abraham Sharp, one of his calculators:

... my hopes that there might be a restitution of the inequality in Saturn and Jupiter, after 2 revolutions of Saturn and 5 of Jupiter, are vanished; and the doctrine of gravitation and its effects are not as yet so perfectly understood as we imagined.

Astronomers in Paris – G. F. Maraldi (Maraldi I), Jacques Cassini, Le Monnier – encountered similar difficulties: no values for the mean motions would satisfy the observations of all ages, nor did the variations follow a discernible pattern.

In April 1746 Jacques Cassini (Cassini II, 1677–1756), who had previously been a partisan of planetary vortices, presented to the Academy a memoir aiming to show that a long-term acceleration of Jupiter and deceleration of Saturn could arise from the gravitational attractions of the two planets. The argument is qualitative, depending on the non-concentric configuration of the orbits which suggests an incomplete compensation of the accelerations and decelerations before and after conjunction. Cassini recognized that a full demonstration would require deriving the continuous action of the perturbative forces in all parts of the orbits. He assumed the anomaly to have a period of 84 000 years, the time for the two lines of apsides, advancing at their mean rates, to move out of and back into coincidence.

The possibility of secular variations in the mean motions of Jupiter and Saturn was not mentioned in the setting of the prize problem for the Paris contest of 1748. It was to assume more importance afterwards, as will become apparent in later chapters.

The contest of 1748 was significant because it engaged mathematicians trained in the Leibnizian calculus in the search for a way to deal with the problem of perturbation. In the essay that Daniel Bernoulli (1700–82) submitted (we possess only the draft), he remarks that, with the advent of Newton's system, astronomy has become inseparably connected with mechanics. Consequently

it is natural that what remains for us to discover cannot fail to be extremely difficult. Such above all is the question proposed by the illustrious Academy of Sciences for the prize of the year 1748. It is so to such an extent that the great Newton himself, foreseeing no doubt the terrible calculations that the solution demands, did not wish to undertake it except by way of some general reflections...,

The prize in 1748 went to Euler; Daniel Bernoulli's essay received the *proxime accessit* (runner-up). But the commission was not satisfied. Euler had failed to give the derivation of his differential equations (this he had done earlier in a memoir submitted to the Berlin Academy in 1747); his application of his solution to observations spanning the years from 1582 to 1745 left errors as high as 8′ and 9′. Daniel Bernoulli had not even attempted observational comparisons. The commission therefore proposed the same topic for the contest of 1750; Euler was again one of the contestants, but no prize was awarded. Then once more, for the contest of 1752, the commission proposed the same problem. This time Euler was the winner, Rudjer J. Bošković (1711–87) received the *proxime accessit*, while Daniel Bernoulli's submission in this year went unmentioned. In none of the essays was there an attempt at an empirical test. Euler excused his failure to carry out such a test on the ground that the available observations were insufficiently accurate.

What influence did these various prize essays have on later developments? Bernoulli's were never published. Bošković caused his essay to be published in Rome in 1756, but neither he nor anyone else appears to have undertaken to apply its theory to planetary observations. (Certain features of both Bernoulli's and Bošković's essays may have influenced Clairaut in his later calculation of the return of Halley's Comet; see the appendix to Chapter 19.) The theories of Euler's two essays contained calculative errors and lacked important terms. On the other hand, they contained seminal ideas on which much of the further development of celestial mechanics would turn. The essay of 1748 was published in Paris in 1749, and on its basis Lagrange would make important discoveries. The essay of 1752 was not published until 1769, but then the disagreement of its results with those of Lagrange triggered Laplace's entry into the fray.

Advances in celestial mechanics during the late 1740s were not the work of Euler alone; beginning in 1746 Clairaut and d'Alembert had taken up the struggle with the three-body problem. Both were members of the prize commission. From the moment the topic for the prize of 1748 was agreed upon, each of them, unbeknownst to the other, set out to derive and solve differential equations representing the motion of a planet or satellite subject to perturbation. As members of the Paris Academy they were ineligible to enter the contest. Their motivation was explained a year later in a letter of the Marquise du Châtelet to François Jacquier: "M. Clairaut and M. d'Alembert are after the system of the world, understandably they do not wish to be forestalled by the essays for the prize".

By late 1746 both Clairaut and d'Alembert had formulated differential equations and a general procedure for their resolution. D'Alembert sent his formulation to the Berlin Academy; Clairaut deposited his in a sealed envelope with the Secretary of the Paris Academy. During 1747 there were more sealed envelopes, from d'Alembert as well as Clairaut. The race came out into the open in the summer, when both of them read memoirs on the subject to the Academy. These appeared in the Academy's *Mémoires* for 1745, published in 1749. More mature accounts of their respective procedures were given in Clairaut's *Théorie de la lune*, which won the prize of the St Petersburg Academy for 1750 and was published in 1752, and in Vol. 1 of d'Alembert's *Recherches sur différens points importans du système du monde*, which also deals with the lunar theory and was published in 1754.

In our account of these several works, it will become apparent that the lunar and planetary problems differ in an important respect. In the case of the Moon the distance of the perturbing body (the Sun) varies only slightly as the Moon circles the Earth, and the variations can be approximated accurately in an expression of a few terms. In the case of Saturn perturbed by Jupiter the distance between the two planets varies by a factor of more than 3 from opposition to conjunction, and because the orbits are not concentric, the force exerted by Jupiter on Saturn at conjunction can vary from one part of Saturn's orbit to another by a factor of more than 1.6. As we shall see, it was Euler who proposed the artifice whereby these wild variations were finally to be coped with successfully.

Abandoned initiatives: Daniel Bernoulli and Bošković on the perturbations of Saturn and Jupiter

The difficulty of the problem of planetary perturbation can be seen in the laborious struggles of Daniel Bernoulli and Bošković to cope with it. The starting-point is necessarily a false supposition, since the perturbations of, say, Saturn by Jupiter depend on the changing distance between the two planets, hence on the perturbations. Bernoulli set out from two false suppositions: that the Sun remains at rest, and that the orbits the planets would follow if unperturbed would be concentric with the Sun. The first of these suppositions he defended as Keplerian, and compatible with Newton's system if the Sun's inertial mass were effectively infinite, and its gravitational mass of the magnitude Newton assigns to it. In his essay of 1752 he withdrew this supposition as unacceptable, excused it in the earlier essay as imposed by lack of time, and proceeded to compute the motions of the Sun about its common centre of gravity with Jupiter and Saturn, thus rectifying the earlier omission. The second supposition he admitted, already in his first essay, to be a source of inaccuracy, since the orbits, although departing only very slightly from circularity, are decidedly not concentric.

But now the mathematical difficulties began. Bernoulli derived a second-order differential equation for the departure (α), in the radial direction, of Saturn from its concentric circle. The equation was only approximately true under the initial assumptions. The independent variable was the heliocentric angle (σ) of elongation between the two planets, but the equation also contained the expressions $1/z$ and $1/z^3$, z being the linear distance between Jupiter and Saturn. To express these factors in terms of σ, Bernoulli first calculated the numerical values of $1/z$ and $1/z^3$ at 10° intervals starting with conjunction and going to opposition, then derived interpolation formulas for the intermediate values in terms of powers of σ for the six 30° intervals from conjunction to opposition. To extend the interpolation formulas over wider intervals, he found, would lead to unacceptable inaccuracies. He was thus led to six differential equations for α, which he solved by the method of undetermined coefficients, obtaining polynomial expressions for α in terms of powers of σ multiplied by numerical coefficients for each of the six intervals. Any error was cumulative, since the accuracy of the expressions for the later 30° intervals depended on that for the earlier intervals.

At 180° of elongation, it emerged, $d\alpha$ was not zero, for Saturn at this point was moving away from the Sun. The laborious calculations needed to be continued, since no sign of periodicity had emerged. "After all the fatigues I had surmounted and all the success I had had in my investigation," Bernoulli remarked, "I would rather have abandoned my enterprise than undertake these new calculations." Fortunately he thought of an artifice that enabled him to avoid the drudgery.

The values of α after 180° of elongation would be symmetrical with those before if the tangent were slightly turned at this point so as to make an equal angle on the other side of the radius vector. Such a change would in effect shift Saturn's circle into an eccentric position with respect to the Sun, and by subtracting out the changes in radial distance that such a shift entails, Bernoulli was able to determine the values of α for succeeding intervals. The accompanying diagram (Figure 20.2) shows his graph of α from first conjunction (C′) to fourth conjunction (C″″). The numbers in the graph, when divided by 211, represent fractions of the radius of the circle initially assumed for Saturn. The dotted lines have to do with the artifice just described. At the fourth conjunction, about 59 or 60 years after the first, both α and $d\alpha$ are nearly zero, and Bernoulli believed he had found a fundamental periodicity. "Nature", he averred, "tends in all its operations to harmonious periods." With α known, he could compute Saturn's orbital velocity.

That the initial assumption of concentric orbits was wrong, Bernoulli was quite aware. (He did not realize, however, that if he chose non-concentric orbits to start with, the periodicity he believed he had found would disappear.) He thought of redoing his investigation with more realistic initial conditions, but the labour appeared daunting. As he wrote to Euler in 1749,

> *I have much less hope than ever of deriving the irregularities of Saturn from mechanical principles; but I can assure you that I see sufficiently into this subject to hope that with equal effort I could bring forth as much as anyone else. An exact solution is impossible and all approximations so dangerous that it would require an unsurpassable effort to determine the irregularities with sufficient exactitude and certainty.*

And he went on to express wonder at Euler's complete confidence in having solved the problem with the utmost precision.

Bošković's procedure, unlike Bernoulili's, involved no differential equations or series approximations. Newtonian in style, it relied on geometry and ratios of infinitesimals. Even astronomers untutored in the integral calculus, Bošković boasted, could follow it.

At each moment the perturbed planet was to be conceived as moving in the elliptical orbit in which it would continue if the perturbing force were just then removed. The changes in this ellipse due to the perturbing force, during the time the planet traversed a tiny arc of its path, were then to be computed by means of formulas which Bošković derived geometrically using infinitesimals. These formulas gave the variations in transverse axis, eccentricity, aphelion, and rate of description of areas caused by infinitesimal changes in the planet's speed and direction. Each formula involved the ratio of the perturbing force to the solar force for the momentary locations of the two planets, the sine or cosine of the angle between the directions of the tangent and perturbing force, and in addition sines or cosines of certain angles, and certain ratios of lengths, in the unperturbed ellipse.

To obtain the changes in the elliptical elements over a finite time, Bošković proposed treating the daily variations as infinitesimal (Saturn moves about 2′, Jupiter about 5′ per day), and adding them up graphically. The abscissas would be lengths of the planet's path; the ordinates would be the diurnal variations of the element whose change was sought. The net area enclosed by the graph was to be approximated using formulas from Cotes's *Harmonia mensurarum* (1722), which Bošković had come to know from a French commen-

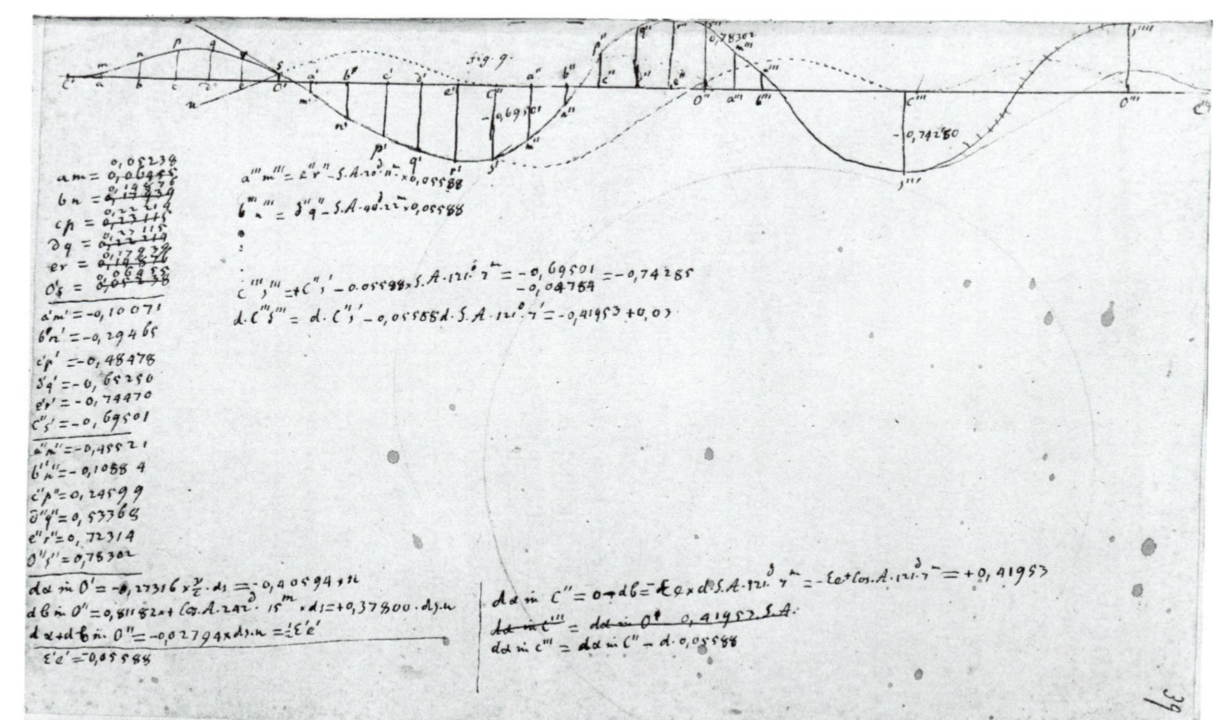

20.2. Daniel Bernoulli's diagram for the departure in the radial direction of Saturn from its concentric circle.

tary by Charles Walmesley. Bošković proposed computing the variations over a period of 60 years, after which time he expected the pattern of variations to repeat very nearly. He suggested various calculational shortcuts; but the labour, we may be sure, would have been immense.

Bošković wrote his *De inaequalitatibus quas Saturnus et Jupiter sibi mutuo videnter inducere* during the years 1750–52, while on expedition with Christopher Maire to measure a degree of the meridian in the papal states. The essay contains neither illustration nor sample of the vast programme of calculations it envisages. Bošković excused the omission as due to the pressure of his other work, and to uncertainty as to the mass of Jupiter, which from data given by Cassini II would come out one-fifth greater than Newton's estimate.

Both Bernoulli and Bošković recognized that there was a problem in determining the initial orbit to which the calculated perturbations were to be applied. Boščović proposed first determining orbital elements from three observations by a procedure then standard. The elements thus obtained were initially to be taken as applying at the time of one of the observations, say the first. The times to the other positions observed were then to be corrected by subtracting out the effect of perturbation. Finally, with the original data and the corrected times, new orbital elements were to be computed, and these, Bošković held, could be assumed to be almost exactly applicable at the time of the first observation. Bernoulli seems to have had in mind some similar procedure of "correcting" the observations by subtracting out the effect of perturbation. Neither author had in mind a way of coping with observational error. It would be Euler who first proposed a technique whereby the orbital elements could be corrected for the effects of both perturbation and observational error.

Euler on the inequalities of Saturn: the first try

It was the Paris contest of 1748 that led to Euler's first published memoirs on the perturbational problem: his "Recherches sur le mouvement des corps célestes en générale", presented to the Berlin Academy in June 1747 and published in the *Mémoires de Berlin* in 1749, and his "Recherches sur la question des inégalités du mouvement de Saturne et de Jupiter", the prize-winning essay of 1748, which was published (as a book) in Paris in 1749. In the first of these Euler derived the differential equations used in the second, which despite its title concerned Saturn's inequalities alone. In the present section we focus on the problem of the inequalities in longitude as dealt with in these two essays.

In both essays, Euler calls the exactitude of the inverse-square law into doubt, while nevertheless basing the mathematical development on this law. The reader is referred to Chapter 16 for an account of how Euler's doubt was eventually put to rest.

Euler began his development by applying Newton's second law in rectangular coordinates:

$$2\frac{d^2x}{dt^2}=\frac{X}{m},\quad 2\frac{d^2y}{dt^2}=\frac{Y}{m},\quad 2\frac{d^2z}{dt^2}=\frac{Z}{m},\tag{1}$$

where m is the mass of the perturbed planet, and X, Y, Z are the components of force acting on it in the coordinate directions. (The factor 2 is due to Euler's choice of units, and eventually drops out of the calculation.) Taking the equations of motion in rectangular coordinates as starting-point is nowadays a standard procedure, but in Euler's use of it here we have its first appearance. He identified the x–y plane with the orbital plane of the perturbing planet (Jupiter); z thus measured the small latitudinal departures that Saturn makes from this plane.

Next, Euler transformed the first two equations of (1) into the polar coordinates used by astronomers by setting $x=r\cos\phi$, $y=r\sin\phi$; also, he replaced X/m by $(-P\cos\phi+Q\sin\phi)$, and Y/m by $(-P\sin\phi-Q\cos\phi)$, where P and Q are accelerations directed so as to diminish, respectively, r and ϕ. The differential equations of motion in the x–y plane thus became:

$$d^2r-rd\phi^2=-P\frac{dt^2}{2},\tag{2}$$

$$2drd\phi-rd^2\phi=-Q\frac{dt^2}{2}.\tag{3}$$

The third equation of (1) also got transformed, but we postpone our discussion of this step to the next section.

If the perturbed planet was to be Saturn, then the radial acceleration P and the transradial acceleration Q had to be expressed in terms of the attractions due to the Sun and Jupiter. Here account had also be taken of the forces exerted by Jupiter and Saturn on the Sun, in order that the Sun could be regarded as at rest; hence the forces on it had to be transferred to Saturn with their directions reversed. Figure 20.3 shows the geometry involved. Saturn was thus considered to be acted upon by four forces:

two forces in the direction $S\odot$, yielding

$$\frac{(\odot+S)\cos^2\psi}{r^2};$$

a force in the direction SJ, $=J/v^2$;
a force in the direction HR, $=J/r'^2$.

By taking components of these forces in the direction of the radius $H\odot$ and in the transradial direction HN, Euler obtained the following expressions for P and Q:

$$P=\frac{(\odot+S)\cos^3\psi}{r^2}+\frac{Jr}{v^3}+\frac{J\cos(\phi'-\phi)}{r'^2}-\frac{Jr'\cos(\phi'-\phi)}{v^3},$$

$$Q=\frac{J\sin(\phi'-\phi)}{r'^2}-\frac{Jr'\sin(\phi'-\phi)}{v^3}.$$

To replace $dt^2/2$ in equations (1) and (2), Euler set the inverse-square force of the Sun on Jupiter at its mean distance a' equal to Jupiter's mean centripetal acceleration: $\odot/a'^2=2a'dM'^2/dt^2$, where M' is Jupiter's mean anomaly; hence $dt^2/2=a'^3dM'^2/\odot$. Next, he replaced the ratios of the planetary to the solar masses by single letters: $S/\odot=n$, $J/\odot=n'$. (In the foregoing we have for clarity's sake replaced Euler's symbols by indicial notation, which was first introduced by Lagrange.) Equations (2) and (3) thus took the form

$$d^2r-rd\phi^2=-a^3dM^2\left[\frac{(1+n)\cos^3\phi}{r^2}+\frac{n'r}{v^3}+\frac{n'\cos(\phi'-\phi)}{r'^2}-\frac{n'r'\cos(\phi'-\phi)}{v^3}\right], \quad (2')$$

$$2drd\phi+rd^2\phi=-n'a'^3dM'^2\sin(\phi'-\phi)\left[\frac{1}{r'^2}-\frac{r'}{v^3}\right]. \quad (3')$$

These differential equations provided an exact expression of the inverse-square forces, but in solving them Euler found it necessary to resort to approximation. Specifically, he replaced the variables r, r', ϕ, ϕ', and v in the right-hand members by approximate expressions in terms of M', the mean anomaly of Jupiter. Having exact formulas to start from, he was in a better position than a Bernoulli or a Bošković to keep track of different orders of approximation.

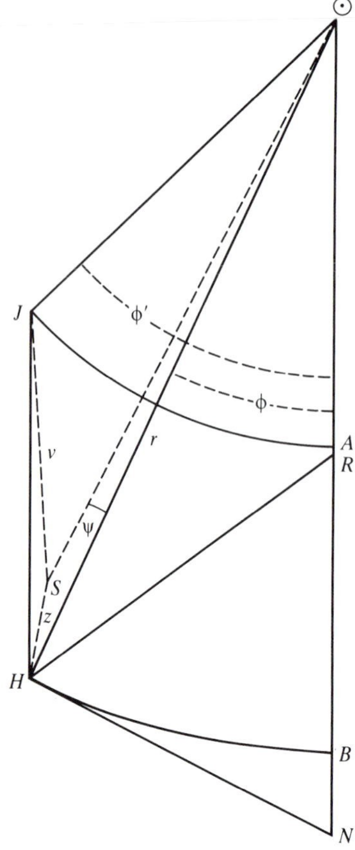

20.3. Euler's figure (with the lettering modified for clarity) to illustrate his theory of the perturbations of Saturn by Jupiter:
$\odot$ = Sun's position (and also its mass)
J = Jupiter's position (and also its mass)
S = Saturn's position (and also its mass)
$r=\odot H$ = the curtate solar distance of Saturn (the projection of Saturn's radius vector on the plane of Jupiter's orbit)
v = distance between Jupiter and Saturn
$r'=\odot J$ = radius vector of Jupiter
ϕ = longitude of Saturn
ϕ' = longitude of Jupiter
ψ = latitude of Saturn (from plane of Jupiter's orbit)
$z=SH$ = distance from Saturn to plane of Jupiter's orbit
HR = line drawn through H parallel to $\odot J$
HN = line drawn through H at right angles to $\odot H$

In a first approximation, r and r' could be identified with a and a', the mean solar distances of Saturn and Jupiter. In effect the planets were here being conceived to move in circles concentric to the Sun, hence uniformly except insofar as they perturbed each other's motions; ϕ and ϕ' became identical with the mean motions of the two planets, so that $\phi' = M' + k'$, and $\phi = k''\phi'$, where k', k'' are constants.

What about v, the distance between the two planets? Its only occurrence was as the inverse cube, v^{-3}, which could be expressed as

$$[a^2 + a'^2 - 2aa' \cos(\phi' - \phi)]^{-\frac{3}{2}},$$

or

$$[a^2 + a'^2]^{-2/3}[1 - g \cos(\phi' - \phi)]^{-\frac{3}{2}},$$

where

$$g = \frac{2aa'}{a^2 + a'^2}.$$

If a were much greater or less than a', as in the case of the Moon perturbed by the Sun where the ratio of the two distances is 389:1, v^{-3} could be accurately approximated by two or three terms of a Taylor expansion:

$$v^{-3} = [a^2 + a'^2]^{-3/2}\left[1 + \frac{3}{2}g \cos(\phi' - \phi) + \ldots\right].$$

In the case of planets perturbing planets, on the contrary, a and a' were much closer to each other in size, and the Taylor series converged much less rapidly. Thus for Jupiter and Saturn, $a'/a = 0.545$, and the series would have to be continued to some sixty terms to yield the same precision as that given by three terms in the case of the Moon. An additional difficulty was that when the expansion was carried past the second term, it contained powers of $\cos(\phi' - \phi)$, which would have to be transformed one-by-one into integrable form.

The difficulty presented by the variation in distance between the two planets, we recall, had led Daniel Bernoulli to resort to elaborate interpolations and successive integrations in numbers, each dependent on the preceding. For a time Euler thought it might be necessary to turn (like Bošković later) to numerical quadratures. But at length he thought of a remedy.

For $\phi' - \phi$, let us write θ. The powers of $\cos\theta$ in the Taylor expansion, Euler discovered, could be replaced by expressions in terms of cosines of multiples of θ (he was the first to develop these expressions). Thus he transformed the Taylor series into something quite new, a trigonometric series: $A + B\cos\theta + C\cos 2\theta + \ldots$. Here the terms were immediately integrable.

But now a new difficulty emerged: each of the coefficients A, B, C,... was itself an infinite series. To meet this difficulty, Euler first showed – by logarithmic differentiation with respect to θ of his general formula for the series,

$$s = (1 - g\cos\theta)^{-\mu}$$

– that each coefficient after the first two was expressible in terms of the two preceding coefficients. Then he provided two different procedures for approximating A and B, one of them an approximation to Fourier integrals by numerical integration. We have here the opening chapter in the pre-history of Fourier analysis.

With the introduction of a trigonometric series for v^{-3}, the right-hand members of (2′) and (3′) assumed integrable form, and Euler proceeded to integrate the equations by the method of undetermined coefficients. The calculation was straightforward, and Euler's result may be written:

$$r = a[1 + 0.000\,854\,2\cos\theta + 0.000\,144\,9\cos 2\theta + 0.000\,033\,5\cos 3\theta + 0.000\,010\,5\cos 4\theta + \ldots]; \qquad (4)$$

$$\phi = \text{const} + k''M' + 4''\sin\theta - 32''\sin 2\theta - 7''\sin 3\theta - 2''\sin 4\theta - \ldots. \qquad (5)$$

Daniel Bernoulli, in a letter to Euler of January 1750 and in his essay for the contest of 1752, objected that this result could not be correct, because it makes r and ϕ functions of θ alone, whereas his own calculations had shown that r is significantly different at successive conjunctions, going for instance from a at the first conjunction to $a[1 - 0.003\,29]$ at the second. He accused Euler of paralogism. The relations between Bernoulli and Euler had by 1750 become strained; whether Bernoulli ever received an answer is unclear. Attempting to reply for Euler we would say: any assault on the perturbational problem involves assuming approximate values to obtain other, more closely approximating values of the same variables. Euler's procedure thus far was such as necessarily to yield results periodic with θ, but a further stage of approximation was to follow. Would these successive steps converge to the result implied by Newton's law? Euler in his essay of 1748 appeared confident that the series he was using were convergent, although he had no proof of this (tests and proofs of convergence were not to be introduced till the next century). Later on, as we shall see, he would become doubtful of the reliability of trigonometric series for computing perturbations when the apparent convergence was slow.

In two further stages of approximation Euler took into account the orbital eccentricities, first Saturn's, then Jupiter's; in each case he treated the orbit of the second planet as circular and non-eccentric. To express the motion of a planet on an eccentric orbit, he used Kepler's eccentric anomaly. Thus for Saturn,

$$r = a(1 + e \cos E),$$

where e is the eccentricity and E the eccentric anomaly. E is related to the mean anomaly M by Kepler's equation

$$M = E + e \sin E,$$

and to the true anomaly ϕ by

$$d\phi = \frac{a}{r}\sqrt{(1 - e^2)}\, dE.$$

There were corresponding equations for Jupiter (in our notation, the letters are the same but distinguished by prime marks). The inverse cube of v had to be developed further than before to take account of the variability of r and r'; this entailed developing new terms in the series for v^{-3}, involving the determination of the coefficients in the trigonometric series

$$(1 - g \cos\theta)^{-5/2} = P + Q\cos\theta + R\cos 2\theta + S\cos 3\theta + T\cos 4\theta + \ldots.$$

The result of integrating the equations that had been elaborated in this way was that the terms obtained in the previous calculation turned up again, but in addition there were new terms proportional to the first powers of the eccentricities. In particular, Euler found two especially large terms:

$$-257'' \sin(\theta - E) - 243'' \sin(2\theta - E').$$

Unfortunately, the two coefficients are wrong: there were errors of sign in Euler's derivation. (In the text as given in *Leonhardi Euleri opera omnia series II* Vol. 25, the mistakes occur on p. 80, where they are indicated by the editor, and on p. 100, where R'' should be, not $+12.63013$, but -12.63013.) When these errors are corrected, the two coefficients become $-182''$ and $+413''$, respectively, which are in close agreement with the values later obtained by Laplace and his successors.

A different difficulty emerged in the term in the radius vector proportional to $\cos(\theta - E')$: the coefficient proved to be a fraction with zero for denominator. Euler took this to be an indication that, in addition to the terms he had already assumed for the integral, there was another term proportional to the time, or to E' understood now as an angle increasing indefinitely. Thus in the longitude ϕ he concluded that there were two terms with argument $(\theta - E')$, namely

$$-Q\, E'\cos(\theta - E') - T\sin(\theta - E');$$

the constants Q and T, indeterminate in his derivation, would have to be obtained by a comparison with observations. The presence of a term with coefficient proportional to E' would explain why astronomers had obtained results for the period of Saturn differing by as much as three or four days. It meant that all the empirical constants would in time need to be re-determined: no eternally valid formula was attainable.

There was another possible interpretation which Euler here mentioned but did not pursue. In obtaining the terms with argument $(\theta - E')$ he had assumed the line of apsides of Jupiter's orbit to be fixed. But he had already found that, when Saturn's orbit is taken to be eccentric, its orbital apse moves owing to the perturbations caused by Jupiter. By symmetry the apse of Jupiter's orbit as perturbed by Saturn must also be in motion. But if the Jovian apse moves, the coefficient of $\cos(\theta - E')$ no longer has zero for denominator. D'Alembert, writing to Euler after the contest was finished, asked why he had not adopted this alternative interpretation. Eventually Euler did so: in his prize essay of 1752 the changes in orbital apse and eccentricity of the two planets were taken as simultaneous and mutually implicated.

In a supplement to his memoir of 1748, Euler observed that the angle $(\theta - E')$ is given (if terms of the order of the eccentricities are ignored) by $-[E + (A - A')]$, where A' and A are the longitudes of the two aphelia, and $(A - A')$ is an angle of about 78° which increases slowly over the centuries. It follows that $\cos(\theta - E')$ is approximately replaceable by $\sin E$, and the term containing $\cos(\theta - E')$

may thus be viewed as contributing to Saturn's equation of centre. The gradual change in the angle $(A-A')$ implies (when proper account is taken of signs) that Saturn's equation of centre is slowly decreasing. To the question of the secular variations in the orbital elements, Euler was to return in his memoir of 1752.

In the main body of his memoir of 1748, however, Euler set himself the task of determining the constants Q and T by means of observations. There were other constants, as well, to be empirically determined. In 1740 Euler had presented to the St Petersburg Academy a memoir on the differential correction of orbital elements, and he now invoked this method to correct the orbital elements of Saturn as given in Cassini's tables, since (in his words) "the inequalities caused by Jupiter have been there enveloped in the eccentricity and position of the orbit of Saturn". All told, there were eight unknowns to be determined by the observations; eight observations and eight simultaneous equations would have sufficed for the determination. Recognizing, however, that the observations inevitably contained error, Euler introduced the use of multiple 'equations of condition' to bring redundant data to bear on the determination of the constants.

This is an important step – a first major step towards bringing planetary astronomy under the aegis of statistics. Euler was here proposing to resolve a problem the difficulty of which Bernoulli and Bošković had failed to recognize: the problem of identifying the unperturbed orbital motion to which the perturbations were to be applied. Euler's solution was to fit both the theoretically derived perturbations and the empirically determined orbital elements to multiple observations, by adjusting the constants of doubtful accuracy so as to minimize the errors.

The method of least squares had not yet been invented, and the procedure that Euler used in its stead was crude: he sought to add equations together in such a way as to render the coefficient of one unknown large, and the coefficients of the remaining unknowns small. And the success of his attempt was limited: for 36 observed oppositions from Tycho's time to 1745 he reduced the error in the theory to about 3′.3 on the average. A few of the earlier observations contained errors larger than this, but the chief reason he could not do better was that he had not derived, and indeed not suspected the existence of, the sizeable inequalities in the motion of Saturn that are proportional to the squares and cubes of the eccentricities. Laplace would be the first to derive these inequalities, and in a test of his theory against 43 observations of Saturn from the period 1582–1786, using basically the same statistical procedure as Euler (but see Chapter 27), would succeed in lowering the errors to about 50″ on the average. The agreement became still better after J.-B. J. Delambre had corrected the observations for aberration and nutation.

During the 1750s Tobias Mayer deployed the method of equations of condition more successfully than Euler (see Chapter 18). According to Delambre, Mayer derived the method from Euler, and Laplace had it from Mayer's writings.

Euler on the secular inequalities

We return now to the third equation of (1), namely

$$2\frac{d^2z}{dt^2}=\frac{Z}{m}.$$

This equation, we recall, determines the perturbations in Saturn's small latitudinal departures from the plane of Jupiter's orbit. In his memoir of 1748, Euler undertook to transform this equation into a result more in accord with astronomical usage. Astronomers since Kepler had viewed each planet as moving in an orbital plane inclined to the ecliptic, its position being specified in terms of the inclination and longitude of the boreal node (the point at which the planet moves from south to north of the ecliptic). The variable z is expressible in terms of the inclination ρ and the longitude of the node π by

$$z=r\sin(\phi-\pi)\tan\rho. \qquad (6)$$

If all perturbations were absent, both ρ and π would be constant, and in forming the differential dz we would consider z to depend solely on the variables r and ϕ. But if perturbations are present, ρ and π become variable, and the differential dz comes to depend on $d\rho$ and $d\pi$ as well as dr and $d\phi$. Yet even in this case, Euler stated, it is permissible to treat dz as depending only on dr and $d\phi$, for ρ and π are changing much more slowly than r and ϕ, and can be regarded as constant during the time dt that the variation dz occurs. Forming dz in both ways, Euler set the two results equal and so obtained

$$d(\ln \tan \rho) = \frac{d\pi}{\tan(\phi - \pi)}. \quad (7)$$

A second differentiation of (6), followed by substitution of the result in the third equation of (1), then yielded an expression for $d\pi$ in terms of the force components P, Q, Z.

We have here the basic step in the method of variation of orbital elements: the formula valid in the two-body case is assumed to be valid in the three-body case, provided that the constants of integration – the orbital elements used to specify the size, shape, and position of the orbit in the two-body case – are taken as variable. At each moment the planet or satellite is imagined as moving on an ellipse; but the ellipse itself is changing position, size, and shape, owing to the perturbations.

An outcome of Euler's investigation of $d\rho$ and $d\pi$ in the case of Saturn (with the plane of reference being taken as Jupiter's orbital plane) was that the node of Saturn's orbit steadily regressed. A similar result must hold for every planet: the nodes of its orbit on the orbital plane of any perturbing planet must undergo a steady motion. In particular this applied to the nodes of the ecliptic on the orbital planes of the planets other than the Earth. Euler undertook a systematic investigation of the resulting motion of the ecliptic in a memoir completed in 1754 and published in 1756, "De la variation de la latitude des étoiles fixes et de l'obliquité de l'écliptique". The question whether the obliquity of the ecliptic was decreasing had long been disputed. Euler demonstrated that the perturbationally induced motion of the nodes implied a secular diminution. His value for it, which depended on hypothesized masses for Mercury, Venus, and Mars, was 47".5 per century, very close to the present-day value.

Euler's prize essay of 1752 applied the idea of mutually induced changes in orbital elements to the perturbations in longitude of Jupiter and Saturn. This time, no *arcs de cercle* emerged. Instead, the apse of each planet was progressing, and each planet was inducing eccentricity in the orbit of the other, so that the observed eccentricity in each case was a conflation of a proper eccentricity and an induced eccentricity. Thus for the *radii vectores* of the two planets Euler obtained expressions of the form

$$r' = a'[1 + e'\cos(\phi' - A') + ke\cos(\phi' - A)],$$
$$r = a[1 + e\cos(\phi - A) + k'e'\cos(\phi - A')],$$

where A, A' were now the longitudes of the two *proper* aphelia; e, e' were the two *proper* eccentricities; and k, k' were two constants for each of which Euler obtained (after neglecting many smaller terms and simplifying) a quadratic equation. Unfortunately, Euler's computation was riddled with error; the two quadratic equations were really one, and the correct values for k, k' were of opposite sign, whereas Euler's mistaken values were both positive. In what amounts to an addition of vectors, Euler showed that the squares of the *apparent* (that is, observationally determined) eccentricities were

$$f^2 = e^2 + k'^2e'^2 + 2k'ee'\cos(A - A'),$$
$$f'^2 = e'^2 + k^2e^2 + 2kee'\cos(A - A').$$

The angle $(A - A')$ was approximately 78° and increasing, and Euler therefore concluded that the apparent eccentricities of both orbits were decreasing. But in fact Saturn's orbital eccentricity is decreasing while that of Jupiter is increasing (for k is really negative).

Another of Euler's inferences, based on the change in the angle $(A - A')$, was that the mean motions of both planets were slowly accelerating. Euler's memoir of 1752 was not published until 1769, and in the meantime Lagrange had published a deduction purporting to show that the mean motion of Jupiter was accelerating while that of Saturn was decelerating, in agreement with observations. The contradiction between the two results was to lead Laplace to re-do the derivation, with an outcome to be reported in Chapter 22.

Euler employed the variation of orbital elements in several later works, including the *Additamentum* to his *Theoria motus lunae* of 1753 and his memoir on the planetary perturbations of the Earth's motion, *Investigatio perturbationum quibus planetarum motus ob actionem eorum mutuam afficiuntur*, which won the prize of the Paris Academy for 1756. But there was one orbital element whose variation he did not examine, namely the epoch. It was left to Lagrange to complete the method by including the variation of this element in the analysis.

Clairaut's procedure for determining perturbations

Clairaut's method, as described in his *Théorie de la lune* of 1752 as well as in earlier papers, is different from Euler's, and worthy of a separate account. It lent itself with a certain clarity to Clairaut's discovery of the full motion of the lunar apse. (For an account of this discovery the reader is referred to Chapter 16.) It was the method that Lalande chose to incorporate in his *Astronomie* (editions in 1764, 1771, and 1792), the chief text from which European astronomers learned astronomy in the last third of the eighteenth century. In carrying out his integration of the equations of motion, Clairaut introduced a number of innovations that were put to use by Euler, Lagrange, and Laplace.

Clairaut began by supposing the orbit of the perturbed body to lie in a single plane; thus he left out of account the perturbations in latitude. His attempt to reintroduce the latitudinal perturbations at a later stage entailed an illogicality, for which d'Alembert took him to task. The initial simplification allowed him to reduce the problem to the following two equations, which differ little from the Eulerian equations we labelled (2) and (3) above:

$$r\,\mathrm{d}^2\phi + 2\,\mathrm{d}r\,\mathrm{d}\phi = Q\,\mathrm{d}t^2; \tag{8}$$

$$r\,\mathrm{d}\phi^2 - \mathrm{d}^2 r = \left(\frac{M}{r^2} + P\right)\mathrm{d}t^2, \tag{9}$$

where M represents the sum of the masses of the central and the perturbed body.

Next, in a series of moves that won Euler's admiration, Clairaut obtained an integrated form of (8). After multiplying by r, he found in a first integration

$$r^2\frac{\mathrm{d}\phi}{\mathrm{d}t} = f + \int Qr\,\mathrm{d}t, \tag{10}$$

where f is a constant of integration. To obtain an expression for the right-hand member of (10) from which $\mathrm{d}t$ is absent, he multiplied the equation by $Qr\,\mathrm{d}t$, integrated, completed the square, and extracted the square root:

$$f + \int Qr\,\mathrm{d}t = [f^2 + 2\int Qr^3\,\mathrm{d}\phi]^{\frac{1}{2}}. \tag{11}$$

Substitution of (11) into (10) then gave

$$\mathrm{d}t = \frac{r^2\mathrm{d}\phi}{[f^2 + 2\int Qr^3\,\mathrm{d}\phi]^{\frac{1}{2}}} = \frac{r^2\mathrm{d}\phi}{f\sqrt{1+2\rho}}, \tag{12}$$

where

$$\rho = \frac{1}{f^2}\int Qr^3\,\mathrm{d}\phi.$$

Equation (12) permits the elimination of $\mathrm{d}t$ from equation (9), yielding an equation for the orbit in which r is a function of ϕ alone. By twice integrating the latter equation, Clairaut then obtained his solution in the form

$$\frac{f^2}{Mr} = 1 - g\sin\phi - c\cos\phi + \sin\phi\int\Omega\cos\phi\cdot\mathrm{d}\phi - \cos\phi\int\Omega\sin\phi\cdot\mathrm{d}\phi, \tag{13}$$

where g and c are constants and

$$\Omega = \frac{\dfrac{Pr^2}{M} + \left(\dfrac{Qr}{M}\right)\left(\dfrac{\mathrm{d}r}{\mathrm{d}\phi}\right) - \left(\dfrac{2}{f^2}\right)\displaystyle\int Qr^3\mathrm{d}\phi}{1 + \left(\dfrac{2}{f^2}\right)\displaystyle\int Qr^3\mathrm{d}\phi}. \tag{14}$$

Equation (13), when the terms involving Ω are deleted, reduces to the equation of a conic section:

$$\frac{p}{r} = 1 - c\cos(\phi - \alpha), \tag{15}$$

where the parameter $p = f^2/M$, the eccentricity $c' = [g^2 + c^2]^{\frac{1}{2}}$, and $\tan\alpha = g/c$.

Suppose now that Ω can be expressed in the form

$$A\cos n\phi + B\cos p\phi + \ldots,$$

where n, p, etc. are numerical factors multiplying the angle ϕ. Clairaut showed that in this case the final two terms of (13) will be expressible in the form

$$+\frac{A}{n^2-1}[\cos\phi - \cos n\phi]$$
$$+\frac{B}{p^2-1}[\cos\phi - \cos p\phi]$$
$$+\ldots.$$

To reduce Ω to the required form, it was necessary to replace r where it occurred in (14) by an approximate expression in terms of ϕ. It would be a mistake, Clairaut urged, to use formula (15), which represents the fixed ellipse resulting when the perturbing forces are absent, since the observations teach us that the line of apsides of the orbit rotates about the central body: in a few revolutions the formula would be inaccurate. Instead, Clairaut proposed using

$$\frac{k}{r} = 1 - e\cos m\phi, \tag{16}$$

which represented a rotating ellipse. Here k, e, and m were constants yet to be determined, and $m\phi$ was the real anomaly, completing 360° when the planet or satellite returned to its apse. Equation (16) would be found to be justified, Clairaut said, if the larger terms of equation (13), once the substitutions were carried out, proved to have the form of (16), permitting k, e, and m to be identified in terms of the other constants of the theory, while the remaining terms were relatively small by comparison. This turned out to be the case.

Besides r, the formula (14) for Ω contained the perturbing forces P and Q, and these, too, needed to be expressed as functions of ϕ. The resulting expression for Ω contained products of sines and cosines of several multiples of ϕ; using trigonometric identities Clairaut converted the products into sums and differences. With Ω thus transformed, the integrations required in the last two terms of (13) were easily performed.

After r had been determined in terms of ϕ, the time corresponding to a given value of r could be determined by integrating equation (12).

Such is the outline of Clairaut's procedure. It involved approximations at every stage. In order to abbreviate the calculation, Clairaut introduced numerical substitutions appropriate to the lunar case before completing his derivation, and so failed to attain a fully algebraic formulation of his theory. His reliance on empirical results to help resolve the differential equations – particularly his use of equation (16) as qualitatively justified by observations – was arguably objectionable from a mathematician's standpoint, and d'Alembert did not fail to urge the objection. Clairaut's rebuttal was that his method was easier to apply than the methods of d'Alembert and Euler; and such followers as Lalande agreed.

D'Alembert's procedure for determining perturbations

D'Alembert's analysis, as presented in his mature lunar theory of 1754, was no doubt superior to Clairaut's from a logical point of view; it avoided the faults that d'Alembert had found in Clairaut's theory. D'Alembert took the ecliptic as his basic plane of reference. He described the Moon's motion by means of the projection of the orbital motion onto the plane of the ecliptic, together with the momentary inclination of the Moon's orbital plane to the ecliptic, and position of the line of nodes. He warned that the ecliptic could not be identified as a plane through the Earth's orbit since, owing to the perturbations in the Earth's motion caused by the Moon, this orbit was not in fact confined to a single plane.

The warning was apropos, but in several writings d'Alembert mistakenly identified the ecliptic as the plane in which the Earth is moving when the Moon, the Sun, and the Earth's line of motion are all in the same plane; and he deduced from this identification that the Sun would be seen to depart from the ecliptic by as much as 15″. An anonymous and largely laudatory review of the second part of his *Recherches sur différens points importans du système du monde* (1754) in the *Journal des sçavans* for December 1754 (by Clairaut?) urged that the ecliptic be identified rather as the plane traced out by the centre of gravity of the Earth and the Moon; the apparent departure of the Sun from this plane is never more than 1″. This definition is in fact more in accord with the ecliptic that astronomers determine observationally, but d'Alembert refused to accept the criticism.

For the equation of the projection of the orbit onto the ecliptic, d'Alembert obtained

$$\frac{\mathrm{d}^2u}{\mathrm{d}\phi^2}+u-\frac{1}{u^2g^2}\left[\frac{S-\left(\frac{Q}{u}\right)\left(\frac{\mathrm{d}u}{\mathrm{d}\phi}\right)}{1+\frac{2}{g^2h^2}\int\left(\frac{Q}{u^3}\right)\mathrm{d}\phi}\right]=0, \quad (17)$$

where $u=1/r$, the reciprocal of the projection of the radius vector onto the ecliptic; S is the sum of all the radial forces, that is, M/r^2+P; g is the initial velocity; and h is the sine of the angle that the Moon's initial direction of motion forms with r.

In preparing to solve (17) to a first order of approximation, d'Alembert replaced u by $K+u'$, where K was the mean reciprocal radius vector, and u', because of the near-circularity of the orbit, was a variable that would always remain small. The substitution made it possible to put (17) in the form

$$\frac{\mathrm{d}^2u'}{\mathrm{d}\phi^2}+N^2u'+R=0, \quad (18)$$

where R is a function of u', $\mathrm{d}u'/\mathrm{d}\phi$, and sines and cosines of multiples of ϕ, and N^2 is a constant differing from 1 by a small quantity of the order of magnitude of the perturbing force.

The reason that N^2 is not identically 1, as substitution of $K+u'$ for u in the second term of (17) might lead one to expect, is that terms linear in u' emerge from the third term of (17). Thus amidst the terms giving his first approximation to the third term of (17), d'Alembert found $n'^2/2g^2u^3$, where n' is the mean rate of motion of the Sun, and on substitution of $K+u'$ for u, this became

$$\frac{n'^2}{2g^2(K+u')^3}=\frac{n'^2}{2g^2K^3}\frac{1-3u'}{K},$$

which contains a term linear in u'.

In his integration of equation (18), d'Alembert employed an exponential integrating factor with complex exponent; he then converted the complex exponential terms into sines and cosines. His first-order solution for u' thus took the form

$$u'=D\cos N\phi-\frac{H}{N^2} \qquad (19)$$
$$+\frac{H}{N^2}\cos N\phi+Km'^2\cos 2(1-m')\phi\ldots.$$

Here H is a constant of integration, m' is the ratio of the Sun's mean motion to the Moon's mean motion, and N is given by $(1-3m'^2/4+\ldots)$. In contrast to Clairaut, d'Alembert has here derived the fact of the apsidal motion from the differential equation, rather than assuming it as an observational result.

In his memoir of 1747, d'Alembert had recognized that N as just calculated gives but half of the observed apsidal motion, but in opposition to Clairaut had supposed the discrepancy to be caused by some non-gravitational force, say magnetism, rather than indicating a need to revise the inverse-square law. In his mature lunar theory of 1754, d'Alembert carefully distinguished the orders of magnitude involved in successive approximations. Equation (19) gives a first-order approximation to u', from which a first-order value of $du'/d\phi$ can be calculated. When these values of u' and $du'/d\phi$ are substituted into the term R of (18) and the integration is carried out, there results a new and more exact expression for u'. The process can be repeated as many times as may be necessary to attain the same precision in the calculations as is found in the observations. D'Alembert carried the iterations far enough to obtain the first four terms in the motion of the apogee, namely 1° 30′ 37″, 1° 3′ 21″, 23′ 30″, 5′ 5″, which add to 3° 2′ 33″, just 1′ 4″ less than the observational value. He gave the algebraic expression for the second term in N, namely $-\frac{225}{32}m'^3$, and formulas from which the algebraic expressions for the next two terms could be determined. Here as elsewhere he carried the algebraic articulation of the theory much farther than Clairaut.

Although the result for the apsidal motion appears to confirm the Newtonian law, d'Alembert warned that it is impossible to demonstrate rigorously the exactitude of the formulas he deduced. Further steps in the iterative process might fail to be convergent. Whether the resulting tables were empirically reliable could be asserted only after a long series of observational tests. D'Alembert himself provided a detailed comparison between his own and the Newtonian tables, as given in P.-C. Le Monnier's *Institutions astronomiques*. He objected to the tables of Tobias Mayer because Mayer provided no explanation of their theoretical and observational basis; the empirical success of such tables, he complained, told nothing about the adequacy of the inverse-square law. In giving a detailed algebraic development of Newton's theory, with clearly defined stages of approximation, d'Alembert prepared the way for an exact test as well as for more precise approximations. But the research programme thus laid out does not appear to have been pursued. As was pointed out in Chapter 18, the lunar tables that came to be generally accepted during the late eighteenth century were those of Tobias Mayer, the obscurity of their theoretical basis notwithstanding.

Later contributions to perturbation theory by Clairaut, d'Alembert, and Euler

As we saw in Chapter 18, Clairaut applied his analytic procedure to the determination of the planetary and lunar perturbations of the Earth's motion, and his results were incorporated in Lacaille's solar tables. In 1757–58, aided by J.-J. L. de Lalande and Mme Nicole-Reine Lepaute, Clairaut used numerical integration to compute the return of Halley's Comet (see Chapter 19). Between 1758 and 1762 he defended himself in an acrimonious controversy initiated by d'Alembert over both this computation and his method of determining lunar and planetary perturbations. He died in 1765 at age 52.

D'Alembert's later memoirs on the lunar theory, the precession of the equinoxes, the shape of the Earth, and other topics relevant to celestial mechanics, were published in his *Opuscules mathématiques* (eight volumes, 1761–80). These memoirs generally presuppose d'Alembert's choice of symbols in earlier writings, and so are not easy reading. They do not appear to contain major discoveries, as d'Alembert himself was admitting in the last volumes. But Lagrange read them all, and claimed to profit from them.

Euler, in contrast, continued to make contributions of considerable importance. In two memoirs completed in 1762 and 1763, and printed in the volumes of the St Petersburg memoirs for 1764 and 1765, he gave the first treatment of what has since come to be known as "the restricted problem of three bodies". Two of the bodies are assumed to be much more massive than the third, which is thus taken to have no perturbing effect; consequently the first two bodies move in exact accord with the solution of the two-body problem. Euler established that there are collinear solutions, in which the massless 'Moon' stays in constant opposition or conjunction with the 'Sun', or moves in small oscillations about the point of opposition or conjunction. (This discovery is sometimes attributed to Lagrange, who published the result only in 1772.)

Euler further showed that, if the 'Sun' and 'Earth' of the restricted three-body problem be identified with our Sun and Earth, then the 'Moon' could be no more than four times the distance of our actual Moon from the Earth, and remain in constant conjunction or opposition. Were its relative distance (compared with the Earth–Sun distance) greater, it would cease to follow the laws of satellites. But – unless its distance were to become much greater – it would not yet orbit the Sun in the manner of a planet, but rather move in so complicated a manner as to defy mathematical analysis. It was Euler's conviction, asserted more than once, that the Creator in designing the universe had taken into account the weakness of the human intellect, and made the celestial bodies either planets or satellites, so that none of them fell into the incomprehensible in-between class.

During the early 1760s Euler became distrustful of the trigonometrical series he had introduced a decade and a half before. In the case of several planets these series did not appear to be very convergent, and Euler feared that the neglected terms might add up to a significant quantity. He therefore proposed (in the Berlin *Mémoires* for 1763, published in 1770) to follow Clairaut's procedure in dealing with Halley's Comet, and use numerical integration; in contrast to Clairaut, however, he would apply the numerical integration directly to the differential equations. The position and velocity of a planet having been determined observationally for some moment of time, the increments to the x, y, and z position coordinates and velocity components were to be calculated from the differentials during successive small intervals of time, taking account of all the forces to which the planet was subject. The error could be expected to increase as the calculation was continued, so that at some point it would be necessary to determine a new starting-point from observation. The initial position and velocity were to be deduced from a sequence of observations by means of the calculus of finite differences – a method later applied by Laplace to determine positions and velocities of comets.

Euler believed the method to be capable of high accuracy. As he pointed out, it was applicable whether the perturbing bodies were one or many.

In the early 1770s, in studies of both lunar and planetary perturbations, Euler introduced a coordinate frame rotating with the mean speed of the perturbed body. Thus in the new lunar theory which he constructed during these years, and for which he received prizes from the Paris Academy in 1770 and 1772, a line rotated about the Earth in the plane of the ecliptic with the Moon's mean speed. The vertical distance from the Moon to the plane of the ecliptic gave the z-coordinate; a line dropped from the foot of this vertical at right angles to the rotating line gave the y-coordinate; and the distance so determined along the rotating line was set equal to $1 + x$, the Moon's mean radius vector being normalized to equal 1. The variations in x, y, z to be determined from the differential equations in these variables were in this way made very small. During the next century, rotating coordinates became standard in treatments of the three-body problem; but they were generally made to rotate with the mean speed of the perturbing rather than that of the perturbed body.

For the prize contests of 1770 and 1772, the Paris Academy has asked the contestants

to perfect the methods on which the lunar theory was founded, and in this way to determine the equations of this satellite that were still uncertain, and in particular to examine whether by this theory one could account for the secular equation in the movement of the Moon.

The gradual acceleration of the Moon's mean motion since ancient times, first detected by Halley, was by this time a well-established fact. But Euler in his deductions found no secular change in the mean movement of the Moon, and – believing his method to be a net from which no detectable inequality could escape – concluded that no such change could be produced by the forces of gravitational attraction. In his prize memoir of 1772 he concluded: "there no longer remains any doubt that this secular equation ... is the effect of the resistance of the milieu in which the planets move". He added, however, that lunar tables could not be made accurate to within less than 30 arcseconds without taking the planetary perturbations of the Moon into account.

In a memoir of 1771 on the perturbations of the Earth due to Venus, besides employing rotating coordinates, Euler espoused the idea of plotting the functions to be integrated at 5° intervals of the angle between the *radii vectores* of Venus and the Earth, and performing the integrations graphically, so as to avoid reliance on trigonometric series. By this time Euler was totally blind, and the calculations were performed by Anders Johan Lexell, a colleague at the St Petersburg Academy. The table of perturbations that emerged proved to be in sharp disagreement with the tables then in use, those of N.-L. de Lacaille and Mayer. Laplace, at the urging of Lalande, wrote to Euler and Lexell to ask their opinion as to the cause of the discrepancy. Lexell, as he reports in a memoir of 1779, discovered that his calculation contained a significant error, and on performing the whole series of quadratures over again, this time at 1° intervals, obtained results in agreement with those in the earlier tables.

Euler, like d'Alembert, died in 1783, two years before Laplace's announcement that he had at last derived from Newton's law the anomalies in the motions of Saturn and Jupiter that had puzzled the members of the prize commission of the Paris Academy thirty-nine years before. Two years later he would claim to have similarly derived the secular acceleration of the Moon – a claim that would turn out to be only partially justified, as we shall see in Chapter 22.

Further reading

Alfred Gautier, *Essai historique sur le problème des trois corps* (Paris, 1817)

Victor J. Katz, The calculus of trigonometric functions, *Historia Mathematica*, vol. 14 (1987), 311–24

F. Tisserand, *Traité de mécanique céleste*, vol. 3: *Exposé de l'ensemble des théories relatives au mouvement de la Lune* (Paris, 1894)

Curtis Wilson, Perturbations and solar tables from Lacaille to Delambre: the rapprochement of observation and theory, *Archive for History of Exact Sciences*, vol. 22 (1980), 53–304

Curtis Wilson, The great inequality of Jupiter and Saturn: from Kepler to Laplace, *Archive for History of Exact Sciences*, vol. 33 (1985), 15–290

21

The work of Lagrange in celestial mechanics

CURTIS WILSON

Joseph Louis Lagrange was born on 25 January 1736 into a well-to-do French–Italian family of Turin, Italy. His father, who was treasurer to the King of Sardinia, lost the family fortune in speculation, so that the young Lagrange was forced to think of embarking upon a career. In later life Lagrange remarked that, if he had been rich, he probably would not have devoted himself to mathematics. In his teens, the chance reading of an essay by Edmond Halley aroused his interest; he was soon awakened to his own extraordinary talent for mathematical analysis. At 18 he sent Leonhard Euler an account of his first important invention, a new mode of analysis which Euler later named the calculus of variations. By age 19 he was teaching mathematics in the artillery school of Turin. By his thirtieth year he had produced a prodigious body of work, and earned the profound respect of Jean le Rond d'Alembert in Paris – an arbiter of fate for up-and-coming mathematicians – as well as of Euler in Berlin. Already he had won two of the contests of the Paris Academy of Science (later, he would garner three more of these lucrative prizes).

In his native city Lagrange helped found a Royal Academy of Sciences; however, it received only meagre financial support, and Turin remained a scientific backwater. In 1766 Euler was leaving Berlin to return to St Petersburg, and through his recommendation, and more especially that of d'Alembert, Frederick II of Prussia was led to designate Lagrange as Euler's successor in the Berlin Academy, heading the mathematical–physical section. The greatest king in Europe, Frederick announced, wished to have the greatest mathematician in Europe at his court. Lagrange remained in Berlin for twenty-one years, but in 1787, on the death of Frederick II, chose to accept Louis XVI's invitation to come to Paris. He spent the last twenty-six years of his life there as a member of the Academy of Sciences and its successor, the Institut National. His death came on 10 April 1813, before he had completed revisions for the second edition of Vol. 2 of his masterwork, the *Mécanique analytique*.

Displacements from one city to another altered little the even tenor of Lagrange's daily life. It regularly included a solitary afternoon walk and a morning and evening of study and writing. Unlike d'Alembert who loved controversy and Euler who could be irascible, Lagrange was a peace-loving man; Frederick II described his character as "full of sweetness and modesty". However, the Parisian mathematician Jean-Charles Borda, himself a stubborn man, remarked after an argument with Lagrange that he had never known anyone more obstinate.

Lagrange's mathematical memoirs had an elegance often commented on. According to the mathematician J. B. J. Fourier, writing in 1829,

> *All [Lagrange's] compositions are remarkable for a singular elegance, for the symmetry of the forms and the generality of the methods, and, if one may so speak, the perfection of analytic style.*

Lagrange's contributions to celestial mechanics were of the first importance. He studied intensively the earlier attempts to deduce the consequences of Newton's gravitational law analytically, penetrated the difficulties to which they led, and showed how these difficulties could be overcome. He perfected the Eulerian methods, giving them an elegant form. Many of the algorithms he introduced came to be widely adopted. His solutions supported and inspired a vision of a stable solar system in which all perturbations could be expected to be oscillatory and confined within narrow bounds. His achievement prepared the way for much later

21.1. Joseph Louis Lagrange

work, above all that of Pierre-Simon Laplace (1749–1827). In Laplace's mathematical development, Lagrange's influence was pervasive. In a letter to Lagrange of November 1778, Laplace wrote:

No one reads you with more pleasure than I, because no geometer [i.e., mathematician] appears to me to have carried to as high a point as you all the parts that go to make a great analyst. Permit me this avowal of my gratitude and respect, since it is principally by an assiduous reading of your excellent works that I have formed myself.

That Lagrange and Laplace felt themselves to be rivals we cannot doubt, but the rivalry never became open or ugly. Lagrange once wrote to Laplace: "If you are not jealous because of your success, I am not less so because of my character."

In reviewing Lagrange's contributions to celestial mechanics, we shall mainly follow a chronological order. However, in some cases he returned to a topic after a lapse of years to modify his original approach or conclusions, and in these cases we shall take up consecutively his several assaults on the same problem.

The libration of the Moon

Lagrange's first memoir on celestial mechanics was the prize-winning essay of the Paris Academy's contest of 1764 concerning the libration of the Moon. The task imposed by the Academy was to explain (a) "the physical reason why the Moon always presents to us approximately the same face", and (b) "how to determine by observation and theory whether the Moon's axis is subject to a movement like that which affects the Earth's axis and causes the precession of the equinoxes and the nutation". Lagrange resolved the first of these problems convincingly, but his result for the second failed to account for the appearances. Sixteen years later, however, he returned to the question and achieved a satisfactory resolution.

Isaac Newton had dealt with the Moon's libration in Props. 17 and 38 of Bk III of the *Principia*. Proposition 17 accounted for the Moon's *optical* libration in longitude. Suppose the Moon's rotation about its axis to be uniform, with a period equal to its orbital period about the Earth. The orbital motion, being eccentric and in accord with the law of areas, is not quite angularly uniform about the Earth's centre. The consequence is that between perigee and apogee the Moon will appear to librate about 6° to one side or the other from the face it shows at perigee and apogee.

Besides the optical libration, Newton hypothesized that there would be a *physical* libration, as he explained in Prop. 38 and its corollary. If the Moon had been originally fluid, the Earth's attraction would raise a lunar tide with a height which he computed to be 93 feet. (The result is too small, principally because Newton's value for the mass ratio of Moon to Earth was too large.) The figure of the Moon would thus be a spheroid, whose greatest diameter when extended would pass through the centre of the Earth, and exceed the diameter in the direction of its orbital motion by 186 feet. (In fact, the excess is about 0.25 miles or 0.4 km.) "Such a figure," Newton claimed, "the Moon possesses, and must have had from the beginning." In consequence of this shape, the Moon would be in equilibrium only if its long axis passed through the Earth; if displaced from this orientation it would librate to either side, the period of the oscillations being very long because of the weakness of the forces exciting them. These qualitative conclusions were to be corroborated in Lagrange's analysis.

Gian Domenico Cassini (Cassini I, 1625–1712) was the first to give a precise description of the lunar libration, basing it on measurements of the positions of the lunar maria within the visible face of the Moon; his conclusions – but not his observations – were published posthumously in the *Mémoires* of the Paris Academy for 1721. The Moon's equator, he found, was constantly inclined to the ecliptic at an angle of 2°.5, and to the plane of its orbit at an angle of 7°.5; the axis of rotation was always in a plane parallel to the great circle of the celestial sphere passing through the poles of the orbit and those of the ecliptic. Further, he found the Moon's period of rotation to be just equal to its Draconitic or nodical month, namely 27 days, 5 hours and a fraction, in which the Moon returns to the node of its orbit with the ecliptic.

These conclusions, with some modification, were confirmed by Johann Tobias Mayer (1723–62) in an treatise published in the *Kosmographische Nachrichten* of Nuremberg for 1750. In deriving his results, Mayer used numerous observations of the positions of the lunar markings, and analyzed them by means of Euler's method of equations of condition (see p. 101 in Chapter 20). He found the inclination of the lunar equator to the ecliptic to be 1° 29′ rather than 2° 30′, and the intersection of these two planes to be always parallel, or nearly so, to the line of nodes of the lunar orbit. Like Cassini, he asserted a precise equality between the Moon's period of axial rotation and the Draconitic month.

In the formulation of the topic for the contest of 1764, d'Alembert no doubt had a hand. In 1756 he had presented to the Paris Academy a memoir on the precession and nutation of a spheroid with dissimilar meridians (that is, an ellipsoid with three different axes); this was published in the Paris *Mémoires* for 1754. He applied his formulas to the Moon in the second volume of his *Opuscules mathématiques*, published in 1761; here he derived values for the precession and nutation of the lunar axis, but they did not appear to be in agreement with observations. Lagrange was familiar with both of d'Alembert's memoirs, as well as the memoirs giving the observational results of Cassini and Mayer, when he wrote his prize essay of 1764.

In this essay Lagrange started from the principle that he would later make fundamental in his *Mécanique analytique* (1788): d'Alembert's principle of dynamics, as formulated by means of the principle of virtual work. According to d'Alembert's principle, a dynamical system may be viewed as a system of forces in static equilibrium, if the products 'mass × acceleration' for each particle, taken negatively, are added to the other forces. According to the principle of virtual work (Lagrange called it the principle of virtual velocities), if to a system of bodies in static equilibrium we give a small movement, and then form the product of each force by the component of displacement in the direction of the force, the sum of these products will be zero.

Lagrange was here formulating his dynamical principle for the first time, and claiming that it comprehended the solution of all problems concerning the motion of bodies. In symbolizing the virtual displacements he used the special differential operator 'δ', which he had introduced in his first letter to Euler on the calculus of variations. An important feature of the new calculus was the commutativity of the operator 'δ' with the ordinary differential operator 'd', employed in symbolizing real velocities and accelerations.

Applying his dynamical principle to the case of the Moon attracted gravitationally by the Earth and the Sun, Lagrange obtained the formula

$$\frac{1}{dt^2}\int \alpha[d^2X\delta X + d^2Y\delta Y + d^2Z\delta Z] \qquad \text{(A)}$$
$$+T\int\left(\frac{\alpha}{R^2}\right)\delta R + S\int\left(\frac{\alpha}{R'^2}\right)\delta R' = 0.$$

Here X, Y, Z are rectangular coordinates of a particle α of the Moon's mass; R and R' are the distances of the particle from the Earth and the Sun, and T and S are the masses of the Earth and the Sun; δX, δY, δZ, δR, $\delta R'$ are the variations in the coordinates and distances brought about by a small 'virtual' motion in the system; and the integrals are taken with respect to all particles α in the body of the Moon.

Lagrange's next step was to transform (A) by a change of variables. Replacing X, Y, and Z, he introduced six new variables, three of them specifying the Moon's orientation with respect to the ecliptic, and three specifying the position of an arbitrary mass-particle of the Moon with respect to axes fixed in the Moon's body. The variations of the first three accounted between them for all possible shifts in the position of the Moon's body about its centre. The expressions by which they were multiplied in the final form of (A) could be set separately equal to zero, yielding three differential equations by which the Moon's motion about its centre was determined.

In the transformation of (A) just described, Lagrange assumed that if the external forces due to the Earth and Sun were absent, the Moon would rotate uniformly about its polar axis. To show that this assumption was not very restrictive he proved that in every rigid body there are three mutually perpendicular axes about which the centrifugal forces cancel, so that the body can rotate without wobbling. Euler had informed Lagrange of his discovery of this result in a letter of October 1759, but Lagrange evidently developed his proof of it independently. Euler's proof, first presented to the Berlin Academy in July 1758, was not published till 1765, when it appeared in both the Berlin *Mémoires* for 1758 and his *Theoria motus corporum solidorum*.

For one of the second-order equations he had derived from (A), Lagrange was able to obtain an approximate solution giving the libration of the Moon in longitude:

$$\theta = -a \sin mV + C \sin V\sqrt{\frac{3M}{H}} + \left(\frac{N}{2M}\right)\left\{1 - \cos V\sqrt{\frac{3M}{H}}\right\}. \qquad (1)$$

Here θ is the angular distance between the apparent centre of the Moon, and a 'prime meridian' of the Moon, which Lagrange chose so that its plane contained the radius vector from Earth to Moon when the Moon's mean place coincided with its true place. V is the Moon's mean motion in longitude, and m is a number slightly less than 1, such that mV is the Moon's mean anomaly. The term $(-a \sin mV)$ is the main term in the Moon's equation of centre, and thus represents the main term in the optical libration. C is an arbitrary constant. H, M, and N are constants determining the Moon's shape and distribution of mass, M and N being very small relative to H, with N (the product of inertia about the polar axis) possibly equal to zero, and M/H giving the ellipticity of the Moon's equator.

The last two terms of the equation for θ represent the physical libration, which will have as period the fraction $(3M/H)^{\frac{1}{2}}$ of the Moon's sidereal period. Because C is arbitrary, Lagrange was able to conclude that the Moon's initial velocity of rotation need not have been exactly equal to its mean motion in longitude, provided only that the difference was relatively small, and that M was not zero or negative. On the assumption of the Moon's homogeneity, Lagrange found for the physical libration a period of about 119 months.

The second and third equations derived from (A) were also second-order differential equations, but the coefficients of the terms containing the second-order differentials were relatively small, and Lagrange decided that these terms could be dropped. Integration of the resulting first-order equations yielded formulas for the inclination of the lunar equator to the ecliptic, and for the longitude of the descending node of the lunar equator where it intersects a plane through the Moon's centre parallel to the ecliptic. Cassini and Mayer, as we have seen, claimed that the longitude of the descending node of the lunar equator was coincident, or nearly so, with the longitude of the ascending node of the lunar orbit; but Lagrange could find in his analysis no reason for this coincidence.

He returned to the problem of the lunar libration in 1780, in a memoir presented to the Berlin Academy. In this second assault he made use of several new techniques.

Once more he took as starting-point his fundamental dynamical principle. But by 1780 he had discovered an algorithm which expedites the derivation of the differential equations of motion. Let the generalized coordinates adopted in formulating the problem (they should be as few as possible, and independent of one another) be ϕ, ϕ', . . . , and let T and V be the functions of these coordinates that we now know as the kinetic and potential energy. Then the differential equation for any one of the variables ϕ will be given by

$$\mathrm{d}\frac{\delta T}{\delta \mathrm{d}\phi} - \frac{\delta T}{\delta\phi} + \frac{\delta V}{\delta\phi} = 0. \qquad \text{(B)}$$

The operator 'δ' here signifies partial differentiation.

For the coordinates of the centre of gravity of the Moon, Lagrange adopted the rotating coordinate system used by Euler in his *Nouvelle théorie de la lune* of 1772: the curtate radius vector, rotating with the Moon's mean speed in the plane of the ecliptic, is $1+x$, where the mean radius vector has been normalized to equal 1; y is at right angles to the radius vector in the plane of the ecliptic; and z is at right angles to the plane of the ecliptic. The variables x, y, z are advantageous because they remain very small, so that their powers and products of more than two dimensions can be neglected. For Lagrange there was the additional advantage that Euler had already calculated their values "with great precision", insofar as they depended on the action of the Earth and the Sun on the Moon, regarded as a point. (In his calculation, Euler had not taken into account the non-sphericity of the Moon, an omission that Lagrange corrected in the present memoir, but without finding any notable effect.)

The most important innovation in the memoir of 1780 was a new choice of variables for the determination of the motions of the Moon's body about its center of gravity. Let ω be the inclination of the Moon's equator to the ecliptic; ψ the longitude of the ascending node of the lunar equator with a plane through the Moon's centre parallel to the ecliptic; ϕ the angular distance of a prime meridian on the Moon from this node. For two of his new variables, Lagrange introduced

$$s=\tan\left(\frac{\omega}{2}\right)\cdot\sin\phi,$$

$$u=\tan\left(\frac{\omega}{2}\right)\cdot\cos\phi.$$

Since according to the observations ω remains less than about 2°, s and u will remain small. An effect of the new definitions is to avoid explicit occurrence of trigonometric functions in the differential equations. Lagrange had used a similar trick in his memoir of 1774 on the secular equations of the nodes and inclinations, to be dealt with in a later section.

To obtain his final new variable, Lagrange set $\theta=\phi+\psi$, and then noted that θ would be equal to $180°+t$, where t is the mean motion of the Moon, if there were no physical libration. Then if r is the physical libration, $\theta=\phi+\psi=180°+t+r$, and r will always remain very small. The motions of the Moon about its centre will be completely determined in terms of r, s, and u.

The three second-order differential equations that resulted from Lagrange's algorithm were linear in r, s, and u, and involved as constants the moments and products of inertia of the Moon about its polar axis and two axes in the plane of its equator, one of them being directed toward the Earth, and the other at right angles thereto. The equations showed that the products of inertia must be very small or zero, and Lagrange set them equal to zero, thereby adopting the hypothesis that the three axes mentioned were natural axes of rotation.

Lagrange's solution for r showed that it must be oscillatory, for the only alternative was that it would contain a term proportional to t, and so increase without bound, contrary to observation. Here as in the memoir of 1764, Lagrange saw the physical libration, however small, as explaining how the Moon's mean rotational and orbital velocities would have become equal, if not created exactly so at the beginning.

The main achievement of the memoir of 1780 was to obtain solutions for s and u while taking account of the second-order differentials. In a discussion of these solutions, Lagrange established that the angular distance of the Moon from the mean ascending node of its orbit must be equal to the mean angular distance of the Moon's prime meridian from the node of the lunar equator. Thus the equality that had appeared as an arbitrary fact in his earlier memoir was at last shown to be a consequence of the theory.

Perturbations in orbital motion: two memoirs of 1766

Lagrange's second memoir on celestial mechanics was his *Recherches sur les inégalités des satellites de Jupiter*, which won the Paris Academy's prize of 1766. This extensive memoir shows Lagrange in full command of the earlier work on celestial mechanics of Alexis-Claude Clairaut (1713–65), Euler, and d'Alembert, and arriving independently at the conclusion that Euler had set forth in his still unpublished memoir of 1752 on the inequalities of Jupiter and Saturn: the discovery that the orbital elements of a perturbed planet, as determined by observation, include major components deriving from perturbation.

We first take notice of the symmetry that Lagrange gave to the equations of motion. Let r be the curtate radius vector in the plane of reference (the orbital plane of Jupiter); u its reciprocal; ϕ, the longitude measured in this plane; p, the tangent of the latitude; F, the force that Jupiter exerts on a satellite at unit distance; R, Q, and P the components of the perturbing force in the three coordinate directions; and c an arbitrary constant. Lagrange's equations for u and p are:

$$\frac{\mathrm{d}^2u}{\mathrm{d}\phi^2}+u-\frac{F(1+p^2)^{3/2}+Rr^2+Qr\left(\frac{\mathrm{d}r}{\mathrm{d}\phi}\right)}{c^2+2\int Qr^3\mathrm{d}\phi}=0, \quad (2)$$

$$\frac{\mathrm{d}^2p}{\mathrm{d}\phi^2}+p-\frac{r^3\left[P-pR+Q\left(\frac{\mathrm{d}p}{\mathrm{d}\phi}\right)\right]}{c^2+2\int Qr^3\mathrm{d}\phi}=0. \quad (3)$$

If R, Q, and P are set equal to zero, the solutions of these equations become

$$u-\frac{F}{c^2}=\rho\cos(\phi-\alpha), \quad (4)$$

the equation of an ellipse of which c^2/F is the semi-parameter, $\rho c^2/F$ the eccentricity, and α the longitude of the inferior apse; and

$$p=\lambda\sin(\phi-\epsilon), \quad (5)$$

where λ is the tangent of the inclination and ϵ the longitude of the node. But even if the perturbing forces are not zero, Lagrange observed, these solutions will still be valid if ρ, α, λ, and ϵ are allowed to vary. Euler had given differential equations for the variations of the inclination and node; Lagrange set them forth alongside the entirely parallel differential equations for the variations of the eccentricity and aphelion. But in determining the perturbations in the present memoir, he did not make use of this idea (it would be the major theme of his later work in perturbation theory). Instead, he employed the method of 'absolute perturbations', in which the perturbations are regarded as independent of the orbital elements.

For this purpose, Lagrange set

$$r=a(1+nx), \quad \phi=\mu t+ny, \quad p=nz, \quad (6)$$

where a and μ are the mean values of r and $\mathrm{d}\phi/\mathrm{d}t$, and n is a very small coefficient. These expressions were substituted into the equations of motion, yielding differential equations in x, y, and z. Terms involving n to a higher power than the second were discarded.

Next, Lagrange developed expressions for the perturbing forces R, Q, and P, due to the action of the Sun and the mutual actions of the satellites. These force components involved the inverse third and fifth powers of the distances between the satellites, which Euler had first shown how to approximate as trigonometric series. Lagrange gave an elegant method of deriving the coefficients of these series by way of De Moivre's theorem,

$$(\cos\theta+\mathrm{i}\sin\theta)^n=\cos n\theta+\mathrm{i}\sin n\theta.$$

The resulting expressions for R, Q, and P were series, to be substituted into the differential equations for x, y, and z.

The first step in obtaining solutions to the differential equations was to solve the simpler equations obtained when all the terms involving the factor n were deleted. Lagrange carried out this step for the first satellite; since he was using what amounts to indicial notation (he used 'traits' or prime marks, which we here replace by subscripts), the solutions for the other satellites could at once be obtained by an appropriate change in indices. The reduced equations in x and z had the general form

$$\frac{\mathrm{d}^2u}{\mathrm{d}t^2}+M^2u+T=0, \quad (7)$$

where M is a constant and T is a function composed of sines and cosines of multiples of the time t. Lagrange gave the solution in exactly the form used by Clairaut; thus the terms

$$B\cos pt+b\sin pt$$

in T led to the following terms in the solution:

$$\frac{B}{p^2-M^2}[\cos pt-\cos Mt] \quad (8)$$
$$+\frac{b}{p^2-M^2}[\sin pt-b\sin Mt].$$

The solution would be invalid, of course, if $p=M$.

In the first approximation Lagrange found for x_1 an expression of the following form:

$$\begin{aligned} x_1=\ &\epsilon_1\cos(M_1t+\omega_1) \quad (9)\\ &-K_{21}\cos(\mu_2-\mu_1)t-K_{22}\cos 2(\mu_2-\mu_1)t\ldots\\ &-K_{31}\cos(\mu_3-\mu_1)t-\ldots\\ &-K_{41}\cos(\mu_4-\mu_1)t-\ldots\\ &-K_5\cos 2(m-\mu_1)t.\end{aligned}$$

Here the first term, multiplied by na, gives the changing part of the first satellite's radius vector due to its elliptical motion, and the remaining terms give the perturbations due to the actions of the three other satellites and the Sun. The *K*s are constants, and μ_1, μ_2, μ_3, μ_4, and m are the mean motions of the satellites and the Sun about Jupiter. In the first approximation, M_1 proves to be equal to μ_1, which means that the line of apsides, in this approximation, remains fixed.

The corresponding expression for y_1 is

$$y_1 = -\frac{2\mu_1\epsilon_1}{M_1}\sin(M_1t+\omega_1) \tag{10}$$
$$+k_{21}\sin(\mu_2-\mu_1)t-k_{22}\sin 2(\mu_2+\mu_1)t\ldots$$
$$+k_{31}\sin(\mu_3-\mu_1)t-\ldots$$
$$+k_{41}\sin(\mu_4-\mu_1)t-\ldots$$
$$+k_5\sin 2(m-\mu_1)t.$$

The first term, multiplied by n, gives the main term of the equation of centre, and the remaining terms represent perturbations in longitude.

The solution of the third equation proved to be

$$p_1 = nz_1 = n\lambda_1\sin(N_1t+\eta_1), \tag{11}$$

where λ_1, and η_1, are arbitrary constants, and in the first approximation $N_1=\mu_1$, so that the nodes are fixed.

These first-approximation solutions proved to be all that was needed to predict the anomalies so far detected in the satellite motions. The Swedish astronomer Pehr Wilhelm Wargentin (1717–83), on the basis of a long series of careful timings of the eclipses of the satellites as they passed through Jupiter's shadow, and comparisons between his own observations and those of earlier astronomers, had found that the uniform circular motions previously ascribed to the first three satellites did not quite accurately hold, but were modified in each case by a single anomaly. Thus in the case of the first satellite Wargentin found a sinusoidal anomaly with a period of $437^d\ 19^h\ 41^m$ and a maximum value of 3.5 minutes of time. For this inequality to be accounted for by the theory, it had to be derivable from the formula for y_1.

Now in this formula we have first of all the term giving the equation of centre, but its coefficient contains an arbitrary constant, so that its size can only be determined by observation. Of the other terms, there is one that dwarfs all the others, the term with argument $2(\mu_2-\mu_1)t$. This term is especially large because (a) the coefficients of the terms in the solution have denominators of the form p^2-M^2, p being the coefficient of t and M being here identical with μ_1; and (b) the mean motion of the first satellite is very nearly twice that of the second (more precisely, $\mu_1{:}\mu_2=2.0075{:}1$). Thus the denominator of the term with argument $2(\mu_2-\mu_1)t$ is especially small:

$$2^2(\mu_2-\mu_1)^2-\mu_1{}^2=\mu_1{}^2\left[4\left(1-\frac{1}{2.0075}\right)^2-1\right]$$
$$=0.00749\mu_1{}^2.$$

Therefore the coefficient is especially large.

The period of the inequality with argument $2(\mu_2-\mu_1)$ is 1.762 731 77 days, while the synodic period of the first satellite with respect to the Sun is 1.769 860 526 days. (These values are derived from Wargentin's determinations of the synodic periods of the satellites, that is, their periods from one eclipse to the next. For convenience of calculation we have put them in decimal form.) Hence the period P in which the inequality returns to the same phase in eclipses will be given by

$$\frac{1}{P}=\frac{1}{1.76273177}-\frac{1}{1.769860526},$$

whence $P=437.63447$ days, very nearly the Q period found by Wargentin.

In the case of the second satellite, Lagrange's theory showed two perturbational terms far surpassing the others in size, the terms with arguments $2(\mu_3-\mu_2)t$ and $(\mu_1-\mu_2)$. The ratio $\mu_2{:}\mu_3$, like the ratio $\mu_1{:}\mu_2$, is just slightly in excess of 2:1, and as a result the coefficients of both terms are made large by small denominators. The two inequalities have very nearly the same period, and the period in which they return to the same phase with respect to eclipses of the second satellite is about 437.67 days, very nearly the same as that found for the first satellite. This result agreed with Wargentin's observational finding. In effect the theory yields a single inequality for the second satellite, for as Lagrange deduced from the observations, $2(\mu_3-\mu_2)t=180°-(\mu_1-\mu_2)t$, so that the sines of the two arguments are equal. Two decades later Laplace would undertake to explain *why* this special relation, first pointed out by Lagrange, must hold.

The theory of the third satellite contained a single sizeable perturbational term, with the argument $(\mu_2-\mu_3)t$, leading to an inequality in eclipses with a period, once again, of some 437.67 days. Wargentin had detected its presence, but had been unable to satisfy himself with regard to its quantity.

Having accounted for the inequalities detected by Wargentin, Lagrange turned in the remainder of his memoir to the further development of the theory. In the second stage of the method of absolute perturbations, the first-stage solutions are substituted back into the differential equations in the terms multiplied by the small coefficient n, and the differential equations are then solved once more. This process, Lagrange showed, results necessarily in terms in which the factor n does not appear – terms therefore belonging to the first-order solution.

Thus in the differential equation for x_1 there is a term of the form $nA_1x_2\cos(\mu_2-\mu_1)t$, where A_1 is a constant and x_2 in the first approximation is given by $e_2\cos(M_2t+\omega_2)$. For the second-stage approximation the constant M_2 takes the form $\mu_2(1-n\beta_2/2)$. With these substitutions, there emerges in the differential equation a term of the form

$$\frac{n}{2}A_1e_2\cos\left[\left(\mu_1-\frac{n\beta_2\mu_2}{2}\right)t+\omega_2\right],$$

and in the solution this leads to the term

$$\frac{\frac{n}{2}A_1e_2\cos\left[\left(\mu_1-\frac{n\beta_2\mu_2}{2}\right)t+\omega_2\right]}{\left[\mu_1-\frac{n\beta_2\mu_2}{2}\right]^2-\left[\mu_1-\frac{n\beta_1\mu_1}{2}\right]^2}$$
$$\approx\frac{A_1e_2\cos\left[\left(\mu_1-\frac{n\beta_2\mu_2}{2}\right)t+\omega_2\right]}{2\mu_1(\beta_1\mu_1-\beta_2\mu_2)},$$

in which n does not appear, and which therefore belongs to the first order in x_1.

There is worse to come. An entirely similar term is found in x_2, and when this is substituted for x_2 in $nA_1x_2\cos(\mu_2-\mu_1)t$, there emerges in the solution a term with zero for denominator, implying that x_1 contains a term proportional to t; indeed, an infinity of such terms lurks ready to emerge. The same is true for y_1 and z_1. Thus, the procedure of successive approximations in the method of absolute perturbations leads unavoidably to the proliferation of *arcs de cercle*, angles increasing proportionally to the time. Lagrange regarded this as the most important discovery of the present memoir. As he wrote to d'Alembert, "[it] appears to me entirely new and of a very great importance in the theory of the planets".

In fact, the discovery was not new; the same insight lay at the basis of Euler's memoir of 1752 on the inequalities of Jupiter and Saturn, which was published only in 1769. Euler, too, had seen that the *arcs de cercle* emerging from the successive approximations of the method of absolute perturbations did not really belong to the solution, and that a different method was necessary to gather into a single finite expression all the variations of a given order, without permitting their exfoliation into an infinite power series in t, the time.

To the achieving of such a solution, Lagrange devoted a long algebraic development, in which his favourite device of integration by parts plays a prominent role. He introduced new variables, in which the reciprocal relations whereby terms contribute to one another through integration became symbolically explicit; six new differential equations emerged; and an integration was effected by means of integrating factors. The solution was similar in form to Euler's, but free of the algebraic errors by which Euler's had been bedevilled. Thus as the first-order result for x_1 Lagrange obtained

$$\begin{aligned}x_1=e_{11}\cos&\left[\left(\mu_1-\frac{n\rho_1}{2}\right)t+\omega_1\right]\\+e_{12}\cos&\left[\left(\mu_1-\frac{n\rho_2}{2}\right)t+\omega_2\right]\\+e_{13}\cos&\left(\left[\mu_1-\frac{n\rho_3}{2}\right)t+\omega_3\right]\\+e_{14}\cos&\left[\left(\mu_1-\frac{n\rho_4}{2}\right)t+\omega_4\right].\end{aligned}$$

Similarly for y_1 he found

$$\begin{aligned}y_1=-2e_{11}\sin&\left[\left(\mu_1-\frac{n\rho_1}{2}\right)t+\omega_1\right]\\-2e_{12}\sin&\left[\left(\mu_1-\frac{n\rho_2}{2}\right)t+\omega_2\right]\\-2e_{13}\sin&\left[\left(\mu_1-\frac{n\rho_3}{2}\right)t+\omega_3\right]\\-2e_{14}\sin&\left[\left(\mu_1-\frac{n\rho_4}{2}\right)t+\omega_4\right].\end{aligned}$$

Here the constants ρ_1, ρ_2, ρ_3, and ρ_4 are the solutions of a biquadratic equation. The expressions for x_1 and y_1 when multiplied by n give the variations in the radius vector and the equation of centre of the first satellite. As in Euler's formulation, we can distinguish a 'proper' eccentricity and equation of centre and 'adventitious' eccentricities and equations of centre; the 'proper' being represented by the first term in each of the two expressions, and the 'adventitious' by the last three terms, which can be regarded as having their source in the perturbing satellites. The expressions for x and y in the cases of the second, third, and fourth satellites have an exactly parallel form.

Entirely analogous results emerge for z. They show that the first-order variation in the tangent of the latitude of any of the satellites may be determined by imagining four planes passing through the centre of Jupiter, of which the first retrogrades with constant inclination on the plane of the orbit of Jupiter, the second moves in the same manner on the first, the third on the second, and the fourth, which will be the plane of the orbit of the satellite, on the third.

In the equations for x, y, and z are contained implicitly the secular equations of the eccentricities, aphelia, nodes, and inclinations of the satellites. To find expressions for these, it would only be necessary to combine the four terms of each expression into a single term, in such a way that x and z take the forms

$$x = D\cos(Mt - w),$$
$$z = G\sin(Lt - k).$$

Here D and G will be found to involve a slow sinusoidal change, and w and k to involve a slow, steady motion with a sinusoidal variation superimposed. In other words, the eccentricities and inclinations of the orbits oscillate about mean values; the apses progress and the nodes regress, but with added oscillatory motions.

Lagrange did not carry out the calculation of these secular variations in his memoir on the satellites of Jupiter, but immediately after completing this memoir he applied the same theory to the interactions of Jupiter and Saturn, carrying the approximation to terms of the order of n^2, and deriving the secular variations of the eccentricities, aphelia, inclinations, and nodes that each of these planets causes in the other. His analysis was published in the midst of a long memoir, "Solution de differents problèmes de calcul intégral", in Vol. 3 of the *Miscellanea Taurinensia* (1762–65). His results for the secular variations of these elements were very close to those that Laplace would give in his *Mécanique céleste*, except that Laplace found somewhat smaller values for Jupiter, because of using a smaller value for the mass of Saturn (Lagrange used the Newtonian value, $\frac{1}{3021}$ of the mass of the Sun, while Laplace reduced this to $\frac{1}{3512}$, which is close to the value now accepted).

In another respect, Lagrange obtained a result with which Laplace was soon to take issue: he found secular variations in the mean motions of the two planets. In effect, the differential equation for y (where ny represents what is variable in the longitudinal motion) contains terms proportional to x^2 and z^2, and from these can be derived, in the solution for y, terms proportional to the square of the time. For the increasing inequality in Jupiter's motion in longitude, Lagrange found $+2''.7402\ N^2$, where N is the number of revolutions since epoch. For Saturn the result was $-14''.2218\ N^2$. These results were empirically plausible, since they showed an acceleration for Jupiter and a deceleration for Saturn, in agreement with what astronomers had been finding from Kepler's time to the 1760s. It was Laplace's discovery (to be discussed in the next chapter) that the mean motions, as well as the mean *radii vectores*, were immune to secular variation; the second-order approximation that had led to Lagrange's result was inadequate.

The prize essays of 1772 and 1774: the problem of three bodies and the secular equation of the Moon

For its contest of 1768, the Paris Academy set the problem of "perfecting the methods on which the lunar theory is founded, in order to determine precisely the equations of this planet that are as yet uncertain, and in particular to examine whether a reason can be derived from this theory for the secular equation of the Moon". D'Alembert early in 1766 encouraged Lagrange to compete, but Lagrange, who was just then entering upon his new duties as head of the mathematical section of the Berlin Academy, declined. No prize was awarded in 1768, and the same problem was set for the contest of 1770, with the offer of a double prize.

In March 1768 Lagrange wrote to d'Alembert that he was at work on the three-body problem, and might have an essay ready for submission in the contest of 1770. In the preceding year, stimulated by Euler's memoirs on the subject, he had written two memoirs on the restricted problem, in which a body is attracted by two fixed centres; these were published in Vol. 4 of the *Miscellanea Taurinensia* (1766–69). Once more, however, the deadline for entry into the contest passed without his submitting an essay. Half the prize was awarded to Leonhard Euler and his son Johann Albrecht Euler, for the lunar theory with rotating coordinates described in the preceding chapter; the remaining half was reserved to be joined to the prize of 1772, when the same problem was to be posed for a third time. The prize commission was evidently not altogether satisfied with Euler's theory, which, despite its innovative method, revealed nothing strikingly new.

The *Essai sur le problème des trois corps* which Lagrange submitted in the contest of 1772, and which won half the prize (the other half being awarded to Euler for a more thorough development of his lunar theory), attacked the three-body problem in a way never before tried. It set out to determine the motions of the bodies relative to each other solely in terms of the distances between them. By means of the integral of *forces vives* and the three integrals expressing the conservation of angular momentum, Lagrange reduced the problem to the solution of seven differential equations: three giving the second derivatives of the squares of the three distances in terms of (a) these distances, (b) the masses of the bodies, and (c) certain auxiliary variables defined in terms of the distances and masses, and the remaining four equations giving relations between the auxiliary variables. Assuming these equations solved, so that the three mutual distances are expressed as functions of the time, Lagrange showed that the coordinates of the bodies with respect to a fixed frame of reference could be found without further integration.

Lagrange's derivation reduced the system of differential equations from the twelfth to the seventh order; later investigators carried the reduction to the sixth order by elimination of the nodes. (It has been claimed, however, that such a reduction is implicit in Lagrange's analysis.) Lagrange failed to give his equations the symmetry they could have had. He defined the three relative velocities as those of bodies B and C with respect to A, and of body C with respect to body B, instead of choosing the velocity of B with respect to A, that of C with respect to B, and that of A with respect to C, in rotational order. This asymmetry in his equations resulted from his being preoccupied with their application to the case of the Moon.

In the second chapter of his essay, Lagrange dealt with the cases in which his equations admitted of closed solutions, without approximation. He found two such cases, in both of which the distances between the bodies were required to be either constant or in fixed ratios to each other. The bodies might fall in a straight line, with one of the distances equal to the sum of the other two; it was the solution that Euler had earlier found for the restricted problem, in which two of the bodies are in Keplerian motion while the third has negligible mass. Or the three distances between the bodies could be equal, so as to form an equilateral triangle. In both cases the motions were confined to a single plane. If the distances were supposed variable but in fixed ratios, the orbits of any two about the third proved to be conic sections.

Lagrange believed these solutions to be inapplicable in "the system of the world", but in the present century one of the cases has been found to be exemplified in the Trojan asteroids, two groups of asteroids with the same period about the Sun as Jupiter, and occupying positions to the east and west of Jupiter so as to form equilateral triangles with the Sun and Jupiter at the other vertices. The first of these asteroids was discovered photographically by Max Wolf in 1906.

In the last two chapters of his essay, Lagrange showed how his theory could be applied to the motion of the Moon. In an initial step he introduced the supposition that the distances of A (the Earth) and B (the Moon) from C (the Sun) were much larger than the distance between A and B; the two larger distances he set equal to R/i and R'/i, where i is a very small coefficient. He then obtained an approximative solution of the differential equations good to terms of the ith order. He did not, however, attempt a numerical calculation of the coefficients of the terms thus found.

For its contest of 1774, the Paris Academy posed a double question: first, how one might be assured, in the calculation of the lunar motions, that no detectable error results from the quantities neglected; second, whether, taking account if necessary of the actions of the planets and the nonsphericity of the Earth and the Moon, one could explain, on the basis of the theory of gravitation alone, why the Moon appears to have a secular equation in its longitudinal motion. In his prize-winning essay, Lagrange dealt only with the second question, arguing that neither the non-sphericity of the Earth nor that of the Moon could account for a secular equation of the magnitude claimed by the astronomers.

The existence of a continual acceleration in the Moon's motion had first been suggested by Halley in the *Philosophical Transactions* for 1693; in a comparison of modern eclipses of the Sun or Moon with Babylonian observations of such eclipses on the one hand, and Arab observations on the other, he had found that a single value of the mean motion would not serve. In the *Philosophical Transactions* for 1749 Richard Dunthorne examined a number of ancient observations of eclipses and found an acceleration such as to yield an increment of 10 arc-seconds in the first century. Tobias Mayer in his first lunar tables of 1753 gave the increment as 7″, but in his second set of tables published posthumously in London in 1770 this number was increased to 9″. Lalande in a memoir of 1757 put the number at 10″.

In effect, the astronomers were assuming the Moon's mean motion to be given by an expression of the form $Z+iZ^2$, where Z is proportional to the time and i is a very small coefficient. As Lagrange pointed out, an apparent secular acceleration could also result from a sinusoidal term of the form $\sin(A+\mu Z)$, where A is a constant and μ is a very small coefficient. Indeed, as Lagrange's analysis showed, the only way in which the theory of gravitation could account for the supposed secular acceleration would be by means of such a sinusoidal term. And in this case an approximation of the form $Z+iZ^2$ would eventually fail. The question to be considered, then, was whether the longitudinal component of the perturbational forces acting on the Moon could produce a sinusoidal term of the form $\sin(A+\mu Z)$.

A noteworthy feature of Lagrange's analysis in the memoir of 1774 was his use of a potential function in deriving the perturbative forces in given directions. Let F be the force between two bodies A and B, and s the distance between them; then, Lagrange pointed out, the component of the force in the coordinate direction x would be $-F.\partial s/\partial x$. Suppose the point-masses of which the Earth is made up to be dm, dm', dm'',..., at distances s, s', s'',..., respectively, from the centre of the Moon. Then the gravitational forces exerted by these point-masses on the Moon will be dm/s^2, dm'/s'^2, dm''/s''^2, etc., which may be expressed by $-\partial(dm/s)/\partial s$, $-\partial(dm'/s')/\partial s'$, $-\partial(dm''/s'')/\partial s''$, etc., and the total force exerted by all the point-masses will be the sum of these terms, or $-(\partial/\partial s)\int(dm/s)$. Let $\int dm/s=V$. It follows that the total force on the Moon in the coordinate direction x will be $(\partial V/\partial s)(\partial s/\partial x)=\partial V/\partial x$. This result holds for any coordinate, whatever the coordinate system. We have here Lagrange's earliest utilization of the potential function in celestial mechanics.

In searching for a term of the form $\sin(A+\mu Z)$, Lagrange first observed that all the perturbational terms in the longitude could be expressed in the form

$$K\sin(m\xi+ns+p\eta+qz+r\psi),$$

where m, n, p, q, and r are positive or negative integers or zero, and

ξ = the anomaly of the Sun,
s = the anomaly of the Moon,
η = the angular distance of the Moon from the Sun,
z = the longitude of the Moon counted from the equinox,
ψ = the angular distance of the prime meridian of the Earth from the equinoctial colure (the great circle through the equinoxes and the celestial poles).

The problem thus reduced to finding a linear combination of the foregoing variables yielding a nearly constant angle. According to Lagrange, the only combination meeting the requirement was $z-\xi-\eta$, along with its multiples; this combination being equal to the longitude of the Sun's apogee, which moves only about 1° 50′ per century. Lagrange therefore went on to identify, among the terms arising from the solution of the differential equations, all those containing a sine or cosine of this angle or its double or triple. The coefficients of these terms were in part dependent on constants expressing the shape and mass-distribution of the Earth.

The final step of the analysis consisted in evaluating these coefficients. If the northern and southern hemispheres of the Earth were assumed to be exactly alike, the coefficients either vanished or yielded a result far too small to account for the supposed secular equation of the Moon. If the northern and southern hemispheres were assumed to be dissimilar, an apparent secular equation of the magnitude Mayer had assigned could only be derived if the dissimilarity were supposed very large – larger than would be compatible with empirical determinations of the Earth's shape or with the theory of the precession of the equinoxes and nutation of the Earth's axis.

As for the non-spherical shape of the Moon, Lagrange claimed without further calculation that this also was incapable of producing a secular equation in the longitude of the size required. For if the quantities in his formulas pertaining to the Earth were reinterpreted as applying to the Moon, no term containing a sine of the form $\sin(A+\mu Z)$ emerged: no combination of angular arguments would produce a nearly constant angle. Six years later, at the end of his second memoir on the libration of the Moon, Lagrange carried out the calculation in detail and found a term of the kind required, namely a term proportional to the physical libration of the Moon; but its coefficient turned out to be too small by several orders of magnitude to account for Mayer's secular equation.

Could the resistance of an interplanetary medium or aether be responsible for the Moon's secular equation? Euler in his prize-winning memoir of 1772 had pronounced it to be the only possible cause, but Lagrange thought the existence of a detectable aethereal resistance doubtful: it was not confirmed in the case of most of the planets, and was even contradicted by Saturn's apparent secular deceleration (the effect of such resistance being to cause the planet to lose potential energy, hence to come closer to the Sun, hence to have a shorter period).

Lagrange devoted the final sections of his memoir of 1774 to a discussion of the seven observations, from 720 BC to AD 1478, from which Dunthorne had derived his value for the secular acceleration. Although in a general way these observations fitted a curve of acceleration, Lagrange regarded them as too uncertain or vaguely reported to give assurance of the existence of a secular acceleration, in particular one proportional to the square of the time. The best course for the time being, he urged, would be to reject the secular equation entirely, while retaining the mean movement established by Mayer, which agreed with observations since the time of Tycho without any adjustment for secular acceleration.

This expression of scepticism brought a protest from the Parisian astronomers, particularly Pierre-Charles Le Monnier. But Lagrange, as he explained in a letter to d'Alembert of May 1774, wanted to provoke the astronomers into providing better proofs for their secular equations. There were several responses to the challenge, one in the *Philosophical transactions* for 1780, another in the Berlin *Mémoires* for 1782. The observations by Ibn Yūnus of the eclipses of AD 977 and 978 provided a strong verification, which Laplace, among others, regarded as settling the question.

The source of the secular change appeared to have been found in 1787, when Laplace announced that the apparent secular acceleration of the Moon was caused by the secular diminution in the eccentricity of the Earth's orbit. The truth was, however, that Laplace's explanation gave only part of the cause, the other part being a backward drag on the Earth's rotation due to the Moon's gravitational action on the Earth's tides. But this second factor came to be recognized only in the 1860s.

Lagrange had already, in 1783, noticed that secular changes in the eccentricities and inclinations of a perturbing planet could produce a secular change in the mean motion of a perturbed planet, and he had carried out the calculation of this effect for the mutual perturbations of Jupiter and Saturn, but found it negligible. In 1792 he applied the same formulas to the calculation of the secular equation of the Moon, and like Laplace found it to be a little over 10″ for the first century. But Lagrange's derivation, like Laplace's, contained an error – the omission of negative terms reducing the coefficient – which would first be pointed out by John Couch Adams in 1854. (See on this topic the penultimate section of Chapter 28 below.)

A new approach to the secular equations of the nodes and inclinations, eccentricities and aphelia: 1774–75

In June 1774 Lagrange wrote to d'Alembert:

I have been occupied of late with the solution of this problem: Given different planes that pass through the same fixed point, and each of which moves at the same time on each of the others while maintaining the same inclination, but causing the line of nodes to retrograde with a given uniform movement, to find the position of the planes at the end of any time. *What gave me the idea was the quarrel between MM. [Joseph-Jérôme Lefrançais] de la Lande and [Jean-Sylvain] Bailly as to the discovery of the cause of the variations in the inclinations of the satellites of Jupiter. It seemed to me that it was necessary to consider the question from a more exact point of view than anyone had yet done, and I have found that it presented difficulties that made it worthy of the attention of geometers, independently of the use it can have in astronomy. When there are only two mobile planes, I can give the complete solution of the problem; but if I suppose a greater number, I fall into formulas that are absolutely intractable. However, I have found a particular method for treating the case of as many planes as one wishes, the only restriction being that the mutual inclinations must be very small, and also the movements of the nodes, which is the case with the planetary orbits. If you find this matter interesting enough, I could compose a memoir on it for your Academy, provided there would not be any indiscretion on my part in imposing on it too often my feeble productions.*

Euler, we recall, was the first to derive differential equations for the variations of the node and inclination of the orbit of one planet perturbed by another. In 1755 he had used these equations to deduce the variation in the position of the plane of the Earth's orbit due to perturbation by the other planets. Here he deduced the differential effect produced by each perturbing planet singly, then added the results together, without attempting an integration. Lalande in 1757 applied the same procedure to the five known planets other than the Earth, and Bailly then applied it to the Galilean satellites of Jupiter. The results would presumably be valid for hundreds of years. Lagrange, by contrast, was seeking formulas valid for all time.

In a memoir presented to the Berlin Academy in June 1774, Lagrange showed that when the problem is confined to two planets the inclinations of the two orbits with respect to a fixed plane oscillate between determinable bounds, and the nodes of the orbits on this same plane either retrograde steadily or oscillate between bounds, depending on the relative value of certain constants. If three or more planets were interacting, the solution turned out to depend on elliptic integrals, and so was not obtainable by known methods. However, if the mutual inclinations of the orbits taken two by two were very small, it became possible to represent the problem by a set of linear differential equations of the first order. The application of such a set of differential equations to the determination of the secular variations of the nodes and inclinations of the six then known planets was the subject of a memoir which Lagrange sent to Paris in October 1774.

Lagrange's derivation of the differential equations turned on a substitution of new variables whereby the equations were freed from sines and cosines. For a given planet let the integrals of area in rectangular coordinates be

$$R=\frac{xdy-ydx}{dt},\quad Q=\frac{xdz-zdx}{dt},\quad P=\frac{ydz-zdy}{dt}. \tag{12}$$

These expressions will be variable or constant, depending on whether perturbation is present or absent. The line of nodes in the xy plane will form an angle χ with the x-axis such that $\tan\chi=P/Q$, and the tangent λ of the instantaneous inclination of the orbit to the xy plane will be $[P^2+Q^2]^{1/2}/R$. The new variables that Lagrange introduced were

$$s=\lambda\sin\chi=\frac{P}{R},\quad u=\lambda\cos\chi=\frac{Q}{R}. \tag{13}$$

Thus

$$\tan\chi=\frac{s}{u},\quad \lambda=\sqrt{s^2+u^2}.$$

If the orbits are assumed to be and remain circular, and all terms that depend explicitly on the positions of the planets in their orbits are discarded, the differential equations take the form

$$0=\frac{ds}{dt}+(0,1)(u-u')+(0,2)(u-u'')+\ldots, \tag{14}$$

$$0=\frac{du}{dt}-(0,1)(s-s')-(0,2)(s-s'')-\ldots,$$

$$0=\frac{ds'}{dt}+(1,0)(u'-u)+(1,2)(u'-u'')+\ldots,$$

$$0=\frac{du'}{dt}-(1,0)(s'-s)-(1,2)(s'-s'')-\ldots,$$

$$0=\ldots.$$

Here prime marks distinguish the variables referring to different planets, and the expressions (0,1), (0,2),..., (1,0), (1,2),..., etc. are constants depending on the masses and *radii vectores* of the planets, and on the coefficients of the trigonometric series representing the inverse cubes of the distances between perturbing and perturbed planets.

The solution of the system, Lagrange showed, is

$$\begin{aligned} s &= A \sin \alpha + B \sin(bt+\beta) + C \sin(ct+\gamma) + \ldots, \qquad (15)\\ u &= A \cos \alpha + B \cos(bt+\beta) + C \cos(ct+\gamma) + \ldots,\\ s' &= A \sin \alpha + B' \sin(bt+\beta) + C' \sin(ct+\gamma) + \ldots,\\ u' &= A \cos \alpha + B' \cos(bt+\beta) + C' \cos(ct+\gamma) + \ldots,\\ &\ldots. \end{aligned}$$

Here t is the only variable, and the values of the other letters can be determined on the basis of observation, although the process increases in complication with the number of planets.

In applying this solution to the six known planets, Lagrange observed that the constants (0,1) and (1,0), where '0' refers to Jupiter and '1' to Saturn, are – because of the larger masses of these two planets – larger by several orders of magnitude than the corresponding constants referring to other pairs of planets. He therefore neglected the effects of the other planets on these two, and took account only of their effects on each other. So for Jupiter he found

$$\tan \chi = \frac{s}{u} = \frac{A \sin \alpha + B \sin(bt+\beta)}{A \cos \alpha + B \cos(bt+\beta)}, \qquad (16)$$

$$\lambda = \sqrt{s^2+u^2} = \sqrt{A^2+B^2+2AB \cos(bt+\beta-\alpha)}, \qquad (17)$$

and the solution for Saturn is the same except that B is replaced by B'. On evaluating the various constants, Lagrange found that λ and λ', the tangents of the inclinations of the two orbits to the ecliptic, undergo oscillations with a period of 51 150 years; Jupiter's orbit is at maximum inclination when Saturn's is at a minimum, and *vice versa*; the ranges of oscillation are 45′ 3″ for Jupiter and 1° 45′ 51″ for Saturn. The motion of the nodes is also libratory with the same period of 51 150 years; the mean positions of the nodes of the two orbits coincide and are fixed with respect to the stars, and the range of libration is 26° 7′ for Jupiter and 64° 8′ for Saturn.

For each of the remaining four planets, Lagrange went on to derive the secular variations of the variables s_i and u_i, but, because of the complexity of the equations, did not attempt to determine whether for these planets the nodes and inclinations are confined to bounded ranges.

The procedure just indicated for determining the secular variations of the nodes and inclinations is applicable, *mutatis mutandis*, to the secular variations of the aphelia and eccentricities. Laplace read Lagrange's memoir immediately it was received by the Paris Academy, and recognized this possibility; in mid-December 1774 he registered with the Academy a memoir developing the idea. It was published in 1775 in Part I of the Paris *Mémoires* for 1772, some three years before the publication of the memoir Lagrange had submitted in October. But Laplace appended to his memoir an extract of a letter from Lagrange of April 1775, which shows that Lagrange had had precisely the same idea as Laplace.

In the quoted extract, Lagrange started from the solution of the three-body problem given in Clairaut's *Théorie de la lune*:

$$\frac{f^2}{Mr} = 1 - \sin u\, [g - \textstyle\int \Omega \cdot \cos u \cdot du] - \cos u\, [c + \textstyle\int \Omega \cdot \sin u \cdot du],$$

where f, M, g, and c are constants, r is the planet's radius vector and u its longitude, and Ω is a function of the perturbing forces. He then set

$$\begin{aligned} g - \textstyle\int \Omega \cdot \cos u \cdot du &= e \sin I = x,\\ c + \textstyle\int \Omega \cdot \sin u \cdot du &= e \cos I = y, \end{aligned}$$

so that the original equation became

$$\frac{f^2}{Mr} = 1 - e \cos(u - I) = 1 - x \sin u - y \cos u,$$

which is the equation of an ellipse of which e is the eccentricity and I the longitude of the aphelion. The equations

$$\begin{aligned} dx &= -\Omega \cdot \cos u \cdot du,\\ dy &= +\Omega \cdot \sin u \cdot du, \end{aligned}$$

will then yield the secular variations of the variables x and y, if Ω and u are expressed in terms of x, y, x', y',..., and t, and only those terms are retained in which x, y, x', y', etc. are linear and multiplied by constant coefficients. Since

$$e^2 = x^2 + y^2$$

and

$$\tan I = \frac{x}{y},$$

the secular variations of e and I are obtainable from those for x and y.

As we have seen, Lagrange's new way of determining the secular variations of the nodes and inclinations, aphelia and eccentricities, showed that these variations could be libratory and periodic. Lagrange continued to classify the perturbational variations as either 'secular' or 'periodic', but the true basis of the distinction could no longer be periodicity. The variations called 'secular', besides being slow, were expressible by formulas in which the orbital positions of the planets did not appear, whereas the variations called 'periodic', besides being short-term, depended explicitly on the configuration of the planets in their orbits.

That the nodes and aphelia of the planets move slowly and unidirectionally had been a premiss of Johannes Kepler's *Rudolphine Tables*, and despite the attempt of Thomas Streete (1622–89) and Nicolaus Mercator (*c.* 1619–87) to challenge this assumption had come to be generally accepted by eighteenth-century astronomers. Steady motion in these orbital parameters did not imperil the stability of the solar system. With the inclinations and eccentricities the case was different: if their variations were not periodic and confined within narrow bounds, the solar system would collapse. The question of the system's stability would soon be addressed, first by Lagrange and then by Laplace.

Lagrange on planetary perturbations, 1775–85

Lagrange, in his letter of April 1775, while thanking Laplace for sending manuscript copies of his memoirs on the secular variations, appeared to relinquish to him the entire subject:

Long ago I proposed to myself to take up once more my old work on the theory of Jupiter and Saturn, to push it further and to apply it to the other planets .... But as I see that you have yourself undertaken this research, I willingly renounce it, and I am very happy that you have relieved me of the necessity of undertaking this work, for I am persuaded that the sciences can only gain much thereby.

In thus proposing to cede the topic to his younger rival, Lagrange was probably moved by a sense of Laplace's ambition, as well as by a recognition of the importance of his discoveries, in particular his demonstration that the mean *radii vectores* were immune to secular variation.

But a month and a half later, writing to d'Alembert, Lagrange reclaimed the topic:

I am now at the task of giving a complete theory of the variations of the elements of the planets due to their mutual action. What M. de la Place has done on this matter has pleased me a great deal, and I flatter myself that he will not be displeased with me for not keeping the sort of promise I made to abandon it entirely to him. I have not been able to resist the desire to occupy myself with it again, but I am not the less delighted that he should also work at it on his side. I am even very eager to read his further researches on this subject, but I beg him to send me nothing in manuscript, only the printed memoirs. I beg you to tell him this ....

During the course of 1775 Lagrange completed and read to the Berlin Academy two long memoirs on the secular variations of the eccentricities, aphelia, inclinations, and nodes, but as he told Laplace in a letter of May 1776, he did not expect to have them published, "above all as I know that you are occupied with the subject .... I have always proposed to make this application, but I expect that your work will now make it unnecessary ...,"

If this was a challenge, Laplace did not rise to it. In the years from 1775 to 1785, his mathematical investigations, insofar as they had to do with the planets, were without direct bearing on the secular inequalities. They dealt with gravitational attraction of spheroids, tidal oscillations of the oceans and atmosphere, and the precession of the equinoxes. As he explained in February 1783 – he was responding to Lagrange's announcement that he was now revising his two memoirs for publication –

Our uncertainty as to the masses of the planets and the derangements they undergo due to comets, made me renounce work on the memoir that I was preparing on the variations of the eccentricities and aphelia .... I do not doubt that you will shed light on a matter so interesting.

The two parts of Lagrange's revised treatise on the secular variations were published in 1783 and 1784, in the Berlin *Mémoires* for the years 1781 and 1782. The first of these parts contains a classic presentation of the solution of the equations of motion by the variation of orbital parameters. A major feature of it was the use of a potential function, from which the components of the total perturbing force in the coordinate directions were derived by partial differentiation. This function, later dubbed by Laplace *the perturbing function*, had made its appearance in two earlier memoirs, the first presented to the Berlin Academy in October 1776, the other in October 1777; both were published in 1779. Before turning to the treatise on the secular variations, we review what these earlier memoirs accomplished.

The first of the memoirs was entitled "Sur l'altération des moyen mouvements des planètes", and was no doubt a response to Laplace's memoir on the same subject, which had been presented to the Paris Academy in 1773 and published in the *Savants étrangers* in 1776. Laplace demonstrated that the mean motions of the planets were immune to secular variation by showing that the constants determining the secular variation of a planet's mean radius vector cancelled one another out, up to the third dimensions of the eccentricities and inclinations. Lagrange wanted to avoid the latter restriction, and to obtain a proof that was *a priori* in the sense of showing not just *that* the result followed from the theory but *why*. His demonstration, like Laplace's, neglects the squares and products of the perturbing forces.

Let the equations of motion for a single planet in rectangular coordinates be

$$0=\frac{d^2x}{dt^2}+\frac{Fx}{r^3}+X,\quad 0=\frac{d^2y}{dt^2}+\frac{Fy}{r^3}+Y,\quad 0=\frac{d^2z}{dt^2}+\frac{Fz}{r^3}+Z, \tag{18}$$

where F/r^2 is the force on the planet due to the Sun, and X, Y, Z are the components of the total perturbative force in the coordinate directions. If X, Y, and Z were zero, the equations would become

$$0=\frac{d^2x}{dt^2}+\frac{Fx}{r^3},\quad 0=\frac{d^2y}{dt^2}+\frac{Fy}{r^3},\quad 0=\frac{d^2z}{dt^2}+\frac{Fz}{r^3}, \tag{19}$$

which are integrable, their complete integrals giving the values of x, y, z in terms of the time and six arbitrary constants of integration. The arbitrary constants determine the size, shape, and orientation of the conic-section orbit in which the planet moves.

Let each of the three integral equations obtained from (19) be differentiated with respect to time; three first-order differential equations result, and these in combination with the three integral equations yield six differential equations of the first order, each containing but one constant of integration. Let any one of these equations be $V=k$, where k is the constant, and V is a function of x, y, z, dx/dt, dy/dt, and dz/dt. Differentiating, we have $dV=0$, which must be identical with one of the original differential equations (18). If in dV we put in place of the differentials $d(dx/dt)$, $d(dy/dt)$, $d(dz/dt)$ their values – namely $-(Fx/r^3)dt$, $-(Fy/r^3)dt$, $-(Fz/r^3)dt$ – the expression becomes identically zero.

But now let us take account of the perturbative forces X, Y, Z in equations (18). The values of the differentials $d(dx/dt)$, $d(dy/dt)$, $d(dz/dt)$ will now be $-(Fx/r^3+X)dt$, $-(Fy/r^3+Y)dt$, $-(Fz/r^3+Z)dt$, and dV will no longer be equal to zero. Thus k will no longer be a constant, and as Lagrange showed, we shall have

$$dk=-\left[\frac{\partial V}{\partial(dx/dt)}X+\frac{\partial V}{\partial(dy/dt)}Y+\frac{\partial V}{\partial(dz/dt)}Z\right]dt. \tag{20}$$

Equation (20) is a recipe for finding the variation of the orbital parameter k.

To find the variation of the orbit's major axis, $2a$, Lagrange made use of an equation obtainable from equations (19) by integration, called the integral of *forces vives*:

$$\frac{1}{\sqrt{x^2+y^2+z^2}}-\frac{dx^2+dy^2+dz^2}{2F\,dt^2}=\frac{1}{2a}. \tag{21}$$

Here V is given by the left-hand member, and $k=1/2a$. Using equation (20) to determine dk, Lagrange found

$$d\left(\frac{1}{2a}\right)=\frac{Xdx+Ydy+Zdz}{F}. \tag{22}$$

It is at this point that Lagrange introduced the perturbing function, to which he gave the symbol Ω. (Clairaut, we recall, had given this symbol to a function similar in gathering together all the perturbing forces.) Let there be perturbing planets with masses T', T'', etc., coordinates (x', y', z'), $(x'',$

y'', z''), etc., and distances from the perturbed planet v', v'', etc. Further, let

$$\Omega = T'\left[\frac{xx'+yy'+zz'}{r'^3} - \frac{1}{v'}\right] + T''\left[\frac{xx''+yy''+zz''}{r''^3} - \frac{1}{v''}\right] + \ldots \quad (23)$$

If we treat the Sun as at rest, transferring to the perturbed planet the actions on the Sun of T', T'', etc., but with changed sign, then it will follow that the components X, Y, Z of the perturbing force are given by $\partial\Omega/\partial x$, $\partial\Omega/\partial y$, $\partial\Omega/\partial z$, and the right-hand member of (22) becomes $\mathbf{d}\Omega/F = \mathbf{d}\Omega/(S+T)$, where the boldface '**d**' means differentiation with respect only to the variables pertaining to the perturbed planet, and S and T are the masses of the Sun and perturbed planet, respectively.

Let θ, θ', θ'', ... be the mean motions of the planets T, T', T'', ... during the time t. Because the eccentricities of the planetary orbits are small, the variables $x, y, z; x', y', z'; x'', y'', z''$; etc. can all be expressed by means of series of sines and cosines of the angles θ, θ', θ'', ..., and their multiples. The distances between T and the perturbing planets can be developed as series of sines and cosines of the angles $(\theta-\theta')$, $(\theta-\theta'')$, etc. Hence the expression $\Omega/(S+T)$ can be reduced to a series of terms of the form

$$M\begin{Bmatrix}\sin\\ \cos\end{Bmatrix}(m\theta + n\theta' + p\theta'' + \ldots),$$

where M is a quantity depending on the orbital elements, and m, n, p, ..., are positive or negative whole numbers or zero.

To obtain the derivative $\mathbf{d}\Omega/(S+T)$ it is only necessary to differentiate with respect to θ; the result will be a series of terms of the form

$$\pm mM\begin{Bmatrix}\sin\\ \cos\end{Bmatrix}(m\theta + n\theta' + p\theta'' + \ldots),$$

Finally, to obtain the integral of this series, and hence the value of $1/2a$, Lagrange made, in accordance with Kepler's third law, the substitutions

$$\theta' = \theta\left[\frac{a}{a'}\right]^{3/2}, \quad \theta'' = \theta\left[\frac{a}{a''}\right]^{3/2}, \ldots,$$

where a', a'', ... are the semi-major axes of the orbits of the perturbing planets. The required integral is then given by terms of the form

$$\frac{\pm mM\begin{Bmatrix}\sin\\ \cos\end{Bmatrix}[(m+n(a/a')^{3/2}+p(a/a'')^{3/2}+\ldots]\theta}{m+n(a/a')^{3/2}+p(a/a'')^{3/2}+\ldots} \quad (24)$$

Lagrange's conclusion was that $1/2a$, being given by a series of terms of form (24), would be subject only to periodic variations, provided that the denominator of (24) is never zero for any particular set of integers m, n, p, ... "But it is easy to be convinced that this case cannot occur in our system, where the values of $a^{3/2}$, $a'^{3/2}$, $a''^{3/2}$, ... are incommensurable with one another."

Lagrange's confidence in his conclusion seems rather astonishing: how could he claim to know that numbers, to which the only access was empirical, were incommensurable? But lack of acuity as to the meaning of incommensurability appears to have been endemic among the mathematicians. Laplace, equally ready with Lagrange to assume the incommensurability of the mean motions, was enthusiastic about Lagrange's argument: "The felicitous application of the beautiful method that you have explained at the beginning of your memoir," he wrote, "... joined to the elegance and the simplicity of your analysis, has given me a pleasure that I cannot put into words." And he went on to infer (erroneously, as he himself would later discover) that the apparent secular inequalities in the motions of Jupiter and Saturn could not be due to their mutual interaction.

A few years later Lagrange discovered that his argument required modification: the coefficient M in (24) could be subject to secular change owing to secular change in orbital elements other than the major axes.

The second memoir published in 1779 used the perturbing function to derive the integrals of motion for a system of bodies interacting in accordance with the inverse-square law of gravitation. Here Lagrange took the centre of gravity of the system as the origin of coordinates, and treated all bodies of the system equivalently. As a consequence the perturbing function took the simple form of a series of terms each having a product of two masses for numerator, and the distance between the two masses for denominator. The proofs that the centre of gravity of the system moves uniformly, and that the angular momentum and *forces vives* of the system are conserved, are remarkably simple and straightforward.

We saw earlier that Lagrange, in his prize-winning memoir of 1774 on the secular equation of the Moon, had used a potential function in deriving the gravitational forces on the Moon due to the non-spherical Earth. Using a potential function in determining the interactions of a system of point-masses was no doubt an obvious extension of the idea. But the importance of this step was nevertheless great. According to Laplace in Vol. 5 of the *Mécanique céleste*, the introduction of this function into celestial mechanics was, because of its utility, "a veritable discovery".

We return now to the treatise of 1783–84 on the secular variations. Its first part is a purely algebraic development, yielding general formulas for the variations of the orbital elements of a planet, and then (by deletion of sinusoidal terms dependent on the planetary positions) arriving at formulas for the variations called secular. Lagrange was here aiming to provide a complete and unified account, using the perturbing function and the procedures he had developed in earlier memoirs.

Once again he showed the mean solar distances to be immune to secular variation. Turning to the two pairs of elements, eccentricity and aphelion, inclination and node, he once again substituted new variables, arriving in both cases at two equations for each planet:

$$x = A\sin(at+\alpha) + B\sin(bt+\beta) + C\sin(ct+\gamma) + \dots,$$
$$y = A\cos(at+\alpha) + B\cos(bt+\beta) + C\cos(ct+\gamma) + \dots. \quad (25)$$

From x and y the variations of the inclination and node, or those of the eccentricity and aphelion, can be calculated. Thus the eccentricity is given by the square root of the sum of the squares of two expressions of the form of (25), and the same is true of the inclination. In (25) the letters $A, B, C, \dots, \alpha, \beta, \gamma, \dots$ are to be determined from the values of the orbital elements found observationally at a given epoch. The letters $a, b, c, \dots$ are the roots of an equation of degree n, where n is the number of interacting planets. If the roots of this equation are all real and unequal, the solution will be oscillatory; if any two roots are equal, the solution will contain terms proportional to the time; and if any of the roots contain $\sqrt{(-1)}$ as a factor, the solution will contain exponential functions of the time. Only in the first case will the variations of the inclinations and eccentricities be bounded.

In the second part of his treatise, Lagrange applied his formulas to the six known planets. Once again, he treated the system of Jupiter and Saturn separately, as being largely uninfluenced by the smaller planets. The solutions for the eccentricities and orbital inclinations of the two planets proved to be oscillatory and confined to narrow bounds. Lagrange concluded:

... it follows that the system of Saturn and Jupiter, insofar as one regards it as independent of the other planets, which is always permitted as we have shown, is of itself in a stable and permanent state, at least if we make abstraction of the action of every foreign cause, such as that of a comet, or of a resistant medium in which the planets move ....

Of the unexplained derangements in the motions of Jupiter and Saturn – Jupiter's apparent secular acceleration and Saturn's apparent secular deceleration – Lagrange was aware. He was asserting the stability of the Sun–Jupiter–Saturn system only in respect to gravitational interactions.

Among the remaining four planets, the masses of Mercury, Venus, and Mars were unknown. Lagrange cleared this obstacle by applying to these planets a relation he found among the densities of the three planets of known mass, Earth, Jupiter, and Saturn (the relation does not in fact apply). For the four planets he obtained eight simultaneous equations of the form of (25), and carried out the detailed process of determining the constants in these equations. The biquadratic equation for the coefficients of the time in the arguments of the sines and cosines proved to have four real and unequal roots. Lagrange concluded:

Thus we are already assured that these expressions cannot contain "arcs de cercle", and that consequently their exactitude will not be limited to a finite time, but will hold for all time.

Uncertainty as to the planetary masses, Lagrange admitted, might lead one to doubt this conclusion; he proposed in the future to seek a way of removing the doubt. (He was to be anticipated by Laplace.) Meanwhile, using his hypothetical masses, he computed upper bounds for the eccentricities and inclinations of the four inner planets, and found them all small. He restated his vision of the solar system at the end of the treatise:

the planets, in virtue of their mutual attraction, change insensibly the form and position of their orbits, but without ever going outside certain limits. The major axes remain unalterable; at least the theory of gravitation implies only alterations that are periodic and dependent on the positions of the planets ....

His long treatise on the secular inequalities completed, Lagrange next turned to the periodic inequalities. His treatise on this subject was published in two parts in 1785 and 1786, in the Berlin *Mémoires* for 1783 and 1784. The aim of the work was to develop the periodic inequalities by the same method of variation of orbital parameters used for the secular inequalities. The mathematical analysis in the new treatise was a natural sequel to that in the earlier one. There the secular inequalities were determined by deleting, from the differential equations for the variations of the orbital elements, all sinusoidal terms; the new deduction required that the deleted terms be restored. Given the earlier integrations for the secular variations, the differential equations with the newly included terms proved to be easily integrable, yielding the periodic contributions by themselves.

Given the periodic changes in the orbital parameters, Lagrange then proceeded by means of partial derivatives to deduce the resulting periodic changes in the perturbed planet's longitude, radius vector, and tangent of the latitude. Thus if the planet's true longitude is q, and q depends on orbital parameters p, x, y, s, u, whose values independent of periodic variation are $\bar{p}$, $\bar{x}$, $\bar{y}$, $\bar{s}$, $\bar{u}$, and whose periodic variations are $\tilde{\omega}$, ξ, ψ, σ, and υ, then the periodic correction in the longitude will be

$$\left(\frac{\partial q}{\partial \bar{p}}\right)\tilde{\omega}+\left(\frac{\partial q}{\partial \bar{x}}\right)\xi+\left(\frac{\partial q}{\partial \bar{y}}\right)\psi+\left(\frac{\partial q}{\partial \bar{s}}\right)\sigma+\left(\frac{\partial q}{\partial \bar{u}}\right)\upsilon.$$

Analogous expressions hold for the radius vector and tangent of the latitude.

In the second part of the treatise, Lagrange applied the formulas thus derived to all the six then known planets, obtaining numerical values for the periodic inequalities of these planets independent of the eccentricities and inclinations. On the whole, the results were not new, but Lagrange's systematic derivation gave them authority. Moreover, the same method could be extended to the derivation of the terms proportional to the eccentricities and inclinations and their powers and products. Lagrange had himself planned to extend his theory to terms proportional to the first powers of the eccentricities and inclinations, but found himself anticipated by Nicolaus Claude Duval-le-Roi (d. 1810), whose derivations and results for perturbations of this order were published in the Berlin *Mémoires* for 1792 and 1793. Earlier, in the Berlin *Mémoires* for 1787, Duval-le-Roi had published the extension of Lagrange's theory to the newly discovered planet Herschel [i.e., Uranus].

More important for later developments, however, were certain theoretical features in the first part of Lagrange's treatise on the periodic variations. At the start, he introduced a change in the analysis whereby, in his earlier treatise on the secular variations, he had obtained the relation between true longitude q and mean longitude p. In the earlier treatise he had derived the differential equation

$$\begin{aligned}&dq+\beta\sin q\cdot dq+\gamma\cos q\cdot dq\\&\quad+\delta\sin 2q\cdot dq+\epsilon\cos 2q\cdot dq+\ldots=dp,\end{aligned}\qquad(26)$$

where β and γ are of the first order in the eccentricities and inclinations, δ and ϵ of the second order, and so on. Because of the perturbations, the coefficients β, γ, δ, ϵ,... had to be regarded as variables. Thus, in the earlier treatise, using integration by parts, Lagrange had derived an approximate integral of (26) in the form

$$\begin{aligned}&q-(\beta)\cos q+(\gamma)\sin q\\&\quad-(\delta)\cos 2q+(\epsilon)\sin 2q-\ldots=p,\end{aligned}\qquad(27)$$

where (β), (γ), (δ), (ϵ), ... were given in terms of β, γ, δ, ϵ,... and their derivatives with respect to q. By reversal of series, it was then possible to obtain q as a function of p.

In his new treatment of this topic, Lagrange proposed to keep the left-hand side of (27) in the form it would have if the orbital elements were constant, then apply to the value of p the corrections resulting from the variability of these elements. Thus, setting

$$\begin{aligned}d\Sigma=&-\cos q\cdot d\beta+\sin q\cdot d\gamma\\&-\tfrac{1}{2}\cos 2q\cdot d\delta+\tfrac{1}{2}\sin 2q\cdot d\epsilon\\&-\ldots,\end{aligned}\qquad(28)$$

and adding this equation to (26), he obtained after integration

$$q-\beta\cos q+\gamma\sin q$$
$$-\tfrac{1}{2}\delta\cos 2q+\tfrac{1}{2}\epsilon\sin 2q-\ldots=p+\Sigma. \quad (29)$$

In the differential $d(p+\Sigma)$ of this equation, dp represents the part of the variation of the mean motion that relates to the mean distance; thus we shall have $dp=dt/a^{3/2}$, where a is the semi-major axis of the ellipse, and is subject only to periodic inequalities which are functions of the orbital positions of the perturbed and perturbing planets. On the other hand, $d\Sigma$ is the variation of the mean motion due to variation of the orbital elements other than the mean distance or semi-major axis. It was to be related, Lagrange proposed, to the epoch of the mean movement, "which in invariable orbits contains the sixth arbitrary constant of the integrals, and so constitutes the sixth element of the elliptic movement."

Thus all six of the orbital elements had now been included in Lagrange's theory of the variation of orbital elements. And he had pinpointed a new source of possible apparent secular variation in the mean motion: secular variation in Σ. He proceeded to show that, to the first power of the eccentricities and inclinations and to the first powers of the perturbing forces, Σ is free of secular variation, but that if the approximation is carried to the second dimensions of the eccentricities and inclinations, there emerge terms in $d\Sigma$ that are proportional to the time. Lagrange devoted a special memoir to these secular variations; like his "Théorie des variations périodiques (Première partie)", it appeared in the volume of the Berlin *Mémoires* for 1783, published in 1785. Here he made the application to Jupiter and Saturn, but found the resulting variations far too small to account for the supposed secular variations in the mean motions of these planets. Nearly a decade later he would use the same formulas to deduce, as Laplace had done before him, the secular equation of the Moon.

The fact was that the apparent secular inequalities of Jupiter and Saturn were not really secular, in the sense of being independent of the configuration of the two planets. They were to be found, not in $d\Sigma$, but in dp; and there Laplace found them in 1785, after reading Lagrange's "Théorie des variations périodiques (Première partie)". And this memoir contains a suggestion that may have played a role in Laplace's discovery; it is nothing less than a description of the very procedure Laplace used, but applied to a different quantity.

Lagrange gives an explicit method for determining the terms of dp that are independent of the eccentricities and inclinations – an approximation which, he says, "suffices in most cases". Then he adds: "The calculations will not be more difficult, but only a little longer, when one wishes to take account also of the terms due to the eccentricities and inclinations." In discussing the terms of $d\Sigma$ that are proportional to the eccentricities and inclinations or to their powers and products, he offers the following suggestion about abbreviating the labour:

Because of the smallness of the terms involved, it will suffice to take account of those which will be augmented a great deal by integration; and it is clear that these are the terms which will contain sines and cosines of angles of which the variation is very small in relation to that of the angle p. *If, for example,* π *designates one of these angles, and* $d\pi=\nu\,dp$, ν *being a very small coefficient, the first integration will introduce into the denominator the coefficient* ν, *and the second integration will introduce the square of* ν. *Since the angle* π *can only be composed of multiples of the angles* p, p′, p″,... *joined together by the signs + or −, it is only by the known relations of the mean motions of the planets that one will be able to judge in each case the value of the coefficient* ν*; there will thus be required a particular examination for each planet of which one wishes to calculate the perturbations.*

If in the case of Jupiter and Saturn Lagrange had carried out the particular examination he recommends in relation, not to $d\Sigma$, but to dp, it would have been he rather than Laplace who discovered the source of the unexplained divagations in the mean motions of these planets. But although not uninterested in applying his theories to observations, his gaze was primarily directed toward achieving unity and clarity in the theory. And in the case at hand his focus was upon the inclusion of the sixth orbital element – the epoch – in the theory of variation of orbital elements, and on showing that there could be secular variation in the mean motion owing to secular variation in elements other than the mean distance.

Lagrange on the perturbations of comets

As the subject of its prize contest of 1778 the Paris Academy of Sciences proposed the theory of perturbations of comets, with particular application to the comet that Halley had hypothesized to account for cometary apparitions in 1532 and 1661. (The two apparitions are now recognized to have been due to two different comets.) In his winning essay Lagrange developed the theory of cometary perturbations by the method of variation of orbital elements; Laplace was to follow this method in his *Mécanique céleste*, with evident reliance on the Lagrangian formulations. For configurations in which the solar distance of the comet was either very large or very small relative to the solar distance of the perturbing planet, Lagrange derived analytical approximations permitting determination of the perturbations by formal rather than numerical integration; these formulas, too, would be followed by Laplace. As for the application to the supposed comet of 1532 and 1661, Lagrange indicated how it could be done, but left the calculations for others. He mentioned predecessors without naming them; he was undoubtedly referring to Clairaut in his *Théorie du mouvement des comètes*, and to d'Alembert in his "Théorie des comètes".

Lagrange's *Mécanique analytique*

Lagrange's *Mécanique analytique*, first published in 1788, was his *chef-d'œuvre*, the work from which later students of mechanics imbibed the Lagrangian formulations. Here Lagrange aimed to present the whole of mechanics as a rational science, expounded analytically. The theory and the art of solving mechanical problems were to be reduced to general formulas. Geometrical constructions were to play no part. As Lagrange wrote in the preface to the first edition,

One will find no diagrams in this work. The methods that I expound require neither geometrical nor mechanical constructions or reasonings, but only algebraic operations ordered in a regular and uniform development. Those who love analysis will be pleased to see mechanics become a new branch of it, and will be grateful to me for having thus extended its domain.

The work is divided into two volumes, one on statics and the other on dynamics. In the first edition, the subsection on celestial mechanics presents the Lagrangian theory of perturbation as we have seen it emerge in the earlier memoirs; there is little here that will appear particularly novel to a reader of the preceding pages. But in the second edition, which Lagrange was preparing during his final years, and of which the first volume was published in 1811, and the second posthumously in 1816, the case is otherwise. Amidst other augmentations and reformulations, volume 2 of this edition contains a totally new section on the variation of arbitrary constants in the solution of dynamical problems. Here appear the concept and the theorem for which Lagrange is chiefly recognized in the later literature on celestial mechanics: the concept of 'Lagrange brackets', and the theorem according to which the arbitrary constants, rendered variable by perturbation, have time-derivatives given by functions of (a) these same constants and (b) the partial derivatives of the perturbing function with respect to these constants.

When Lagrange came to Paris in 1787, it appears that he was mentally exhausted, his interest in celestial mechanics depleted. Aside from his application, in 1792, of a formula he had derived earlier to the derivation of the Moon's secular equation, he wrote nothing on celestial mechanics till 1808. What revived his interest in that year was the presentation to the Institute, on 20 June 1808, of a memoir by Siméon-Denis Poisson (1781–1840), a student of both Laplace and Lagrange. The memoir was a proof that the immunity of the mean solar distances of the planets to secular variation held when the approximation was extended to the squares and products of the planetary masses.

Poisson's demonstration depended on the formulas for elliptical motion. The result, Lagrange conjectured, must be an analytical consequence of the form of the differential equations and of the conditions of variability of the constants, hence a more general truth, not limited in its application to the special solution represented by elliptical motion.

In a memoir read in two segments on 22 August and 12 September 1808, Lagrange proved the correctness of his conjecture. As in his memoir of 1776 on the variation of the major axis, he took the equations of motion of the unperturbed planet to be six differential equations of the first order, each containing a single arbitrary constant together with the coordinates x_i of the planet and the components x_i' of the velocity of the planet in the coordinate directions. Let the arbitrary constants be a, b, c, f, g, h, and let the bracket (a,b) be defined by

$$\sum_{i=1}^{3}\left[\frac{\partial x_i}{\partial a}\frac{\partial x_i'}{\partial b}-\frac{\partial x_i}{\partial b}\frac{\partial x_i'}{\partial a}\right],$$

where the summation extends to all three spatial coordinates. There are parallel definitions for $(a,c), \ldots, (b,c), \ldots, (g,h)$. Once perturbing forces are introduced, there will be a perturbing function Ω, and the constants become parameters subject to variation. Lagrange showed that the partial derivatives of the perturbing function with respect to these parameters are given as linear functions of the time-derivatives of the parameters each multiplied by a bracket. Thus

$$\frac{\partial \Omega}{\partial a}\,\mathrm{d}t=(a,b)\mathrm{d}b+(a,c)\mathrm{d}c+(a,f)\mathrm{d}f + (a,g)\mathrm{d}g+(a,h)\mathrm{d}h.$$

There are corresponding equations for the partial derivatives of Ω with respect to b, c, f, g, and h. Hence it is possible to solve for the time-derivatives of the parameters, $\mathrm{d}a/\mathrm{d}t$, $\mathrm{d}b/\mathrm{d}t$, etc.; they will be given by expressions in terms of the brackets and the partial derivatives of Ω, in which the time t does not explicitly figure.

Now let Ω be developed as a series in sines and cosines of angles proportional to the time. By discarding, in the new expressions for the time-derivatives of the orbital parameters, all terms except those independent of the time, one obtains immediately the equations of the secular variations of these parameters, good to the squares and products of the planetary masses. In the particular case of $\mathrm{d}a/\mathrm{d}t$, the variation of the major axis, Lagrange found essentially the same formulas as Poisson, showing the immunity of the major axis to secular variation.

The proof of invariability thus obtained by Poisson and Lagrange was limited insofar as it referred only to terms arising from the variation of the parameters of the *perturbed* planet. Would the invariability subsist if account were taken of the variation of the parameters of the *perturbing* planets? To extend the proof in this way, Lagrange used a coordinate system with centre not in the Sun but in the centre of gravity of the solar system; in such a coordinate system, the perturbing function takes a symmetric form with respect to perturbed and perturbing planets, and the extension of the proof becomes straightforward. Once more, the major axis proved immune to secular variation.

In two further memoirs, read on 13 March 1809 and 19 February 1810, Lagrange generalized his theorem to apply to any mechanical system of interacting bodies, where the equations of motion are integrable when abstraction is made of certain 'perturbing' forces, and account can be taken of these perturbing forces by permitting the constants of integration to vary. He then inserted a unified account of the general theorem in a new subsection, the fifth, of Vol. 2 of the second edition of his *Mécanique analytique*. His working out of this theorem, although marred here and there by minor errors, is a major achievement. So the aged Lagrange, in his mid-seventies, generalizing, completing symmetries, put in place the capstone of his masterwork.

Because of its generality and elegance, Lagrange's variational formulation of the principles of mechanics is still widely used; it is employed, for instance, in the development of relativistic gravitational theory. To Lagrange we owe the introduction of the concept of potential in the formulation of gravitational interactions; the use of this concept, and in particular of the special case of it known as the perturbing function, has been a regular feature of works on celestial mechanics ever since. The expression of the time-rates of change of the orbital elements in terms of 'Lagrange brackets' is also standard in such works. The memoirs of Lagrange's final years form an essential antecedent for the emergence of the Hamiltonian formulation of mechanics.

In Lagrange's own time, a primary result of his work in celestial mechanics was its impact on Laplace. Laplace's first great *coup* in celestial mechanics, his discovery of the immunity of the mean solar distances to secular variation up to the third order in the eccentricities and inclinations, emerged from a comparison of Lagrange's and Euler's differing computations of the secular variations of the orbital elements of Jupiter and Saturn. Lagrange's memoir on the satellites of Jupiter formulated the problem that Laplace resolved in his own study of the interactions of these satellites; and the idea of the pendulum-like oscillation that Laplace showed to be the guarantor of stability in these interactions was probably inspired by Lagrange's establishment of the reality of the physical libration of the Moon. Laplace's discovery in 1785 of the great inequality of Jupiter and Saturn may well have been triggered by an assiduous reading of the memoirs of Lagrange that were published in the same year. Lagrange's vision of a stable solar system, subject only to minor oscillations in form, was adopted and promulgated by Laplace, who made it the theme of his *Exposition du système du monde*.

Neither Lagrange nor Laplace can be looked to today for the kind of mathematical rigour that would be developed after them, by such mathematicians as A.-L. Cauchy and K. T. W. Weierstrass. Lagrange's concern for symmetric form may appear to us to be at times almost a fetish; Laplace's sharp eye for results may seem opportunistic. But in the interactive relation between the two men, opposing qualities complemented each other, and accounted for the main advances of celestial mechanics during the last forty years of the eighteenth century.

Further reading

Oeuvres de Lagrange (ed. J. A. Serret; 14 vols., Paris, 1867–92)

Alfred Gautier, *Essai historique sur le problème des trois corps* (Paris, 1817)

Curtis Wilson, Perturbations and solar tables from Lacaille to Delambre, . . . , Part II, *Archive for History of Exact Sciences*, vol. 22 (1980), 189–243

Curtis Wilson, The great inequality of Jupiter and Saturn: from Kepler to Laplace, *Archive for History of Exact Sciences*, vol. 33 (1985), 15–290

22

Laplace

BRUNO MORANDO

Pierre-Simon Laplace was born on 23 March 1749 at Beaumont-en-Auge in Lower Normandy, a region that produces apple cider and cheeses, among them the famous Camembert. Of Laplace's family little is known. In his later life Laplace, having become Count and then Marquis de Laplace, rich and celebrated, scarcely ever mentioned relatives that he no doubt regarded as unworthy of the glory he had acquired. His father was a cultivator of the land, but not as poor as has sometimes been suggested; his mother, born Marie-Anne Sochon, was the daughter of a rich farmer. The young Pierre-Simon, for his primary and secondary education, attended the school run by the Benedictines of Saint-Maur at Beaumont, where one of his uncles, the Abbé Louis Laplace, had taught. Destined for the priesthood, in 1766 he entered upon the required preparatory studies at the University of Caen. But one of his professors, Pierre Le Canu, interested him in mathematics, and in 1768 sent him to Paris with a letter of recommendation to Jean le Rond d'Alembert (1717–83), doyen of French mathematicians and a powerful influence within the Paris Academy of Sciences. D'Alembert, it is said, was unimpressed by the letter of recommendation; but happily a mathematical essay that the young man brought him a few days later won his admiring approval, and he arranged for Laplace to be given the post of professor of mathematics at the École Militaire.

Thus began Laplace's scientific career; it would be a long and brilliant one, lasting nearly six decades. After initially seeking, without success, for a position at the Academy of Berlin, in 1773 he was named an associate member for mechanics in the Paris Academy of Sciences. In 1784 he became examiner for the Royal Artillery Corps, succeeding the illustrious Étienne Bezout. And in 1785 he became a full member of the Academy of Sciences, hence recipient of a pension. In March 1788 he married Charlotte de Courty by whom he had a son, who would study at the École Polytechnique and then become a general in the army, and a daughter who died at the age of 21 while bringing into the world a daughter, the future Marquise de Colbert.

Laplace's work in mathematical astronomy to 1789

The first of the mathematical memoirs that Laplace submitted to the Paris Academy of Sciences was registered by the Academy's secretary on 28 March 1770, five days after Laplace's twenty-first birthday. More than a hundred such would follow, and the last would be entered in the register on 8 September 1823, when Laplace was 74. The memoirs of the first few years dealt with a variety of mathematical topics: application of the integral calculus to the study of finite differences, use of finite differences in the study of probability, recurrent series as applied to probability, investigation of particular solutions of differential equations, the probability of causes (Bayes's rule). Astronomical questions also made their appearance early: the motion of the nodes of the planetary orbits, variation of the plane of the ecliptic, determination of the lunar orbit, perturbations of the planets caused by the motions of their satellites, the secular equations of the planets, etc.

Laplace's first published memoir on astronomy

In 1776, in the volume of the Paris Academy's *Savants étrangers* for the year 1773, there appeared Laplace's "Sur le principe de la gravitation universelle, et sur les inégalités séculaires des planètes qui en dépendent" – the first of his astronomical memoirs to be published. It began with a presentation of "the general equations of motion of a body

22.1. Pierre-Simon de Laplace.

of any figure, acted upon by any forces"; here Laplace showed himself in full command of the theory of the motion of rigid bodies as developed by d'Alembert and Leonhard Euler (see Chapter 17 above).

The memoir next turned to "an examination of the principle of universal gravitation". According to Laplace, Newton in his application of this principle to the celestial bodies made four assumptions, which were now generally accepted: (1) the attraction exerted on a body varies directly as its mass and inversely as the square of the distance from the attracting body; (2) the attractive force of a body is the sum of the attractions of each of its parts; (3) the force is propagated instantaneously; and (4) it acts in the same way on bodies at rest and in movement.

In the discussion that followed Laplace examined these assumptions both speculatively and critically. He suggested a possible metaphysical justification for the inverse-square law in the fact that only this law permits the planetary paths to remain geometrically similar to themselves under proportional shrinkage or magnification of all distances and velocities. He examined the evidence for the second assumption, urging that the phenomena of precession and nutation required its acceptance. He also showed that if, contrary to the third and fourth assumptions, the speed of propagation of gravitation was finite, and its effect on bodies dependent on their relative speeds, then there were mathematical consequences to be reckoned with. Daniel Bernoulli, in his prize-winning memoir of 1740 on the tides, had imagined that as much as two days might be required for the attractive force to travel from the Moon to the Earth. Laplace showed that the hypothesis of a finite speed would account for the secular acceleration of the Moon, provided that the velocity of propagation were supposed more than six million times the speed of light. The planets also, under this hypothesis, would be subject to secular accelerations. The idea is presented as a conjecture for the consideration of philosophers; it would be acceptable, Laplace says, only if it were known that the Moon's secular acceleration could not be explained by more ordinary assumptions. With other mathematicians of the time he saw this acceleration as a problem demanding solution. He examined the suggestion that the friction of the trade winds might slow the rate of the Earth's diurnal rotation, and so make the Moon appear to accelerate; but concluded that no such effect could result. As possible explanations he also considered, but dismissed, the frictional effect of an interplanetary aether on the Earth's motion, and the Sun's loss of mass due to the emission of light, causing the Earth to recede into a larger orbit.

In the second half of the memoir, Laplace turned to the derivation of the secular inequalities of the planets, and quickly arrived at a new and important result. He began by pointing out that, of all the inequalities, the secular ones are the most important; the planets in each of their revolutions move very nearly in ellipses, and the perturbational effect most important to determine is the change in the orbital elements of these ellipses. And among the secular inequalities, the most crucial is that of the mean movement; but, Laplace urged, it had not been determined with the precision that its importance demanded. Euler, in his memoir of 1752 on the irregularities of Jupiter and Saturn, had concluded that both Jupiter and Saturn were accelerating at the present epoch. Joseph Louis Lagrange (1736–1813), in a treatise published in 1766 in the *Mémoires de Turin*, had determined Jupiter to be accelerating and Saturn decelerating, a result in agreement with observations, but still, according to Laplace, incorrect: the derivable effects were negligible. In a footnote, Laplace stated that after reading his memoir to the Academy, he had been able to prove that the secular inequalities of the mean movement were identically zero.

Laplace's derivation had certain novel procedural and notational features that were to remain characteristically his in his later work on planetary perturbations. His differential equations for the longitude (ϕ), radius vector (r), and tangent (s) of the latitude of a perturbed planet were essentially Lagrangian. But into them Laplace proceeded to introduce the differential operator δ, borrowed from Lagrange's formulation of the calculus of variations. He initially designated the masses of the perturbed and perturbing planets by δm and $\delta m'$, respectively, thereby indicating their smallness relative to the Sun's mass (denoted by S). He followed Lagrange in distinguishing the variables referring to the second planet by prime marks. Deleting from the differential equations for the longitude and radius vector the quantities involving the factor $\delta m'$, he obtained the following equations:

$$\frac{d\phi}{dt}=\frac{c}{r^2}, \tag{1}$$

$$0=\frac{d^2r}{dt^2}-\frac{c^2}{r^3}+\frac{S+\delta m}{r^2}, \tag{2}$$

where c is a constant of integration. These equations when integrated yield

$$\phi=nt+A'-2\alpha e\sin(nt+\epsilon) + \tfrac{5}{4}\alpha^2e^2\sin 2(nt+\epsilon) + \dots \tag{3}$$

and

$$r=a\left[1+\frac{\alpha^2e^2}{2}+\alpha e\cos(nt+\epsilon) - \frac{\alpha^2e^2}{2}\cos 2(nt+\epsilon)+\dots\right], \tag{4}$$

which are the equations for motion in an ellipse, in which a is the semi-major axis, αe the eccentricity, n the planet's rate of mean motion, A' the planet's mean longitude when $t=0$, and ϵ the angle by which the planet has advanced beyond its aphelion at the same moment. (Later, in the *Mécanique céleste*, Laplace would shift the starting-point for anomaly from aphelion to perihelion, a change which is effected simply by changing the sign of e in the preceding equations, and which allows for a uniform treatment of cometary and planetary motions.) In the foregoing equations the tangent s of the latitude of the perturbed planet is assumed to be so small that s^2 can be neglected. The symbol α was the tag Laplace used to keep track systematically of orders of magnitude.

With a view to re-introducing into his equations the terms involving $\delta m'$, Laplace next differentiated (3) and (4) using the operator δ, and taking advantage of the fact that this operator commutes with the standard differential operator 'd'; then he added to the resulting equations the terms from the original equations of motion involving $\delta m'$. He thus obtained

$$\frac{d\delta\phi}{dt}=-\frac{2c}{r^3}\delta r - \frac{\delta m'}{r^2}\int r dt \sin(\phi'-\phi)\left[\frac{1}{r^2(1+s^2)^{3/2}}-\frac{r'}{v^3}\right], \tag{5}$$

$$0=\frac{d^2\delta r}{dt^2}+\frac{3c^2}{r^4}\delta r-\frac{2(S+\delta m)}{r^3}\delta r + \text{terms involving } \delta m'. \tag{6}$$

Here v denotes the distance between perturbing and perturbed planets. The variations of r and ϕ

distinguished by the operator δ are those due to the perturbations caused by $\delta m'$.

In a first step towards solving (5) and (6), Laplace replaced the terms containing $\delta m'$ as a factor by expressions involving undetermined coefficients A, B, C, D, and including only such terms as contributed to the secular variations up to the degree of approximation he was aiming to achieve. Thus the terms of (6) containing $\delta m'$ were replaced by

$$\begin{aligned} &a\frac{\delta m'}{a^3}A + \alpha^2 a\frac{\delta m'}{a^3}Bnt \\ &\quad + \alpha a\frac{\delta m'}{a^3}C\cos(nt+\epsilon) \\ &\quad + \alpha a\frac{\delta m'}{a^3}D\sin(nt+\epsilon), \end{aligned}$$

and the last term of (5) was replaced by

$$-\alpha^2\frac{\delta m'}{a^3}\frac{a^2}{2c}Bnt.$$

For the quantity $(S+\delta m)/a^3$ Laplace substituted n^2, in accordance with Kepler's third law, and for the quantity $\delta m'/S$, $\delta\mu'$. Then, for $\delta\phi$ and δr in (5) and (6), he substituted the expressions

$$\delta\phi = \delta\mu' gnt + \alpha^2\delta\mu' hn^2t^2, \qquad (7)$$

and

$$\begin{aligned} \delta r = a\delta\mu'[&L + \alpha Pnt\cos(nt+\epsilon) \\ &+ \alpha Qnt\sin(nt+\epsilon) \\ &+ \alpha^2 Knt]. \end{aligned} \qquad (8)$$

By comparing the coefficients of t, t^2, $\cos(nt+\epsilon)$, and $\sin(nt+\epsilon)$, he then determined g, h, L, P, Q and K in terms of A, B, C, and D. In this way he obtained expressions for the secular increase in the equation of centre, the movement of the aphelion, and the acceleration of the mean movement in terms of A, B, C, and D. Thus the acceleration of the mean movement took the form

$$\frac{3}{2}\pi\alpha^2\delta\mu'(B - eD)\cdot i\cdot 360, \qquad (9)$$

where i is the number of revolutions.

The final step in evaluating the expressions for the secular variations was the determination of the values of A, B, C, and D in terms of the trigonometric series expressing the inverse third, fifth and seventh powers of the distance between perturbing and perturbed planets. Letting $a'/a = z$ be the ratio of the semimajor axes of the two planets, and θ the heliocentric angle between their *radii vectores*, Laplace set

$$\begin{aligned} (1-2z\cos\theta+z^2)^{-\frac{3}{2}} &= b + b_1\cos\theta + b_2\cos 2\theta + \ldots, \\ (1-2z\cos\theta+z^2)^{-\frac{5}{2}} &= b' + b'_1\cos\theta + b'_2\cos 2\theta + \ldots, \\ (1-2z\cos\theta+z^2)^{-\frac{7}{2}} &= b'' + b''_1\cos\theta + b''_2\cos 2\theta + \ldots. \end{aligned} \qquad (10)$$

(We have here an early form of the double indicial notation that he would make standard for the coefficients of the trigonometric series.) The acceleration of the mean motion turned out to involve the factor

$$\begin{aligned} \{&3[b - b' + \tfrac{1}{2}(b'_2 - b_2)] \\ &+ \tfrac{1}{4}z[7(b'_1 - b'_3) - 5(b''_1 - b''_3)] \\ &- z^2[3(b' - \tfrac{1}{2}b'_2) - \tfrac{5}{4}(5b'' - 2b''_2 - \tfrac{1}{2}b''_4)] \\ &- \tfrac{5}{4}z^3(b''_1 - b''_3)\}. \end{aligned} \qquad (11)$$

At first, in applying his formula to Jupiter and Saturn, Laplace substituted numerical values for the *b*s. On finding the result to be nearly zero, he then made use of the relations first developed by Euler whereby all these coefficients can be expressed in terms of the first two, b and b_1, and showed that the above factor reduced identically to zero.

The result meant that, at least to the order of $\alpha^4\delta\mu'$, secular accelerations or decelerations in the motions of the planets – that is, terms in the mean motion proportional to t^2, and independent of the positions of the planets – did not occur. Laplace drew the conclusion that the apparent secular acceleration of Jupiter and secular deceleration of Saturn must be due to causes other than their mutual gravitation; they might be caused, he suggested, by the actions of comets. Neither he, nor Lagrange when he learned of Laplace's new result, considered the possibility that the observed alterations in the mean motions of Jupiter and Saturn might be produced by long-term *periodic* inequalities – oscillations in longitude depending on the positions of the planets, but with periods much longer than those of the planets themselves.

Our sketch of the course of Laplace's derivation, besides exhibiting notational features that were to remain distinctively Laplacian, illustrates characteristics of his mathematical style: while gifted with a keen sense for subtle consequences of both mathematical assumptions and empirical data, he almost always set his sights on some final practical or numerical result to be obtained, and frequently took the short way through a tangle without fully explaining his processes. In contrast to Lagrange, he was little moved by the 'beauty' of a theory, as a quality independent of the theory's power to yield practical results. To the determination of order of approximation in the calculation of approximate numerical results, on the other hand, he brought a new rigour.

Secular inequalities of the eccentricities, aphelia, inclinations, and nodes

As reported in the preceding chapter, Lagrange in October 1774 presented to the Paris Academy a memoir giving a new method for determination of the secular variations of the nodes and inclinations of the planetary orbits; it would be published in 1778. (Memoirs published in the *Mémoires de l'Académie Royale des Sciences de Paris* will hereinafter be designated by *MARS*, followed first by the year of the volume in question and then by the year of publication; and similarly for other publication series. Thus the memoir of Lagrange just cited appeared in *MARS* 1774/78.) The stroke of genius in Lagrange's memoir consisted in the introduction of new variables $s=\theta \sin \omega$, $u=\theta \cos \omega$, where θ is the tangent of the inclination and ω is the angle between the line of nodes and some chosen reference line; there were corresponding variables for each of the planets. The differential equations for the secular part of the variables s and u of the several planets, independent of short-term oscillations, then proved to be linear and of the first order in s and u. Laplace read this memoir immediately it reached the Paris Academy, and realized that the same procedure was applicable to the aphelia and eccentricities. He proceeded to register his elaboration of this idea with the Academy's secretary, and it was published in Pt I of *MARS* 1772/75, three years before Lagrange's essay appeared. Before it was printed, however, Laplace received from Lagrange a letter of 10 April 1775 outlining precisely the same application, and this he appended to the end of his memoir. Laplace's new variables were of the form $x=e \sin L$ and $y=e \cos L$, where e is the eccentricity of the planetary orbit and L the longitude of the aphelion; and in analogy with the case of the nodes and inclinations, he derived first-order linear differential equations for the secular variations of x and y.

Further consideration of this Lagrangian procedure led Laplace to a new method for integrating differential equations by approximation – a method of eliminating the *arcs de cercle* or terms proportional to the time which, as Lagrange had shown in his prize-winning memoir of 1766 on the satellites of Jupiter (see the preceding chapter), emerge inevitably in the solution of the differential equations when they are solved by the ordinary process of successive approximation. Laplace presented this method in Pt II of *MARS* 1772/76. The gist of the method consisted in replacing the eccentricity, aphelion, inclination, and node by the new-style variables, and then showing that, wherever in the solution of the differential equations the time t appeared outside the arguments of sines and cosines, it could be eliminated by taking into account the secular variations of the new variables. From the latter it was then a matter of simple algebra to deduce the secular variations of the eccentricites, aphelia, inclinations, and nodes.

Having explained a general method for the determination of all perturbations, both periodic and secular, Laplace concluded: "It would now remain to apply the preceding theory to the different planets; but the already excessive length of this memoir obliges me to put off these applications to another time ...". As reported in the previous chapter, Lagrange, who was expecting this application to be soon forthcoming, in his letter to Laplace of 10 April 1775 went so far as to say that he was ceding the entire subject to his young rival. A month and a half later, however, he withdrew this offer, suggesting that in their further work on planetary perturbations he and Laplace engage in a friendly rivalry, exchanging only the published memoirs. Lagrange himself proceeded to develop his own general treatment of perturbations by the method of variation of orbital parameters, with application to all the known planets; the resulting memoirs would finally begin to appear in published form in 1783. Laplace, by contrast, published nothing on this subject for a decade; as he explained in a letter to Lagrange of 10 February 1783, he was deterred from pursuing the topic by uncertainties as to the masses of the planets and as to the extent of the derangements they underwent due to the gravitational action of comets.

Tides, precession of the equinoxes, nutation

During the decade from 1775 to 1785, Laplace turned his attention to the gravitational action of *extended* bodies: the attraction of spheroids, the tidal motions of the oceans and atmosphere, the precession of the equinoxes and the nutation of the Earth's axis. One concern was undoubtedly an unresolved difficulty concerning the shape of the Earth: Isaac Newton had deduced, for a homogeneous fluid spheroid rotating with the speed of the Earth's diurnal rotation, an ellipticity of $\frac{1}{230}$; the several geodetic measurements made in Lapland, Peru, France, and the Cape of Good Hope implied an ellipticity greater than $\frac{1}{230}$, while universal gravitation together with plausible assumptions appeared to imply that the ellipticity should be less (see Chapter 15). But this difficulty was as yet unresolvable, and Laplace saw that there were prior problems that needed to be addressed, and first of all the theory of the tides and the role of the fluid parts of the Earth in relation to the actions causing precession and nutation.

Beginning with Newton, the earlier analysts investigating the tides had left the Earth's axial rotation out of account; they had imagined a stationary Earth covered by a thin layer of fluid and subject to the attraction of a single celestial body (Moon or Sun); the fluid would assume an equilibrium shape under the attraction, and if the Moon or Sun were then set into a real or apparent motion of revolution about the Earth, the resulting bulge of fluid would be dragged round the Earth without change of form. This analysis implied that the two full tides occurring each day would differ considerably in height; observations indicated, on the contrary, that the difference was very slight. The traditional analysis failed to take into account another fact: in the Earth's diurnal rotation the linear velocities eastward were different at different latitudes, and the resulting relative displacements of ocean water were of the same order of magnitude as those caused by the attractions of the Moon and the Sun. Moreover, given the inclinations of the Moon's and Sun's paths to the Earth's equator, the tidal distributions in the southern and northern hemispheres must differ, and the effect of the asymmetry on the precession and nutation of the Earth's axis needed to be evaluated.

In a three-part memoir that appeared in *MARS* 1775/78 and 1776/79 and ran to over 200 pages, Laplace grappled with these issues. He showed that, when all factors of the same order of magnitude were taken into account, consecutive high tides would be of nearly the same height if and only if the sea were assumed to be of uniform depth, and further that the slight difference between consecutive high tides was proportional to the influence of the tides on the ratio of nutation to precession. In a further memoir "Sur la précession des équinoxes" (*MARS* 1777/80), he corroborated the result for the precession by a simplified analysis using d'Arcy's principle (conservation of angular momentum); here he announced the theorem:

> *If the Earth be supposed an ellipsoid of revolution covered by the sea, the fluidity of the water in no way interferes with the attraction of the Sun and the Moon on precession and nutation, so that this effect is just the same as if the sea formed a solid mass with the Earth.*

Because d'Arcy's principle applies to all interactions including those involving friction, shock, and turbulence, Laplace now claimed that his analysis could be extended to "the case in nature", in which the shape of the Earth and the depth of the sea are very irregular.

Determination of cometary orbits

Another topic that Laplace concerned himself with in this period was the determination of cometary orbits. In 1771 R. J. Bošković had presented to the Paris Academy a memoir on the determination of a cometary orbit from three observations close to one another in time; he assumed that, during the interval of the observations, the motions of the comet and the Earth were rectilinear and uniform. The memoir appeared in the *Savants étrangers* in 1774, and on reading it Laplace saw at once that the method neglected quantities of the same order as those entering into the supposed solution. He delivered a devastating critique from the floor of the Academy; Bošković, believing himself to have been personally affronted, made a reply. A commission appointed to adjudicate the controversy concluded in its report of June 1776 that Laplace's critique was analytically correct, but that the manner in which it had been couched was unnecessarily abrasive.

In 1781 Laplace returned to the problem of determining cometary orbits and undertook to deal with it in a new way. The motion of a body in the solar system, perturbations aside, is entirely determined if one knows six initial conditions of motion, for instance the three spatial coordinates and the three components of velocity in relation to the Sun at a given 'initial' instant of the motion. Three observations of a planet or comet on the celestial sphere at three different instants give six quantities (right ascensions and declinations or ecliptic longitudes and latitudes), from which the elements of the planet's or comet's heliocentric orbit can be determined. The problem is a difficult one. Newton and others, particularly Lagrange and A.-P. Dionis du Séjour, had contributed to a partial solution of it.

To begin with, Laplace proposed deriving, for the mean instant of the observations, a geocentric longitude and latitude, together with their first and second time-derivatives, from multiple observations (preferably many) by way of the calculus of finite differences. Euler had suggested a similar method for comparing lunar observations with theory in a memoir published in *Mémoires de Berlin* 1763/70. The advantage was that the further analysis could be made rigorous, while less reliance was placed on individual observations, which were in any case necessarily imprecise. For the mean instant of the observations the comet's heliocentric coordinates and components of velocity in relation to the Sun could then be deduced, provided that one knew the geocentric coordinates of the Sun (as given by ephemerides) and the distance of the comet from the Earth at the given instant. Laplace determined the latter quantity as the root of an equation of the seventh degree (or of the sixth degree if the orbit was supposed parabolic). To resolve this equation he used an iterative method, beginning with an approximate value of the distance sought.

In 1784 the Abbé A.-G. Pingré in his *Cométographie* pronounced Laplace's method to be the best available. Together with the later methods of H. W. M. Olbers and C. F. Gauss it has continued in favour with astronomers (see Chapter 25). It has been improved by Henri Poincaré and A. O. Leuschner.

Laplace played a role in the determination of the orbit of the planet Uranus. This planet, discovered by William Herschel on 13 March 1781, was at first taken for a comet (see Chapter 24), but the efforts of P.-F.-A. Méchain, A. J. Lexell, Bošković, and in fact of Laplace himself to determine parabolic elements proved unsuccessful. J.-B.-G. Bochart de Saron, Laplace's friend and protector, appears to have been the first to show that the heliocentric distance of the newly discovered 'star' was at least 14 AU, and hence that one had to do with a planet. On 22 January 1783 Laplace presented a memoir to the Academy giving orbital elements for the planet. In Vol. 3 of the *Mécanique céleste* he would give a set of more precise elements, and calculate the perturbations due to Jupiter and Saturn.

The attractions of spheroids

In late 1782 or early 1783 Laplace was appointed (along with Étienne Bezout and d'Alembert) to a committee for the review of a memoir by Adrien-Marie Legendre (1752–1833) on the attraction of homogeneous spheroids of revolution. Legendre's aim was to prove the theorem:

If the attraction of a solid of revolution is known for every external point on the prolongation of its axis, it is known for every external point. With the origin of coordinates placed at the centre of the spheroid, the radial component of the force of attraction on an external point is given by an integral over all the points within the spheroid. If r is the radius vector of the external point, r' the radius vector of an arbitrary internal point, and γ the angle between them, then the integrand can be expressed as a function of r'/r and of $\cos \gamma$.

Legendre showed how this function could be expanded as an infinite series in even powers of r'/r, with coefficients P_2, P_4, ... (now called Legendre polynomials) that were rational integral functions of $\cos \gamma$. At a suggestion from Laplace, he went on to obtain the potential function for the spheroid in terms of Legendre polynomials, and with its aid evaluated the component of the attraction at right angles to the radius vector.

Stimulated by Legendre's work, Laplace took up the problem on his own account, generalizing it and making some applications to the figure of the planets, in Pt II of *Théorie du mouvement et de la figure elliptique des planètes* (1784), and in *MARS* 1782/85 and 1783/86. To know the attraction exerted by an spheroid (taking this to be a body bounded by any surface expressible by a single equation in spherical coordinates), it was only necessary to know the corresponding potential function, and this, Laplace now asserted, was subject to a fundamental partial differential equation – the equation now known as Laplace's equation or the potential equation. In rectangular coordinates, it is

$$\frac{\partial^2 V}{\partial x^2}+\frac{\partial^2 V}{\partial y^2}+\frac{\partial^2 V}{\partial z^2}=0. \qquad (12)$$

In his early memoirs on the equation, however, Laplace gave it in the more complicated form it has in spherical coordinates, without informing the reader whence it was derived. To solve it he proposed an expansion in series of the form

$$V=\frac{U_0}{r}+\frac{U_1}{r^2}+\frac{U_2}{r^3}+\ldots; \qquad (13)$$

the U_n, he then showed, can be expressed as integrals in terms of the Legendre polynomials. This result enabled him to obtain a series expression for the potential of any spheroid differing little from a sphere. Armed with this general theory, Laplace returned to the problem of tidal oscillations in *MARS* 1776/79, and showed that equilibrium conditions entailed periodicity, and further that the equilibrium would be stable only if the layer of fluid covering the spheroid were less dense than the underlying spheroid. In *MARS* 1783/86 he brought the theory into relation with the available data on the Earth's figure, and in *MARS* 1787/89 he made a tentative application to the rings of Saturn.

The observations of the London optician and instrument-maker James Short (1710–68) had shown that the two rings of Saturn distinguished by G. D. Cassini (Cassini I) were actually multiple, each consisting of many rings. Laplace initially treated each of these many rings as a thin stratum of liquid, its cross-section by any plane through Saturn's axis being a narrow ellipse; he then sought the conditions of equilibrium of this stratum. He took the duration of the rotation of each ring to be equal to that of a satellite at the same distance from Saturn's centre. Finally, he showed that for the equilibrium to be stable the separate rings must be irregular in shape, their centres of gravity differing from their centres of figure. Proof that the rings cannot be fluid, but must be ensembles of tiny satellites, would have to await James Clerk Maxwell's essay of 1856 "On the stability of the motion of Saturn's rings".

A new attack on planetary perturbations

On 23 November 1785 Laplace read to the Academy a memoir "Sur les inégalités séculaires des planètes"; it marks his return, after the lapse of a decade, to the problem of the mutual perturbations of the planets. The memoir as finally printed (in *MARS* 1784/87) contained a sketch of a new theory concerning the mutual perturbations of the satellites of Jupiter, and to the title was added the phrase "et des satellites"; but the presence of this section in the original memoir may be doubted, for in its original form the memoir may have dealt with only two subjects: the anomalies in the mean motions of Jupiter and Saturn, and a proof of the stability of the solar system.

Laplace's new attack on the perturbational problem was manifoldly influenced by memoirs of Lagrange published between 1779 and 1785. In these Lagrange had introduced the perturbing function as applied to the planets conceived as mass-points; had derived with its aid all the known integrals of motion; had articulated an argument for the stability of the solar system; and in the first part of his "Théorie des variations périodiques", published in 1785, had made a suggestion about locating perturbational terms that would be made large by small divisors, owing to the near-commensurability of the mean motions of the planets concerned (see Chapter 21). All these features played a role in Laplace's new assault; and in fact Lagrange's suggestion about small divisors may have triggered Laplace's discovery of the source of the puzzling anomalies in the mean motions of Jupiter and Saturn.

At the start Laplace reviewed what he and Lagrange had proved with regard to the immunity of the mean motions to secular change from mutual gravitational action; his own proof was valid up to the third powers of the eccentricities and inclinations, while Lagrange's proof was valid independent of the eccentricities and inclinations, on the assumption that the mean motions of the planets were incommensurable, "as is the case for the planets in our system". But, Laplace went on, there were perceptible variations in the revolutions of Jupiter and Saturn; Edmond Halley had found, for a period of 2000 years, 3° 49′ of accelerational advance in Jupiter's motion, and 9° 16′ of retardation for Saturn.

These alterations, Laplace next argued, were "very probably" an effect of the mutual action of Jupiter and Saturn. For from the integral of *forces vives*, by neglecting all terms of the third order in the masses (m^3), and terms of the order m^2 that were either periodic or constant, he had deduced the relation

$$\frac{m}{a}+\frac{m'}{a'}+\frac{m''}{a''}+\ldots=\text{const}, \tag{14}$$

where $m, m', m'', \ldots$ are the planetary masses and a, a', a'' the semi-major axes of the orbits. Because the masses of Jupiter and Saturn were very large relative to the masses of the other planets, all terms in the foregoing relation other than those pertaining to these two planets could be neglected, and it followed that if Jupiter's mean solar distance (a) diminished, Saturn's mean solar distance (a') would increase. From Kepler's third law one could derive that

$$mn^{\frac{2}{3}}+m'n'^{\frac{2}{3}}=\text{const}, \tag{15}$$

where n and n' are Jupiter's and Saturn's mean motions. In mutual action affecting the mean motions it followed that

$$\delta n' = -\frac{m}{m'}\left(\frac{a}{a'}\right)^{1/2}\delta n, \tag{16}$$

or with the values Laplace accepted for the masses and mean distances, $\delta n' = -2.33\,\delta n$. The ratio of Halley's values for the secular changes, $-9°\,16'/3°\,49'$, is -2.43, nearly the same.

It is therefore very probable that the observed variations in the movements of Jupiter and Saturn are an effect of their mutual action, and since it is established that this action can produce no inequality that either increases constantly or is of very long period and independent of the situation of the planets, and since it can only cause inequalities dependent on their mutual configuration, it is natural to think that there exists in their theory a considerable inequality of this kind, of which the period is very long.

The mean motions of Jupiter and Saturn, Laplace now observed, were nearly as 5 to 2; "whence I concluded that the terms which, in the differential equations of motion of these planets, have for argument five times the mean longitude of Saturn minus two times the mean longitude of Jupiter, could become sizeable because of the integrations, though they are multiplied by the cubes and products of three dimensions of the eccentricities and inclinations of the orbits." If λ is the mean longitude of Jupiter and λ' the mean longitude of Saturn, then the quantity

$$V=5\lambda'-2\lambda \tag{17}$$

varies very little with time and the terms in sin V in the longitudes of Jupiter and Saturn cause inequalities having a period of about 850 years and large amplitudes which account for the accelerations and retardations observed in the motions of these two planets. In a preliminary formulation Laplace gave the inequality in the longitude of Jupiter as

$$20'\sin(5n't-2nt+49°\,8'\,40''),$$

and that in the longitude of Saturn as

$$46'\,50''\,\sin(5n't-2nt+49°\,8'\,40''),$$

where the zero of time is taken as the beginning of 1700; a later, more accurate derivation produced somewhat larger coefficients, 21.1′ for Jupiter and 49.0′ for Saturn.

Laplace's discovery of this inequality not only resolved an enigma but demonstrated the importance of resonant terms in celestial mechanics (that is, terms involving near-commensurabilities in the mean motions). C. G. J. Jacobi would later give an elegant method for isolating these resonant terms in the perturbing function, which was published in *Astronomische Nachrichten* in 1848.

It is a remarkable fact that the "great inequality of Jupiter and Saturn", as it is called, reached its null point toward the end of the sixteenth century, so that in the observations of Tycho Brahe employed by Johannes Kepler the displacements of Jupiter and Saturn due to this inequality were close to zero. Kepler's values for the mean motions of these planets, obtained from a comparison of Tycho's observations with ancient observations, were therefore very close to the correct values. Throughout the seventeenth and most of the eighteenth centuries Jupiter's average motion was found to be more rapid and Saturn's slower than Kepler's numbers implied. Then in 1773 Johann Heinrich Lambert showed that with respect to the seventeenth-century observations of Johannes

Hevelius the motion of Jupiter was now slowing and that of Saturn accelerating: the apparent acceleration of Jupiter and deceleration of Saturn were rounding off. The displacements in longitude caused by this inequality reached their maxima toward the end of the eighteenth century, then commenced to decrease.

In the second part of the memoir Laplace undertook to provide what he believed to be a definitive answer to the question of the stability of the solar system. Given the immunity of the mean distances to secular change (the Laplacian and Lagrangian proofs had in fact established this immunity only to the order of the first powers of the perturbing masses), there remained the problem of secular changes in the eccentricities and inclinations; if an orbital eccentricity were to increase beyond the value 1, the motion would cease to be periodic. Laplace now gave an argument to show that neither the orbital eccentricities nor the tangents of the orbital inclinations contained *arcs de cercle* or exponential terms that could lead to indefinite increase; ergo the solar system was stable. Given the fact that all the planets turn in the same direction round the Sun, he derived from the conservation of angular momentum, on neglecting terms of the order of m^2, the two equations

$$\text{const} = m(a)^{\frac{1}{2}}(e)^2 + m'(a')^{\frac{1}{2}}(e')^2 + \ldots, \quad (18)$$

$$\text{const} = m(a)^{\frac{1}{2}}(\theta)^2 + m'(a')^{\frac{1}{2}}(\theta')^2 + \ldots, \quad (19)$$

where $e, e', \ldots$ are the eccentricities, and $\theta, \theta', \ldots$ are the tangents of the inclinations. These equations, he showed, could not hold if the es and θs contained terms that would cause them to increase indefinitely with time.

In 1808 Siméon-Denis Poisson, a protégé of Lagrange and Laplace, would demonstrate that the immunity of the mean distances to secular change held to the second order in the perturbing masses. Yet the general problem of stability is a more difficult one than Lagrange, Laplace, and Poisson envisaged; it remains today a subject of active research (see the appendix to Chapter 28).

Laplace next devoted himself to a systematic and thorough development of the mutual perturbations of Jupiter and Saturn; the resulting "Théorie de Jupiter et de Saturne" was read in two parts on 10 May and 15 July 1786, and published in *MARS* 1785/88.

In an extended initial section Laplace developed the equations of motion into forms convenient for the calculations that he now knew to be necessary. Thus for the perturbations of the radius vector he obtained

$$\delta r = \frac{\left\{\begin{array}{l} + a\cdot\cos v.\int ndt\cdot r\cdot\sin v\cdot\{2\int DR + r\,\partial R/\partial r\} \\ - a\cdot\sin v\cdot\int ndt\cdot r\cdot\cos v\cdot\{2\int DR + r\,\partial R/\partial r\} \end{array}\right\}}{\sqrt{1-e^2}}, \quad (20)$$

where v is the true anomaly, R is the perturbing function, and 'D' means differentiation with respect only to the variables pertaining to the perturbed planet. Similarly, for the perturbations in longitude he obtained

$$\delta v = \frac{\left\{\begin{array}{c} (2r\cdot d\delta r + \delta r\cdot dr)/a^2 ndt \\ + 3a\int ndt\int DR + 2a\int ndt\cdot r\,\partial R/\partial r \end{array}\right\}}{\sqrt{1-e^2}}. \quad (21)$$

The first of these equations is closely parallel to the perturbative part of Clairaut's equation for the radius vector, and like Clairaut's equation has the merit of permitting expeditious application. but the perturbing function R is now understood as being developable systematically as an infinite series in accordance with increasing powers of the eccentricity and tangent of the orbital inclination; Laplace reduces this development to an essentially mechanical decision procedure, allowing him to calculate the value of any term in the expansion. For the inequalities independent of the squares and higher powers of the eccentricities and inclinations, a complete derivation was feasible, and Laplace proceeded to supply it; but for the higher powers such a derivation would have been costly in time and labour, and here he resorted to a species of sharpshooting, picking out the terms in R farther down the line that would become large enough to be detectable on integration, and deriving the inequalities resulting from these terms by themselves. In this way he obtained not only the "great inequality" of Jupiter and Saturn, proportional to $\sin(5n't - 2nt + \text{const})$, with coefficients for the two planets dependent on the third order in the eccentricities and inclinations, but also several terms proportional to the squares and products of two dimensions in the eccentricities, for instance, in the longitude of Saturn the term $-13'\ 16''$ $\sin(2nt - 4n't - 2°\ 27'\ 4'')$; this and other oscillations of the same order in the longitude had probably caused as much perplexity for his predecessors as the great inequality itself.

The derivation of these inequalities presupposed values for the eccentricities, aphelia, inclinations, nodes, and mean longitudes at epoch; all which must be determined from observations. But there is a logical circle here, of which Laplace was clearly aware: to determine the orbital elements, it was necessary to have good observations from which the perturbations had been subtracted. Thus "the determination of the inequalities of Jupiter and Saturn and that of the elements of their orbits depend reciprocally on one another, and we can come to know them well only by successive approximation." Laplace proceeded to use equations of condition in the manner of Tobias Mayer to determine differential corrections to the orbital elements; for a description of his procedure see Chapter 26 below.

On 10 May 1786, the day on which Laplace presented the first part of his "Théorie de Jupiter et de Saturne", Jean-Baptiste Joseph Delambre (1749–1822), an astronomer who had recently made a name for himself in revising the solar tables of N.-L. de Lacaille and in detecting an error in the accepted tables of Mercury, offered to Laplace his services in the correction and refinement of the data – the observed oppositions – on which any theory of Jupiter and Saturn must rest. All the original unreduced observations had to be re-examined and assessed, and corrected for aberration, nutation, and faulty assumptions as to the positions of stars. Then, on the basis of Laplace's theory and making use of his method of equations of condition, the orbital elements had to be re-determined, and the results embodied in new tables of the two planets. This work, Delambre tells us, "required nine months of the most constant and assiduous application". Delambre would similarly undertake the observational refinement of Laplace's theories of Uranus and of the Galilean satellites of Jupiter.

Long-term inequalities of the satellites of Jupiter

Lagrange, in his memoir of 1766 on the inequalities of the Galilean satellites of Jupiter, had found certain sinusoidal terms in the longitudes of the first three satellites to be especially large, because of near-commensurabilities between the mean motions of these satellites. Thus if the mean motions for the first, second, and third satellites are n, n', and n'', then $n{:}n' = 2.0073{:}1$ and $n'{:}n'' = 2.0147{:}1$. For the second satellite two large terms resulted, one with the argument $(n-n')t$ and the other with the argument $2(n'-n'')t$, and the curious fact was that these two inequalities appeared to have precisely the same period, for Pehr Wargentin in his observational study of the satellite had found only one inequality. Therefore, to within the precision of the observations, the *difference* between the mean motions of the first and second satellites was exactly *twice the difference* between the mean motions of the second and the third:

$$n-n' = 2(n'-n''), \quad \text{or } n-3n'+2n''=0. \tag{22}$$

Lagrange had further noticed that (to use Laplace's symbols)

$$(n-3n'+2n'')t+\epsilon-3\epsilon'+2\epsilon''=180°, \tag{23}$$

where ϵ, ϵ', and ϵ'' are the longitudes of the three satellites at epoch. At some point in the mid-1780s Laplace set out to try to explain these equations on the basis of gravitation. Inspiration for his inquiry came from Lagrange's work on the libration of the Moon.

The explanation that he found was sketched in *MARS* 1784/87, then given with full detail in *MARS* 1788/91. Let $s=n-3n'+2n''$, and $V=(n-3n'+2n'')t+\epsilon-3\epsilon'+2\epsilon''$. From an examination of the mutual perturbations dependent on the squares and products of the masses Laplace obtained two differential equations having to do with variations in the mean motions, namely

$$\frac{ds}{dt} = \alpha n^2 \sin V, \tag{24}$$

where α represents a complicated function, and

$$\frac{dV}{dt} = s. \tag{25}$$

From (24) and (25) it followed that

$$\frac{d^2V}{dt^2} = \alpha n^2 \sin V, \tag{26}$$

whence

$$\pm\frac{dV}{dt} = \sqrt{\lambda - 2\alpha n^2 \cos V}. \tag{27}$$

Here λ is a constant of integration. The case in which α is positive and $\lambda < 2\alpha n^2$ required that V be periodic with a value oscillating around 180°, and this appeared to be the case in nature, the oscillations being very small; their amplitude and zero-point could be determined, if at all, only observationally.

A practical consequence was that, in the construction of tables, equations (22) and (23) could and should be taken as exact. On a more theoretical note, one could conclude that, even if originally the distances of the satellites from the centre of Jupiter had not been in the ratios now observed, the pendular oscillations implied by (27) would have brought them into that relation. It was a case entirely analogous to that which Newton had conjectured and Lagrange had derived for the physical libration of the Moon (see Chapter 21).

The secular equation of the Moon

Besides the anomalies in the mean motions of Jupiter and Saturn, so triumphantly accounted for on the basis of Newtonian theory in Laplace's "Théorie de Jupiter et de Saturne" (*MARS* 1785/88), there was another major unexplained anomaly in the solar system: the secular equation of the Moon. It had been a puzzle since Halley first drew attention to it in 1693. In 1749 Richard Dunthorne, from an examination of ancient and modern lunar and solar eclipses, had found the rate of acceleration to be 10″ per century per century; Johann Tobias Mayer in his first lunar tables of 1753 put it at 7″, but in the revised version of these tables published in London in 1770 it was raised to 9″. D'Alembert and Euler had shown that this acceleration could not arise from direct perturbations due to the Sun and planets. Lagrange, in a prize-winning memoir of 1774, had proved that it could not be caused by the spheroidal shapes of the Earth and Moon, and had raised doubts about the reality of the acceleration. (In his *Mécanique céleste* Laplace would claim that it was placed beyond doubt by ancient eclipses, and those observed in AD 977 and 978 by Ibn Yūnus.) As we have seen, Laplace in his first published astronomical memoir pointed out that the assumption of a finite speed of propagation of the gravitational force would account for it; but he regarded this assumption as doubtful. After completing his "Théorie de Jupiter et de Saturne" in August 1786, he set as his next challenge the answering of this question: could the anomalous acceleration of the Moon be accounted for on the basis of a strictly Newtonian theory?

On 19 December 1787 Laplace read to the Academy a first draft of his resolution of the question; the completed memoir "Sur l'équation séculaire de la lune" was published in *MARS* 1786/88. The acceleration, he argued, was the result of an indirect perturbation. Owing to planetary perturbations, the eccentricity of the Earth's orbit is subject to an oscillation of very long period; in the present age of the world (as he had proved in an earlier memoir) it is diminishing. The diminution causes a slight decrease in the average radial component of the Sun's perturbing force on the Moon (the perturbing force being the difference between the Sun's force on the Earth and its force on the Moon); as a consequence the Moon is less attracted away from the Earth, and the increased net force toward the Earth causes it to revolve more rapidly. After a few hundred thousand years the eccentricity will cease to diminish and begin to increase, and then the Moon's motion will be subject to a corresponding deceleration.

Laplace's explanation, however, does not completely resolve the problem. As John Couch Adams was to point out in 1854, the Laplacian explanation accounts for only about half the Moon's observed secular acceleration; the effect of the diminishing eccentricity on the tangential components of the Sun's perturbing force, which Laplace failed to include in his calculation, subtracts from the total effect. The remaining acceleration is due to the fact that the scale of mean time, in accordance with which all the celestial motions were timed in Laplace's day, is not uniform because of the slowing of the Earth's rotation caused by the friction of the tides. Laplace, in his earliest astronomical memoir, we recall, had considered and then dismissed the possibility that the secular acceleration of the Moon could be an illusion resulting from frictional slowing of the Earth's rotation. The idea would not be taken up again until the 1860s, and an atomic clock capable of establishing its validity would not be available till the twentieth century.

Revolution, Consulate, Empire, and Restoration

The Revolution, and more especially the Terror, turned upside down the lives of Frenchmen generally, including scientists. Initially the revolutionaries took an antipathetic view toward science, especially physical and mathematical science; the academies, seen as elitist, were suppressed by a decree of the Convention of 8 August 1793. A number of scientists, friends of Laplace like A.-L. Lavoisier, J.-S. Bailly, and Bochart de Saron who had been president of the Parlement and one of Laplace's first protectors in the Academy, were guillotined. Laplace had been appointed to the Commission on Weights and Measures which was charged with the establishment of the metric system; but he was removed from this commission by the decree of 3 nivôse An II (23 December 1793), on suspicion of lacking "republican virtues and the hatred of kings". He took refuge at Melun, southeast of Paris and, there pursued his work, writing his *Exposition du système du monde* and the first volumes of his *Mécanique céleste*. He was assisted by the young Alexis Bouvard (1767–1843), who became his faithful collaborator and verified the calculations of the *Mécanique céleste*.

Often quoted is the stupid remark of a member of the Convention on the occasion of Lavoisier's death: "The Republic has no need of *savants*". Anti-scientific rhetoric was abandoned, however, with the Thermidorean reaction and rise of the first Directory, for the new leaders recognized the importance of scientific research for a great nation – a nation, moreover, engaged in a long and difficult war with half of Europe. The law of 7 messidor An III (25 June 1795) created the Bureau des Longitudes, charged with calculating and publishing ephemerides, directing the Paris Observatory, and perfecting the theories of celestial mechanics; among the members originally appointed to it were Lagrange and Laplace. The École Polytechnique, created in December 1794, received on 1 September 1795 the name by which it has been known ever since; the École Normale had its beginnings in January 1795, but then underwent temporary eclipse till 1812. The academies were re-created in the form of the five classes of the Institut de France by the law of 3 brumaire An IV (25 October 1795), and on 29 brumaire An IV (20 November 1795) Laplace and Lagrange were named members of the mathematical section of the class of physical and mathematical sciences.

The Republican Calendar had been adopted by the Convention in 1793, with 22 September 1792 designated as the beginning of its first year. The group that had drawn it up, under the direction of the Committee of Public Instruction, had included Lagrange but not Laplace; on one occasion, however, the latter had served as a consultant. On 22 fructidor An XIII (9 September 1805) it was Laplace, now a senator, who presented the report advising a return to the Gregorian Calendar, to be effective on 1 January 1806. In the first two editions of the *Exposition du système du monde* (1796 and 1799), Laplace presented and explained the Republican Calendar; in the third edition (1808) he did not mention it, but confined himself to praising the Gregorian Calendar.

After the *coup d'état* of 18 brumaire An VIII (9 November 1799), by which Napoleon came to power, Laplace was named Minister of the Interior, only to be replaced six weeks later by Lucien Bonaparte. In 1803 he became Chancellor of the Senate. Honours accumulated: Mme Laplace became Dame of Honour in attendance on the Princess Elisa, Napoleon's sister, and Laplace became Grand Officer of the Legion of Honour and Count of the Empire in 1808.

In 1814, with France assaulted by enemies on all frontiers, the Senate, Laplace included, voted the end of the Empire and Napoleonic power. During the Hundred Days (from Napoleon's return in March 1815 to Waterloo, 18 June 1815), Laplace avoided rallying to the Bonaparte cause; thus, after Waterloo and the definitive fall of the Empire, he remained in favour with Louis XVIII. He received the Grand Cross of the Legion of Honour, and was given the title Marquis. In 1816 he was made a member of the Académie Française, and here he is said to have played an ultraconservative role, refusing even to support freedom of the press. For this attitude – the docile acceptance of the regimes that succeeded one another in France during this troubled time, without any sign of having made critical distinction among them – Laplace has been much blamed. He has been defended, on the other hand, as the scientist totally preoccupied with science; could one not argue that it was proper for him to accept the honours that successive governments, whatever their character might be, owed him as the greatest scientist of his nation and time?

Laplace always showed much solicitude for the young scientists who came to him for counsel and support. In 1806 he acquired a country house at Arcueil near Paris, and with the chemist Claude Berthollet, a colleague in the Academy of Sciences who had a house nearby, he created the "Société d'Arcueil". It aimed to foster the association of illustrious elder statesmen of science, like Laplace and Berthollet themselves, with younger scientists who could benefit from the knowledge, experience, and counsel of such elders. Among those who profited from the society were J. A. Chaptal, J. L. Gay-Lussac, Alexander von Humboldt, Jean-Baptiste Biot, François Arago, and S.-D. Poisson.

Laplace's work in celestial mechanics after 1789

In 1793, while Laplace was taking refuge from the Revolution at Melun, he began his *Exposition du système du monde*; the first edition appeared in 1796, and the sixth, which was posthumous but in large part revised by the author before his death, in 1835. In the interim there had appeared four other editions, in 1799, 1808, 1813, and 1824. This work is a very special one among Laplace's writings: it is a general exposition of all the astronomical knowledge of the time in which there appears not a single mathematical formula, and – what is perhaps stranger yet – not a single geometrical figure. The exposition is nevertheless perfectly clear. According to Arago in his "Notice" on Laplace, "This work, written with a noble simplicity, an exquisite propriety of expression, a scrupulous correctness, is classed today, by universal consent, among the beautiful monuments of the French language."

The work contains five books. The first concerns the apparent movements of the celestial bodies (the diurnal motions, the apparent motions of planets and satellites, etc.). The second is entitled "On the real motions of the celestial bodies", and describes the heliocentric orbits of the planets and the laws of planetary motion. The third is dedicated to mechanics (forces, the motion of a mass-point under the action of forces, etc.). In Bk IV, entitled "On the theory of universal gravitation", we are given an exposition of Laplace's own discoveries, for instance those having to do with the secular inequalities and the stability of the solar system. Book V gives a history of astronomy from Antiquity to Newton, and concludes with a chapter entitled "Considerations on the system of the world and on the future progress of astronomy", in which Laplace draws attention to the structure of the solar system, to the fact that the orbits are nearly circular and little inclined to one another, and to the stability of the observed motions, which he believes he has definitely demonstrated. One finds here a scarcely veiled rejection of direct action by the Deity as an explanation for astronomical phenomena. "These phenomena and some others similarly explained, lead us to believe that everything depends on these laws [the primordial laws of nature] by relations more or less hidden, but of which it is wiser to admit ignorance, rather than to substitute imaginary causes solely in order to quieten our uneasiness about the origin of the things that interest us."

Note 7 at the very end of Bk V has made the *Exposition du système du monde* especially celebrated. It contains the hypothesis on the formation of the solar system known as Laplace's nebular hypothesis. Georges-Louis Leclerc, Comte de Buffon, here cited by Laplace, had suggested that the matter that would later form the planets had been dragged out of the Sun by the attraction of a passing comet. A very similar idea was to be suggested in 1920 by Sir James Jeans, who proposed that matter had been dragged out of the Sun as the result of tidal action due to the near passage of a star.

Laplace rejected Buffon's hypothesis because it failed to explain why the direction of revolution of the satellites around their primary planets is the same as that of the planets around the Sun, or why the direction of rotation of the planets about their axes is again the same, or above all why the planetary orbits are so little eccentric. He proposed instead that a vast nebula in rotation had cooled and condensed to form the Sun at its centre and, by the agglomeration of the remaining matter at different distances from the Sun, the planets. (Immanuel Kant had proposed a similar hypothesis in 1755, but Laplace probably did not know of it.) In the successive editions of the *Exposition du système du monde* Laplace revised his formulation of the hypothesis and presented it with increasing confidence. He was aware that William Herschel, in his examination of the nebulae, had persuaded himself that they were stars in process of formation through condensation. We know today that this is true for a gaseous nebula like the Orion nebula, but false for other nebulae such as the galaxies and the globular clusters. Remarkably, Laplace cited the example of the Pleiades, suggesting that this group had been created in the middle of a "nebula with several nuclei" – an hypothesis completely in

TRAITÉ

DE

MÉCANIQUE CÉLESTE,

PAR P. S. LAPLACE,

Membre de l'Institut national de France, et du Bureau des Longitudes.

TOME PREMIER.

DE L'IMPRIMERIE DE CRAPELET.

A PARIS,

Chez J. B. M. DUPRAT, Libraire pour les Mathématiques, quai des Augustins.

AN VII.

22.2. The title-page of the first volume of Laplace's *Traité de mécanique céleste.*

accord with twentieth-century astrophysics. The hypotheses of Buffon and Jeans have been abandoned, while Laplace's nebular hypothesis retains the interest of modern astronomers who strive to establish the theory on a more scientific basis.

The *Traité de mécanique céleste* (Figure 22.2), commonly known simply as the *Mécanique céleste*, is Laplace's most famous work, contributing most to his renown both in France and abroad. For the astronomers of the nineteenth century, it was the chief source from which they derived the foundations of the theories by which they made their century "the golden age of celestial mechanics". If Lagrange's *Mécanique analytique* (1788) had, in some of its theoretical formulations, an equal importance, it was the *Mécanique céleste* that combined theoretical formulation with detailed application in such a way as to serve most effectively as guide and spur to further research; as such it was undoubtedly the most influential work of celestial mechanics since Newton's *Principia*. Numerous translations were made almost immediately, but the most celebrated – the bedside book of American astronomers like Simon Newcomb and G. W. Hill – was the English translation by Nathaniel Bowditch, published in Boston between 1829 and

1839. Bowditch added explanatory notes and redid all the calculations.

For a twentieth-century astronomer the reading of the *Mécanique céleste* is not easy; nor was it easy for Laplace's contemporaries. Bowditch and Biot, who read page-proofs, complained of the absence of numerous intermediary calculations, replaced by "it is easy to see"; to bridge these gaps they were forced to devote long hours of reflection and calculation. F. F. Tisserand, in the preface to the first volume of his *Traité de mécanique céleste* (1889), proposed his own work as an easy introduction to an arduous science, but urged those who wished to penetrate more deeply to turn to Laplace's treatise, "of which all the chapters present, still today, and even to the most accomplished astronomers, varied subjects for fruitful meditation". A century later, the interest of Laplace's text must be admitted to be largely historical.

The work contains five volumes of which the first four appeared between 1798 and 1805 and the fifth in 1825. Laplace's purpose, as formulated in a brief preface, was to bring together into a coherent whole all the theories of celestial mechanics, both his own and those of his predecessors and contemporaries. For celestial mechanics thus unified he had an explicit goal:

Astronomy, considered in the most general manner, is a grand problem in mechanics, in which the elements of the celestial motions are arbitrary constants; its solution depends both on the accuracy of the observations and on the perfection of Analysis, and it is very important to banish all empiricism and to borrow nothing from observation except indispensable data. It is the aim of this work to achieve, as much as may be in my power, this interesting result.

The first volume consists of two books and the second volume of three. These two volumes contain the general theory of the motions and figures of the celestial bodies. Book I presents a general course in mechanics, based on the laws of equilibrium and motion. In Bk II Laplace introduces universal gravitation, with its consequences for the elliptical motion of the centres of gravity of the planets about the Sun and the perturbations of this motion due to mutual planetary attraction. Book III deals with the figures of the heavenly bodies, and includes a comparison of theoretical predictions and observational data relative to the figure of the Earth. Book IV concerns the oscillations of the sea and atmosphere. In Bk V Laplace treats of the rotations of the heavenly bodies about their centres of gravity, and deals in particular with the precession and nutation of the Earth's axis, possible changes that the Earth's rotation might undergo, the libration of the Moon, and the rotation of the rings of Saturn.

The third volume consists of two books: Bk VI which gives the detailed theories of the motions of the several planets, and Bk VII which contains the detailed theory of the Moon's motions. The fourth volume contains three books: Bk VIII deals with the motions of the satellites of Jupiter, Saturn, and Uranus; Bk IX with the theory of comets; and Bk X with miscellaneous topics including astronomical refraction and the possible effect of a resisting medium on planetary motion.

The fifth and last volume is divided into six books, each concerned with one of the major topics dealt with in the earlier books, but giving results that Laplace had published more recently in the *Connaissance des temps* or in the *Mémoires* of the Institute. Each of these books begins with an historical notice, describing the relevant work of Laplace's predecessors along with his own contributions.

As we might expect, the *Mécanique céleste* is to a considerable extent a presentation and elaboration of the results of Laplace's earlier work – on the secular equations and the stability of the solar system; the attraction of spheroids with its consequences for the figure of the Earth, the tides, and the precession and nutation of the Earth's axis; the effect of near-commensurabilities in the motions of planets and satellites in producing inequalities of long period; the libration in the motion of the first three Galilean satellites of Jupiter; and the secular equation of the Moon. But the *Mécanique céleste* also incorporated a number of discoveries and developments dating from after 1789; the following deserve special mention.

The invariable plane

In the late 1790s Laplace made the discovery that there exists in the solar system an invariable plane, about which the planetary orbits, owing to mutual planetary interactions, perpetually oscillate through small angles to either side. In the study of the motions of the planets astronomers ordinarily use the ecliptic – the plane of the Earth's orbit about the Sun – as a plane of reference; but this choice has the disadvantage that because of perturbations the ecliptic does not remain parallel to a fixed plane in a system of absolute coordinates. Laplace defined his invariable plane as follows: it passes through the centre of gravity of the solar system, and is such that if the areal velocity of the projection of each planet onto the plane is multiplied by the planet's mass, then the sum of these products is a maximum. Using the later language of vectors we may say that the invariable plane is the plane perpendicular to the vector sum of the angular momenta of the planets. For the year 1750, Laplace found the ecliptic to be inclined to the invariable plane by an angle of $1° 35' 31''$. Changes in the inclination and position of the node of the ecliptic with respect to the invariable plane are very slow; it is important to determine them, however, whenever long-term variations in the orientations of the planetary orbits are considered.

The figure of the Earth

Laplace had early recognized the disagreement between the gravitational theory of the Earth's figure, as it had been developed by A.-C. Clairaut, and the several measures of a degree of the meridian in different latitudes; the ellipticity apparently implied by the latter was a good deal greater than that implied by the theory. In Bks III and XI of the *Mécanique céleste* Laplace carried the development of the theory further than Clairaut, but with general corroboration of Clairaut's results: the Earth should be an ellipsoid of revolution, very nearly, with layers of increasing density toward the centre, and with an ellipticity considerably less than $\frac{1}{230}$ (the precession and nutation set limits between which the ellipticity must fall: $\frac{1}{304}$ and $\frac{1}{578}$ according to Laplace, $\frac{1}{279}$ and $\frac{1}{578}$ according to Bowditch). How were these conclusions to be reconciled with the observational results? Laplace was undoubtedly one of those who had supported the basing of the metric system on a new measurement of the meridian of France: the surreptitious aim was to resolve the enigma of the Earth's shape. From the re-measurement carried out by Delambre and Méchain during the years from 1792 to 1799, compared with the measured length of the degree in Peru, Laplace in Bk III of the *Mécanique céleste* derived an ellipticity of $\frac{1}{334}$, and in Bk XI an ellipticity of $\frac{1}{308}$. Similar low values were supported by the variation of the length of the seconds-pendulum with latitude, and by two inequalities in the Moon's longitude that were dependent on the Earth's oblateness. Only the measure of the degree of the meridian in Lapland appeared to belie these conclusions; but the re-measurement of that degree by Jons Svanberg in 1802–03 showed that the original result obtained by Maupertuis and his colleagues in the 1730s had erred greatly in excess. (For further details see Chapter 15 above.)

The tides

As we have seen, already by the early 1780s, Laplace had achieved the first truly dynamic theory of the tides; a systematic presentation of the theory is to be found in Bk IV of the *Mécanique céleste*. At the beginning of the twentieth century G. H. Darwin paid homage to Laplace as the first to put in evidence the whole difficulty of the problem, and to demonstrate the fundamental role that the Earth's rotation plays in producing the phenomena of the tides.

Laplace based his theory, he tells us, on the result that "the state of a system of bodies in which the initial conditions of motion have disappeared owing to resistance is periodic, like the forces to which it is subject." At the end of Chapter 3 of Bk IV, he established a formula giving the height of the tide at a given place as a function of the hourly coordinates of the Moon and the Sun and of certain constants characteristic of the place. In the next chapter he proceeded to compare this formula with a series of measurements of the tides made at the port of Brest during the years 1711–16, and in Bk XIII he carried out a further comparison with new measurements made at Brest during the years 1807–22. Laplace's formula is still used to calculate the *Brest-Reference Tide*, which serves as an intermediary in the prediction of tides in all ports situated on the coast of France.

Development of the perturbing function of the planets

In 1774 Lagrange had introduced a potential function in the calculation of the attraction exerted by an extended body, and in memoirs published in *Mémoires de Berlin* 1776/79 and 1777/79, he had introduced a potential function, later called by Laplace "the perturbing function", for the determination of the mutual perturbations of the planets considered as point-masses. The introduction of the perturbing function, according to Laplace, constituted "a veritable discovery". In his "Théorie de Jupiter et de Saturne" he proceeded to employ this function to obtain formulas for the perturbations in radius vector, true anomaly, and latitude; these formulas reappear as formulas (X), (Y), and (Z) of Bk II, Chap. 6 of the *Mécanique céleste* (equations (X) and (Y) are identical with our equations (20) and (21) above). Of them he says:

The formulas (X), (Y), (Z) have the advantage of presenting the perturbations under a finite form. This is very useful in the theory of comets, in which those perturbations cannot be found, except by the [mechanical] quadrature of curves. But the smallness of the eccentricities, and the inclinations of the orbits of the planets to each other, enables us to develop their perturbations in converging series of sines and cosines of angles, increasing in proportion to the time, and we can then arrange them in tables which will answer for an indefinite time.

The development of the perturbations in series entailed that of the perturbing function, which entailed in turn the development in series of the odd inverse powers of the distance between the perturbed and perturbing planets, or

$$(1-2\alpha\cos\theta+\alpha^2)^{-s}=\tfrac{1}{2}\sum_{-\infty}^{+\infty} b_s^{(j)}\cos j\theta, \qquad (28)$$

where α is the ratio of the mean solar distances of the perturbed and perturbing planets, $s=\frac{1}{2},\frac{3}{2},\frac{5}{2},\ldots,$ and

$$b_s^{(-j)}=b_s^{(j)}.$$

The bs, here given in the double indicial notation that Laplace employed in the *Mécanique céleste*, are today called Laplace coefficients. They are functions solely of α. In deriving them, Laplace made use of a device introduced by Lagrange in 1766, whereby the left-hand member of equation (28) is factored into two complex factors, each of which is then expanded by the binomial theorem; the two resulting series are then multiplied together. The resulting product gives the bs as hypergeometric series. Laplace also exploited the Eulerian relations among the bs and their derivatives with respect to α, thus reducing the number of necessary summations of series to two.

Laplace then proceeded to develop the perturbing function to the third order in the eccentricities and inclinations. This development was much utilized by the astronomers of the early nineteenth century, for instance by Bouvard in his theory of Uranus. Some fifty years were to elapse before the development was carried to the sixth order in the eccentricities and inclinations by Philippe de Pontécoulant and Benjamin Peirce, and then to the seventh order by U. J. J. Le Verrier.

The theory of comets

In Bk IX Laplace took up the problem of determining the changes resulting in cometary orbits owing to perturbation by planets. Here, in contrast to the case of planets perturbed by planets, no analytic formulas could be derived that would give the perturbations for all time; it was necessary to have recourse to mechanical quadratures, and starting at a given time, proceed step by step round the cometary orbit. Following Lagrange's prize-winning essay on cometary perturbations (see the preceding chapter), Laplace employed the eccentric anomaly as independent variable, and proposed calculating the variations of the orbital elements – eccentricity, perihelion, mean solar distance and mean motion, epoch, orbital inclination and node of the cometary orbit on the ecliptic – for each degree or two degrees round the orbit. For the required numerical integrations, he developed the appropriate approximating formulas by means of a generating function. Like Clairaut (see Chapter 19 above), in the superior half of the orbit he separated the calculation into two parts, one concerned with the (very small) direct effect of the planet on the comet, the other with the effect of the planet on the Sun; in both cases formal integration could be invoked. And again like Clairaut, he proposed calculating backwards in the last quadrant, from 360° to 270° of eccentric anomaly, when the date corresponding to 360° was known. Clairaut had used a single cometary reference ellipse for the determination of planet–comet distances; Laplace proposed rectifying the reference ellipse after every fourth of a quadrant of eccentric anomaly in the first and fourth quadrants, and at least once in the superior part of the orbit. These procedures were applied by the Baron de Damoiseau, J. K. Burckhardt, O. A. Rosenberger, Pontécoulant and J. W. Lubbock to the calculation of the return of Halley's Comet in 1835, and led to predictions considerably more accurate than the prediction Clairaut had been able to achieve in 1759.

In a second chapter on the cometary perturbation, Laplace considered the case of a comet coming so close to a planet that for a time it could be considered as moving in an unperturbed ellipse round the planet; later, when it had passed out of the 'sphere of influence' of the planet, it could be considered as moving in an ellipse about the Sun. The formulas derived from this conception were applied by Burckhardt, at Laplace's request, to Comet Lexell. This comet had been seen only once, in 1770; from the observations Lexell and Burckhardt concluded that it had a period of 5.6 years, and a solar distance at perihelion of about 0.68 AU. Applying Laplace's formulas to the near approach of this comet to Jupiter in 1767, Burckhardt showed that the comet had previously had a mean solar distance of 13.29 AU, and a solar distance at perihelion of 5.08 AU, so that it would have been invisible from the Earth. Applying the formulas again to the near approach of the comet to Jupiter in 1779, Burckhardt showed that the orbital elements must have been so altered as to make the mean solar distance 6.39 AU, and the solar distance at perihelion 3.33 AU; thus the comet once more became invisible from the Earth.

Laplace concluded Bk IX by assembling evidence to show that the masses of comets are not sufficiently great to perturb the planets detectably; the consideration that had deterred him from pursuing the analysis of mutual planetary perturbations during the decade 1775–85 was thus disposed of.

Astronomical refraction

Light rays from a star on entering the Earth's atmosphere are refracted in such a way that the apparent zenith distance of the star is less than its true zenith distance. The estimation of astronomical refractions had been a concern of astronomers since Tycho's time; again and again the tables drawn up for this purpose had to be refined, to keep pace with the increasing precision of observational instruments. Delambre in 1797 made observational determinations of the refractions from 70° to 90° of zenith distance, and Jean-Charles Borda undertook similar studies. In Chap. 1 of Bk X of the *Mécanique céleste* Laplace derived a formula, which now carries his name, and which gives the astronomical refraction to the third order of the tangent of the zenith distance of the star.

The scientific legacy of Laplace

The work of Laplace largely defined the problems with which celestial mechanics would be concerned during the nineteenth century, and provided many of the methods that the celestial mechanicians of that century would use in dealing with them. The heirs and appropriators of the Laplacian legacy, both in France and abroad, were numerous and brilliant. The works of Bouvard, Damoiseau, Pontécoulant, Le Verrier, C.-E. Delaunay, and A.J.-B. Gaillot in France, of Newcomb and Hill in the United States, of Adams in England and of Peter Andreas Hansen in Germany, were direct developments from the foundation laid in the *Mécanique céleste* (see Chapter 28 below for further detail). We have already insisted on the continuing importance of Laplace's cosmogonic hypothesis as presented in his *Exposition du système du monde*.

Laplace was a proponent of philosophical determinism, and no-one has been more resolute in championing the doctrine, as the following oft-quoted passage testifies:

> *We must therefore envisage the present state of the universe as the effect of its anterior state and as the cause of its future state. An intelligence that knew, for a given instant, all the forces acting in nature and the respective situations of the beings that made it up, if it were in addition vast enough to submit these data to analysis, would embrace in a single formula the motions of the largest bodies of the universe and those of the smallest atom: nothing would be uncertain for it and the future like the past would be present to its eyes.*

The tranquil assurance here evident could hardly have survived unshaken in face of such later developments in science as the emergence of quantum mechanics. Recently it has been shown that, because of sensitivity to tiny errors in the initial conditions, predictions of the future state of the solar system are valid only for durations of a hundred million years or so (see the appendix to Chapter 28). Celestial mechanics as Laplace conceived it nevertheless remains the fundamental tool of the astronomers for calculating ephemerides of the bodies of the solar system.

Almost one hundred years after Newton, on 5 March 1827, Laplace died in Paris, at 108 rue du

Bac, a house which still stands. The comparison with Newton is an interesting one. Newton stood at the origin of a scientific revolution without precedent. He discovered a law applying to the least particle wherever it is found in the universe; at one blow he provided explanations of the most diverse phenomena, and with apparently unrestricted validity. The success of his derivations encouraged the hope that the employment of mathematics as a tool would suffice for deriving all the consequences of universal gravitation. The genius of Laplace, on the other hand, consisted in wielding mathematical devices in such a way as to discover and demonstrate a very considerable number of these consequences. He definitively placed the law of universal gravitation in the rank of the great scientific truths by showing that the phenomena that appeared to contradict it or limit its domain (e.g. the great inequality of Jupiter and Saturn, the secular acceleration of the Moon, the motions of the satellites of Jupiter) were, on the contrary, explained by it.

Despite his achievement, Laplace is relatively little known to the general public. Is it, as has suggested Jacques Merleau-Ponty, because Laplace was not the originator of a spectacular revolution but the hero of "normal science"?

Further reading

Article on Laplace, in *Dictionary of Scientific Biography*, vol. 15 (New York, 1978), 273–403

H. Andoyer, *L'oeuvre scientifique de Laplace* (Paris, 1929)

B. A. Vorontsov-Veliaminov, *Laplace* (Moscow, 1985; in Russian)

François Arago, Laplace, in J. A. Barral (ed.), *Oeuvres de François Arago*, Vol. 3 (Paris, 1859)

Oeuvres complètes de Laplace (14 vols, Paris, 1878–1912)

PART VII

Observational astronomy and the application of theory in the late eighteenth and early nineteenth centuries

23

Measuring solar parallax: the Venus transits of 1761 and 1769 and their nineteenth-century sequels

ALBERT VAN HELDEN

Since the time of Copernicus, the astronomical unit – the mean distance between the Earth and the Sun – has been the fundamental unit of our solar system. Kepler's third law, published in 1619, made it an easy matter to express with great precision the distances of the planets from the Sun in terms of this astronomical unit. It was, however, extremely difficult accurately to express the distance from the Earth to the Sun in terms of precisely defined earthly distance units. In order to express the sizes and distances of all bodies within the solar system in terms of common terrestrial measures, astronomers had to bridge this gap between cosmic and earthly dimensions. For this, at least one distance in the heavens had to be measured in earthly terms.

Important as this task was for cosmology, measuring at least one celestial distance was crucial to predictive astronomy because precise knowledge of the astronomical unit, or of the angle from which it is determined (the solar parallax, or the angle subtended by the Earth's radius as seen from the centre of the Sun*) is fundamental to planetary theory. In the Copernican universe, the motions of the planets must be referred to the Sun, and therefore the Earth's motion around the Sun must be known accurately. Determining this motion requires accurate measurements of the Sun's positions, which are obtained by correcting the raw measurements downwards for atmospheric refraction and upwards for solar parallax. Since historically these two corrections were interdependent, it was impossible to construct reliable refraction tables until the solar parallax was known with acceptable accuracy. The uncertainties in the Sun's altitudes resulting from inadequate refraction and parallax corrections introduced errors in the obliquity of the ecliptic and the eccentricity of the Earth's orbit. Errors in the Earth's orbit led to systematic errors in the calculated orbits of the other planets, so that their calculated positions deviated from their observed positions. Moreover, in the physics of Newton, without an accurate knowledge of solar parallax, it was impossible precisely to determine the Earth's mass relative to the Sun's, a ratio essential in determining the perturbations of the Earth on the other planets. Finally, there is a connection between solar parallax and the parallactic inequality of the Moon, an essential component in lunar theory: one is needed to determine the other.

Now much of this was academic at the time of Copernicus: planetary theory had not progressed much beyond Ptolemy and measuring instruments perhaps not at all. Over the next two centuries, however, both theory and measurement increased immeasurably in sophistication and precision, and by the beginning of the eighteenth century, after the elimination of many other errors and uncertainties, the accurate determination of solar parallax had become a central problem in technical astronomy.

The quest for the astronomical unit had begun in the middle of the sixteenth century, and it was to take several centuries before the problem would reach a reasonably satisfactory conclusion. By the end of the seventeenth century, telescopic observations had resulted in a consensus figure of 10 to 12 arc-seconds for solar parallax, corresponding to a solar distance of some 20 500 Earth radii or about 80 million miles. This figure was, however, the result not of direct measurements of a parallax, but rather of a combination of factors. The assumption that solar parallax was less than about 12″ led to better refraction tables. That figure also made the

*Unless otherwise noted, in this chapter 'solar parallax' stands for the mean horizontal solar parallax.

Earth, a planet with a moon, larger than Venus, a planet without a moon, while making Mercury, a primary planet, larger than the Moon, a secondary planet (see Chapter 7 in Volume 2A).

The centre piece in this new consensus was the measurement of the parallax of Mars made by French and English astronomers in 1672, when a favourable opposition brought Mars unusually close to the Earth. A close examination of these measurements shows, however, that at that time Mars's parallax still lay within the error margin of the measuring instruments. In retrospect, therefore, the consensus was not based on a positive measurement.

Although Gian Domenico Cassini (Cassini I) and John Flamsteed, the architects of the new consensus, at times cautioned their audiences against assigning too great a precision to their measurements, they firmly believed that they had measured Mars's parallax in 1672 and had confirmed it by subsequent measurements. For several generations measurements of Mars's parallax during favourable oppositions remained the preferred method of determining solar parallax. In 1685 Francesco Bianchini used it to arrive at a solar parallax of about 15″; in 1704 and 1719 Cassini's nephew Giacomo Filippo Maraldi concluded that solar parallax was about 10″; in 1719 James Pound and his nephew James Bradley (1693–1762) reported a value of between 9″ and 12″; in 1736 Cassini's son Jacques (Cassini II) obtained a result of 11″ to 15″; and in 1751 the abbé Nicolas-Louis de Lacaille (1713–62) obtained a value of 10″ from his measurements of the diurnal parallaxes of Mars and Venus, made at the Cape of Good Hope. In view of the improvements in instruments and observing techniques between 1670 and 1750, if Cassini, Flamsteed, and their successors had actually measured Mars's parallax, one might have expected that during those years the Sun's parallax would have been confined to increasingly narrow limits around the actual value of 8″.8. Since, however, the values reported above show the opposite tendency, we can see that Mars's parallax was still lost in the error margin of the instruments.

From the beginning, one astronomer, Edmond Halley (1656–1742) refused to lend his name to the consensus. Acting on a suggestion by James Gregory, Halley had observed Mercury's transit of 1677 in St Helena, where he was mapping the southern skies. He was the first to observe both the planet's ingress onto the solar disk and its egress, measuring their times as accurately as he could. Only one observer in Europe – Jean Charles Gallet in Avignon – had made similar measurements: a comparison of the two observations yielded a solar parallax of 45″, a value already considered much too large by influential astronomers. Although Halley did not put much stock in this result because Mercury's parallax is only slightly greater than the Sun's, he did think the method to be useful, especially if applied to transits of Venus. His subsequent role as an assistant to Flamsteed in several measurements of Mars's parallax led him to conclude that this method was useless. In his *Catalogus stellarum australium* of 1678 he wrote:

There remains but one observation by which one can resolve the problem of the Sun's distance from the Earth, and that advantage is reserved for the astronomers of the following century, to wit, when Venus will pass across the disk of the Sun, which will occur only in the year 1761 on 26 May [Julian]. For if the parallax of Venus on the Sun is then observed by this method I have just explained, it will be almost three times greater than the Sun's, and the observations required for this are the easiest of all, so that through this phenomenon men can instruct themselves of all they could wish on that occasion.

Over the next decade, one of Halley's research projects was the improvement of the elements of Venus and Mercury, building on the work of Johannes Kepler, Jeremiah Horrocks, and Thomas Streete. His own daylight measurements of Mercury's positions showed that Streete's elements were more accurate than those of Kepler. In 1691 Halley published his results in a paper in the *Philosophical Transactions* entitled "An Astronomical dissertation on the Visible Conjunctions of the Inferior Planets with the Sun," the first paper ever devoted entirely to the subject of transits. He gave the inclination of Venus's orbit to the ecliptic as 3° 23′ (modern value for 1700, 3° 23′ 28″) and the location of the ascending node as 14° 18′ in the sign of Gemini (modern value 13° 59′ Gemini). He calculated that in eight sidereal years Venus moved 1° 30′ $28\frac{1}{4}$″ more than thirteen revolutions, in 235 years 42′ 21″ less than 382 revolutions, and in 243 years 48′ 8″ more than 395 revolu-

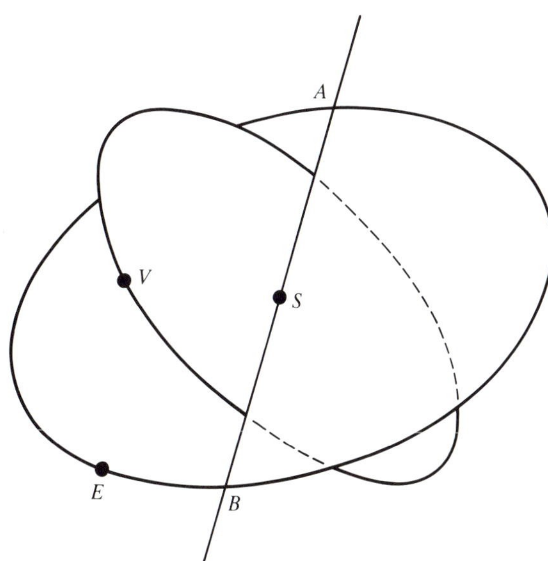

23.1. Diagram showing the orbits around the Sun (*S*) of Venus (*V*) and the Earth (*E*). Transits of Venus are observed when Venus and the Earth are simultaneously in direction *SA* or simultaneously in direction *SB*.

tions. Transits of Venus happen when the planet is at or very near one of its nodes and the line from the centre of the Sun through that node passes through the Earth (Figure 23.1). Since the Sun's radius is about 16′, the planet may be a little distance removed from its node and still appear on the Sun's disk. The greatest duration of a Venus transit occurs when the planet passes over the centre of the Sun's disk: Halley gave this time as 7 hours 56 minutes. Transits at the ascending node occur early in December and those at the descending node early in June.

With this and similar information for Mercury, the dates of past and future transits could easily be calculated. Halley gave a list of 29 transits of Mercury between 1615 and 1789, and seventeen transits of Venus between 918 and 2004. He correctly gave the June transits of Venus of 1518, 1526, 1761 and 1769 and the December transits of 1631 and 1639 (the last of which had been observed by Horrocks and William Crabtree). For those farther removed from his time period, his predictions were not as good. His list therefore does not show the rigid sequence of alternating pairs of June and December transits, the transits of each pair being separated by 8 years, and the successive pairs by 235 years. (Thus, June transits occurred in 1518 and 1526, 1761 and 1769, and will occur in 2004 and 2012; December transits occurred in 1631 and 1639, and 1874 and 1882.) The important thing was, however, that he was correct in his prediction of the next set of Venus transits: they were to happen in June 1761 (already predicted by Kepler) and 1769 (not predicted by Kepler).

Halley concluded his article with the statement that the chief use of Venus transits was the determination of solar parallax, a quantity astronomers had investigated in vain by various methods and with the most delicate instruments. During a transit, one could measure the time lapse between ingress and egress (using the moments of internal contact) accurately to a second of time, or 15 arcseconds, by means of even a mediocre telescope and a pendulum clock that runs accurately for six or eight hours. How, from two such observations, made simultaneously at different locations, the solar parallax and distances could be determined to one part in 500 he promised to explain in a future paper.

Although Halley's more senior colleagues, Newton and Flamsteed, did not react publicly to this paper, some of his younger ones did. In his *Astronomiae physicae & geometricae elementa* of 1702, David Gregory repeated his uncle James Gregory's remarks about the use of transits in determining solar parallax, and stated that the world would have to wait until the Venus transit of 1761 for an accurate determination. In his *Praelectiones astronomicae* of 1707 William Whiston drew on Halley's 1691 paper to advocate the use of Venus transits. Both books went through several English and Latin editions.

Halley himself returned to the subject in 1716. In that year he published in the *Philosophical Transactions* a paper entitled "A Singular Method by which the Parallax of the Sun or its Distance from the Earth can be Determined Securely, by means of Observing Venus in the Sun." In this paper, Halley's most elaborate treatment of the subject, he reviewed the history of this problem and gave some harmonic reasons why the Sun's parallax should be taken as about $12\frac{1}{2}''$ (this would make the Earth larger than Venus and Mercury larger than the Moon) until the observations he was about to advocate could be made. The measurements that had been made up to then, he insisted, were useless because they often produced a negative solar parallax.

Halley related how in 1677, in St Helena, he had been able to observe the ingress and egress of Mercury: he had been able to time the transit of Mercury "without an error of one second" by taking the internal contacts as his reference points. The reason was that the moments when a thin filament of the Sun's light first appeared between the Sun's limb and the dark body of the planet at ingress (first internal contact), and disappeared at egress (last internal contact) could, in Halley's judgment, be timed very accurately. If Mercury were not so close to the Sun, transits of this planet would be ideal for determining parallax, for they happen frequently. It remained therefore to use the much less frequent transits of Venus, whose parallax is four times greater than the Sun's. Differences in the measured times of a Venus transit as observed from different places on Earth should allow a determination of the Sun's parallax to within a small part of an arc-second. For such observations the only instruments needed were common but good telescopes and clocks, and all that was required of the observers besides trustworthiness and diligence was a modicum of skill in astronomical matters. Nor was it necessary scrupulously to measure the latitude of the locations of the observations but only to record the local times accurately.

The length of Venus's path across the Sun was a function of the planet's observed latitude on the Sun and ranged from a maximum equal to the Sun's diameter to zero when the planet merely grazed the Sun. By measuring the times of ingress and egress, an observer could determine the length of time Venus took to cross the Sun and therefore its path. The angular separation between two paths observed at two observing stations was the measure of the parallactic displacement of Venus on the Sun caused by the separation between the stations (Figure 23.2). Since the ratio of the distances of Venus and the Sun was known, the parallaxes of Venus and the Sun could be calculated.

In London, the transit of 5 June 1761 would begin before sunrise, but observers in more northern latitudes, Norway and perhaps the Shetlands, would be able to observe both ingress and egress. Since at the midpoint of the transit the Sun would

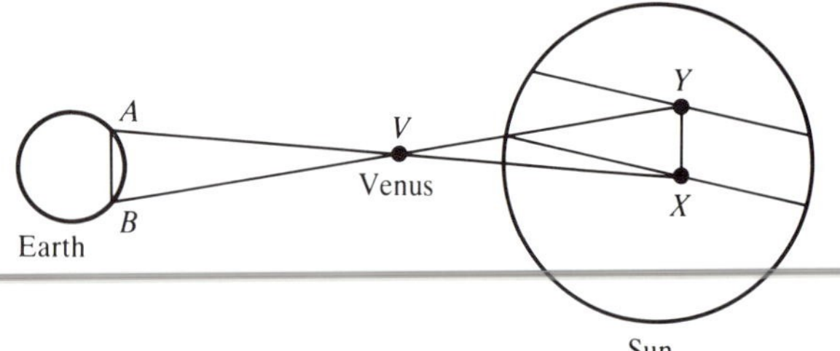

23.2. Halley's method of deriving the distance of the Sun from observations of transits of Venus. The method depends, first, on consideration of two (virtually) similar triangles each with vertex at the planet (*V*), one triangle having as base a pair of observing locations on Earth (*A* and *B*) and the other having as base the projections of Venus on the Sun as seen from these locations (*X* and *Y*). Second, from Kepler's third law we obtain the ratio between the Earth–Sun distance (*AX*, say) and the Venus–Sun distance (*VX*), and hence the ratio between *AV* and *VX*, which by similar triangles is the same as that between *AB* and *XY*. Third, we use the known distance in miles between *A* and *B* to obtain the distance in miles separating *X* and *Y*. Multiplying this by the angular diameter of the Sun divided by the angular separation of *X* and *Y*, we obtain the linear diameter of the Sun in miles. By comparing this linear diameter with the Sun's angular diameter, we obtain the Sun's distance from Earth.

The drawing is not to scale, and the distance *XY* is greatly exaggerated. It is therefore important that terrestrial observating locations *A* and *B* are chosen that are very different in latitude; even so, *XY* is typically only 1/50th of the Sun's disk. For the same reason transits of Venus are chosen in preference to those of Mercury, for Venus is nearer to Earth when in transit and so the separation between *X* and *Y* is greater.

In the method that Halley proposed, the angle *XY* is obtained by observing from both *A* and *B* the times taken for the complete transit of the planet across the Sun, for (if we extract from the observations the effects of the Earth's diurnal and annual movements during transit) the times the transit takes are proportional to the lengths of the lines across the Sun traced out by *X* and *Y*, and knowledge of these lengths allows us to locate these lines on the Sun.

be directly overhead near the northern shore of the Bay of Bengal, the entire transit would be visible there as well. It would also be useful to have an observation at a point where the midpoint of the transit occurred at midnight, and this would be the case in the Hudson Bay colony, where ingress would happen shortly before sunset on 4 June and egress shortly after sunrise on 5 June. Halley also advocated observations in Madeira and Benkulen (on the western shore of Sumatra). All these were areas controlled by the British. If the French wished to make observations, it would be useful to have observers in Pondicherry on the western shore of the Bay of Bengal; if the Dutch had a desire to advance the science of the heavens in this matter, the observatory of their celebrated city of Batavia in the Dutch East Indies would be a good observing station. Halley wished this phenomenon to be witnessed by many observers in different locations, so that out of a consensus a more trustworthy result would emerge, but also to ensure that the effort would not be foiled by cloudy skies:

Therefore to the curious investigators of the stars to whom (when we shall have ended our lives) these observations are reserved, we recommend again and again that, bearing in mind this warning of ours, they set themselves to carrying out this observation actively and with all their powers. And we wish them luck and pray above all that they be not deprived of the desired spectacle by the untimely gloom of cloudy skies, and that at last the magnitudes of the celestial orbs, confined within narrower limits, will bestow upon them eternal glory and fame.

The paper ended with a geometric analysis of the 1761 transit as seen from London, and a brief mention of the 1769 transit.

Halley's two papers were the basic canon of transits. Since the seventeenth-century attempts to measure the parallax of Mars were, in retrospect, failures, Halley's establishment of the science of transits on a solid footing, educating his fellow astronomers on how parallaxes could be measured by means of Venus transits, is perhaps his most important contribution to astronomy.

If Halley eschewed the use of Mercury transits, concentrating his attention entirely on Venus, others were not so quick to abandon the more frequent transits of the smaller planet. The transit of this planet predicted by Halley for 1723 would be partly visible in western Europe. Whiston published a broadsheet in which he showed the paths of Mercury and Venus across the Sun in their transits from the sixteenth to the end of the eighteenth century, and he urged astronomers to observe the transit of Mercury of 1723 with the hope of determining solar parallax. Indeed, Whiston argued that transits of Mercury were more useful for determining solar parallax than transits of Venus. Observations of this particular transit in England and France, however, produced no useful results, and as a consequence astronomers, especially in England, began to focus more on the upcoming transits of Venus. In France, the astronomical establishment led by the Cassini and Maraldi families clung for some time to their belief in the accuracy of measurements of the parallax of Mars. A new generation of French astronomers, however, led by Joseph-Nicolas Delisle (1688–1768), became champions of the transit method.

After a review of the literature in 1723, Delisle had become convinced that measurements of the parallax of Mars were hopeless and therefore turned his attention to transits. He proposed a slightly different method of observation. Halley's method meant that the times of both ingress and egress had to be measured. But, as Delisle pointed out, local weather conditions might allow only one of these events to be observed. In some places one of the events might occur when the Sun was below the horizon. In such cases, Delisle pointed out, a measurement of the exact moment of ingress *or* egress would suffice, provided that the observer could accurately determine the longitude of his observing station. The time difference between the same event at different places was a function of the difference in path length, and so would produce the desired result as well. In practice, Delisle's method was more practical for it meant that observations could be made in more locations – more practical, that is, provided that observers could make accurate measurements of their longitude. He used this method in his observation of the transit of Mercury of 1723, with disappointing results. On a visit to London, in 1724, however, Delisle visited Halley and was reinforced in his conviction that transits, especially those of Venus, were the only practical way of determining parallax.

During his tenure as director of the observatory of the Russian Imperial Academy of Sciences in St Petersburg (1725–47), Delisle retained his interest in this problem, and through his wide correspondence he kept transit observations before the astronomical community. Delisle gathered information from correspondents on the relationship between the length (i.e., power) of the telescope and the duration of the transit, a knotty technical problem that had to be solved before useful results could be extracted from a comparison of observations. Upon his return to France he became the central figure in the preparations for the transit of Venus of 1761.

Two events were important in those preparations: Lacaille's observations at the Cape of Good Hope in 1751 and the transit of Mercury of 1753. Lacaille went to the tip of Africa to map the southern stars. Before he left he asked European astronomers to measure the positions of Mars and Venus, so that from comparisons with his measurements at the Cape the parallaxes of those planets could be determined. As it turned out, the results were disappointing, but in advocating this method Lacaille pointed out several weaknesses of the transit method. He argued that the accuracy of one part in 500 mentioned by Halley was impossible to attain. First, in those places where ingress and/or egress occurred when the Sun was close to the horizon, the uncertainty of the Sun's limb caused by changing refractions would make it impossible to time internal contacts accurately. Second, in instruments of high magnification, the Sun moves very rapidly through the field of view, and this would make the timing of internal contacts difficult. Since Mercury moves across the Sun $1\frac{1}{2}$ times as fast as Venus, its internal contacts were easier to determine with precision than those of Venus, yet during the Mercury transit of 1743 the most skilled astronomers using excellent instruments differed on the moment of internal contact by as much as 40 seconds. Obviously, one should not put all one's hopes on the upcoming transits of Venus: parallax investigations should be conducted whenever an opportunity arose. Events would prove Lacaille's comments justified.

The Mercury transit of 1753 promised to be the most favourable for parallax investigations, because the planet would pass very near the centre of the Sun. If there was any hope of determining solar parallax by means of transits of this planet, this would be the occasion. In preparation for this transit, Delisle reviewed Halley's planetary tables, published posthumously in 1749, and made several corrections for Mercury. He urged his many correspondents to prepare for the observation, and in the winter preceding the event he published an *avertissement* on this subject. Here he instructed observers on how to make the measurement, specifying his own method instead of Halley's, and included a figure illustrating the different paths of Mercury across the Sun predicted by the tables of Streete (1661), Philippe de La Hire (1702), Jacques Cassini (1740), and Halley as corrected by Delisle. Delisle saw to it that the little tract reached the far corners of Europe and that instructions reached the New World as well. Through his English colleagues he hoped, too, to enrol observers in the East Indies. As the event drew nearer he also prepared a world map on which the visibility of the transit was depicted for easy reference.

Observations of the Mercury transit of 1753 failed to produce a satisfactory measure of solar parallax, but the event served almost as a dress rehearsal for the impending Venus transit of 1761. Here (and in Lacaille's expedition to South Africa) we see the efforts at marshalling an international body of observers through printed tracts and private correspondences, the correction of tables in order to predict the event more exactly, and the circulation of diagrams and instructions. In fact, a number of the key observers of the Venus transits learned how to observe transits at this time.

In preparing for the Venus transit of 1761, the first in order of business was the correction of the tables. The Mercury transit had confirmed that in spite of the many transits of that planet that had been observed since the first such observation by Gassendi in 1631, there were still important errors in its tables. The best tables were those of Jacques Cassini and Halley, but Delisle's calculations based on corrections of Halley's tables had still predicted the entry of Mercury on the Sun 17 minutes late. There was every reason to believe that the tables of Venus (of which only one transit had been observed, in 1639) contained greater errors. In 1753 the young astronomer Guillaume-Joseph-Hyacinthe-Jean-Baptiste le Gentil de La Galaisière (1725–92) compared the predictions of different tables (chief among them those of Jacques Cassini and Halley), and produced a diagram showing the paths of Venus across the Sun's disk in 1761 and 1769 according to the different tables. It was clear that there was much uncertainty concerning the times and paths of the transits. This uncertainty made it difficult to choose the best locations for observing stations.

The corrections occupied Delisle and his assistants for the next several years, and, finally, early in May 1760 – only thirteen months before the

1761 transit – his world map, showing the visibility of the transit, and his memoir of instructions were published by the Paris Academy of Sciences. Delisle's memoir showed that observations made from Port Nelson in the Hudson Bay Colony, as advocated by Halley, would not be useful for the determination of solar parallax and he therefore advocated a station farther north – at a latitude of at least 65°. The optimum locations for viewing the entire transit were in India and the East Indies. The shortest apparent duration of the total transit would be in Tobolsk, Siberia, while the longest would be in Batavia in the Dutch East Indies. But Delisle also stressed the importance of stations where only ingress or egress would be visible, suggesting Yakoutsk, Kamchatka, the Cape of Good Hope, and the island of St Helena for this purpose. He further urged that observations be made in Peking, Macao, Archangel, Torneo, and St Petersburg. Observers were advised to pay close attention to the characteristics of their telescopes. Copies of Delisle's memoir and map were sent to all regions of Europe, including St Petersburg, Constantinople, and Stockholm.

Interest in the Venus transit had grown in France since 1753, often extending to the popular press, and calls for expeditions were heard frequently. Acting on his own initiative, le Gentil found funding for an expedition to Pondicherry and left early in 1760. Shortly after Delisle presented his memoir to the Academy, late in April of that year, Joseph-Jérôme Lefrançais de Lalande (1732–1807), who was to become the central figure in the Venus transits after Delisle's death in 1768, read a memoir that ended with the following call to arms:

The occasion presented to us by this celebrated phenomenon is one of those precious moments of which the benefit, if we let it escape, will not be compensated later – neither by the efforts of genius nor by dint of hard work, nor by the munificence of the greatest kings. It is a moment that the past century envies us and which in the future will be, I dare say, an injury to the memory of those who will have neglected it.

The Academy now took official action and weighed various possibilities. There were no locations in French hands on the west coast of Africa, and cooperation with the Portuguese and Dutch would take time to arrange. In the meantime, it was decided to send Alexandre-Gui Pingré (1711–96) to the island of Rodrigue in the Indian Ocean, in the shipping lanes of the French East India Company. On the invitation of the St Petersburg Academy, J.-B. Chappe d'Auteroche was to go to Tobolsk in Siberia. Le Gentil was already underway to Pondicherry, and César-François Cassini de Thury (Cassini III, 1714–84) was to observe in Vienna. Efforts to organize a joint expedition with the Dutch to Batavia were scrapped when it became apparent that the British would send an expedition to Benkulen.

Events in England had taken a somewhat different course. Since 1723, transits of Mercury had been assiduously observed by British astronomers, but, now following the lead of Halley, they had used them only to improve the elements of that planet. Although British scientists kept themselves informed of the events in France, preparations for the transit of 1761 did not begin until the Royal Society discussed Delisle's memoir in June 1760. That same month the Council of the Society took charge of the observations and quickly decided to send expeditions to St Helena and Benkulen. Bradley, the Astronomer Royal, supplied the Society with a list of instruments needed: "A Reflecting Telescope of Two Foot with Dollond's Micrometer, Mr Dollond's Refracting Telescope of Ten Feet; a Quadrant of the radius of Eight Inches; and a Clock or time piece".

Because of the lateness of British preparations, the expedition to Benkulen was under time-pressure from the start. Charles Mason (1728–86) and Jeremiah Dixon (1733–79) left Portsmouth early in December 1760, on board the *Seahorse*, a warship provided by the admiralty. Britain and France were engaged in the Seven Years War, and although both Crowns had instructed their navies not to interfere with the expeditions sent out by their respective enemies, such orders were difficult to enforce. Before it could get out of the English Channel, the *Seahorse* was attacked by a French frigate and after a violent battle had to return to port for repairs. The ship did not put out to sea again until 3 February 1761. As had been foreseen by Mason and Dixon, this was too late to reach the East Indies in time, and when the *Seahorse* reached the Cape of Good Hope on 27 April, they decided to make the observation there. The expedition to St Helena under Nevil Maskelyne (1732–1811) had, in the meantime, left England on 17 January 1761 aboard the East Indiaman *Prince Henry* and arrived in St Helena without trouble.

Europeans had thus sent out six expeditions to faraway parts: the French sent le Gentil to Pondicherry, Pingré to Rodrigue (where he arrived on 28 April 1761), and Chappe to Tobolsk (where he arrived on 19 April 1761), while Cassini de Thury went to Vienna; the English sent Maskelyne to St Helena and Mason and Dixon to Benkulen (although as described above they succeeded in getting only as far as the Cape of Good Hope). As the event approached, the Harvard professor John Winthrop (1714–79) organized an expedition to St Johns in Newfoundland, paid for by the government of the Massachusetts colony.

At the island of Rodrigue, it was overcast on 6 June, the day of the transit, and Pingré and his assistant Denis Thuillier had only brief glimpses of the transit, not including ingress and egress. But the astronomers did make some useful measurements of Venus's positions on the Sun, noting the exact times. Chappe was able to observe the entire transit, and his measurements were of the greatest importance. He also observed an eclipse of the Sun on 3 June. Le Gentil had the worst luck. After many complications and delays, he arrived near Pondicherry aboard a ship of the French fleet, only to find that this post had been captured by the British. The fleet turned back toward Île de France (Mauritius), and le Gentil observed the transit from the deck of the ship. He was not able to make useful measurements.

Mason and Dixon, at the Cape of Good Hope, observed the entire transit, and both were able to measure the moments of ingress and egress, differing 4 and 2 seconds respectively in their times. They were also able to make observations of Jupiter's satellites for longitude purposes and determined the latitude of their observatory accurately. Their expedition was, therefore, a great success. Maskelyne and Robert Waddington, on the other hand, had bad luck, for on St Helena the sky was overcast on 6 June. During brief periods of visibility they were able to make some quick measurements of Venus's position on the Sun, but on the whole their expedition was unsuccessful.

At St Johns in Newfoundland, only the egress, shortly after sunrise, was visible. The sky was clear and Winthrop was able to make five measurements of Venus's positions on the Sun and, most importantly, time the moment of egress exactly. After some difficulties he was able also to make an observation of an occultation from which the longitude of his station could be determined.

Besides these expeditions, the Europeans organized observations in a number of places closer to home. The Swedish and Danish academies were particularly active in this respect, and the Jesuits also observed at a number of stations.

In all, more than 120 observations were made, mostly in Europe, but also in Peking, Calcutta, Madras, and Constantinople. Most of these observations were, of course, made by amateurs with little experience in precise astronomical measurements. The best cadres of observers were to be found in France, Sweden, and Britain.

The data obtained from all these observations were not unproblematic. First, the quality of the successful observations varied greatly, and calculators had to choose which data to favour and which to ignore, choices that were, in retrospect, not always the best. Sometimes supposed errors in the observations were even "corrected". In the absence of accepted statistical methods of dealing with such a large and uneven body of data, there was no standard of data reduction to which practitioners could appeal for unanimity. Second, Halley's prediction of an accuracy of one part in 500 was vitiated by the "black drop" effect. The moment of internal contact was defined as the first or last continuous filament of sunlight between the dark body of the planet and the Sun's limb. In practice, the planet seemed to adhere to the Sun's limb as though attached to it by a sticky substance (see Figure 23.3). The duration of this effect varied from observer to observer and depended on several factors. Third, several observers saw a luminous ring around the planet that made the moment of internal contact even more difficult to time. This was the first evidence that Venus has an atmosphere, a fact confirmed a few years later by William Herschel. Finally, in cases where only the ingress or egress could be observed it was crucial to have exact longitude measurements. In most cases there was an unacceptable uncertainty in these measurements. Lalande admitted that even the longitude difference between the established observatories in Paris and Greenwich was uncertain to 20 arc-seconds.

23.3. One of the problems encountered in observations of transits of Venus: the "black drop" effect when Venus is near internal contact. The black silhouette of the planet then appears to be joined to the limb of the Sun by a ligament, and this makes the moment of internal contact observationally ill-defined.

As a result, calculations of the solar parallax based on the transit observations ranged from 8″.28 to 10″.60, without any immediate sign of consensus. In this respect, then, the expeditions of 1761 were a disappointment. On the other hand, many valuable lessons had been learned that could be used in observing the transit of 1769, when, it was hoped, there would be peace between the French and British. Moreover, the expeditions did much more than observe the transit: their longitude observations made possible corrections to maps and charts, and their studies of local geology, flora, and fauna, brought back a wealth of important information that is useful to this day (Figure 23.4).

The arguments about the value of the solar parallax kept the importance of Venus transits before a wide audience, and therefore preparations for the 1769 transit began early. Le Gentil, who had been foiled by the war, decided to stay in the East, and proceeded to Manila in 1767. The French Academy, however, advised him to proceed to Pondicherry, now that the war with the British was over. Le Gentil arrived there in March 1768, fourteen months before the transit. An expedition, under the leadership of Pingré, left France on 9 December 1768. Its primary function was to test chronometers for the purpose of finding longitude at sea. This expedition observed the transit from Cap-François, Saint Domingue.

Chappe proposed to go to the South Sea islands, but Spanish cooperation for this project could not be secured. A proposal to make the observation from the California peninsula was, however, accepted by the Spanish Crown, and a joint French–Spanish expedition was organized. Chappe departed from Paris in September 1768, set sail from Cadiz a month later, and arrived in Vera Cruz on 8 March 1769. It took more than two months more to cross Mexico and traverse the sea between the mainland and the peninsula, where they arrived on 16 May, three weeks before the transit. Although the French–Spanish team was successful in their observation, all died from an epidemic, with the exception of Vincente de Doz, who brought the observations back to Europe.

The British, likewise, took the opportunity to send an expedition whose goals were wider than the observation of the Venus transit. The *Endeavour* sailed from Plymouth on 26 August 1768, under the command of James Cook. Charles Green was on board to make the astronomical observations, while Joseph Banks went along to study the natural history of the places visited. Cook was to explore the South Pacific, test the chronometers of John Harrison, and make the observation of the transit from the newly discovered island of Tahiti. Under the guidance of the Royal Society, the British Crown also sent expeditions to the northern part of Norway and County Donegal in northwest Ireland.

In the British colonies in North America, efforts were made to organize expeditions to the West, but funds could not be found. Nevertheless, the transit of 1769 was observed at no fewer than nineteen stations in the colonies, the most celebrated of which were those of Winthrop at Harvard, Benjamin West in Providence, Rhode Island, and David Rittenhouse in Norristown, Pennsylvania.

Father Maximilian Hell, SJ, was invited by the Danish/Norwegian Crown to observe from the island of Vardö off the northeastern shore of Norway (Figure 23.5). Having obtained permission from his superiors and the Hapsburg Crown, Hell headed a successful expedition whose results were for some time mistrusted because of a delay in publication. Lalande's suspicion led J. F. Encke, in the 1820s, to reject Hell's observations. In 1835 Karl Ludwig von Littrow published what he thought was direct proof that Hell had falsified his data. Hell's results were rehabilitated by Simon

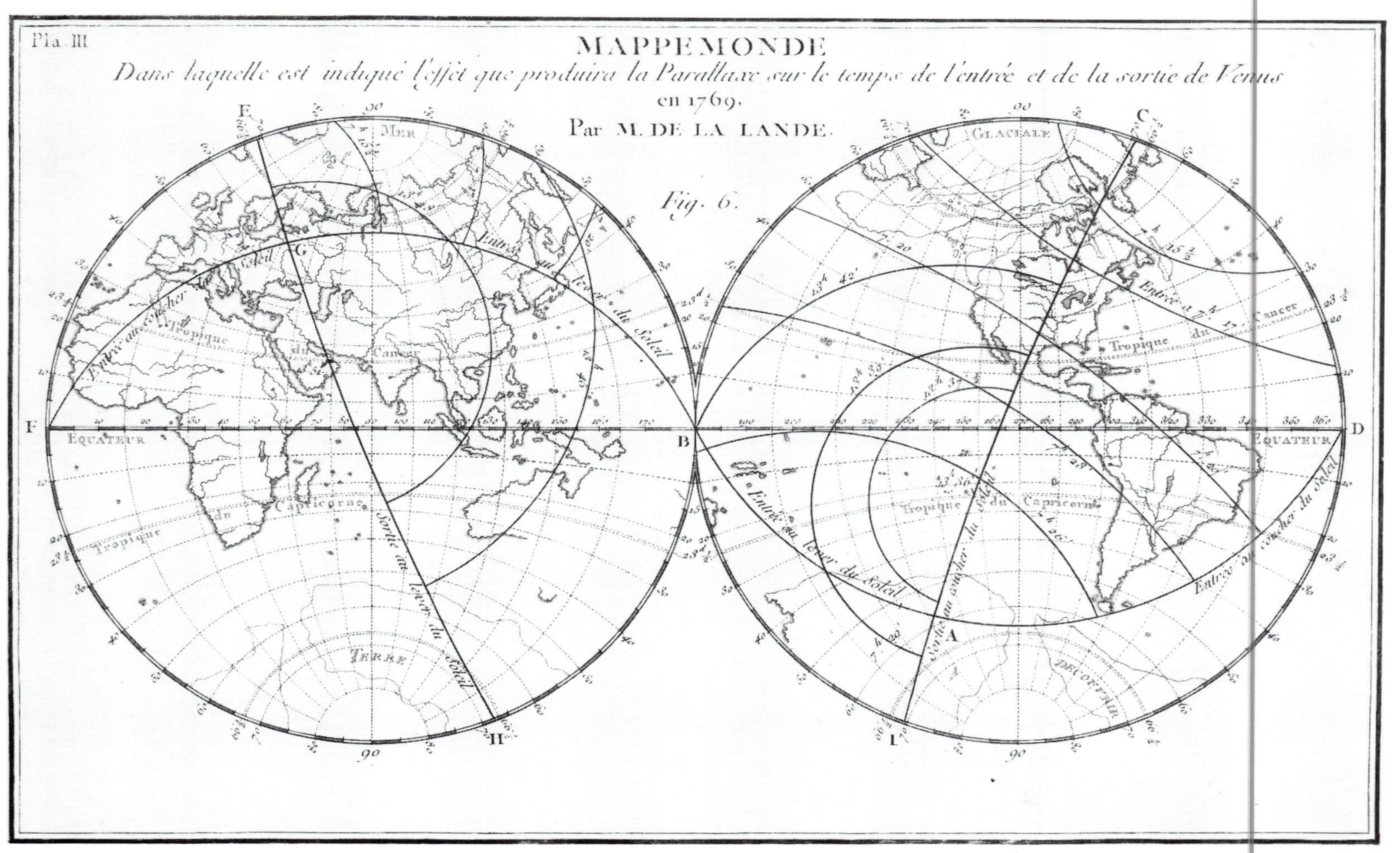

23.4. Lalande's world map for the 1769 transit of Venus, showing the times when the planet will be seen entering and leaving the Sun.

23.5. Maximilian Hell observing the transit of 1769 on the island of Vardö.

Newcomb (1835–1909) in 1883 after an examination of his journal of observations.

In Russia, with the enthusiastic backing of Catherine the Great, the St Petersburg Academy coordinated a number of expeditions, equipped with instruments ordered from James Short in England. Observing stations were established as far east as Yakutsk (east longitude 130°), in the Urals (east longitude 55° and 60°), near the Caspian sea, and on the Kola peninsula between the Arctic Ocean and the White Sea.

As in the previous transit, the Swedes organized a large and coherent effort, making certain that observations were made from a number of locations in Swedish territory.

Altogether, this transit was prepared for by 151 observers at 77 stations, from Yakutsk, Manila, Peking, and Batavia in the East, to Baja California and Tahiti in the West. The British counted no fewer than 69 observers, the French 34, and the Russians 13. Of all the experiences, Le Gentil's is the most poignant. Having been foiled by the war in 1761, and having spent the intervening eight years in the East, in 1769 he was foiled by the weather.

Calculations of the resulting solar parallax began immediately but were not definitive until the results from the expeditions to faraway places arrived. Thus, in his memoir of 1770 Pingré calculated solar parallax to be 8″.88 ± 0″.05 but upon further consideration concluded (in 1772) "that the horizontal parallax of the Sun at its median distances is very nearly eight seconds and eight tenths." Lalande, who on the basis of the 1761 transits had used 9″.0 in his influential *Astronomie* of 1764, put the value between 8″.5 and 8″.75 in his memoir of 1770. The next year he narrowed the range to 8″.55 – 8″.63: "thus, in taking in round numbers 8″.6 for the middle latitudes, like the one of Paris, one cannot differ sensibly from the truth; the observations made in 1769 cannot suffice for removing that small degree of incertitude of a twelfth of a second." From the English observations Maskelyne found a parallax of 8″.8; Anders Planmann of the Swedish Academy arrived at a figure of 8″.43, but upon reconsidering agreed with Lalande on a value of 8″.50. At the St Petersburg Academy, Anders Johan Lexell based his calculations on Leonhard Euler's theoretical treatment of the determination of solar parallax by means of transits, and found a value of 8″.68 in his memoir of 1771 but reduced that value to 8″.63 the next year.

There was, therefore, an agreement considerably closer than in the case of the transit of 1761. Yet, nagging doubts remained. The two transits had clearly shown Halley's claim, that transits of Venus could reveal solar parallax accurate to one part in 500, to be a mirage: the best observations still yielded a range of some 0″.4, or about one part in 20. The limitation was due to several sorts of problem. First, the longitudes of many of the far-flung places were not known with a sufficient degree of accuracy. This was a problem that might, over time, disappear. Second, there was the problem of the black drop effect and the luminous rings around Venus. Here one had to choose between the moment of internal tangency or the first and last appearance of the bright filament between the dark body of the planet and the limb of the Sun. But all these estimates depended on the condition of the atmosphere as well as the power and quality of the telescope used. This problem was much less tractable. Last, there was the problem of various inaccuracies introduced by human factors: the skills of the observers varied greatly. Those who assembled the observations and calculated the parallax had to decide how much weight to give to each observation, a difficult task in the absence of a formal theory of error.

The rough consensus of a value around 8″.6 held steady for about eighty years. In his *Exposition du système du monde* (1796) Pierre-Simon Laplace derived a solar parallax of 8″.6 from the parallactic inequality of the Moon, thus confirming Lalande's findings. Laplace concluded: "It is very remarkable that an astronomer without leaving his observatory, by merely comparing his observations with analysis, has been enabled to determine with accuracy the magnitude and figure of the Earth and its distances from the Sun and Moon, elements, the knowledge of which has been the fruit of long and troublesome voyages in both hemispheres." And in his influential *Histoire de l'astronomie du XVIIIe siècle* (1827) Jean-Baptiste Joseph Delambre, upon reviewing the results of the Venus transits, concluded that the solar parallax could be held to be 8″.6 in round numbers.

As better statistical methods of evaluating data became available, astronomers reviewed the results of the Venus transits. In memoirs dated 1822 and 1824 Encke subjected the data to C. F. Gauss's method of least squares (1809) and concluded that the mean horizontal solar parallax was 8″.5776 ± 0″.0370. In 1835 Encke further refined his figure to 8″.57116 ± 0″.0371. It seemed that the accuracy of the measure, now one part in 200, was approaching Halley's prediction after all. Encke's results were very influential: astronomers felt supremely confident of the accuracy of this, and other, fundamental constants. The consensus fell apart, however, in the middle of the century.

In his researches on lunar theory, Peter Andreas Hansen derived from the parallactic equation a solar parallax of 8″.916 (published in 1857 and 1863), a value considerably larger than Encke's. At about the same time Urbain Jean Joseph Le Verrier studied the Earth's perturbations on Mars and Venus and from the Earth's mass found a solar parallax of 8″.95. These values were supported by the results of Jean Bernard Léon Foucault's laboratory determinations of the speed of light, which, by way of the aberration of light, produced a solar parallax of 8″.86. In the meantime, the greater accuracy of nineteenth-century measuring instruments promised that perhaps the parallax of Mars could now be measured more accurately. At the favorable opposition of 1862, a number of observers in widely separated observatories measured the planet's declination, and their results yielded values of solar parallax averaging around 8″.95.

These developments led to new reappraisals of the observations of 1761 and 1769, with a view to improving on them in the upcoming Venus transits of 1874 and 1882. In his doctoral dissertation published in 1864, Karl Rudolph Powalky went back to the data and by eliminating a large number of observations teased out a value of 8″.83. Four years later Edward James Stone took the observations of five stations where both ingress and egress were visible in 1769 and produced a value of 8″.91. But neither of these analyses was entirely satisfying.

In preparing for the transit of 1874, several different approaches were taken. Astronomers were generally in agreement that new techniques and better measuring instruments held out great hope of improving on the results of the observations of the previous century. The Germans, under the leadership of Hansen and A. J. G. F. von Auwers, pinned their hopes on heliometers, with which they equipped all ten of their expeditions. The English expeditions, coordinated by the Astronomer Royal, George Biddell Airy, used both visual and photographic techniques (Figure 23.6). Because heliometers were not available for this purpose in the United States, the Americans, under Newcomb, used traditional instruments to observe contacts but also equipped each expedition with a photoheliostat (a forty-foot horizontal instrument in which the image of the Sun was projected by a rotating mirror). The French also availed themselves of a version of this instrument. Expeditions were also organized by Russia and other European countries.

The entire course of the transit could be observed only in East Asia and the Indian and Pacific Oceans. Information was gathered on weather conditions before the choices of observing stations were made. Altogether the Europeans sent over fifty expeditions to these areas. Travel by ship was now relatively easy and, since the European powers were at peace with each other, safe. Further, equipment had been standardized, especially by the Germans and Americans, and observers and photographers had been trained. (The Americans went so far as to construct an artificial transit machine on which their observers could practise.) In view of all these preparations, astronomers were optimistic about the results.

Although the full reports took decades to complete, preliminary results began to appear within a few years. They were very disappointing. Airy found 8″.76 for the solar parallax, Stone 8″.88, G. L. Tupman 8″.81, Auwers 8″.88 (heliometer measurements), and D. D. Todd 8″.88 (photoheliostat measurements). This range was much larger than had been hoped. As a result, astronomers now realized that transit observations were inadequate for this purpose, and there was decidedly less enthusiasm for observing the transit of 1882 (although a number of observations were made).

There was now a large body of data from four transits, and these were subjected to thorough analyses. Newcomb (no great enthusiast for this method) reviewed the eighteenth-century observations, incorporating up-to-date determinations of the longitudes of the observing stations. His effort, published in 1891, produced a solar parallax of 8″.79 with a mean error of 0″.051 and a probable error of 0″.034. In *The Elements of the Four Inner Planets and the Fundamental Constants of Astronomy* of 1895, Newcomb incorporated the results of the 1874 and 1882 observations. From the measurements of Venus's distance from the centre of the Sun, made by means of heliometers by German observers and by means of photographs by American observers, Newcomb derived a solar parallax of 8″.857 ± 0″.016. He then compared the observed contacts of Venus with the Sun's limb from all four Venus transits, arriving at a parallax of 8″.794 ± 0″.023.

Astronomers in the eighteenth century may have been very satisfied with a solar parallax accurate to one part in several hundred, but their successors a century later were not, and they looked for alternative means of determining that important constant. Measurements of Mars at opposition held out promise, and David Gill (1843–1914) even went to Ascension Island to determine the diurnal parallax of the planet by means of a

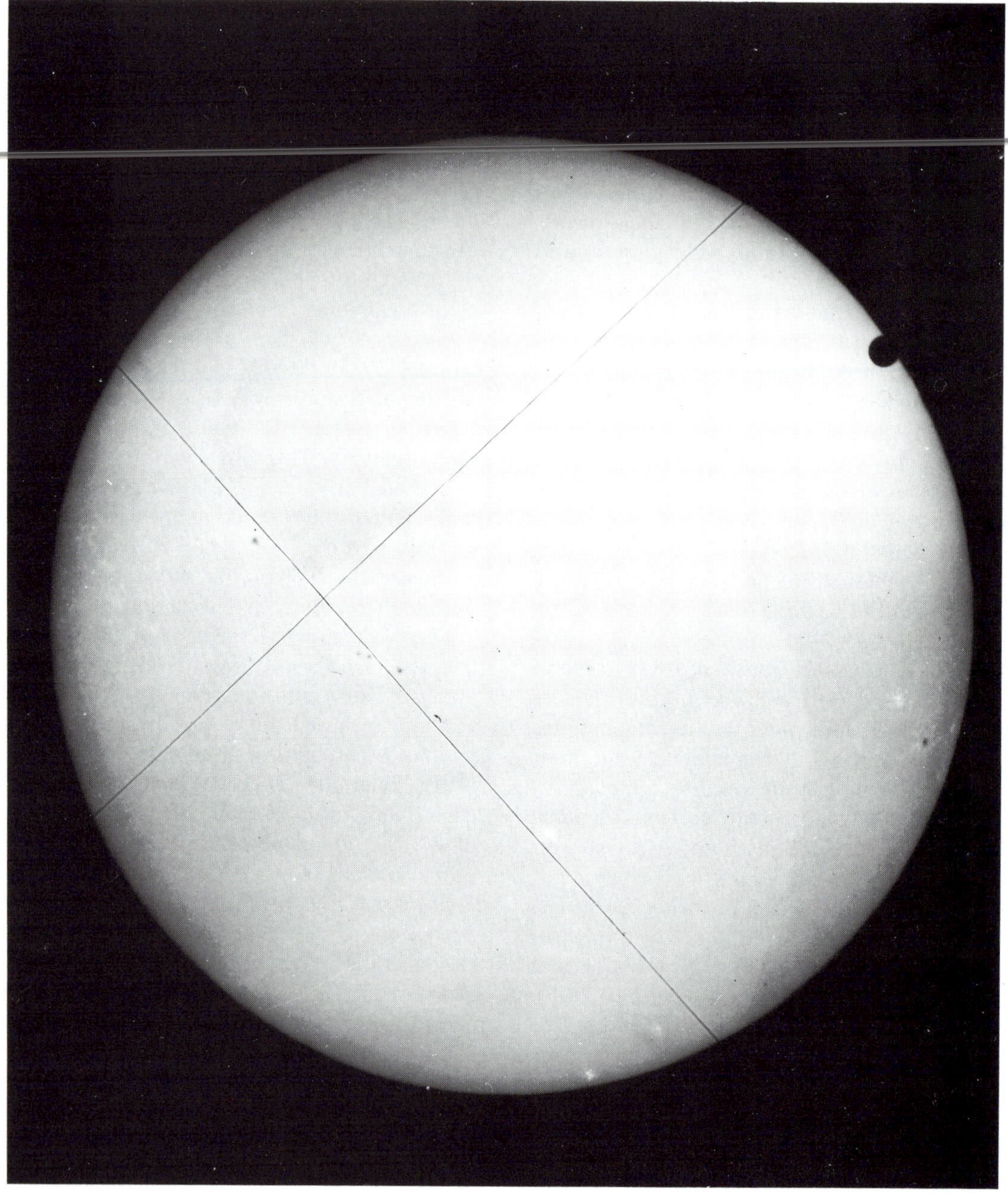

23.6. A photoheliograph of Venus near the solar limb during the transit of 1874, taken by the Greenwich Observatory team on Honolulu.

heliometer during the perihelion opposition of 1877, deriving a solar parallax of 8″.78. Although Gill's result was judged reasonably accurate, it was felt that the use of Mars for such measurements was of limited value because of the inevitable inaccuracy involved in centring the image of a star on the planet's disk in a heliometer or centring the threads on the disk in a transit instrument. Newcomb said of this method that "we have here a cause so mixed up with personal error in making the observations that the objective and subjective effects cannot be completely separated."

There were, perhaps, alternatives to Mars. In 1872 Johann Gottfried Galle had proposed that instead of Mars a planetary body so small that it showed no disk be used. The minor planets proposed were Iris, Victoria, and Sappho, whose nearest approaches to the Earth were all about 0.85 astronomical units – much greater than Mars's distance of 0.38 AU at its perihelion oppositions. In 1888–89 observatories in Europe and the southern hemisphere made measurements of the positions of these bodies, and from them Gill derived a solar parallax of 8″.802, a value that agreed very well with the parallax figure of 8″.80 ± 0″.01, derived from the new measurements of the speed of light by Newcomb and Albert Abraham Michelson around the same time.

The minor-planet method received an unexpected boost when, in 1898, Karl Gustav Witt discovered asteroid no. 433, later named Eros. Computations of its orbit showed that its perihelion lay well within the orbit of Mars and approached the Earth's orbit closely. At its opposition of 1900–01 it would approach to within 0.27 AU from the Earth, and during its opposition of 1930–31 to within 0.17 AU. Both visual and photographic measurements were made during the first opposition, resulting in solar parallax values of 8″.806 ± 0″.004 and 8″.807 ± 0″.003. The reduction of the data of the 1930–31 opposition, published by Harold Spencer Jones in 1942, yielded the value 8″.790 ± 0″.001. The confidence limits of these measurements were thus almost an order of magnitude better than those of the value derived by Newcomb from the observations of Venus transits. Yet, the difference between the 1900–01 and 1930–31 measurements is considerably greater than the calculated error margin of either one.

Several conclusions can be drawn from the astronomers' great struggle to determine the astronomical unit to within acceptable error margins. An analysis of seventeenth-century measurements of the parallax of Mars shows that this method of determining solar parallax was not satisfactory. A not uncharitable interpretation of these measurements is that Cassini I and Flamsteed did little more than determine the error margin of their instruments. Yet the resulting partial agreement that solar parallax was 10″–12″ fitted in rather well with other considerations such as refraction. From 1672 to 1761 this rough consensus was confirmed time and again by measurements of the parallax of Mars and transits of Mercury.

The observations made during the Venus transits of 1761 and 1769 were of widely varying quality and could support solar parallaxes anywhere from 8″.5 to over 9″.0. The trend from 1770 to 1850 was toward a lower solar parallax, from Lalande's 8″.6 to Encke's 8″.57. Then, because values for the solar parallax derived by means of other methods tended to argue for a larger value, astronomers returned to the same body of data and duly teased out a slightly larger value. In so doing, they made ever better corrections for the longitudes of the observing stations and subjected the thus improved data to ever more sophisticated statistical analyses. Yet, even in the work of Newcomb, there is a residual degree of arbitrariness that serves to bring the final value closer to the value desired at that particular time. In retrospect it is difficult to escape the conclusion that, in spite of the increased sophistication of the analyses from Lalande to Encke and Newcomb, the values of solar parallax obtained from the transit observations of 1761 and 1769 have an accuracy no better than ± 0″.2.

Measurements of important constants of nature are not made in isolation. The very importance of these constants ensures that there will be converging pressures toward a value that is consistent with the limits imposed by several related approaches to the same problem. Thus, Encke's value was rejected when Hansen, Foucault, and Le Verrier arrived at values fully 0″.3 larger by three different methods. Astronomers promptly derived such larger values from the 1761 and 1769 data.

Observations in astronomy are, therefore, not one-time affairs that serve for a while as the best and are then replaced by better ones. Observations have a rather long life and their raw data are analysed time and again in the hope that they will yield results that are more in agreement with other and more recent observations. The observations of the transits of Venus of 1761 and 1769 provide a good example of this phenomenon.

In 1976 the International Astronomical Union (IAU) adopted 8″.794 148 as its value of the mean horizontal solar parallax, corresponding to a mean solar distance of $1.495\,978\,70 \times 10^{11}$ metres and an equatorial radius of the Earth of $6.378\,140 \times 10^{6}$ metres. The range of error is ±0″.000 007. Since 1984 the IAU astronomical constants have been the basis for all the national ephemerides.

The modern system of sizes and distances was first arrived at in rough outline by astronomers at the end of the seventeenth century. It was then fine-tuned by subsequent measurements in which the observations of the Venus transits of 1761 and 1769 were central. These measurements were made at enormous cost, measured in terms of money as well as manpower. The Venus expeditions were a great achievement in international scientific cooperation, something that had never been tried on this scale. Since then such cooperative projects among astronomers have become routine. In the eighteenth century scientific expeditions were anything but routine: astronomers underwent great hardships and some lost their lives. Le Gentil spent more than eleven years of his life thousands of miles away from home and missed both transits. When he finally returned, his heirs were dividing his estate among themselves in the mistaken belief that he was dead. Chappe and all his assistants save one died of an epidemic on the Lower California peninsula. The magnitude of such sacrifices and the great financial cost of such expeditions give us a measure of the power of the spirit of the Enlightenment.

Further reading

R. d'E. Atkinson, The Eros parallax, 1930–31, *Journal for the History of Astronomy*, vol. 13 (1982), 77–83

A. J. Meadows, The transit of Venus in 1874, *Nature*, vol. 250 (1974), 749–52

Simon Newcomb, Discussion of observations of the transits of Venus in 1761 and 1769, *Astronomical Papers prepared for the Use of the American Ephemeris and Nautical Almanac*, vol. 2 (1890), 295–405

Simon Newcomb, *The Elements of the Four Inner Planets and the Fundamental Constants of Astronomy* (Washington, 1895)

Simon Newcomb, On Hell's alleged falsification of his observations of the transit of Venus in 1769, *Monthly Notices of the Royal Astronomical Society*, vol. 43 (1883), 371–81

Albert Van Helden, *Measuring the Universe: Cosmic Dimensions from Aristarchus to Halley* (Chicago, 1985)

Harry Woolf, *The Transits of Venus: A Study of Eighteenth-Century Science* (Princeton, NJ, 1959)

24

The discovery of Uranus, the Titius–Bode law, and the asteroids

MICHAEL HOSKIN

The Mars–Jupiter 'gap' in the planetary system

The traditional Ptolemaic astronomy that the Middle Ages inherited from Antiquity placed the Earth at the centre of the cosmos. As a result not only the distances, but even the order of the planets was uncertain; for observers on a supposedly central Earth were (it seemed) handicapped when it came to determining cosmic dimensions.

All this changed with the publication in 1543 of Nicholas Copernicus's *De revolutionibus*, which located the Sun at the centre of the cosmos. If Copernicus was right, then the Earth-based observer was in fact on a planet circling the Sun (at a mean distance later termed 'the astronomical unit' or AU); and earlier astronomers had been misled by the Earth's changes of position when interpreting their observations of the apparent movements of other planets. It now seemed that the circular movements with a period of one year that occurred in the traditional geometric models of the planets were no more than the reflection of the terrestrial motion; if so, then the radius of each of these circles should now be equated with the astronomical unit, which thus became a yardstick for a true planetary system.

The system is portrayed in outline for the first time in the famous diagram in Bk I of *De revolutionibus*. But Bk I is cosmological and essentially qualitative, and Copernicus's diagram is not drawn to scale. Had it been to scale, the reader could not have failed to note the disparately large gap between the (mean) orbits of Mars and Jupiter. This anomaly was to jar the Platonic sensibilities of the young Johannes Kepler (1571–1630), who in the closing years of the sixteenth century set himself to understand the motives that had led the Divine Geometer to construct the universe in the way he had chosen. In the Preface to his *Mysterium cosmographicum* (1596) Kepler tells us that he tried an approach

> *of striking boldness. Between Jupiter and Mars I placed a new planet, and likewise another between Venus and Mercury, two new planets that perhaps we would not see on account of their tiny size; and I assigned periodic times to them. For I reckoned that in this way I should produce some equality between the ratios, as the ratios between the pairs would be respectively reduced in the direction of the Sun and increased in that of the fixed stars.... Yet the interposition of a single planet was insufficient for the enormous gap between Jupiter and Mars; for the ratio of Jupiter to the new planet remained greater than is the ratio of Saturn to Jupiter.*

Eventually Kepler found suitable motivation for the Divine Geometer in a quite different approach, the nesting of spheres and regular solids (see Chapter 5 in Volume 2A). But such a solution did not commend itself to the generations that followed. Isaac Newton (1642–1727) in 1692, when pressed by the theologian Richard Bentley, suggested that God had located the massive planets Jupiter and Saturn at a great distance from the lesser planets so that their gravitational attraction should not disrupt the system:

> *... Jupiter and Saturn, as they are rarer than the rest, so they are vastly greater, and contain a far greater Quantity of Matter, and have many Satellites about them; which Qualifications surely arose not from their being placed at so great a Distance from the Sun, but were rather the Cause why the Creator placed them at a great Distance. For by their gravitational Powers they disturb one another's Motions very sensibly, ... and had they been placed much nearer to the Sun and to one another, they would by the same Powers have caused a considerable Disturbance in the whole System.*

Two of the cosmological speculators of the mid-eighteenth century likewise found the explanation of the gap in the sheer size of the outer planets. Immanual Kant (1724–1804) declared in the

Eighth Section of Pt II of his *Allgemeine Naturgeschichte und Theorie des Himmels* (1755) that the gap "is worthy of the greatest of all planets, namely, of that which has more mass than all the others together [i.e. Jupiter]". Johann Heinrich Lambert (1728–77) was as committed as Newton had been to stability in the large-scale structure of the universe, but he was prepared to accept the Mars–Jupiter gap as the consequence of the powerful attraction of Jupiter. In the first of his *Cosmologische Briefe* (1761), he asks: "And who knows whether already planets are missing that have departed from the huge space between Mars and Jupiter? Is it then true of celestial bodies as well as of the Earth, that the stronger harass the weaker, and are Jupiter and Saturn destined to plunder forever?"

Was the gap between Mars and Jupiter real (and therefore requiring an explanation) or merely apparent, and occupied by planets as yet undiscovered? Astronomers had been taught a lesson by the telescopic discoveries of the seventeenth century and were alive to the possibility of discoveries yet to come. William Whiston, for example, Newton's successor at Cambridge and one of the most influential writers of semipopular treatises in astronomy, repeatedly speaks of the 'known' planets, thereby reminding his readers that there may be some as yet unknown. However, most speculations about undiscovered planets concentrated on regions where planets would inevitably be hard to see: a planet within the orbit of Mercury, it was pointed out, might well remain lost to our sight in the glare of sunlight, while a planet beyond Saturn would be only faintly illuminated by the Sun and therefore difficult to detect. Only occasionally did a writer follow the young Kepler and posit a planet to fill the Mars–Jupiter gap. One who was later reputed to have done so in the early decades of the eighteenth century was the Scottish mathematician Colin Maclaurin. Another, from the next generation, was the amateur astronomer and maverick theologian Thomas Wright of Durham (1711–86). A speculation of his in Letter I of the *Second Thoughts* that remained in manuscript until our own day (and so had no influence on the history of astronomy) is nevertheless so remarkable as to merit citation:

That comets are capable of distroying such worlds as may chance to fall in their way, is, from their vast magnitude, velocity, firey substance, not at all to be doubted, and it is more than probable from the great and unoccupied distance betwixt ye planet Mars and Jupiter some world may have met with such a final dissolution.

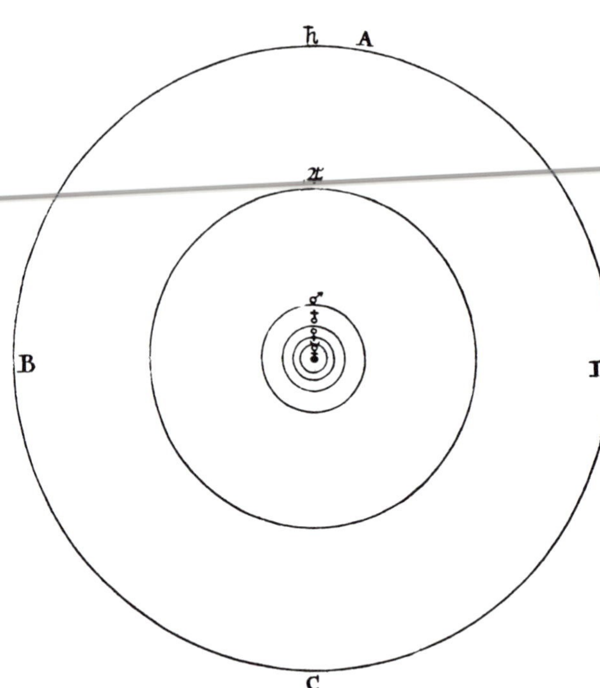

24.1. The orbits of the then-known planets of the solar system (Figure 1 of Plate 1 in David Gregory's *Elements of Astronomy*). The diagram illustrates how disporportionate was the apparent gap between Mars and Jupiter. The planet symbols are ♄ (Saturn), ♃ (Jupiter), ♂ (Mars), ♁ (Earth), ♀ (Venus), and ☿ (Mercury).

The formulation of the Titius–Bode law

Most writers, however, concerned themselves with the known planets, and astronomical treatises would list their distances from the Sun. David Gregory, professor at Oxford and disciple and confidant of Newton, in his *Astronomiae elementa* (1702) (Figure 24.1) expresses these distances in proportional numbers:

... supposing the distance of the Earth from the Sun to be divided into ten equal Parts, of these the distance of Mercury will be about four, of Venus seven, of Mars fifteen, of Jupiter fifty two, and that of Saturn ninety five.

Gregory's work quickly became a standard text, the original Latin edition being followed by an English translation in 1715 and second editions in both languages in 1726. The sentence quoted is prominently placed in Prop. I of Section 1 of Bk I, and it came to the notice of the German philosophical popularizer Christian Wolff (1679–1754), who incorporated a translation of it in Section 85 of his *Vernünfftige Gedanken von den Absichten der natürlichen Dinge*, which appeared in 1724 and went through several editions. There it was read by Johann Daniel Titius of Wittenberg (1729–96), and when in 1766 Titius put out a German translation of the *Contemplation de la Nature* (Figure 24.2) that had been published two years earlier by the distinguished French natural philosopher Charles Bonnet, he drew on these quantities for an (unsigned) interpolation he chose to make in Chap. 4 of Pt I of Bonnet's text. Titius had evidently noticed that if Gregory's numbers were slightly modified, so that Mars was assigned 16 instead of 15, and Saturn 100 instead of 95, then the resulting sequence fell into an arithmetical pattern, whereby each number was equal to 4 plus a suitable multiple of 3. Bonnet's text was now made to read:

Take note of the distances of the planets from one another, and recognize that almost all are separated from one another in a proportion that matches their bodily magnitudes. Divide the distance from the Sun to Saturn into 100 parts. Then Mercury is separated by 4 such parts from the Sun, Venus by 4 + 3 = 7 of the same, the Earth by 4 + 6 = 10, Mars by 4 + 12 = 16. But note that between Mars and Jupiter there occurs a departure from this very exact progression. After Mars there is a gap of 4 + 24 = 28 such parts, but thus far no planet or satellite has been sighted there. But should the Lord Architect have left this space empty? Never! Let us therefore confidently assume that this space belongs without doubt to the as yet undiscovered satellites of Mars [!]; let us add that perhaps Jupiter also has several around itself that are not yet visible in any telescope. Next to this for us still unexplored space comes Jupiter's sphere of influence at 4 + 48 = 52 parts, and Saturn's at 4 + 96 = 100 parts. What a remarkable relationship!

Titius published a second edition of his translation, with the relationship now more appropriately located in a translator's footnote, in 1772, just as the young Johann Elert Bode (1747–1826) was putting the finishing touches to the second edition of his introduction to astronomy, *Anleitung zur Kenntniss des gestirnten Himmels*, which he had published when he was only nineteen and which was to be in print for nearly a century. Bode rejected the nonsense about the satellites of Mars, but he was attracted by the arithmetical relationship, and inserted it (in this edition, without acknowledgement) as a footnote to his text:

This latter point appears to follow in particular from the remarkable relationship that the six known major planets follow in their distances from the Sun. Call the distance from the Sun to Saturn 100, then Mercury is separated from the Sun by 4 such parts; Venus 4 + 3 = 7; the Earth 4 + 6 = 10; and Mars 4 + 12 = 16. But now comes a gap in this very orderly progression. After Mars there follows a gap of 4 + 24 = 28 parts, where up to now no planet is seen. Can one believe that the Creator of the Universe has left this position empty? Certainly not. From here we come to the distance of Jupiter by 4 + 48 = 52 parts and finally to that of Saturn by 4 + 96 = 100.

Bode was convinced that a primary planet lay undiscovered in the Mars–Jupiter gap, at a distance from the Sun of some 28 such units. Within a few years, his conviction was to receive a wholly unexpected boost.

The discovery of Uranus

On 13 March 1781, in the fashionable English resort of Bath, the Hanoverian-born organist and amateur astronomer William Herschel (1738–1822) was devoting his spare time to what he termed his "second review of the heavens". Self-taught in astronomy, and equipped with home-made reflectors of previously unattained optical quality, Herschel was fascinated by problems of stellar astronomy, in contrast to the professionals of his day whose main concern was the solar system. Herschel was systematically using a 7-ft (2.1-m) reflector of 6.2-inch (16-cm) aperture (Figure 24.3) to examine the stars down to magnitude 8, partly to familiarize himself with the heavens, partly to collect a store of double stars that might be suitable subjects for Galileo's method of measuring annual parallax. That evening he was studying stars in Taurus, when he came across one that – so excellent was his mirror, and so experienced was he as an observer – he instantly recognized as anomalous: "In the quartile near ζ Tauri the lowest of two is a curious either Nebulous Star

und dem Weltgebäude überhaupt. 7

nämlich die Nebenplaneten jährlich um ihren Hauptplaneten, wie Trabanten, herumlaufen.

Venus und die Erde haben jegliche ihren Trabanten. Mit der Zeit wird man ohne Zweifel auch einen um den Mars entdecken. Jupiter hat ihrer viere, Saturn fünfe, nebst einem Ringe, oder einer leuchtenden Atmosphäre, welche die Stelle vieler kleinen Monden zu vertreten scheint. Da er beynahe dreyhundert Millionen Meilen von der Sonne entfernt ist, so würde er ein zu schwaches Licht bekommen, wenn nicht seine Trabanten und sein Ring dasselbe zurückwürfen und vermehrten.

Wir kennen siebzehn Planeten, die unser Sonnensystem ausmachen helfen; aber wir sind nicht versichert, daß ihrer nicht noch mehrere vorhanden sind. Ihre Anzahl ist seit Erfindung der Fernröhre sehr gewachsen; vielleicht wird sie noch mehr wachsen, wenn wir noch vollkommenere Werkzeuge, noch fleißigere und glücklichere Bemerker bekommen. Der Trabante der Venus, der im vorigen Jahrhunderte nur auf einen Augenblick gesehen, seit kurzem aber aufs neue erblicket worden, verkündiget der Sternkunde noch manche neue Entdeckungen.

Gebet einmal auf die Weiten der Planeten von einander Achtung; und nehmet wahr, daß sie fast alle in der Proportion von einander entfernt sind, wie ihre körperliche Größen zunehmen. Gebet der Distanz von der Sonne bis zum Saturn 100 Theile, so ist Mercurius 4 solcher Theile von der Sonne entfernt: Venus 4 + 3 = 7 derselben; die Erde 4 + 6 = 10; Mars 4 + 12 = 16. Aber sehet, vom Mars bis zum Jupiter kömmt eine Abweichung von dieser so genauen Progression vor. Vom Mars folgt ein Raum von 4 + 24 = 28 solcher Theile, darinn weder ein Haupt- noch ein Nebenplanete zur Zeit gesehen wird. Und der Bauherr sollte diesen Raum ledig gelassen haben?

A 4 Nimmer-

8 Von Gott

Nimmermehr! lasset uns zuversichtlich setzen, daß dieser Raum sonder Zweifel den bisher noch unentdeckten Trabanten des Mars zugehöre, laßt uns hinzuthun, daß vielleicht auch Jupiter noch etliche um sich habe, die bis itzt noch mit keinem Glase gesehen werden. Von diesem, uns unbekannten Raume erhebt sich Jupiters Wirkungskreis in 4 + 48 = 52; und Saturnus seiner, in 4 + 96 = 100 solcher Theile. Welches bewundernswürdige Verhältniß!

Es war der heutigen Sternkunde vorbehalten, nicht nur unsern Himmel mit neuen Planeten zu bereichern, sondern auch die Gränzen unsers Sonnenwirbels viel weiter hinauszusetzen. Die Kometen, welche, ihres betrüglichen Anblickes halber, ihres Schweifes, ihres haarigten Kernes, ihrer den Planeten oft entgegen gesetzten und von ihnen verschiedenen Richtung, ihres Erscheinens und Verschwindens wegen, für Erscheinungen gehalten wurden, die eine erzürnte Macht in der Luft angezündet hatte; diese Kometen sind zu planetischen Körpern geworden, deren lange Laufbahnen unsre Sternkundige berechnen, ihre entfernte Rückkehren vorhersagen, und ihren Ort, ihre Annäherungen und Entfernungen bestimmen. Vierzig dieser Körper erkennen anitzt schon die Herrschaft unsrer Sonne, und die Bahnen, welche einige von ihnen um dieselbige beschreiben, sind so sehr ausgedehnt, daß sie solche, erst nach einer langen Reihe von Jahren, oder wohl gar in vielen Jahrhunderten, einmal durchlaufen.

Gleichergestalt war es ein Vorrecht der neuern Sternkenntniß, zu zeigen, daß diese Kometen vermuthlich diejenigen Wandelsterne sind, wodurch die unzählichen Systeme so vieler Sonnen zusammen hängen, und die das eigentliche Verbindungsglied in der gesammten Kette der Sterngebäude abgeben. Denn wozu wäre der große Raum nöthig, der vom Saturn bis zum nächsten Fixsterne

24.2. Pages 7 and 8 of Titius's German translation of Bonnet's *Contemplation de la nature*, with the statement of the numerical relationship between the planetary distances interpolated by Titius into Bonnet's text.

24.3. The home-made 7-ft (2-m) reflector with which Herschel discovered Uranus, from the drawing by his friend and ally William Watson.

or perhaps a Comet", he noted (Figure 24.4). Four days later he returned to the object and found that it had moved: it was indeed a member of the solar system.

Interestingly, it did not occur to Herschel that his "curious" object might be a planet, perhaps because while he knew very well that no new planet had been discovered since the dawn of history, he was unaware of the widespread opinion among astronomers that there were nevertheless planets to be found. Nevil Maskelyne, the Astronomer Royal, was within days writing to a mutual friend to tell him of Herschel's "comet or new planet", and on 23 April he wrote to Herschel:

I am to acknowledge my obligations to you for the communication of your discovery of the present Comet, or planet, I don't know which to call it. It is as likely to be a regular planet moving in an orbit nearly circular round the sun as a Comet moving in a very excentric ellipsis.

Maskelyne in fact had had difficulty in identifying the object to which Herschel was referring; for whereas the organist with his home-made reflector had seen at a glance that this was no ordinary star, the Astronomer Royal with the professionally-built instruments of Greenwich Observatory had been able to find nothing unusual in that region of the sky and had been forced to identify Herschel's object by its movement. Thomas Hornsby, the professor at Oxford, had done no better.

Meanwhile, Herschel himself was running into difficulties. His micrometric observations of the angular diameter of the object in the month following discovery seemed to imply that it had nearly

March 12. 5h 45' in the morning

Mars seems to be all over bright but the air is so frosty & undulating that it is possible there may be spots without my being able to distinguish them. Nr 4. 20ft

53' I am pretty sure there is no spot on Mars

---

the Shadow of Saturn's ring lays at the left upon the ring

Tuesday March 13

Pollux is follow'd by 3 small stars at abt 2' and 3' distance.

α as usual. p H

---

in the quartile near ζ Tauri the lowest of two is a curious either Nebulous star or perhaps a Comet.

~~preceding the star that precedes ν Geminorum double about 30"~~

a small star follows the Comet at 2/3 of the field's distance

~~Σ 2. 33.~~

24.4. The page of Herschel's journal with his discovery of Uranus, thought by him at first glance to be "a curious either Nebulous Star or perhaps a Comet".

doubled in apparent diameter and so was rapidly approaching the Earth. In fact these changes were illusory; and their publication was to cast a question-mark over Herschel's reliability. But the unknown amateur's discovery of a planet moving so slowly that its change of position in a single night was almost imperceptible, astonished astronomers throughout Europe and gave notice that an observer of exceptional ability had appeared on the scene.

Meanwhile, the question of the nature of the object remained unresolved. The first orbits were computed on the assumption that the object was a comet. P.-F.-A. Méchain derived a perihelion distance of 0.46 AU and a perihelion date of 23 May 1781. In contrast, Anders Johan Lexell, professor of mathematics at St Petersburg and a specialist in cometary orbits, who chanced to be in England at the time, estimated a perihelion distance of 16 AU, and a perihelion date of 10 April 1789. But within weeks it was clear that no cometary orbit would suffice. Who first gave a satisfactory demonstration that the orbit was nearly circular – that is, planetary – and at a distance about twice that of Saturn is not clear, but Lexell's later claim to priority in the derivation of the elements of a circlar orbit may well have been justified. He found the radius of the orbit to be 18.93 AU and its motion no more than some 4° 20′ per annum, both excellent values: the body was a primary planet of the solar system. The discovery – and some tactful lobbying by his friends – soon earned Herschel royal patronage that would enable him to give up music and devote himself to astronomy, and he was happy in return to name his planet Georgium Sidus in honour of King George III; but astronomers preferred to follow tradition and adopted instead Bode's suggestion of Uranus, since in mythology Uranus was father to Saturn as Saturn was father to Jupiter.

The search for a planet between Mars and Jupiter

Gregory had taken the mean Earth–Sun distance (1 AU) as equal to 10 units, and on this scale had assigned the satisfactory value of 95 units to the distance of Saturn from the Sun. Titius and Bode had forced an arithmetical relationship out of the data by making two modifications to Gregory's numbers, one of which was to increase the 95 to 100; and they had then used this figure of 100 to define the units in which they were now working. But even when this "fudging" is seen for what it is, the fit between the next term in the series, $4+192=196$, and what was found to be the mean distance of Uranus is remarkable. For within a very few years of the planet's discovery, students of its orbit were taking into account pre-discovery observations of the planet, and even the perturbations by other planets, and nearly all agreed that the mean distance of Uranus was between 19.0 and 19.2 AU, with a gradual convergence towards the upper end of that range – a fit good enough to convince not only Bode but Baron Franz Xaver von Zach (1754–1832), the court astronomer at Gotha, of the validity of the Titius–Bode relationship and so of the existence of a planet between Mars and Jupiter. In 1787 Zach undertook a solo search for the planet, equipping himself for the purpose with a catalogue of zodiacal stars arranged by right ascension; but without success. The autumn of 1799 found him visiting astronomers in Celle, Bremen and Lilienthal, and (as he later explained to readers of his journal *Monatliche Correspondenz*) it was there that the idea of a cooperative attack on the problem emerged:

> *It was the opinion of these men of discernment, that to get onto the trail of this so-long-hidden planet, it cannot be a matter for one or two astronomers to scrutinize the entire Zodiac down to the telescopic stars.*

It took another year for the collaboration – probably without precedent in the history of science – to become a reality. On 21 September 1800 Zach met in Lilienthal with five other astronomers, among them J. H. Schröter, the chief magistrate of Lilienthal, whose world-famous collection of instruments included a Herschel reflector of 27-ft (8.1-m) focal length, C. L. Harding (1765–1834), who was employed by Schröter, and H. W. M. Olbers (1758–1840), a physician from nearby Bremen and long-time collaborator with Schröter. They judged that even six observers were too few for the task ahead, and they decided to coopt practising astronomers from throughout Europe, to make the numbers up to twenty-four. Schröter was to be president of the team and Zach secretary. They divided the entire Zodiac into twenty-four zones each of 15° in longitude and extending some 7° or 8° north and south of the ecliptic in latitude. The zones were allocated to the members by lot. Each member was to draw up a star chart for his zone, extending to the smallest telescopic stars,

> *and through repeated examination of the sky was to confirm the unchanging state of his district, or the presence of each wandering foreign guest. Through such a strictly organized policing of the heavens, divided into twenty-four sections, we hoped eventually to track down this planet, which had so long escaped our scrutiny – supposing, that is, that it existed and could be seen.*

The discovery of Ceres

Zach accordingly set out to recruit the team that was to police the Zodiac. One of those chosen was, naturally, Giuseppe Piazzi (1746–1826) of Palermo in Sicily. As a young man Piazzi had joined the Theatine Order, and had then taught mathematics in a number of Italian cities. In 1780 he had accepted an invitation to take the chair of higher mathematics at the Academy of Palermo.

Although himself inexperienced in astronomy, he had succeeded in convincing his royal patron of the desirability of an astronomical observatory in Palermo, which would give access to stars further south than were visible from any existing European observatory. Piazzi was given leave to travel to northern Europe, to learn the craft of the astronomical observer from such disparate figures as Maskelyne and Herschel, and to purchase instruments of quality. Piazzi succeeded in persuading Jesse Ramsden, greatest of the instrument-makers, to undertake a 5-ft (1.5-m) vertical circle of unique design, and – what was much more difficult – to complete it, a feat that required Piazzi to be present in Ramsden's workshop and literally breathing down his neck (Figure 24.5).

In Palermo, Piazzi found himself privileged to possess this masterpiece of eighteenth-century technology and a site worthy of it. He accordingly set to work to undertake a star catalogue of greater accuracy than any that had gone before (see Volume 3).

The beginning of 1801 found Piazzi patiently at work on the star catalogue. The dramatic events that then unfolded were described by him a few months later in a pamphlet entitled *Risultati delle osservazioni della nuova stella*, which we may quote from the translation made in England for Maskelyne:

... on the evening of the 1st of January of the current year [1801], together with several other stars, I sought for the 87th of the Catalogue of the Zodiacal stars of Mr La Caille. I then found it was preceded by another, which, according to my custom, I observed likewise, as it did not impede the principal observation. The light was a little faint, and of the colour of Jupiter, but similar to many others which generally are reckoned of the eighth magnitude. Therefore I had no doubt of its being any other than a fixed star. In the evening of the 2d I repeated my observations, and having found that it did not correspond either in time or in distance from the zenith with the former observation, I began to entertain some doubts of its accuracy. I conceived afterwards a great suspicion that it might be a new star. The evening of the third, my suspicion was converted into certainty, being assured it was not a fixed star. Nevertheless before I made it known, I waited 'till the evening of the 4th, when I had the satisfaction to see it had moved at the same rate as on the preceding days. From the fourth to the tenth the sky was cloudy. In the evening of the 10th it appeared to me in the Telescope, accompanied by four others, nearly of the same magnitude. In the uncertainty which was the new one, I observed them all, as exactly as possible, and having compared these observations with the others which I made in the evening of the 11th, by its motion I easily distinguished my star from the others. Mean while however I greatly wished to see it out of the Meridian, to examine and to contemplate it more at leisure. But with all my labour, and that of my assistant D. Niccola Cacciatore and [of] D. Niccola Carioti belonging to this Royal Chapel both enjoying a sharp sight, and very expert in the knowledge of the heavens, neither with the night Telescope, nor with another acromatic one of 4 inches aperture, was it possible to distinguish it from many others among which it was moving. I was therefore obliged to content myself with seeing it on the meridian, and for the short time of two minutes, that is to say the time it employed in traversing the field of the Telescope; other observations, which were [being made] at the same time, not permitting the instrument to be moved from its position.

In the mean time, in order to render the observations more certain, while I was observing with the Circle, D. Niccola Carioti observed with the transit instrument. The sky was so hazy, and often cloudy, that the observations were interrupted 'till the 11th of February; when the star having approached so near the Sun, it was not possible to see it any longer at its passage over the meridian. I intended to search for it, out of it [the meridian], by means of the Azimuth; but having fallen ill on the thirteenth of February, I was not able to make any further observations. These, however, which have been made, though they are not at the necessary distance from one another in order to assure us of the true course which the star describes in the heavens, are, notwithstanding, sufficient in my opinion, to make us know the nature of the same, as one may collect from the results, which I have deduced from them.

Piazzi had in fact measured the position of the object on a total of 24 nights between 1 January and 11 February, though some positions were marked as "doubtful" or even "very uncertain". On 24 January, Piazzi had announced his discovery in three letters to fellow astronomers. One was to his friend, Barnaba Oriani of Milan; in it, Piazzi confides that

I have announced this star as a comet, but since it is not accompanied by any nebulosity and, further, since its

24.5. The circle by Ramsden that Piazzi used in compiling his star catalogues and with which he detected the motion of what proved to be the first known asteroid. In this drawing, two of the four supporting arches have been omitted in the interests of clarity.

movement is so slow and rather uniform, it has occurred to me several times that it might be something better than a comet. But I have been careful not to advance this supposition to the public.

To Bode and J.-J. L. de Lalande, however, he limited himself to the cautious claim that the object was merely a comet, though making it clear that the 'comet' had no nebulosity or tail.

When after 11 February he could no longer see the object, he set to work to investigate its orbit, though such mathematical investigations were not his strength. He began with the assumption that it was indeed a comet, and fitted a parabola to three of the observations to see if the orbit would account for the others. It did not. A second attempt with a different group of three observations likewise failed:

From the parabolic hypothesis I passed then to the circular; and having made a few suppositions, I found two radii, 2.7067 and 2.6862; with each of which all the observations were represented a great deal better than any parabola. The planets describing ellipses more or less eccentric, and not circles, it is to be believed that ours will not deviate from this rule. In an ellipsis I should then have continued my calculations; but as the arch observed is very small, the results would be very uncertain, and the labour long and painful. I have therefore preferred the circle....

The agreement of the observed longitudes with the calculated ones in the circular hypothesis, its motion in the Zodiac, from which it only departs a little way in the greatest latitudes, and its position between Mars and Jupiter, leave no doubt that this new star is a true planet....

It is easy to believe that not only Oriani, but more especially Bode, were, in Piazzi's words, "instantly of the opinion that it was a new planet; and settled nearly the same elements of its orbit, as I have done". One can imagine the German's delight that the hoped-for planet had been found, even if the carefully-laid plans of the celestial police had played no part in the discovery.

But now Piazzi was beginning to have doubts. His first estimate of the size of the object was based on the fact that it was almost, but not quite, covered by one of the wires of his telescope, and he concluded that it was larger than the Earth. However, it would seem that in the hazy nights that followed, the true (and much smaller) size of the object became more evident to the Palermo astronomer, who began to think that the object was diminishing in size and therefore moving rapidly away, so that it must be a comet after all:

As after the 23rd [of January] the star began sensibly to diminish in size and brightness, uncertain whether it was to be attributed to its rapid receding from the Earth, or rather to the state of the atmosphere, which became after that time still more dark and hazy, I began to doubt of its nature, so as even to believe it was a Comet and not a planet.

Eventually, in April, after illness had prevented him from making further progress in the investigation of the object's orbit, Piazzi sent his complete observations to Oriani, Bode, and Lalande in Paris. Like Herschel's, his discovery had been an unexpected – but not therefore accidental – byproduct of a lengthy observational campaign in stellar astronomy; like Herschel, he had been deceived into suspecting the object was a comet; and like Herschel, he had been forced in the end to entrust his discovery to the mathematicians. Nor did the similarities end there; for just as Herschel's attempt to name his planet in honour of his royal patron was rejected by astronomers, so Piazzi's expressed wish that his object – should it ever be recovered – be named Ceres Ferdinandea, Ceres for the patron goddess of Sicily and Ferdinandea for Piazzi's royal patron, has only partially been honoured.

Ceres had been observed for only a fragment of its orbit, a geocentric arc of a mere 3°, and to recover it when it at length emerged from the glare of the Sun seemed beyond the present capacity of mathematicians. Piazzi's own opinion was that recovery depended upon the identification of some earlier sighting when the object had been seen but taken for a star. But fate had provided a brilliant new mathematical talent in the person of C. F. Gauss (1777–1855), and Gauss's analysis of the orbit (see Chapter 25) enabled Zach at Gotha to recover the object on 31 December 1801, a year after its first discovery; the following evening, at Bremen, Olbers also saw it.

Gauss's analysis assigned to Ceres a mean distance from the Sun of 2.767 AU. When this figure was scaled up by a factor of 10 or so, to match the units of the Titius–Bode relation, the fit with the distance predicted for the 'missing' planet –

$4 + 3.8 = 28$ – was almost perfect: Piazzi's planet, it seemed, matched the description of the missing body sought by the celestial police. The relation – 'law' would by now be not too strong a term – had received its second triumphant vindication within a couple of decades.

It seemed too good to be true, as indeed it was. William Herschel found to his surprise that he could barely perceive a planetary disk in Ceres, and his preliminary conclusion, announced to the Royal Society in mid-February, 1802, was "that the real diameter of the new planet, remarkable as it may appear, is less than five eights of the diameter of our moon", a surprisingly low estimate but one that he was soon to revise sharply downwards.

Worse was to follow. At the end of March, Olbers, who had become very familiar with the region of the sky where the reappearance of Ceres had been predicted, noticed there a small star that formed an equilateral triangle with two stars that were in the catalogues. He was convinced this third star had newly appeared on the scene, perhaps because it was a variable now at maximum; but in hours he could see that it had moved. Within a month Gauss had calculated its mean distance, 2.670 AU, or almost the same as Ceres; its orbit had an eccentricity even greater than that of Mercury, and was inclined to the ecliptic by an astonishing 34°. Meanwhile, Herschel was taking micrometer measurements of both bodies, and concluded that Ceres had a diameter of less than 162 miles (261 km), while the new object, which Olbers named Pallas, was even smaller, with a diameter of less than 111 miles (178 km): underestimates, as we now know, but serving to highlight the contrast between these objects and the known planets. Herschel did not yet know the elements that Gauss had calculated for Pallas, but he was well aware that the orbits of the two bodies were sharply inclined to the ecliptic. "If bodies of this kind", he drily remarked in a paper read to the Royal Society early in May, "were to be admitted into the order of planets, we should be obliged to give up the zodiac; for, by extending it to them, should a few more of these stars be discovered, still farther and farther deviating from the path of the earth, which is not unlikely, we might soon be obliged to convert the whole firmament into zodiac...." Such bodies could not properly be termed planets; since they were not comets either, they deserved a name of their own, and he proposed the descriptive term 'asteroid' – though this was naturally unwelcome to Piazzi who in consequence would cease to be the discoverer of a planet.

Confronted with not one but two bodies in the gap between Mars and Jupiter, Olbers suggested that they were in fact fragments of a full-size planet that had once occupied the gap – thereby unknowingly reviving the unpublished speculation of Thomas Wright. If this was so, then they might indeed have orbits of various inclinations and eccentricities, but these orbits would all have similar mean distances from the Sun and, more importantly, would all intersect in two positions in opposite parts of the sky, one of which would be the position of the planet at the moment of fragmentation. The two known orbits served roughly to define these two points of intersection, and Harding agreed to compile charts of the faint stars in these regions. In September 1804, while at work on this task, he noticed an object that proved to be a third asteroid, which he named Juno. Its mean distance was again 2.670, while its eccentricity was once more far greater than that of any planet.

Olbers persevered in the search, but not until 1807 was he rewarded by the discovery of a fourth asteroid, for which Gauss, at Olbers's invitation, selected the name Vesta.

Ceres is in fact much the largest of the asteroids, with a diameter now estimated as about 1000km. Pallas and Vesta are the only others with diameter in excess of 500km. The rest are even smaller, and none is as bright as eighth magnitude, so it is not surprising that astronomers wearied of the search. It was revived by a German amateur of extraordinary dedication, an ex-postmaster named Hencke, who began to search around 1830 and was at last rewarded with his first success in 1845, and his second in 1847. By this time the star charts sponsored by the Berlin Academy of Sciences that were to prove decisive in the discovery of Neptune (Chapter 28) were becoming available, and private observers now had the data on the basis of which they could prosecute an effective search. Later that year, J. R. Hind, who was employed as astronomer at the observatory of George Bishop at Regent's Park in London, found two more asteroids, and the next year another was found by Andrew Graham, astronomer at Edward J. Cooper's observatory at

Markree in Ireland. Since then, the number of asteroids known has grown constantly; and as a result of the application of photography, asteroids are today known in their thousands.

The Titius–Bode 'law' had been rescued by Olbers's suggestion that the planet in the Mars–Jupiter gap had fragmented; and it survived to play a key role in the discovery of Neptune. But asteroids were being found whose orbits showed they could not be fragments of a single planet; and the discovery of Neptune, which refused to conform to the 'law' in any simple manner, brought the relationship into disrepute – though it continues to exert a fascination on astronomers of the planetary system.

Further reading

A. F. O'D. Alexander, William Herschel discovers the 'Georgian Planet', 13 March 1781, Chap. 1 in his *The Planet Uranus: a History of Observation, Theory and Discovery* (London, 1965)

Eric G. Forbes, Gauss and the discovery of Ceres, *Journal for the History of Astronomy*, vol. 2 (1971), 195–99

Robert Grant, *History of Physical Astronomy* (London, 1852), 237–43

Stanley L. Jaki, The early history of the Titius–Bode Law, *American Journal of Physics*, vol. 40 (1972), 1014–23

Michael Martin Nieto, *The Titius–Bode Law of Planetary Distances* (Oxford, 1972)

R. Porter *et al.*, History of the discovery of Uranus, in Garry Hunt (ed.), *Uranus and the Outer Planets* (Cambridge, 1982), 21–89

Simon Schaffer, Uranus and the establishment of Herschel's astronomy, *Journal for the History of Astronomy*, vol. 12 (1981), 11–26

25

Eighteenth- and nineteenth-century developments in the theory and practice of orbit determination

BRIAN G. MARSDEN

The eighteenth century

In 1705 Edmond Halley published the results of his investigation of the orbits of twenty-four comets observed up to the end of the seventeenth century (see Chapter 19). As to how he carried out this investigation, we know only that he adapted the graphical method developed by Isaac Newton in his *Principia* for the determination of cometary orbits, and (as Halley puts it) "attempted to bring the same method to an arithmetical calculation", substituting arithmetic for at least part of Newton's graphical construction. Halley gave the first explicit statement of the cubic equation relating the time interval from perihelion passage in a parabolic orbit to the perihelion distance and the tangent of half the true anomaly. He also supplied a table to aid its solution and attempted a generalization to an elliptical orbit.

After the dramatic achievements of Newton and Halley, there was surprisingly little work of either a theoretical or practical nature on the determination of cometary orbits for several decades. James Bradley determined a parabolic orbit for a comet observed in 1723, and Pierre Bouguer (1698–1758), in a piece published in the *Mémoires* of the Paris Academy of Sciences for 1733, computed a hyperbolic orbit for the comet of 1729. While from the point of view of accuracy a step backwards from the results of Newton and Halley, Bouguer's work is of interest because it represented the first attempt at an orbital solution in algebraic terms. Noting that the orbit of the comet is confined to a plane, Bouguer made use of what in vector notation would be written

$$\mathbf{r}_0 = c_1\mathbf{r}_1 + c_3\mathbf{r}_3, \qquad (1)$$

where $\mathbf{r}_1$, $\mathbf{r}_0$, and $\mathbf{r}_3$ are the position vectors from the Sun to the comet at three observations times t_1, t_0, and t_3, and c_1 and c_3 are scalar constants. He combined Equation (1) with a geometrical condition at each t_i ($i = 1, 0, 3$), namely,

$$\mathbf{r}_i = \rho_i\hat{\boldsymbol{\rho}}_i - \mathbf{R}_i, \qquad (2)$$

where ρ_i is the distance from the observer to the comet, $\hat{\boldsymbol{\rho}}_i$ is the unit vector from the observer to the comet, and $\mathbf{R}_i$ is the vector from the observer to the Sun, to obtain

$$c_1\rho_1\hat{\boldsymbol{\rho}}_1 - \rho_0\hat{\boldsymbol{\rho}}_0 + c_3\rho_3\hat{\boldsymbol{\rho}}_3 = c_1\mathbf{R}_1 - \mathbf{R}_0 + c_3\mathbf{R}_3. \qquad (3)$$

The vector quantities in Equation (3) are knowable from observations and solar tables, and if the scalars c_1 and c_3 are also known, one can – at least in principle – solve for the three unknown distances ρ_i. Substitution of the ρ_i into Equation (2) then yields points – only two of the three actually being required – on the heliocentric orbit and eventually the orbital elements. For c_1 and c_3 Bouguer adopted the time ratios

$$c_1^0 = \frac{t_3 - t_0}{t_3 - t_1}, \quad c_3^0 = \frac{t_0 - t_1}{t_3 - t_1}, \qquad (4)$$

considering these to be adequate approximations if the time span covered by the observations is small.

For complete accuracy, c_1 and c_3 should be the ratios of the areas of triangles formed by the Sun and respective positions of the comet. The numerator triangle for c_1 involves the comet's positions at t_0 and t_3 and that for c_3 those at t_1 and t_0; in each case the denominator triangle involves the positions at t_1 and t_3. By defining them instead as the ratios of time differences, Bouguer was – as a consequence of Kepler's second law – replacing the *triangles* by the corresponding *sectors* of the comet's orbit. The correct expressions for c_1 and c_1 are therefore

$$c_1 = \frac{t_3 - t_0}{t_3 - t_1}\frac{y_{13}}{y_{03}}, \quad c_3 = \frac{t_0 - t_1}{t_3 - t_1}\frac{y_{13}}{y_{10}}, \qquad (5)$$

where each y_{ij} is the ratio of the areas of the sector

and the triangle involving the Sun and the positions of the comet at t_i and t_j.

Nicolaas Stryuck, in a work of 1740 in Dutch, and Giovanni Domenico Maraldi, in the Paris *Mémoires* for 1743, strongly criticized Bouguer's work, largely because of his audacity in claiming a hyperbolic orbit for the comet of 1729. Interest in the subject of orbit determination became widespread during the 1740s, as these and several other astronomers, notably Eustacho Zanotti and Pierre-Charles Le Monnier, began to compute orbits whenever new comets were discovered. Cornelis Douwes and Nicolas-Louis de Lacaille also worked on the orbits of comets observed in previous centuries. Various processes of trial and error were used, involving in particular the adjustment of assumed values for the comet's distance from the Earth at two observations so that the time interval between those observations was consistent with parabolic motion, followed by the selection of the particular parabola that satisfied the comet's observed longitude at a third observation. A method of orbit determination proposed by the Dalmatian Jesuit Rudjer Josip Bošković (in his *De determinanda orbita planetae* of 1749) utilized a geometrical construction based on that used to determine the reflection of light by a spherical mirror.

The first real progress in understanding the orbit-determination process from an analytical point of view was made by Leonhard Euler (1707–83), in his *Theoria motuum planetarum et cometarum* of 1744. Euler appreciated that it was necessary to do something about the y_{ij}, the three sector-triangle ratios in Equations (5). For rather small arcs of an orbit, these ratios are clearly slightly larger than unity, and Euler established for the t_1-t_3 configuration, for example, the approximation

$$y_{13}=1+\tfrac{1}{6}\frac{(\tau_3-\tau_1)^2}{r_1^{\frac{3}{2}}r_3^{\frac{3}{2}}}, \tag{6}$$

where τ_1 and τ_3 are the time differences t_1-t_0 and t_3-t_0 specified in 'canonical' units of approximately 58.13 days (that is, such that the Earth's revolution period is approximately 2π).

Euler utilized Equation (6) (and its companion equations) in practice by noting that it means that a small fraction, $-\frac{1}{2}\tau_1\tau_3/r_0^3$, of the comet's radius vector at t_0 is isolated by the chord joining the comet's positions at the first and third observations. He was thereby able to derive values for ρ_1 and ρ_3 from assumed values for ρ_0. Since eighteenth-century astrometry left a lot to be desired, Euler generally made use of a fourth observation well separated in time from the first three, and the accuracy with which this fourth observation was represented would enable him to select the most satisfactory result. Like Bouguer, Euler settled for any type of conic, and in the case of the comet of 1742 he obtained an ill-defined ellipse with a revolution period of only 42 years.

The fact that Euler made no attempt to force a parabolic solution is particularly surprising in view of his demonstration of the celebrated equation

$$(r_1+r_3+s_{13})^{\frac{3}{2}}-(r_1+r_3-s_{13})^{\frac{3}{2}}=6(\tau_3-\tau_1), \tag{7}$$

where s_{13} is the length of the chord joining the points at t_1 and t_3 in an exactly parabolic orbit. Although Euler established this equation in 1743, he does not seem ever to have put it to practical use. Johann Heinrich Lambert (1728–77), in his *Insigniores orbitae cometarum proprietates* of 1761, rediscovered and generalized it to the case of an elliptical orbit. As was pointed out by Joseph Louis Lagrange (1736–1813), Euler's equation was in fact implicit in Newton's work.

Lambert's other important contribution to orbit literature, published in the Berlin *Mémoires* for 1771, was his demonstration that the apparent trajectory of a comet is convex or concave towards the Sun according as the comet is farther from or closer to the Sun than the Earth. Lambert's theorem, as this is usually called, can be expressed by means of the equation

$$\rho_0=\rho_0^{\circ}\left(1-\frac{\gamma_0}{r_0^3}\right), \tag{8}$$

where $\gamma_0=R_0^3$, the cube of the distance from the observer to the Sun, and ρ_0° is positive or negative according as r_0 is greater or less than R_0. Bouguer's process effectively supposes that $\gamma_0=0$, that is, that $\rho_0=\rho_0^{\circ}$, and it was precisely because the comet of 1729 had such a large r_0 (greater than $4R_0$ even at perihelion) that he obtained a tolerably satisfactory result in this case.

The earliest practical use of Lambert's theorem in orbit determination is usually attributed to Lagrange, who in the Berlin *Mémoires* for 1778 used this *dynamical* relationship between ρ_0 and r_0 in conjunction with the obvious *geometrical* relationship,

$$r_0^2 = \rho_0^2 + 2\rho_0 R_0 \cos\epsilon_0 + R_0^2, \tag{9}$$

where ϵ_0 is the object's elongation from the opposition point, to obtain an eighth-degree equation for r_0. Since $r_0 = R_0$ (that is, $\rho_0 = 0$, yielding the observer's own motion) is clearly a solution, the equation can be reduced to the seventh degree.

Lagrange also gave the Taylor-series developments of what are nowadays called the 'f and g functions' that allow the heliocentric position vector $\mathbf{r}_i$ at time t_i to be expressed in terms of the heliocentric position and velocity vectors $\mathbf{r}_0$ and $\mathbf{v}_0$, namely,

$$\mathbf{r}_i = f_i \mathbf{r}_0 + g_i \mathbf{v}_0. \tag{10}$$

Although the series may not converge, the first two terms are

$$f_i = 1 - \frac{1}{2}\frac{\tau_i^2}{r_0^3}, \quad g_i = \tau_i\left(1 - \frac{1}{6}\frac{\tau_i^2}{r_0^3}\right), \tag{11}$$

and subsequent terms, which depend also on the radial velocity, rapidly increase in complexity. No practical use seems to have been made of the f and g functions before well into the nineteenth century. We note that the constants c_1 and c_3 of Equation (1) are related to the f and g functions by

$$c_1 = \frac{g_3}{f_1 g_3 - g_1 f_3}, \quad c_3 = -\frac{g_1}{f_1 g_3 - g_1 f_3}. \tag{12}$$

Lagrange did not present his work in a very practical form, however, and it was Pierre-Simon Laplace (1749–1827) who in the Paris *Mémoires* for 1780 widely publicized the first complete analytical orbit-determination procedure that had a good degree of practicality. In order to set up Lambert's theorem, Laplace utilized Equation (2) and its first and second time derivatives (at time t_0), with then the substitution for the second derivatives of both $\mathbf{r}_0$ and $\mathbf{R}_0$ in terms of $\mathbf{r}_0$ and $\mathbf{R}_0$ themselves from the Newtonian equations for gravitational motion. Although this idea is conceptually very simple, there exists the practical difficulty of the computation of the first and second time derivatives of $\hat{\rho}_0$. In principle, this can be accomplished from only three observations of the comet's position projected on to the sky, but the use of several additional observations is desirable in practice. An added complication is that the observer is also assumed to be travelling on an exact Keplerian orbit about the Sun; for the results to be strictly accurate the observed positions should therefore be reduced to the centre of the Earth, and this requires prior knowledge of the distances one is attempting to determine. There is one special convenience to Laplace's method, namely, that the computed heliocentric velocity vector can be constrained so that the comet's orbital energy is precisely zero, as is the case in an exactly parabolic orbit.

In spite of its problems, the simplicity and convenience of Laplace's method resulted in its immediate popularity. Alexandre-Gui Pingré, the leading practitioner of orbital computations during the late eighteenth century, in the second volume of his *Cométographie* (1784) gave an extensive critique of methods of orbit determination and concluded that Laplace's was the best. The quick and widespread acceptance of Laplace's method was unfortunate, for it caused a significant concept that had just been presented in the Paris *Mémoires* for 1779 by the distinguished amateur A.-P. Dionis du Séjour (1734–94) to be largely overlooked and certainly misinterpreted.

Dionis du Séjour's method starts with the elimination of the middle terms in both the left-hand and right-hand sides of the vector Equation (3) and then division by the resulting coefficient of ρ_1. This yields an expression of the form

$$\rho_1 + M_{13}\rho_3 = N_{13}. \tag{13}$$

Eliminating instead the last terms on each side, one obtains

$$\rho_1 + M_{10}\rho_0 = N_{10}. \tag{14}$$

Here the quantities M_{13}, M_{10}, N_{13} and N_{10} depend, in particular, on c_1 and c_3. Dionis du Séjour followed Bouguer and approximated c_1 and c_3 by c_1° and c_3°. In so doing, he recognized that the same approximations apply equally well to the orbit of the Earth, in which case the right-hand side of the vector Equation (3), and hence also the N_{13} and N_{10} of Equations (13) and (14), vanish. Though not in fact mentioned by Dionis du Séjour, it is this point that shows the fallacy of Bouguer's method, for Equation (3) can therefore be satisfactorily solved only for the *ratios* of the ρ_i, not for the ρ_i individually. Nevertheless, Dionis du Séjour went on to eliminate c_1 and c_3 from Equation (1), and to substitute from Equation (2) and then from Equations (13) and (14) with the zero right-hand sides, obtaining thereby a cubic equation in ρ_1, which he reduced to a quadratic by taking the ecliptic as the reference plane. In reality, this procedure is not significantly different from that of Bouguer, and it can therefore be criticized for the same reasons.

The real significance of Dionis du Séjour's work is that he went on to consider that the square of the length of the chord joining the first and third points on the orbit can be written

$$s_{13}^2 = 4\frac{(\tau_3-\tau_1)^2}{r_1+r_3}\left[1+\frac{1}{3}\frac{(\tau_3-\tau_1)^2}{(r_1+r_3)^3}\right]. \qquad (15)$$

This is an approximation to Equation (7), Euler's equation, recast in a form that is more convenient to use. With the help of Equations (2) and (13) – again with $N_{13}=0$ – Dionis du Séjour obtained a *second* quadratic equation for ρ_1. By analogy with the *vis viva* integral in the Laplace method, this second quadratic equation provides the additional dynamical constraint necessary to ensure that the orbit is a parabola. This was the missing link in the orbit-determination process, utilized by Newton but not by Euler, even though the latter had written it down in exact algebraic form in Equation (7).

Pingré accorded Dionis du Séjour's work some publicity in his aforementioned critique, but his numerical examples made use of only the first quadratic equation in ρ_1, not the second. He justified his action with the unfortunate comment that "this will almost always suffice", adding that "the observational circumstances almost always make it possible to decide which root one must choose".

The pinnacle of orbit-determination achievement during the eighteenth century was therefore the publication by the Bremen physician H. W. M. Olbers (1758–1840) in 1797 of an essay to which, with a fair amount of justification, he gave the title *Abhandlung über die leichteste und bequemste Methode die Bahn eines Cometen zu berechnen* (Treatise on the easiest and most convenient method of computing the path of a comet). Before introducing the method, Olbers gave a summary of earlier work on the subject. This included a detailed description of Dionis du Séjour's method, much as is given above, although Dionis du Séjour's use of Equation (15) is mentioned only in passing. Very properly dismissing Dionis du Séjour's first quadratic equation as insufficiently accurate, Olbers suggested that the user should simply adopt $\rho_1=1$ AU and calculate the corresponding r_1, r_3, and s_{13} from Equation (13) with $N_{13}=0$ and the obvious geometrical expressions. Substitution into Equation (7) then yields a value for the time interval $\tau_3-\tau_1$ that is to be compared with the observed value. The correct value of ρ_1 can be established by interpolation (and extrapolation) among the results of further trials.

Olbers formalized the interpolation process into what is frequently known as the "method of the variation of geocentric distances". This involves the examination of two small quantities, for example the differences between the observed and computed values of $\tau_3-\tau_1$ and between the observed and computed ecliptic longitudes at the second observation, that correspond, not only to some initially assumed values of ρ_1 and ρ_3, but also to the effect of changing ρ_1 and ρ_3 in turn by small increments. By elementary calculus, improved values for ρ_1 and ρ_3 then readily follow from the two increments and the three sets of residuals.

The popularity – even today – of Olbers's method is due to the careful and enthusiastic exposition of it by Olbers, Johann Franz Encke (1791–1865), and numerous later authorities. Olbers acknowledged that the linear two-variable interpolation procedure had previously been employed by Laplace in modifying a comet's perihelion time and perihelion distance to fit the observed longitude and latitude at some time. That all the other basic features of Olbers's method had earlier been espoused by Dionis du Séjour seems to have escaped attention for almost a century, until this fact was pointed out in 1883 by W. Fabritius.

Gauss and the *Theoria motus*

As the nineteenth century dawned there were thus essentially two workable methods for the determination of parabolic orbits for new comets – those broadly attributed to Laplace and Olbers. Although Laplace's method could be utilized for elliptical orbits, the generally poor determination of the necessary derivatives caused difficulties in practice.

Before the nineteenth century there had been little reason for the direct computation of general orbits. The rather small departure of Halley's Comet from parabolic motion could be measured directly from the observed interval between successive passages through perihelion, just as the sizes of the orbits of the known planets were derived by Kepler's third law from observations covering innumerable revolutions. The only significant exception was afforded by the first comet of 1770, for which Anders Johan Lexell, working at St Petersburg under the guidance of Euler, eventually succeeded in obtaining a viable elliptical solution (given in the *Acta* of the St Petersburg Academy for 1777). Even the discovery of Uranus in 1781 caused no difficulty, for when a parabolic solution was found to be untenable a simple circular solution proved to be quite adequate to prevent this slow-moving planet from becoming lost.

Giuseppe Piazzi's discovery of Ceres in 1801 brought to a head the need for a satisfactory method for determining a general elliptical orbit. The discoverer made twenty-four positional observations of Ceres during the course of the first six weeks of the year, and although Johann Karl Burckhardt at Paris made an attempt (published in the *Monatliche Correspondenz* later that year) at an elliptical solution, it was clear that neither it nor the various available parabolic and circular solutions would yield a prediction accurate enough to guarantee the recovery of this object – believed in mid-1801 to be the single, long-sought-for planet between Mars and Jupiter – when it reappeared in a very different part of the sky following its conjunction with the Sun. As is well known, the mathematical prodigy Carl Friedrich Gauss (1777–1855) (Figure 25.1) came dramatically to the rescue, and Ceres was reobserved at the end of the year within half a degree of his prediction.

The procedures utilized by Gauss in 1801 were extensively modified and perfected during the eight years that elapsed before he published the celebrated *Theoria motus*. As to just what Gauss did in 1801, misconceptions have persisted; it has been asserted, for instance, that "we do not know how Gauss actually computed the orbit of Ceres", and also that the methods he used were based on least squares. Clues in the *Theoria motus* – as well as in contemporary publications and documents – show that Gauss was quite reluctant to use least squares in the orbit-determination process.

Much of the third section of the second book of the *Theoria motus* is devoted to the method of least squares and the theory of the distribution of errors. Gauss noted that he had used the method as early as 1795, and the earliest least-squares computation extant seems to be his four-variable application in 1799 to the equation of time. A full six-variable application to the adjustment of an orbit involves more work than even a human supercomputer such as Gauss would undertake lightly, and most of the adjustment he made was simply a two-variable interpolation procedure like the method of

25.1. Carl Friedrich Gauss at the age of 63.

the variation of geocentric distances. The most extensive section of the *Theoria motus* is devoted to a discussion of possible choices of variables P and Q, say, for which initial values are to be adopted, and then of the small quantities, P' and Q', that depend on them. The discussion includes a generalization to the use of three independent sets of P and Q, rather than the variation of P and Q separately for the second and third sets.

The presentation of Gauss's Ceres predictions by Franz Xaver von Zach in the *Monatliche Correspondenz* in 1801 shows that Gauss made separate three-observation solutions from Piazzi's observations on 2 and 22 January and 11 February, and on 1 and 21 January and 11 February. From Gauss's notebooks it is clear that Gauss used an interpolation procedure in which P and Q were taken to be Ceres's orbital inclination I and longitude of the ascending node Ω. If these are defined with respect to the origin and reference plane in which x_i, y_i, and z_i are the planet's heliocentric coordinates (at time t_i), it follows that

$$\begin{aligned} r_i \cos u_i &= x_i \cos \Omega + y_i \sin \Omega \\ r_i \sin u_i &= -x_i \sin \Omega \cos I + y_i \cos \Omega \cos I + z_i \sin I \\ 0 &= x_i \sin \Omega \sin I - y_i \cos \Omega \sin I + z_i \cos I, \end{aligned} \quad (16)$$

where u_i is the argument of latitude at t_i, or the angle subtended at the Sun by the arc from the ascending node to the planet. On substitution from Equation (2), it follows that the third of Equations (16) can be used to yield ρ_i, and the first two can then be used to give r_i and u_i. The area of the triangle encompassing the Sun and the positions of the planet at the first and third observations then follows from r_1 and r_3 and the angle $u_3 - u_1$. With the help of the sector–triangle ratio y_{13} Gauss converted this to the sector area, which is equal to $2(\tau_3 - \tau_1)\sqrt{p}$, p being the semilatus rectum of the orbit, permitting the computation of p and then the remaining orbital elements. From these elements he could derive dynamical values for r_0 and u_0 and take for the functions P' and Q' required in the adjustment process the differences between these and the geometrical values computed from Equations (16).

Adopting values of approximately 11° and 81° for I and Ω, respectively, and taking the first and third observations to be those of 1 January and 11 February, Gauss's trial computation gave $r_1 = 2.695\,106$ AU, $r_3 = 2.664\,372$ AU, $u_3 - u_1 = 9°.268\,87$, whence Equation (6) indicated $y_{13} = 1.004\,286\,41$; and with $\tau_3 - \tau_1 = 0.703\,478$ this meant that $p = 2.726\,286\,5$ AU. Aware that Equation (6) was an inadequate approximation for the sector–triangle ratio, Euler, in collaboration with Lexell, had refined it in their *Recherches et calculs sur la vraie orbite elliptique de la comète de l'an 1769 et son temps périodique*, published in St Petersburg in 1770. Application of the improved formula to Ceres gave $y_{13} = 1.004\,299\,43$, $p = 2.726\,357\,2$ AU. Recognizing that p would still be good to only six significant figures, Gauss developed and utilized in 1801 an improved procedure that is essentially that given as Formulas I–V of Article 86 of the *Theoria motus*; for Ceres this yielded $y_{13} = 1.004\,299\,82$, $p = 2.726\,359\,3$ AU. The Euler–Lexell formula is given by Gauss as Formula VI.

Around the end of 1805, evidently in the course of the investigation of the hypergeometric function that he published in 1813, Gauss solved the sector–triangle problem completely and exactly. His elegant solution occupies Articles 88–105 of the *Theoria motus* and gives in the Ceres case $y_{13} = 1.004\,299\,85$, $p = 2.726\,359\,5$ AU, which are correct to the last figure stated.

What is normally termed "Gauss's method" of orbit determination is contained in Articles 131–163 of the *Theoria Motus*, and except for his continued use of the extended variation of distances procedure, it is very different from the method he utilized in his 1801 work on Ceres. As in the Dionis du Séjour – Olbers method, the basic Gauss orbit-determination method starts from the vector Equation (3). This time, however, it is the first and third terms on the left-hand side that are eliminated, yielding an expression of the form

$$\rho_0 = S_0 - c_1 S_1 - c_3 S_3. \qquad (17)$$

By adopting as his basic variables

$$P = \frac{c_3}{c_1}, \quad Q = 2r_0^3(c_1 + c_3 - 1), \qquad (18)$$

Gauss showed that Equation (17) can be written in the form of Equation (8), with

$$\rho_0^{\circ} = S_0 - S, \quad \gamma_0 = \tfrac{1}{2}\frac{SQ}{S_0 - S}, \qquad (19)$$

where

$$S = \frac{S_1 + PS_3}{1 + P}. \qquad (20)$$

Gauss's first approximations for P and Q were

$$P = -\frac{\tau_1}{\tau_3}, \quad Q = -\tau_1\tau_3, \qquad (21)$$

that for P being obvious from Equation (4), while that for Q requires development from Equations (11) and (12) and in the terms of higher order the assumption that $\tau_1 = -\tau_3$. For P' and Q' Gauss simply took the difference between the improved and the initial values of P and Q (or rather their logarithms). The computation could then be repeated with P and Q as the improved values. In a short-arc computation two or three trials might be all that would be required to reduce P' and Q' to zero, but if more than three trials were necessary, use could be made of the generalized interpolation process mentioned earlier.

By utilizing Equation (8) Gauss was effectively acknowledging Lambert's theorem, and indeed, for short-arc orbits $\gamma_0 \sim R_0^3$. Gauss's approach to the problem is much more general, however, and it is applicable over much longer arcs, where γ_0 may be very different from R_0^3. Rather than attempt to solve the equivalent of Lagrange's eighth-degree equation for r_0 (which, contrary to popular belief, does not necessarily include the solution $r_0 = R_0$), Gauss ingeniously transformed it into what is sometimes called Gauss's equation,

$$\alpha \sin^4\beta_0 = \sin(\beta_0 - \chi), \qquad (22)$$

where the single unknown is the object's phase angle β_0 (that is, the angle at the object in the Sun–Earth–object triangle). Of the possible values of β_0 the one of interest, at least for an object reasonably near opposition, is the one that is less than ϵ_0, the object's elongation from the opposition point.

The *Theoria motus* contains an enormous amount of insight into both the two-body problem and the treatment of observations, and it is still worth serious scrutiny almost two centuries after its publication. Concerned about the indeterminacy that can all too easily arise in a practical orbit solution from three observations, Gauss provided a modified procedure that would utilize four observations. To specify the constant k that converts time intervals in days to those in the convenient canonical units may seem a trivial point, but this introduced a simplification not present in eighteenth-century literature. Gauss's numerical value for k, namely 0.017 202 098 95, was based on the values, now obsolete, that he used for the various physical quantities from which it is derived. Yet the same value continues to be used today, for the practice, having proven to be immensely useful, has become traditional; thus Gauss's number is now taken as yielding the definition of the astronomical unit. Gauss's clever procedure for handling Kepler's equation (and its hyperbolic equivalent) in nearly parabolic orbits by means of a particularly efficient iterative expansion about the exact parabolic case is another beautiful illustration of what has been described as "the perspicacity that marked his genius", and it is still very useful today.

Gauss addressed at length various possibilities for incorporating into the computations the effects of aberration and parallax. In favouring the idea of removing the annual aberration from the observed positions and predating the observed times, and in hinting at the use of topocentric rather than geocentric solar coordinates, he was more than a century ahead of almost all other workers in the field.

Other nineteenth-century contributions

Since Gauss obtained such a complete and elegant solution to the problem of determining an orbit from three observations, and since he also discussed at length the rationale for the incorporation of additional observations, any subsequent contribution to the orbit-determination process is necessarily an anticlimax. Much of this later work was designed to simplify the practical computation of an orbit by the primitive means then available.

In 1806 the Paris mathematician Adrien-Marie Legendre (1752–1833) published his *Nouvelles méthodes pour la détermination des orbites des comètes*. This work is of interest because it includes the earliest published account of the method of least squares, even though there is no consideration of the theory of errors. Like Gauss, Legendre did not advocate a rigorous least-squares solution from a handful of observations, but he did recommend a simplified computation along these lines. Legendre's approach was to calculate the effect on the computed ecliptic longitudes and latitudes of small incremental changes in the initially assumed elements. In modern parlance, he set down the linearized equations of condition and approximated the various partial derivatives by simple differences. To simplify the solution he assumed that the equations corresponding to one of the observations would be satisfied exactly. This meant that the corrections to two of the orbital elements would be expressed in terms of those to the remaining elements. The actual least-squares computation was therefore restricted to the computation of the corrections to the remaining elements, which in the case of an exactly parabolic orbit would imply only three unknowns.

The method Legendre used to obtain his initial set of parabolic elements is also of interest because, although it was not particularly accurate, it seems to have been the earliest method that utilized the f and g expressions in a direct manner. From Equations (10) and (2) it follows that the components of the velocity v_0 can be expressed as a linear combination of ρ_0 and, say, ρ_1. These can be modified by the use of Equations (11) and (14). On calculating the kinetic energy in a parabolic orbit and equating it to the potential energy one can write $1/r_0$ as a quadratic expression in ρ_0, which can then be solved together with Equation (9) to obtain r_0 and ρ_0 and thence the first approximation to the orbit.

A useful modification of Olbers's method was made by James Ivory in 1814, while Encke later improved it, at least for logarithmic computation – by reformulating Equation (7) along the lines of

Equation (15) but in terms of a rapidly convergent power series.

Pursuing Legendre's work, J. W. Lubbock in 1830 noted that $1/r_0$ in a parabolic orbit is also proportional to the second derivative of r_0^2. Equating the expansions for the two, he was able to obtain ρ_0 from a single quadratic equation. However, Lubbock's approximation is valid only when $\tau_1 = -\tau_3$. As already remarked, this assumption of equal time intervals also appears in Gauss's first approximation for P and Q. In the Gauss procedure, however, these initial values are then rigorously corrected. In any case, Encke showed in 1851 that Gauss's initial approximation can be generalized simply by replacing the second of Equations (21) by

$$Q = -\tau_1\tau_3\left(1+\tfrac{1}{3}\frac{P-1}{P+1}\frac{S_1-S_3}{S}\right). \qquad (23)$$

Encke also produced a three-term expansion for the logarithm of the sector–triangle ratio, and he examined at length the practical solution of Equation (22). He noted that there can be no meaningful solutions for β_0 unless α and χ are contained within specific limits and that there may be two meaningful solutions when $\epsilon_0 > 63°.4$.

In 1876 Friedrich Tietjen formulated a differential procedure for solving Equation (22), and he expressed the sector–triangle ratio with the aid of an auxiliary trigonometric relationship. A popular procedure introduced by P. A. Hansen in 1863 developed the sector–triangle ratio as a continued fraction.

The best-known late-nineteenth-century contribution to the subject of orbit determination is that made in 1889 by Josiah Willard Gibbs (1839–1903), professor of mathematical physics at Yale, in his "On the determination of elliptic orbits from three complete observations", which appeared in the *Memoirs* of the National Academy of Sciences. The 'Gibbs ratios' yield c_1 and c_3 in terms of the heliocentric distances and the time differences expanded out to the fourth power. Of more lasting value, however, was Gibbs's introduction of vectors. Gauss was supreme in both his rigorous treatment of the dynamics of the situation and his statement of the relevant equations in a form convenient for logarithmic computation; if there is to be any adverse criticism of Gauss, it must concern his treatment of the geometry, which his brilliant algebraic manipulations could render quite obscure. By recasting much of Gauss's work in vector form Gibbs paved the way for twentieth-century thought on the subject.

Gibbs is sometimes credited with the introduction of the unit vectors **P** and **Q** directed from the Sun to the perihelion point of an orbit and to the point in the orbit 90° ahead of perihelion. The use of these particular vectors (which can alternatively be regarded as two rows in a 3 × 3 matrix) represented a great conceptual advance in the problem of transferring the position and velocity components of an orbiting object from one reference system to another. H. Buchholz incorporated their use by Gibbs in his revision of 1899 of the *Theoretische Astronomie* of Wilhelm Klinkerfues, but otherwise they are not discussed in any of the standard nineteenth-century texts – or even in the authoritative *Die Bahnbestimmung der Himmelskörper* (1906) of Julius Bauschinger. However, J. C. Watson in his *Theoretical Astronomy* of 1868, in recognizing that equatorial values of the position components in an elliptical orbit can be conveniently specified in terms of the eccentric anomaly, came close to using the unit vectors, and Klinkerfues's original edition of 1871 briefly mentions them without explanation. The vectors were in fact originally given in the *Annales de l'Observatoire de Paris* in 1855 by U. J. J. Le Verrier, who expressed them in terms of Ω, I, and the argument of perihelion in the case where the ecliptic is taken as the reference plane. The corresponding expressions with respect to the equator were finally given by C. E. Adams in 1922. Emphasis on logarithmic computation undoubtedly precluded much early use of Equation (10), and the 'closed expressions' for f and g in terms of differences in eccentric anomaly were first published at a surprisingly late date – namely by Franz Kühnert in 1879.

Laplace's method continued to have many adherents throughout the nineteenth century, among them G. B. Airy and James Challis. The still older methods of Lagrange, Newton, and Lambert were revived, respectively, by P. de Pontécoulant in 1829, Emile Plantamour in 1839, and J. Glauser in 1889. In the late 1840s the end of the hiatus in the discovery of new minor planets yielded increased interest in orbit determination, and there were particularly extensive writings during the next decade by A. L. Cauchy, W. C. Goetze, J. A. Grunert, A. de Gasparis, and Y. Villarceau. A. Cayley, in a memoir of 1871, included consideration of the curves described by the pole of an orbit as a function of variations in some of the elements.

Concern about multiple solutions and indeterminacy led to further developments on orbit determination from four observations, notably by Theodor von Oppolzer in his *Lehrbuch zur Bahnbestimmung der Kometen und Planeten* of 1870, and by Adolf Berberich in a paper of 1902. As for fitting orbits to more extensive series of observations, the four discoveries of minor planets early in the century rapidly yielded a wealth of raw material. The study of the motion of Pallas was particularly difficult, owing to high values of both inclination and eccentricity. It was in his *Disquisitio de elementis ellipticis Palladis* of 1810 that Gauss made his principal practical application of least squares to orbital improvement. Here, however, he simplified matters by separating the corrections to I and Ω from those to eccentricity and longitude of perihelion, and he calculated some partial derivatives analytically and others by approximating them by differences. Utilizing single equations of condition for the mean anomaly and for the heliocentric latitude at each opposition, he linked six oppositions during 1803–09 in three groups of four. In a separate memoir, Gauss made a step-by-step integration of the effect of planetary perturbations on Pallas up to 1811 and included a seventh opposition in his least-squares solution for the orbit.

In the mean time Friedrich Wilhelm Bessel (1784–1846) in his *Untersuchungen über die scheinbare und wahre Bahn des in Jahre 1807 erschienenen grossen Kometen* of 1810 had made a six-unknown least-squares determination of the orbit of the comet of 1807, using six normal places representing 70 observations spanning six months and also allowing for perturbations. Since the orbit of this comet was nearly parabolic, Bessel had to modify the equations given by Gauss for the differential effect of adjustments to the elements of an elliptical orbit on geocentric ecliptic longitude and latitude. For this purpose he used his own expansion of near-parabolic motion about the parabolic case. While less inspired than Gauss's near-parabolic procedure, Bessel's form is in many respects more convenient for this purpose; his expansion was modified and extended by F. F. E. Brünnow in 1858. In a memoir of 1816 on the orbit of Olbers's comet of 1815, Bessel first set down and utilized equations for differential correction in terms of right ascension and declination.

By the 1830s several astronomers were regularly making least-squares orbit improvements. Thomas Clausen in 1831 wrote the equations of condition in terms of small rotations about the axes, and Le Verrier in 1846 expressed concern at the need to calculate partial derivatives to as many as five significant figures when the initial uncertainty of an orbit was rather large. He showed that only two figures were needed if one selected for the six orbital 'elements' the geocentric longitudes and latitudes corresponding to three of the normal places, although he did not indicate how he calculated these partial derivatives.

A method of orbit improvement given by M. A. Kowalski in 1859 separated the corrections involving the plane of the orbit from those connected with the object's position in this plane, but his paper did not become widely known. Tietjen in 1877 presented a rather similar method that became particularly popular. The most thorough treatment of the differential-correction process was published by Eduard Schönfeld in the *Astronomische Nachrichten* for 1885. This involved the simultaneous solution for corrections to all six elements, but the partial derivatives of the right ascension and declination with respect to the elements were derived in an eminently convenient and systematic manner, both for an ellipse of low to moderate eccentricity and for the near-parabolic case by the Gauss procedure. Expressions for variations of orbital elements were also set down in the 1880s by J.-C.-R. Radau in 1887 and by L. Schulhof in 1889. H. Kloock, in a paper of 1893, revived Le Verrier's point about the precision of the partial derivatives. P. Harzer in 1896 also indirectly addressed this problem by introducing a differential-correction process into his adaptation of the Laplace method. Harzer used the f and g series to approximate the partial derivatives required when the corrections are made to the position and velocity components at t_0. Later, Henri Poincaré and A. O. Leuschner carried out to even greater lengths procedures for correcting the Laplace first approximation.

Twentieth-century postscript

Following the introduction of mechanical calculating machines, C. Veithen in 1912 and Gerald Merton in 1925 independently modified the Gauss method so that the use of Equations (18)–(22) was bypassed, an expression having the form of Equation (8) being obtained directly from Equation (17) by means of the 'Encke' approximation that led to Equation (23). The Veithen–Merton method, as well as a Lagrange-type method given by Alexander Wilkens in 1919 and a version of Olbers's method presented by Tadeusz Banachiewicz in 1925, were featured in G. Stracke's textbook of 1929, *Bahnbestimmung der Planeten und Kometen.*

F. R. Moulton in 1914 proposed a method of orbit determination in which Equation (2) was substituted into the left-hand side of Equation (10), but he was unable to give a satisfactory procedure for solving the resulting nine equations. This matter was independently rectified in a paper by the Finnish astronomer Y. Väisälä in 1939, and in the unpublished Harvard University Ph.D. dissertation of L. E. Cunningham (1946).

These authors continued to stress the use of approximations; and the Gibbs ratios, the Hansen continued fraction, the Encke power series for Euler's equation, and the derivation of perhaps five or more terms in the f and g series, are still given prominence in recent texts. In fact it is preferable to use Gauss's rigorous hypergeometric-function solution for the sector–triangle ratios, as well as the closed expressions, rather than the series development as in Equations (11), for f and g.

In 1993 R. H. Gooding published a completely new method of orbit determination based on his sophisticated solution of Lambert's elliptical generalization of Equation (7). Gooding's method has proven particularly useful in applications where the orbiting body might make several complete revolutions between observations.

In cases where more than three observations are to be included, the 'standard' least-squares differential-correction development is nowadays considered to be that given by W. Eckert and D. Brouwer in a paper of 1937. A modification by P. Herget in 1939 of the differential-correction procedure in terms of $\mathbf{r}_0$ and $\mathbf{v}_0$ allowed the use of the closed expressions for f and g. With modern computers, however, it is often convenient simply to revert to Legendre's procedure and approximate *all* the partial derivatives used in an orbital differential correction. A full least-squares solution is performed, of course, and the procedure can be particularly effective when perturbations are also included.

Further reading

The most important contributions to the subject of orbit determination during the eighteenth and nineteenth centuries are listed by K. Zelbr in "Bahnbestimmung der Planeten und Kometen", in W. Valentiner's *Handwörterbuch der Astronomie*, vol. 1 (Breslau, 1897), 568–73, and by R. Radau in "Bibliographie relative au calcul des orbites", *Bulletin astronomique*, vol. 16 (1899), 427–45. A bibliography of the literature from 1900 to 1929 is given by G. Stracke in his *Bahnbestimmung der Planeten und Kometen* (Berlin, 1929), 352–61. For the more recent literature, the reader can turn to the references listed in B. G. Marsden, "Initial orbit determination: the pragmatist's point of view", *Astronomical Journal*, vol. 90 (1985), 1541–5, and "The computation of orbits in indeterminate and uncertain cases", *Astronomical Journal*, vol. 102 (1991), 1539–52.

26

The introduction of statistical reasoning into astronomy: from Newton to Poincaré

OSCAR SHEYNIN

A feeling for the opposition of randomness and order is presumably older than philosophy if, as Robert Burns put it, "the best laid schemes of mice and men gang aft agley". Chance was a key concept for the pre-Socratic atomists; Aristotle gave an important if strictly limited place to the role of chance. According to some atomists, order arose out of purely chance events, while Aristotle claimed that chance pertained only to the terrestrial region, where it was the result of an intersection of two or more causally determined sequences, each with its own final cause or *telos*. "What happens everywhere and in every case," Aristotle asserted (*De caelo* 289b), "is no matter of chance." Thus the fact that all the stars in their diurnal revolutions, some on greater and others on smaller circles, move with the same angular speed, could not be the effect of chance, but indicates that they are all moved on and by the same sphere.

In medieval and early modern Europe the almost universal acceptance of the doctrine of divine creation meant that any fundamental randomness in the world was denied; and the atomism of the ancients was accordingly anathematized. Johannes Kepler (1571–1630) in a work on the new star of 1604 called randomness "an idol, ... an abuse of God ... and of the world created by Him". Yet when in his teaching on eclipses in the *Rudolphine Tables* (1627) he could not obtain good agreement between observations and his theory, Kepler supposed the discrepancies to be due to small, accidental physical variations *extra ordinem*. Insofar as the interactions of bodies were "physical" or "smelled of matter", he allowed that they might depart slightly from the patterns imposed by the Creator-God.

In the mid-seventeenth century the French priest, astronomer, and philosopher Pierre Gassendi (1592–1655) gained a following for an atomism reconciled with the doctrine of divine providence; and Isaac Newton (1642–1727) was one of those who accepted the new atomism or corpuscularianism divorced from the rule of chance. Yet Newton, projecting a view of the world as ruled by universal laws, was forced to recognize that an initially imposed order could be subject to decay: a pattern of initial conditions could deteriorate owing to the subsequent interactions of bodies.

Order was predominant, but it was mixed with irregularity or randomness. Up to the middle of the eighteenth century, the desire to discover definite rules for discriminating between randomness and divine design determined the very development of the theory of probability. The English mathematician (of French Huguenot extraction) Abraham de Moivre (1667–1754), in dedicating the first edition of his *Doctrine of Chances* (1718) to Newton, declared his intention of giving a

> *Method of calculating the Effects of Chance ... and thereby fixing certain Rules, for estimating how far some sorts of Events may rather be owing to Design than Chance ... so as to excite in others a desire of ... learning from your [i.e., Newton's] philosophy how to collect, by a just Calculation, the Evidences of exquisite Wisdom and Design, which appear in the Phenomena of Nature throughout the Universe.*

That there was a Newtonian 'philosophy' concerning proofs of design is shown also by a letter which the Rev. William Derham wrote to John Conduitt in 1733, in which he speaks of

> *a peculiar sort of Proof wch Sr Is: mentioned in some discource wch he and I had soon after I published my Astro-Theology [1715]. He said that there were 3 things in the Motions of the Heavenly Bodies, that were plain evidences of Omnipotence and Wise Counsel. 1. That the Motion imprest upon those Globes was*

Lateral, or in a Direction perpendicular to their Radii, not along them or parallel with them. 2. That the Motions of them tend the same way. 3. That their orbits have all the same inclination.

Thus the near-circularity of the planetary orbits, the uniform direction in the motion of all the planets (counterclockwise as seen from the ecliptic's north pole), and the smallness of all the orbital inclinations to the ecliptic, were, in Newton's opinion, evidences of design.

Earlier, the main fields to which probability theory had been applied were population statistics and games of chance. Now, under Newton's influence, probabilistic reasonings would be applied to the "wonderful uniformity in the planetary system". Even Pierre Simon Laplace (1749–1827), despite his celebrated avoidance of the divine, would continue the trend, proclaiming nature's lawfulness as contrasted with the disorder of randomness. With Daniel Bernoulli and Laplace, the reasonings became explicit calculations.

The inclinations of the planetary orbits

In 1735 Daniel Bernoulli (1700–82), mathematician, physicist, and physician of Basel, argued that there must exist a general cause for the fact that all the planetary orbits lie nearly in the same plane. Here is his reasoning: Let the inclination of the orbit of planet i to the ecliptic be a_i, where $0° < a_i < 90°$, and $i = 1, 2, \ldots, 5$ (only 5 planets were considered since the sixth known planet, the Earth, was by definition in the ecliptic). Further, let A be any number of degrees less than 90. If the values of the a_i were independent and uniformly distributed, the probability that $a_i < A$ for any of the five planets would be $A/90$, and the probability that all were less than A would be $(A/90)^5$. But the latter probability becomes negligible if A is much smaller than 90°, as it is in fact. Consequently, the quantities a_i cannot be independent, and must therefore be the result of a general cause.

Actually, Bernoulli had supposed that the unperturbed situation was one in which all the planetary orbits would be in the same plane, and that the small inclinations of the several planetary orbits to the ecliptic were occasioned by random disturbances of some kind. He also believed that a random event having a sufficiently small probability of occurrence was effectively impossible – a principle that underlies most applications of probability theory to this day.

The direction of motion of planets and satellites

In 1776 Laplace, like Newton before him, argued that the movement of all the planets and of their satellites (fifteen such bodies in all were known at this date) in the same direction was inconsistent with the assumption that their directions were the effect of chance. Indeed, he claimed that under the latter assumption the probability of the observed uniformity of direction would be only 2^{-15}, and thus negligible.

Laplace's assumption that direct motion has a probability of $\frac{1}{2}$ was in fact arbitrary, although he did not recognize it to be so. If we suppose that the probability of direct motion is 0.9 (say), so that retrograde motion has a probability 0.1, the probability of direct motion of all the bodies would be $(0.9)^{15} \approx 0.2$; and thus a random occurrence of the phenomenon would not be ruled out. The formation and testing of statistical hypotheses is a more delicate operation than Laplace implied. Laplace's reasoning here was to serve as underpinning for his celebrated *nebular hypothesis*, first proposed in 1796, to explain the origin of the solar system (see Chapter 22).

As later discoveries were to show, not all bodies in the solar system revolve in the same direction. In 1787 William Herschel discovered two satellites of Uranus (Titania and Oberon) and, in 1797, he ascertained that they move in the retrograde direction. In 1851 William Lassell discovered two more satellites of Uranus (Ariel and Umbriel), which are also retrograde. However, the orbital planes of all four were in line with the planet's equator, and inclined by 98° to the ecliptic, so that with a slight change in angle the motions would have been described as direct. Triton, a satellite of Neptune discovered by Lassell in 1846, proved to have a retrograde motion, with an orbit inclined at 35° to the ecliptic. Of more recently discovered satellites, four of Jupiter's and one of Saturn's have retrograde motions, on orbits inclined to the ecliptic at various angles less than 35°.

Comets

In 1776 Laplace undertook a mathematical study of the inclinations of cometary orbits to the ecliptic. A senior colleague in the Paris Academy, Achille-

Pierre Dionis du Séjour (1734–94), had posed the question whether the cause that produced the planetary motions – whatever it might be – also produced the phenomena of comets. He had calculated the mean inclination of the 63 known cometary orbits and found it to be 46° 16′. The departure from 45° seemed so small as to be insignificant. Moreover, the ratio of forward to retrograde motions was about 5:4, not very different from 1:1. Dionis du Séjour concluded that the planetary system constituted a distinctive causal ordering, whereas the cometary orbits followed a principle of indifference. Laplace wanted to calculate the degree of certainty of this conclusion.

Assuming the comets to have been originally projected randomly into space, what would be the probabilities, Laplace asked, for the mean inclination of their orbits, and the ratio of direct to retrograde revolutions, to fall within given limits? If the mean inclination were $45° + \alpha$, and on the assumption of a uniform random distribution (there were very large odds that $|\alpha|$ should have a small value), then it could be concluded with *vraisemblance* that some particular cause accounted for the comets moving in one particular plane. Taking up in succession the cases of 2, 3, 4, . . . , 12 comets, Laplace constructed the curves representing the probability for the mean inclination to have any value between 0° and 90°. By determining the fraction of the total area under a curve falling between given limits, he could compute the probability that the mean fell within these limits. Thus, the odds for the mean inclination of 12 comets to fall between 37°.5 and 52°.5, he found, were 678 to 322. Was there any evidence, he asked, for a cause tending to make the comets move in the ecliptic? The mean inclination of the 12 most recently discovered comets was 42° 31′; the odds for this mean to exceed 42° 31′ seemed to him too small to reject the independence of the inclinations. He did not go on to apply his reasoning to all 63 of the comets for which orbits had been computed. The calculations failed to provide a conclusive answer as to the dependence or independence of the inclinations. Laplace renewed his attack on the problem in 1810 and again in 1812, but again without achieving a convincing analysis.

In 1792 the German astronomer J. E. Bode claimed that the inclinations of cometary orbits were distributed with near-uniformity. He compared a strictly uniform distribution (the inclinations being ranked as to size and differing by a constant angle) with the actual distribution, and maintained that the differences were negligible. However, he was unable to offer a quantitative estimate of the degree of fit.

Clustering of stars: the work of John Michell

The Rev. John Michell (*c.* 1724–93) was the first astronomer to attempt to calculate the probability of two stars being within a given apparent angular distance of one another. This topic belongs in principle to Volume 3, but we include it here because of the techniques employed.

Michell's first and chief foray into this subject was made in "An Inquiry into the probable Parallax, and Magnitude of the fixed Stars . . .", which appeared in the *Philosophical Transactions* of the Royal Society of London in 1767. Earlier Michell had been a fellow of Queens' College, Cambridge, and had written on artificial magnets, the causes of earthquakes, and topics in surveying and navigation. In 1764 he resigned his fellowship for marriage and an ecclesiastical living, and from 1767 onwards held a rectorship in Thornhill, Yorkshire.

The paper of 1767 had as its announced purpose the estimation of an upper limit for the stellar parallax that astronomers might hope to discover. The method Michell used was not new with him, but had been suggested earlier by James Gregory. Michell assumed the size and intrinsic brightness of the fixed stars to be equal to the size and brightness of the Sun; the differences, he granted, might in fact be very great, but an error of 1000 in size (volume) would alter the distance by a factor of only 10. From Saturn's solar distance and apparent diameter, Michell deduced that – supposing Saturn to reflect all the solar light it receives – the Sun would have to be removed to 220 000 times its present distance from us to send back only as much light as does Saturn; at such a distance the Sun's whole parallax on the diameter of the Earth's orbit would be less than 2″. (If Saturn reflects only $\frac{1}{4}$ or $\frac{1}{6}$ of the light it receives from the Sun, as seems likely, the distance of the Sun would have to be increased in the ratio of 2 or 2.5 to 1.) Now Sirius has an apparent magnitude comparable with that of Saturn. Michell therefore concluded that the parallax of Sirius, assuming this star to be of the same size and intrinsic brightness as the Sun, was no greater than 1″. From rough experiments he judged the light received from the faintest naked-eye stars to be between $\frac{1}{400}$th and $\frac{1}{1000}$th of that from Sirius, and so inferred that the parallax of these stars, supposing them to be of the same size and intrinsic brightness as the Sun, was between 2‴ and 3‴.

We come now to Michell's application of statistical reasoning to the distribution of stars in space. If the stars had been scattered by mere chance, the probability that any particular star should be within 1° of any other given star, according to Michell, would be given by a fraction with numerator equal to the area of a circle of 1° radius, and denominator equal to the area of the whole celestial sphere; the result is $p=\frac{1}{13131}$. The probability that these same stars would *not* be within 1° of one another, Michell then reasoned, is $1-p=\frac{13130}{13131}$; and if there were n stars as bright as the stars in question, the probability that none of them was within 1° of the given star would be $(1-p)^n$. Finally, Michell inferred (erroneously) that the probability of no two such stars being within 1° of each other was $[(1-p)^n]^n$.

On this basis Michell concluded that the probability of close optical doubles was very slight. Thus he calculated that the odds against any two stars being within as small a distance of one another as the two stars of β Capricorni, assuming there to be 230 stars equal to these in magnitude, was 80 to 1. The odds against six stars being within as small a distance of one another as are the six brightest stars of the Pleiades he found to be about 500000 to 1.

From these and similar calculations, Michell drew the conclusion that

the stars are really collected together in clusters in some places, where they form a kind of systems, whilst in others there are either few or none of them, to whatever cause this may be owing ....

From the contrast between deduced probability and observed frequency, Michell inferred that attractive forces – presumably gravitational – are operative among the stars called 'fixed'. This was the first attempt to provide a detailed argument for gravitational action beyond the solar system. When in 1779 William Herschel embarked on a search for double stars to be used in parallax measures, Michell reiterated his point in a second paper, which was published in the *Philosophical Transactions* for 1784:

The very great number of stars that have been discovered to be double, triple, &c. particularly by Mr Herschel, if we apply the doctrine of chances, ... cannot leave a doubt with any one, who is properly aware of the forces of those arguments, that by far the greatest part, if not all of them, are systems of stars so near to each other, as probably to be liable to be affected sensibly by their mutual gravitation: and it is therefore not unlikely, that the periods of the revolutions of some of these about their principals (the smaller ones ... to be considered as satellites to the other) may some time or other be discovered.

Michell's argument was to be favourably reviewed by Laplace, John Herschel, and Wilhelm Struve, and came to be widely accepted. However, as pointed out by the Edinburgh physicist and geologist J. D. Forbes (1809–68) in two articles of 1849 and 1850, Michell's calculations assume that a random distribution of stars would be a *uniform* distribution, whereas a uniform distribution was quite unlikely as the result of chance. Because of this and other errors in the calculation, Forbes concluded that Michell's mathematical argument had proved nothing at all. The mathematician George Boole (1815–64) took a largely similar stand in an article of 1850.

A correct calculation of the required probabilities can be based on Poisson's distribution. Let the expected number of stars situated not farther than 1° apart be $a=pn$, where p and n have the same significations as before. Then the probability that there is at least one optical double is

$$P=1-e^{-a}(1+a).$$

For $n=5000$, we find that $a=0.3808$ and $P=0.056$. This is a small though not negligible probability, considerably smaller than the several hundreds of optical doubles discovered by William Herschel would imply. Thus Michell's final conclusion was correct in making the probability of star clusters smaller than would be deduced from their actual frequency.

The practical result of Michell's work was to convince people of the essential truth of what he said, and he was vindicated observationally by William Herschel. Re-examining his double stars soon after the turn of the century, Herschel found that some had indeed orbited around each other. By about 1830 several astronomers (including William's son John Herschel) had enough evidence to show that the orbits were Keplerian and the law therefore indeed gravitational.

Sunspots

The first telescopic observation of a sunspot was made by Thomas Harriot in 1610. Galileo exhibited sunspots to observers in Rome in the spring of 1611, and it was by means of sunspots that he showed that the Sun rotates about an axis passing through its centre. In addition, he successfully separated the regular movement of sunspots due to the Sun's rotation from the random component, that is, their proper motions relative to the Sun's disk. His estimate of the period of rotation of the Sun, namely one lunar month, was a good first approximation. Actually, the sidereal rotation period varies with latitude, being about 25 days at the Sun's equator, 27.5 days at latitude +45°, and as much as 33 days at latitude +80°: the surface of the Sun is not solid.

The first to suspect periodicity in the number of sunspots may have been the Danish astronomer Peder Nielsen Horrebow (1679–1764), who entered a remark about it in his diary. William Herschel in 1801, drawing on data for the period 1650–1717, attempted to determine the relationship between sunspots and the price of wheat, the latter being, presumably, an index of the weather.

In 1844, after studying the occurrence of sunspots for about eighteen years, the German amateur observer Samuel Heinrich Schwabe (1789–1875) announced that their number varies with a period of about ten years; but the future alone, he prudently added, would show whether this period was constant. His yearly data included the number of days of observation, the number of days when no sunspots were seen at all, and the number of groups of sunspots. Schwabe's work appears to have passed unnoticed until Alexander von Humboldt described it in his *Kosmos* (1850).

In 1859 the astronomer Johann Rudolf Wolf (1816–93) of Zurich collected all observations of sunspots beginning with the middle of the eighteenth century, determined the epochs of their extreme numbers, and so derived an estimate of the periodicity of their occurrence, $T=11.1$ years. Twenty years later he put on record observations covering 120 years and again calculated the period, testing, in the course of doing so, nineteen hypotheses ($T=9$ years 6 months; 9 years 8 months; 9 years 10 months; ...; 12 years 6 months). He considered the deviations of the mean yearly data from the general mean number of sunspots and applied as a test the range of the deviations or, alternatively, the root of the sum of their squares divided by the number of deviations. He concluded that there were two periods: $T_1=10$ years, and $T_2=11.3$ years, their least common multiple (the single common period), as he himself noticed, being 170 years.

Today it is generally accepted that there is an approximate period of about 11 years, but a strict periodicity is denied.

The influence of sunspots on terrestrial magnetism and climate was discovered in the middle years of the nineteenth century. In 1840 Johann von Lamont (1805–79) established a magnetic observatory at Bogenhausen near Munich, and ten years later he announced that the Earth's magnetic field is subject to variation. Wolf later correlated this variation with the cycle of sunspot activity, but without supplying a quantitative measure of the correlation, such measures being as yet unknown.

In 1874 the Scots meteorologist Charles Meldrum announced that

not only the number of cyclones, but their duration, extent, and energy were also much greater in the [years of maximum] than in the [years of minimum sunspot frequency], and ... there is a strong probability that this cyclonic fluctuation has been coincident with a similar fluctuation of the rainfall over the globe generally.

At about the same time the English astrophysicist J. N. Lockyer voiced a similar opinion regarding cyclones, while in 1880 the head of the Meteorological Department of India, the Englishman H. F. Blanford, pointed out a connection between atmospheric pressure and sunspots. But all these authors, lacking as they did a mathematical theory of correlation, were able to give only qualitative comparisons.

The founder of the theory of correlation was the London gentleman-scientist Francis Galton, in a paper on "Co-relations and their measurement, chiefly from anthropometric data", read to the Royal Society in December 1888. The theory was developed mathematically by the Irish statistician Francis Ysidro Edgeworth during the 1890s, and made widely known in its biological applications by Karl Pearson (1857–1936) of University College, London. Between 1908 and 1910 Pearson, partly in collaboration with the statistician, astronomer, and medical researcher Julia Bell, published a number of papers in astronomical journals, applying correlation theory and the Pearsonian test for goodness of fit. They were concerned with the connection between the colours of stars and their spectral classes and magnitudes.

The asteroids

Because of their large number, the asteroids or minor planets constitute the most representative instance of a statistical population in the solar system. The first four were discovered between 1801 and 1807 (see Chapter 24). Since the asteroids were small, astronomers were convinced there must be more. To facilitate their discovery, the Berlin Academy sponsored the construction of star maps of the zodiacal zone containing all stars down to the ninth magnitude. A fifth asteroid was discovered in 1845, and, with the aid of the new star maps, from 1847 onward at least one was discovered each year. By 1852 the number had risen to 20, and by 1870 it reached 110. Since 1891 most of the discoveries have been made photographically (see Chapter 28), and several thousands have been catalogued.

In the 1860s Simon Newcomb (1835–1909), astronomer of the US Naval Observatory in Washington, DC, and a leading celestial mechanician of his day, devoted a number of papers to the study of the asteroids with a view to testing for group properties – characteristics that might apply to asteroids yet to be discovered, or that might indicate a common origin. Thus in 1862, from the Lagrangian formulas for the secular variations of the orbital elements of these bodies, and assuming uniform distribution of certain constants dependent on the totality of the perturbations, he derived probabilities for the perihelia and nodes to fall into each of the four quadrants of the ecliptic. The perihelia, he found, fell within the quadrant containing the perihelion of Jupiter, and the nodes within the quadrant containing the node of Jupiter, with greater probability than in the remaining quadrants – an expected result, since Jupiter is the greatest perturber of the motions of asteroids. In 1869 he compared the actual distribution of perihelia and nodes with a uniform distribution, and attempted to investigate the possibility that the actual distribution had diverged from a uniform distribution; but he lacked the statistical modes of analysis requisite for such a task.

In 1900 Newcomb returned to the asteroids to test whether the 'Kirkwood gaps' might have arisen by chance. In 1857, and again in 1866, the American mathematician Daniel Kirkwood (1814–95) had remarked that the periods of revolution of the then known asteroids were not uniformly distributed about their mean of 4.7 years; there were gaps at $\frac{1}{3}$, $\frac{2}{5}$, and $\frac{2}{7}$ of Jupiter's period. Later Kirkwood discovered other, somewhat less distinct gaps at $\frac{1}{2}$, $\frac{3}{5}$, $\frac{4}{7}$, $\frac{5}{8}$, $\frac{3}{7}$, $\frac{5}{9}$, $\frac{7}{11}$, and $\frac{4}{9}$ of Jupiter's period. Kirkwood's explanation was that the perturbational forces due to Jupiter would not permit such commensurabilities to exist; an asteroid in one of the gaps would be "pumped" out of it by repeated application of similar forces similarly applied, or its orbit made so eccentric that it would become subject to collisions.

Newcomb selected 354 asteroids with mean motions varying from 600″ to 1000″ per day (the periods thus varied from 3.55 to 5.91 years), and divided them into forty groups according as the mean motions fell into the intervals 600″–610″, 610″–620″, . . ., 990″–1000″. Assuming a binomial distribution of the 'population' within the forty groups, Newcomb computed what he called 'the probable number' of groups having x planets each (it is more properly called the mean number). Then, from a qualitative comparison between the observed and probable [mean] numbers of groups with x planets each, he concluded – in agreement with Kirkwood – that the inequalities of distribution could not have arisen in a group of asteroids whose original periods or mean distances had been distributed uniformly.

Not only were the orbital parameters of the asteroids studied statistically, but even the very existence of any as yet undiscovered asteroids was proposed as a subject for statistical inference. The great French mathematician and mathematical physicist Jules Henri Poincaré (1854–1912) in his

Calcul des probabilités (1896, with a revised and expanded version in 1912) undertook to estimate the entire number (N) of asteroids. Suppose that M of these are known and that during a certain year n asteroids were observed, m of which had been known before, so that $N \approx Mn/m$. Let $p = n/N$ be the probability that during the same year a certain (existing) asteroid was observed; ω_i, the prior probability that i asteroids exist (Poincaré assumes that $\omega_i = \text{const}$); and p_i, the probability that n asteroids were observed given that i of them existed ($i \geq n$). In this notation, and writing $q = 1 - p$, Poincaré found

$$p_i = \frac{i!}{n!(i-n)!} p^n q^{i-n},$$

$$\sum_{i=n}^{\infty} p_i = p^n + \frac{n+1}{1!} p^n q + \frac{(n+1)(n+2)}{2!} p^n q^2 + \ldots$$

$$= p^n/(1-q)^{n+1} = 1/p.$$

The probability that there are N asteroids is then pp_N and the expected or mean value of N is

$$E(N) = \frac{n+q}{p}.$$

In his *Calcul des probabilités* and in other writings published at the beginning of this century Poincaré attempted to explain the notion of randomness, and in this connection he once again considered an example concerning the asteroids. He did not provide detailed derivations, and from a methodological point of view his analysis was imperfect; nevertheless, it constituted an appreciable contribution to probability theory and its philosophy.

Why, Poincaré asked himself, are the ecliptic longitudes of the asteroids uniformly distributed? The solar distances of these bodies differ from one another, though often by only small amounts; hence their mean motions do not coincide. If one supposes that they once constituted a single body, over millions of years they would, because of the slight differences in their mean motions, come to be scattered across the entire ecliptic.

Poincaré's proof is too intricate to reproduce here, but it is worth noting that he was dealing with a random process of a kind that later came to be included in the province of probability theory, and that his proof closely approaches that of a celebrated theorem due to Hermann Weyl in 1916 on the distribution of the fractional parts of irrational numbers on the unit interval.

Poincaré offered several explanations of randomness and chance. One was instability of motion, such as may occur in unstable equilibrium, and another was the existence of complicated causes, as in the great number of collisions among gaseous molecules, leading to a relatively uniform distribution. He also mentioned the intersection of chains of events as an explanation of chance.

Conclusion

The examples of statistical reasoning given in the present chapter belong to the prehistory of mathematical statistics. Only in the twentieth century would quantitative tests of correlation and of statistical significance become part of the standard arsenal of investigative techniques. During the period considered here, the steps taken in applying probabilistic reasoning to astronomical phenomena were initial ones, often innovative but sometimes flawed because of the inadequacy of the available analytical tools.

In one area, however, probabilistic reasoning applied to astronomy led to a result of signal and lasting importance: the theory of observational error. It is to the development of this theory that we turn in the following chapter.

Further reading

O. B. Sheynin, Newton and the classical theory of probability, *Archive for History of Exact Sciences*, vol. 7 (1971), 217–43

O. B. Sheynin, On the history of statistical method in astronomy, *Archive for History of Exact Sciences*, vol. 29 (1984), 151–99

O. B. Sheynin, H. Poincaré's work on probability, *Archive for History of Exact Sciences*, vol. 42 (1991), 137–71

27

Astronomy and the theory of errors: from the method of averages to the method of least squares

F. SCHMEIDLER

with additions by Oscar Sheynin

The problem of how best to employ observations in obtaining information about unknown quantities is of fundamental importance for astronomy. Ancient and medieval astronomers, in determining any particular quantity, appear often to have relied upon the smallest number of observations necessary to determine the quantity in question. Thus Ptolemy in the *Almagest* frequently reported only a single observation for the determination of a parameter. In such cases, however, it may be doubted whether he is recounting all the data on which he relies, or merely giving an illustration of how the determination is to be made.

In the centuries after Ptolemy, the accuracy of the observations that he reported remained almost entirely unquestioned. But the problem of observational error was not one that astronomers ignored. Thus al-Bīrūnī (AD 973–p. 1050), in a work on the correction of the distances between cities, says that random errors are inevitable "because celestial observation is a very delicate matter". In one case he chooses a particular value "because it is close to the average between the smaller amount ... and the larger amount, and because the indirect method ... produces an amount which is not far from that amount and [thus] corroborates it". Although referring to the arithmetic mean, he does not adopt it as his general rule; in each case he chooses a common-sense value for the constant sought.

Nicholas Copernicus (1473–1543) maintained toward the observations reported by Ptolemy an attitude of pious acceptance. Tycho Brahe (1546–1601), however, early came to recognize that the predictions of both the Ptolemaic and Copernican theories were often in error by several degrees (see Chapter 1 in Volume 2A), and that the only remedy was an *instauratio*, a renewal of astronomy on the basis of a regular series of accurate observations. But Tycho continued the practice of selecting from among the available observations the one he believed, on whatever grounds, to be the most suitable or trustworthy for the purpose in hand.

Johannes Kepler (1571–1630), when adjusting observations in his *Astronomia nova*, appears to have used the arithmetic mean, and also (in Chapter 10, Volume 2A) a *medium ex aequo et bono*, in which the observations seem to have been weighted according to their presumed likelihood of being in error. But during the seventeenth century such practices did not become general. The Danzig astronomer Johannes Hevelius (1611–87) used the one observation of the position of a star that he considered the best, even when he had made many measurements. And according to Francis Baily, the nineteenth-century biographer of John Flamsteed (1646–1719), England's first Astronomer Royal, Flamsteed

> *does not appear to have taken the mean of several observations for a more correct result.... [Where] more than one observation of a star has been reduced, he has generally assumed that result which seemed to him most satisfactory at the time, without any regard to the rest.*

In the middle years of the eighteenth century, two astronomers stand out as having made a regular practice of using the mean. One was James Bradley, the third Astronomer Royal. According to his own report, "When several observations have been taken of the same star within a few days of each other, I have either set down the mean result, or that observation which best agrees with it." The other, Nicolas-Louis de Lacaille, likewise regularly employed the arithmetic mean in determining observed positions as well as constants such as orbital elements derived from such positions (see Chapter 18).

Early use of equations of condition

Not till the eighteenth century did astronomers and mathematicians undertake systematic studies of how unknown quantities ought to be determined from multiple data. Suppose that a certain number of unknowns, say m, are to be determined, and further, that n observations, where $n > m$, have been made which are algebraically dependent on the unknowns. Then each observation constitutes a condition that must be fulfilled at least approximately by the solution. The equations by which the observed values and the unknowns are connected are in most cases linear; if not, they can usually be linearized. The simplest case is that in which a single unknown quantity can be measured directly; and in this case the arithmetic mean of the several measured values came to be considered the best value – the one on which to rely. During the eighteenth century several attempts were made to refine and generalize the use of the arithmetic mean for more complicated problems. These methods were gradually improved, and finally replaced by the method of least squares which was invented around 1800 and in the following decades came into general use.

The first mathematician to investigate how unknown quantities should be determined when the number of equations exceeds the number of unknowns was the ingenious Roger Cotes (1682–1716), editor of the second edition of Newton's *Principia*, and an assiduous student of Newton's methods. He considered only the case in which a single unknown quantity, x, is to be determined from n equations of condition

$$a_i x = l_i \quad (i = 1, 2, \ldots, n).$$

From each of these equations, the value of the unknown could be calculated as $x = l_i/a_i$. According to Cotes, among these results the more reliable are those for which the coefficients a_i (all taken as positive) have the larger values. Cotes therefore concluded that the best value would be given, not by the mean of the values l_i/a_i, but by the formula

$$x = \frac{l_i + l_2 + \ldots + l_n}{a_1 + a_2 + \ldots + a_n}.$$

The formula in effect gives a weighted mean, with larger weights for the values l_i/a_i having the larger a_i. Cotes's researches on the subject are contained in his *Harmonia mensurarum*, which was published posthumously in 1722.

How should one proceed if there is more than one unknown? This question was attacked some decades later by the Swiss mathematician Leonhard Euler (1707–83), and by a young cartographer and astronomer in Nuremberg, Johann Tobias Mayer (1723–62), who would later become director of the observatory in Göttingen. Their general idea was to subdivide the equations of condition into as many subsets as there were unknowns to be determined, in such a way that one or two of the unknowns in each subset had relatively large coefficients, while the others had relatively small coefficients; the terms involving the latter unknowns were then neglected, and the resulting equations solved for the unknowns they contained. After the first unknowns were determined, the remaining unknowns could be found consecutively by a similar procedure. This is the so-called 'method of averages'.

Euler's application of the method occurs in his "Recherches sur la question des inégalités du mouvement de Saturne et de Jupiter", which was the winning essay in the Paris Academy's contest of 1748, and was published in 1749. After deriving the perturbational terms in the longitude of Saturn due to the attraction of Jupiter, Euler undertook to compare his theory with some 95 heliocentric longitudes of Saturn reported by Jacques Cassini (Cassini II) in his *Élemens d'astronomie* of 1740; they covered the period from 1582 to Euler's own time. To make the comparison, it was necessary to have values for Saturn's orbital elements – epoch, mean motion, aphelion, and maximum equation of centre; and these can be determined only empirically. Provisionally, Euler adopted Cassini's values for these parameters, but recognized that they stood in need of correction, since "the inequalities caused by the action of Jupiter are there enveloped in the eccentricity and position of the orbit of Saturn". In addition, his theory included (mistakenly) a perturbational term proportional to the time, and Euler, being unable to derive its coefficient from the theory of gravitation, hoped to fix its value from the observations.

All in all, there were eight unknowns, and Euler showed that the differences between the observed longitudes of Saturn and the theoretical values could be represented by linear equations of condition in these eight unknowns. Five of the unknowns were multiplied by sines or cosines – purely periodical coefficients. In order to separate the unknowns, Euler selected pairs of observations 59 years apart, this time interval being equal, very nearly, to two sidereal periods of Saturn and five of Jupiter; in such paired observations the purely periodical coefficients were nearly equal. Subtracting one of the paired equations of condition from the other, Euler obtained an equation in which the terms with the periodic coefficients were negligible; from three such equations he could determine the unknowns having non-periodical coefficients. By other combinations of the equations he then sought to determine the remaining unknowns, and to reduce the maximum error of his theory as much as possible. But the result of these manipulations was only partially successful, because his theory contained incorrect perturbational terms and lacked other terms it should have contained.

Mayer, in undertaking to determine the libration of the Moon, made use of a procedure that was similar in principle, but different in some details. He had at his disposal twenty-seven observations of the angular distance between the crater Manilius and the apparent centre of the lunar disk. These angular distances could be expressed as linear functions of three unknowns: the inclination of the Moon's equator to the ecliptic, the node of the lunar equator on a plane parallel to the ecliptic through the Moon's centre, and the selenographic latitude of the crater Manilius. There were thus twenty-seven equations of condition for three unknowns. To reduce these equations to three, Mayer first added together the nine equations in which one of the unknowns had the largest positive coefficients; then he formed the sum of the nine equations in which the same unknown had the largest negative coefficients; and finally he added together the remaining nine equations, in which the coefficients of this unknown were comparatively small in absolute value. Thus he obtained three equations from which the three unknowns could be directly determined.

Mayer's memoir on the libration of the Moon was published in 1750 in the *Kosmographische Nachrichten und Sammlungen auf das Jahr 1748*. From Mayer's earliest letter to Euler, written in July 1751, we know that Mayer had previously made a thorough study of Euler's prize-winning essay on the inequalities of Jupiter and Saturn (see Chapter 18); but whether he learned his method of combining equations of condition from Euler's essay cannot be settled definitively. In any case his application of the method was more successful than Euler's, since he was not applying a mistaken theory. His procedure was reported in detail and with high praise in the second and third editions of J.-J. L. de Lalande's *Astronomie* (1771 and 1793), the best known textbook of astronomy during this period.

The method of Bošković

Euler's and Mayer's way of combining equations of condition produced reasonable results, but in an important respect it was unsatisfactory. The aim of forming subsets in such a way that particular coefficients were maximal in each subset guided the computation, but there was more than one way in which this aim could be realized. It was therefore possible that, with different combinations of the equations, the solutions would have been different.

It was because of this deficiency in the method that the Croatian polymath Rudjer J. Bošković (1711–87) developed a new method for combining equations of condition. At an early age Bošković had entered the Jesuit order, and studied at the Collegio Romano in Rome, where he became conversant with Newton's *Opticks* and *Principia*, and with a wide range of mathematical disciplines. In 1750 he and a fellow Jesuit, Christopher Maire, were ordered by Pope Benedict XIV to measure the length of an arc of the meridian in the Papal States, and to carry out other operations for the correction of the map of these territories. During the following three years the two of them carried out triangulations, and made determinations of the height of the celestial pole, over an arc of 2° of the meridian in central Italy. In 1755 they published an account of their results in Latin, and here Bošković did not yet make use of his new method for dealing with equations of condition. However, a short summary of this memoir appeared in the *Memoriae de Bononiensi Scientiarum et Artium Instituto atque Accademia* for 1757, and into this Bošković inserted a description of his method. He described it again in his commentary on a Latin poem concerning Newtonian natural philosophy by Benedict Stay, published in 1760. A French translation of the original work of 1755 appeared in 1770 under the title *Voyage astronomique et géographique dans l'État d'Église*, and the description of the method was here given as an appendix. Bošković did not mention the work of earlier authors on this problem, but from his essay on the perturbations of Jupiter and Saturn, which won the *proxime accessit* in the Paris Academy's contest of 1752, we know that he had read Euler's prize-winning essay of 1748.

Bošković wished to compare his result for the length of a degree of the meridian with four other determinations that had been made at different geographical latitudes. His aim was to derive a value for the flattening of the Earth: on the assumption that the Earth is a flattened spheroid, the degrees of the meridian lengthen as one proceeds from the equator towards either geographical pole (see Chapter 15). Bošković sought to represent the five determinations of a degree of the meridian by a function meeting three conditions:

(a) The difference between the length of a degree of the meridian at latitude ϕ_i and the length of a degree of the meridian at the equator should be proportional to the square of the sine of the latitude, or equivalently, to the versed sine of double the latitude (this condition holds for an ellipsoid of revolution). Thus each measurement of a degree of latitude, D_i, should fit, very nearly, a linear equation of the form

$$D_i - D_0 - (1 - x\cos 2\phi_i) = 0, \qquad (1)$$

where D_0 is the (unknown) length of a degree of the meridian at the equator, and x is another unknown dependent on the parameters of the Earth's spheroid. Evidently not all five equations of form (1) could be satisfied exactly, and the best values of D_0 and x would presumably be such that the differences between the left-hand members of Equations (1) and zero (the 'residuals', ΔD_i) were in some fashion minimized.

(b) The sum of the positive residuals should be equal to the sum of the negative residuals, or

$$\sum_{i=1}^{n} \Delta D_i = 0. \qquad (2)$$

(c) The sum of the residuals taken without regard to their algebraic signs should be a minimum, or

$$\sum_{i=1}^{n} |\Delta D_i| = \text{min}. \qquad (3)$$

Bošković's solution of this problem was mainly geometrical. He plotted the lengths of the five meridian arcs against the corresponding values of the versed sines of 2ϕ. His goal was then to fit a straight line to these points in the "best" way. First he determined the 'centre of gravity' of the points, that is, the point with coordinates given by the mean value of the versed sine of the double latitude and the mean value of the measured degrees. Because of Equation (2), the straight line would have to pass through this point. Finally, by trial and error Bošković adjusted the slope of the line in such a way as to satisfy Equation (3).

As we shall see, Laplace was later to give a purely analytic formulation of this method, and Gauss would enunciate a fundamental objection to it.

Early investigations concerning the arithmetic mean and the law of errors

In the long run, progress in minimizing the effect of observational errors would depend on theoretical investigations of a more philosophical character. Galileo in the Third Day of his *Dialogue Concerning the Two Chief World Systems* had formulated some of the basic propositions of a theory of observational errors: their inevitability, the equal probability of positive and negative errors, the accumulation of observations close to the true value, and the advisability of rejecting outliers. The first attempts to develop a theory of errors mathematically date from the middle years of the eighteenth century; they concerned the use of the arithmetic mean, and the question of the statistical distribution of errors of observation about the mean.

In the *Philosophical Transactions* for 1756 there appeared a paper by Thomas Simpson (1710–61), a mathematics teacher in a military academy, entitled "An attempt to show the advantage arising by taking the mean of a number of observations in practical astronomy"; a revised version of the paper was published in the following year in Simpson's *Miscellaneous Tracts*. The main merit of the paper was the idea that there is a law of the frequency of errors of different amounts. In the original version Simpson considered two different discrete laws of the distribution of astronomical errors. In the one, the errors

$$-v, \ldots, -3, -2, -1, 0, 1, 2, 3, \ldots, v$$

were supposed equally probable; in the other, the probabilities of these same errors were supposed proportional to

$$1, \ldots, v-2, v-1, v, v+1, v, v-1, v-2, \ldots, 1,$$

hence outlining an isosceles triangle when plotted against the errors. For each of these distributions, Simpson provided generating functions for determining the probability that the sum of the errors of n observations should equal a given integer m. He was here giving the first application of the theory of probability to the investigation of observational errors. In both of the error distributions that Simpson considered, the errors of positive and negative sign occurred with equal frequency, and there was an upper limit to the size of the errors; under these conditions, the arithmetic mean of several observations could be expected to be a good approximation to the true value. Simpson's immediate aim was to refute the opinion that one careful observation was as reliable as the mean of a large number of observations. In the revised version of the paper he considered a continuous distribution along with the discrete distributions.

Johann Heinrich Lambert (1728–77) was another contributor to the theory of errors. The son of a poor tailor, he was entirely self-educated, making intensive studies in many areas of mathematics and philosophy. In his *Photometria* of 1760, which was devoted to laying the foundations for the measurement of the intensity of light, he described the properties of observational errors, remarking that the numbers of errors of opposite sign tend to equality as the number of observations is increased. He proved that an extreme observation should be rejected when it is considerably separated from the rest, because in this way the error of the arithmetic mean is reduced. For a continuous unimodal frequency curve, $\phi(x-x_0)$, he enunciated the principle that x_0, the "most probable" value of x, be determined by the condition that

$$\phi(x_1-x_0)\phi(x_2-x_0)\ldots\phi(x_n-x_0)=\max.$$

(This is now referred to as the condition of maximum likelihood.) In most cases, he stated, x_0 would not differ from the arithmetic mean.

In later works, Lambert carried further his discussion of these matters, introducing the term 'theory of errors', and undertaking to develop this theory as a distinct discipline.

Although the arithmetic mean of several observations had come to be considered the most reasonable approximation to the true value, the first attempt to justify this assumption appeared in 1756–57 (Simpson). Then, in 1776, the mathematician Joseph Louis Lagrange (1736–1813), then head of the mathematical–physical section of the Berlin Academy, published in the *Miscellanea Taurinensia* his memoir "Sur l'utilité de la méthode de prendre le milieu entre les résultats de plusieurs observations". Here he went beyond Simpson in considering many different frequency distributions and applying generating functions to the determination of the probability of errors in each case. For the cases considered he was thus able to compute the most probable value of the error committed by taking the arithmetic mean for the true value. He then turned to the general problem of deriving the most probable correction that must be applied to the arithmetic mean if the law of the frequency of the errors is known, and showed that the arithmetic mean is the most probable value of the unknown if the law of errors is an even function and unimodal. He did not propose a formula for the law of errors.

Further development of the method of averages

The theoretical investigations described in the preceding section had no immediate effect on the practice of astronomers. For them, the method of averages remained the accepted procedure for dealing with equations of condition. Bošković's proposal that the sum of the absolute values of the residuals should be minimized was not generally followed (although Laplace would make use of it in the study of the Earth's figure: see below). But certain refinements in applying the method of averages deserve notice.

In 1787, Pierre-Simon Laplace (1749–1827) presented to the Paris Academy of Sciences his memoir on the inequalities of Jupiter and Saturn – the memoir in which, for the first time, the positions of these planets since antiquity became derivable from the theory of gravitation with only a small margin of error. In fitting the constants of his theory to observations, Laplace used a refinement of Mayer's method. Comparing his theory with twenty-four observations, he compiled twenty-four equations of condition, each having the form

$$\Delta w + a_{i1}\Delta x + a_{i2}\Delta y + \ldots = r_i$$

where Δw, Δx, Δy, ... are differential corrections to the initially assumed values of the constants, the a_{i1}, a_{i2}, etc. are sines or cosines of arguments proportional to the time of the *i*th observation, and r_i is the 'residual', that is, the difference between the observed position and the position as predicted by means of the initially assumed values of the constants. Unlike Mayer, who had added his equations of condition together in disjoint groups, Laplace combined the same equations together in several different ways, always with a view to maximizing the coefficients. Thus he added all equations in which the coefficient of one particular differential correction was large, either positive or negative, but where the coefficient was negative, he multiplied the equation through by -1 before summation. The aggregation chosen in each case was dependent on *all* the equations. Repeating the same procedure for each term, he obtained as many equations as unknowns, in each of which one term was predominant in comparison with the others. It was then easy to determine the values of the unknowns satisfying the equations.

Evidently the method would fail if, in the equations of condition, some unknowns were to have only positive (or only negative) coefficients; in this case it would be necessary to subtract the arithmetic mean of the equations of condition from each of these equations. In applications to celestial mechanics, however, where the unknowns are normally multiplied by certain values of trigonometrical functions, such a case would almost never arise.

During the late 1780s and 1790s, Jean-Baptiste Joseph Delambre (1749–1822) worked on the construction of tables of the Sun, Moon, Jupiter, Saturn, Uranus, and the satellites of Jupiter; in all cases he followed Laplace's procedure in fitting constants to observations, although he did not mention Laplace. The procedure was adopted by other astronomers, and became the standard practice in the early years of the nineteenth century. Even after the invention of the method of least squares it was still used, mainly by English authors, for instance the Astronomer Royal G. B. Airy (1801–92) and his collaborators. The latest case of its use appears to have been in the correction of B. A. von Lindenau's tables of Venus carried out by James Breen, an assistant at the Greenwich and later at the Cambridge Observatory; the corrected tables were published in 1849.

Laplace's elaboration of Bošković's method

In a memoir of 1789, and again in the third book of the second volume of the *Mécanique céleste*, published in 1799, Laplace treated the problem of the Earth's figure. First he derived a theoretical shape for the Earth, assuming that its parts had been originally fluid and subject to both centrifugal and gravitational forces. Then he undertook to compare this theoretical shape with observational determinations of lengths of degrees of the meridian, and of degrees of longitude, in different latitudes. For this purpose he had recourse to Bošković's method; that is, he stipulated that the algebraic sum of the residuals should be zero, and that their sum taken without regard to sign should be a minimum. In the memoir of 1789 Laplace attributed the method to Bošković; in the *Mécanique céleste*, Bošković's name went unmentioned.

To determine the values of the unknowns meeting the stipulated conditions, Bošković had used a geometrical construction. Laplace, in contrast, proceeded analytically, giving a long and tedious description of the algebraic steps of the process, and also providing a proof that his formulas yielded the required result.

The invention of the method of least squares

In a manuscript of the mid-1750s Simpson had posed, but did not further discuss, the following problem: to find a point *N* as the intersection of four lines drawn from four given points, in such a way that the sum of the squares of the distances from *N* to the four given points was a minimum. In 1778, in commenting on a memoir of Daniel Bernoulli, Euler suggested the principle of determining a measured value by, practically speaking, minimizing the sum of the squares of the errors, but did not pursue the implications of the principle. In the last decade of the eighteenth century and first decade of the nineteenth, a number of mathematicians independently hit upon the idea of determining the unknowns of a set of equations of condition by minimizing the sum of the squares of the residuals. This *method of least squares*, it was assumed, would yield the most probable values of the unknowns.

It seems the first to hit on the method in this form may have been Daniel Huber, who was born in 1768 in Basle, Switzerland, and became professor of mathematics at the university there in 1791. According to an obituary by Peter Merian in the *Verhandlungen der allgemeinen schweizerischen Gesellschaft für die gesamten Naturwissenschaften* (1830), Huber invented the method of least squares, but published nothing about it:

*Already in his early days (*frühen Zeiten*), through his own reflection, he discovered the method – later made known by Gauss and Legendre – of least squares, for the obtaining of the most probable results of a series of observations.*

Merian gives no date for Huber's invention of the method, but "frühen Zeiten" suggests it was before 1800.

In 1806, Adrien-Marie Legendre (1752–1833) published a book entitled *Nouvelles méthodes pour la détermination des orbites des comètes*; in an appendix bearing the date 6 March 1805, he expounded the method of least squares. This method, he stated, was in his opinion the most general and most accurate way of determining several unknowns from a larger number of equations of condition. He also showed that, in the case of a single unknown, minimization of the squares of the residuals implies the rule of the arithmetic mean. Further justification he did not supply.

In 1808, Robert Adrain (1775–1843), a teacher of mathematics and principal of an academy in Reading, Pennsylvania, and later professor of mathematics at Queen's College (now Rutgers) in New Jersey, and at Columbia College, New York, published a paper in which he derived the normal law of errors and used it to establish the principle of least squares in the same way as Gauss was to do in 1809. The paper appeared in *The Analyst*, a short-lived journal of which Adrain himself was the publisher. Perhaps he was aware of Laplace's earlier work on error distributions; whether at this point he had read Legendre's memoir is unclear (his library contained the memoir). He included a number of applications of the principle of least squares to problems of navigation and land surveying. In 1818 in the *Transactions of the American Philosophical Society* he applied the same principle to the determination of the size and shape of the Earth by the adjusting of meridian arc measurements and pendulum observations. However, his work was little known in Europe before 1872, when J. W. L. Glaisher published a review of it in the *Memoirs of the Royal Astronomical Society*.

In 1809, in the second book of his *Theoria motus corporum coelestium*, the mathematician, physicist, and astronomer Carl Friedrich Gauss (1777–1855) published a probabilistic argument for the method of least squares. He remarked that he had used the method already in 1795. His reasoning, briefly outlined, went as follows. Let V_i ($i=1, 2, \ldots, \mu$) be functions of the ν quantities p, q, r, s, etc., to be determined by observation, with $\nu<\mu$. Further, let M_i be the values of the V_i given by observation. Since the observations contain error, we cannot suppose $V_i=M_i$ exactly, but any system of values of p, q, r, s, etc. is possible which gives values of the functions V_i-M_i within the limits of observational error. Gauss's aim was to pick the most probable of these systems.

Suppose there is no prior reason to regard one observation as less reliable than another, and let the probability of an error $v_i=M_i-V_i$ in an observation be $\phi(v_i)$. Gauss assumed that $\phi(v_i)$ is a maximum for $v_i=0$, is equal for opposite values of v_i, and vanishes for v_i sufficiently large. For ease of calculation, he proposed to treat ϕ as a continuous function. Using the (Bayesian) rule of inverse probability, he argued that the probability for a particular set of values of the V_i on the assumption of given values of the M_i would be equal to the product

$$\prod_{i=1}^{\mu}\phi(M_i-V_i)=\prod_{i=1}^{\mu}\phi(v_i)=\Omega. \tag{4}$$

He then proposed to maximize (4) by setting the partial derivatives of Ω with respect to the ν unknowns equal to zero:

$$\frac{\partial\Omega}{\partial p}=0, \quad \frac{\partial\Omega}{\partial q}=0, \text{ etc.} \tag{5}$$

Equations (5), divided through by Ω, become

$$\sum_{i=1}^{\mu}\frac{1}{\phi(v_i)}\cdot\frac{\partial\phi(v_i)}{\partial v_i}\cdot\frac{\partial v_i}{\partial p}=0, \tag{6}$$

with similar equations for q, r, s, etc. The system of equations (6) will have a determinate solution if the form of the function ϕ is known.

To determine ϕ, Gauss considered a direct measurement of a single quantity, so that all the V_i reduced to a single constant p. Moreover, in this case it was generally assumed that the arithmetic mean of the measurements yielded the most probable value, and on this assumption Gauss showed that (6) implied that

$$\phi(v) = K \exp\left(\frac{kv^2}{2}\right).$$

In order that $\phi(v)$ should have a maximum when v is 0, $k/2$ must be negative, and Gauss accordingly set it equal to $-h^2$. He introduced the normalizing factor $h/\sqrt{\pi}$, and set

$$\phi(v) = \frac{h}{\sqrt{\pi}} \exp(-h^2v^2), \tag{7}$$

which brings it about that

$$\int_{-\infty}^{\infty} \phi(v)\, dv = 1,$$

so that the total probability is 1. As Gauss explained, the number h is a measure of the precision of observations. Expression (7) has come to be called the normal (or "Gaussian") density function.

The product (4) now takes the form

$$\Omega = \left(\frac{h}{\sqrt{\pi}}\right)^{\mu} \exp\left[-h^2\left(\sum v_i^2\right)\right], \tag{8}$$

and it follows that for (8) to be a maximum, $\sum v_i^2$ must be a minimum. Hence, Gauss concluded,

> that will be the most probable system of values of the unknown quantities p, q, r, s, etc. in which the sum of the squares of the differences between the observed and computed values of the functions V, V', V'', etc. is a minimum, *if the same degree of accuracy is to be presumed in all the observations.*

The principle can be generalized to the case in which the observations are of unequal precision, by making the sum $\sum h_i^2 v_i^2$ a minimum. In a final remark on the logical status of the method, Gauss affirmed that

> *this principle, which promises to be of most frequent use in all applications of mathematics to natural philosophy, must everywhere be considered as an axiom with the same propriety as the arithmetical mean of several observed values of the same quantity is adopted as the most probable value.*

Gauss further asserted that the alternative principle of Bošković and Laplace – the method of seeking a minimum of the sum of the absolute values of the errors – is unsatisfactory because it implies that ν of the μ equations are exactly satisfied, while the remaining equations influence the result only insofar as they help to determine the choice of the equations to be exactly satisfied. This surprising result can in fact be proved, although Gauss did not prove it.

The justification for the method of least squares given by Gauss in his *Theoria motus* of 1809 remained popular over a long period, appearing in textbook presentations throughout the nineteenth and into the twentieth century. Gauss himself, however, soon came to reject it.

Since the method of least squares was discovered by no fewer than four scientists working – it appears – independently of one another, the question of priority arises. The first to publish the method was undoubtedly Legendre. Gauss no doubt knew the method about ten years earlier, and Huber may have found it even some years before Gauss, but there is no documentation for the discoveries before 1800. Among the four scientists who derived the method of least squares, only Gauss published a proof of it.

Later derivations of the method of least squares

In a memoir read to the Paris Academy in April 1810 Laplace undertook to derive a theorem, called since 1920 the *central limit theorem*, for the sum of identically distributed random variables. According to this theorem, such sums will, if the number of terms is large, be approximately normally distributed. A rigorous proof was obtained only in the twentieth century.

Only after writing his memoir did Laplace encounter Gauss's treatment of error theory in the *Theoria motus*. His reaction was swift: he wrote a *Supplement* to his memoir in which he gave a new rationale, different from Gauss's, for the choice of a normal distribution of errors, hence a different argument for the method of least squares. Suppose the number of observations of a quantity is large; let these observations be split into several groups, and take the mean in each group. Then in accord with the central limit theorem, Laplace assumed that these means would be normally distributed. The rest of his derivation, like Gauss's, involved maximization of the likelihood function.

In his *Théorie analytique des probabilités* (first edition, 1812), Laplace gave the following criticism of Gauss's derivation:

M. Gauss ... sought to derive this method [of least squares] from the Theory of Probability, by showing that the same law of the errors of observations which yields the rule of the arithmetic mean among several observations – a rule followed generally by observers – implies equally the rule of least squares of the errors of the observations .... But, as nothing proves that the first of these rules gives the most advantageous result, the same uncertainty exists in relation to the second. The investigation of the most advantageous way of forming final equations is without doubt one of the most useful in the Theory of Probability; its importance in physics and astronomy has led me to concern myself with it.

Laplace did not here note that his own formulas presuppose a large number of observations, whereas Gauss in his derivation made no such requirement.

In his *Essai philosophique sur les probabilités* (1814), Laplace disavowed all methods of adjusting observations used prior to the emergence of the method of least squares, including Bošković's method which he had previously held in high regard. In still later work it appears that he came to regard the normal distribution not merely as a limit but as the actual distribution of errors of directly measured quantities.

Meanwhile, Gauss responded to Laplace's criticism by abandoning the argument for the method of least squares given in his *Theoria motus*, and rethinking the subject statistically. No longer did he consider it appropriate to describe the arithmetic mean as in general "the most probable value"; thus in a letter of 1831 to Encke he urged:

... the problem of finding the most probable value ... intrinsically presumes knowledge of the error law, and it leads to the arithmetic mean only *when the error law has the form* e^{-kkxx}. *To put it abstractly, the probability of the "most probable" value is still only infinitesimal ... and for that reason has little practical interest – much less than the value that on the average makes the error least harmful. For this reason (as well as for other reasons ...), I have settled upon this latter principle, which should not be confused with the first one.*

Gauss's new justification for the method of least squares appeared in his *Theoria combinationis* of 1823. He supposed the density function $\phi(x)$ of the errors to be unimodal and in most cases even. Hence,

$$\int_{-\infty}^{\infty} x\phi(x)\,dx = 0. \tag{9}$$

Then, as a measure of precision, he introduced m, "the mean error to be feared", or (simply) "the mean error", where

$$m^2 = \int_{-\infty}^{\infty} f(x)\phi(x)\,dx = \int_{-\infty}^{\infty} x^2\phi(x)\,dx. \tag{10}$$

In present-day terminology, (9) gives the 'mean value' or 'expectation of x, written $E(x)$, and (10) gives the 'variance', written $E(x^2)$.

The cornerstone of Gauss's new theory was the principle of minimal variance. The choice of an *integral* measure of precision or of error, as in (10), was, according to Gauss, expedient: it avoided the resort to maximized probabilities that were merely infinitesimal. But the selection of $f(x)$ in (10), he allowed, was arbitrary, except for the reasonable restriction that this function should increase more rapidly than the error itself.

Gauss's derivation involves finding multipliers for each of the equations of condition such that the "mean error to be feared" will be a minimum. His somewhat ponderous solution of this problem is tantamount to the solution by the method of least squares. In sum, he shows that the method of least squares leads to minimal variance and is thus justified.

In the *Theoria combinationis* Gauss articulated his objections both to his own earlier deduction of the least squares principle, and to Laplace's deduction of it by way of the central limit theorem, as follows:

The first is totally dependent on the hypothetical form of the probability of the errors; and as soon as that is rejected, the values of the unknown quantities found by the method of least squares really are not the most probable ones any more, not even in the simplest case of the arithmetic mean.... The second ... leaves us completely unenlightened as to what we should do for a moderate number of observations....

He expressed the hope that friends of mathematics would enjoy seeing how

the new justification given here reveals the method of least squares as the most suitable combination of the observations in general, not just approximately but in mathematical rigour, no matter what the function for the probability of the errors is, and no matter whether the number of observations is large or small.

Whereas in the *Theoria motus* he had characterized the least-squares solution as *maxime probabile*, in the *Theoria combinationis* he described it as *maxime plausibile*.

The new substantiation was not widely accepted during the nineteenth century. Even during Gauss's lifetime the so-called 'theory of elementary errors' came to the fore. According to this theory, the error of each observation is the result of a large number of tiny, component ('elementary') errors. Therefore – so the proponents of this theory argued – the central limit theorem (in one version or another) leads to the normality of the errors, and, consequently, to a Laplacian-type substantiation of the method of least squares.

Astronomers tended to assume that if the errors had many causes whose influence was of the same order of magnitude, the law of distribution would be approximately normal. Their opinion on this point was strengthened by findings in other branches of natural science, notably physics: in 1860 James Clerk Maxwell published his celebrated theorem stating that the velocities of gas molecules at equilibrium are normally distributed.

Gauss's substantiation of least squares in the *Theoria combinationis* was revived in 1899 by the Russian mathematician A. A. Markov, and in the twentieth century has come to be generally accepted.

Application of the method

Gauss was probably the first to make decisive and important applications of the method of least squares. He continued in later years to be assiduous in applying the method to astronomic, geodetical, and magnetic measurements – so much so that friends protested at the extent to which he involved himself in the operations for the triangulation of Hanover. He was a master of experimental science, an expert in the analysis of instrumental errors. The inauguration by F. W. Bessel (1784–1846) of a new era of precision in astrometry owed much to Gauss's teaching and inspiration.

Laplace also was a virtuoso in the application of the theory of errors, particularly in the detection of small effects comparable in magnitude with errors of observation. The following quotation from his *Théorie analytique des probabilités* indicates the importance such applications could have in the celestial mechanics of his day:

This [lunar] inequality, although indicated by the observations, had been neglected by most astronomers, because it did not appear to result from the theory of universal gravitation. But, when I had submitted its existence to the calculus of probability, it appeared to me to be indicated with such a strong probability that I felt obliged to investigate its cause.

The great achievements of celestial mechanics in the nineteenth century – the discovery of Neptune, the construction of planetary tables by Le Verrier and Newcomb (see Chapter 28) – were all to involve arduous and painstaking application of the method of least squares.

Further reading

E. Czuber, *Theorie der Beobachtungsfehler* (Leipzig, 1892)

O. B. Sheynin, articles in *Archive for History of Exact Sciences*, vols 7 (1971), 9 (1972), 11 (1973), 12 (1974), 16 (1976), 17 (1977), 20 (1979), 26 (1982), 29 (1984), 31 (1984)

Stephen M. Stigler, *The History of Statistics* (Cambridge, Mass., 1986)

I. Todhunter, *A History of the Mathematical Theory of Probability* (Cambridge, 1865)

PART VIII

The development of theory during the nineteenth century

28

The golden age of celestial mechanics

BRUNO MORANDO

with an appendix on the stability of the solar system by Jacques Laskar

Introduction

The purpose of this chapter is to give an account of the developments in celestial mechanics during the period that extends roughly from the death of Laplace in 1827 to that of Poincaré in 1912. Unquestionably, this period is the golden age of celestial mechanics, and the names that we shall meet with, Hamilton, Le Verrier, Jacobi, Hill, and so forth, and, of course, that of Poincaré himself, bear witness to the fact. However, if we compare this period to the one that preceded it, from Newton to Laplace, we get the impression that it was wholly dedicated to the perserving and skilful exploitation of the splendid achievements of the prior century. Hamilton and Jacobi perfected the analytical mechanics that Euler and Lagrange had founded. Delaunay, Le Verrier, and Newcomb calculated the perturbations of the motions of the Moon and planets by applying methods that derive directly from the works of Euler, d'Alembert, Clairaut, Lagrange, Laplace.... Indeed, Poincaré himself, in the introduction to his *Méthodes nouvelles de la mécanique céleste*, defined with accuracy the goal that this science set for itself in the nineteenth century:

The ultimate goal of celestial mechanics is to resolve the great question whether Newton's law by itself accounts for all the astronomical phenomena; the sole means of doing so is to make observations as precise as possible and then to compare them with the results of calculation. The calculation can only be approximate, and, moreover, it would serve no purpose to calculate more decimal places than the observations can make known. It is thus useless to require more precision from the calculation than from the observations; but one must not, on the other hand, require less.

It would be impossible to improve on this definition of the goals pursued by the celestial mechanicians of the nineteenth century; and it would be ungracious to reproach them for having followed a course approved by such an authority.

Yet the celestial mechanics of that epoch appeared in its own time, and still appears to us today, as a scientific monument of the highest importance. At first its methods consisted in the pure and simple application of methods already known. But they were then improved upon decisively, and from the progress achieved other branches of science were to benefit; this is the case, for instance, with the methods of resolving canonical systems as in the work of Delaunay. Nor should the results obtained be underestimated. The comparisons of observation with calculation of which Poincaré spoke ended by supplying an answer to the question he posed apropos the ultimate goal of celestial mechanics. The discovery by Le Verrier of an advance that he could not explain in the perihelion of Mercury led to the conclusion that the law of gravitation by itself fails to account for all the astronomical phenomena. This is a fact of the first importance, leading to the revolution, at once scientific and philosophical, that is the General Theory of Relativity. The nineteenth-century scientists had an unclouded faith in their science; but they took it as their task to accumulate materials for the theorists of the future, rather than to build a finished structure.

Finally we must not forget that through a large part of the nineteenth century celestial mechanics was the whole of astronomy. Fraunhofer's discovery of the solar spectrum dates from 1814, and the ensuing experiments of Kirchhoff from 1859, but the prodigious development of astrophysics to which they led came only later. For those who lived under Louis-Philippe and Queen Victoria, the discovery of Neptune was comparable in importance to the theory of black holes for us today, a triumphant achievement of the astronomers. And these astronomers, Le Verrier and Adams, were practitioners of celestial mechanics.

Celestial mechanics at this epoch was devoted almost exclusively to the problem of the movements in the solar system – the particular case of the *n*-body problem that Nature had placed most insistently before the astronomers. There were a few exceptions: the work of Edouard Roche (1820–83) on the tidal stability of celestial bodies under the action of the mutual attraction of the particles composing them, leading to his introduction in 1849 of the notion of 'Roche's limit'; the work of G. H. Darwin on the secular acceleration of the Moon; the work of Laplace and of Poincaré on the tides. Here we shall confine ourselves to what the astronomers were primarily concerned with: to formulate and solve the equations of the *n*-body problem for the particular cases where the bodies are planets or satellites of planets.

Equations and solutions

Isaac Newton, in his earliest enunciation of the law of universal gravitation, when considering the multiple attractions influencing the motion of each planet, expressed the opinion that "to define these motions by exact laws admitting of easy calculation exceeds, if I am not mistaken, the force of any human mind". In the second half of the eighteenth century Joseph Louis Lagrange (1736–1813) and Pierre-Simon Laplace (1749–1827) addressed the problem of the complete integration of the equations of motion for *n* bodies attracting one another two by two according to Newton's law, and their studies quickly showed the problem to be insusceptible to a general solution. Not till the very year of Poincaré's death, 1912, was there decisive progress on the theoretical level. This was due to the Finn Karl Sundmann (1878–1949). But Sundmann's method, which consists in developing the coordinates of the bodies, and also the time, in very slowly convergent series in terms of one and the same variable, is hardly applicable in practice. The nineteenth-century theorists, like those of today, struggled to cope with the problem in a variety of ways.

Approximate solutions of the equations of motion can be sought in terms of developments in series. This has been the chief mode of attack for the bodies of the solar system; it requires on the one hand that analytical mechanics be perfected, and on the other that the use of series be justified through the study of their convergence. Another approach is to investigate whether there exist functions of the coordinates and velocities that remain constant over the course of the motion, the 'first integrals' as they are called. Given such integrals it becomes possible to delimit the domains within which the motions can occur. A third mode of attack is to study the configuration that a system of material points had at a time infinitely far in the past, or that it will have at a time infinitely far in the future.

The works of Hamilton and Jacobi

In his *Mécanique analytique* (1788) Lagrange established general equations of mechanics' ('the Lagrangian equations') involving a function of the coordinates and the velocities called the 'Lagrangian'. Also, he established another system of equations especially suited to celestial mechanics. These 'equations of Lagrange' require that the body considered move very nearly in the Kepler-motion that Newton had shown to be the solution of the two-body problem. Such a motion is characterized by six constants determined observationally. In the perturbed motion Lagrange supposed these constants to be augmented by functions of the time having small amplitude or varying with extreme slowness. To determine these functions, he introduced a function of the positions of the mass-points called the *disturbing* or *perturbing function*. This function has played a fundamental role in the celestial mechanics of the nineteenth and twentieth centuries. It is associated with the notion of intermediary orbit (an orbit the departures from which are considered as perturbations) and with the method of variation of constants. For Lagrange the intermediary orbit was the Keplerian ellipse, but as we shall see later on, Hill utilized an intermediary orbit of a different kind to resolve the problem of the motion of the Moon.

Two mathematicians, Jacobi and Hamilton, introduced new concepts and new variables, permitting the Lagrangian equations to be transformed into systems that could be treated more advantageously. Carl Gustav Jacob Jacobi (1804–51), a German mathematician known especially for his studies of the elliptic functions, introduced into mechanics a set of differential equations called canonical equations. The six variables that characterize every problem of mechanics are here associated two by two in three pairs of variables 'canonically conjugated'; the system of differential equations they obey has an especially simple form.

Jacobi investigated the conditions under which a change of variables in such a system would preserve its canonical character. An appropriate change of variables can be a powerful means of resolving a system of differential equations, if it is possible to choose the new variables in such a way as to simplify the system. In his *Vorlesungen über Dynamik* (1866), Jacobi studied the motion of two bodies by a method now called 'Jacobian', which reduces the problem to that of finding a particular solution of a partial differential equation. He identified certain canonical variables, 'Jacobian variables', that are constant in the two-body problem but which by the method of variation of constants can be utilized in the n-body problem. Also, inquiring into the conditions under which a canonical system remains canonical under a change of variables, he showed that certain functional determinants, 'Lagrange's brackets' and also 'Poisson's brackets', must have a particular form. Poincaré later returned to this question and found an extremely simple condition, called Poincaré's condition of canonicity, which reduces to the statement that a certain differential form is a total differential.

The works of Jacobi are closely related to those of Sir William Rowan Hamilton, who was born in Dublin in 1805, was elected Andrews professor of astronomy and Astronomer Royal of Ireland at the age of twenty-two, and died at Dunsink (site of Dublin's Trinity College Observatory) in 1865. An eminent mathematician, founder of the algebraic theory of complex numbers and the theory of quaternions, Hamilton also made major contributions to mechanics. He showed that, among all the possible movements of a material point between two instants t_0 and t_1, the real movement is that which minimizes the integral of a certain function between these two instants. This is called 'the principle of least action' or 'Hamilton's principle'; it leads quite naturally to a system of canonical equations, the right-hand members of which are the partial derivatives with respect to the variables of a function called the 'Hamiltonian'.

If, following Hamilton and Jacobi, we write the equations of celestial mechanics in canonical form, it is then necessary to resolve two difficult problems: to find how to express the Hamiltonian as a function of the canonical variables, permitting us to calculate the partial derivatives appearing in the right-hand members of the equations; and to find a method of integrating the equations.

Development of the disturbing function

To express the Hamiltonian of the motion as a function of the required variables, it is sufficient to express the disturbing function – defined initially in terms of the coordinates of the bodies in the system under consideration – as a function of these same variables. This is possible only if we represent the disturbing function as a sum of a number of terms, in which certain of the variables appear under cosine signs, while the conjugate variables appear in the coefficients of these cosines. To the analytic development of the disturbing function, there were many contributors during the nineteenth century. Laplace had introduced the functions called 'Laplace coefficients' to express as a trigonometric series the reciprocal of the distance between two planets, and the odd powers of this reciprocal. F. W. Bessel (1784–1846) was the first to make use of 'Bessel functions' in a treatise on the development of the disturbing function published in Berlin in 1824. In 1831 Peter A. Hansen – of whom we shall have more to say later – in his *Untersuchung über die gegenseitigen Störungen des Jupiters und Saturns* introduced certain functions, called Hansen's coefficients, which permit the development of the *radii vectores* of the planets, or any powers of these, as trigonometric series in the mean anomaly, the coefficients being each an entire series in the eccentricity. We note also the important role played by Legendre polynomials in all these questions; they had been introduced in Adrien-Marie Legendre's treatise on the *Figure des planètes* of 1782.

To Laplace is due the first expansion of the disturbing function for the planets to the third order of the eccentricities and inclinations. Pontécoulant carried it out to the sixth order in his *Théorie analytique du système du monde* (1829–46). Benjamin Peirce also published a development to the sixth order in 1849 in the *Astronomical Journal*. Le Verrier pushed the development to the seventh order in the first volume of the *Annales de l'Observatoire de Paris* (1855). In the development of the disturbing function, different angular variables – the variables fixing the position of the orbit in space and the position of the planet on the orbit – may be used. Hansen, in investigating the motion of the minor planet Egeria in 1859, employed the eccentric anomaly in place of the mean anomaly. George W. Hill, for his theory of Jupiter and Saturn in 1890, utilized the eccentric anomaly of one of the planets and the mean anomaly of the other. Simon Newcomb (1835–1909) in 1891 and 1895 gave developments to the sixth order, using first the eccentric anomalies of the four interior planets (Mercury, Venus, Earth, and Mars) then their mean anomalies.

Resolution of the equations of motion

Given the canonical equations and an expansion of the disturbing function to a certain order of precision, how does one obtain a solution of corresponding precision? In the case of the theory of a planet, one can seek an expansion in powers of the masses of the perturbing planets, as these are small relative to the Sun's mass; in the case of the Moon, the small parameter employed is the ratio of the Earth–Moon distance to the Earth–Sun distance, or a related quantity, the ratio of the mean motions. The coefficients of the powers of the small parameters are sums of a large number of periodic terms – sums it is not easy to obtain.

One of the methods devised consists in carrying out a change of variables on the canonical system, in such a way as to simplify it. Thus Delaunay in his theory of the Moon (to be described in a later section) reduces the Hamiltonian to an isolated periodic term, which is a linear combination of three angular variables and the time. This linear combination is taken as the sole angular variable of a new canonical system which is then integrable. This operation is repeated by taking successively the different periodic terms of the Hamiltonian. The method was greatly admired by Hill who in 1907 generalized its application in his memoir "On the extension of Delaunay's method in the lunar theory to the general problem of planetary motions". An important modification of Delaunay's method was made in 1916 by Hugo von Zeipel in his article "Recherches sur le mouvement des petites planètes", which appeared (in French) in a Swedish journal. The modification consists in introducing a generative function in order to effect one of Delaunay's operations, and simultaneously in eliminating several periodic terms even though they are not multiples of the same argument.

All these methods miscarry if the Hamiltonian contains resonant terms, that is, combinations of angular arguments in which the coefficient of the time is zero or nearly zero. This situation was studied by Hill and by von Zeipel in the memoirs referred to above, and also by Poincaré in Vol. 2 of his *Méthodes nouvelles de la mécanique céleste*, published in 1893.

A canonical system can be integrated if the variables are separable, but unhappily this is not the case for systems describing the motion of n bodies. One known case that is integrable – aside, of course, from the two-body problem – is that of a point attracted by two fixed centres; the solution was given by C. V. L. Charlier in Vol. 1 of his *Mechanik des Himmels*, which was published in Leipzig in 1902. The orbits obtained can be employed as intermediary orbits in more general problems of celestial mechanics, but this was not attempted, it appears, until much later, around 1960, when the device was made use of in the study of the trajectories of artificial satellites.

As stated earlier, however, the principal preoccupation of the celestial mechanicians of the nineteenth century – they were mainly astronomers rather than mathematicians – was to construct ephemerides that would give the positions of planets and satellites in as near agreement as possible with the observations. We shall accordingly pass in review all, or nearly all, the theories of the Moon and of the planets and satellites developed during this period – theories that are characteristic of the epoch, and have moreover continued to serve as the basis for the construction of ephemerides into the third quarter of the twentieth century. But first we shall recount the circumstances in which the most spectacular result of the century was obtained: the discovery of Neptune.

The discovery of Neptune

If there is a single outstanding event in the history of celestial mechanics in the nineteenth century, it is surely the discovery of Neptune. The difficulty of the problem to be resolved and the boldness required in addressing it, the excitement produced by the telescopic discovery of the planet very close to the predicted position and, last but not least, the

28.1. Urbain Jean Joseph Le Verrier.

piquancy added to all this by the controversy between France and England over priority in the discovery – there you have what made the event a sensational affair, widely echoed in the daily press of the time.

The principal protagonists in this drama were two astronomers very different in character, but not in their passion for scientific research.

Urbain Jean Joseph Le Verrier (Figure 28.1) was born at St Lô (département de la Manche) in 1811. As student at the École Polytechnique, then as *élève-ingénieur* of the state tobacco company, he pursued an interest in chemistry. But when a position as *répétiteur d'astronomie* became vacant at the École Polytechnique in 1837, he accepted it, devoting himself henceforth to celestial mechanics and astronomy. In 1839 he published a memoir on the secular variations of the planets. Made famous by his discovery of Neptune, in 1854 he was appointed head of the Paris Observatory, and

28.2. John Couch Adams (engraving by Samuel Cousins from the painting by Thomas Mogford, 1851).

remained in that position till 1870, then occupied it again from 1874 until his death in 1877. Despite grave difficulties, owing in large measure to his authoritarianism and difficult disposition, he was able to effect a lasting transformation in the observatory and to turn it into a scientific institution of the first rank (see Volume 4A, p. 116). The principal object of his own scientific writings was the theory of the planets, of which we shall have more to say in a later section. Le Verrier was an ambitious, autocratic man, avid for honours. He played an important role in the inner councils of the Academy of Sciences, and also in politics, for he served as deputy, senator, and president of the General Council of the Département de la Manche.

A quite different man was John Couch Adams (1819–92) (Figure 28.2). In 1839 he was awarded a scholarship to enter St John's College, Cambridge. In 1850 he obtained a teaching post at the University of St Andrews, and in 1859 he was named Professor of Astronomy and Geometry at Cambridge. In 1861 he succeeded James Challis as director of the Cambridge Observatory, and organized there a series of meridian observations. After his work on the perturbations of Uranus, he devoted himself to the lunar theory and, in particular, brought to light an error committed by Laplace and, following him, Plana (more will be said of this in a later section). Later he devoted himself largely to mathematics and even calculated Euler's constant (which occurs in the summation of harmonic series) to 236 decimals! Self-effacing and timid, and also concerned about the cost of keeping up appearances consonant with the title, Adams in 1847 refused a knighthood from Queen Victoria.

The planet Uranus had been discovered by William Herschel in 1781, but it was then realized that earlier observations of it had been made by astronomers who took it to be a fixed star. By 1820 there were thus available for analysis some forty years of meridian observations, together with nineteen observations made between 1690 and 1781 by John Flamsteed, P.-C. Le Monnier, James Bradley, and Johann Tobias Mayer. It was Alexis Bouvard (1767–1843), long an assistant of Laplace and the one who had carried out the detailed calculations of his *Mécanique céleste*, who undertook the analysis. In 1808 he had published tables of Jupiter and Saturn, and for years he had supervised the calculation of the ephemerides of the *Annuaire du Bureau des Longitudes*. In 1821 he undertook to recalculate the tables for the motion of Jupiter and Saturn, and to add thereto tables of the motion of Uranus, following the methods developed by Laplace in the *Mécanique céleste* to take account of perturbations. He found it impossible to represent the observations with the expected accuracy, the discrepancy between theory and observation being in some cases as high as 65″. The error, he decided, must lie in the older observations, but on discarding all those made before 1781, he found that errors of about 30″ still remained. He wrote: "... I leave to the future the task of finding out whether the difficulty ... in fact has its source in the inaccuracy of the older observations, or derives from some foreign and hitherto unperceived action to which the planet is subject." But the discrepancies between the theory and the observations continued to increase and reached nearly 2′ in 1845 (see Figure 28.3), despite Bouvard's having discarded the older observations. Beginning in 1835, G. B. Airy, then Eugene Bouvard (nephew to Alexis), François Arago, Bessel, and Sir John Herschel entered into discussions of this subject, and proposed to search for a hypothetical perturbing planet.

Other hypotheses also made their appearance. The discrepancies, it was proposed, could be due to the resistance of the aether, or to a large and unobserved satellite of Uranus, or to a comet that had perturbed the planet's progress. It was even proposed that at such a great distance from the Sun the law of universal gravitation might no longer be exact. Le Verrier in a memoir of October 1846 – the final memoir that he devoted to the discovery of Neptune – disposed of these hypotheses. The resistance of the aether would have manifested itself elsewhere in the solar system and would have produced observable perturbations in bodies of low density. The perturbations due to a satellite of Uranus would have been of short period, contrary to what was observed. Finally, the encounter of a comet with Uranus would have produced its effect during a limited time between the epoch of the older observations and that of the observations used by Bouvard; why then should the discrepancies continue to grow? As for the suggested failure of Newton's law, many, including Le Verrier, simply did not wish to believe in it. There thus remained the hypothesis of the unknown planet, but it was unthinkable to search for it at random. As Le Verrier remarked much later:

How were the astronomical observers to discover, in the immense extent of the sky, the physical cause of the perturbations of Uranus, unless their task could be provided with limits, and their search confined within a determinate region? Who among them would have undertaken to seek a telescopic star successively in the twelve signs of the zodiac? It was thus necessary to begin by showing that the observations should be concentrated within a small number of degrees. Only thus could one expect that the vigils of the observers would not be in vain, and that in the not distant future physical astronomy would be enriched with a star of which theoretical astronomy had previously revealed the existence and fixed the position.

Adams's interest in the divagations of Uranus arose earlier than Le Verrier's. Already in July 1841, while still an undergraduate, he had formed the design of investigating the perturbations of Uranus to see whether they might be caused by a hypothetical perturbing planet. On completing his degree in 1843 he set to work by reviewing Bouvard's tables, correcting errors and pushing further than Bouvard the calculation of the perturbations

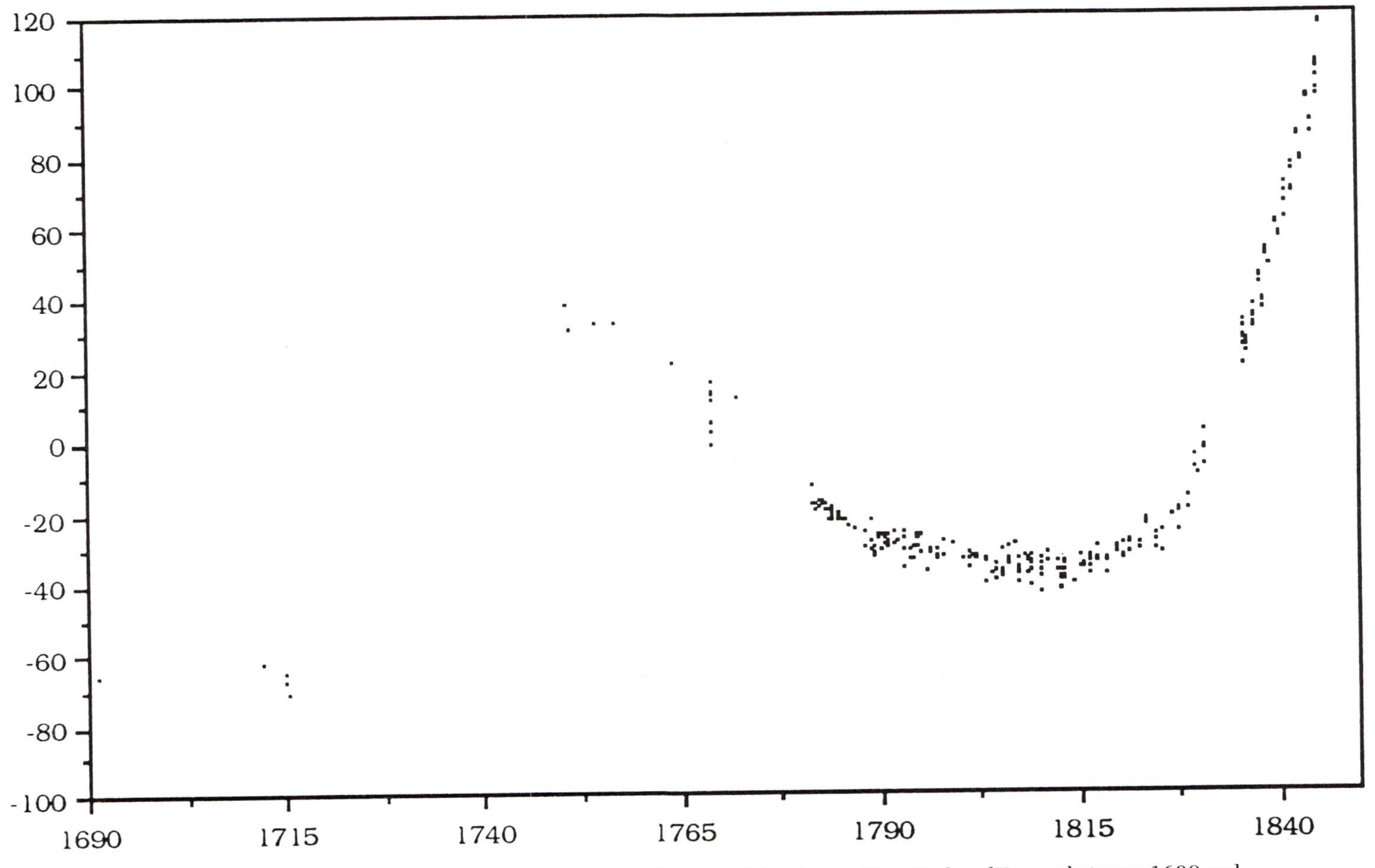

28.3. The discrepancy (in arc-seconds) between the predicted and the observed longitudes of Uranus between 1690 and 1845.

due to known planets. Then he made the following two hypotheses, both of them reasonable: (1) the unknown planet was in the ecliptic, since the latitudinal anomalies in the movement of Uranus were negligible; and (2) the semi-major axis of its orbit was double that of the orbit of Uranus, therefore some thirty-eight astronomical units – a hypothesis in agreement with the empirical relation called 'the Titius–Bode law' (see Chapter 24). Using successive approximations and varying the parameters of the problem, Adams by October 1845 succeeded in determining a set of orbital elements for the unknown planet that led to acceptably small values for the residuals of the observations. He predicted the position of the planet in the heavens, and in September 1845, armed with a letter of introduction from James Challis, the director of the Cambridge Observatory, made an unannounced call on Airy, the Astronomer Royal. As Airy was at this time away on business, Adams returned on a day in late October, and twice made a further attempt to see the Astronomer Royal, both times without success; but before leaving he left a three-page letter for Airy, specifying the elements of his hypothetical planet and its position.

Airy responded on 5 November, requesting the source of the value that Adams ascribed to the radius vector – he did not regard the Titius–Bode law as an adequate basis, and in any case did not want to involve the Greenwich Observatory in what he considered a merely theoretical investigation. Adams judged it futile to reply, and the matter rested there. Thus the honour of the discovery was lost to Adams and to England. The works of Adams on the problem of the hypothetical planet were later published in an appendix to the *Nautical Almanac* for 1851. The theoretical position of the planet at the moment of Galle's discovery of it (of which more will be said below), as calculated from Adams's orbital elements, differs from the observed position by 2° 27′.

Throughout this time Le Verrier was working assiduously. In the summer of 1845 Arago had convinced him of the importance of resolving the problem, and had persuaded him to abandon for the time being the research he had undertaken on comets. Unlike Adams, Le Verrier published successive accounts of his work as it progressed. On 10 November 1845 he presented to the Academy of Sciences a "Premier mémoire sur la théorie d'Uranus"; this was followed on 1 June 1846 by a memoir entitled "Recherches sur les mouvements d'Uranus" and on 31 August by a third memoir "Sur la planète qui produit les anomalies observées dans le mouvement d'Uranus. Détermination de sa masse, de son orbite et de sa position actuelle". After the discovery he read to the Academy of Sciences on 5 October 1846 a final memoir entitled "Sur la planète qui produit les anomalies observées dans le mouvement d'Uranus. Cinquième et dernière partie, relative à la détermination de la position du plan de l'orbite". He then combined all these into a single work, which was published in the *Connaissance des temps pour 1849* under the title "Recherches sur les mouvements de la planète Herschel (dite Uranus)". In a piquant note at the bottom of the first page he justified the title in the following fashion: "In my further publications, I shall consider it as a strict duty to make the name *Uranus* disappear completely, and to refer to the planet henceforth only by the name *Herschel.* I regret keenly that the printing of this memoir is already so far advanced that it is not possible for me to bring it into conformity with this decision, which I shall observe religiously in the future." In calling Uranus "Herschel" he hoped that Neptune would be named after himself. More on this later.

Le Verrier began his investigation by developing, in two different ways, the perturbing function for the motion of Uranus and then, after integration of the second members of the equations, the inequalities of the orbital elements of Uranus as subject to the attractions of Jupiter and Saturn. One of the methods consisted in utilizing the literal development of the perturbing function given by Laplace in his *Mécanique céleste*, and calculating the partial derivatives; the other consisted in calculating the developments numerically by giving particular values to the eccentric anomalies of the interacting planets. The use of two different methods was requisite, Le Verrier urged, because "the importance of the question I am examining, and the nature of the results I shall obtain, require that I omit nothing that may serve to convince astronomers that my conclusions are in agreement with the truth".

He then compared with observations the results given by his theory, forming equations of condition for all the available observations from 22 December 1690 to 27 December 1844. He found residuals as large as 20″.5 – unacceptably high. It was at this point that he introduced the hypothesis of an unknown perturbing planet. Like Adams, he supposed it to be in the ecliptic, with a solar distance double that of Uranus; thus the unknowns remaining were the mass of the planet, the eccentricity and longitude of the perihelion of its orbit, and its longitude at epoch. Actually, in place of the eccentricity and longitude of the perihelion, he used two other variables that Laplace had introduced, namely h', the product of the eccentricity and the sine of the longitude of the perihelion, and l', the product of the eccentricity and the cosine of the longitude of the perihelion. He calculated the perturbations as a function of these unknowns and introduced them into the equations of condition where, except in the case of the mean longitude at epoch, he could give them a simple form. He then expressed m' (the unknown planet's mass), $m'h'$, $m'l'$ as functions of the mean longitude at epoch ϵ'. In order that m' not be negative, it was necessary that ϵ' fall within certain limits. The next step was thus to give a value to ϵ' within these limits such as to reduce the residuals of the observations to a minimum. Finally, he improved the precision of the results by utilizing observations of Uranus at its quadratures between 1781 and 1845 – observations that he had set aside at an earlier stage while retaining only the observations made in the neighbourhood of opposition, for which the equations of condition were easier to form. He also varied the semi-major axis, which he had taken at first as double that of Uranus.

Thus on 31 August 1846 he presented to the Academy of Sciences the following results: mass, $\frac{1}{9322}$ times the mass of the Sun; semi-major axis, 36.1539 astronomical units; eccentricity, 0.10761; longitude of the perihelion on 1 January 1847, 284° 45′ 8″; heliocentric longitude of the planet on the same date, 326° 32′.

It remained to find the planet. This was not as simple as it might seem. Most observers, particularly in France, appear to have been sceptical and unwilling to commit themselves. It has often been claimed that Le Verrier turned to Johan Gottfried Galle (1812–1910), astronomer at the Berlin Observatory, because he knew that the Berlin Observatory had at its disposal the good star charts of Bremiker, but there is no evidence for this; Le Verrier makes no mention of it in his letter to Galle, and as we shall see, Galle himself did not think of referring to the star chart until the last moment. Galle had sent his doctoral thesis to Le Verrier a short time before, and so was known to him. Le Verrier's letter, dated 18 September, arrived on 23 September and, on the evening of that very day, aided by the young astronomer Heinrich Louis d'Arrest, Galle turned the excellent Fraunhofer telescope of 9-inch (23-cm) aperture toward the indicated region of the heavens; but he did not find the disk of 3 arc-second diameter that, according to Le Verrier, the planet ought to present. There is nothing astonishing in this: as the theoretical resolving power of his instrument was 0″.5, a particularly clear sky would have been required in order to see a disk so small; a magnification of at least 300 to 400 would have been required, and this would have limited the field to several minutes of arc – very little when it is a matter of finding an unknown object!

In was at this point that d'Arrest suggested comparing the field with a chart. Rummaging through the drawers they came on the Berlin Academy's Star Atlas charts, which had been compiled by Carl Bremiker (1804–77), inspector for the minister of commerce of Prussia, and a collaborator in the calculations of the ephemerides of the *Berliner Astronomisches Jahrbuch*. These excellent charts had been printed late in 1845 but had not yet been distributed to foreign observatories. Since the hypothetical planet was within the constellation of Capricorn, it was the chart Hora XXI, corresponding to the zone of 21 hours of right ascension, that they took for the comparison. Galle returned to the telescope and at length found, some 52′ from the position predicted by Le Verrier, a star of magnitude 8 which did not appear on Bremiker's chart. The following day the star had shifted; it was indeed the new planet (Figure 28.4).

On 25 September Galle wrote to Le Verrier a letter that has remained famous: "Monsieur, the planet of which you indicated the position really exists [réellement existe] ...." (Amusingly for us, in citing this letter in an appendix to his final memoir, Le Verrier corrected Galle's French by making him say: "existe réellement".)

The news of the discovery of a new planet resounded throughout Europe like a clap of thunder. The journalists and the cartoonists seized upon it. Airy, meanwhile, had not remained entir-

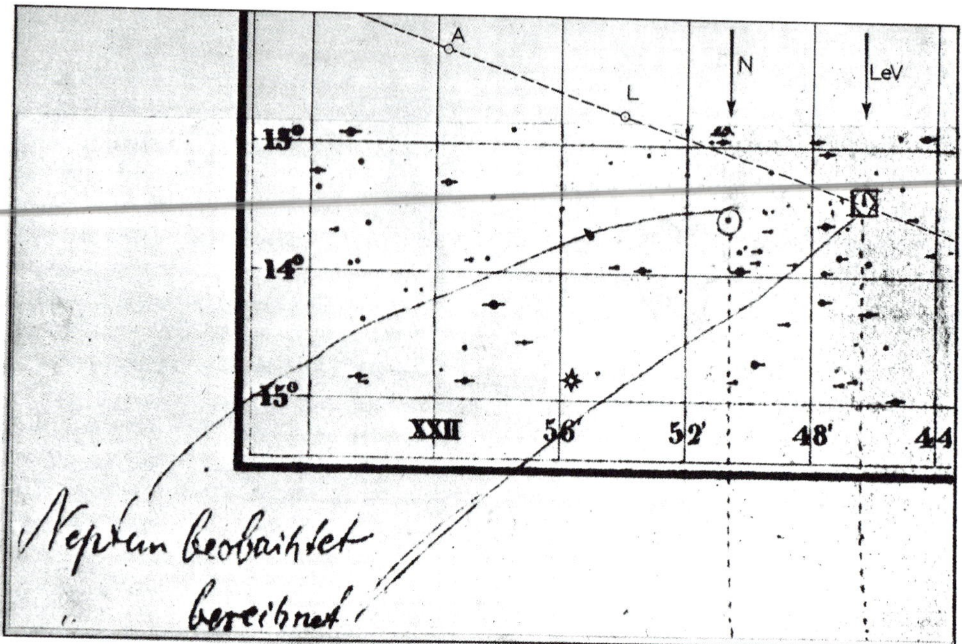

28.4. A section of the newly completed chart Hora XXI used by Galle and d'Arrest to locate the planet predicted by Le Verrier. (The ecliptic and various letters have been added.) The observed position of Neptune is indicated by N, the prediction of Le Verrier by LeV, and that of Adams by A. The handwritten notes are said to be by Galle.

ely inactive since receiving Adams's note. He had read and admired Le Verrier's publications, and on 29 June 1846, at a meeting of the Board of Visitors of Greenwich Observatory, he made known his desire that an observational search be instituted. If we believe what Challis wrote in 1847 in a letter to the editor of the *Astronomische Nachrichten*, it was Le Verrier's publication of 1 June 1846 that persuaded Airy that a search should be begun:

M. Le Verrier by an investigation published in June last, obtained almost precisely the same heliocentric longitude which Mr Adams had arrived at. This coincidence from two independent sources very naturally inspired confidence in the theoretical deductions, and accordingly Mr Airy shortly after suggested to me the employing of the Northumberland Telescope of this Observatory in a systematic search after the planet.

Challis observed at the Cambridge Observatory beginning on 29 July; he did not have very precise charts and was reduced to comparing from one day to another the relative positions of the stars of the field with the aim of discovering a proper motion in some one of them. On 4 and again on 21 August he had Neptune in the field of view and noted carefully its position without realizing that it was the sought-for planet. On 29 September, six days after the discovery that Galle had made on the basis of Le Verrier's prediction, he was again seeking the planet.

The history of the discovery of Neptune must be completed by some account of the controversies to which it gave rise. The name *Neptune* that was in due course adopted did not, of course, please Le Verrier who had hoped to give his own name to the planet. We have seen him playing the saint in attempting to impose the name *Herschel* on the planet Uranus, obviously not without a hidden motive. Arago agreed to support him and promised never to refer to the new planet other than as Le Verrier's. But the names *Janus* and *Oceanus* had been proposed, and Le Verrier himself claimed that the Bureau des Longitudes had chosen the name Neptune – which it does not appear to have done if we are to believe the recorded minutes of its meetings.

Although it does not strictly pertain to the history of celestial mechanics, we must say a word about the controversy that arose between France and England concerning priority of discovery.

Before Le Verrier had even begun his investigation, Adams had completed calculations that would have led to the discovery, but these calculations remained in a drawer, while Le Verrier published the results of his investigations and informed an observer where exactly the planet was to be searched for. A letter from Sir John Herschel to the London journal *The Athaeneum* of 3 October 1846 opened the controversy. The role played by Airy was much criticized: his negligence or rather his prejudice against the work of the young and little known astronomer that Adams then was deprived England of a great discovery. In the issues of the *Astronomische Nachrichten* beginning with no. 585 (1847) one can read Airy's attempts to justify himself.

Today it is generally agreed that equal merit should be assigned to Adams and to Le Verrier. The problem that they were able to resolve, each in his own way, was of great difficulty and importance; its solution constitutes a landmark in the history of science. The two antagonists themselves became good friends after their meeting in Oxford in 1847.

Another controversy, more technical but likewise dictated by envy, arose concerning the circumstances of the discovery. It became possible to determine the true orbit of Neptune on the basis of observations – all the better after it was found that Michel de Lalande (nephew of Jérôme) had twice observed that planet, while taking it for a star, on 8 and 10 May 1795. The discovery of the satellite Triton by William Lassell in 1846 made it possible, in addition, to calculate the mass of the new planet with precision. It was then found that between the reality and the results obtained respectively by Adams and Le Verrier there were notable differences (Figure 28.5). The mass of Neptune is 17 times that of the Earth; Le Verrier had found it 32 times, Adams 45 times. The semi-major axis of the orbit is 30.11 astronomical units, while Le Verrier and Adams, taking as their point of departure the value 38 given by Bode's law, had found it much greater. But this error was compensated by the large eccentricity attributed to the orbit (0.1 in place of 0.01), which brought it about that in the neighbourhood of the conjunction of Uranus and Neptune of 1821, the time at which the perturbations would be especially great, the three orbital arcs (the real orbit, Le Verrier's orbit, Adams's orbit) coincided very nearly, the errors in the distance being compensated by the errors in the mass.

The controversy on this point, which seems futile today, was quite the rage for a while, above all due to the impetus given it by a certain Jacques Babinet, a member of the Academy of Sciences, who went so far as to say that Adams and Le Verrier had discovered nothing at all. Babinet even asserted that there were two perturbing planets, one being a certain Hyperion of which, by means of some ridiculously abridged calculations, he claimed to supply the orbital elements!

The discovery of Neptune by pure calculation was the grand triumph of celestial mechanics as founded on Newton's law of gravitation. The discovery was scientifically important, and the circumstances under which it occurred gave it a spectacular character, such as to make a deep impression on the public imagination. Yet Le Verrier himself, as we shall see later, was the first to put his finger on what would prove a failure of the theory of gravitation, when he studied the motion of the perihelion of Mercury. He did not, however, dare to doubt this law to whose triumph he had contributed so much.

The theories of the movement of the Moon

The Moon is close to the Earth and is of relatively low brightness; hence its displacement among the stars is rapid and easily observable. This explains why a good many of the idiosyncracies of its motion, such as the rotation of the line of nodes, the equation of centre, and the evection, became known to Hipparchus, Ptolemy, and the Arabic astronomers. Tycho Brahe observed further inequalities, but it was of course necessary to await Newton and those who pursued celestial mechanics in the eighteenth century before seeing the emergence of mathematical theories of the motion based on the law of universal gravitation.

To what point had the theory advanced at the time of Laplace's death? We recall that Alexis-Claude Clairaut in 1752 published a theory of which the precision with respect to the Moon's position was of the order of 1′.5. The theories of Jean le Rond d'Alembert and Tobias Mayer and the two theories of Leonhard Euler had a comparable precision. The theory that Laplace published in 1802 had a precision of 0′.5, and also the merit of partly explaining the secular acceleration (we

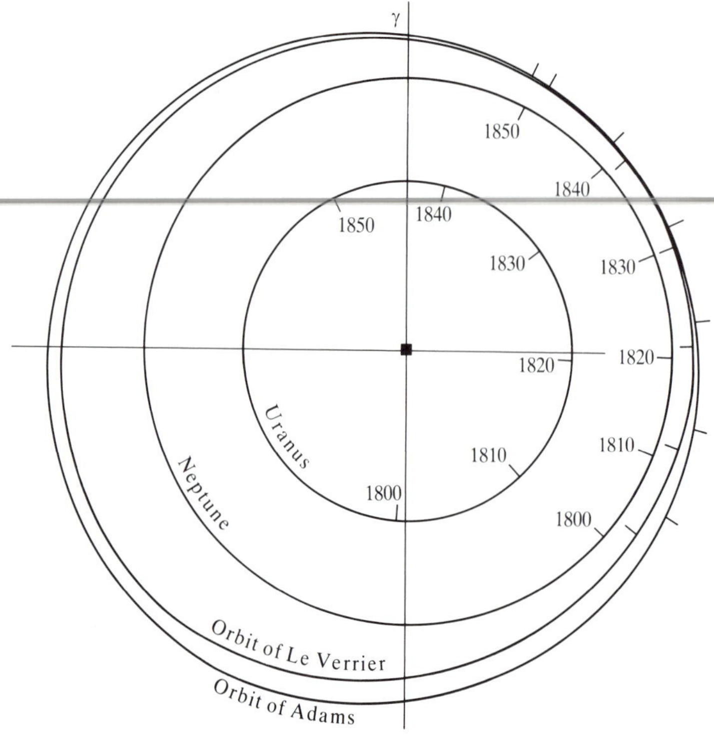

Le Verrier:	Semi-major axis 36.2 a.u. Perihelion 285° Eccentricity 0.11
Adams:	299° 0.12 [earlier 0.16] 38.4 a.u. [earlier 316°]
Neptune:	30.1 a.u. (almost perfect circle)
Uranus:	between 17.3 and 20.1 a.u. Eccentricity 0.05 Perihelion 170°

28.5. The observed positions of Uranus and Neptune, and the positions of Neptune expected on the basis of the calculations of Adams and Le Verrier.

return to the latter subject in a later paragraph). The tables published by J. T. Bürg in 1806 and those published by J. K. Burckhardt in 1812 were based directly on Laplace's theory. It is starting from this date that we shall study the post-Laplacian lunar theories.

The theories of the Moon between Laplace and Delaunay

In 1820 the Academy of Sciences of Paris set as a prize problem the motion of the Moon. Two theories were entered in the competition, one by Damoiseau and the other by Plana.

Marie Charles, Baron de Damoiseau (1768–1846), an officer of artillery who had emigrated in 1792, returned to France after the Restoration and became director of the observatory of the École Militaire in Paris. His lunar theory, first published in 1824, attempted no more than to push to a higher order of precision the kind of calculation instituted by Laplace.

The Italian Giovanni Plana (1781–1864) had been a student of Lagrange at the École Polytechnique. From 1811 until his death he was professor of astronomy at the University of Turin. His lunar theory, which was developed in literal notation throughout, was published at Turin in 1832. Like Damoiseau's it was based on the theory of Laplace.

The theory of Siméon-Denis Poisson (1781–1840) was published in 1833. It is not a complete theory of the Moon's motion, for Poisson was unable to carry very far the calculations to which it led him. He supposed in a first approximation that the semi-major axis, eccentricity, and inclination of the orbit were constants and that the mean longitude of the Moon, that of the perigee and that of the node were linear functions of the time. These quantities were substituted in the second members of equations which, on being developed, were integrated term by term. He thus obtained better approximations for the semi-major axis, eccentricity, etc., and these being substituted into the second members, the whole process was then repeated. The calculations, given the limited means available at the time, quickly became too complicated; much later the method was to be usefully employed in certain problems of celestial mechanics, once the drudgery of the calculations could be turned over to electronic computers.

Next to be mentioned is the theory of Pontécoulant. Philippe, Comte de Pontécoulant (1795–1874), a scientist of great merit of whom there will be more to say when we come to the theories of Jupiter and Saturn, published his first memoir on the Moon's motions in 1837. His theory dates from 1846, but revisions were published in the *Monthly Notices of the Royal Astronomical Society* for 1860, and in the *Comptes rendus* of the Academy of Sciences of Paris for 1862. The variables that Pontécoulant used were the longitude, latitude, and radius vector of the Moon; already in the approximate orbit with which he began he introduced the movement of the node and that of the perigee. His theory is entirely literal, that is, he does not assume as known *a priori* the numerical values of any of the variables of the problem.

The most decisive progress was brought about by Peter A. Hansen (1795–1874), one of the great German theorists of nineteenth-century astronomy. At first a watchmaker, he learned French, Latin, and mathematics during his leisure hours. His learning and enterprise brought him to the attention of the Danish astronomer H. C. Schumacher, founder of the *Astronomische Nachrichten*, who had engaged him as a calculator; so Hansen began his career as a researcher. In 1825 he succeeded Encke as director of the observatory of the Duke of Mecklembourg at Seeberg near Gotha, and here he remained until his death. His theory of the Moon was published in 1838 at Gotha under the title *Fundamenta nova investigationis orbitae verae quam Luna perlustrat*. Modifications were published in 1861 and 1864. Hansen himself put the theory in tables which were published in London in 1857 under the French title *Tables de la Lune construites d'après le principe newtonien de la gravitation universelle*. These tables were employed from 1862 onwards for the calculation of ephemerides. As intermediary orbit Hansen chose an ellipse of fixed dimensions, located in the mobile plane of the real orbit, and with uniformly turning perigee. He then calculated the perturbations of the mean anomaly and of the radius vector in this ellipse. The procedure was an advance over the method used and extolled by Laplace in his planetary theories (Hansen employed the same procedure to calculate the perturbations of Saturn and Jupiter).

The theory of Delaunay

Charles Delaunay (1816–72) constructed the most precise and complete of all the literal theories of the Moon's motion, a prodigious piece of work requiring twenty years to finish; it was published in 1860 and 1867 in two huge volumes of the *Mémoires* of the Academy of Sciences. His method was first explained in 1846 in a "Mémoire sur une nouvelle méthode pour la détermination du mouvement de la Lune", then generalized nine years later in "Sur une méthode d'intégration applicable au calcul des perturbations des planètes et de leurs satellites". After Delaunay's death, Airy and later Henri Andoyer (1862–1929) introduced certain corrections into the theory, which was put into tables by R. Radau. It was used for the lunar ephemerides of *La Connaissance des temps* from 1915 to 1925, but was then supplanted by the theory of Hill–Brown (to be dealt with below). The truth is that Delaunay's theory, admirable as it is, cannot, because of the tremendous calculations it entails, lead to suffciently precise lunar ephemerides.

Delaunay had studied at the École Polytechnique from 1834 to 1836, receiving the Laplace prize which consisted of the complete works of Laplace; the award led to his deciding on mathematics as a career. A mining engineer to begin with, he wrote a dissertation on mathematics and became a professor of the Faculty of Sciences of the University of Paris. Besides his theory of the Moon he published calculations on the second-order terms in the motion of Uranus, and an article on the secular acceleration of the Moon, of which we shall have more to say in the section devoted to this topic.

He was a member of the Royal Society and of the Academy of Sciences; the Academy's *Procès-ver-*

baux re-echo with the strife between him and Le Verrier, whose authoritarianism he opposed. He replaced Le Verrier as director of the Paris Observatory when the latter was removed from the position in 1870, but he died in 1872, drowned at the harbour of Cherbourg while on a promenade with his cousin.

We have already referred to Delaunay's method in the section on the problem of *n* bodies. He employed a canonical system in variables called 'Delaunay variables', and by 57 successive changes of variable eliminated the most important terms of the disturbing function, then eliminated the less important terms by simplified operations. In the end he obtained the development in literal form of the longitude, latitude, and parallax of the Moon to the seventh order in relation to the small parameters (ratio of the mean motions of the Sun and the Moon, eccentricities of the terrestrial and lunar orbits, inclination of the lunar orbit). Because of the slow convergence of certain terms, however, he found it necessary to add complementary terms, extrapolated and calculated in numerical form (some of these terms exceed 1″ of arc).

If we accept the comparisons made by F. F. Tisserand and Newcomb, we can judge the precision of the lunar theories reviewed so far by saying that the theories of Damoiseau and Pontécoulant had errors of the order of 4″; those of Hansen and Delaunay, errors of the order of 1″.

The theory of Hill–Brown

The works of Hill were the point of departure for a theory of the Moon's motion constructed by E. W. Brown. The tables that Brown derived from the theory were published in 1919 and served as the basis for the British and American lunar ephemerides until 1959, when they were refined by W. J. Eckert.

George W. Hill (1838–1914) (Figure 28.6) was one of the great American astronomers of the latter part of the nineteenth century. He was raised in the countryside at a time when transportation was slow and uncertain, and it was by the good luck of his going to Rutgers College and having Theodore Strong as a mathematics teacher there that he was put in the way of reading Nathaniel Bowditch's English translation of Laplace's *Mécanique céleste*. He went on to read in French the works of Poisson, Pontécoulant, Lagrange, Legendre, and others. Also through Strong's influence he became an admirer of Euler. In 1861 he began to work for the *American Ephemeris and Nautical Almanac* in Cambridge (Massachusetts).

28.6. G. W. Hill.

When in 1877 Newcomb became director of the Nautical Almanac Office, he transferred the operation to Washington, and, undertaking to remake the theories and tables of the motions of the Moon and planets, he turned to Hill for assistance.

The new method that Hill proposed for the study of the lunar motion was published in 1877 in a memoir entitled "On the part of the motion of the lunar perigee which is a function of the mean motion of the Sun and Moon", and then in a second memoir of 1878 entitled "Researches in the lunar theory". Euler's influence on Hill is seen here in his utilizing, as had Euler in 1772, a system of rotating axes. In setting out the problem Hill made the following simplifications: the Sun describes, round the centre of mass of the Earth–Moon system, a circular orbit with uniform motion; the perturbing function of the Moon's movement is limited to terms independent of the ratio of the semi-major axis of the lunar orbit to the semi-major axis of the terrestrial orbit. The latter condition reduces to saying that the radius of the circular orbit described by the Sun is infinite.

In his system of rotating axes Hill placed the axis

of abscissas in the direction of the Sun. He obtained a differential equation of which he sought a particular solution, dependent solely on the ratio m of the mean motion of the Sun to the mean motion of the Moon. The numerical value of this ratio, being very well known owing to numerous observations from Antiquity onwards, could be introduced at the very start into the equations of the curve that Hill found as a solution for his differential equation. The curve, called the intermediate orbit of Hill, is a closed oval centred at the origin in the plane of the rotating axes. When we later encounter Poincaré's theory of periodic orbits, we shall see that Hill's intermediate orbit is a particular case of Poincaré's periodic orbits of the first kind. Starting from the intermediate orbit of Hill, one can introduce the perturbations that depend on the eccentricity of the Moon's orbit, then those that depend on its inclination to the plane of the Earth's orbit about the Sun.

Ernest W. Brown (1866–1938) was born in England and died in the United States, to which he had emigrated in 1891 and where, as a professor at Yale University, he ended his career. At the instigation of G. H. Darwin, Brown attacked the problem of the Moon's motion, using Hill's method. The theory was completed in 1908; the tables derived from it were published in 1919. Hill had calculated all the terms of which the coefficients were greater than a hundredth of an arc-second, but the precision of the theory was in fact less. In particular, Hill had encountered inequalities in the longitude that would only be explained through recognizing the non-uniformity of a scale of time based on the rotation of the Earth.

The satellites of the Planets

The two satellites of Mars, Phobos and Deimos, were discovered by Asaph Hall in 1877. They are very small and very close to the planet, so that to observe them was difficult. Since these satellites move very nearly in the plane of Mars's equator, which is itself inclined by 27° to the plane of the orbit of Mars about the Sun, the question arises whether this arrangement is stable. The question was investigated by Adams in 1879. He, and Tisserand after him, showed that whatever reasonable value one adopts for the flattening of Mars, the orbits of the satellites remain very close to the planet's equatorial plane.

"The determination of the motions of the Galilean satellites of Jupiter", wrote Tisserand in 1896, "constitutes one of the most beautiful problems of celestial mechanics." These four bodies revolve round Jupiter while attracting each other, just as do the planets in revolving about the Sun, but like the Moon in its motion round the Earth they are perturbed by the Sun. Hence their motions present simultaneously a problem of the planetary type, with all its inherent difficulties, and a problem of the lunar type, which also has its own special difficulties. It is also necessary to take account of the perturbations due to the flattening of Jupiter and, above all, to resolve the problems posed by the near-commensurability between the periods of certain of the satellites. For if we designate by n, n', n'' the mean motions of the first three Galilean satellites Io, Europa, and Ganymede, we have almost exactly $n - 3n' + 2n'' = 0$. Laplace showed that this relation is always true. It implies the presence of small divisors which are sources of great difficulty in the theory.

The first true theory of the motion of the Galilean satellites was the one constructed by Laplace in 1788, before the start of the period with which we are concerned (see Chapter 22 above). The second truly complete theory appeared at the very end of this period. It is due to Sampson who published tables in 1910, while his theory itself appeared only in 1921.

Ralph A. Sampson had been, like Adams, a student at St John's College, Cambridge. In 1893 he was named professor of mathematics at the University of Durham; in 1910 he became Astronomer Royal of Scotland and professor of astronomy at the University of Edinburgh. His theory is semi-analytic, that is, the series representing the solution are trigonometric series of which the coefficients are given in numerical form. The method employed derives from that of Hansen. Sampson's theory has recently been corrected and modified, and is still the basis for the ephemerides published today; newer theories now in preparation will eventually replace it.

In the century and more separating the dates of appearance of the theories of Laplace and Sampson, there is little of note in the development of the theory of the Galilean satellites – nothing, indeed, producing a detectable improvement in the precision of ephemerides. J.-B. J. Delambre in 1817, and again Damoiseau in 1836, published tables based on Laplace's theory; but they are scarcely superior to the earlier tables constructed empirically on the basis of analyses of long series of observations. Souillart in 1880 and Tisserand in 1896 introduced important supplements to Laplace's theory; in particular, they modified the values of the principal terms of resonance. Cyrille Souillart (1828–98), after his studies at the École Normal Superieure in Paris, was a lyceum professor, then professor at the University of Nancy from 1867, finally at the University of Lille from 1874. François Félix Tisserand (1845–96), director of the Paris Observatory from 1893 to 1895, is best known for his remarkable *Traité de mécanique céleste*, which appeared in four volumes between 1888 and 1896, and which for astronomers and students became and remained the replacement for Laplace's treatise practically down to the present day. It is in the fourth volume of this work, after having expounded Laplace's and Souillart's contributions to the theory of the Galilean satellites, that he presents his own results.

Of the nine satellites of Saturn known in the nineteenth century, seven had already been discovered by the turn of the century. Hyperion was discovered simultaneously in 1848 by G. P. and W. C. Bond at Harvard and by Lassell in Liverpool in 1848, and Phoebe in 1898 by W. H. Pickering at Harvard. After the works of Laplace on the perturbations of the plane of the orbit of the satellite Japet, we should make special mention of the results published by Tisserand on this same satellite and on the rings of Saturn in 1880, and above all the works of Newcomb on the perturbations of Hyperion by Titan (1891) and those of Hermann Struve on Tethys, Mimas, Dione, and Enceladus. We shall return to Newcomb in the section on planetary theories. As for Hermann Struve, he was a member of the famous line of astronomers that began with Wilhelm Struve (1793–1864), founder of the Pulkovo Observatory, and was in fact the latter's grandson. Himself an astronomer at Pulkovo Observatory for a time, he was subsequently named director of the observatory of the University of Königsberg in 1895, and director of the observatory of Berlin–Babelsberg in 1904.

As the satellites of Saturn are bodies of low brightness, the observations of them are few and of poor quality. The observations do show, however, that the ratio of the mean motion of Tethys to that of Mimas, and the ratio of the mean motion of Dione to that of Enceladus, are both close to $\frac{1}{2}$, and that the ratio of the mean motion of Hyperion to that of Titan is close to $\frac{3}{4}$. As is well known in celestial mechanics, such small whole-number ratios induce terms of resonance which are a source of difficulty in the theories. Hermann Struve showed that the conjunctions of Enceladus and Dione are always approximately in the direction of the point of the orbit of Enceladus closest to Saturn (the perisaturn) and that they oscillate round this point; further, that the conjunctions of Mimas and Tethys oscillate with a period of 68 years and an amplitude of 45° round the middle of the arc of Saturn's equator comprised between the ascending nodes of the two orbits.

Newcomb, for his part, studied the effects due to the resonance between the periods of Titan and Hyperion, and showed that the point of Hyperion's orbit farthest from Saturn (the aposaturn) has a movement of libration round the middle point of the arc in which the conjunctions of Titan and Hyperion occur.

Two of the satellites of Uranus, Titania and Oberon, were discovered by William Herschel, the discoverer of Uranus itself; two others, Ariel and Umbriel, were discovered by Lassell in 1851. During the period that concerns us there was little progress in the theory of these satellites. Newcomb showed that they move approximately in the same plane, inclined at 98° to the plane of the ecliptic.

In 1846, shortly after the spectacular discovery of Neptune, Lassell discovered its satellite Triton (the second satellite, Nereid, was discovered in 1949). This discovery, as previously remarked, made it possible to calculate the mass of Neptune. Between 1848 and 1892 the longitude of the node of the satellite's orbit advanced by 7°; the advance was attributed to the flattening of Neptune. Newcomb and Tisserand published several works devoted to this question.

The planetary theories

Various works

Before treating of the complete (indeed, magnificent) theories of Le Verrier and Newcomb, we shall review the results obtained in planetary theory during the nineteenth century, except for those having to do with the perturbing function, which were dealt with in an earlier section.

Laplace in 1773, and in a more complete way

Lagrange in 1776, showed that the semi-major axis of the orbit of a planet perturbed by other planets was subject only to periodic variations, and not to secular variations (perturbations, that is, proportional to the time or to the powers of the time), which would have had the long-term effect of dispersing the solar system. This result, however, was demonstrated only for perturbations of the first order with respect to the planetary masses. In 1808 Poisson extended the theorem to the second order of the masses, and Lagrange attempted to give a simpler demonstration; his calculations, however, were mistaken. In 1875–76 Tisserand gave a correct demonstration of Poisson's theorem. S.C. Haretu, however, in the thesis that he defended at the University of Paris in 1878, showed that among the terms of third order with respect to the masses there are some that are secular – a result confirmed in 1889 by D. Egenitis. But this result is somewhat artificial, in that the secular terms appear as such only for reasons of ease of calculation. Newcomb, in fact, showed in 1876 that the orbital elements of a planet can be represented by series in which time is the variable, and which, while satisfying the differential equations of motion, are purely periodic. The question that remained concerned the convergence of these series – a question resolved in the negative, as we shall see later, by Poincaré.

In general, therefore, it appeared appropriate to calculate the secular inequalities of the orbital elements, basing the calculation on the mean value of the perturbing function with respect to all the terms of short period. In 1818 Carl Friedrich Gauss (1777–1855) showed how to resolve this problem by taking the mass of each planet as distributed in a sort of ring along the length of its orbit, the density of the ring at each point being less in the measure in which the planet's speed at this point is greater. He then showed how to determine the deformations of such a ring when subjected to the attraction of other such rings. Applications of this method were published by Hill in 1882 and by P. J. O. Callandreau in 1885.

Before taking up the planetary theories of Le Verrier and Newcomb, we should mention a method of development of such theories proposed by Hansen in 1829 and again in 1859; it is similar to the method he employed for the lunar theory. It permits what no one had known how to do previously, the application of all the perturbations, whether of long or short period, to the mean longitude, or, what comes to the same thing, to the mean anomaly; the earlier practice, stemming from Laplace, had been to apply the perturbations of long period to the mean longitude and those of short period to the true longitude. The new calculation of the true anomaly, by starting from a mean anomaly containing all the perturbations, thus furnished at one stroke the true longitude as affected by all the perturbations; there remained only a few corrections to be applied to the radius vector and the latitude. To determine these, Hansen considered an ellipse of fixed dimensions situated in the osculating plane at the given instant, that is, in the plane containing the radius vector and velocity vector at this instant; he then obtained by means of a single function the perturbations of this ellipse in the plane. Also original in his procedure is a special mode of calculating certain integrals with respect to the time, whereby he avoids determining the result as the difference of two large numbers – a determination in which precision is inevitably lost. For this purpose he initially uses one symbol for the time in the functions to be integrated, and another symbol for its occurrences elsewhere; after the integration he returns to a single symbol for all occurrences.

The planetary theories of Le Verrier

Le Verrier devoted his scientific efforts almost exclusively to the construction of extremely precise theories of the motion of the planets – theories which served for the calculation of ephemerides of these bodies in the *Connaissance des temps* up to 1984.

In 1839 Le Verrier presented to the Academy of Sciences a memoir "Sur les variations séculaires des orbites des planètes". In 1843 there appeared a second memoir of his entitled "Détermination nouvelle de l'orbite de Mercure et de ses perturbations". From 1844 to 1847 he presented several memoirs on the periodic comets, in particular on Comet Lexell and Comet Faye – studies which he interrupted in order to resolve the problem of the perturbations of Uranus. We have already cited the memoirs that he published on the latter subject.

Having decided to devote himself to the study of the motions of the planets, he established his plan of work in July 1849, and in the same year published a study of the perturbing function in the first volume of the *Annales de l'Observatoire de Paris*. This publication, founded at Le Verrier's initiative, carried in subsequent volumes the results of all his further work on the planets. A little later the *Astronomical Papers of the American Ephemeris* were to play the same role for Newcomb.

The construction of planetary theories took place in stages: first, it was necessary to obtain a solution to the equations of motion, and in Le Verrier's time, to translate this solution into tables, since other means for directly calculating the positions were lacking; second, the positions of the planet under consideration had to be calculated for all the dates of the available observations, and the orbital elements then modified in such a way as to reduce the discrepancies between these positions and those furnished by observation; finally, if it proved impossible to reduce the discrepancies to the level of the observational error by this route, it would become necessary to modify the values of certain of the constants entering into the problem, in particular the planetary masses. It could happen that this expedient of last resort would itself prove insufficient (we shall see an example), and it then became necessary to make new hypotheses as to the causes of such phenomena. If one remembers the limited methods of calculation available at the time, the magnitude of the task will be apparent.

In 1852, Le Verrier published the results of the reduction of the 9000 meridian observations of the planets made by Bradley between 1750 and 1762. In 1858 he published his theory of the Sun's movement. Here he introduced perturbations of the second order with respect to the perturbing masses; the mean longitude of the Sun for 1850 thus received a correction of 5″. In addition he found it necessary to diminish the mass adopted for Mars by 10% and to augment that of the Earth by 10%. We know today, in the light of modern theories based on numerical integration, that the errors in Le Verrier's theory can go as high as 1″.5.

Le Verrier dealt with the theory of Mercury's motion in a volume published in 1859. Here he met with the unexplained advance in the perihelion, to which we shall devote a special section. In 1861 he published his theory of the motion of Venus. One finds here a new inequality discovered by Airy as well as an inequality of the second order due to the Earth and Mars. In developing this theory Le Verrier was led to diminish the mass then adopted for Mercury and to augment the Earth's mass. For determining the constants of integration he employed, besides the meridian observations made from 1751 to 1857, two transits of Venus across the disk of the Sun observed in 1761 and 1769. We know today that the errors in the position of Venus calculated from Le Verrier's theory can go as high as 19″.

In this same year he published his theory of the motion of Mars, based on meridian observations made between 1751 and 1858, and on a conjunction of the planet with the stars ψ_1, ψ_2, ψ_3 of Aquarius, which had been observed in 1672 in France by Jean Picard and at Cayenne by Jean Richer with a view to calculating the parallax of the Sun and thus its distance from the Earth. Le Verrier was led to introduce into the longitude of Mars a hitherto unrecognized term of amplitude 1″.5 and of period 40 years, and two previously unknown inequalities of the second order, one due to the Earth and Venus and the other to the Earth and Jupiter. His memoirs on the theories of the Sun, Mercury, Venus, and Mars won him in 1868 the gold medal of the Royal Astronomical Society.

The problem of the motions of Jupiter, Saturn, Uranus, and Neptune is much more difficult, and Le Verrier was unable to complete his treatment of these planets prior to his death. In November 1872 there appeared a first "Mémoire sur les théories des quatre planètes supérieures" which dealt with the secular terms for these planets. Memoirs dealing with the individual planets appeared in the next two years. But at the time the available means of calculation made it impossible to construct ephemerides by summing trigonometric series, as is done today with the aid of electronic computers. It was necessary to form tables, generally of double entry, and to interpolate in computing the several inequalities at a given instant. The tables for Jupiter appeared in 1874, those for Saturn in 1875, those for Uranus in 1876. The tables for Neptune were completed by A. J.-B. Gaillot and presented by him in October 1877, a month after Le Verrier's death.

The theories of the motions of Jupiter and Saturn are difficult partly because of the magnitude of the perturbing forces, the two planets being massive and far from the Sun, and partly because the ratio of the mean motion of Jupiter to that of Saturn is almost exactly 5:2. As a consequence there is a 'great inequality' in the longitudes of the two planets, of which the period is close to 900 years and the amplitude about 20′ for Jupiter and 50′ for Saturn. C. G. J. Jacobi found a method for isolating the corresponding terms of resonance in the perturbing function, and published this result in 1849. Le Verrier's theory represented the observations of Jupiter made between 1750 and 1869 with an error in longitude less than 1″, and, after certain corrections were introduced by Gaillot, the observations of Saturn made from 1750 to 1890 with an error in longitude less than 3″ in absolute value. In subsequent decades these maxima did not increase; in the 1970s the errors were still found to be always approximately the same.

The planetary theories of Newcomb and Hill

We group under this heading the works of Newcomb on the four planets closest to the Sun, and those of Hill on Jupiter and Saturn tackled by Hansen's method. These planetary theories served as the basis of the *American Ephemeris* until 1984. To such a degree were they the product of the will and direction of Newcomb that they are often referred to by his name alone.

Simon Newcomb (1835–1909) (Figure 28.7), like many Anglo-Saxon astronomers of his time, learned celestial mechanics from Bowditch's English translation of Laplace's *Mécanique céleste*. In 1857 he was appointed to the Nautical Almanac Office in Cambridge, Massachusetts, and starting in 1863 he carried out determinations of the right ascensions of stars with the meridian circle of the Naval Observatory in Washington DC. With a view to improving the constants of the lunar theory, he travelled to the Paris Observatory in 1871, and there located observations of good quality going back as far as 1672. His visit was during the revolutionary insurrection known as the Paris Commune. Delaunay, as director of the observatory, presided over a deserted and menaced institution; on 23 April he wrote to his wife:

All the astronomers have left, and so too the servants, the gardener, and the porter . . . , I live almost alone in the building; it has not been invaded and probably will not be. It has been sufficient for me to hoist the red flag in place of the tricolor, following the order given me by the Paris Commune. I have with me here an American astronomer with his wife. He works all day copying in our library, especially such documents from among our older manuscripts as can be of interest in his work.

28.7. Simon Newcomb.

In 1877 Newcomb was named superintendent of the Nautical Almanac Office, which by then had moved to Washington; he remained in charge until his retirement in 1897, and thereafter continued to work at the Almanac Office as a scientific adviser.

Why undertake to construct new planetary theories at the moment when Le Verrier's theories appeared? First of all, because the empirical basis had been augmented with observations of good quality. But there was a more profound reason: in order to make the theory of each planet fit the observations, Le Verrier had been led to attribute to one and the same planet different masses in the different planetary theories in which it figured as a perturbing body, and to employ similarly varying values for other constants entering into the theories. Newcomb, in the preface of his memoir entitled "The elements of the four inner planets and the fundamental constants of astronomy", which appeared in 1895, expressed his reaction to this inconsistency as follows:

The diversity in the adopted values of the elements and constants of astronomy is productive of inconvenience to all who are engaged in investigations based upon these quantities, and injurious to the precision and symmetry of much of our astronomical work .... On taking charge of the work of preparing the American Ephemeris *in 1877 the writer was so strongly impressed with the inconvenience arising from this source that he deemed it advisable to devote all the force which he could spare to the work of deriving improved values of the fundamental elements and embodying them in new tables of the celestial motions.*

Newcomb was thus the first to concern himself with the construction of a coherent system of astronomical constants. After him this idea slowly made progress and ended by being erected into a dogma by the International Astronomical Union, which endeavoured to supply astronomers with such a system. It can seem astonishing that a goal so logical did not come to be adopted more easily. But this goal is much more difficult to realize than initially one might suppose – at the present time perhaps more than ever.

Newcomb's earliest work on planetary theory concerned the development of the disturbing function. Newcomb undertook to generalize the results that Laplace had obtained for circular orbits in introducing the 'Laplace coefficients' – certain functions of the ratios of the radii of these orbits – for elliptical orbits. To do this he created 'Newcomb's operators', which operate on the Laplace coefficients, and he tabulated the functions obtained to the eighth order in the eccentricity. These results appeared in 1891.

He then attacked the problem of the motions of Mercury, Venus, Mars, and Earth, employing 62000 meridian observations, which he had to reduce, as well as the data obtained from observations of the transits of Mercury and Venus across the solar disk. Like Le Verrier, he too found himself confronted with the unaccounted-for advance in the perihelion of Mercury; the explanations that he proposed will be described later on. The memoirs devoted to these planets, "The elements of the four inner planets, and the fundamental constants of Astronomy", appeared in 1895. The corresponding tables appeared in 1898.

Using the method of Hansen previously described, Hill during this time worked on the theory of the motions of Jupiter and Saturn; the result appeared in 1890. The same volume also contains an exposition of Hansen's method that was clearer than that supplied by Hansen himself! The corresponding tables appeared in 1898 along with tables for Uranus and Neptune, but Hill is not very explicit as to the theory on the basis of which the latter tables were constructed; he refers merely to an "investigation as yet unpublished". It seems it has never appeared.

What can be said as to the precision of Newcomb's and Hill's theories? In the case of the four inner planets, the maximal error appears to be 3″ for Venus; in the case of the outer planets the errors are of the order of a second. Newcomb's and Hill's theories are therefore superior to those of Le Verrier and Gaillot.

Other works

The asteroids

On 1 January 1801, Guiseppe Piazzi of Palermo discovered in the constellation of Taurus a faint star which was not listed in the catalogue of N.-L. de Lacaille: it was the asteroid, or minor planet, Ceres (see Chapter 24). Gauss calculated its orbit and found the semi-major axis to be about 2.8 astronomical units, thus falling in the space left between Mars and Jupiter by the empirical law of Titius–Bode. In March 1802 H. W. M. Olbers discovered Pallas, then in 1804 K. L. Harding discovered Juno and in 1807 Olbers discovered Vesta. By 1860 sixty-four asteroids (a term coined by William Herschel) had been discovered, by 1891 some three hundred.

It was at this point that Max Wolf proposed a method that consists in photographing a region of the sky for several hours, while compensating for the Earth's rotation by rotating the telescope in the opposite sense. The stars appear on the photographic plate as points, but the asteroids, because of their proper motions, leave a rectilinear track. With the aid of this method, the total number of asteroids identified had been increased by 1912 to eight hundred.

In 1772 Lagrange had shown that if an equilateral triangle is formed in the plane of the relative orbits of two bodies, with the bodies occupying vertices, then a body at the third vertex will be in a position of stable equilibrium. Remarkably, starting in 1906, asteroids began to be discovered that occupied precisely such positions in relation to the Sun and Jupiter. These are the 'Trojan' planets, so-called because they have been named after the heroes of the Trojan War: Achilles, Hector, Priam, etc. Their orbits round the Sun are similar to Jupiter's, but they are located in two groups 60° to the east and west of Jupiter.

The problem of the motion of the asteroids, as compared with the corresponding problem for the major planets, presents a number of special aspects. The orbital eccentricities are large and can reach 0.4; the inclinations of the orbital planes to the ecliptic go as high as 35° (the case of Pallas); the principal perturbing planet is Jupiter, the most massive planet in the solar system; the ratios of the semi-major axes of the orbits of the asteroids to that of the orbit of Jupiter are between 0.4 and 0.8 (at least for the asteroids known at the beginning of the twentieth century). Finally, between the mean motions of Jupiter and certain asteroids there are relations of near-commensurability, which introduce resonances, phenomena mentioned earlier in our discussion of 'the great inequality' of Jupiter and Saturn.

Here the classical methods are not very effective, although Damoiseau attempted to apply them to Ceres and Juno (*Addition* to *La Connaissance des temps pour 1846*), and H. J. A. Perrotin used them with success for Vesta (*Annales de l'Observatoire de Toulouse*, 1880). Gauss, in a letter to Bessel, describes a theory he had constructed for Pallas, which contained 800 inequalities and which he intended to compare with a calculation by quadratures (that is, a numerical integration). However, this theory was not published during Gauss's lifetime.

Le Verrier had found that a resonance term of period 800 years in the motion of Pallas had an amplitude of 895″. A.-L. Cauchy (1789–1857), although a mathematician rather than an astronomer, was charged by the Academy of Sciences with verifying this result, and in six weeks invented an ingenious method which he published in 1845 in the *Comptes rendus*.

Earlier we mentioned Hansen's method and indicated its principal characteristics. This method was employed by F. F. E. Brunnöw for the motion of the asteroids Iris and Flora, by L. Becker for that of Amphitrite, and by O. L. Lesser for the motions of Pomona, Metis, etc. G. Leveau applied the method to the construction of a very complete theory of the motion of Vesta which was published between 1880 and 1892.

Another important method is that due to Hugo Gyldén, director of the observatory of Stockholm, which appeared in Swedish in 1874. An exposition is given in Vol. 4 of Tisserand's *Traité de mécanique céleste*. Gyldén made it his aim to avoid the appearance of secular terms; also, he took the true anomaly as independent variable instead of the time. He was led to a linear differential equation of the second order which he resolved; in it the second derivative of a function is equal to the product of this function and a periodic function of the variable. Poincaré was greatly interested in Gyldén's method, and devoted to it an important part of Vol. 2 of his *Méthodes nouvelles de la mécanique céleste*, which appeared in 1893. This method was applied by J. O. Backlund in 1875 to the motion of the asteroid Iphigenia, and by Callandreau in 1882 to the motion of the asteroid Hera.

An interesting peculiarity of the ensemble of the asteroids is that the distribution of the semi-major axes of their orbits is not uniform round the mean value of 2.8 astronomical units; there are lacunae, that is, zones empty of asteroids. These zones correspond to values of the semi-major axis for which the asteroid's period is resonant with the period of Jupiter. The American Daniel Kirkwood (1814–95), a professor of mathematics first at Delaware College, later at Washington and Jefferson College in Pennsylvania, and at the University of Indiana, was the first to draw attention to the existence of these lacunae, now called 'Kirkwood gaps' (see Chapter 26). In 1860 he gave examples of commensurability between the asteroids and Jupiter in a note to the *Astronomical Journal*, then proved the existence of the lacunae in an article entitled "On the meteors' which appeared in 1867 in the *Proceedings of the American Association for the Advancement of Science for 1866*. He later made a comparison between these lacunae and the spaces between the rings of Saturn, such as 'the division of Cassini' between ring *A* and ring *B*.

The explanation of this phenomenon is very difficult and has continued to be a subject of investigation down to the present day. Among the numerous memoirs devoted to it in the nineteenth century we mention in particular that of Callandreau, published in 1896, and that of K. Bohlin, which appeared in 1888 and gives a method for treating resonances that Poincaré later studied. In 1895 Bohlin applied his method to the asteroids whose mean motion is triple that of Jupiter.

Finally, a word concerning hypotheses as to the origin of the asteroids. It was tempting to think that these small bodies, so well placed according to the law of Titius–Bode, had arisen from the break-up of a large planet. This idea was suggested by Olbers after the discovery of Pallas. But as more and more new objects were discovered the hypothesis ceased to be tenable, and as described in Chapter 26, Newcomb in the late nineteenth century showed that the orbits then known belied it.

The appearances of Halley's Comet in 1835 and 1910

The return of Halley's Comet in 1759 (see Chapter 19) confirmed Newton's hypothesis that comets are part of the solar system like planets, but with orbits that can be very elongated and very steeply inclined to the ecliptic; with this, comets lost something of their mystery, and it seemed possible that the problem of their motions could be resolved by celestial mechanics. However, as we have seen, all the methods elaborated for constructing literal or semi-numerical theories of the motions of the planets, Moon, and satellites of the planets presuppose that the orbital eccentricities and inclinations are small, for they employ series developments in terms of these small parameters. Since the eccentricities and inclinations of cometary orbits are large, the number of terms of the series that would have to be retained becomes exceedingly, indeed prohibitively, large. Worse yet, the convergence of some of the series employed is no longer assured when the values of the eccentricities and inclinations become too large. Thus the developments in series of the functions of the radius vector and of the true anomaly that enter into the disturbing function fail to converge for all values of the time unless the eccentricity is below the critical value 0.662 743. Similarly, if a comet has an orbit of which the semi-major axis is equal to 3 astronomical units, then the inverse of the distance from the comet to Jupiter, which figures in the disturbing function for the perturbations caused by Jupiter, can be developed in a converging series only if the inclination is less than about 30°; and this limit of inclination becomes smaller in the measure in which the orbit of the comet comes closer to that of Jupiter.

For this reason the problem of the motion of comets is almost always treated by the numerical integration of equations. The different methods used and their history are described in Chapter 29 below, and we limit ourselves here to citing the most important of the works on comets to which these methods led.

(i) In June 1770 Messier discovered a comet whose orbit was studied by A. Lexell (1740–84), member of the Academy of Sciences of St Petersburg; it was henceforth called Comet Lexell. The orbit proved to be an ellipse with a semi-major axis equal to 3 astronomical units, so that the cometary period was 5 years and 7 months. With such a short period, the comet should have been discovered much earlier; moreover, it was lost some years after its discovery. It was then shown, in particular by Burckhardt working at Laplace's instigation, that passages of the comet very close to Jupiter had twice profoundly modified the orbit. This posed a peculiar problem which, by the techniques of numerical calculation available at the time, could not be satisfactorily resolved. Laplace, here following an idea of d'Alembert, then introduced the notion of 'sphere of activity'. In a certain part of the trajectory of a comet perturbed by Jupiter, one considers the motion as heliocentric and proceeds to determine the perturbations due to Jupiter. But close to Jupiter, within its 'sphere of activity', it is more advantageous to consider the motion of the comet as jovicentric, and to calculate the perturbations due to the Sun. Le Verrier in 1857 applied this method to Comet Lexell in his memoir "Sur la théorie de la comète périodique de 1770".

(ii) An interesting problem was posed and resolved by Tisserand in 1889, using the rule now known as 'Tisserand's criterion'. The problem is to know whether or not two periodic comets which appeared at two different epochs are the same body, the orbital elements of which were modified by a passage close to Jupiter. Employing a first integral of the restricted three-body problem found by Jacobi, Tisserand showed that the two comets are the same provided that

$$\frac{1}{2a}+\sqrt{\frac{a(1-e^2)}{a'^3}}\cos i=\text{const},$$

where a is the semi-major axis of the cometary orbit, a' that of the orbit of Jupiter, e the eccentricity of the cometary orbit and i the inclination of the plane of the cometary orbit to the plane of the orbit of Jupiter. The eccentricity of Jupiter's orbit is neglected, but in 1892 Callandreau showed how to take account of the first power of this eccentricity.

The influence of Jupiter on a great many comets besides Lexell's was recognized and studied from the middle of the century onwards. D'Arrest showed in 1857 that in 1842 Comet Brorsen had approached Jupiter with the result that its perihelion distance was diminished by half. Similarly in 1890 R. Lehmann-Filhés studied the effects of the passage close to Jupiter in 1875 of Comet Wolf, discovered in 1884. The problem of the capture of parabolic comets by Jupiter, that is, of the transformation of a parabolic into an elliptical orbit owing to perturbations by Jupiter, was also broached by Tisserand in Vol. 4 of his *Traité de mécanique céleste*. He there cites a memoir of 1891 by H. A. Newton (not to be confused with Isaac Newton!) in which it is shown that the probability of the capture of a parabolic orbit by Jupiter is very small.

(iii) The fact that non-gravitational forces play a significant role in the motions of many comets was first brought into prominence by Encke. Johan Franz Encke (1791–1865) began his career in astronomy in a post that Gauss obtained for him at the observatory of Seeberg near Gotha. In 1825 he became director of the Berlin Observatory, and in 1844 he was named a professor at the University of Berlin. He is known for his work on the first four asteroids to be discovered, and also as founder of the *Berliner Astronomisches Jahrbuch*, providing ephemerides which he published in collaboration with J. P. Wolfers from 1830 until his death.

On 26 November 1818 J. L. Pons discovered a comet that remained observable until mid-January 1819. Encke determined its orbit, and found it to be periodic with a period of 3.5 years. More important, he showed that the same comet had been seen by Pierre-François-André Méchain, Charles Messier, and J.-D. Cassini (Cassini IV) in 1786, by Caroline Herschel in 1795, and by Pons, J. S. G. Huth, and Alexis Bouvard in 1805. He predicted its return in 1822 and again in 1825. The comet, which was named Comet Encke, was the second comet whose periodicity had been established with certainty in the years since the return of Halley's Comet in 1759. It was noticed that its period diminished constantly (and inexplicably) by about two and a half hours from one perihelion passage to the next. Encke suggested that the friction of a resistant medium could account for the phenomenon: a force of friction opposed to the velocity, while diminishing the eccentricity of the orbit, also diminishes the semi-major axis and hence the period – a problem that was to be studied by A. V. Bäcklund in 1894. On 28 June 1851 at the Leipzig Observatory d'Arrest discovered the comet that bears his name, another comet of short period. It was later shown (in 1939) that it, too, has an acceleration, but that the hypothesis of a resistant medium fails to account for it. Not until 1950, with the appearance of Fred Whipple's model, which takes the nucleus of a comet to be a "dirty snowball" formed of a mixture of ice and dust, was it possible to give a satisfactory account of the non-gravitational forces.

(iv) The period with which we are concerned saw two returns of the celebrated Halley's Comet, in 1835 and in 1910.

Among those calculating predictions of the passage of 1835 we meet once more the two competitors in the prize contest on the lunar theory, Damoiseau and Pontécoulant. Damoiseau in 1820 predicted the comet's return for 17 November 1835. He took account of the perturbations by Jupiter, Saturn, and Uranus (a planet unknown at the time of the previous passage in 1759) over the period extending from 1682 to 1835. He then added the perturbations due to the Earth in a second memoir which appeared in 1829; this brought the predicted date of passage to the evening of 4 November. Pontécoulant, with the same hypotheses as Damoiseau, published three memoirs, which appeared between 1830 and 1835. His final prediction for the passage was 12 November at 10 p.m. Others who tackled the problem were J. W. H. Lehman, who predicted perihelion would be on 26 November, and O. A. Rosenberger, who took account of all the planets then known and gave 12 November as the date of the perihelion passage. The comet was rediscovered on 6 August by Father Étienne Dumouchel of the Collegio Romano in Rome, and the perihelion passage occurred on 16 November at 10 a.m.

The prediction of the passage of 1910 was undertaken by Pontécoulant in 1864. Taking account only of perturbations by Jupiter, Saturn, and Uranus, he calculated that the comet would pass its perihelion on 24 May 1910 at 9 p.m. Later, however, P. H. Cowell and A. C. D. Crommelin in England found faults in Pontécoulant's memoir, and re-did all the calculations, which were the subject of numerous articles in the *Monthly Notices of the Royal Astronomical Society* in 1907 and 1908. Here they employed the method always used hitherto, which consists in finding by numerical integration the variations of the orbital elements; and they took account of all the planets from Venus to Neptune. Later they undertook a new prediction, employing a method that Cowell had introduced in 1908 for studying the motion of the eighth satellite of Jupiter; it consists in carrying out a numerical integration to determine, not the orbital elements, but the rectangular coordinates. According to their new calculation, the perihelion passage should occur on 17 April. However, by the time this publication appeared, the comet had already been rediscovered by Max Wolf at Heidelberg on a photograph taken on 11 September 1909, and so it was already known that the passage would occur on 20 April. The difference comes from the non-gravitational forces, which once more had been left out of account. The return of 1986 would be predicted, before the rediscovery, to within a few hours.

Two "small difficulties"

At the beginning of this chapter we quoted Poincaré's assertion that the goal of celestial mechanics is to decide whether Newton's law is true. Having completed our survey of the nineteenth-century achievements, let us see how and in what measure the question was answered. Among those who participated in the progress we have surveyed, we generally find that the response was a confident one. Thus, in the preface to the fourth and last volume of his *Traité de mécanique céleste* Tisserand stated: "everything advances with admirable consistency except for one or two small difficulties over which our successors will no doubt triumph".

We now turn to the two most significant of these "small difficulties". We have "triumphed" today over the second of them by taking into account the effects of friction due to the tides, hence due to forces ultimately of gravitational origin, and over the first, not indeed by rejecting the law of universal gravitation, but by taking it as a very good local approximation in the solar system to the behaviour of space-time according to the general theory of relativity.

The advance of the perihelion of Mercury

The Earth passes through the line in which the orbit of the planet Mercury intersects the ecliptic (the line of nodes) in the months of November and May. If at these times Mercury is at inferior conjunction, one can observe a transit of this planet across the disk of the Sun. These observations are easy to carry out and yield, among other results, a very precise determination of the planet's position. Le Verrier employed nine November transits from 1677 to 1848 and five May transits from 1753 to 1845. But he found it impossible to account for these observations theoretically; it was necessary to postulate that the perihelion of the orbit of Mercury rotated round the Sun not 527″ per century as the theory of planetary perturbations implied, but 565″ per century. There was a discrepancy of 38″ that he could not explain.

The first idea that came to mind was to increase the mass of Venus so as to obtain a suitable mean motion for Mercury's perihelion, but for this it would have been necessary to increase the Venusian mass by a tenth, an inadmissible increase: for the mass of Venus contributes very significantly to the secular variation in the obliquity of the ecliptic, and an augmentation by 10% would have led to variations in the obliquity completely incompatible with precise observations carried out since Bradley's time.

There remained the hypothesis of an unknown planet or group of planets, which by disturbing the motion of Mercury would constitute the cause of the anomaly. Such a hypothesis was well suited to seduce the discoverer of Neptune, and he undertook the calculations necessary to confirm it – calculations which he explained and commented on in 1859.

First he showed that the perturbing body must be located between the Sun and Mercury; if exterior to the orbit of Mercury it would have caused empirically inadmissible perturbations in the motions of the Earth and Venus (in the case of Venus, whose theory he had not yet completed, this conclusion was in part based on hypotheses).

Next he established a relation connecting the mass of the unknown planet with the radius vector of its circumsolar orbit (assumed to be circular). This relation, as one would expect, indicated that the closer the planet was to the Sun the greater must be its mass, if its action was to cause the observed effect on the motion of the perihelion of Mercury. An examination of this relation shows that at a distance of 0.17 astronomical units the planet would have a mass equal to Mercury's and be 10° from the Sun at its greatest elongation; it would thus be easily visible, even to the naked eye.

These figures were later revised because Le Verrier was found to have given Mercury a mass too large by a factor of two. If, as Tisserand shows in Vol. 4 of his *Traité de mécanique céleste*, a more accurate mass is taken for Mercury, then the unknown planet, if assigned a mass equal to Mercury's, would be at a distance of 0.21 astronomical units from the Sun, and its greatest elongation would be 13°, so that it would be even more easy to observe. If it were farther from the Sun, its mass, although smaller, would yet be sufficiently great that it could be easily observed, either in its greatest elongation or at the moment of a transit across the solar disk. If it were closer to the Sun than 0.21 astronomical units, its mass would become unreasonably large: for example, at 0.12 astronomical units from the Sun the mass would be six times that of Mercury, while the greatest elongation would still be 7°. Le Verrier, obliged to discard this solution, hypothesized that a ring of small bodies circulating between the Sun and Mercury could have the same effect, the reduced size of the individual bodies being the reason for their not having been observed.

Meanwhile a French amateur astronomer, E. M. Lescarbault, who lived at Orgères in the department of Eure-et-Loir, believed that he had observed on 26 March 1859 the transit of a planet across the disk of the Sun. He had noted the times and places of the contacts and had clocked the transit as lasting 1 hour 17 minutes. Informed of the observation nine months later in a letter from Lescarbault, Le Verrier visited him. From their discussion he concluded that the planet had a period of 19.7 days, and he determined the position of the plane of the orbit. From Lescarbault's estimate of the apparent diameter of the planet, and assuming a density equal to Mercury's, he found that its mass would be $\frac{1}{17}$ that of Mercury. The fact that the greatest elongation was 8° explained why the planet had not hitherto been observed. But the supposed discovery left the problem of the advance of Mercury's perihelion quite unresolved. The planet was given the name Vulcan, the merit of its discovery being attributed to Le Verrier. (Although partially effaced, the planet Vulcan can still be seen in a representation of the solar system engraved on the pedestal of the statue of Le Verrier erected in 1888 in the north court of the Paris Observatory.) No other observation of Vulcan has ever been made and it is now accepted that it does not exist.

It was clearly a possibility that numerous other observations of spots could be due to small bodies passing in front of the Sun, and in 1856–59 the Swiss astronomer Johann Rudolf Wolf drew up a list of these. Le Verrier sought to isolate among all these reports those that could relate to Vulcan, and even attempted predictions of transits for 1877 and 1882 (he died five years before the second of these dates). Neither of these transits was observed. Accordingly, Tisserand in 1896 concluded that Le Verrier's hypothesis to account for the advance of Mercury's perihelion should be abandoned.

Newcomb, in his reconstruction of the theories of the planetary movements, corrected the mass of Mercury, diminishing it by half. Also, he sought explanations for the anomalous part of the advance of its perihelion, which he put at 41″ per century rather than the 38″ found by Le Verrier. To begin with, he made clear that meridian observations are unsuitable for determining the secular variations of the orbital elements of the inner planets, and that without observations of transits Le Verrier's determination of the advance of Mercury's perihelion would have been impossible. He re-examined the hypothesis of an intro-Mercurial body and, like Le Verrier, concluded that the existence of a single planet in this region and of such mass as to produce the observed motion of Mercury's perihelion is to be excluded. As to the hypothesis of a hitherto unperceived ring, he thought it improbable: given the large mass that would need to be assigned to the ring, it would reflect too much light to remain undetected.

Another attempt to explain the anomalous advance of Mercury's perihelion was made by Hall in 1895. Hall's idea was that the exponent of the inverse of the distance in the law of universal gravitation was perhaps not 2 but a slightly different number. Taking this exponent as unknown, he calculated what value it would need to have in order to account for a 41″ advance of the perihelion of Mercury per century. His result was 2.000 000 151. However, it was then necessary to see what consequences this change, slight as it was, would have for the orbital elements of the other planets. All went well for the perihelia of Venus and the Earth, but an advance of the node of Venus's orbit that Newcomb had found, some 10″ per century, remained unexplained (according to present-day theory, the advance was largely an artefact). In the case of the Moon, Hall's hypothesis proved inadequate to account for certain anomalies which were in fact due to the inadequacies of the assumed measure of time.

It was in late 1915 that Albert Einstein published four articles in Berlin, including "Erklärung der Perihelbewegung des Merkur aus dem allgemeinen Relativistät Theorie". Einstein calculated the extra advance of Mercury's perihelion in the general theory of relativity and found it equal to 43″ per century. He wrote: "I have found an important confirmation of the Relativity Theory which explains qualitatively and quantitatively, without any special hypothesis, the secular precession of the orbit of Mercury discovered by Le Verrier." This result was taken as a brilliant proof of the exactitude of the new theory, just as the discovery of Neptune 70 years earlier had been taken as a brilliant proof of the exactitude of the law of universal gravitation.

The secular acceleration of the Moon

In the 1690s Edmond Halley found discrepancies between the observed and the calculated times for certain ancient eclipses, and suggested that the Moon might be subject to an acceleration. In the mid-eighteenth century, studies by Richard Dunthorne, Tobias Mayer, and J.-J. L. de Lalande established that the mean motion of the Moon increases by about 20″ per century.

In 1787 Laplace claimed to have accounted completely for the phenomenon by attributing it to a secular variation in the eccentricity of the terrestrial orbit caused by planetary perturbations. This result was confirmed in 1820 in the theories of both Damoiseau and Plana. The latter, in his literal theory, gave an expression for the secular acceleration as a function of m, the ratio of the mean motion of the Sun to that of the Moon. This expression contained terms in m^2, m^4, and m^5.

In 1854 there appeared in the *Monthly Notices of the Royal Astronomical Society* an article by Adams, who showed that Plana, by failing to take account of the transverse component of the perturbation, had committed an error in calculating the secular acceleration and that the coefficient of the term in m^4, which is negative, must be more than tripled in absolute value. This implied a secular acceleration of 16″ instead of the previously deduced 21″ per century per century. Thus, a few years after the discovery of Neptune, a new subject of Franco–British quarrel arose in which Adams, the co-discoverer of Neptune, was once more involved!

Pontécoulant bestrode this new war-horse and in 1860 published in the *Monthly Notices of the Royal Astronomical Society* some not very convincing calculations that claimed to prove that Adams was mistaken. In the very same issue Adams replied, analysing lucidly the calculations of both Plana and Pontécoulant and showing that there was indeed an error of principle. The quarrel abated, more especially as Delaunay, for his theory of the Moon, made the same calculations to the eighth order of m and found the same coefficient of the term in m^4 as had Adams. Finally, once everyone was in agreement, the theory was found to imply a secular acceleration of 12″ per century per century, while the analysis of the ancient observations yielded an acceleration about twice as large.

The answer to this dilemma is well known today: the longitude of the Moon was expressed as a function of the time as measured on the scale of Mean Solar Time, and this scale assumes that the rotation of the Earth round its axis is perfectly constant. If, on the contrary, the Earth is slackening in its rotation, the Moon will appear to accelerate. The question of the uniformity of the Earth's rotation had been posed by Immanuel Kant in 1754, and Laplace himself had remarked that a deceleration in the Earth's rotation would give the illusion of an acceleration in the movement of the Moon.

Delaunay in 1865 published an article in which he hypothesized that a slackening in the rotation of the Earth due to the friction of the tides was causing an apparent acceleration of the Moon. Indeed, he took the fact of the Moon's acceleration as proof that the Earth was slowing down, and proposed to evaluate its deceleration by measuring the Moon's acceleration. G. H. Darwin published in 1880 a memoir on the evolution of the Earth–Moon system, and there established the principles of a theory of the effects of the tides on the Earth's rotation. Poincaré cited this memoir in his *Leçons sur les hypothèses cosmogoniques*; but at the time of his death the dogma of the invariability of the Earth's rotation had not yet been definitively rejected by astronomers.

28.8. Henri Poincaré.

The work of Poincaré

Poincaré's name has appeared more than once in the preceding pages; indeed, many of the paths explored by nineteenth-century celestial mechanicians led to works on these same subjects by this mathematician of genius.

Born at Nancy on 29 April 1854, Henri Poincaré (Figure 28.8) belonged to a solid bourgeois family that had long been established in Lorrain. His father was a professor of medicine at the University of Nancy, and one of his cousins, Raymond Poincaré, a man of politics, was to serve as President of the French Republic from 1913 to 1920. A brilliant student, especially in mathematics, and winner in a general competition, Poincaré entered the École Polytechnique in 1873, and then the École des Mines. He occupied a post as engineer of mines while preparing a doctoral thesis which he defended in 1879. At first professor at the University of Caen, in 1881 he was appointed professor in the Faculty of Sciences in Paris, in which position he remained until his death in 1912. He was elected to the Academy of Sciences in 1887 and to the French Academy in 1908.

Poincaré's principal discoveries in celestial mechanics were published in the following works: "Mémoire sur le problème des trois corps et les équations de la dynamique", for which he received in 1889 the first prize in an international competition on the *n*-body problem that had been proposed by Oscar II, King of Sweden; the three volumes of *Les méthodes nouvelles de la mécanique céleste*, which appeared between 1892 and 1899; and the three volumes of *Leçons de mécanique céleste*, which appeared between 1905 and 1910. It is impossible here to enter into the details, which are necessarily very technical, of so considerable a body of work; we shall rather seek to sketch in outline the principal results he obtained.

We know that the equations of motion of n bodies have ten independent first integrals – ten functions of the coordinates and the velocities that remain constant during the motion. Six of these first integrals express the fact that the centre of mass of the system of bodies has a uniform, rectilinear motion in space; three others express the fact that the moment of momentum of the set of bodies is a fixed vector; and the tenth is the theorem of *vis viva*, which states that the sum of the kinetic energy of the ensemble of points of the system (due to the velocities of these points) and the potential energy arising from the gravitational attractions is equal to a constant. These ten first integrals make possible a resolution of the problem of two bodies. Poincaré showed that, whatever the number of bodies, there exist only ten first integrals – or rather, only ten *uniform* first integrals, the only sort that would be utilizable.

In 1878 Hill had found a particular intermediate orbit for the motion of the Moon. It is a closed curve when referred to a plane that rotates uniformly, and is called a periodic orbit in phase-space or simply a periodic orbit. Poincaré generalized this result in magisterial fashion by establishing that there are three families of periodic orbit in the three-body problem provided that two of the bodies are of small mass compared to the third (the case of the solar system). He demonstrated the importance of these orbits by showing how one can construct orbits close to periodic orbits and to asymptotic orbits, that is, orbits which approximate as closely as one pleases to a periodic orbit as the time tends to $+\infty$ or $-\infty$. He showed further that there can be doubly asymptotic orbits, orbits that approximate to periodic orbits as the time tends both towards $+\infty$ and towards $-\infty$. By means of what he called 'characteristic exponents' he showed that, close to any given periodic solution, there are nearby periodic solutions. Using integral invariants, he demonstrated that 'recurrent trajectories', that is, trajectories such that if they pass through a given region of phase space once do so an infinite number of times, are infinitely more numerous than non-recurrent ones.

Poincaré justified the investigation of periodic orbits as follows:

It seems at first that [this] can be of no interest in practice. In fact there is a zero probability that the initial conditions of motion should be precisely those corresponding to a periodic solution. But it can happen that they differ very little, and this occurs just in the case where the old methods are no longer applicable. One can then advantageously choose the periodic solution as a first approximation, as intermediate orbit, *to use the term introduced by Gyldén.*

Later on in the same memoir we meet the following formulation, which has often been quoted:

Moreover, what makes these periodic solutions so precious to us is that they are, so to speak, the sole breach by which we can attempt to penetrate into a stronghold hitherto believed impregnable.

As we have seen, the celestial mechanicians, in seeking solutions of the differential equations of motion of the celestial bodies, employed developments with respect to the small parameters of the problem, for example the masses; successive powers of these parameters figured as factors in the coefficients of the terms of trigonometric series, the arguments being functions of the time. But the methods of integration employed in practice sometimes made the time appear outside the arguments of sine and cosine, either in the form of 'secular terms', powers of the time, or in the form of 'mixed secular terms' or 'Poisson terms', products of a monomial of the time and a periodic function. Newcomb had shown that theoretically we can obtain the solution as a series of terms periodic in the time, but without any assurance as to the convergence of these series. Poincaré demonstrated that such series are not uniformly convergent, but that they permit us to have solutions approximating as closely as we please to the real solution during a given interval of time.

In addition to this work on the *n*-body problem, Poincaré carried out investigations of two other aspects of celestial mechanics. In 1885 he published in the *Acta mathematica* a memoir on the figures of equilibrium of fluids in uniform rotation. Colin Maclaurin had found particular solutions of this problem in the form of ellipsoids of revolution, and Jacobi had found another family formed of ellipsoids with three unequal axes. Poincaré found the most general figure answering to the question, to which he gave the name "figure piriforme" (pear-shaped figure); he believed it to be a figure of stable equilibrium. But A. M. Lyapunov and after him James Jeans showed that it is unstable, and ends by breaking into two distinct masses.

In 1892 Poincaré began an investigation into the problem of the tides, and this led him to discover solutions of integro-differential equations of Fredholm's type. In this problem the form of the coasts and the configurations of the sea bottom impose limiting conditions that make the solution of the equations especially complex. Poincaré returned to the problem in his *Leçons de mécanique céleste*; but his investigation of it was never to be completed.

Finally, we should mention the *Leçons sur les hypothèses cosmogoniques*, based on the course given by Poincaré in 1910 at the Sorbonne. It is curious to observe a genius like Poincaré giving himself so much trouble to expound theories, often merely confused, that had been born in the heads of incompetent dilletantes. Yet this work is pleasant to read and full of ingenious insights. We find there at the turn of a page the famous virial theorem which Jeans was later to exploit in founding stellar dynamics.

Such, rapidly surveyed, is the immense work of Poincaré in the domain of celestial mechanics. It is an achievement so concentrated and so profound that it is far from having been assimilated by those present-day investigators who continue to concern themselves – and profitably so – with a more classical celestial mechanics, and who are well aware that the exploitation of Poincaré's ideas, for instance those relative to periodic orbits, can supply the key to many problems.

Conclusion

The death of Poincaré in 1912 marks the end of the triumphal epoch of classical celestial mechanics. An astrophysics whose conquests were becoming ever more numerous and spectacular was thereafter to be the focus of attraction for students. It began to seem impossible to improve on the work of Le Verrier, Hill, Newcomb, and Brown, and for several decades the observers themselves ceased to make observations of the positions of planets and satellites, to the great regret of present-day astronomers. Almost alone J. F. Chazy undertook to make the connection between classical celestial mechanics and general relativity, and William de Sitter re-did the theory of the Galilean satellites.

In 1960, at last, there came a renewal of celestial mechanics – an extremely fruitful one, impelled forward by the prospect of the exploration of space, and aided by the advent of powerful computers. Yet, as we look forward to the moment when present-day investigators will have assimilated the ideas of Poincaré and other modern mathematicians, and vanquished the difficulties involved in bringing celestial mechanics completely under the aegis of general relativity, it is interesting to note that the brilliant results obtained recently are direct prolongations of the achievement of the "golden age"

Further reading

The work of many of the authors referred to appeared in publications that are not widely accessible but may be found in the libraries of nearly all observatories and universities. This is so, at least, for the following: *Annales de l'Observatoire de Paris, Monthly Notices of the Royal Astronomical Society, Astronomical Papers of the American Ephemeris, Astronomical Journal, Comptes rendus de l'Académie des Sciences, Astronomische Nachrichten.*

Tisserand's *Traité de mécanique céleste*, which was published by Gauthier–Villars in Paris between 1888 and 1896, contains expositions of methods and theories that are often clearer than those given by the original authors; this is the case, for instance, for Hansen's theory. The first two volumes of Tisserand's work were reissued by Gauthier–Villars in 1960.

For the discovery of Neptune, we refer the reader to *The Discovery of Neptune* by Morton Grosser, first published in 1962 and republished in paperback in 1979; it contains an extensive bibliography.

There is is also much to be learned from the relevant entries in the *Dictionary of Scientific Biography*, in particular those on Jacobi, Hamilton, Le Verrier, and Poincaré.

Appendix: The stability of the solar system from Laplace to the present

JACQUES LASKAR

As we have seen in Chapter 22, Pierre-Simon Laplace (1749–1827) established, to his own satisfaction and that of his contemporaries, the stability of the solar system: it was subject only to small, pendulum-like oscillations of short or long period. Before turning to the nineteenth-century developments, we review the nature of the relevant Laplacian proofs.

Motion in accordance with Kepler's three laws can be considered as motion of order zero; this is the motion that the planets would have if their masses were nil, so that they would not perturb each other's motions. Laplace's work consisted in describing the approximation of order 1. He studied the variations of the orbital elements of the ellipses (of order zero) due to perturbations caused by the planets taken as moving on the orbits of order zero. The types of inequality arising in the approximation of order 1, according to Laplace, were two: (1) *short-period inequalities*, which depend on the situations of the bodies either with respect to each other or with respect to their aphelia, and which have periods of only a few years; and (2) *secular inequalities*, which alter the orbital elements almost imperceptibly in each revolution, but end after millions of years by changing entirely the shapes and positions of the orbits. The latter inequalities are those that put in question the stability of the solar system.

An initial question was whether there were secular terms in the semi-major axes or, equivalently, in the periods or mean motions of the planets (according to Kepler's third law, the mean motion n of a planet is related to the semi-major axis a by the equation $n^2a^3 = \text{const}$). A change in the semi-major axes of the planetary orbits could have disastrous consequences, for a planet might then break loose from the system or two planets collide.

Laplace showed that, to an approximation carried to the third order in the eccentricities, there existed no secular terms in the semi-major axes or in the mean motions of the planets. Lagrange then extended this result to all powers of the eccentricities. According to Laplace, "M. de la Grange has extended [this result] to an unlimited time, in showing by an ingenious and simple analysis that the mean solar distances of the planets are immutable." But this statement exaggerates: the approximation carried out by Laplace and J. L. Lagrange takes into account only terms of the first order with respect to the planetary masses – a point to which we shall return.

These proofs were apparently contradicted by an observational result: Edmond Halley – here rediscovering an anomaly already detected by Kepler in 1625 – in the 1690s had found Jupiter to be accelerating, and Saturn decelerating, at rates such that the displacements in 2000 years amounted to $+3° 49'$ for Jupiter and $-9° 16'$ for Saturn.

Laplace resolved the discrepancy. He showed that the apparent variations in the mean motions of Jupiter and Saturn arose from an inequality involving the positions of both planets, but having a period sufficiently long to resemble a secular inequality. The inequality depended on the combination of longitudes $2\lambda_J - 5\lambda_S$; its period was about 900 years. His complete theory of the couple Jupiter–Saturn proved to be in excellent agreement with both ancient and modern observations. He drew the important corollary that the masses of comets must be very small, for otherwise they would have perturbed the motion of Saturn. Thus the new theory confirmed the constancy of the mean semi-major axes, and established Newton's law as the ultimate formula for the precise description of the motion of the celestial bodies in the solar system.

But constancy of the semi-major axes was not

sufficient for stability. For if the eccentricities of two orbits could vary considerably, these orbits could intersect, even though their semi-major axes remained constant. Laplace therefore turned to the problem of the eccentricities and inclinations.

For greater simplicity, we use complex notation, similar to that introduced by Simon Newcomb; it differs in nothing essential from the notation used by Laplace. We set

$$z = e \exp(\sqrt{-1}\tilde{\omega}),$$
$$\zeta = \sin(i/2)\exp(\sqrt{-1}\,\Omega),$$

where e, i, $\tilde{\omega}$, and Ω are the eccentricity, inclination, longitude of the perihelion and longitude of the node. Laplace showed that, if one considers terms of the first order in relation to the masses, and of the first degree in the eccentricities and inclinations, the solutions for the variables z_i and ζ_i of the different planets are given by a system of linear differential equations with constant coefficients:

$$\frac{d}{dt}\begin{bmatrix} z_1 \\ \cdot \\ \cdot \\ \cdot \\ z_k \\ \zeta_1 \\ \cdot \\ \cdot \\ \cdot \\ \zeta_k \end{bmatrix} = \sqrt{-1}\begin{bmatrix} A & 0 \\ 0 & B \end{bmatrix}\begin{bmatrix} z_1 \\ \cdot \\ \cdot \\ \cdot \\ z_k \\ \zeta_1 \\ \cdot \\ \cdot \\ \cdot \\ \zeta_k \end{bmatrix}$$

where A and B are real matrices with constant coefficients. The two systems in eccentricity and inclination are therefore decoupled, and their solutions are combinations of complex exponentials of the form

$$z = \sum_{j=1}^{k} \alpha_j \exp(\sqrt{-1}g_j t),$$
$$\zeta = \sum_{j=1}^{k} \beta_j \exp(\sqrt{-1}s_j t),$$

where α_j, β_j are constant values, the g_j are the eigenvalues of the matrix A, and the s_j the eigenvalues of the matrix B.

The whole problem of the stability of the solar system thus reduced to the calculation of the eigenvalues of the two real matrices A and B; these were 7×7 matrices in Laplace's time, since Neptune had not yet been discovered. If the eigenvalues were all real and distinct, then the solutions were quasi-periodic, and the eccentricities and inclinations were subject only to periodic variations about their mean values. But if one of the eigenvalues had a non-zero imaginary part, then the solutions contained an exponential instability, and an eccentricity could become very great, so that the possibility arose of two planets colliding. The possibility of collision also arose if two of the eigenvalues were identical.

In Laplace's time, the numerical calculation of the eigenvalues of a 7×7 matrix was not an easy matter. Laplace avoided this problem by finding a very simple demonstration showing all the eigenvalues of these matrices to be real and distinct – a demonstration valid, moreover, for any number of planets revolving about the Sun in the same direction. It is based on the constancy of the angular momentum. The total angular momentum of the system, if we neglect terms of order 2 with respect to the masses, is

$$C = \sum_{i=1}^{k} m_i \sqrt{a_i(1-e_i^2)}\cos i_i.$$

Noting that the semi-major axes are constant, and neglecting terms of degree 4 in the eccentricities and inclinations, Laplace derived the result that

$$\sum_{i=1}^{k} m_i \sqrt{a_i}\left(\frac{1}{2}e_i^2 + 2\sin^2\frac{i_i}{2}\right) = \text{const.}$$

But to this approximation, the eccentricities are independent of the inclinations. Laplace could thus write

$$\sum_{i=1}^{k} m_i \sqrt{a_i} e_i^2 = \text{const},$$
$$\sum_{i=1}^{k} m_i \sqrt{a_i}\sin^2\frac{i_i}{2} = \text{const.}$$

These equations must hold for all time; hence there could be neither a term proportional to the time nor a real exponential in the solutions, and Laplace therefore concluded that the eigenvalues of the matrices A and B were all real and distinct.

In sum, Laplace had shown that (1) the semi-major axes of the planets were constant, except for periodic variations of small amplitude and relatively short period; (2) the eccentricities and inclinations were subject only to periodic variations about their mean values. The latter variations included both short-period oscillations dependent on the positions of the planets, and secular variations of the orbits with periods ranging from about 50 000 to some millions of years.

Laplace's calculations, however, involved two fundamental approximations. He carried out the developments only to the first order in the masses, neglecting terms that depended on the squares and products of the planetary masses. And while the invariance of the semi-major axes was proved for all powers of the eccentricities and inclinations, the result concerning the smallness and quasi-periodicity of the variations in the eccentricities and inclinations was obtained only by neglecting terms of degree greater than the second. Because of these approximations, the duration of the validity of Laplace's proofs was limited. Although conscious of the dependence of his proofs on approximations, Laplace appears to have underestimated their importance.

The calculations of Le Verrier: a new formulation of the problem of the stability of the solar system

Urbain Jean Joseph Le Verrier (1811–77), besides playing a major role in the discovery of Neptune and carrying out the immense task of constructing new and more precise theories of all the planets, also pursued, in Vol. 2 of the *Annales de l'Observatoire de Paris* (1856), the study of the stability of the solar system. Having calculated the perturbing function between two planets to degree 7 in the eccentricities and degree 6 in the mutual inclination of the planets, he returned to Laplace's calculation of the secular variations of the eccentricities and inclinations, and pushed it to a higher order of approximation. Laplace had carried out the calculation only for the quadratic part of the perturbing function, and only for the planets Jupiter and Saturn. Le Verrier pushed the calculations to degree 4, and took all the planets into account. He found that in this case Laplace's beautiful demonstration of the stability of the solar system no longer held, or rather, no longer had the same implications.

Laplace had employed the conservation of angular momentum, truncated at degree 2 in the eccentricities and inclination, to obtain

$$\sum_{i=1}^{k} m_i \sqrt{a_i} e_i^2 = \text{const.} \qquad (1)$$

This relation puts a limit on possible values of the eccentricities, and hence leads to the conclusion that the eigenvalues of the secular system are all pure imaginaries; but the limit obtained is of practical interest only if the masses are of approximately the same value, as in the case of Jupiter and Saturn. As Le Verrier put the matter,

The eccentricity e will always remain very small if the planet m constitutes a considerable part of the sum of the masses of the system of planets. But no analogous conclusion can be drawn with regard to the planets whose masses are a small fraction of the total mass of the system; only the complete integration of the equations can show, with respect to these planets, whether their eccentricities will remain confined within narrow limits.

Thus the existence of a quasi-periodic solution for the secular system no longer sufficed to establish the stability of the solar system; in addition it was necessary that the amplitudes of the coefficients not be too large – a condition not guaranteed by the conservation of angular momentum. Le Verrier thus reformulated the conditions for stability:

The conditions necessary for the stability of our planetary system, relative to the eccentricities, are of two kinds: the first have to do with the nature of the roots of the equation in [g_i and s_i as above]; the second with the absolute magnitude of the coefficients, N, $N_1, \ldots, N'$, N'_1[α_j and β_j as above], etc.

Le Verrier also raised the question of the convergence of the series approximations he employed. The approximations carried out *a priori*, he knew, had a meaning only if the eccentricities and inclinations remained small:

Only by calculating the coefficients N, N′, ... can we know whether any of them will be large, and whether, as a consequence, any of the eccentricities can increase in such a way as to make the series – on which this entire analysis is based – but slowly convergent or even divergent. For the equation [(1) above], in which the value of the constant in the right-hand member is given by actual observations, shows us that the eccentricities of planets of large mass cannot increase beyond rather narrow limits; but it tells us nothing about the limits of the orbital eccentricities of planets of small mass.

Le Verrier therefore launched into a calculation of the eigenvalues; it consisted in resolving an equation of degree 7 (he omitted Neptune, initially, from the account). The task was extremely complicated, and involved successive approximations, with a decoupling of the systems of the planets. He obtained a complete solution for the linear secular system of the seven planets, and so was able to derive maximal values for the eccentricities and inclinations, and thus to show that these when calculated by the secular equations of the first order remained "eternally small'. What was new in his study was the consideration of the bodies of small mass, in particular the interior planets. For them, this result was obtained, Le Verrier emphasized, "only for the relations of the major axes that we have considered: we do not know what the result would be for other mean distances of the planets."

Le Verrier also laid the basis for a study of the secular resonances of the minor planets (asteroids).

There exists, for example, a position between Jupiter and the Sun such that if one placed there a small mass, in an orbit initially but little inclined to the orbit of Jupiter, this small mass could depart from its original orbit, and attain a large inclination relative to the plane of the orbit of Jupiter, owing to the action of Jupiter and Saturn. It is remarkable that this position is located approximately at a distance double that of the Earth from the Sun, that is, at the lower limit of the zone where all the minor planets have so far been found.

For the solar distance of the position of resonance Le Verrier found 1.977 AU, a value very close to that accepted today. Later, Le Verrier also suggested that the large inclination of Mercury's orbit could be due to secular resonances.

Le Verrier thus pushed the study of the linear system much further than Laplace. In a yet more important development, he raised the question of the influence of non-linear terms, that is, terms of degree 3 in the secular equations, or of degree 4 in the perturbing function.

Laplace's analysis had in effect yielded only the first term of an infinite series. This term was no doubt the most important part of the solution; but, did the later terms have only a negligible effect? Le Verrier posed the question of convergence in the manner of an astronomer who examines only the early terms of the series to see if they decrease with sufficient rapidity: he did not raise the problem of *mathematical convergence*. He then recognized that, for some initial conditions, the series failed to converge; he considered also that Laplace's approximation was insufficiently precise to yield a solution valid for a very long time; thus it was necessary to have regard to terms of higher degree, namely of degree 3 in the eccentricities and inclinations.

Here is the programme Le Verrier set for himself:

We propose to determine whether, by the method of successive approximations, the integrals do in fact develop in series sufficiently convergent so that one can answer the question about the stability of the solar system; and, in the second place, to give to these integrals all the accuracy possible in the present state of our knowledge of the masses of the planets.

We shall give an indication of Le Verrier's calculations in a simple example illustrating the problem of 'small divisors' in celestial mechanics. Suppose that we have already carried out Laplace's first reduction, and let us consider only a single non-linear term in one of the equations:

$$\frac{dz_j}{dt} = i(c_j z_j + z_1 z_2 z_3)$$

with $z_j(0) = A_j$, and where $z_1 z_2 z_3$ is a term of degree 3. If one considers only the linear part, one obtains by integration

$$z_j(t) = A_j e^{ic_j t}.$$

We seek a solution for small eccentricities e_i; in this case $z_1 z_2 z_3$ is very small, and we can assume that this term leads to only a small correction in the solution of the linear part. We introduce a small formal parameter, say ϵ, in carrying out a change of variables $z_j = \sqrt{\epsilon} z_j$ (the square root makes things prettier). By changing also the initial conditions $A_j = \sqrt{\epsilon} A_j$, we then obtain

$$\frac{dz_j}{dt} = i(c_j z_j + \epsilon z_1 z_2 z_3)$$

with the initial condition $z_j(0) = A_j$. We seek a solution in the form of an integral series in ϵ:

$$z_j = z_j^{(0)} + \epsilon z_j^{(1)} + \epsilon^2 z_j^{(2)} + \ldots.$$

There are many possibilities for the choice of the initial conditions of the $z_j^{(0)}$, and to simplify, we choose $z_j^{(0)}(0) = A_j$, $z_j^{(i)}(0) = 0$ for $i \neq 0$. By carrying out the development to the order 1 in ϵ and identifying terms in the same powers of ϵ, we find

$$\frac{dz_j^{(0)}}{dt} = ic_j z_j^{(0)}; \quad z_j^{(0)}(0) = A_j$$

$$\frac{dz_j^{(1)}}{dt} = i(c_j z_j^{(1)} + z_1^{(0)} z_2^{(0)} z_3^{(0)}); \quad z_j^{(1)}(0) = 0.$$

The whole problem reduces to the resolution of the last equation with

$$z_j^{(0)}(t) = A_j e^{ic_j t}.$$

This gives, if we denote $A_1 A_2 A_3$ by B,

$$\frac{dz_j^{(1)}}{dt} = i(c_j z_j^{(1)} + Be^{i(c_1+c_2+c_3)t}).$$

This last equation is resolved easily by the method of variation of constants introduced by Lagrange, yielding

$$z_j^{(1)}(t) = \{iB\int e^{i(c_1+c_2+c_3-c_j)t}dt + C\}e^{ic_j t},$$

where C is a constant of integration. If $c_1+c_2+c_3-c_j$ is not zero, we obtain

$$z_j^{(1)}(t) = \frac{1}{c_1+c_2+c_3-c_j}(Be^{i(c_1+c_2+c_3)t} - Be^{ic_j t}),$$

for we have $z_j^{(1)}(0)=0$. We see that the terms of degree 3 add a term of frequency $c_1+c_2+c_3$ and correct the amplitudes of the linear part. No difficulties arise unless the combination of frequencies $c_1+c_2+c_3-c_j$ is very small, in which case the effect of the degree 3 can become greater than that of the linear part, strongly contradicting a hypothetical convergence. This is the problem of the small divisors in celestial mechanics. Le Verrier did not face this problem for the moment because he had a more serious problem to deal with. For in the part of degree 3 there also exist terms for which the combination of frequencies is of the form $c_j+c_k-c_k-c_j=0$. Such a term is said to be resonant, and one then has

$$z_j^{(1)}(t) = iBte^{ic_j t}.$$

Thus a mixed term appears in the solution, with an amplitude Bt which increases with time; if this term corresponds to a physical reality, the stability of the solar system is compromised.

Le Verrier proceeded to eliminate the mixed term:

We have taken as point of departure the rigorous integrals obtained by keeping only the terms of the first order, and it is by varying the arbitrary constants introduced into these equations by integration that we have been able to take into account the terms of the third order. This procedure produced arcs de cercle *outside the sine and cosine signs; but we can eliminate them by changing appropriately the values of the arguments introduced in the first approximation.*

He introduced here the premisses of the method that Poincaré would call "the method of Lindstedt" for the resolution of such systems. The solutions causing secular terms to enter were (if we take account only of these terms) of the form

$$z = Ae^{ict} + iBte^{ict}.$$

Le Verrier eliminated the mixed terms in Bte^{ict} by changing slightly the frequency of precession c, which amounts to considering $1+itB/A$ as the beginning of a power series in $e^{i\delta ct}$ ($e^{i\delta ct}=1+i\delta ct\ldots$), where $\delta c = B/A$ is the correction of the frequency. He found corrections amounting to 0".3231 per year for the largest frequency g_6 related to the Jupiter–Saturn couple ($g_s = 22''.427$). A modification this large surprised him, for he knew that Laplace and Lagrange had thought their formulas to be valid for an infinite time: "[these] illustrious mathematicians ... were far from thinking that the terms of the third order could introduce into the arguments such large corrections".

To give an order of magnitude, one can say that an uncertainty of 0".3 per year in one of the frequencies of precession limits the duration of the validity of the solution to about three million years. But Le Verrier's judgment of his predecessors should perhaps be tempered, since the age of the Earth admitted in Laplace's time attained this value only in the boldest speculations of Georges-Louis Leclerc, Comte de Buffon (in print the latter, concerned perhaps to avoid shocking his readers, went only so far as to propose a duration of 75 000 years).

These corrections, however, were small in relation to the values already determined, and only the value of g_6 was notably changed. Le Verrier therefore believed that at least for the exterior planets, whose masses were well known, these values were very close to the true values of the frequencies:

We limit ourselves therefore to concluding from the preceding calculations that the corrections of the arguments are very small in relation to the arguments themselves, so that the series in which the integrals are developed are regarded as convergent.

Le Verrier appears to have been persuaded that further corrections would be much smaller, and that his solution would be utilizable for a time comparable to the age attributed to the Earth – some tens of millions of years in Le Verrier's time.

By means of this change, one can answer, at any epoch whatever, for the accuracy of the results, to within the limits fixed by means of the formulas we have given for taking account of the uncertainty of the values assigned for the masses. However, we make abstraction of the influence of the constant terms in the perturbing function, that depend on the higher powers of the masses, with which we are not concerned here.

Le Verrier was well advised to enter this *caveat*, for George W. Hill in 1897 showed that the contribution of the second order in the masses increased the value of g_6 by about 5″ per year because of the quasi-resonant term $2\lambda_J - 5\lambda_S$ between Jupiter and Saturn.

Finally, Le Verrier posed the question of the existence of small divisors in the system of interior planets – an especially important question because some of the masses were poorly known, and an admissible change of mass could make a divisor very small. The lack of determination of these masses, however, prevented him from settling the matter, and all he could do was to appeal to the mathematicians:

It thus appears impossible, by the method of successive approximations, to decide whether, in virtue of the terms of the second approximation, the system composed of Mercury, Venus, Earth, and Mars will enjoy stability indefinitely; and one must hope that the mathematicians, by the integration of the differential equations, will give a means of removing this difficulty, which could very well be only a matter of form.

Poincaré: the response of mathematics

Laplace had established the stability of the solar system in taking into account the linear approximation of the equations of its motion. Le Verrier, we have seen, was not so optimistic. He showed that the problems of small divisors met with in the system of interior planets prevented him from drawing a conclusion as to the stability of the solar system, and he asked the mathematicians to find new methods for integrating the differential equations of the motion, so as to be able to resolve this important problem.

Le Verrier had examined the influence of the non-linear terms in the development of the equations of the secular motion of the planets; his method was to investigate solutions in the form of series depending on a small parameter. He did not pose the problem of the convergence of these series as a mathematician would do it today. For him, a series converged, or rather converged sufficiently, if the successive approximations introduced successively smaller corrections. He thought that he would thus obtain a formula which would approach the true motion of the planets. For Le Verrier, this convergence did not hold universally, but depended on initial conditions. He knew that for certain of these initial conditions there emerged very small divisors that compromised the convergence of the series. The implied resonances could correspond to real instabilities, such as that of the orbits of minor planets with semi-major axes less than about 2 AU. But he was unable to reach systematic results, and his methods, despite improvements later introduced by Newcomb and Lindstedt, could not be extended easily to higher orders. Astronomy had arrived at an impasse; hence Le Verrier's appeal to the mathematicians.

The response to the question posed by Le Verrier was given some years later by Henri Poincaré (1854–1912). In the three volumes of his *Méthodes nouvelles de la mécanique céleste* (1892–99), Poincaré took an altogether new approach, and transformed celestial mechanics totally. He undertook to simplify the equations in order to study them in their greatest generality. For this purpose, he took advantage of the canonical formalism introduced by C. G. J. Jacobi (1804–51) and W. R. Hamilton (1805–65). He proposed to the astronomers a set of variables, very close to the variables that they had used since Laplace, but permitting a considerable simplification in the writing of the equations of motion, which thus became

$$\begin{cases} \dfrac{dI_i}{dt} = \dfrac{\partial H(I,\theta)}{\partial \theta_i} \\ \dfrac{d\theta_i}{dt} = -\dfrac{\partial H(I,\theta)}{\partial I_i} \end{cases} \qquad (2)$$

where (I_i, θ_i) are called canonical variables. The variables I_i are related to the semi-major axes, eccentricities, or inclinations of the planets, while the angular variables θ_i represent the longitudes of the planets, perihelia and nodes. The function $H(I,\theta)$, or Hamiltonian, represents the energy of the system. In the case of the solar system, this function takes the special form

$$H(I,\theta) = H_0(I) + \epsilon H_1(I,\theta), \qquad (3)$$

where $H_0(I)$ represents the energy of the planets orbiting round the Sun when one neglects their interactions, while $\epsilon H_1(I,\theta)$ represents the mutual interaction of the planets which is much smaller (the small parameter ϵ is of the order of the planetary masses). If one neglects the interaction of the planets, the second term is suppressed, and the equations can be integrated very simply because the variables of action I_i are then constant and the angles θ_i are linear functions of the time ($\theta_i = \nu_i t + \theta_{i0}$). In this case the motion can be considered as compounded of uniform circular motions with radii I_i and angles θ_i; the number of couples (I_i, θ_i) is called the number of degrees of freedom of the system. In the case of such a system with two degrees of freedom, the motions will be represented by curves on a torus. One sees here the power of this formalisation, for to seek to integrate these equations of motion now reduces to finding transformations which reduce the motion (which can appear complex) to the products of uniform circular motions. Poincaré showed that all the perturbational methods available to Laplace and Le Verrier reduced to seeking changes of variables in the form of series in the small parameter ϵ

$$\begin{cases} I' = I + \epsilon S_1(I,\theta) + \epsilon^2 S_2(I,\theta) + \epsilon^3 S_3(I,\theta) + \dots \\ \theta' = \theta + \epsilon T_1(I,\theta) + \epsilon^2 T_2(I,\theta) + \epsilon^3 T_3(I,\theta) + \dots \end{cases} \quad (4)$$

so as to keep the simple form of Equations (2), while seeking to transform the Hamiltonian $H(I,\theta)$ into a new Hamiltonian which depends only on the variables of action (I_i). The solutions would then be combinations of uniform circular motions, corresponding in the original variables to quasi-periodic motions, combinations of periodic motions.

The simplification introduced by Poincaré permitted him to respond to Le Verrier's plea, but his response was negative. For in his memoir "Sur le problème des trois corps et les équations de la dynamique", he showed the impossibility of integrating the equations of motion of three bodies (the Sun, Jupiter, and Saturn, for instance) as one could integrate the equations for two bodies. The only thing that could be done was to seek approximate solutions with the help of perturbational methods.

Poincaré showed that the "methods of successive approximation" used by astronomers could be extended to all orders. But he also showed that the series of Lindstedt, which in general defined these changes of variables and which made it possible to obtain quasi-periodic solutions, were divergent, because of the arbitrarily small divisors that always appeared in the expression of them. The result was that, contrary to Le Verrier's hope, it was impossible to find series that converged towards quasi-periodic solutions for the whole domain of initial conditions.

These results profoundly disturbed the astronomers of the time, for they had been persuaded that their series would converge provided that they took certain precautions. But Poincaré left them some hope, for he knew that their methods permitted them to calculate the motions of the planets with precision for a finite time; he proceeded to clarify the implications of his discoveries. He made precise the difference in meaning of the expression 'convergent series' for the astronomers and for the mathematicians, and laid the basis of the formal calculation that could be made with divergent series:

To take a simple example, let us consider the two series that have for general term

$$\frac{1000^n}{1\cdot 2\cdot 3\cdot \ldots \cdot n}, \quad \frac{1\cdot 2\cdot 3\cdot \ldots \cdot n}{1000^n}.$$

The mathematicians will say that the first series converges, and even that it converges rapidly, because the millionth term is much smaller than the 999 999th; but they regard the second as divergent, because the general term can increase beyond all limits.

The astronomers, on the contrary, regard the first series as divergent, because the first 1000 terms show an increase; and the second as convergent, because the first 1000 terms show a decrease, and this decrease is at first very rapid.

The two rules are both legitimate: the first, in theoretical investigations; the second, in numerical applications.

The series used by the astronomers are thus in general divergent, but they can serve to approximate the motion of the planets for a certain time, which can be long, but not infinite. Poincaré did not seem to think that his results could have much practical importance, except in the study of the stability of the solar system:

The terms of these series, in fact, decrease very rapidly at first, and then begin to increase; but since the astronomers stop with the first terms of the series, long before these terms have ceased to decrease, the approximation is sufficient in practice. The divergence of these expansions would be inconvenient only if one wished to use them to establish rigorously certain results, for instance the stability of the solar system.

By "the stability of the solar system" Poincaré here meant its stability over an infinite time, which is very different from its stability in a practical sense, that is, for an interval of time comparable to its expected duration.

Le Verrier had reformulated the question of stability by pointing out that it was necessary to take account of terms of higher degree than those considered by Laplace and Lagrange; Poincaré was still more demanding, in requiring the convergence of series. Possibly, under certain conditions, the series used by the astronomers might be convergent, but he regarded this as improbable.

Postscript: the modern results

Around 1960, the mathematicians A. N. Kolmogorov, V. I. Arnold, and J. Moser demonstrated that, contrary to Poincaré's intuition, certain initial conditions can lead to convergent series and thus to quasi-periodic solutions. But these conditions required very small values for the planetary masses, eccentricities, and inclinations, and could not be applied directly to our solar system.

In 1988 the Americans G. Sussman and J. Wisdom at MIT, using a computerized numerical integration of the equations of motion for the exterior planets (Jupiter, Saturn, Uranus, Neptune, and Pluto) over a duration of 875 million years, found Pluto's motion to be chaotic. That is, this motion proved to be extremely sensitive to initial conditions: the uncertainty in the initial conditions increases by a factor of 3 every 20 million years, so that prediction beyond about 400 million years becomes impossible. Because Pluto's mass is relatively small ($\frac{1}{130\,000\,000}$ of the solar mass), no macroscopic instability results for the rest of the solar system; the motion of the large planets remains very regular.

A little afterwards, using a very different method, J. Laskar obtained similar results for the interior planets (Mercury, Venus, Earth, and Mars). Because of the very rapid orbital motion of these planets, direct numerical integration over comparable intervals of time was not within the reach of computers. To obtain results for the whole of the solar system, it was first necessary – using methods similar to those of Laplace and Le Verrier – to transform the Newtonian equations into a new and much larger system (it contained 150 000 terms), which no longer represented the motions of the planets but rather the mean motions of their orbits. It was then possible, in only a few hours of calculation, to integrate numerically the motion of the solar system over 200 million years. The outcome was surprising. For the large planets, the result was a regular motion, as given also by direct numerical integration, but for the interior planets the behaviour of the trajectories was chaotic. Initial uncertainties were found to increase by a factor of 3 every 5 million years, preventing all prediction beyond about 100 million years. An error of 15 metres in the initial conditions produced an error of only 150 metres at the end of 10 million years, but an error of 150 000 000 kilometres after 100 million years.

More recently, the origin of this chaotic motion has been identified: it comes from resonances between the periods of precession of the orbits of Mars and the Earth on the one hand, and between those of Mercury, Venus, and Jupiter on the other. In the neighbourhood of such resonances, there exist regions in which the dynamics is complex and highly sensitive to initial conditions.

Thus celestial mechanics, which Laplace erected as the model *par excellence* of predictable science, has shown its limits. A new formulation of the problem of stability is imposed on us today. The solar system, we have shown, is unstable; it is a matter now of knowing with precision the effects of the instabilities for a time comparable to the system's age. To do this, it is necessary to study globally all the neighbouring trajectories, and thus follow the way opened a hundred years ago by Henri Poincaré. A better knowledge of the ensemble of these motions will not permit us to predict whether a catastrophic event, such as a sudden increase in the Earth's orbital eccentricity, will actually happen in the next billion years; it will authorise us only to say whether such an event, within such a period of time, is possible or not.

Further reading

P. S. Laplace, *Exposition du système du monde* (five different editions) (Paris, 1796–1835)

J. Laskar, A numerical experiment on the chaotic behaviour of the solar system, *Nature*, vol. 338 (1989), 237–8

J. Laskar, The chaotic behaviour of the solar system: A numerical estimate of the size of the chaotic zones, *Icarus*, vol. 88 (1990), 266–91

J. Laskar, La stabilité du système solaire, in *Chaos et déterminisme*, ed. by A. Dahan Dalmedico *et al.* (Paris, 1992), 170–211.

U. J. J. Le Verrier, *Annales de l'Observatoire de Paris*, vol. 2 (Paris, 1856)

Henri Poincaré, *Méthodes nouvelles de la mécanique céleste* (Paris, 1892–99)

G. J. Sussman & J. Wisdom, Numerical evidence that the motion of Pluto is chaotic, *Science*, vol. 241 (1988), 433–7

PART IX

The application of celestial mechanics to the solar system to the end of the nineteenth century

29

Three centuries of lunar and planetary ephemerides and tables

BRUNO MORANDO

Almanacs giving predictions of notable celestial events for the successive months of the year have an origin lost in Antiquity. Ephemerides – tables of day-to-day or month-to-month positions of the Moon and planets – were already being constructed in the fourth century BC in Seleucid Babylon. But a new era in lunar and planetary prediction began in the late sixteenth century with Tycho Brahe's observational work; on the basis of the large body of accurate observations that Tycho compiled, Johannes Kepler was able to discover his laws of planetary motion, and so to develop unprecedentedly accurate planetary tables. Later in the century, the introduction of new and more precise instruments of observation – the filar micrometer, the pendulum clock, and telescopic sights applied to graduated arcs – led to further increments in predictive accuracy. Finally, in the 1750s, the law of universal gravitation that Isaac Newton had announced in 1687 at last began to have an effect upon prediction, as the approximate solution of differential equations was brought to bear upon the problem (see Chapter 18). Successive refinements in theory, as well as in instruments and techniques of observation, would characterize the further history of lunar and planetary prediction up to the present day.

These increments in predictive accuracy had as their consequence the appearance of ephemerides of an increasing accuracy and precision. Simple traditional calendars and almanacs – often filled with absurd information – continued to appear, but where accuracy of prediction mattered, as it crucially did in navigation and astronomy, the more accurate ephemerides were increasingly in demand. This trend culminated in the emergence of the great national nautical almanacs.

The new ephemerides fell into two classes. On the one hand, there were *annual* ephemerides, their tables giving the day-by-day coordinates of the bodies of the solar system over an entire year. Their form, admitting of easy interpolation, varied little from year to year. They were constructed at first by isolated astronomers, then by national bureaux such as the Bureau des Longitudes and the British and American Nautical Almanac Offices.

On the other hand, there were tables of various astronomical data, collections of planetary ephemerides or of star positions, which covered periods of time other than the year. These were intended for particular clienteles, often mariners. They became very numerous during the eighteenth century, so that it would be difficult to make an exhaustive list, but we shall be citing some of them.

But the word 'tables' in astronomy has yet another important meaning. The theories of celestial mechanics give the positions of the planets and the Moon in the form of series composed of a large number of terms, usually trigonometrical; in accordance with an astronomical tradition that predated celestial mechanics but persisted into the nineteenth century, the individual terms were called 'equations'. The summation of these series for each body, and for each date of a projected ephemeris, entailed calculations that were – given the means of computation available – truly horrendous. To simplify the calculations, astronomers derived, on the basis of the theory, tables of single or double entry, each one yielding either a single term or several terms of a series by linear interpolation; the several 'equations' had then only to be added together. Until the advent of electronic computers in the mid-twentieth century, these tables played an enormous role. We shall devote a section to the most important of them, and to the astronomers who compiled them.

Astronomical ephemerides and tables

Ephemerides before the national nautical almanacs

As we have remarked, the extraordinary progress in astronomy that began with the work of Tycho Brahe and Kepler led to ephemerides of a new order of accuracy. But, as also mentioned, the new ephemerides were successors in a tradition that stretched back to Antiquity. Collections of diverse predictions arranged according to the calendar had long been in common use. They were called *almanacs* (a term of Arabic origin that appeared in Medieval Latin by the thirteenth century) or *ephemerides* (a name of Greek origin that was already used for the *Ephemerides* of Johannes Regiomontanus published in 1474). The almanacs tended to contain instruction of all sorts, mixed with astronomical predictions such as those of eclipses and phases of the Moon. The positions of the planets and Moon, whether in almanacs or ephemerides, were calculated from tables based in part on empirical data and in part on purely kinematic theories employing epicycles and deferents to describe the motions. The most widely used of these tables up to the seventeenth century were the *Alfonsine Tables*, traditionally associated with the thirteenth-century King of Castile, Alfonso X, and the *Prutenic Tables*, prepared by Erasmus Reinhold on the basis of Nicholas Copernicus's *De revolutionibus* and published in 1551.

The advent of printing with movable type gave rise to numerous printed ephemerides. As early as 1471, a wealthy inhabitant of Nuremberg, Bernhard Walther, founded a printing-house for the express purpose of publishing ephemerides, and engaged Regiomontanus who computed ephemerides for the years 1475–1506. Subsequently, a succession of astronomers took up the task: Johann Stöffler, Cyprian Leovitius, Joannes Stadius, and others.

A forward leap in predictive accuracy came with the introduction of Kepler's elliptical orbit and areal rule, incorporated in his *Tabulae Rudolphinae (Rudolphine Tables)*, which were published at Ulm in 1627. Kepler himself calculated and published ephemerides for the years 1617–37. These were continued in Danzig by Lorenz Eichstadt and then by Johann Hecker, who computed positions for the period from 1666 to 1680.

When no Danzig successor to Hecker appeared, the astronomers of the newly established Paris Observatory decided that Hecker's ephemerides, which had been reprinted in Paris, should be continued. So appeared the first ephemerides worthy of the title *Connaissance des temps*: a regular annual publication, exact, simple to employ, and conceived specially for the use of those most urgently concerned, astronomers and navigators.

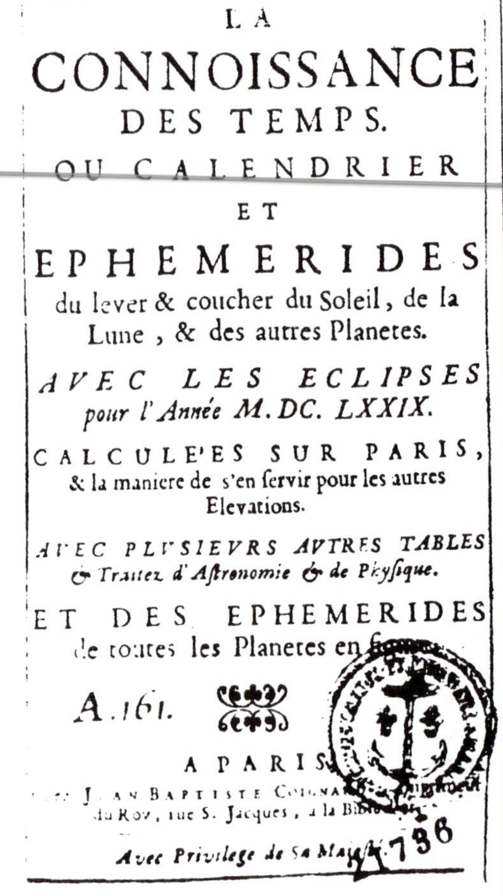
LA
CONNOISSANCE
DES TEMPS.
OU CALENDRIER
ET
EPHEMERIDES
du lever & coucher du Soleil, de la Lune, & des autres Planetes.
AVEC LES ECLIPSES pour l'Année M.DC. LXXIX.
CALCULE'ES SUR PARIS, & la maniere de s'en servir pour les autres Elevations.
AVEC PLUSIEURS AUTRES TABLES & Traitez d'Astronomie & de Physique.
ET DES EPHEMERIDES de toutes les Planetes en
A.161.
A PARIS,
Jean Baptiste Coignard
du Roy, rue S. Jacques, à la Bi
Avec Privilege de Sa Maj

29.1. Title-page of the first volume of the *Connaissance des temps*.

The "Connaissance des temps" and "The Nautical Almanac"

The title of the new publication (see Figure 29.1) may be translated as follows:

The Knowledge of Times or calendar and ephemerides of the rising and setting of the Sun, Moon, and the other planets. With the eclipses for the year 1679 calculated for Paris and the way to use them for other elevations [of the celestial pole]. With several other tables, and treatises on astronomy and physics, and ephemerides of all the planets in figures.

In the preface addressed to the King we read that the ephemeris has been "purged of all the ridiculous things with which works of this sort have been filled till now". The preface is signed only with three stars (***), and the 'privilege' (authorization for publication necessary at the time) states: "It is permitted to Sieur *** to cause to be printed by whatever printer he shall choose ... etc." It was long believed that the person so designated was Jean Picard (1620–82), but it appears that it was rather Joachim Dalencé (*c.* 1640–1707). Of Dalencé we know very little: he purchased a telescope in England in 1668; became the intermediary between Henry Oldenburg, Secretary of the Royal Society, and Christiaan Huygens; and in 1685 settled in the Low Countries where he wrote a treatise on the magnet in 1687 and proposed a thermometer whose points of reference were to be the melting points of ice and butter!

Although the *Connaissance des temps* was directed primarily to astronomers, the general public was not altogether forgotten: according to the preface preceding the tables the first part contained "everything that has been considered useful and necessary to the public, and so easy to apply that the least intelligent can make use of it".

Dalencé retained the publication privilege till 1685, but Picard seems to have carried out many of the calculations, leading A.-G. Pingré to believe that he had been the founder of the ephemeris. The privilege next passed to Jean Le Fèvre (1652?–1706) and after him to Jacques Lieutaud (*c.* 1660–1733), then in 1702 was vested in the Academy of Sciences. The volume for 1702 appeared under the title: *Connaissance des temps pour l'année 1702 au méridien de Paris publiée par l'ordre de l'Académie Royale des Sciences et calculée par M. Lieutaud de la même académie.* Lieutaud continued the calculations for the years through 1726. Among the astronomers who were subsequently charged with the publication were Joseph-Jérôme Lefrançais de Lalande (1732–1807), for the years from 1760 to 1776, and Pierre-François-André Méchain (1744–1804), for the years from 1788 to 1795. The Bureau des Longitudes at its creation in 1795 was assigned the responsibility of publishing the *Connaissance des temps*; two centuries later it continues to carry out this task.

Other ephemerides appeared: the Bolognese in 1715, the Viennese in 1757, the Milanese in 1775, those of Berlin in 1776. But the most renowned publication of this kind was the *Nautical Almanac* (Figure 29.2) founded in England in 1767 by Nevil Maskelyne (1732–1811).

Maskelyne, appointed Astronomer Royal in 1765, was a firm defender of the method of lunar distances for the determination of longitude at sea. For this purpose he proposed using the lunar tables of Johann Tobias Mayer (1723–62). Mayer first published his tables in 1753 in Göttingen, but he subsequently refined them. Neither their theoretical basis nor the mathematical methods employed in them were especially novel (see Chapter 18). But in fitting the theory to observations, Mayer was eminently skilful, and consequently the tables were very accurate. After Mayer's death, his widow submitted the amended tables to the British Board of Longitude, which was charged with improving the determination of longitude at sea. In October 1765 Maskelyne persuaded the Board that Mayer's tables should be officially endorsed, and proposed the creation of the *Nautical Almanac*. In the preface to its first volume the series is announced as follows:

The Commissioners of Longitude, in pursuance of the powers vested in them by a late Act of Parliament, present the Publick with the NAUTICAL ALMANAC and ASTRONOMICAL EPHEMERIS for the year 1767, to be continued annually; a Work which must greatly contribute to the Improvement of Astronomy, Geography, and Navigation.

The preface (of which Maskelyne was the author) also claims that Mayer had so perfected the tables of the Moon that one could hope to determine the longitude at sea with an accuracy of the order of a degree.

For a long time the Americans made use of this same almanac, but in 1849 a Nautical Almanac Office was created by act of the Congress of the United States. The first volume of *The American Ephemeris and Nautical Almanac*, for the year 1855, was published in 1852. Since 1981 the two publications have been united under the title *The Astronomical Almanac*.

THE

NAUTICAL ALMANAC

AND

ASTRONOMICAL EPHEMERIS,

FOR THE YEAR 1767.

Publiſhed by ORDER of the

COMMISSIONERS OF LONGITUDE.

LONDON:

Printed by W. RICHARDSON and S. CLARK,
PRINTERS;
AND SOLD BY
J. NOURSE, in the Strand, and Meſſ. MOUNT and PAGE,
on Tower-Hill,
Bookſellers to the ſaid COMMISSIONERS.
M DCC LXVI.

29.2. Title-page of the first volume of the *Nautical Almanac*.

Other tables and ephemerides

As indicated earlier, besides the regular annual ephemerides there appeared numerous tables for periods longer than the year, intended for special clienteles. Eustachio Manfredi, an astronomer at Bologna, published ephemerides for successive decades from 1715 to 1750. His work was carried on till 1786 by Eustachio Zanotti, Petronio Matteucci, and others. Apropos of these ephemerides the Baron Franz Xaver von Zach (1754–1832) remarked in 1819 (on p. 474 of Vol. 2 of his *Correspondance astronomique*):

It is especially to be remarked that it was these ephemerides that constituted the science of the Jesuits in China. There they used them, we know, as a vehicle for introducing and propagating the Christian religion among that idolatrous and ignorant people, whose great knowledge in astronomy they have always so ridiculously and deceitfully praised.... Manfredi's ephemerides were then distributed to the four quarters of the world and a prodigious number of them were sold in all the missions.

The Abbé Nicolas-Louis de Lacaille published ephemerides in three volumes for the years from 1745 to 1774. Then Lalande published three volumes for the years from 1775 to 1804. Another famous producer of ephemerides was Father Giovanni Inghirami (1779–1851), professor of astronomy at the Ximenes Institute of Florence and director of its observatory. He calculated ephemerides of the two brightest planets, Venus and Jupiter, for the years 1820 to 1824; they were intended for the use of navigators, and were published in Zach's *Correspondance astronomique*.

In 1786 Zach had entered the service of the Duke of Saxe-Coburg, who had erected for him an observatory on the Seeberg near Gotha; Zach was director of this observatory until 1806, and there trained a number of astronomers who would later produce planetary tables. In 1809 he himself published in Florence his *Tables abrégées et portatives du Soleil calculée pour le méridien de Paris sur les observations les plus récentes d'après la théorie de M. Laplace*. These tables are followed by tables of the Moon ... *d'après la théorie de M. Laplace et d'après les constantes de M. Bürg*. The idea of publishing ephemerides in a small volume, which has become a preoccupation of the almanac offices in the late twentieth century, is thus not new; Zach, in his preface, pointed out its advantages and disadvantages:

In the older tables of the Sun it was believed sufficient to give four or five equations of perturbation; the new theories require twenty-two. Astronomers, navigators, topographical engineers, amateur astronomers who, either because of their responsibilities or their interests, travel a great deal, find themselves in the necessity and embarrassment of having to go abroad with whole libraries.... One must concede, however, that the advantage of abbreviated tables can be purchased only at the price of lengthening the calculations. For there are only two possibilities: either the volume of the tables must be increased, in order to abbreviate the calculations, or the calculations must be increased, in order to diminish the size and number of the tables.

Tables for the construction of ephemerides

Tables employed to the middle of the nineteenth century

In France a law of 1795 assigned to the Bureau des Longitudes the perfecting of astronomical tables. In 1806 the Bureau published *Tables du soleil* by Delambre and *Tables de la lune* by Bürg. These tables are described in the *Connaissance des temps* for 1808.

Jean-Baptiste-Joseph Delambre (1749–1822) had established the basis of his solar tables in the *Mémoires* of the Academy of Berlin for 1784 and 1785; he then calculated them in 1790 using 300 observations of Maskelyne, and they were published in the third edition of Lalande's *Astronomie* in 1792. In the revised version published in 1806 under the aegis of the Bureau des Longitudes, Delambre had recourse to a much larger number of observations: 718 observations made by James Bradley from 1750 to 1755, 700 new observations made by Maskelyne, and some observations of Alexis Bouvard. According to Delambre, "All these calculations have resulted in only 1″.7 to be subtracted from the mean longitude in 1800 and 3″ to be subtracted from the place of the apogee; the equation of centre remains the same to within a few tenths of an arc-second, but the mean motion has undergone a greater diminution." The constant of precession previously accepted, he found, needed to be reduced by 15″ per century. He was also able to correct the masses of Venus, Mars, and the Moon.

Johan Tobias Bürg (1766–1834), after serving as a calculator at the observatory of Seeberg under the direction of Zach, became a professor at the University of Vienna, then an astronomer at the university's observatory. The lunar tables of Bürg derived from Tobias Mayer's tables, which had been improved by Mason. Charles Mason (1728–86) had been Bradley's assistant at the Royal Greenwich Observatory; after observing the transits of Venus across the Sun in 1761 and 1769, he worked on the *Nautical Almanac*. In 1778, making use of 1200 unpublished observations of Bradley, he published his *Lunar Tables in Longitude and Latitude according to the Newtonian Laws of Gravity*; a second, improved edition appeared in 1780. Bürg found some new 'equations', but despite the fact that he had at his disposal 2000 new observations of Maskelyne, he modified but little the values of Mayer's and Mason's coefficients. In 1806 the Bureau des Longitudes instituted a prize of 6000 francs for the best lunar tables, and this prize was awarded to Bürg after his tables had been compared with P.-S. Laplace's lunar theory as published in the third volume of the *Mécanique céleste*.

In 1812 the Bureau des Longitudes published Burckhardt's *Tables de la lune*. Johan Karl Burckhardt (1773–1825), born in Leipzig, had learned astronomy under the tutelage of Zach. In 1797 Lalande induced him to come to Paris; he became an adjunct member of the Bureau des Longitudes in 1799, then a full member in 1817. His tables were based on Laplace's theory, but many of the coefficients of perturbational terms were still (as in Mayer's and Bürg's tables) adjusted to fit the observations. The Bureau des Longitudes had appointed a special commission consisting of Laplace, Delambre, Bouvard, S.-D. Poisson, and D. F. J. Arago to study Burckhardt's tables; it concluded that they were much more accurate than Bürg's, and the Bureau des Longitudes thereupon adopted them.

In 1824 and 1828 there appeared the *Tables de la lune formées par la seule théorie de l'attraction* by Marie Charles, Baron de Damoiseau. They were derived from the theory entered by Damoiseau in the Paris Academy's contest of 1820, a contest in which Giovanni Plana also had presented a theory of the Moon's motions.

In 1808 Bouvard, who had been Laplace's assistant and had carried out the calculations for the *Mécanique céleste*, published tables of Jupiter and Saturn based on Laplace's theory. They were laid out in the same form as Delambre's solar tables, and were based on observations made at opposition between 1747 and 1804 by Bradley, Maskelyne, Lacaille, and the astronomers of the Paris Observatory. In 1821 Bouvard published new tables of Jupiter and Saturn, utilizing 126 observations of oppositions and quadratures of Jupiter, and 129 of Saturn, made between 1747 and 1814. He justified this new publication not only by the greater number of observations brought to bear but more importantly by the fact that since the publication of his first tables in 1808 Laplace had corrected an error in the sign of the terms of the fifth order in the "great inequality".

To the tables of Jupiter and Saturn of 1821 Bouvard added tables of Uranus. He of course employed the observations made since William Herschel's discovery of the planet in 1781, but he also used seventeen earlier observations, made by astronomers who had taken Uranus for a fixed star: Flamsteed in 1690, 1712 and 1715; Bradley in 1753; Mayer in 1756; and P.-C. Le Monnier who sighted the planet twelve times, including eight times in 1769 alone. Laplace had applied Laplacian theory to this planet to reach a result he considered satisfactory. Barnaba Oriani (1752–1832), director from 1802 of an observatory established by the Jesuits at Milan, and dubbed a count and senator of the realm of Italy by Napoleon, had published (among numerous other astronomical tables) tables for Uranus that appeared to satisfy all the observations of this planet, including the older ones. But Bouvard's tables of 1821 showed an irreducible discrepancy between the older and the more recent observations. This discrepancy, we know, would only increase, and was to result in the discovery of Neptune (see Chapter 28).

Bernhard von Lindenau (1779–1854), after working at the observatory of Seeberg near Gotha under the direction of Zach, became its director in 1808, and remained in this position till 1818 when he was succeeded by Encke. For a time he served as editor-in-chief of the *Monatliche Correspondenz zur Beförderung der Erd- und Himmelskunde*, a journal founded by Zach in 1800; then in 1816 he founded with J. G. F. von Bohnenberger the *Zeitschrift für Astronomie und verwandte Wissenschaften*. His activities were many and various. His astronomical work dealt principally with the problem of improving the constants of aberration and nutation, and with the construction of planetary tables. His tables were based on Laplace's theory: *Tabulae Veneris novae et correctae ex theoria cl. de Laplace et ex observationibus recentissimis in specula astronomica Seebergensi habitatis erutae* (Gotha, 1810), tables of Mars with an analogous title (Eisenberg, 1811), and *Investigatio novae orbitae a Mercurio circa Solem descriptae cum tabulis planetae* (Gotha, 1813). Lindenau's tables were employed in the *Nautical Almanac* from 1834 to 1863 for Mercury, to 1864 for Venus, and to 1865 for Mars.

The solar and planetary tables of Le Verrier, Newcomb, and Hill

The solar and planetary tables employed in ephemerides during the first half of the nineteenth century were all more or less directly the outcome of Laplace's work as set forth in the *Mécanique céleste*. Urbain Jean Joseph Le Verrier (1811–77) was the first to make a new beginning and to undertake to treat the motions of all the planets (see Chapter 28). Le Verrier's theory and tables of the Sun appeared in Vol. 4 of the *Annales de l'Observatoire de Paris* (1858), his theory and tables of Mercury in Vol. 5 (1859), and his theory and tables of Venus and Mars in Vol. 6 (1861). In Vol. 12 (1876) his tables of the motions of Jupiter and Saturn "based on comparison of the theory with the observations" appeared. The theories of the motions of Uranus and Neptune appeared in Vol. 13 (1876), and the corresponding tables in Vol. 14 (1877), some months after Le Verrier's death.

A. J.-B. Gaillot (1834–1921), Le Verrier's sole collaborator from 1861, played an important role in the improvement of these tables. In 1873 he became director of the Bureau of Computation of the Paris Observatory; in 1897 he became assistant to the director of the observatory; and in 1908 he was made a *Correspondant* of the Academy of Sciences. The discrepancy between the observed positions of Saturn and the positions calculated by Le Verrier had reached 4″. Gaillot attributed this discrepancy to that fact that Le Verrier had neglected the terms of the third order in the masses, and undertook to recompute the tables, not by the method of Le Verrier which had led to inextricable calculations, but by means of a 'method of interpolation'. Thus Gaillot established his *Tables rectifiées du mouvement de Saturne*, published in 1904 in Vol. 24 of the *Annales de l'Observatoire de Paris*. In Vol. 28 (1910) he published his *Tables nouvelles des mouvements d'Uranus et de Neptune*, in which he followed Le Verrier's method but re-did the entire calculation in order to take account of modified values of the perturbing masses. Finally, Gaillot rounded off this very considerable body of work with his *Tables rectifiées du mouvement de Jupiter*, which were published in 1913 in Vol. 31 of the *Annales de l'Observatoire de Paris*. He justified these new tables, like those for Saturn, on the grounds that, Le Verrier having neglected the terms of the third order in the masses, the departures of the calculated from the observed positions varied periodically in an unacceptable way.

The tables of Le Verrier and Gaillot were employed in the computation of ephemerides for the *Connaissance des temps* until 1984, and up to the same date, tables derived from the theories of Le Verrier were the basis of the ephemerides published in the *Nautical Almanac*; the errors in position being of the order of 1″ to 3″ for the large planets, but reaching 19″ for Venus. *The American Ephemeris and Nautical Almanac* did not employ Le Verrier's tables, although Joseph Winlock (1826–75), who worked in the Nautical Almanac Office of the US Naval Observatory from 1852 and was superintendant of this office from 1858 to 1859 and from 1861 to his death in 1875, established tables of the motions of Mercury founded on Le Verrier's theory, and these were the basis of the ephemeris of this planet in *The American Ephemeris* from 1855 to 1899.

The new theories constructed in the United States under the direction of Simon Newcomb (1835–1909) led to the establishment of tables which appeared in the *Astronomical Papers of the American Ephemeris* (hereinafter abbreviated as *Astronomical Papers*). The tables of Jupiter and Saturn were constructed by George William Hill on the basis of his theory which employed the method of Peter Andreas Hansen (1795–1874). They were published in Vol. 7, parts 1 and 2, of the *Astronomical Papers* (1895). Provisional tables constructed on the basis of observations made before 1830 had been published in Vol. 4 of the *Astronomical Papers*; the tables of 1895 employed observations extending to 1888.

Parts 3 and 4 of Vol. 7 contain the tables of

Uranus and Neptune, respectively. They were the work of Newcomb. After having established the tables of Neptune, then the last known planet of the solar system, Newcomb investigated whether the residues in the observed longitudes might not betray the action of an unknown planet. He found these residues very high, but he could not discern in them any systematic character enabling him to exploit his hypothesis. In 1855 Marian A. Kowalski, a professor at the University of Kazan, had published a theory of Neptune based on his doctoral dissertation, which he had defended in 1852. Tables deduced from this theory were employed in *The American Ephemeris* till 1870. Newcomb used Kowalski's theory of Neptune for his global theory of the planets.

Newcomb's tables of the Sun, Mercury, Venus, and Mars were introduced into *The American Ephemeris* beginning in 1900. His tables of Uranus had been utilized for the ephemerides beginning in 1877, and his tables of Neptune beginning in 1870 (therefore before their publication in *Astronomical Papers*).

The tables of the Moon after 1850

In 1853 the lunar tables of Benjamin Peirce (1809–80) appeared. While still a student at Harvard, Peirce had found his vocation in astronomy when Nathaniel Bowditch asked him to re-read the proofs of the famous Bowditch translation of Laplace's *Mécanique céleste*. His tables of the Moon were based on Plana's theory, with corrections by G. B. Airy and M. F. Longstreth, and two inequalities of long period due to Venus that Hansen had calculated, and also the secular variations of the mean motion and perigee found by Hansen.

Through his works on the Moon published in 1838 and in 1862–64, Hansen made possible a considerable improvement in the precision of the lunar theory. In 1857 he published tables based on his theory, and these were employed from 1862 in the *Nautical Almanac*. Corrections to the right ascension and declination, which Newcomb had calculated, were introduced in 1883, and corrections in the parallax and semi-diameter were introduced in 1897. These tables were employed for the ephemerides until 1922, when they were replaced by Brown's tables based on the theory of Hill–Brown.

Charles-Eugène Delaunay's purely analytic theory of the motion of the Moon was published in the *Mémoires* of the Academy of Sciences, Vols. 28 (1860) and 29 (1867). In presenting Vol. 29 to the Academy, Delaunay stated that he proposed to publish in a third volume the solution of various accessory questions. There followed his "Expressions numériques des trois coordonnées de la lune" in the *Connaissance des temps* for 1869; and various additional notes appeared up to his accidental death in 1872.

But Delaunay had not awaited completion of his theoretical researches to initiate preparation of tables based on his theory. The first sheets, calculated with the support of the Bureau des Longitudes, were presented by Victor Puiseux in 1873 after Delaunay's death. No further progress was then made till 1878, when F. F. Tisserand supervised the resumption of calculations, responsibility for which he assigned to Leopold Schulhof. Progress, however, was slow. Hansen's tables of 1857 were believed to be very exact, and were of a convenient form; an incentive was thus lacking for the construction of other tables. In 1880 Newcomb and J. Meier published in *Astronomical Papers*, Vol. 1, part 2, a work entitled "Transformation of Hansen's lunar theory, compared with the theory of Delaunay", in which they showed that the two theories were in excellent agreement, at least for the century 1750–1850, which was the period of the observations used by Hansen. A note due to Tisserand, "Sur l'état actuel de la théorie de la Lune", published in *Bulletin astronomique* in 1891, and numerous corrections effected by Henri Andoyer, gave impetus to the carrying through of the calculations for the lunar tables derived from Delaunay's theory.

These tables, completed under the direction of E. Radau, appeared in 1911 in the *Annales du Bureau des Longitudes*. Brown's theory, a completion of Hill's lunar theory, had been published in 1908, although not yet put into tables; and Radau, in a long introduction to Delaunay's tables, gave grounds for concluding that tables derived from Brown's theory would prove more accurate. This judgement was later confirmed; the tables of Delaunay–Radau were employed in the *Connaissance des temps* from 1915 to 1925, and were then supplanted by those of Brown.

The theory of Hill–Brown, which took for its starting-point an intermediate orbit calculated in a system of rotating coordinates (see Chapter 28), thus showed itself superior to Delaunay's theory, in particular because the introduction from the beginning of the well-established numerical value of the ratio of the mean motions of the Sun and Moon avoided severe problems of convergence. This theory was put into tables by Brown himself, assisted by H. B. Hedrick. The tables were published in six volumes in 1919 under the title *Tables of the Motion of the Moon*, and were employed for the construction of ephemerides from 1923 to 1959. The decision was then made to renounce tables and, starting in 1960, to employ directly Brown's series, as corrected and completed in the *Improved Lunar Ephemeris*.

Conclusion

The year 1960 can be considered as marking the end of a period stretching over nearly three centuries, which saw the appearance of ever more carefully worked-out ephemerides, derived from tables which in the eighteenth century had come to be deduced from analytic or semi-analytic theories. The advent of electronic computers made it possible, first to program the existing theories, then directly to construct much more accurate ephemerides, using either numerical integrations or new theories that depended on algebraic manipulations previously unthinkable. In 1984 the ephemerides based on the works of Le Verrier, Gaillot, Hill, Newcomb, and Brown at last disappeared altogether from the *Connaissance des temps* and *The Nautical Almanac*.

To be sure, these publications, the form of which has greatly changed, will continue to exist; the number of useful ephemerides they contain has increased. But henceforth the venerable tables that were used to construct them and which cost so much effort to so many illustrious astronomers will repose on the highest and dustiest shelves of our libraries for eternity.

30

Satellite ephemerides to 1900

YOSHIHIDE KOZAI

By the end of the nineteenth century, the telescope had added to the Moon some twenty-one satellites of primary planets of the solar system. The first four had been the 'Galilean' satellites of Jupiter, discovered by Galileo in 1610 (see Chapter 9 in Volume 2A); the last was the Saturnian satellite Phoebe, the first satellite to be discovered photographically (by W. H. Pickering in 1898). These twenty-one satellites, each with its currently accepted name, period and orbital radius, along with the discoverer and year of discovery, are listed in Table 30.1.

Before the introduction of photographic methods astronomers made position observations of these satellites by measuring the satellite's distance from the planet's centre (or its nearest limb), and determining the angle between the meridian and the line joining the two components (called the 'position angle' of the line), by means of a heliometer or micrometer; or by similarly measuring the mutual distance between two satellites and the position angle. During the nineteenth century the mean error in the distance measurement was reduced to about 0″.5 when measurements were made with respect to the centre of the planet, while for the mutual distances of satellites the r.m.s. error could reach as low as 0″.1 in the best cases. For the Galilean satellites of Jupiter, observations of the eclipses, transits and occultations by Jupiter could provide even more accurate position information.

An early motive for careful observation of the Galilean satellites was to develop ephemerides for use in determining geographical longitudes. With the publication in 1687 of Newton's *Principia*, another motive emerged, namely, determination of the masses of the primaries as fractions of the Sun's mass by way of Kepler's third law – values for these masses being crucial to the determination of mutual planetary perturbations. Yet another motive was potent from the first days of telescopic observation, an interest in satellite systems for their intrinsic beauty and for the analogy they supplied to the solar system as a whole.

The orbital period of any satellite is shorter – usually much shorter – than that of its primary; thus the motions of the satellites can be traced over many orbital periods. Therefore, without the need for any analytical theory, it was usually possible to derive fairly accurate expressions for the motions from observations. Ephemerides of the satellites could then be computed by extrapolation.

Perturbations

In some cases, observation when extended over a few years, besides confirming mean motions, epochs, orbital inclinations and the longitudes of pericentres and nodes, yielded with considerable accuracy the secular motions and the amplitudes and periods of several long-period perturbations. Thus in 1719 James Bradley completed a study of the Galilean satellites of Jupiter, in which he detected a small inequality in the eclipses of the first two satellites, with a period of 437 days, and also discovered the eccentricity of the orbit of the fourth satellite. In the 1740s Pehr Wilhelm Wargentin detected, in addition to the aforementioned inequalities of the first two satellites, a similar inequality in the motion of the third satellite.

The theoretical question was then posed: were these inequalities derivable from Newton's law of gravitation? In a prize essay of 1766 Joseph Louis Lagrange succeeded in deriving the perturbational inequalities that Bradley and Wargentin had detected, but failed to account for a peculiar relation among these inequalities (see Chapter 21). Another foray into theoretical explanation was made by Jean-Sylvain Bailly, using the equations of Alexis-Claude Clairaut's lunar theory. But the first to put the analytical theory of the motion of

Table 30.1

Satellite	Period (days)	Distance from planet (kms)	Discoverer	Year of discovery
Mars (a = *1.52 AU*, P = *1.88 years)**				
Phobos	0.319	9 378	A. Hall	1877
Deimos	1.262	23 459	A. Hall	1877
Jupiter (a = *5.20 AU*, P = *11.86 years)*				
Io	1.769	422 000	Galileo	1610
Europa	3.551	671 000	Galileo	1610
Ganymede	7.155	1 070 000	Galileo	1610
Callisto	16.589	1 883 000	Galileo	1610
Amalthea	0.498	181 000	E. E. Barnard	1892
Saturn (a = *9.55 AU*, P = *29.46 years)*				
Mimas	0.942	185 400	W. Herschel	1789
Enceladus	1.370	238 200	W. Herschel	1789
Tethys	1.888	294 600	Cassini I	1684
Dione	2.737	377 400	Cassini I	1684
Rhea	4.518	526 800	Cassini I	1672
Titan	15.945	1 200 000	C. Huygens	1655
Hyperion	21.277	1 482 000	W. C. Bond, W. Lassell	1848
Iapetus	79.330	3 558 000	Cassini I	1671
Phoebe	550.480	12 960 000	W. H. Pickering	1898
Uranus (a = *19.22 AU*, P = *84.02 years)*				
Ariel	2.520	191 800	W. Lassell	1851
Umbriel	4.144	267 300	W. Lassell	1851
Titania	8.706	438 700	W. Herschel	1787
Oberon	13.463	568 600	W. Herschel	1787
Neptune (a = *30.11 AU*, P = *164.77 years)*				
Triton	5.877	353 600	W. Lassell	1846

Note:
**a* is the mean solar distance of the planet in the astronomical unit, which is 1.496 × 108 km; *P* is the planet's period.

satellites on an adequate mathematical basis was Pierre-Simon Laplace; his final account of the theory appears in Vol. 4 of his *Traité de mécanique céleste* (1805). A refinement of the theory is given in Vol. 4 of François Félix Tisserand's work of the same title (1896).

Celestial mechanics, here as elsewhere, undertook to compute positions with an accuracy equal to that of the observations. Among satellites, different cases may be distinguished according to whether the solar perturbation or the non-sphericity of the planet has a preponderant effect. The solar perturbing force is proportional to $(n'/n)^2$, where n' is the mean motion of the planet about the Sun, and n is that of the satellite about the planet. The dynamical oblateness factor is measured by J_2/a^2, where J_2 is the second-order zonal harmonic for the planet and a is the mean distance from the satellite to the planet's centre. The latter factor is evidently greater for satellites close to their primaries, while the solar factor is larger for satellites with smaller mean motions, hence farther from the primary.

Computation of periodic perturbations, as required for planets, is usually unnecessary for satellites. However, in the case of satellites for which the solar perturbations are appreciable, long-period perturbations must be included. In the case of a satellite for which the dynamical oblateness factor is dominant over the solar factor, the inclination of the satellite's orbital plane to the planet's equator is nearly constant; in the contrary case, the inclination of the two orbital planes is nearly constant. For intermediate cases, the pole of the orbital plane moves around a point between the pole of the planet's equator and that of its orbital plane.

The pericentre and the ascending node move slowly with a rate proportional to the oblateness factor or the solar action. If the eccentricity and the inclination are very small, they move with the same rate but in opposite directions. From the secular motions of the longitudes of the pericentres and the nodes, the dynamical oblateness can be determined.

There are several pairs of satellites whose motions are especially interesting from a dynamical point of view. These are pairs for which simple relations hold between their mean motions, namely, the inner three of the Galilean satellites, and the Saturnian satellites Mimas–Tethys, Enceladus–Dione, and Hyperion–Titan. Since Rhea is very close to Titan, a giant satellite, its motion suffers a large perturbation due to Titan. By determining amplitudes and/or periods of certain of the perturbation terms, the masses of several of the disturbing satellites can be derived. The interactions in many cases involve pendulum-like librations in one parameter or another, providing case-studies in mechanical resonance and stability. The relative shortness of satellite periods as compared with planetary periods provides a magnification of the time scale, making satellite systems of special interest for the study of secular and long-period perturbations.

Ephemerides

In their satellite ephemerides, the earlier editions of the *Nautical Almanac* and *Connaissance des temps* gave only predictions of eclipses and diagrams for configurations of the Galilean satellites of Jupiter. From 1855, the first year for which it was issued, to 1881, the *American Ephemeris* gave ephemerides only for these four satellites; in 1882 it added diagrams for the configurations of these satellites, and at the same time introduced ephemerides of the epochs of elongations of the satellites of Mars, Saturn, Uranus and Neptune as well as diagrams of their apparent orbits. In 1896 the *Connaissance des temps* introduced ephemerides of the epochs of elongations for the satellites of Saturn, Uranus and Neptune, and in 1899 it added several tables useful for computing ephemerides of these satellites. In 1899 the *Nautical Almanac* also introduced diagrams of the apparent orbits at the time of opposition for the satellites of Mars, Saturn, Uranus and Neptune. Ephemerides for the satellites were computed annually by Albert Marth and published in *Monthly Notices of the Royal Astronomical Society* and in *Astronomische Nachrichten*: for Uranus from 1870, for Saturn from 1873, for Neptune from 1878, and for Mars from 1883.

Mars

The two satellites of Mars were discovered in 1877 by Asaph Hall (1829–1907) at the US Naval Observatory, using the 26-inch (66-cm) refractor built shortly before by Alvan Clark & Sons. Hall first observed the outer satellite on 11 August, while he was measuring positions of white spots on Mars; but cloudy weather prevented its certain recognition as a satellite that night. On 16 August he observed it again, and established its motion over a period of two hours, during which it moved over 30 arc-seconds. The following night he discovered the inner satellite. The discoveries were confirmed by the Clarks using the 26-inch refractor they then had under construction for the Leander McCormick Observatory, and by E. C. Pickering and his assistants at Harvard. On 19 August the Smithsonian Institution was informed, and from there the news was transmitted worldwide.

The orbital elements of the satellites (and the mass of Mars itself) were computed by Simon Newcomb using the data of ten days of observations, and published in *Astronomische Nachrichten*. Early the following year Hall named the satellites Phobos and Deimos, thereby adopting a suggestion emanating from Britain.

The satellites were again observed at the US Naval Observatory in 1879, 1892, 1894 and 1896. In 1894 W. W. Campbell made numerous micrometric measurements with the 36-inch (91-cm) refractor of Lick Observatory. Meanwhile, the satellites came under the scrutiny of Karl Hermann Struve (1854–1920) and others at Pulkovo in 1894 and 1896.

For the two Martian satellites, the dynamical oblateness effects are dominant over the solar perturbations, and therefore the inclinations to the equator (0°.96 and 1°.73) are almost constant. The eccentricities, 0.017 and 0.003, are also small. Hence their motions are relatively simple and without large perturbations. However, the longitudes of the pericentres and the ascending nodes move secularly with rates of 158° per year for Phobos and 6°.37 for Deimos. In the twentieth century the mean motions of the satellites have been found to be subject to secular accelerations.

Jupiter

Amalthea

The fifth satellite of Jupiter, Amalthea, was discovered on 9 September 1892 by the noted observer, E. E. Barnard, using the Lick Observatory 36-inch refractor. Its existence was confirmed the following month by Ormond Stone of the Leander McCormick Observatory, and by Taylor Reed of Princeton. Barnard himself made extensive observations from first discovery through to the end of 1894, and meanwhile Struve observed it at Pulkovo from October 1893 to January 1895. The observations of Barnard and Struve were analysed by Fritz Cohn of Königsberg to determine the orbital elements of the satellite and the mass of Jupiter itself. In addition, he was able to determine the dynamical oblateness of Jupiter from the annual secular motion of the ascending node, which was derivable from observations.

The satellite's eccentricity (0.003) and inclination with respect to the equator of Jupiter (0°.4) are small. The secular motions of the longitudes of the pericentre and the node are 914° per year, and are mainly due to the oblateness of Jupiter. Thus the motion can be expressed in very simple form, except for a secular acceleration term in the mean longitude, which was discovered in this century.

The Galilean satellites

In the eighteenth and nineteenth centuries many observations were made of the Galilean satellites of Jupiter. Most of these observations were of eclipses of the satellites by the parent planet, and they were usually published in *Monthly Notices of the Royal Astronomical Society* or in *Astronomische Nachrichten*. Among the mean motions of the inner three satellites there is a remarkable relation. The mean motion of Io, n, is a little more than twice as large as that of Europa, n', which in turn is a little more than twice as large as that of Ganymede, n''. Moreover, the following relation:

$$n - 2n' = n' - 2n''$$

almost exactly holds, so that

$$n - 3n' + 2n'' = 0. \qquad (1)$$

In fact, if one forms the expression $\Theta = \lambda - 3\lambda' + 2\lambda''$, where λ, λ' and λ'' express the mean longitudes of the satellites, then Θ is found to librate around 180° with a very small amplitude and a period of six years. The near constancy of this relation, which had been recognized by Lagrange, was first explained by Laplace, and so it is termed 'the Laplace relation'.

Laplace rounded out the theory of motion of the four satellites by determining from observations the mass and the dynamical oblateness of Jupiter and the masses of the satellites. J.-B. J. Delambre published in 1817 a table for computing positions of the Galilean satellites based on Laplace's theory and on observations made between 1662 and 1802. W. S. B. Woolhouse published new tables in the *Nautical Almanac* for 1835, and other tables were published by C. T. de Damoiseau the following year. In the *Nautical Almanac* for 1881 John Couch Adams extended Damoiseau's tables to later dates; these extended tables were then used in the preparation of the ephemerides of eclipses and the drawings of configurations of the satellites that appeared in the *Connaissance des temps* and the *Nautical Almanac*. A more complete theory was developed by Cyrille Souillart of Lille and published in *Memoirs of the Royal Astronomical Society* in 1880; Souillart's theory is the chief basis for the account of the motions of the Galilean satellites in Tisserand's *Traité de mécanique céleste*.

The eccentricities of the orbits of the Galilean satellites and their inclinations to the equator of Jupiter are very small: for Io, Europa, Ganymede and Callisto, they are 0.0000, 0.0003, 0.0015 and 0.0075, and 0°.03, 0°.47, 0°.18 and 0°.27 respectively. Therefore, in spite of the commensurabilities, the inequalities in the mean longitudes with the arguments $2\lambda' - \lambda$ and $2\lambda'' - \lambda'$ are small, and were barely detectable in observations. However, the terms with such arguments produce appreciable perturbations in the eccentricities and the longitudes of the pericentres for the three inner of the four satellites. In consequence, the following terms appear in the expressions for the true longitudes of these satellites:

$$\begin{aligned} &25'.9 \sin 2(\lambda-\lambda'), \\ &-61'.5 \sin (\lambda-\lambda'), \\ &-3'.8 \sin (\lambda'-\lambda''). \end{aligned} \tag{2}$$

Even in the eighteenth century, the effects represented by these terms could be detected in the timing of eclipses. From the period of the inequality of each satellite, together with the synodic period of that satellite with respect to the Sun, one can compute the period in which the inequality returns to the same phase in eclipses; for all three satellites this proves to be 437.6 days (see Chapter 21). The amplitudes of the inequalities, expressed in Equations (2) as maximum angular deviations in the Keplerian motion about Jupiter, imply differences in the times of onset of eclipses; the greatest maximum time-difference is that for the second satellite and comes to about 14 minutes. These time-differences were the inequalities that Bradley and Wargentin had detected.

Souillart corrected the expressions in Equations (2) by adding terms in the disturbing function with the arguments $4\lambda'-2\lambda-2\varpi$, $4\lambda'-2\lambda-\varpi-\varpi'$, $4\lambda'-2\lambda-2\varpi'$, $4\lambda''-2\lambda'-2\varpi'$, $4\lambda''-2\lambda'-\varpi'-\varpi''$, and $4\lambda''-2\lambda'-2\varpi''$, where ϖ, ϖ' and ϖ'' are the longitudes of the pericentres of the three satellites. The resulting corrections for the amplitudes in Equations (2) were $-91''$, $+186''$ and $-36''$.

When these perturbations are taken into account, it can be shown that the critical argument, $\Theta=\lambda-3\lambda'+2\lambda''$, satisfies the following equation:

$$\frac{d^2\Theta}{dt^2}=-K\left(\frac{m'm''}{a^2}+\frac{9m''m}{a'^2}+\frac{4mm'}{a''^2}\right)\sin\Theta, \tag{3}$$

where K is a positive constant depending on the ratios of the semi-major axes, and m, m' and m'' are the masses of the three satellites, their products two by two indicating that mutual perturbations of the satellites are involved. Since Equation (3) is the equation for a pendulum, it follows that Θ librates; the period of this libration is computed to be about six years, which is of the order of the orbital period divided by the mass of one satellite.

The secular motions of the pericentres and nodes for the four satellites were determined from observations, and from these data the mass of the fourth satellite and the dynamical oblateness of Jupiter were derived. Finally, the equations for the secular perturbations were solved simultaneously as in the case of planetary theory, the mutual perturbations being taken into account, along with the actions due to the Sun and the oblateness of Jupiter, and even the precession torque exerted on Jupiter by the Sun and the satellites, as expressed in the node-inclination equations. In this way, as new observations became available, the masses of the satellites and the mass and oblateness of Jupiter, as well as the orbital elements and their time variations, were improved step by step.

Saturn

Phoebe

Phoebe was one of nine satellites of Saturn known by the end of the nineteenth century (see Table 30.1), and the first to be discovered on photographic plates, the plates in question having been taken on 16 and 18 August 1898 by W. H. Pickering using the 13-inch (33-cm) Bruce photographic telescope at the Boyden Station of Harvard Observatory at Arequipa, Peru. In 1904 A. C. D. Crommelin was able to infer, from the 1898 position and from four positions determined in 1904, that Phoebe's motion is retrograde; the orbital inclination was later found to be 175° with respect to the ecliptic, and the orbital eccentricity to be 0.166. The solar perturbation is rather large and the period of long-period perturbations is long, of the order of ten years; as a result the perturbations are expressed by polynomials with time as the variable, as in planetary theory. The secular motions for the pericentre and node are slow: 0°.27 and 0°.435 per year, respectively. The expressions for the orbital elements are therefore very simple, being in fact linear functions of time.

Iapetus

Iapetus was discovered by G. D. Cassini (Cassini I) at Paris on 25 October 1671, with a small glass of 17-ft (5.2-m) focal length. Soon after, Cassini noticed a difference in brightness of the satellite when in different parts of its orbit – a phenomenon now well known, not only for this satellite but also for several others. He hypothesized that the reflectivity of the satellite was different on opposite sides, and that the period of the satellite's axial rotation was equal to its orbital period, as is the case for the Earth's Moon. These hypotheses have been corroborated by later investigators. Continuing the observations, Cassini was able by 1673 to obtain an approximate value of 80 days for the orbital period of Iapetus.

Reports by Jacques Cassini (Cassini II) of observations of the satellite consist of drawings of the planet and its ring with the relative position of the satellite indicated by a cross or mark; as to time of observation, only the date is supplied. Nevertheless, from these observations it is possible to infer the approximate times of the superior and inferior conjunctions of Iapetus in March and May 1685 and in May 1714, and these are still useful for the determination of the orbital period when employed in combination with observations of later years.

Few observations of Saturnian satellites were made between 1714 and 1787, when attention was again directed to them by J.-J. L. de Lalande in a paper on the motion of Iapetus. Lalande's call for data resulted in observations by P. J. Bernard at the Marseilles Observatory during the opposition of 1787 and by William Herschel during the opposition of 1789. Lalande analysed these observations, and on the basis of the knowledge thus acquired published tables of the motion of Iapetus in the *Connaissance des temps* for 1791 and 1792.

There then followed a forty-year gap, before observations by F. W. Bessel with a heliometer were published in volumes of the *Astronomische Beobachtungen . . . zu Königsberg* in the 1830s. John Herschel, during his stay at the Cape of Good Hope, observed the position angles of the satellite between 1835 and 1837, and another English amateur, the wealthy brewer William Lassell (1799–1880), made a series of observations in 1850. An extended series of observations was made at Madras by W. S. Jacob in 1856–58 and published in the *Memoirs of the Royal Astronomical Society*. Then, in 1874, the US Naval Observatory commenced a series of observations of Iapetus with the 26-inch refractor. In 1885 Hall published an analysis based on observations up to and including those of 1884 in an Appendix to the *Washington Observations*.

The peculiar characteristics of the motion of Iapetus's orbital plane are due to the fact that the oblateness and the solar factors are of the same order of magnitude. As a consequence, the inclinations with respect to the equator and the orbital plane of Saturn – the latter two planes being inclined to one another by 28° – are not constant, but the pole of the orbital plane moves slowly along an ellipse defined by

$$K \cos^2 i + K' \cos^2 i' = \text{const}, \tag{4}$$

where K and K' represent the oblateness and the solar factors, respectively, and i and i' represent the inclinations with respect to the equator and the orbital plane of Saturn. As previously mentioned, if K is dominant, i is nearly constant, while if K' is dominant, i' is nearly constant; but if, as here, K and K' are of the same order, then the motion of the orbital plane is complicated.

This strange motion was noticed by Laplace, who undertook to explain it by a theory of the secular perturbations in his *Traité de mécanique céleste* (Vol. 4, Bk VIII, Chap. 17). However, as the orbital period of Iapetus and, therefore, the secular period of the ascending node are rather long, the inclination to the ecliptic, i, and the longitude of the ascending node, Ω, as well as the longitude of the pericentre, ϖ, can be expressed by linear functions of the time, the constants in the expressions being derived from observations at several oppositions:

$$\begin{aligned} i &= 18^\circ\, 33'\, 39''.5 - 10''.80t, \\ \Omega &= 142^\circ\, 26'\, 41''.4 - 126''.00t, \\ \varpi &= 353^\circ\, 14'\, 56''.5 + 86''.28t, \end{aligned} \tag{5}$$

where t is measured in the unit of a year from the epoch (1880 March 17.0 GMT) and i, Ω and ϖ are referred to the mean equinox of 1880.0.

Hyperion

The faint satellite Hyperion was discovered independently in September 1848 by W. Lassell and by W. C. Bond of the Harvard Observatory. In 1884 Hall pointed out in the Royal Astronomical Society's *Monthly Notices* a peculiar phenomenon in its motion: the pericentre moved in a retrograde direction at the rate of 18° per year, an effect that cannot be explained as due either to the oblateness of Saturn or to the solar action. Newcomb noticed that the critical argument, $\Theta = 4\lambda' - 3\lambda - \varpi'$, where λ' and λ are, respectively, the mean longitudes of Hyperion and Titan and ϖ' is the longitude of the pericentre of Hyperion, librates around 180°

with an amplitude of 36° and a period of 640 days; the following equation holds:

$$\frac{d^2\Theta}{dt^2} = -K\,m \sin\Theta, \qquad (6)$$

where m is the mass of Titan and K is a positive constant.

Because Hyperion's mean motion (n') is to Titan's (n) very nearly as three to four, Θ librates around 180° with a period of the order of the orbital period divided by the square root of m. As a consequence, Hyperion can avoid very close approach to Titan, because any conjunction of the two satellites takes place only near the apocentre of Hyperion (the eccentricity of Hyperion is large, namely 0.109). The secular motion of the pericentre of Hyperion's orbit is not due to the oblateness of Saturn but is equal to $4n' - 3n$ because of the libration of the critical argument.

The motion of Hyperion was studied in the 1880s by Stone, Tisserand, and G. W. Hill, all of whom sought to solve the problem by computing perturbations. Then in 1892 W. S. Eichelberger published in the *Astronomical Journal* a complete analysis of observations at the US Naval Observatory for the period 1875–90, and derived the orbital elements and their time variations as well as the masses of Saturn and Titan.

Titan and Rhea

Titan, the second largest satellite in the entire solar system, was discovered on 25 March 1655 by the Dutch physicist Christiaan Huygens with his 12-ft (3.6-m) refractor, while Rhea was found by Cassini I on 23 December 1672.

As Titan is a bright object, position measurements of it are relatively easy. However, Huygens, and likewise Bernard in 1787, gave only estimates of conjunction times. In 1829 Bessel began observations, later published in *Astronomische Nachrichten*, with the heliometer of the Königsberg Observatory; he was able to determine with accuracy the orbital elements of the satellite and the mass of the planet. Bessel also derived expressions for the solar perturbations of Titan, which are appreciable, and he published tables for computing positions of this satellite.

In 1887 Hall published in the *Washington Observations* an analysis of observations of Titan made at the US Naval Observatory between 1874 and 1884, and determined its orbital elements. The following year H. Struve published his analysis of the orbits of Saturn's satellites including that of Titan. For this his main sources were recent observations by Bessel, Jacob, Newcomb, W. Meyer and himself, but he took into account eighteenth-century observations of conjunctions by Cassini I, Bradley, William Herschel and J. G. Köhler, and even two each by Edmond Halley and Cassini I from the 1680s. He derived the orbital elements and their time variations including the solar and secular perturbations with the difference of the longitudes of the pericentres of Titan and Iapetus as argument, and determined the mass of Iapetus as well as the position of the equator of Saturn from the motion of the orbital plane.

As the orbital elements of Titan had now been well established and are rather stable, and as Titan is a very bright object, positions of other satellites besides Iapetus could thereafter be measured with respect to Titan. Such observations led to repeated improvements in the orbital elements of both Titan and Rhea. (Another satellite that has been adopted as a reference is the inner satellite Tethys, the orbit of which has an extremely small eccentricity, in most computations equatable to zero.)

For Rhea, H. Struve made an analysis similar to the one he had made for Titan. He used observations of conjunctions made by Cassini I in 1689 and 1704 and by Bradley in 1719, as well as more recent ones, and derived the orbital elements and their time variations. But since the orbital eccentricity of Rhea is only 0.0010, he could not determine the longitude of the pericentre precisely, and so for the secular motion of the pericentre he gave the theoretical value due to the oblateness of Saturn and the actions of the Sun and Titan. Later, however, it was found that the eccentricity of Rhea is the forced one due to the perturbation by Titan, and as a consequence the longitude of the pericentre of Rhea moves secularly with the same rate as that of Titan. Although the mean motion of Rhea does not have any special relation with that of Titan, the orbits of the two satellites are very close to each other and Titan is massive, which is why the orbit of Rhea suffers this large perturbation.

The four inner satellites

Despite the enormous length of the aerial telescopes he was using, Cassini I succeeded in discovering Tethys and Dione on 21 March 1684, just before the ring disappeared. Enceladus was in fact seen by William Herschel in 1787, but went unrecognized at the time. On 18 July 1789, when the ring had almost disappeared (the condition most favourable for observing faint inner satellites) he saw it again, but took it to be Tethys. He did not identify it as a new satellite until 28 August, when he was inaugurating his 40-ft (12.4-m) reflector. A few weeks later he discovered Mimas. Since Mimas is very near the ring, and faint, it was very difficult to observe, and no routine observations were possible until the middle of the nineteenth century with even the best instruments. Herschel's 1789 observations of Mimas and Enceladus were analysed in 1836 by Wilhelm Beer and Johann Heinrich Mädler and the orbital elements derived.

Until the 1789 discoveries, the four satellites previously known were referred to by number in order of the increasing size of their orbits. The new discoveries made this scheme obsolete, and the names we now use were later introduced by John Herschel.

Beginning in the middle of the nineteenth century, the positions of the four inner satellites were the subject of micrometer measurements by numerous observers, among them Lassell. In 1887 Hall determined the orbital elements and the mean motions of Rhea and the four inner satellites from observations made at the US Naval Observatory. He also published tables for the computation of the positions of the six inner satellites.

The ratio of the mean motions of Enceladus and Dione is nearly 2 to 1, and the critical argument, $2\lambda' - \lambda - \varpi$, librates around 0° with a small amplitude. The observed eccentricity of Enceladus, 0.0045, is the forced one due to Dione, the free one being much smaller (0.0001); the longitude of the pericentre of Enceladus moves at the rate $2n' - n$. From this resonance arise detectable inequalities in the mean longitudes of the two satellites, observed to be 15′ at maximum for Enceladus and 1′ for Dione; from these empirical constants were derived the masses of the two satellites.

The ratio of the mean motions of Mimas and Tethys is also nearly 2 to 1, and for this pair the critical argument, which librates about a mean value, is $4\lambda' - 2\lambda - \Omega' - \Omega$, where Ω and Ω' are the longitudes of the ascending nodes, and the amplitude and period of the libration are 98° and 72 years respectively. As a consequence of this libration, large inequalities appear in the mean longitudes, the amplitudes being 43°.4 for Mimas and 2°.1 for Tethys. From the observed amplitudes and period of the libration the masses of the two satellites were derived.

H. Struve carried out a complete analysis and theoretical investigation for the two pairs in 1890; for Mimas–Tethys it involved the study of observations going back to William Herschel's of 1789. He also derived the dynamical oblateness of Saturn by using the observed secular motions for the six inner satellites, the masses of most of them being known from other sources. In addition, he determined from satellite observations the position of the equator of Saturn, which is believed to coincide with the ring plane.

Uranus

The outer two satellites of Uranus, now called Titania and Oberon, were discovered by William Herschel in January 1787, nearly six years after his discovery of the planet itself. On 7 February he observed the satellites for nine hours to confirm their nature, and from subsequent observations he determined the mean radii of the orbits, the periods of revolution, and the mass of Uranus. He found also the remarkable fact that the orbital planes of the satellites are almost perpendicular to the ecliptic. Later he announced, but not very confidently, the discovery of four other satellites, which he termed *supplementary*; because of difficulties in measuring their distances from Uranus, he was unable to assign sizes to their orbits. At least some of these 'supplementary satellites' must have been phantoms or incorrectly identified objects.

Herschel continued to make micrometric measurements of the satellites of Uranus until 1798, but thereafter no one attempted to observe them until his son John took up the challenge in 1828. Uranus, having then acquired southern declination, was in an unfavourable position for observation of the satellites, and he was able to observe only the brightest two, Titania and Oberon. He was unable to measure their distances from the planet, even after he repolished his telescope mirror in 1830, but on the basis of his father's observations and his own, he was able to determine accurately their orbital periods.

In 1837 Johann von Lamont of the Royal Observatory at Munich observed the satellites for a month, using a 10.5-inch (22-cm) refractor with a micrometer. He reported that he observed three satellites, one of them only once. Using the orbital sizes and periods of Titania and Oberon, he determined a value for the mass of Uranus; this is smaller, however, than the value accepted today.

In September 1847 Lassell began observing and measuring the positions of the satellites. Besides the two bright satellites discovered by William Herschel, he believed he was seeing one or two inner satellites – satellites that Herschel had probably observed – but he was unable to find any satellite with orbit larger than that of Oberon. He succeeded in obtaining data on one of the faint inner satellites. That same autumn, Otto Struve at Pulkovo observed the satellites with a 15-inch (38-cm) refractor, and succeeded in measuring the size and period of the orbit of one of the inner satellites.

Towards the end of the same year, W. R. Dawes, who had observed with Lassell, undertook an analysis of Lassell's and Struve's observations, showing that they were not consistent if they were of the same satellite, and concluding that the two observers had been observing two different satellites, one with a period of 2.1108 days and the other with a period of 3.9236 days. Because of different reflectivities of the satellites on different sides, and the difference in the times at which Lassell and Struve had been able to observe, each failed to see what the other had seen. Finally, in 1851, Lassell announced the discovery of the two inner satellites, now called Ariel and Umbriel, which his observations of October and November 1851 showed to have periods and mean radii in accordance with Kepler's third law. Both Lassell and Struve later observed all four Uranian satellites extensively.

Soon after the 26-inch refractor came into use at the US Naval Observatory in 1873, Newcomb commenced observations of the satellites of Uranus, and these he continued during the oppositions of 1874 and 1875. From them he determined the orbital elements of the satellites and the mass of the planet. His orbital elements were to be used for the computation of ephemerides for many years. In the early 1880s Hall observed the two outer satellites, in an attempt to improve the accuracy of their orbital elements and hence of the mass and the position of the equator of the planet.

Motions of the four satellites are very simple: all can be taken as moving in circular orbits in the plane of the equator of Uranus, which is inclined to the ecliptic by 98°.

Neptune

On 10 October 1846, only 17 days after the discovery of Neptune by J. G. Galle and H. L. d'Arrest, Lassell glimpsed what he suspected to be a satellite of the planet. It was months afterwards before he could confirm his suspicion; he announced his discovery of Triton, as it was later named, on 8 July 1847. At the same time, he had the impression that he was seeing a ring about the planet. For nearly six years he sought to confirm this impression, but finally concluded that the appearance must be due to faults, and in particular to a flexure, in his 2-ft (61-cm) reflector. The satellite, he early noticed, was much brighter in one half of its path than in the other – a phenomenon explicable, as before mentioned, by supposing the periods of axial rotation and orbital revolution to be equal, and the reflectivities on opposite sides of the satellite to be different. By late 1848 W. C. Bond of Harvard had sufficient observations to allow a determination of the mass of the planet. In 1875 Newcomb was able to derive orbital elements for Triton from an analysis of observations made at the US Naval Observatory over the previous two years.

Triton was the first satellite to be discovered whose orbital motion is retrograde, the inclination of its orbit to Neptune's equator being 160°. (In the twentieth century, five additional planetary satellites have been found to have retrograde motions, namely four outer satellites of Jupiter together with Saturn's Phoebe.) The eccentricity of Triton's orbit is almost zero, while the longitude of the ascending node moves at a rate of 61°.5 per year, mainly owing to the oblateness of Neptune; it is not necessary to take into account any other perturbations. The mass of Triton was later determined from observations of the motion of Neptune around the common centre of mass of planet and satellite.

Conclusion

Through the nineteenth century, observations of planetary satellites and analyses of their motions were rewarded by the knowledge they supplied of the planetary masses, a prerequisite for the calculation of planetary perturbations. An additional reward was acquaintance with the variety and beauty of the satellite systems, their mechanical resonances and librations and such special features as the retrograde motion of Triton. In the twentieth century, particularly since space probes have begun to supply new and more accurate data, a lively interest has arisen in the evolution of satellite systems, as depending on tidal interactions of the primaries and satellites; it is suggested that studies of the evolution of the regular satellite systems of Jupiter, Saturn and Uranus may hold the key to the origin of the solar system. In the late twentieth century, satellites of the outer planets have come to play a role in the navigation of space missions, with resulting new demands on the accuracy of satellite ephemerides.

ILLUSTRATIONS: ACKNOWLEDGEMENTS AND SOURCES

14.1 I. Newton, *Principia*, 3rd edn (London, 1726), p. 384.
14.2 *Ibid.*, p. 377.
14.3 From G. W. Leibniz, Tentamen de motuum coelestium causis, *Acta eruditorum*, Feb. 1689, 82–96.
14.7 From *Histoire et mémoires de l'Académie royale des Sciences*, 1736, Plate 9, Fig. 3.
15.1 Collection of Owen Gingerich.
15.2 R. Outhier, *Journal d'un voyage au nord en 1736 et 1737* (Paris, 1744), p. 96.
15.3 P. Bouguer, *Figure de la terre* (Paris, 1749), Planche VII, courtesy of the Institute of Astronomy, Cambridge University.
16.1 I. Newton, *Principia*, 3rd edn (London, 1726), p. 134.
16.2 Courtesy of Académie des Sciences, Paris
17.1 From *Philosophical Transactions*, vol. 45 (1748), p. 16.
17.2 Courtesy of Académie des Sciences, Paris.
18.1 N.-L. de Lacaille, *Astronomiae fundamenta* (Paris, 1757), title-page.
18.2 N.-L. de Lacaille, *Tabulae solares* (Paris, 1758), title-page.
18.3 From Göttingen *Comentarii*, vol. 3 (1754), p. 393, by permission of the Syndics of Cambridge University Library.
19.1 From *Philosophical Transactions*, vol. 24 (1705), p. 1886, by permission of the Syndics of Cambridge University Library.
19.2 From *Philosophical Transactions*, vol. 49 (1755), p. 350.
19.3 From *Hollandsche Maatschappij der Wetenschappen Verhandelingen*, vol. 2 (1755), 288.
19.4 Broadside by B. Martin, courtesy of the British Library.
19.5 From *Mémoires de l'Académie royale des Sciences*, 1760, pp. 385–6, by permission of the Syndics of Cambridge University Library.
19.6 From *Mémoires de l'Académie royale des Sciences*, 1760, plate II, facing p. 465, by permission of the Syndics of Cambridge University Library.
20.1 Courtesy of Académie des Sciences, Paris.
20.2 Courtesy of Öffentliche Bibliothek der Universität Basel.
21.1 Courtesy of Académie des Sciences, Paris.
22.1 Courtesy of Académie des Sciences, Paris.
22.2 P.-S. de Laplace, *Traité de mécanique céleste*, vol. 1 (Paris, 1799), title-page, by permission of the Syndics of Cambridge University Library.
23.1 Redrawn from Harry Woolf, *The Transits of Venus* (Princeton, NJ, 1959), p. 18.
23.2 Redrawn from H. H. Turner, *Astronomical Discovery* (London, 1904), p. 29.
23.3 Simon Newcomb, *Popular Astronomy* (London, 1878), p. 179.
23.4 From *Mémoires de l'Académie royale des Sciences*, 1757, plate III, preceding p. 251, by permission of the Syndics of Cambridge University Library.
23.5 Engraving by J. G. Kaid, 1771, after a portrait from life by W. Pohl, courtesy of I. Bernard Cohen.
23.6 Courtesy of the Royal Greenwich Observatory Archives, Cambridge, and by permission of the Syndics of Cambridge University Library.
24.1 D. Gregory, *The Elements of Physical and Geometrical Astronomy*, 2nd English edn (London, 1726), plate 1.
24.2 *Beobachtung über die Natur vom Herrn Karl Bonnet* (Leipzig, 1766), trans. by J. D. Titius, pp. 7–8.
24.3 From the drawing by W. Watson, RAS MS Herschel W.5/5, no. 3, courtesy of the Royal Astronomical Society.
24.4 W. Herschel, *Collected Scientific Papers* (London, 1912), vol. 1, plate A.
24.5 W. Pearson, *An Introduction to Practical Astronomy*, vol. 2 (London, 1829), plate XVIII.
25.1 Lithograph by E. Ritmüller after the painting by C. A. Jensen, courtesy of the Mary Lea Shane Archives of Lick Observatory.
28.1 Courtesy of Académie des Sciences, Paris.
28.2 Engraving by Samuel Cousins from the painting by Thomas Mogford, 1851, courtesy of the Institute of Astronomy, Cambridge University.
28.4 From *Vistas in Astronomy*, vol. 3 (1960), p. 44.
28.6 G. W. Hill, *Collected Mathematical Works*, vol. 1 (Washington, D.C., 1905), frontispiece.
28.7 S. Newcomb, *Reminiscences of an Astronomer* (London, 1903), frontispiece, courtesy of the Institute of Astronomy, Cambridge University.
28.8 From the photograph of Henri Poincaré in his study, Viollet Collection
29.1 *La Connoissance des temps*, vol. 1 (Paris, 1679), title-page, courtesy of the Royal Greenwich Observatory Archives, Cambridge, and by permission of the Syndics of Cambridge University Library.
29.2 *Nautical Almanac*, vol. 1 (London, 1766), title-page, courtesy of Institute of Astronomy, Cambridge University.

COMBINED INDEX TO PARTS 2A AND 2B

(Page references to Part 2A are italicized, G refers to the Glossary of Part 2A.)